AA002393

Proceedings

THE 23RD IEEE INTERNATIONAL SYMPOSIUM ON DEFECT AND FAULT-TOLERANCE IN VLSI SYSTEMS

DFT 2008

23rd IEEE International Symposium on Defect and Fault Tolerance in VLSI Systems

(DFT 2008)

Boston, Massachusetts
1-3 October 2008

IEEE Catalog Number: CFP08078-POD
ISBN: 978-0-7695-3365-0

Proceedings

THE 23RD IEEE INTERNATIONAL SYMPOSIUM ON DEFECT AND FAULT-TOLERANCE IN VLSI SYSTEMS

Boston, Massachusetts
1–3 October 2008

Sponsored by
The IEEE Computer Society Test Technology Technical Council
The IEEE Computer Society Technical Committee on Fault Tolerant Computing

Edited by
Cristiana Bolchini, Yong-Bin Kim, Dimitris Gizopoulos, and Mohammad Tehranipoor

Los Alamitos, California

Washington • Tokyo

Copyright © 2008 by The Institute of Electrical and Electronics Engineers, Inc.
All rights reserved.

Copyright and Reprint Permissions: Abstracting is permitted with credit to the source. Libraries may photocopy beyond the limits of US copyright law, for private use of patrons, those articles in this volume that carry a code at the bottom of the first page, provided that the per-copy fee indicated in the code is paid through the Copyright Clearance Center, 222 Rosewood Drive, Danvers, MA 01923.

Other copying, reprint, or republication requests should be addressed to: IEEE Copyrights Manager, IEEE Service Center, 445 Hoes Lane, P.O. Box 133, Piscataway, NJ 08855-1331.

The papers in this book comprise the proceedings of the meeting mentioned on the cover and title page. They reflect the authors' opinions and, in the interests of timely dissemination, are published as presented and without change. Their inclusion in this publication does not necessarily constitute endorsement by the editors, the IEEE Computer Society, or the Institute of Electrical and Electronics Engineers, Inc.

IEEE Computer Society Order Number P3365
BMS Part Number CFP08078-PRT
ISBN 978-0-7695-3365-0
ISSN Number 1550-5774

Additional copies may be ordered from:

IEEE Computer Society	IEEE Service Center	IEEE Computer Society
Customer Service Center	445 Hoes Lane	Asia/Pacific Office
10662 Los Vaqueros Circle	P.O. Box 1331	Watanabe Bldg., 1-4-2
P.O. Box 3014	Piscataway, NJ 08855-1331	Minami-Aoyama
Los Alamitos, CA 90720-1314	Tel: + 1 732 981 0060	Minato-ku, Tokyo 107-0062
Tel: + 1 800 272 6657	Fax: + 1 732 981 9667	JAPAN
Fax: + 1 714 821 4641	http://shop.ieee.org/store/	Tel: + 81 3 3408 3118
http://computer.org/cspress	customer-service@ieee.org	Fax: + 81 3 3408 3553
csbooks@computer.org		tokyo.ofc@computer.org

Individual paper REPRINTS may be ordered at: reprints@computer.org

Editorial production by Lisa O'Conner

Cover art production by Joe Daigle/Studio Productions

Printed in the United States of America by The Printing House

Conference Publishing Services

http://www.computer.org/proceedings/

Table of Contents

IEEE INTERNATIONAL SYMPOSIUM ON DEFECT AND FAULT TOLERANCE IN VLSI SYSTEMS

DFT 2008

Message from the Symposium Chairs..x
Organizing Committee..xii
Program Committee...xii

Keynote Talk

The Evolving Role of Test ... it is now a "Value Add" Operation 3
   Phil Nigh, IBM

Session 1: Defect and Fault Tolerance

Using TMR Architectures for Yield Improvement 7
   J. Vial, A. Bosio, P. Girard, C. Landrault, S. Pravossoudovitch, and A. Virazel
Module Grouping for Defect Tolerance in Nanoscale Memory..................................16
   Yoonjae Huh and Yoon-Hwa Choi
Coping with Obsolescence of Processor Cores in Critical Applications..........................24
   Francesco Abate and Massimo Violante
A Low-Power Safety Mode for Variation Tolerant Systems-on-Chip...............................33
   David Wolpert and Paul Ampadu

Session 2: Dependability Analysis and Evaluation

Built-In Self-Diagnostics for a NoC-Based Reconfigurable IC for Dependable
Beamforming Applications 45
   Oscar J. Kuiken, Xiao Zhang, and Hans G. Kerkhoff
Network Fault Model for Dependability Assessment of Networked Embedded Systems 54
   Franco Fummi, Davide Quaglia, and Francesco Stefanni
Obtaining Microprocessor Vulnerability Factor Using Formal Methods............................63
   Syed Z. Shazli and Mehdi B. Tahoori
System Reliabilities when Using Triple Modular Redundancy in Quantum-Dot
Cellular Automata 72
   Timothy J. Dysart and Peter M. Kogge

Invited Talk

Error Detection and Tolerance for Scaled Electronic Technologies 83
Kartik Mohanram, Rice University

Session 3: Hot Topics

Hardware Trojan Detection and Isolation Using Current Integration
and Localized Current Analysis 87
Xiaoxiao Wang, Hassan Salmani, Mohammad Tehranipoor, and Jim Plusquellic

Built-In Proactive Tuning System for Circuit Aging Resilience 96
Nimay Shah, Rupak Samanta, Ming Zhang, Jiang Hu, and Duncan Walker

Exploring Density-Reliability Tradeoffs on Nanoscale Substrates: When Do Smaller
Less Reliable Devices Make Sense? 105
Andrey Zykov and Gustavo de Veciana

Impact of Technology and Voltage Scaling on the Soft Error Susceptibility
in Nanoscale CMOS 114
Vikas Chandra and Robert Aitken

Session 4: Design for Testability

Enhancing Silicon Debug via Periodic Monitoring 125
Joon-Sung Yang and Nur A. Touba

A Digital BIST for Phase-Locked Loops 134
Kevin Sliech and Martin Margala

On Optimizing Fault Coverage, Pattern Count, and ATPG Run Time Using a Hybrid
Single-Capture Scheme for Testing Scan Designs 143
Shianling Wu, Laung-Terng Wang, Zhigang Jiang, Jiayong Song, Boryau Sheu,
Xiaoqing Wen, Michael S. Hsiao, James C.-M. Li, Jiun-Lang Huang, and Ravi Apte

Analyzing the Impact of Fault-Tolerant BIST for VLSI Design 152
W. Robert Daasch, Saurabh Jain, and David Armbrust

Invited Talk

Targeting "Zero DPPM" – Can We Ever Get There? 163
Nilanjan Mukherjee, Mentor Graphics

Session 5: Posters

A BIST Technique for Crosstalk Noise Detection in FPGAs 167
Waleed K. Al-Assadi and Sindhu Kakarla

A Fault Tolerance Aware Synthesis Methodology for Threshold
Logic Gate Networks 176
Manoj Kumar Goparaju, Ashok Kumar Palaniswamy, and Spyros Tragoudas

A Framework to Evaluate the Trade-Off among AVF, Performance and Area
of Soft Error Tolerant Microprocessors 184
Rui Gong, Kui Dai, and Zhiying Wang

A Power Efficient Masking Technique for Design of Robust Embedded Systems
against SEUs and SETs 193
Mahdi Fazeli and Seyed Ghassem Miremadi

Can Knowledge Regarding the Presence of Countermeasures Against Fault Attacks
Simplify Power Attacks on Cryptographic Devices? 202
*Francesco Regazzoni, Thomas Eisenbarth, Luca Breveglieri, Paolo Ienne,
and Israel Koren*

Modeling and Evaluation of Threshold Defect Tolerance 211
Zachary Patitz and Nohpill Park

Defect Tolerance for a Capacitance Based Nanoscale Biosensor 220
Glenn H. Chapman and Vijay K. Jain

Fault Detection of Bloom Filters for Defect Maps..229
Jae-Young Choi and Yoon-Hwa Choi

Fault Tolerant Schemes for QCA Systems...236
Xiaojun Ma and Fabrizio Lombardi

On Reducing Circuit Malfunctions Caused by Soft Errors 245
Ilia Polian, Sudhakar M. Reddy, Irith Pomeranz, Xun Tang, and Bernd Becker

Realization of L2 Cache Defect Tolerance Using Multi-bit ECC 254
Hongbin Sun, Nanning Zheng, and Tong Zhang

Selective Hardening of NanoPLA Circuits 263
Ilia Polian and Wenjing Rao

Soft Error Hardened FF Capable of Detecting Wide Error Pulse 272
Shuangyu Ruan, Kazuteru Namba, and Hideo Ito

XOR-Based Low Cost Checkers for Combinational Logic 281
Carlos Arthur Lang Lisboa and Luigi Carro

Minimization of CTS of k-CNOT Circuits for SSF and MSF Model 290
Muhammad Ibrahim, Ahsan Raja Chowdhury, and Hafiz Md. Hasan Babu

Keynote Talk

Architectural Vulnerability Factor (or, Does a Soft Error Matter?) 301
Shubu Mukherjee, Intel

Session 6: Reliability and Fault Tolerance

Automatic Detection of In-Field Defect Growth in Image Sensors 305
Jenny Leung, Glenn H. Chapman, Israel Koren, and Zahava Koren

Material Fatigue and Reliability of MEMS Accelerometers...314
Xingguo Xiong, Yu-Liang Wu, and Wen-Ben Jone

Fault-Tolerance with Graceful Degradation in Quality: A Design Methodology
and its Application to Digital Signal Processing Systems...323
Nilanjan Banerjee, Charles Augustine, and Kaushik Roy

Design Space Exploration for the Design of Reliable SRAM-Based FPGA Systems 332
Cristiana Bolchini and Antonio Miele

Session 7: Error Detection and Correction (1)

A Low Cost Scheme for Reducing Silent Data Corruption in Large Arithmetic Circuits 343
Abhisek Pan, James W. Tschanz, and Sandip Kundu

Adaptive Error Control for NoC Switch-to-Switch Links in a Variable Noise Environment 352
Qiaoyan Yu and Paul Ampadu

Arbitrary Error Detection in Combinational Circuits by Using Partitioning 361
Osnat Keren, Ilya Levin, Vladimir Ostrovsky, and Beni Abramov

Error Detect Logic Resulting in Faster Address Generate and Decode for Caches 370
Prashant D. Joshi

Invited Talk

A Case Study of ATPG Delay Path Performance Based on Measured Power Rail Integrity 381
Zahi Abuhamdeh, Transwitch

Session 8: Testing Techniques

ATPG Heuristics Dependant Observation Point Insertion for Enhanced Compaction
and Data Volume Reduction 385
*Santiago Remersaro, Janusz Rajski, Thomas Rinderknecht, Sudhakar M. Reddy,
and Irith Pomeranz*

Detection of Transistor Stuck-Open Faults in Asynchronous Inputs of Scan Cells 394
*Fan Yang, Sreejit Chakravarty, Narendra Devta-Prasanna, Sudhakar M. Reddy,
and Irith Pomeranz*

Efficient Determination of Fault Criticality for Manufacturing Test Set Optimization 403
Yiwen Shi, Kellie DiPalma, and Jennifer Dworak

Core Test Wrapper Design to Reduce Test Application Time for Modular SoC Testing 412
Hyunbean Yi and Sandip Kundu

Session 9: Panel

Zero Defects: How Can We Get There?
Organizer: M. Tehranipoor, University of Connecticut

Invited Talk

Computing at the Nanoscale 423
John E. Savage, Brown University

Session 10: Error Detection and Correction (2)

A Generalized Approach for the Use of Convolutional Coding in SEU Mitigation 427
    Laura Frigerio, Matteo Alan Radaelli, and Fabio Salice

A Novel Error Detection and Correction Technique for RNS Based FIR Filters 436
    S. Pontarelli, G.C. Cardarilli, M. Re, and A. Salsano

An Asymmetric Checkpointing and Rollback Error Recovery Scheme
for Embedded Processors 445
    Hamed Tabkhi, Seyed Ghassem Miremadi, and Alireza Ejlali

Design and Evaluation of a Timestamp-Based Concurrent Error Detection Method (CED)
in a Modern Microprocessor Controller 454
    Michail Maniatakos, Naghmeh Karimi, Yiorgos Makris, Abhijit Jas,
    and Chandra Tirumurti

Session 11: Testing for Timing and Parametric Failures

Novel On-Chip Clock Jitter Measurement Scheme for High Performance Microprocessors 465
    Cecilia Metra, Martin Omaña, TM Mak, Asifur Rahman, and Simon Tam

Prioritization of Paths for Diagnosis 474
    Rajsekhar Adapa and Spyros Tragoudas

Delay Fault Testability on Two-Rail Logic Circuits 482
    Kazuteru Namba and Hideo Ito

Diagnosis of Analog Circuits by Using Multiple Transistors and Data Sampling 491
    Yukiya Miura and Jiro Kato

Invited Talk

Design for Test Challenges of High Performance/Low Power Microprocessors 503
    Kamran Zarrineh, AMD

Invited Talk

Defect-Tolerant Hybrid CMOS/Nanoelectronic Circuits 504
    Konstantin K. Likharev, Stony Brook University

Session 12: Emerging Technologies

A Statistical Model for Assessing the Fault Tolerance of Variable Switching Currents
for a 1Gb Spin Transfer Torque Magnetoresistive Random Access Memory....................................507
    Yoshiaki Asao, Masayoshi Iwayama, Kenji Tsuchida, Akihiro Nitayama,
    Hiroaki Yoda, Hisanori Aikawa, Sumio Ikegawa, and Tatsuya Kishi

A Tile-Based Error Model for Forward Growth of DNA Self-Assembly 516
    Masoud Hashempour, Zahra Mashreghian Arani, and Fabrizio Lombardi

Checkpointing of Rectilinear Growth in DNA Self-Assembly 525
    Stephen Frechette, Yong Bin Kim, and Fabrizio Lombardi

Fabrication Variations and Defect Tolerance for Nanomagnet-Based QCA....................................534
    Michael Niemier, Michael Crocker, and X. Sharon Hu

Author Index 543

Message from the Symposium Chairs

Welcome to Boston and the 2008 International Symposium on Defect and Fault Tolerance in VLSI Systems, sponsored by the IEEE Computer Society Technical Council on Test Technology and on Dependable Computing and Fault Tolerance.

This is the 23rd in a series of productive technical meetings bringing together top researcher and developers from the academic world, research laboratories and companies from numerous countries. This year the authors come from around the world: Bangladesh, Brazil, Canada, China, Korea, France, Germany, Netherlands, Iran, Israel, Italy, Japan, Switzerland, Taiwan - R.O.C., UK and the USA, hosts of the event.

The technical program covers traditional DFT topics such as defect analysis and test, error detection and fault tolerance, reliability analysis and evaluation, diagnosis. Moreover, additional topics are gaining importance and attention, including nanotechnology defect and fault modeling and tolerance techniques, System-on-Chip architectures with reliability properties, Quantum-Dot Cellular Automata. The papers propose both methodological solutions and innovative techniques to tackle the numerous issues defect and faults raise.

Beginning last year, DFT has introduced Best Paper Awards: it has been awarded to the paper "A Scalable Framework for Defect Isolation of DNA Self-assembled Networks", authored by Masaru Fukushi, Susumu Horiguchi, Luke Demoracski and Fabrizio Lombardi. This year there are several good papers candidate for the award.

Many people have contributed to this event; first of all we would like to thank the authors for their manuscripts, the Program Committee members for the time devoted to reading the papers and providing valuable technical feedback.

In the next few days we will embark in an exciting technical path, located in the Boston atmosphere; we hope that to old DFT habitués as well as to new participants it will offer a fruitful experience.

General Chairs

Cristiana Bolchini   Yong-Bin Kim

Program Chairs

Dimitris Gizopoulos   Mohammad Tehranipoor

Organizing Committee

General co-chairs
Cristiana Bolchini, *Politecnico di Milano, Italy*
Yong-Bin Kim, *Northeastern University, USA*

Program co-chairs
Dimitris Gizopoulos, *University of Piraeus, Greece*
Mohammad Tehranipoor, *University of Connecticut, USA*

Publicity chair
Marco Ottavi, *Advanced Micro Devices Inc., USA*

Local arrangements chair
Harry Chen, *MediaTek Inc., USA*

Program Committee

S. Chakravarty, *LSI Logic*
M. Choi, *University of Missouri Rolla*
Y. Choi, *Hongik University*
G. Chapman, *Simon Fraser University*
R. Datta, *Texas Instruments*
M. Favalli, *University of Ferrara*
J. Figueras, *Universitat Politècnica de Catalunya*
E. Fujiwara, *Tokyo Institute of Technology*
M. Fukushi, *Tohoku University*
S. Horiguchi, *Tohoku University*
C. Huang, *National Tsing Hua University*
H. Ito, *Chiba University*
A. Jas, *Intel*
N. Jha, *Princeton University*
W. Jone, *University of Cincinnati*
I. Koren, *University of Massachusetts, Amherst*
R. Leveugle, *TIMA Laboratory*
J. Lo, *University of Rhode Island*
F. Lombardi, *Northeastern University*
Y. Makris, *Yale University*
M. Margala, *University of Massachusetts, Lowell*
I. Markov, *University of Michigan*
C. Metra, *University of Bologna*
Z. Navabi, *Worcester Polytechnic Institute*
N. Nicolici, *McMaster University*
N. Park, *Oklahoma State University*
A. Paschalis, *University of Athens*
Z. Peng, *Linkoping University*
W. Pleskacz, *Warsaw University of Technology*
S. Pontarelli, *University of Rome "Tor Vergata"*
J. Prashant, *Intel*
M. Rebaudengo, *Politecnico di Torino*
S. Reddy, *University of Iowa*
F. Salice, *Politecnico di Milano*
A. Salsano, *University of Rome "Tor Vergata"*
D. Sciuto, *Politecnico di Milano*
S. Shukla, *Virginia Tech*
J. Teixeira, *INESC-ID Lisboa*
C. Thibeault, *École de Technologie Supérieure*
N. Touba, *University of Texas at Austin*
S. Tragoudas, *Southern Illinois University*
R. Velazco, *TIMA Laboratory*
M. Violante, *Politecnico di Torino*
H. Walker, *Texas A&M University*
L. Wang, *University of Connecticut*
X. Wen, *Kyushu Institute of Technology*
K. Zarrineh, *AMD*

KEYNOTE TALK

The Evolving Role of Test ... it is now a "Value Add" Operation

Phil Nigh, *IBM*

The role of IC testing is changing – from being viewed as mainly a non-value (cost) operation – to one which provides additional value to products. Historically, the role of Test was to separate the good (passing) devices from faulty (failing) ones. The emerging role of Test is to provide feedback to product designers (performance, functionality, power) and to the fab (variability, sources of yield loss, defect characterization). Test is also the focal point for product dispositioning; applying unique methods to maximize yield while still achieving high levels of Quality & Reliability.

This talk will focus on the key trends driving these changes to Test — and the emerging methods that are enabling Test to be a "Value Add" operation. Industry examples will be shown where Test has provided unique insight to IC process performance and defect behavior using these methods.

Speaker Bio – Phil Nigh is with Test Strategy & Development, IBM Server & Technology Group. Phil got his Ph.D. from Carnegie Mellon University (1990) and has been with IBM for 25 years.

SESSION 1
DEFECT AND FAULT TOLERANCE

IEEE International Symposium on Defect and Fault Tolerance of VLSI Systems

Using TMR Architectures for Yield Improvement

J. Vial   A. Bosio   P. Girard   C. Landrault   S. Pravossoudovitch   A. Virazel

Laboratoire d'Informatique, de Robotique et de Microélectronique de Montpellier
Université Montpellier II / CNRS – 161, rue Ada 34932 Montpellier – France
Email: {vial, bosio, girard, landraul, pravo, virazel}@lirmm.fr

Abstract

With the technology entering the nano dimension, manufacturing processes are less and less reliable, thus drastically impacting the yield. A possible solution to alleviate this problem in the future could consist in using fault tolerant architectures to tolerate manufacturing defects. In this paper, we use the classical Triple Modular Redundancy (TMR) fault tolerant architecture as a case study. Firstly we analyze the conditions that make the use of TMR architectures interesting for yield improvement purpose. In the second part of the paper, we investigate the test requirements for the TMR architecture and we propose a solution for generating test patterns for this type of architecture. Finally, we propose a new manner to implement the TMR architecture that makes it very effective for yield improvement purpose. Experimental results are provided on ISCAS and ITC benchmark circuits to prove the efficiency of the proposed approach in terms of yield improvement with a low area overhead.

1. Introduction

Deep submicron technologies allow hundreds of millions of transistors to be integrated on the same chip. However, the yield of such large chips decreases due to gross imperfections caused by manufacturing defects or process deviations. Even with a well controlled manufacturing process, it becomes more and more difficult to guarantee a chip without any defect. Consequently, the design and test of an electronic system benefiting from technological advances will be even more challenging in the coming years [1].

To increase the yield for future VLSI systems, fault tolerant architectures have been proposed as a potential solution [2]. Fault tolerant architectures are commonly used to tolerate on-line faults, *i.e.* faults that appear during the normal functioning of the system irrespective of their transient or permanent nature [3]. In the near future, fault tolerant architectures could also be used to tolerate permanent defects due to an imperfect manufacturing process.

For example, in the case of memories, fault tolerance architectures are currently used to tolerate manufacturing defects [4]. Extra core-cells are used as spare elements in order to substitute the faulty core-cells indentified during test application. Nevertheless, fault tolerance techniques used in regular array, such as memories, cannot be reused for random logic blocks.

This paper analyzes the potential interest of using fault tolerant architectures to tolerate manufacturing defects in the case of logic circuits. As a case study, we consider the well-known Triple Modular Redundancy (TMR) fault tolerant architecture in order to tolerate manufacturing defects while increasing the yield. We first determine the set of conditions to be satisfied in order to successfully resort to TMR to ramp-up the yield. Then, we propose a test pattern generation solution for TMR architectures in order to evaluate the condition that makes such architectures suitable for yield improvement. The first set of results show that the combinational depth of circuits has a strong impact on the effectiveness of the TMR architecture. We therefore propose to improve the tolerance of TMR architectures by partitioning circuits and adding voters on circuit's cuts. Results show that this improvement of the TMR architecture is very fruitful to improve its tolerance capability with a low overhead in term of silicon area (less than 3% for the biggest circuits).

1550-5774/08 $25.00 © 2008 IEEE
DOI 10.1109/DFT.2008.23

The remaining of the paper is organized as follows. In Section 2, we present the TMR architecture and we analyze its capability to tolerate manufacturing defects. Section 3 details the test strategy targeting the TMR architecture while Section 4 presents the experimental results on ISCAS and ITC benchmark circuits. Section 5 presents the improvement of the TMR architecture to make it able to tolerate more defects. Finally, concluding remarks are given in Section 6.

2. The TMR approach

2.1. Basic principle

A TMR structure is a fault tolerant architecture based on three identical modules performing the same function. Their inputs receiving the same data are tied together, and their outputs feed a majority voter (V) circuit as shown in Figure 1. As a result, the TMR architecture significantly reduces the error probability at the primary outputs of the system. A defective module propagating an erroneous value can be masked thanks to the presence of the two other fault-free modules. In the simplest structure, the voter is the weak point. If a fault appears in the voter, the TMR structure can be possibly faulty. To avoid this type of problem, the voter can be realized in software or with more robust design techniques [5].

2.2. How many defects can be tolerated?

Basically, the TMR architecture can tolerate one defect but, in practice, it can tolerate more than one defect. In fact, if there are two defects, the TMR can function properly depending on the nature and the location of the defects. Two defects are simply tolerated by the structure if their induced errors cannot simultaneously drive the majority voter.

To be not tolerated, two defects must be located in two different modules and then propagate an error towards identical outputs on each module.

In Figure 2, two examples are shown with the same pattern feeding the three modules. The voter has been omitted. Each defect is modeled as a stuck-at fault (f_1 and f_2 respectively). In the case of Figure 2.a, f_1 is propagated towards O1 in the first module and f_2 is propagated towards O2 in the second module. The majority voter receives two correct values and one wrong value. The outputs of the TMR are therefore correct and, consequently, faults f_1 and f_2 are tolerated.

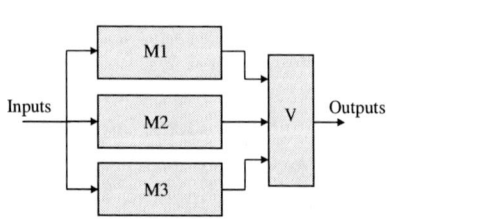

Figure 1. TMR principle

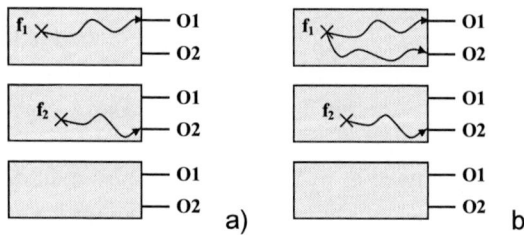

Figure 2. Two stuck-at faults are a) tolerated and b) not tolerated

In the case of Figure 2.b, f_1 is propagated towards O1 and O2 while f_2 is propagated towards O2. The voter receives one wrong value for O1 and two wrong values for O2. So, the value on the second output of the TMR is faulty. Consequently, faults f_1 and f_2 are not tolerated.

As a result, two faults are tolerated when there is no pattern able to propagate errors, coming from the two faults in different modules, toward identical outputs in each module. In the case of more than two defects, multiple defects can be handled by considering all the possible fault couples between them.

2.3. Yield Improvement with TMR architectures

In this sub-section, we investigate the interest of producing the TMR version of a circuit, instead of its single version, in order to tolerate manufacturing defects and consequently improve the yield as presented in [6]. We therefore analyze the conditions that have to be fulfilled to achieve benefits in implementing a TMR architecture instead of a simple non-tolerant architecture which is three times smaller.

For our analysis, we consider that the voter area is negligible (very small compared to the individual area of each of the three modules). Thus, if we triplicate a circuit to implement a TMR structure on a given silicon area S, we can have N TMR structures each having a yield equal to η_{TMR} (η_{TMR} x N fault-free TMR structures) or 3N regular circuits (without redundancy) each having a yield equal to η_c (η_c x 3N fault-free regular circuits). Then, a TMR architecture is worthwhile only if η_{TMR} x N > η_c x 3N with $\eta_{TMR} \leq 1$ and $\eta_c \leq 1$. Consequently:

$$\eta_{TMR} > 3 \times \eta_c \Rightarrow \eta_c \leq 1/3 \tag{1}$$

First of all, due to Eq. 1, it is important to notice that a TMR architecture can be profitable only if $\eta_c \leq 1/3$, i.e. when a low manufacturing yield is expected due to the use of aggressive nanotechnologies.

Let us now compute η_{TMR} and η_c by using the Poisson distribution. It would not be completely accurate to use the Poisson distribution for large circuits due to clustering effects on defects [7, 8]. But for a first and rough evaluation this is reasonable. The Poisson distribution is a discrete probability distribution that defines the probability that a number of manufacturing defects occur in a fixed area if these defects occur with a known probability.

Let X be the number of manufacturing defects. Let λ be the average number of expected defects for a given silicon area. Then, $\lambda = n \times p$ with n being the number of logic gates (or transistors) and p the average defect rate of a gate (or a transistor). Let $P\{X = k\}$ be the probability that the structure has k manufacturing defects. If n is high and p is low, the binomial distribution becomes the Poisson distribution:

$$P\{X = k\} = e^{-\lambda} \times \frac{\lambda^k}{k!} \tag{2}$$

The yield of the circuit without redundancy is denoted as η_c. The presence of a fault makes the entire system faulty. So, η_c is the probability that there is no defect inside the circuit:

$$\eta_c = \underbrace{P\{X = 0\} = e^{-\lambda_c} \times \frac{(\lambda_c)^0}{0!}}_{\text{Probability that there is no defect}} \Leftrightarrow \eta_c = e^{-\lambda_c} \tag{3}$$

Let R be the probability that two defects are tolerated (see Section 2.1). The yield of the TMR structure η_{TMR} is thus given by:

$$\eta_{TMR} = \underbrace{P\{X = 0\}}_{\text{Probability that there is no defect}} + \underbrace{P\{X = 1\}}_{\text{Probability that there is 1 defect}} + R \times \overbrace{P\{X = 2\}}^{\text{Probability that there are 2 defects}} + R^3 \times \underbrace{P\{X = 3\}}_{\text{Probability that there are 3 defects}} + \ldots$$

n defects are equivalent to C_n^2 couples of defects

$$\Leftrightarrow \eta_{TMR} = e^{-\lambda_{TMR}} \times \left(1 + \lambda_{TMR} + R \frac{(\lambda_{TMR})^2}{2!} + R^3 \frac{(\lambda_{TMR})^3}{3!} + \ldots \right)$$

There are three times more gates (or transistors) into a TMR architecture than into a non-redundant circuit. So, by substituting λ_{TMR} by $3\lambda_c$ and with $\eta_c = e^{-\lambda_c} \Rightarrow \lambda_c = -\ln \eta_c$, we obtain:

$$\eta_{TMR} = e^{3\ln\eta_c} \times \left(1 - 3\ln\eta_c + \sum_{i=2}^{\infty} R^{C_i^2} \times \frac{(-3\ln\eta_c)^i}{i!} \right)$$ (4)

A TMR architecture will improve the resulting yield if $\eta_{TMR} > 3\eta_c$ ($\eta_c \leq 1/3$). Figure 3 gives η_{TMR} as a function of η_c for different values of the R probability. The bold dotted line represents the condition $\eta_{TMR} > 3\eta_c$. From Figure 3, it appears that the condition can only be satisfied starting from values of R greater than 92.58% (percentage of tolerated defects) and for a yield η_c lower than 33.33%.

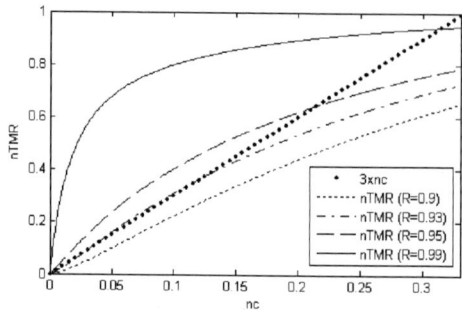

Figure 3. η_{TMR} **related to** η_c

To summarize, implementing a TMR architecture for a given circuit can improve the yield if i) the yield of the non-tolerant circuit is lower than 33.33% and ii) the percentage of tolerated fault pairs using a TMR architecture is greater than 92.58%. A question is still open concerning the meaning of 92.58% value of R. In order to answer this question we will investigate test issues related to TMR architectures.

3. The test of TMR architectures

Testing a TMR architecture depends on its final use. In a classical way of use, the goal is to tolerate potential on-line defects (permanent and/or transient). Every module has to be fault-free after manufacturing. Due to the intrinsic impossibility to test single stuck-at fault, the architecture has to be modified during test [3] as shown in Figure 4.a. In this modified architecture, redundancy is removed and each single stuck-at fault becomes testable. Each module is individually tested.

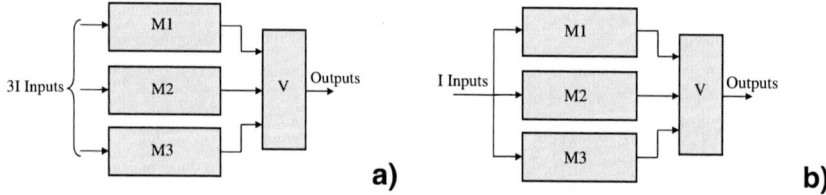

**Figure 4. Test of a) on-line fault tolerant structure
and b) manufacturing fault tolerant structure**

In the proposed way of use, the goal is to increase the yield by tolerating permanent defects due to an imperfect manufacturing process. In this case, the test consists in testing globally the TMR structure in order to determine what are the couples of defects which are tolerated or not (determination of the R value). In this case, the architecture is not modified for test purpose like the one presented in Figure 4.b. Lastly, we remind that the TMR basically tolerates one single defect and optionally two or more defects as discussed previously. For example, if several defects are located in the same module, the TMR still works properly.

To summarize, in the first approach, the question to be answered during the test of the TMR is: *"Are there one or more manufacturing defects in each module?"*. In the second approach, the question becomes: *"Does the circuit pass the test despite the presence of one or more manufacturing defects?"*.

3.1. Fault Model

An ATPG based on single stuck-at fault model generates patterns able to detect all single stuck-at faults. Such patterns are not efficient in the context of multiple stuck-at faults occurring in a redundant TMR architecture. Since the single stuck-at fault is by definition tolerated (untestable) in the TMR architecture, a different fault model has to be considered. This fault model must be testable on primary outputs of the TMR, and also be representative of actual manufacturing defects. We have seen in Figure 2 that a couple of stuck-at faults can be observed (not tolerated) on primary outputs and also can be testable.

In the following, we refer to a couple of stuck-at fault as fault pair [2]. If there are two stuck-at faults in the structure, these two faults are a fault pair. If there are more than two stuck-at faults, these multiple stuck-at faults can be gathered in fault pairs (under the assumption that no masking effect occurs between faults). For example, three stuck-at faults f_1, f_2, and f_3 are equivalent to three fault pairs $\{f_1, f_2\}$, $\{f_1, f_3\}$ and $\{f_2, f_3\}$. In general, n stuck-at faults are equivalent to C_n^2 fault pairs.

3.2. Size of the fault list

Let us consider that each module has n single stuck-at faults. The whole number of stuck-at faults in the three modules is 3n. As $\{f_1, f_2\} = \{f_2, f_1\}$, the total number ψ of fault pairs is:

$$\Psi = C_{3n}^2 = \frac{3n!}{(3n-2)! \times 2!} = \frac{9n^2 - 3n}{2} \tag{5}$$

We have now to consider when the percentage of untestable (tolerated) fault pairs is greater than 92.58% to determine if the TMR architecture is worthwhile for yield improvement. So, we simply generate the test patterns able to detect all the testable fault pairs. The remaining fault pairs are the untestable ones. Since ψ is quadratic to n, the use of a classical ATPG to detect the entire fault pair set will be unfeasible due to a huge CPU time. In the next section, we show how to handle this particular point considering ISCAS and ITC benchmark circuits as case studies.

4. ISCAS and ITC Analysis

4.1. ATPG Procedure

For the analysis, both ISCAS and ITC (combinational part only) benchmark circuits are used to implement TMR architectures by simply cloning each circuit three times. In order to determine the convenience of using the TMR architecture to handle manufacturing defects for these benchmark circuits, we have first to determine the untestable fault pairs. An ATPG targeting the fault pair model has to be run. Since the goal of this work was not to develop an ATPG for fault pairs, we have adapted an ATPG tool targeting single stuck-at faults to make it able to test fault pairs.

Considering that a fault pair is composed of two stuck-at faults (f_1 and f_2), the proposed approach is to inject a permanent fault in one module of the TMR architecture (f_1) by modifying the 'netlist'. Then, an ATPG is run to test all stuck-at faults in presence of the permanent one. This process is repeated until we have injected all stuck-at faults in one module.

The simplicity of this approach is obtained at the cost of high simulation time since n (number of single stuck-at faults in one module) full ATPG runs are needed. To decrease this drawback, we have first reduced the fault pair list to be handled as follows:

- When there is only one faulty module, the outputs of this module are masked by the voter. Thus, all fault pairs which impact only one module are structurally untestable. These fault pairs can be removed from the ATPG fault list.
- Symmetries of the TMR architectures are used to reduce the fault pair list of the ATPG. As each module's inputs are tied together during test, fault pairs are equivalent when their two stuck-at faults have the same location in two different modules. Therefore, equivalent fault pairs are removed from the ATPG fault list.

With the help of these reductions, it can be demonstrated that the size of the fault pair list becomes $(n^2 + n)/2$. The list of fault pairs can be further reduced by determining and removing the fault pairs which are structurally untestable. Let us consider the fault pair $\{f_1, f_2\}$ and their output cones (list of outputs where the fault effect may be propagated) $\{\phi_1, \phi_2\}$. Due to the presence of the voter, if $\phi_1 \cap \phi_2 = \emptyset$, the fault pair $\{f_1, f_2\}$ is structurally untestable and therefore, can be removed from the fault pair list. In practical cases, the number of structurally untestable fault pairs is quite large leading to a huge improvement of the overall ATPG performance. For example, the percentage of structurally untestable fault pairs is 82.86% for circuit c7552, 76.16% for circuit b05 and 88.01% for circuit b13.

4.2. Results

Results are reported in Table 1. The first column lists the circuit name. Second and third columns show the number of stuck-at faults in one module and the number of equivalent fault pairs in the TMR structure respectively. The number of stuck-at faults in one module of the TMR structure varies from 155 (b06) to 7438 (c7552). Consequently, the total number of fault pairs can be very high (over 249 million for c7552).

Circuits	# of stuck-at faults in one module	# fault pairs in TMR	# fault pairs in the ATPG fault list	Reduced ATPG fault list factor	Overall results			
					% untestable fault pairs (R)	% detected fault pairs	Fault efficiency	# patterns
c880	886	3.53M	133k	96.2%	78.41%	19.86%	98.28%	1896
c1908	1812	14.8M	1.30M	91.2%	56.42%	36.43%	92.85%	3852
c2670	2852	36.6M	1.87M	94.9%	75.95%	17.51%	93.46%	4987
c3540	3438	53.2M	4.91M	90.8%	54.09%	23.75%	77.84%	1756
c5315	4970	111M	3.44M	96.9%	93.20%	3.75%	96.95%	642
c6288	6250	176M	18.2M	89.7%	38.02%	54.79%	92.81%	4532
c7552	7438	249M	7.40M	97.0%	84.92%	11.71%	96.63%	8745
b03	382	656k	291k	95.6%	87.93%	12.06%	99.99%	114
b04	1477	9.81M	535k	94.5%	84.32%	12.70%	97.02%	3390
b05	2553	29.3M	1.17M	96.0%	88.65%	8.98%	97.63%	4699
b06	155	108k	7.73k	92.8%	87.51%	12.49%	100%	16
b07	1120	5.64M	399k	92.9%	81.91%	16.00%	97.91%	1635
b08	439	867k	59.1k	93.2%	89.09%	10.78%	99.87%	132
b09	417	782k	57.3k	92.7%	83.07%	16.93%	100%	211
b10	468	985k	63.5k	93.6%	89.39%	10.60%	99.99%	149
b11	1308	7.70M	703k	90.9%	74.50%	21.16%	95.66%	1551
b12	2777	34.7M	857k	97.5%	95.45%	4.09%	99.54%	1274
b13	835	3.14M	59.6k	98.1%	96.94%	3.06%	100%	167

Table 1. ATPG results

The fourth column shows the number of fault pairs in the ATPG fault list. The reduced ATPG fault list factor (in percentage) is given in the fifth column. For example, for the c880 the ATPG can neglect 96.2% of all the fault pairs.

The sixth column summarizes the results after ATPG. The first sub-column presents the percentage of untestable fault pairs. The second one shows the percentage of detected fault pairs. The fault efficiency, defined as the sum between the percentage of untestable fault pairs and the percentage of detected fault pairs, is given in the third sub-column. The last sub-column gives the number of test patterns generated.

After ATPG runs, we know the percentage of untestable fault pairs in every circuit. This percentage has to be greater than 92.58%. From results reported in Table 1, we report that three circuits only have this characteristic (93.2% for c5315, 95.45% for b12, 96.94% for b13). Moreover, it is important to notice that the percentage of untestable fault pairs is higher for ITC'99 circuits than for ISCAS'85 circuits. The smallest combinational depth of ITC'99 circuits could explain this difference.

We have also to notice that the percentage of untestable faults may be higher than shown in Table 1. In fact, when the fault efficiency is not 100%, we do not know if some among the fault pairs left are testable or not.

5. Improvement of the TMR architecture

As shown in the previous section, modifying a circuit using the TMR architecture is suitable only for few circuits. This result comes out from the fact that the TMR architecture does not tolerate enough defects to compensate the overhead of silicon area. Consequently, we have to improve the tolerance of the TMR to make it suitable for yield improvement purpose.

Results reported in Table 1 have shown that ITC'99 benchmark circuits have a probability R (tolerated faults) greater than ISCAS'85 benchmark circuits. ITC circuits have, most of the time, a lower combinational depth compared to ISCAS circuits. Actually, it appears that the lower the combinational depth, the greater is the probability to have an empty intersection of logic cones, and hence a higher tolerance capability (R probability) of the TMR architecture. Then, in the following sub-section we propose to modify the basic TMR architecture by partitioning the circuit in order to decrease the combinational depth.

5.1. Circuit partitioning

In order to reduce the combinational depth, each circuit is partitioned into two or three equivalent blocks so as to increase the tolerance of the TMR architecture. As shown in Figure 6, majority voters are added on circuit's edge cut.

An important feature of partitioning the circuit is that the tolerance of fault pairs increases when the number of partition increases as well. For example, in the case of double TMR (Figure 5.a), circuits (modules) are divided into two equivalent blocks. Each block operates independently and a manufacturing defect in the first block has no impact on the second block. In a basic TMR architecture, the percentage of untestable fault pairs is always greater than 33.33% as two stuck-at faults in the same module are untestable. In a double TMR architecture, if we consider that the first fault f_1 impacts M1', then the fault pair $\{f_1, f_2\}$ can be detected (not tolerated) if and only if f_2 is in M2' or M3'. Conversely, if f_2 is in M1', M1'', M2'' or M3'', the fault pair is necessarily tolerated due to the presence of the voters. Therefore, the percentage of tolerated fault pairs in a double TMR architecture is always greater than 66.66%.

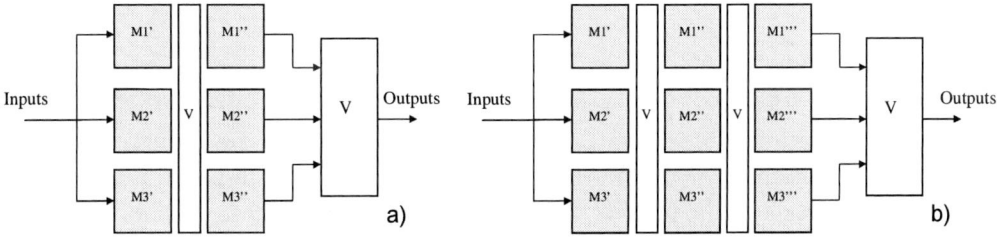

Figure 5. Improvement of the TMR architecture a) double and b) triple TMR architectures

In the case of a triple TMR architecture (see Figure 5.b), circuits (modules) are divided into three equivalent blocks. If f_1 impacts M1', the fault pair $\{f_1, f_2\}$ is necessarily tolerated if f_2 impacts M1', M1'', M2'', M3'', M1''', M2''' and M3'''. Consequently, the percentage of tolerated fault pairs is always higher than 77.77%. More formally, if circuits are partitioned into k blocks, then the percentage of tolerated fault pairs is always higher than 100x(3k-2)/3k.

With this technique, the tolerance of the TMR architecture increases with the increase of the number of partitions. The main drawback of this technique is the requirement of additional voters. In the next sub-section, we show that the overhead of silicon area due to additional voters still remains very low (less that 3% for large benchmark circuits).

The circuit partitioning consists in three steps. First, we transform the circuit into a hypergraph. A hypergraph is a generalization of a graph, where the set of edges is replaced by a set of hyperedges. A hyperedge extends the notion of an edge by allowing more than two vertices to be connected together. The vertices of the hypergraph can be used to represent the gates of the circuit, and the hyperedges can be used to represent the lines connecting these gates.

In a second step, we make the hypergraph partitioning. The proposed technique of circuit partitioning for TMR architecture improvement utilizes sh-METIS, a multilevel hypergraph partitioning algorithm based upon the multilevel paradigm [9].

The final step consists in adding voters in the place of hyper-cuts. The number of added voters is equal to number of hyper-cuts which is optimal using sh-METIS software.

5.2. Experimental results

Results are reported in Table 2. The first column lists the circuit name. The first part of the table (basic TMR) shows the number of voters (#voters). This number is equal to the number of module outputs. The second sub-column recalls the R probability shown in Table 1.

Circuits	Basic TMR		Double TMR			Triple TMR		
	#voters	R (%)	#voters	R (%)	AO (%)	#voters	R (%)	AO (%)
c880	26	78.41%	40	92.58%	3.02%	39	96.47%	2.81%
c1908	25	56.42%	53	89.49%	3.03%	68	96.50%	4.65%
c2670	140	75.95%	160	92.46%	1.40%	179	93.90%	2.72%
c3540	22	54.09%	56	88.36%	1.90%	95	93.78%	4.09%
c5315	123	93.20%	149	94.30%	1.05%	161	96.38%	1.53%
c6288	32	38.02%	49	76.35%	0.58%	64	84.14%	1.10%
c7552	108	84.92%	134	94.40%	0.69%	158	96.22%	1.32%
b03	34	87.93%	43	95.07%	4.25%	47	96.75%	6.14%
b04	74	84.32%	99	93.19%	3.23%	109	95.53%	4.52%
b05	60	88.65%	73	89.09%	1.08%	91	93.30%	2.57%
b06	15	87.51%	22	93.68%	8.64%	26	96.26%	13.58%
b07	57	81.91%	80	91.23%	4.58%	84	94.11%	5.38%
b08	25	89.09%	41	95.28%	7.25%	51	96.75%	11.79%
b09	29	83.07%	40	95.15%	4.96%	44	97.13%	6.77%
b10	23	89.39%	36	94.92%	5.28%	50	97.34%	10.97%
b11	37	74.50%	76	87.75%	6.14%	94	93.91%	8.97%
b12	127	95.45%	139	96.27%	0.85%	167	97.79%	2.84%
b13	63	96.94%	66	97.36%	0.67%	70	97.83%	1.56%

Table 2: Comparisons of basic TMR with double and triple TMR architectures

The next two parts of Table 3 show the double TMR (modules are partitioned into two equivalent blocks) and the triple TMR (modules are partitioned into three equivalent blocks). The first sub-columns show the number of voters in these architectures. This number is equal to the number of module outputs plus the number of hyper-cuts related to the circuit partitioning. The probability R of each circuit has been computed again and is reported in the second sub-columns of each part of Table 2. The last sub-columns show the area overhead

(AO%) of a double/triple TMR architecture compared to a basic TMR architecture (expressed in terms of equivalent gates).

We have shown in Section 2.3 that a TMR architecture can be used for yield improvement if the probability R is greater than 92.58%. With a basic TMR architecture only three circuits have this property. With the results shown in Table 2, we can observe that this number is now equal to 11 (for the double TMR architecture) and 17 (for the triple TMR architecture).

Concerning the area overhead due to the additional voters, we can see that this percentage decreases with the size of the circuit and is almost negligible for the biggest benchmark circuits (<3%). Consequently, the modification of TMR architectures based on circuit partitioning is really of interest for yield improvement purpose.

6. Conclusion

This paper analyses the use of fault-tolerant circuits for yield improvement purpose. It is conceivable to imagine a lower and lower yield for future technologies. Consequently, fault-tolerant circuits could be used to tolerate manufacturing defects. We want to accept circuits that have a correct functioning in spite of the presence of manufacturing defects.

TMR architectures have been studied in this paper. We have computed the necessary conditions that make TMR architectures more attractive compared to non-tolerant circuits. These conditions are i) the yield of the non-tolerant circuit has to be lower than 33.33% and ii) the percentage of tolerated fault pairs using a TMR architecture must be greater than 92.58%. We have performed a set of experiments using ISCAS'85 and ITC'99 benchmark circuits. For each one, an ATPG has been run with a fault pair model. For three circuits, the percentage of untestable fault pairs is greater than 92.58%. Thus, despite a larger silicon area (three times larger) and despite an increasing number of defects (on average three times higher), it is suitable to implement the circuit using the TMR architecture to increase the yield for these circuits. Then, we have proposed to partition the circuit to improve the tolerance of the TMR architecture. This has been done by using hypergraph partitioning with the help of sh-METIS; a software package for partitioning large hyper-graph with a minimal number of hyper-cuts. Results have shown that this improvement of the TMR architecture is very fruitful to improve its tolerance capability with a low overhead in term of silicon area.

7. References

[1] International technology roadmap for semiconductors (ITRS), 2007 edition.
[2] L. Fang and M. S. Hsiao, "Bilateral Testing of Nano-scale Fault Tolerant Circuits", Proc. of IEEE Defect and Fault Tolerance in VLSI Systems, pp. 309-317, 2006.
[3] C. E. Stroud and A. E. Barbour, "Design for Testability and Test Generation for Static Redundancy System Level Fault Tolerant Circuits", Proc. of IEEE International Test Conference, pp. 812-818, 1989.
[4] I. Schanstra, "Semiconductor Manufacturing Process Monitoring using Built-in Self-test for Embedded Memories", Proc of IEEE International Test Conference, pp. 872-881, 1998.
[5] J. M. Cazeaux, D. Rossi and C. Metra, "New High Speed CMOS Self-Checking Voter", Proc. of IEEE International On-Line Testing Symposium, pp. 58-63, 2004.
[6] J. Vial, A. Bosio, P. Girard, C. Landrault, S. Pravossoudovitch and A. Virazel, "Yield Improvement, Fault-Tolerance to the Rescue?", to appear in Proc. of IEEE International On-Line Testing Symposium, 2008.
[7] Y. Gagnon et al., "Are Defect-Tolerant Circuits with Redundancy Really Cost Effective? Complete and Realistic Cost Model", Proc. of IEEE Defect and Fault Tolerance in VLSI Systems, pp. 157-165, 1997.
[8] P. de Gyvez, "Integrated Circuit Manufacturability", IEEE Press, 1999.
[9] G. Karypis et al., "Multilevel Hypergraph Partitioning: Applications in VLSI Domain", Technical Report, Department of Computer Science, University of Minnesota, 1998.

IEEE International Symposium on Defect and Fault Tolerance of VLSI Systems

Module Grouping for Defect Tolerance in Nanoscale Memory

Yoonjae Huh, Yoon-Hwa Choi

Department of Computer Engineering, Hongik University, Seoul, Korea
yhchoi@cs.hongik.ac.kr

Abstract

Designing a nanoscale memory system with defect rate as high as 10% poses a significant challenge. Redundancies at various levels have been employed to tolerate the high defect rates. Multiple crossbar modules that share the same address space can be used to build a simple and robust memory architecture to overcome the defects in the crossbar. In this paper, we presents a module grouping scheme for tolerating defects in a nanoscale memory composed of nano-modules. Redundancy at nano-module level with some degree of flexibility in assigning nano-modules is used to achieve defect tolerance. Computer simulation shows that the proposed scheme can construct a functioning memory with up to 45% reduction in the required number of nano-modules as compared to the existing simple redundancy scheme.

1. Introduction

Molecular electronics has drawn significant attention as a potential alternative to Si-based CMOS devices to fabricate nanoscale memories and logic devices [1][2]. Crossbar architecture to connect molecular switches in a grid [5] is known to be attractive in realizing such circuits. Nanoscale devices produced in a bottom-up manner, however, are expected to have high defect rates. Although the exact defect rate is unknown and could be improved, defect rate as high as 10% has been reported in [3]. Defect-tolerance techniques are thus required to deal with such high defect rates.

Several defect-tolerance techniques for crossbar-based molecular memory and logic arrays have been proposed in [6-10]. Most of the proposed techniques are based on defect-avoidance through reconfiguration with spares. In [9] spare rows and columns with address remapping are used to construct fully functional memories. A probabilistic model to design high yield defect-tolerant nanoscale memories has been presented in [11].

Defect tolerance using redundancy in memory modules, sharing the same address space, has been presented in [8]. It has been claimed to outperform the TMR with reliable majority voting. An improved space efficiency in defect map was achieved by the use of a Bloom filter [12], a randomized data structure with a small false positive probability. Some reductions in the number of nano-modules have been achieved by using augmented crossbars and address remapping [10].

In this paper, we present a module grouping scheme for defect tolerance in a redundant nanoscale memory, where a memory consists of groups of logical modules. Each group has multiple nano-modules to form the required number of logical modules. A small degree of flexibility in module grouping allows a significant reduction in the number of redundant nano-modules. Defect maps and all the supporting logic are assumed to be implemented in CMOS as seen in hybrid technology

1550-5774/08 $25.00 © 2008 IEEE
DOI 10.1109/DFT.2008.47

[13][14][15] to enhance the density and to overcome the technological difficulties in implementing molecular circuits.

The rest of the paper is organized as follows. In Section 2, a redundant nanoscale memory using multiple nano-modules sharing the same address space is briefly introduced. The proposed module grouping scheme for reducing crossbar area overhead is presented in Section 3. Section 4 estimates by computer simulation the memory configurability and the required number of nano-modules to build a functioning memory. Conclusion is given in Section 5.

2. A Redundant Nanoscale Memory

A simple technique for achieving defect tolerance in nanoscale memory is to employ multiple nano-modules that share the same address space as shown in Fig. 1, where Bloom filters are used to build defect maps. The memory system in the figure has 16 modules, each of which consists of three major functional blocks: k nano-modules $N_{1\sim k}$ of $2^m \times 2^m$ cells, k Bloom filters $BF_{1\sim k}$, and a selection logic. The k nano-modules are used to construct a $2^m \times 2^m$ memory module. Each nano-module has its own defect map, implemented in its associated Bloom filter. In Fig. 1, BF_i stores the defect map of the nano-module N_i. In other words, the locations of defective devices in the nano-module N_i are programmed in the Bloom filter BF_i. Each cell in the nano-module can be addressed by the offset of $2m$ bits and a defect-free location will be selected by a combined effort of the k Bloom filters and the selection logic.

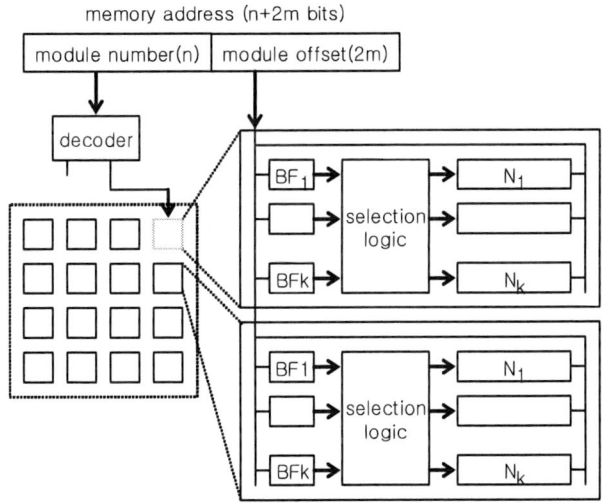

Figure 1. A redundant nanoscale memory

For a given address of $n+2m$ bits the decoder will select one of the 2^n modules and the corresponding k nano-modules $N_{1\sim k}$ will be accessed with the $2m$-bit offset. The k Bloom filters of the selected module are then queried to see if the corresponding locations are defective. If at least one of them shows a 0 (i.e., defect-free), the defects (if exist) can be tolerated and one of the nano-modules with result=0 will be selected by the selection logic and accessed with the $2m$-bit address. This simple redundancy scheme assigns each nano-module to exactly one memory module, requiring an increased number of nano-modules to construct a functioning memory. Although each nano-module has its own Bloom filter in Fig. 1, the hashing units of the Bloom filters can be shared in actual implementation. Detailed structure of the Bloom filters, however, has nothing to do with the memory configurability and the number of nano-modules required, and thus it is beyond the scope of this paper.

3. Module Grouping for Defect Tolerance

In order to improve the utilization of the nano-modules and thus to reduce the number of nano-modules required, we propose a module grouping scheme (MG) as illustrated in Fig. 2, where two modules in Fig. 1 are grouped into one and r nano-modules in each group are used to logically configure two modules for the nanoscale memory. To generalize the module grouping, we say that a memory is composed of multiple groups of modules of the same size and r nano-modules in each group are used to logically construct 2^b memory modules, where b is a small positive integer. In other words, the number of modules to form a group is determined by b. Fig. 2 illustrates the module grouping for b=1. If b is equal to 2, each group will provide 2^b (=4) memory modules.

For an address of $n+2m$ bits, the leftmost n-b bits are the group number, the rightmost $2m$ bits are the module offset, and the middle b bits are the index to select a module in a group. The (n-b)-bit group number will be decoded to select a group (in Fig. 2, the last one in the first row), the corresponding r Bloom filters $BF_{1\sim r}$ and r nano-modules $N_{1\sim r}$ will be accessed with the $2m$-bit offset. The r Bloom filters will first be queried with the offset to find a nano-module to access. The resulting output is an r-bit vector, Z=($z_1, z_2, ..., z_r$). The b-bit index will be used to select the nano-module by finding the location of the corresponding 0. If b=2 and the b-bit index is 01, for example, the nano-module N_i, where z_i is the second 0 in the r-bit vector, will be selected.

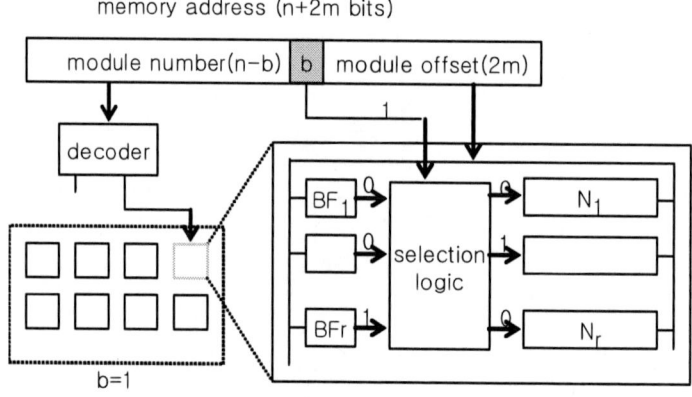

Figure 2. Module grouping for defect tolerance

In Fig 2, the outputs of the r Bloom filters, Z, for the $2m$-bit address (offset) are 00...1, indicating that at least two nano-modules are defect-free at the given address. If the index is 1, the second nano-module N_2 will be selected and accessed, as marked with a 1 in the figure.

Fig. 3 depicts the steps in accessing a memory location for a memory consisting of groups of modules of the same size. Depending on the implementation of the structure, nano-modules can be accessed while the selection logic selects the nano-module to access (i.e., step 3 and step 4 simultaneously), although its usefulness needs to be further investigated.

In the proposed defect tolerance scheme, only the defects in the nano-modules are under consideration. All the Bloom filters and selection logic are to be implemented in the CMOS layer, and they thus are assumed to be highly reliable.

Due to the flexibility in nano-module assignment module grouping can reduce the number of nano-modules required to achieve defect tolerance. As b increases, the number of nano-modules

Given an address of $n+2m$ bits with r Bloom filters $BF_{1\sim r}$ and r nano-modules $N_{1\sim r}$
Step 1: Decode the $(n\text{-}b)$-bit module number and select the corresponding module
Step 2: Compute the bit vector Z by performing $BF_i(\text{offset})$ in parallel for $1 \le i \le r$
Step 3: Select N_j using the index of the middle b bits and bit vector Z as follows
   If z_j is the kth zero in Z from left, select N_j
Step 4: Access N_j with the offset of $2m$ bits

Figure 3. Accessing memory with an address of $n+2m$ bits

decreases, with an increase in the complexity of the selection logic. Since b is expected to be small, the selection logic is simple enough to be practically useful. In addition, it can be shared by multiple groups as long as only one of the groups is active at a time. The circuits for Bloom filters are the same regardless of the value of b. Hence they are not discussed in this paper. In the next section, we will estimate by computer simulation the number of nano-modules required for the proposed module grouping scheme to achieve defect tolerance.

4. Memory Configurability and Number of Nano-Modules Required

As the performance metrics we use the memory configurability and the number of nano-modules required, where the memory configurability is defined as the probability that at least one defect-free cell exists for each address in the memory address space.

Computer simulation is performed to estimate the memory configurability and the number of nano-modules of $2^m \times 2^m$ required to construct a functioning memory for various values of m and defect probability p. For comparison purposes, we have also considered the simple redundancy scheme (SR) presented in Section 2.

In the simulation we assume that all the defects are independent and each memory cell is defective with probabilities 0.01, 0.03, 0.05, and 0.1, respectively. In addition, defects are assumed to occur at crosspoints (i.e., cells) and defects affecting an entire row and/or column are not taken into account. To obtain memory configurability 2×10^4 sample defect patterns are used. The require number of nano-modules to achieve a near perfect configurability (≈ 1) in SR (simple redundancy scheme) for various values of p and m are shown in Table 1. For relatively small p and m, the overhead might be acceptable. For the defect rates under consideration in this paper, however, some reduction in the value of k would be desirable as long as the simplicity of the structure can be maintained.

Table 1. The number of nano-modules k in SR to achieve a near perfect configurability of $2^m \times 2^m$ memory for various values of p and m

p	$m=4$	$m=5$	$m=6$	$m=7$	$m=8$	$m=9$
0.01	3	4	4	4	5	5
0.03	4	5	5	5	6	6
0.05	5	5	6	6	7	7
0.10	6	7	7	8	8	9

In our module grouping scheme (MG) for improving nano-module utilization, 2^b modules are logically constructed within a group by sharing nano-modules in the group. The configurability of

2^b modules (in a group) for various values of p, m, and r when $b=1$ are shown in Table 2. Some increase in r is necessary to maintain near perfect configurability as p and m increase in the given ranges.

Table 2. Configurability for various values of p, m and r in constructing two modules of $2^m \times 2^m$ for MG. (a) $p=0.01$, (b) $p=0.03$, (c) $p=0.05$, (d) $p=0.10$

m	$r=4$	5	6
4	0.999	1.000	1.000
6	0.987	1.000	1.000
8	0.771	0.996	1.000

m	$r=4$	5	6	7	8
4	0.973	0.999	1.000	1.000	1.000
6	0.652	0.984	1.000	1.000	1.000
8	0.001	0.770	0.990	0.999	1.000

m	$r=4$	5	6	7	8	9
4	0.891	0.992	1.000	1.000	1.000	1.000
6	0.138	0.887	0.993	1.000	1.000	1.000
8	0.000	0.140	0.889	0.992	1.000	1.000

m	$k=4$	5	6	7	8	9	10	11
4	0.387	0.893	0.988	0.998	1.000	1.000	1.000	1.000
6	0.000	0.149	0.797	0.976	0.998	0.999	1.000	1.000
8	0.000	0.000	0.027	0.654	0.954	0.994	0.999	1.000

A notable improvement in the number of nano-modules required has been observed even for large m and p. Based on the memory configurabilities of SR and MG we can construct Table 3, where the numbers of nano-modules required for SR and MG when $b=1$ are compared. The reductions in the number of nano-modules for the given parameter values are in the range of 16% to 45%.

Table 3. The number of nano-modules required to construct two modules of $2^m \times 2^m$ each in SR and MG for various values of p and m when $b=1$

p	$m=4$ SR	$m=4$ MG	$m=5$ SR	$m=5$ MG	$m=6$ SR	$m=6$ MG	$m=7$ SR	$m=7$ MG	$m=8$ SR	$m=8$ MG	$m=9$ SR	$m=9$ MG
0.01	6	5	8	5	8	5	8	6	10	6	10	6
0.03	8	6	10	6	10	7	10	7	12	8	12	8
0.05	10	6	10	7	12	7	12	8	14	8	14	9
0.10	12	8	14	9	14	10	16	10	16	11	18	11

We have also performed simulation to estimate the required number of nano-modules to build a group of modules when $b=2$ (i.e., four modules form a group). The configurabilities for various values of m and r when $p=0.05$ are shown in Fig. 4. When the module size m increases, the number of nano-modules for near perfect configurability increases rather slowly unlike that for SR.

As expected, with the small increase in b further reductions in the number of nano-modules have readily been observed. As m increases, the performance of SR becomes poor due to the lack of flexibility in nano-module assignment.

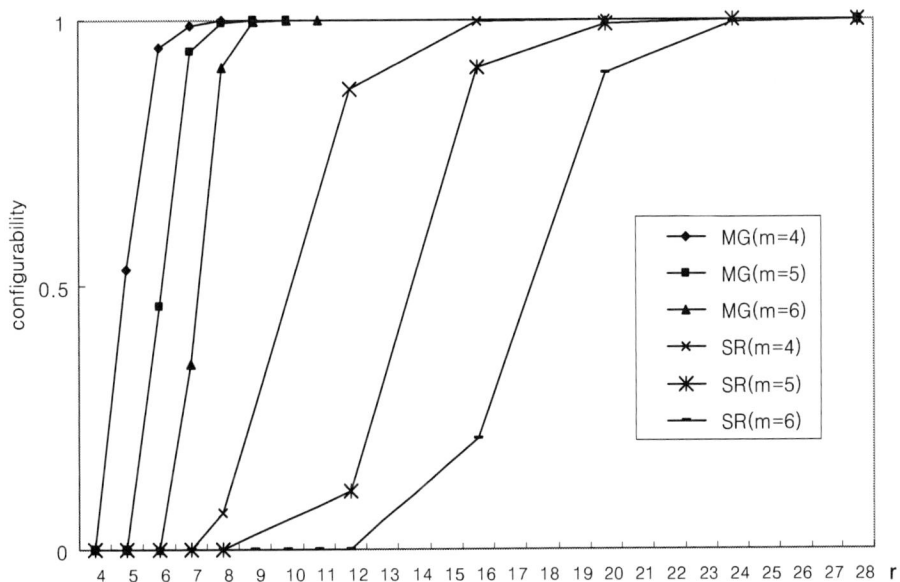

Figure 4. Configurabilities of MR and SR for p=0.05 and b=2

The number of nano-modules required for b=2 is shown in Table 4. The reductions in the number of nano-modules are in the range of 40% to 60% for the chosen values of p and m. As can be seen in the table, the reduction increases notably as b changes from 1 to 2. We, however, suggest to choose $b = 1$ to gain some flexibility while maintaining simplicity in the selection logic, although the selection logic can possibly be shared by multiple groups.

Table 4. The number of nano-modules required to construct four modules of $2^m \times 2^m$ each in SR, $MG(b$=1$)$, $MG(b$=2$)$ for various values of p and m

	m=4			m=6			m=8		
p	SR	MG(b=1)	MG(b=2)	SR	MG(b=1)	MG(b=2)	SR	MG(b=1)	MG(b=2)
0.01	12	10	7	16	10	8	20	12	8
0.03	16	12	8	20	12	9	24	16	10
0.05	20	12	9	24	14	10	28	16	11
0.10	24	16	11	28	20	13	32	22	14

The simulation results simply show the fact that some degree of flexibility in module assignment could reduce the number of nano-modules required. An extensive study, however, is necessary for the defect tolerance scheme to be practically useful since all other important aspects have not been considered. Any further reduction in the number of nano-modules could possibly be made by adding spare rows and columns of memory cells to each nano-module as long as the address decoding logic can support the required flexibility in assigning addresses to memory cells. Some additional reduction in the number of nano-modules will make the selection logic simpler.

Simple redundancy scheme(SR), where at least two defect-free locations are available for each address in the memory address space, can be useful in detecting and tolerating faults during normal operation. Table 3 in the previous section shows that a small increase in the number of nano-modules will provide at least two defect-free cells for each memory address. For a given address of $m+2n$ bits, the selected module after decoding the module number will be accessed with the offset of $2n$ bits. The Bloom filters associated with the module, in that case, will generate at least two 0's regardless of the offset value, allowing parallel write/read. Hence on read at least two nano-modules will provide the requested data. They must be identical unless there is a fault in the corresponding memory cells. Errors generated will be detected by a simple comparison during normal operation. In addition, run-time faults may be tolerated if the faulty cells can be identified.

5. Conclusion

In this paper, we have presented a module grouping scheme for tolerating defects in a redundant nanoscale memory composed of nano-modules. Module grouping has been proposed to give flexibility in assigning nano-modules and thus to enhance the utilization of nano-modules. The resulting reductions in the number of nano-modules to form a fully functional memory module have been up to 45% for the chosen values of defect rate and module size, even when the group size is small. A small increase in the number of nano-modules for simple redundancy scheme can realize duplication of data by writing/reading the data simultaneously, resulting in enhanced error detection and fault tolerance during normal operation. Although we have estimated the crossbar area by counting the number of nano-modules employed, interfacing CMOS and nanoelectronics has to be investigated to obtain more realistic estimates.

References

[1] J.R. Heath and M.A. Ratner, "Molecular electronics," Physics Today, May 2003, pp.43-49.

[2] C.P. Collier et al., "Electronically configurable molecular-based logic gates," Science, 285, pp. 391-394, 1999.

[3] M. Mishra and S.C. Goldstein, "Defect tolerance at the end of the roadmap," Int. Test Conf., 2003, pp. 1201-1211.

[4] Y. Chen el al., "Nanoscale molecular-switch crossbar circuits", *Nanotechnology*, 14(2003), pp. 462-468.

[5] P.J. Kuekes and R.S. Williams, "Defect-tolerant molecular electronics," IEEE Int. Symposium on Circuits and Systems, Phoenix-Scottdale, AZ, May 2002, pp. II-42-44, vol 2.

[6] W. Rao, A. Orailoglu, and R. Karri, "Logic level fault tolerance approaches targeting nano-electronic PLAs," IEEE DATE, April 2007, pp.1-5.

[7] T. Hogg, G. Snider, "Defect-tolerant logic with nanoscale crossbar circuits," JETTA, vol.23, issue2-3, June 2007, pp. 117-129.

[8] G. Wang, W. Gong and R. Kastner, "On the use of Bloom filters for defect maps in nanocomputing," IEEE ICCAD, Nov. 2006, pp.743-746.

[9] M.-H. Lee, Y.K. Kim and Y.-H. Choi, "A defect-tolerant memory architecture for molecular electronics," IEEE Trans. Nanotechnology, 2004, pp. 152-157.

[10] Y.-H. Choi and M.-H. Lee, "A defect-tolerant molecular-based memory architecture," 22nd Int. Symp. DFTS, 2007, pp. 143-151.

[11] G. Venkatasubramanian, P.O. Boykin, and R.J. Figueiredo, "Design of high-yield defect-toleratn self-assembled nanoscale memories," IEEE Nanoarch, 2007, pp. 77-84.

[12] B. Bloom, "Space/time tradeoffs in hash coding with allowable errors," Communications of the ACM 13:7, 1970, pp. 422-426.

[13] M.M. Ziegler and M.R. Stan, "The CMOS/nano interface from a circuit perspective," Int. Symp. Circuits and Systems, May 2003, pp. 904-907.

[14] M.M. Ziegler and M.R. Stan, "CMOS/Nano codesign for corssbar-based molecular electronic systems," IEEE Trans. Nanotechnology, Vol. 2, No.4, Dec. 2003, pp. 217-230.

[15] X. Ma, D.B. Strukov, J.H. Lee, K.K. Likharev, "Afterlife for silicon: CMOS circuit architectures," IEEE Nano 2005, pp. 175-178.

IEEE International Symposium on Defect and Fault Tolerance of VLSI Systems

Coping with obsolescence of processor cores in critical applications

F. Abate, M. Violante

Politecnico di Torino

Torino, Italy

{francesco.abate, massimo.violante}@polito.it

Abstract*

Critical applications are more and more relying on electronic components to provide the services they are designed for. The obsolescence of such electronic components has already been recognized as a critical issue that must be properly addressed. Among the different possibilities, FPGA-based emulation of obsolete digital components seems particularly interesting. This paper proposes an automatic approach for customizing the processor cores before they are implemented on the FPGA of choice, by removing the unnecessary instructions. The benefits stemming from this approach are cost savings thanks to the possibility of adopting smaller, less expensive, FPGA devices, the possibility of integrating both the processor and its companion chip on the same FPGA device, and the possibility of adopting hardware redundancy without incurring in high overheads. The approach has been evaluated on a case study, showing that a customized and hardened processor obtained from our design flow requires the nearly the same area of the original, un-hardened, processor.

1. Introduction

Critical applications are more and more relying on electronic components to provide the services they are designed for. The obsolescence of such electronic components has already been recognized as a critical issue that must be properly addressed. As an example taken from [1] we can consider a stealth bomber that first flew in 1989: by 1996, significant components of the aircraft's defensive management system were obsolete. Repairing the system entailed either redesigning a few circuit boards and replacing other obsolete integrated circuits for US $21 million, or spending $54 million for replacing the whole system.

The problem of component obsolescence is related to the changes that semiconductor industries have experienced in the recent years. 40 years ago the expected lifetime of electronic components was between 20 and 25 years, today is less than 2 years [1]. As a result, critical systems developed in the recent years are likely to become operational after an intense and long validation process when most of the electronic components they employ are already out of the market. Therefore, proper strategies have to be developed to set available the electronic parts needed to maintain the systems fully operations even if they are no longer available on the market.

Different solutions can be adopted for coping with the obsolescence issue [1]: stocking parts to be used when replacement is needed, having the needed electronic parts re-manufactured from the original design upon request, reverse engineer the obsolete components and manufacture new parts on more recent technologies, or redesign the electronic system where the obsolete part was used.

---

* Contact author: Francesco Abate, Dip. Automatica e Informatica, Politecnico di Torino, C.so Duca degli Abruzzi 24, 10129, Torino, ITALY, Email: francesco.abate@polito.it

1550-5774/08 $25.00 © 2008 IEEE
DOI 10.1109/DFT.2008.28

To reduce the costs of solutions, a new approach was introduced in [2] where the authors proposed to use Field Programmable Gate Arrays (FPGAs) to implement the behavior of the electronic component no longer available. The approach is particularly appealing as it is a general solution to the obsolesce problem of digital electronic parts: FPGAs can implement the behavior of practically any digital component, provided that a suitable model of the component's behavior is available. The concept already found its implementation: companies exist that prove Intellectual property (IP) cores under the form of Register Transfer Level (RTL) synthesizable models for processors and companion chips already disappeared from the market, but that are heavily used in critical systems. An example can be found in [3], where the RTL model of a complete 80186 system is set available for implementation on a Xilinx's Spartan 3 FPGA.

By replacing obsolete parts with IP cores mapped on FPGAs, program managers can save enormous amount of moneys as the systems have not to be redesigned, and the software running on the processor can be used without any modification by the FPGA-emulated processors. Despite its potential benefits, the introduction of FPGA-emulated processors to replace obsolete parts may have noticeable costs. Indeed, when critical systems are considered, resources must be spent to guarantee that the FPGA-emulated processors provide the same level of dependability of the obsolete processor it replaces. Obsolete processors based on old manufacturing technologies may indeed be insensitive to some phenomena that are affecting the most recent manufacturing technologies that are likely to be used in modern FPGAs. In particular, radiation-induced soft errors in FPGAs have to be carefully evaluated when selecting the FPGA technology to map the IP core that will replace the obsolete part [4].

When analyzing the problem of ionizing radiations, it is mandatory to first identify the radioactive environment where the system has to be deployed, so to define the type of radiation we have to consider. For the sake of this paper, we will focus on the atmospheric radiation environment, where neutrons are the primary source of soft errors [5].

When analyzing the FPGA technology, we can find three possibilities:
1. Commercial off the shelf (COTS) SRAM-based FPGAs, whose configuration memory and user memory can be affected by neutrons.
2. COTS Flash-based FPGAs, whose configuration memory is insensitive to radiations, while its user memory can be affected.
3. Fuse-Antifuse FPGAs that, being radiation-hardened, are insensitive to radiations.
4. Radiation-hardened SRAM-based FPGAs that, being radiation-hardened, are insensitive to radiations.

Due to their reasonable prices, COTS FPGAs are probably the best candidates for implementing IP cores replacing obsolete parts in atmospheric-based critical systems while radiation-hardened devices, being very expensive, are probably better suited for more aggressive radiation environments (like space).

When COTS FPGAs are selected for implementing processor IP cores to replace obsolete parts in atmospheric-based critical systems, resources must be spent for guaranteeing that neutrons-induced soft errors does not affect the functionalities of the FPGA-emulated processor.

A dependable system can be obtained resorting to hardware and/or software techniques; both approaches introduce some redundancies in the sense of hardware components or code sections that are not necessary to implement the desired functionality. Redundancy guarantees the system against fault effects, but its implies some drawbacks: higher hardware costs or performances loss. Section 2 describes these techniques in detail.

When processor IP cores are considered, innovative hardening solutions that selectively adopt redundancy can be exploited. For instance, a processor instruction set may offer a 32x32 multiply instructions, which is never used in the software the processor has to run. Before implementing the processor in the FPGA of choice, the processor IP can be simplified by removing the unused instructions, thus minimizing the amount of resources on which redundancy has to be applied.

Based on this observation, this paper proposes a new design flow that takes advantage of the customization property offered by IP processor cores to reduce the area overhead that redundancy techniques introduce. Thanks to this approach smaller, less expensive, devices can be used for implementing the FPGA-emulated processors, and both obsolete processors and companion chips can be implemented on the same device. The paper proposes an automatic design flow that, starting from an application source code, analyzes a processor's instructions set, removes all the unused instructions for the considered application, and applies hardware redundancy to those resources that are mandatory to implement the considered application. We evaluated the approach on several applications, considering as target an IP implementing the Intel 8051 processor. The results show that, when error detection is implemented though processor customization and duplication, the resulting dependable hardware requires almost the same resources of the un-hardened original processor core.

It is worthwhile to underline here that being FPGA-emulated processors aiming at replacing obsolete parts, the software the processors will run can be assumed as already available in a stable form, i.e., it is not expected to change during the IP customization process.

The rest of the paper is organized as follows. Section 2 introduces hardware, software and hybrid approaches for hardening embedded systems, underlining how a dependable system implies costs and performances overhead that must be balanced properly. Section 3 describes in details each phase of proposed design flow and in Section 4 we exploit our design flow for building a dependable system based on a processor soft core. Section 5 shows the validity of our approach analyzing the synthesis results on the obtained system. Finally, Section 6 draws some conclusions.

2. Previous works

Several different approaches have been adopted in the past in order to achieve dependable and safe systems. The proposed solutions can be classified in three main categories: hardware, software and hybrid techniques. Each method offers different advantages and disadvantages in terms of costs, performance, and achieved dependability.

Software-implemented techniques aim to protect both the data and the code segment of an embedded program. These methods can be categorized in two main groups: those based on program execution redundancy with check of the results and those based on introducing control code merged into the application program code [6]. N-Version Programming [8] and Recovery Blocks [7] are examples of the first category. N-Version programming is based on software redundancy in the sense of developing N different version of the same software specifications. Even if such a kind of technique allows to achieve an as dependable as simple solution on the base of fault masking concept, the high costs in terms of manual interventions and software development make this solution not always practicable. Also the Recovery Blocks methods, even if offering both active redundancy and passive one, implies too expensive costs. In order to decrease the software developing costs, [9] proposes an automatic method that duplicates both the instructions and the variables involved in the operations and executes some consistence checks. However, the main drawback is the very huge cost in term of memory occupation because both the data and the code segment results almost doubled. The solution proposed in ED^{4}I methods [10] brings at similar consequences. It consists in an error detection system based on instructions duplication so that another program version is executed along with the original one, and finally checking for an eventual mismatch. Thus, the memory overhead implies too much expensive consequences.

About the second category of software techniques, the Algorithm Based Fault Tolerance (ABTF) [11] offers a less expensive solution in terms of memory overhead and resource costs. The software is modified at algorithm level adopting a form of information redundancy. However, it is an as not general solution as strongly related to the implemented algorithm, thus the high specificity represents the main disadvantage.

Hardware-based techniques overcome those limitations proper of software modification and offer system architectures specifically designed for achieving safe and dependable systems. A typical dependable hardware system is based on redundancy, i.e. it includes some components that are as redundant for executing the assigned task as necessary in safety-critical contexts. The fault tolerance techniques based on the redundancy are classified according to the number of replicated components.

The Triple Modular Redundancy (TMR) [12] is based on giving the system outputs as result of a majority vote of three identical components. Thus, it is possible to build a reliable system that masks a single fault combining not reliable components. TMR can be applied at different levels: *processor* level, *register* level and *gate* level. In the first case it is possible to cover both hardware and software faults at very expensive costs. The register and gate level redundancies cover only the faults in the hardware and has the advantage of hardening just a restricted part of the system, and they are useful if the faults probability is not uniformly distributed among the system. An extension of TMR is the NMR where a voting system is built using a greater than three odd number of replicated components. Regardless of the number of copies, the main point of failure of such a kind of a technique is the voter. In fact, it must be fault free otherwise the result of comparison is not trustworthy. Replicating the voter allows to avoid faults both in the processors and in the memory elements and in the voter itself too. However, it presents several drawbacks because of the higher number of interconnections and area resources [14].

A less expensive fault tolerant technique is the Duplication With Comparison (DWC) that consists in duplicated hardware blocks followed by a comparator. Such technique allows error detection, but the comparison by even number of replicated elements cannot establish which is the correct result. Thus, it does not guarantee a consistent state for the system if a recovery procedure is not implemented.

As hardware techniques present a big overhead in terms of area occupation, software ones are general and imply very high memory usage. Hybrid solutions represent a compromise between hardware and software methods. Some examples are Lockstep [15] and Watchdog [13], but the high resources usage and the low portability, especially in the watchdog case, represent a limit that is more and more necessary to overcome. Thus, alternative solutions must be looked for in order to minimize the number of used resources and to find the best compromise between reliability, performance and cost.

The method we proposes tackle these issues and consists in removing from the processor instruction set all the unnecessary instructions for running the application software the system is based on. This solution is possible as we are replacing obsolete parts (i.e., the processor) of an already existing systems, whose application software is already available in a stable form. After the initial customization step, hardware redundancy is exploited for providing the level of dependability the system requires. Thanks to this approach, the software of the system has not to be modified, and significant cost savings can be achieved.

3. Design flow

The following section describes the proposed approach for generating a custom processor that can be inserted in a dependable design with a considerable saving in resources occupation.

In developing the design, we assumed that the complete register-transfer model for the target processor is available and synthesizable (as for example [3]). In fact, all the core details must be provided in order to exploit such information to customize the IP and to build a fault tolerant system. The design flow we developed is shown in Figure 1. It is composed of three phases. The first one (on the left side of the diagram) is the *Instruction Set Extraction*. Starting from the embedded application source code, the binary code is obtained after compiling and linking all the needed library. The resulting binary code is the same that the embedded system has to run during its lifetime. A binary code analyzer selects all the instructions involved in the program execution flow obtaining a smaller instruction sub set. Thus, the Instruction Set Extraction phase yields the minimum list of all the instructions the processor has to execute in order to realize the embedded application task.

The Processor Configuration phase aims to obtain a processor VHDL model free from all the superfluous instructions not need for running the application the processor is tailored for. A parsing operation is performed starting from the reduced instruction sub set combined with a Processor Core Database that contains all the processor VHDL details: for each instruction the list of VHDL statements needed for its decoding, sequencing and executing is provided. Thus, all the component devoted to the instruction management (decoding unit, control unit and arithmetic/logic unit) and the relative VHDL source code is obtained exploiting the processor database information.

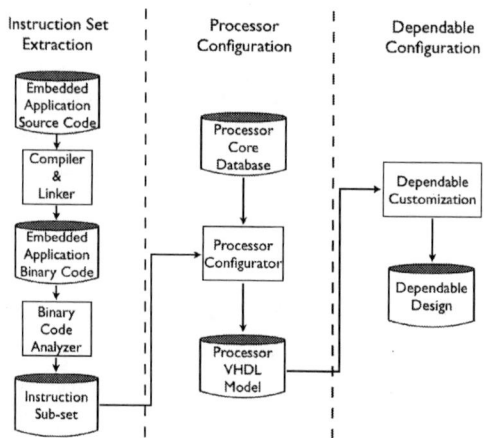

Figure 1: The proposed processor customization flow

Finally, in the most right side of the picture is illustrated the Dependable Configuration, that is the last phase of the proposed flow. A Dependable Design is obtained applying a design customization according to some pre- established techniques. For the sake of this paper, the processor VHDL model resulting from Processor Configuration phase has been duplicated and the two outputs have been connected to a comparator, obtaining the scheme known as duplication-with-comparison (DWC). Thus, at the end of this flow, a safe system is obtained with a considerable saving in terms of area occupation with respect to the cost one may have to pay in case of full processor duplication.

4. A case study

The flow described in the previous section has been applied to a case study: the Intel 8051 soft core processor. Although very simple, this processor core is still used in many embedded systems and is therefore a well-suited case study. All the VHDL components are available in detail in order to make the customization process possible. The macro blocks of the 8051 involved in the design are:

- *RAM* memory containing the program and data variable.
- *ROM* memory containing the application code instructions.
- *Control Unit* (CU) that manage the program flow execution.
- *Arithmetic Logic Unit* (ALU) that performs all the needed computations.
- *Decoder Unit* (DEC) translating the operational code in a specific instruction.

The customization flow has been applied in particular to the ALU, Control Unit and Decoder Unit because their size is strongly related to the number of instructions the program has to perform in order to execute the embedded application task.

Figure 2 shows the scheme of the DWC system obtained from the 8051 design. In the proposed solution we assumed a fault model based on the Single Event Upset. In details, the Control Unit, ALU, and Decode Unit have been duplicated and the output signals have been connected to a comparator that communicates any mismatch. In our approach we assumed that both the RAM and the ROM memory are hardened with a Parity Code technique to prevent common mode failures.

The area savings resulting from our approach depends on the number of instructions involved in the software application the processor is running. If the application task is so specific that requires just a restricted instruction set, then it is possible to build hardened designs that are much smaller than not hardened ones. Nevertheless, even if the software application executes a general-purpose task, using several instructions, it is possible to gain area savings. Another feature of the proposed solution concerns the extensibility field. In fact, as the proposed flow is based on reducing the processor instruction set in order to obtain area savings, the application running on the processor can be upgrade only if it presents the same instruction set.

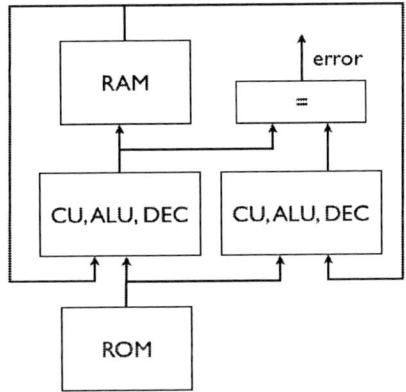

Figure 2: Duplication With Comparison scheme

5. Synthesis results

In this section we report the results coming from the application of the design flow described in section 3. In our analysis we assumed that designers are willing to replace an obsolete Intel 8051 processor using the FPGA-emulated approach used in a ground-based safe application where only fault detection has to be implemented (fault masking is not required). As we adopted a Xilinx's Spartan XA3S500E device for implementing the processor core, processor customization in combination with DWC is exploited for detecting the occurrence of soft errors affecting the processor's memory elements. Moreover, configuration memory checking is used for detecting soft errors in the FPGA's configuration memory.

In the experimental analysis we aim at demonstrating the saving in terms of FPGA resource occupation, and we are interested in observance the effect on the achieved saving of the software application the processor has to execute. The Intel 8051 soft core we considered amounts to about 7,200 lines of synthesizable VHDL code and it implements 106 instructions.

We adopted four different programs as benchmark:

- Matrix multiplication of two 4x4 matrices (M) computes the product of two 4x4 integer matrices.
- Sum of a 13 length vector (S) sums the items of a vector with 13 elements.
- Ascending Sort of a 10 length vector (AS) orders a vector of 10 items in ascending way.
- Finite impulse response filter (F) implements a filtering algorithm over a set of 32 samples.

Table 1 summarizes the main features of the used application, showing for each one the code size and the number of instructions that constitutes the instruction set. For each of them we applied the Instruction Set Extraction methodology described in Section 2, and the binary code has been obtained through the KEIL C complier [16].

All the adopted programs are based on general structures that are commonly used in many applications.

Application Name	Lines of C Code [#]	Instruction Set [#]
M	37	39
S	26	37
AS	33	40
F	62	45

Table 1: The considered applications

To collect figures concerning the silicon area occupation for the different versions of the system we developed, we run the synthesis process obtaining the results showed in Table 2. We analyzed the percentage device occupation calculated as the ratio of the number of used slices with respect to the total number of slices (4.656 for the considered device). Area occupation for data/code memory are not included in our measures. We considered three different design scenarios:

- *Scenario 1*: we synthesized the Intel 8051 soft core without any error detection/correction features. This figure gives the minimum area occupation that designers have to pay when using the considered soft core in their application for the selected implementation technology.
- *Scenario 2*: we applied the DWC error detection scheme to the Intel 8051 implementing all complete instruction set. This figures is the area occupation that have to be paid in case the design flow presented here is not used.

- *Scenario 3*: we applied the DWC error detection scheme to the Intel 8051 tailored to the considered application using the proposed design flow.

	Scenario 1 [%]	Scenario 2 [%]	Scenario 3 [%]
M	48	67	46
S	48	67	46
AS	48	67	46
F	48	67	48

Table 2: Synthesis results for the four workbench

The results we gathered show that, as expected the adoption of the DWC scheme for hardening the whole processor implies a significant area overhead, which is not justified as the considered applications exploits only a subset of the instruction set. As a result of the DWC application to the whole processor, several unused instructions are hardened even if they are not necessary.

Conversely, the use of the proposed design flow allows saving a significant amount of resources by selectively applying hardening only to those instructions that are needed for running the considered applications. The results also show that the cost for the hardened design is nearly the same of the un-hardened design. The proposed approach can thus be used for minimizing effectively the cost for developing safe systems.

As far as error detection is considered, the proposed approach allows detecting all the faults affecting the processor's memory as well as all the faults affecting the combinational and sequential logic of the processor. In our conceptual scheme of Figure 2 only one comparator is reported. In order to prevent common mode errors affecting this module, in the final implementation of the hardened system two comparators will be used.

6. Conclusions

As component obsolescence is becoming more and more relevant due to the very high pace at which components exist from the market, solutions are needed for preserving the correct function of those systems that exploits obsolete parts. FPGA-based emulation can be a solution to the obsolescence of digital parts, as for example processor cores.

In this paper we propose a new design flow aiming at customizing processor cores intended for being emulated using FPGAs to replace obsolete parts. The design flow starting from the analysis of the software application source code obtain a minimal soft core that implements only the instruction needed by the software application, thus minimizing resource occupation. As a result smaller, less expensive FPGA devices can be used for processor emulation. Moreover, redundancy techniques can be exploited to harden the customized processor with negligible hardware overhead. The experiments reported in the paper show that, no matter the considered application, a processor core customized and hardened according to the proposed methodology occupies the same area of the original un-hardened processor.

References

[1]   P. Sandbord, "Trapped on Technology's Trailing Edge", IEEE Spectrum,

[2]   L. Anghel, R. Velazco, S. Saleh, S. Deswaertes, A. El Moucary, "Preliminary Validation of an Approach Dealing with Processor Obsolescence", 18th IEEE International Symposium on Defect and Fault Tolerance in VLSI Systems, p. 493, 2003

[3]   http://www.iwavesystems.com/iW-X86SOC.pdf

[4] P.E. Dodd, L.W. Massengill, "Basic mechanisms and modeling of single-event upset in digital microelectronics", IEEE Trans. on Nuclear Science, vol. 5, pp. 583- 602, 2003

[5] J.L. Barth, C.S. Dyer, E.G. Stassinopoulos, "Space, atmospheric, and terrestrial radiation environments", IEEE Trans. on Nuclear Science, vol. 5, pp. 466- 482, 2003

[6] P. Cheynet, B. Nicolescu, R. Velazco, M. Rebaudengo, M. Sonza Reorda, M.Violante, "Experimentally evaluating an automatic approach for generating transient errors", IEEE Transactions on Nuclear Science, Vol. 47, No. 6, December 2000, pp. 2231-223

[7] B. Randell, "System structure for software fault tolerant," IEEE Trans. Software Eng., vol. 1, pp. 220–232, June 1975.

[8] A. Avizienis, "The N-version approach to fault-tolerant software," IEEE Trans. Software Eng., vol. 11, pp. 1491–1501, Dec. 1985.

[9] P. Cheynet, B. Nicolescu, R. Velazco, M. Rebaudengo, M. Sonza Reorda and M. Violante. "Experimentally evaluating an automatic approach for generating safety-critical software with respect to transient errors", IEEE Transactions on Nuclear Science, Vol. 47, No. 6 (part 3), December 2000, pp. 2231-2236

[10] N. Oh, S. Mitra, E.J. McCluskey, "ED4I: error detection by diverse data and duplicated instructions", IEEE Transactions on Computers, Vol. 51, No. 2 , Feb. 2002, pp. 180-199

[11] K. H. Huang and J. A. Abraham, "Algorithm-based fault tolerance for matrix operations", IEEE Trans. Comput., vol. 33, pp. 518–528, December 1984

[12] Wei Chen; Rui Gong; Kui Dai; Fang Liu; Zhiying Wang, "Two New Space-Time Triple Modular Redundancy Techniques for Improving Fault Tolerance of Computer Systems", Computer and Information Technology, 2006. CIT '06. The Sixth IEEE International Conference on. Sept. 2006 Page(s):175 - 175

[13] M. Namjaoo, E. J. McCluskey, "Watchdog processors and capability checking", in Proceedings of the 12th International Symposium on Fault-Tolerant Computing (FTCS-12), 1982, pp. 245-248.

[14] F. Kastensmidt, L. Sterpone, M. Sonza Reorda, L. Carro, "On the optimal design of Triple Modular Redundancy Logic for SRAM-Based FPGAs", DATE 2005: IEEE Design, Automation and Test in Europe, 2005, pp. 1290-1295.

[15] www.xilinx.com, "PPC405 Lockstep System on ML310", Application Note.

[16] www.keil.com

IEEE International Symposium on Defect and Fault Tolerance of VLSI Systems

A Low-Power Safety Mode for Variation Tolerant Systems-on-Chip

David Wolpert and Paul Ampadu
ECE Dept., University of Rochester, Rochester, NY 14627
wolpert@ece.rochester.edu, ampadu@ece.rochester.edu

Abstract

Process, voltage, and temperature (PVT) variations are difficult to manage in multi-core SoCs, as each core may have different voltage and reliability requirements. Indeed, common implementations of variation-tolerant techniques (e.g. dynamic voltage and frequency scaling) are ineffective in multi-core SoCs because of the large overheads they incur. In this work, we propose a simple low-power safety-mode system that adjusts each core's frequency independently to compensate for PVT variation at runtime. This module combines an on-chip sensor with a safety-mode lookup table to guarantee SoC functionality. The module is simulated in a four-core SoC framework, and allows the system to operate at a frequency corresponding to the average system temperature instead of the worst-case temperature. In the case presented, a room temperature system is capable of operating at an average frequency that is 2X that of the worst-case temperature requirements. The safety mode module consumes an average power of 0.56 µW and has a 3% area overhead.

1. Introduction

PVT variations are exacerbated by technology scaling, making reliability among the most important of the many challenges facing nanoscale system design. To avoid delay errors resulting from these effects, many techniques addressing PVT variation use frequency guardbands. Unfortunately, temperature fluctuations of up to 50°C [1] and supply voltage variations of 22% [2] mean that worst-case guardbands can become quite large, limiting system throughput and reducing the effectiveness of power saving techniques. Adaptive solutions such as adaptive supply voltage scaling (ASV) and adaptive body bias (ABB) have been proposed [3-15] to reclaim some of this lost performance, allowing systems to tolerate extreme worst-case scenarios without sacrificing delay and power under normal conditions. These methods have been used to consider variation in a number of ways, such as restoring performance when variation thresholds are crossed [3], improving yield by managing speed and power trade-offs [4][5], and using runtime variation and process aging data to throttle chip frequency [6]. Adaptive systems have also been studied for reducing power and energy consumption [7][8], as the same techniques used to compensate for variation can also significantly reduce power consumption during low activity or idle periods. Ultra-dynamic voltage scaling systems have been explored as well [9], scaling voltage into the subthreshold region where the impacts of variation are even more significant [16][17].

Adaptive systems can track PVT delay variations using ring oscillators [10], delay lines [11], or critical path replicas [12][13]. Rather than sensing delay changes, look-up table (LUT)-based methods allow for guardbands to be set at each voltage and temperature operating point [6][14]. Previous designs using LUTs have explored compensating for the effects of temperature variation [14] or using voltage droop detectors and temperature sensors to compensate for individual parameter fluctuations

1550-5774/08 $25.00 © 2008 IEEE
DOI 10.1109/DFT.2008.17

[6]. LUT techniques can be modified to include intra-die variations by adjusting the table values to consider worst-case process ranges. These adjustments generally result in lower frequencies and higher voltages than necessary, wasting power and lowering yield as more chips exceed power budget limits. A combination of a delay sensor and guardband LUT has been proposed to minimize the amount of wasted resources [15].

All of the aforementioned techniques guarantee reliability at the expense of large power and complexity overheads. In this paper, we focus on a much simpler safety mode module with a small power and area footprint, allowing it to be duplicated as necessary to ensure reliability in a multi-core system. The safety mode module determines when a core is near delay failure and reduces that core's frequency by half, ensuring correct functionality even under extreme worst-case operating conditions. The remainder of the paper is organized as follows: Section 2 explores the delay impact of PVT variation across a wide range of operating conditions. This information is used in Section 3 to propose a variation-tolerant multi-core framework, with a specific multi-core implementation provided. Variability tolerances and safety mode functionality of the implementation are shown in Section 4. Conclusions are presented in Section 5.

2. PVT Variation

In order to know when to activate the safety mode, it is important to have a detailed model of the variations being tracked. By examining these variations over the entire range of potential operating conditions, a table of the worst-case delay effects for each condition can be populated. This table will be part of the safety mode control system.

Runtime variations in voltage can affect delay according to the delay sensitivity [18]

$$S_{V_{DD}}^{\tau} = \lim_{\Delta V_{DD} \to 0} \frac{\frac{\Delta \tau}{\tau}}{\frac{\Delta V_{DD}}{V_{DD}}} = \frac{V_{DD}}{\tau} \cdot \frac{\partial \tau}{\partial V_{DD}}, \tag{1}$$

which in a ring oscillator simplifies to

$$S_{V_{DD}}^{\tau} = -\frac{\frac{V_{DD}}{V_T} \cdot (\alpha - 1) + 2}{\frac{V_{DD}}{V_T} - 1}. \tag{2}$$

Here τ is delay, ΔV_{DD} and $\Delta \tau$ are the variation in delay and corresponding change in delay, respectively; V_T is the threshold voltage, and α is the alpha-power law exponent [19].

Temperature variation also affects delay through device mobility and threshold voltage [20], according to

$$\mu = \mu_0 (T / T_0)^{\alpha_\mu} \tag{3}$$

and

$$V_T(T) = V_{T0} + \alpha_{V_T} (T - T_0), \tag{4}$$

where T_0 is nominal temperature, μ_0 is mobility at T_0, α_μ is an empirical measure of the temperature dependence on mobility, V_{T0} is nominal threshold voltage, and $\alpha_{VT} = \partial V_T / \partial T$ is the threshold voltage temperature coefficient, also an empirical measure.

To examine the combined effects of runtime variation, drain currents (I_D) of equally-sized, diode-connected, bulk NMOS and PMOS devices from the 90 nm BSIM4 predictive technology model [21] are shown in Fig. 1. The operating conditions chosen include temperatures between -55°C and 125°C (military environmental specification) and voltages

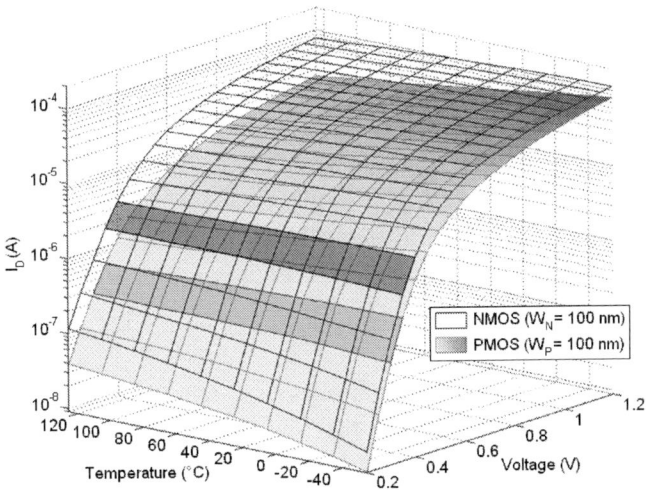

Fig. 1. Temperature and voltage sensitivity of NMOS and PMOS devices.

between 200 mV and 1.2 V. Each plane in Fig. 1 is divided up by a shaded voltage slice, in which temperature variation has the smallest impact on I_D. At voltages larger than the shaded regions, current decreases as temperature increases, termed *normal* temperature dependence. At lower voltages, current increases as temperature increases, termed *reverse* temperature dependence. The reversal of the temperature dependence at low voltages is caused by temperature, voltage, and technology dependencies of threshold voltage and mobility [20][22][23]. These changing trends cause the impact of each type of variation to depend upon the operating point. For example, a ±12°C variation at 0.4 V causes a negligible change in current, while the same variation at 1.2 V can cause current to change by 15%.

Variation-tolerant circuits must be capable of operating over the entire range of voltages and temperatures to which the circuit might be exposed. In addition, they must be able to handle cycle-to-cycle variation and across-chip variation, which can cause communication failures from delay differences between IP cores in a SoC. Models for these types of variation are examined in the remainder of this section.

2.1. Temperature variation

Temperature variation refers to the combination of ambient temperature (climate, cooling systems, etc.) and on-chip temperature (caused by power dissipation) changes during normal operation. To determine the effect of temperature on delay performance, we use an on-chip temperature variation model taken from a chip and package thermal model of an IC [24]. In this model, temperature varies by 24°C across the chip under normal conditions.

2.2. Voltage variation

Two major contributors to voltage variation are fluctuation in the supply load, which can change by up to 700% depending upon input patterns [2] (and also result in simultaneous switching noise [25]), and IR drop, caused by non-zero interconnect resistance. Variations in supply voltage can also arise from substrate noise or interconnect coupling. Voltage variations are particularly important because many adaptive systems control applied voltages [3-15]; any voltage instability could result in inaccurate or undesirable speed and power adjustments. Reductions of voltage variations to 2% V_{DD} have been reported using decoupling capacitors [26]; however, supply variations up to 10% V_{DD} are common, even with well-designed distribution networks and decoupling capacitors [27].

2.3. Process variation

Process variations generally result from lithography limitations, quantum effects in oxides, and variations in doping profiles [28]. In the process variation model used in this work [27], gate length, L, varies by ±30% and the primary threshold voltage coefficient, V_{T0}, varies by ±10%. The V_{T0} parameter change is implemented by adjusting the vth0 parameter in the 90 nm predictive technology model [21]. To reduce the number of variables, intra-die process variation is not considered in this work, though it is certainly an important consideration in robust system design and will be examined in the future.

2.4. Combined effects of variation

The combined effects of runtime variation are shown in Fig. 2a for an 11-stage ring oscillator at 1.2 V and 27°C. The square marker shows the performance of the system prior to the introduction of variation; the circular and triangular markers show the performance impacts of temperature and voltage variations, respectively. As shown, the impact of temperature variation on frequency is nearly as large as the impact of voltage variation at the 1.2 V/27°C nominal point, yet voltage variation causes a far greater fluctuation in power dissipation than temperature variation. The four star markers represent combinations of the variation sources. For example, the lower left star marker represents the system performance at the +12°C/-10% V_{DD} operating point. The lower left and upper right star markers indicate the maximum range of frequencies over which a system must function to tolerate runtime variation at the 1.2 V/27°C operating point, a 26% frequency difference.

In Fig. 2b, process variation is added to the runtime variation analysis. To show the combined effect of PVT variation, the runtime variation data from Fig. 2a is copied as the 'No Process Variations' case in Fig. 2b, and fast and slow process corners are included to show their impact on frequency and power. As shown, the combination of PVT variation increases the frequency range from 26% in Fig. 2a to 45% in Fig. 2b. This 45% variation in delay is a key motivating factor for the need for variation-tolerant safety modes in SoCs.

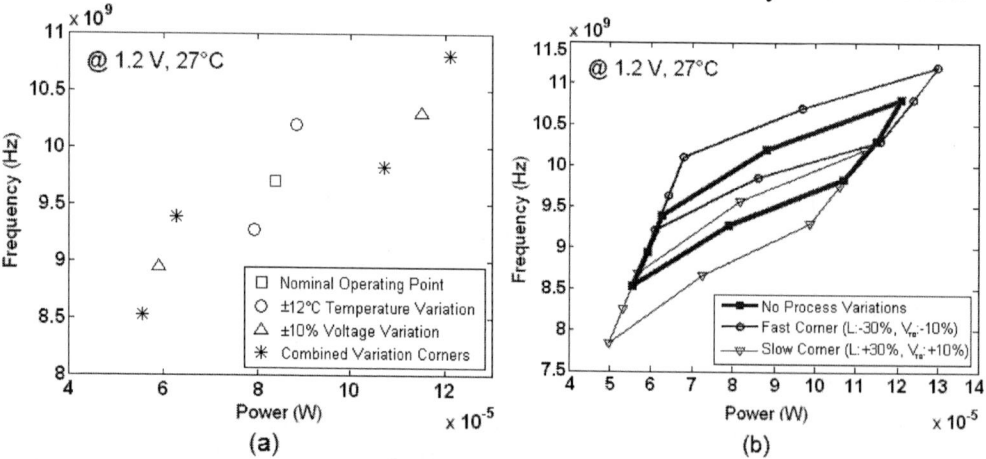

Fig. 2. Runtime variation impacts (a) with no process variation and (b) with fast and slow process corners.

3. Variation-tolerant system design

An important component of any adaptive system that compensates for variation is a sensor to track the runtime operating conditions. Depending on the application environment, this sensor (or multiple sensors, if necessary) may be required to track a wide variety of effects, such as PVT variation, power dissipation, or substrate noise. These effects can result in a variety of problems, including timing failure. The sensors must have some way of quantifying their results to determine when and how to adjust the adaptive system to maintain functionality. One such framework is shown in Fig. 3, which will be described in this section and analyzed in Section 4.

3.1. A multi-core framework with a variation-tolerant safety mode

The safety mode module presented in this paper is capable of halving the frequency in each core individually whenever runtime variations approach levels which could cause system failure, in this case due to delay error. By limiting the per-core overhead to a simple frequency divider and one column of a LUT, this framework is extendable to a large number of cores which would be unmanageable by systems with large per-core overhead requirements such as PLLs or voltage regulators. The purpose of the safety mode module is variation tolerance, not power optimization, and the large frequency margin ensures correct functionality over a wide range of conditions. Future work will examine the power and area trade-offs associated with more than two choices of operating frequency.

The example multi-core framework shown in Fig. 3a includes a sensor, look-up table (LUT), four computation cores, an arbiter for handling the I/O of each core, and system memory. While the arbiter controls the flow of information, the sensor and LUT control the arbiter and core frequencies. Fig. 3b provides a closer look at the safety mode module (the dashed box in Fig. 3a). As shown, the sensor feeds its quantized information to the LUT, which is created at design time (the LUT function can be efficiently implemented using simple gates rather than an addressable array of memory elements). The LUT controls whether or not to implement the safety mode in each core using the *Safe* signals, which choose from a system clock and a frequency divider. The LUT also provides this information to the arbiter, which controls the data rate to each core.

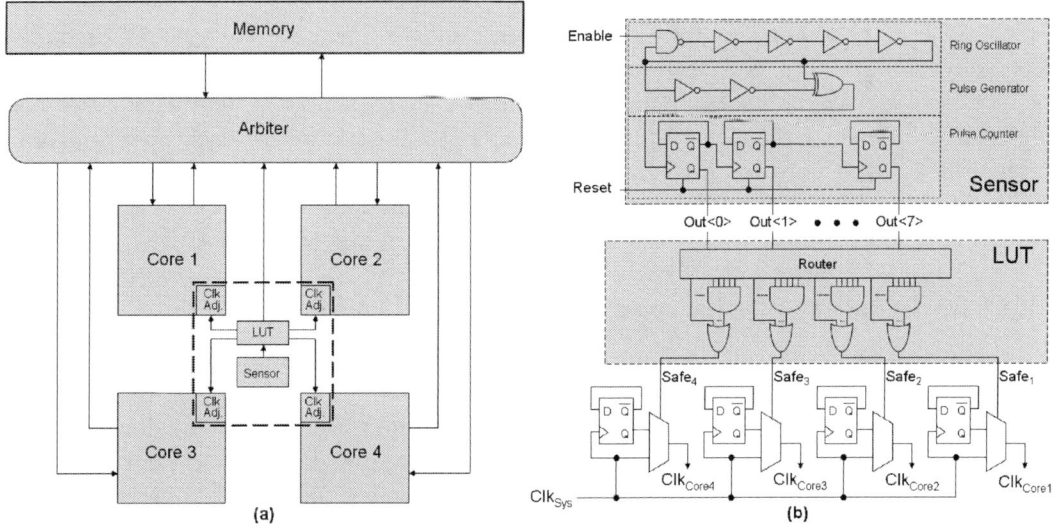

Fig. 3. Multi-core framework for variation-tolerant design. (a) System diagram and (b) safety mode module diagram.

3.2. Runtime variation sensor

The sensor used for the test case, shown in Fig. 3b, is an all-digital delay sensor [29], modified by including a pulse generator (the XOR gate fed by a two inverter buffer) to double the sensor resolution and make it more sensitive to runtime changes in temperature. The sensor can also be used to measure die-to-die process variation, or voltage losses due to IR drop, by recording the sensor output at a specific temperature and comparing that result to other known values (similar to how performance-sensitive ring oscillators are used [30]). The accuracy of the all-digital sensor is not limited by process mismatch and DC offset as in analog sensors [31], though it is susceptible to voltage variations due to changing loads. These voltage variations must be built into the frequency guardband of each unit. Voltage spikes are also problematic due to their short duration, which could result in oscillation between the normal and safety modes. This oscillation could be reduced by adding hysteresis to the system, for example using an incrementer to artificially increase the digital output when the safety mode is triggered, which would avoid this oscillation pattern until the temperature was reduced enough to reset the hysteresis signal.

The sensor is composed of three parts, shown in Fig. 3b. The delay sensing mechanism is an enable-controlled ring oscillator, which will continue to run as long as the enable signal is asserted. By adjusting the pulse width of the enable signal, very small variations in delay can be accumulated across multiple clock cycles using the pulse counter. The oscillator is tapped and fed into a pulse generator, which converts every transition in the ring oscillator into a full pulse, effectively doubling the sensor accuracy. The pulse generator feeds a pulse counter, which tallies up the number of pulses and provides a digital read-out, with each temperature corresponding to an n-bit state. Here, n is determined by the number of pulses the sensor must count, a function of the enable pulse width and the range of operating conditions. If necessary, the capacity of the pulse counter can be doubled by simply adding a flip-flop to the end of the chain. For this application, n is set to 8 bits, and the sensor can achieve a resolution of 4°C when enabled for 30 ns. Reduced sensor accuracies require the safety mode to be asserted at lower temperatures (meaning larger temperature safety margins are needed), at the expense of lost system throughput.

4. Simulation Results

Schematic level simulations were performed in Cadence Spectre using a 90 nm Predictive Technology Model [21]. Each core in this system is a 64-bit adder composed of cascaded four-bit carry lookahead blocks. In order to simulate a more general system where each core might have different frequency requirements and variation tolerances, each core is given a different body bias. Core 1 is given a normal bias ($V_{B,P} = 1$ V, $V_{B,N} = 0$ V), Core 2 is given 0.2 V of forward bias ($V_{B,P} = 0.8$ V, $V_{B,N} = 0.2$ V), Core 3 is given 0.4 V of forward bias ($V_{B,P} = 0.6$ V, $V_{B,N} = 0.4$ V), and Core 4 is given 0.6 V of forward bias ($V_{B,P} = 0.4$ V, $V_{B,N} = 0.6$ V).

4.1. Characterization of the test cores

To illustrate the impact of variation on the system, the temperature response of the four cores is shown with and without process and voltage variation in Fig. 4. As shown, the combination of process and voltage variation can result in a nearly 3x delay degradation at 125°C, and generally increases the temperature dependence of the delay in each core. Note that below 0°C, the large forward bias in Core 4 changes the trend slightly due to a combination of gate leakage and reduced signal swings.

Fig. 4 is also useful as a visualization tool for how the safety net system works, as illustrated by a simple example. By drawing a horizontal line at 333 MHz as shown, it is possible to find the temperatures at which each core can no longer operate at that desired frequency (assuming no additional frequency guardbands are needed). When process and voltage variation are considered, we can see that Core 1 will fail beyond ~30°C, Core 2 will fail at ~47°C, Core 3 will fail at ~64°C, and Core 4 will fail near 80°C. Applying the safety mode module to these cores reduces the operating frequency of each core by 2x (changing the period to 6 ns) when a core approaches its failure point. From Fig. 4, we see that all cores have a delay of <6 ns up to 125°C; thus, using the safety mode system, all four cores are variation-tolerant across the entire range of military specified temperature conditions, albeit with reduced throughput beyond those temperature limits. Without this adaptive capability, system frequency would be limited to the worst case frequency of 167 MHz.

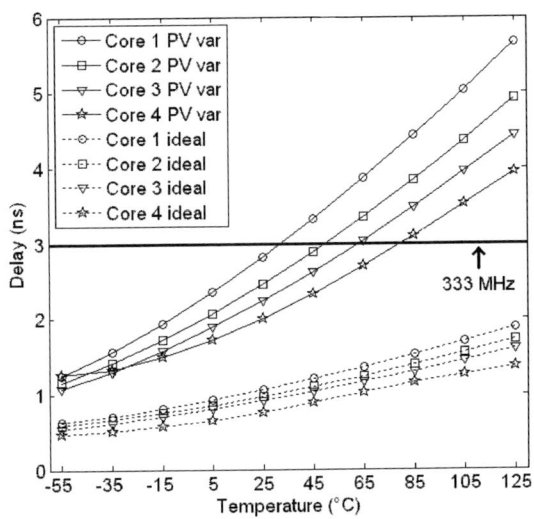

Fig. 4. Temperature response of the four cores with and without process and voltage variation.

4.2. Operation of the safety mode system

The safety mode system operation is shown in Fig. 5 at 27°C with 10% voltage variation and a worst-case process corner. As described in Section 3, the *Enable* signal is asserted over a number of clock cycles to accumulate the delay error in the ring oscillator. After that pulse has completed, the *Readout* signal passes the sensor output to the LUT. Beyond 27°C, Core1 is incapable of operating correctly at the 333 MHz system clock, so the *Safe1* signal is activated, indicating Core1 needs to be put into the low-frequency safety mode. Core2, Core3, and Core4 are all capable of operating correctly at 333 MHz, so the other Safe signals are kept at logic 0. The safety mode is shown to operate correctly on the *Core1Clk* signal, with the operating frequency dropping from 333 MHz to 167 MHz after the 40 ns system latency. The clock signals to the other three cores remain unchanged. In future designs, the first cycle of the safety mode clock will be skipped to avoid the partial clock pulse during transition between modes, shown in Fig. 5 at 40 ns on signal *Core1Clk*.

For our particular system and technology, the power overhead of the safety mode module, including the sensor, LUT, and clock dividers, is 105.1 μW per adaptation at 27°C; however, because temperature changes very slowly (on the order of microseconds) the sensor is left inactive for the majority of runtime, making its power overhead very small compared to the 638.1 μW average power of the combined cores (also measured at 27°C). The total latency of adaptation is 40 ns, including the sensor latency, LUT and clock dividers. Assuming one sample is taken every ten microseconds [31], the average power of the safety mode module is just 0.56 μW (idle leakage power in the module is 0.14μW). The area overhead of the safety mode module (including the sensor and clock adjustment circuitry in each core) is only 3% of the overall multi-core system, making our approach both power- and area-efficient for variation-tolerant multi-core design. This small power and area overhead would also allow multiple sensors to be implemented as necessary to take into account intra-die variation.

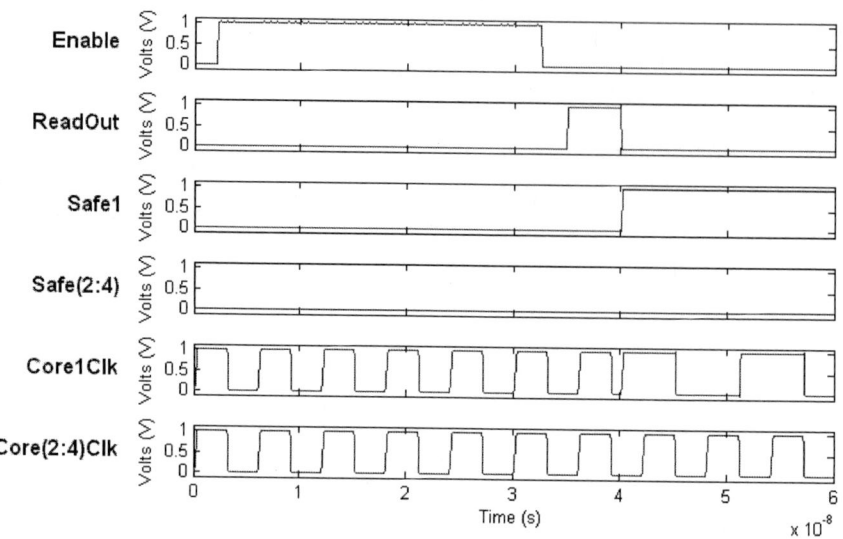

Fig. 5. Safety mode system operation at 27°C.

5. Conclusions and future work

Multi-core systems can be made variation-tolerant without the need for extensive, large-overhead adaptive systems. The proposed safety mode module is able to adapt each core individually for variation tolerance with a 3% area and 0.09% power overhead in the four-core implementation examined. This low overhead makes the approach scaleable for SoCs with a larger number of cores. The individual handling of each core also enables more optimal use of resources—rather than shutting down or scaling large sections of a chip due to overhead limits, only those in danger of failure need to be adjusted. The variation-tolerant multi-core framework offers a number of exciting paths for future work. Intra-die variation will be examined in the future for larger numbers of cores, where single sensors may be unable to capture the effects of local variations. The use of additional frequencies will be examined to reduce the throughput penalty of switching to the safety mode. Sensors can also be combined to address multiple types of faults while still maintaining a low per-core overhead, and the combination of core and interconnect fault-tolerance will be a great challenge as nanoscale design continues to exacerbate the effects of PVT variation.

6. References

[1] S. Borkar, *et al.*, "Parameter variations and impact on circuits and microarchitecture", *Proc. ACM IEEE 40th Design Automation Conference (DAC'03)*, Jun. 2003, pp. 338-342.

[2] N. James, *et al.*, "Comparison of split- versus connected-core supplies in the POWER6™ microprocessor", *Proc. IEEE Int. Solid-State Circuits Conf.*, Feb. 2007, pp. 298-299.

[3] K. Shakeri and J. Meindl, "Temperature variable supply voltage for power reduction", *IEEE Comp. Soc. Ann. Symp. on VLSI*, Apr. 2002, pp. 64-67.

[4] T. Chen and S. Naffziger, "Comparison of adaptive body bias (ABB) and adaptive supply voltage (ASV) for improving delay and leakage under the presence of process variation", *IEEE Trans. Very Large Scale Integr. (VLSI) Syst.*, vol. 5, no. 11, Oct. 2003, pp. 888-899.

[5] J. Tschanz, S. Narendra, R. Nair, and V. De, "Effectiveness of adaptive supply voltage and body bias for reducing impact of parameter variations in low power and high performance microprocessors", *IEEE J. Solid-State Circuits*, vol. 38, no. 5, May 2003, pp. 826-829.

[6] J. Tschanz, *et al.*, "Adaptive frequency and biasing techniques for tolerance to dynamic temperature-voltage variations and aging", *Proc. IEEE Int. Solid-State Circuits Conf.*, Feb. 2007, pp. 292-293.

[7] S. Martin, K. Flautner, T. Mudge, D. Blaauw, "Combined dynamic voltage scaling and adaptive body biasing for lower power microprocessors under dynamic workloads", *IEEE/ACM Int. Conf. on Computer Aided Design*, Nov. 2002, pp. 721-725.

[8] S. Das, *et al.*, "A self-tuning DVS processor using delay-error detection and correction," *IEEE J. Solid-State Circuits*, vol. 41, no. 4, Apr. 2006, pp. 792-804.

[9] B. Calhoun, A. Chandrakasan, "Ultra-dynamic voltage scaling (UDVS) using sub-threshold operation and local voltage dithering", *IEEE J. Solid-State Circuits*, vol. 41, no. 1, Jan. 2006, pp. 238-245.

[10] J. Rabaey, A. Chandrakasan, and B. Nikolic, *Digital Integrated Circuits—A Design Perspective* (Prentice Hall, New Jersey, 1996).

[11] M. Miyazaki, G. Ono, K. Ishibashi, "A 1.2-GIPS/W microprocessor using speed-adaptive threshold-voltage CMOS with forward bias," *IEEE J. Solid-State Circuits*, vol. 37, no. 2, Feb. 2002, pp. 210-217.

[12] J. Kao, M. Miyazaki, A. Chandrakasan, "A 175-mV multiply-accumulate unit using an adaptive supply voltage and body bias architecture", *IEEE J. Solid-State Circuits*, vol. 37, no. 11, Nov. 2002, pp. 1545-1554.

[13] M. Elgebaly, M. Sachdev, "Variation-Aware Adaptive Voltage Scaling System", *IEEE Trans. Very Large Scale Integration (VLSI) Systems*, vol. 15, no. 5, May 2007, pp. 560-571.

[14] T. Gyohten, *et al.*, "An On-Chip Supply-Voltage Control System Considering PVT Variations for Worst-Caseless Lower Voltage SoC Design", *IEICE Trans. Electron.*, vol. E89-C, no. 11, Nov. 2006, pp. 1519-1525.

[15] D. Wolpert, P. Ampadu, "Adaptive delay correction for runtime variation in dynamic voltage scaling systems," *J. Circuits, Systems, and Computers* (to appear).

[16] B. Zhai, S. Hanson, D. Blaauw, D. Sylvester, "Analysis and mitigation of variability in subthreshold design", *Proc. Int. Symp. on Low Power Electronics and Design*, Aug. 2006, pp. 20-25.

[17] S. Hanson, *et al.*, "Ultralow-voltage, minimum-energy CMOS", *IBM J. Res. & Dev.*, vol. 50, no. 4/5, Aug. 2006, pp. 469-490.

[18] M. Alioto, G. Palumbo, "Impact of supply voltage variations on full adder delay: analysis and comparison," *IEEE Trans. Very Large Scale Integration (VLSI) Systems*, vol. 14, no. 12, Dec. 2006, pp. 1322-1335.

[19] T. Sakurai, A. R. Newton, "Alpha-power law MOSFET model and its applications to CMOS inverter delay and other formulas," *IEEE J. Solid-State Circuits*, vol. 25, no. 2, Apr. 1990, pp. 584-594.

[20] I. Filanovsky and A. Allam, "Mutual compensation of mobility and threshold voltage temperature effects with applications in CMOS circuits", *IEEE Trans. Circuits Syst. I, Fundam. Theory Appl.*, vol. 48, no. 7, Jul. 2001, pp. 876-884.

[21] W. Zhao and Y. Cao, "New generation of predictive technology model for sub-45nm design exploration", *Proc. Int. Symp. on Quality Electronic Design*, Mar. 2006.

[22] C. Park, *et al.*, "Reversal of temperature dependence of integrated circuits operating at very low voltages", *Proc. Int. Electron Devices Mtg.*, Dec. 1995, pp. 71-74.

[23] A. Bellaouar, A. Fridi, M. Elmasry, and K. Itoh, "Supply voltage scaling for temperature insensitive CMOS circuit operation", *IEEE Trans. Circuits Syst. II, Analog Digit. Signal Process.*, vol. 45, no. 3, Mar. 1998, pp. 415-417.

[24] W. Huang, E. Humenay, K. Skadron, and M. Stan, "The need for a full-chip and package thermal model for thermally optimized IC designs", *Proc. Int. Symp. on Low Power Electronics and Design*, Aug. 2005, pp. 245-250.

[25] S. Chun, *et al.*, "Modeling of simultaneous switching noise in high speed systems," *IEEE Trans. on Adv. Packaging*, vol. 24, no. 2, May 2001, pp. 132-142.

[26] G. Ji, T. Arabi, and G. Taylor, "Design and validation of a power supply noise reduction technique", *IEEE Trans. on Adv. Packaging*, vol. 28, no. 3, Aug. 2005, pp. 445-448.

[27] Y. Cao, P. Gupta, A.B. Kahng, D. Sylvester, and J. Yang, "Design sensitivities to variability: extrapolations and assessments in nanometer VLSI", *Proc. IEEE Int. ASIC/SOC Conf.*, Sept. 2002, pp. 411-415.

[28] Y. Cao and L. Clark, "Mapping statistical process variations toward circuit performance variability: an analytical modeling approach", *Proc. ACM IEEE 42nd Design Automation Conference (DAC'05)*, Jun. 2005, pp. 658-663.

[29] S. Griffith, "Method and apparatus for measuring the speed of an integrated circuit device," United States Patent #4890270, Dec. 1989.

[30] J. Xiong, V. Zolotov, C. Visweswariah, P. A. Habitz, "Optimal margin computation for at-speed test," *Design, Automation and Test in Europe*, Mar. 2008, pp. 622-627.

[31] L. Luh, J. Choma, J. Draper, H. Chiueh, "A high-speed CMOS on-chip temperature sensor", *Proc. European Solid-State Circuits Conf.*, Sept. 1999, pp. 290-293.

SESSION 2

DEPENDABILITY ANALYSIS AND EVALUATION

Built-In Self-Diagnostics for a NoC-Based Reconfigurable IC for Dependable Beamforming Applications

Oscar J. Kuiken[1], Xiao Zhang[1], and Hans G. Kerkhoff [1]

Testable Design and Test of Integrated Systems Group, CTIT, University of Twente

o.j.kuiken@student.utwente.nl, x.zhang@utwente.nl, H.G.Kerkhoff@utwente.nl

Abstract

Integrated circuits (IC) targeting at the streaming applications for tomorrow are becoming a fast growing market. Applications such as beamforming require mass computing capability on a single chip as well as flexibility to adapt to new algorithms. A reconfigurable IC with many processing tiles based on the Network-on-Chip architecture is considered ideal for such applications as it balances efficiency and flexibility. Due to the highly regular arrangement of the processing tiles connected by the communication network, it is possible to adopt new Design-for-X strategies to improve the dependability of the reconfigurable IC. The communication network can be reused as a test-access mechanism. On-chip deterministic test pattern generators can multicast test-vectors through the network to the cores under test and test responses from multiple cores can be collected and analyzed by a test result evaluator. The faulty core, or functional parts of it, will be labeled and isolated from the whole system by re-mapping the computing resources and thus improve the dependability of the whole system.

1. Introduction

Beamforming is a technique for combining signals that are received at multiple antennas [1]. It is a streaming-data application that requires a massive amount of digital signal processing and is applied in e.g. phased-array radars. For such applications, it is of key importance to ensure the dependability of the platform on which the beamforming application runs.

The Montium reconfigurable processor (Processing Tile, PT) has been developed [2] to perform highly efficient streaming-data applications, such as the beamforming task. In this case, many Montium processors are interconnected by a Network-on-Chip (NoC) to form a General Stream Processor (GSP) on a single chip while run-time mapping techniques are used to ensure optimal allocation of the computing resources. As a result of the high regularity of the NoC as well as the PTs, new Built-In Self-Diagnostics methods can be used to improve the dependability of the beamforming system.

2. The Montium processor and ATE based tests

A Montium processor consists of identical processing tiles [2]. Before investigating Built-In Self-Diagnostics, it is necessary to determine whether the Montium core does not contain any undetectable and ATPG untestable faults. This will provide a reference level for the

---

[1] This research is conducted within the FP7 Cutting edge Reconfigurable ICs for Stream Processing (CRISP) project (ICT-215881) supported by the European Commission.

performance of the on-chip test-pattern generator, as ATE-based test-vectors are considered to have the highest possible quality, that is, they have the highest possible fault coverage using the least number of test-vectors.

When testing the Montium and its network interface (NI), all memories were isolated from the Core-Under-Test (CUT) by means of boundary wrapper cells. The complete design was carried out using Synopsys and its TetraMAX tools in a standard way. All memories in the Montium core have their own BIST structures and are thus not part of the task at hand, that is, to determine the fault coverage of the Montium.

When only looking at stuck-at faults (SA), TetraMAX is able to achieve a fault coverage (FC) of 99.82%, using 2,746 test vectors, each with a length of 23,145 elements (Figure 1). For the different fault classes is referred to [3]. The total Design-for-X (X can be Test (T), Debug (D) and Dependability (DEP)) overhead is 17% [4]. The latter is explained by the large number of reconfiguration registers in the highly-reconfigurable Montium.

```
    Uncollapsed Stuck Fault Summary Report
------------------------------------------------
fault class                      code    #faults
------------------------------    ----   --------
Detected                         DT       683968
Possibly detected                PT            0
Undetectable                     UD          292
ATPG untestable                  AU           85
Not detected                     ND         1135
------------------------------------------------
total faults                              685480
test coverage                              99.82%
------------------------------------------------
            Pattern Summary Report
------------------------------------------------
#internal patterns                           2746
    #basic_scan patterns                     2746
------------------------------------------------
            CPU Usage Summary Report
------------------------------------------------
Total CPU time (sec)                     12802.11
------------------------------------------------
```

Figure 1. TetraMAX results show that ATPG leads to a fault coverage of 99.82%

3. The use of the NoC as a test access mechanism

According to the IEEE 1500 standard testability method for embedded core-based ICs, dedicated test buses are required for embedded core-based testing [5]. As silicon test-area overhead is to be minimized, it is a natural design methodology to reuse as much hardware resources for as many applications as possible. Research on using NoC as a test access mechanism (TAM) has been carried out and it is proven possible to combine the NoC and a modified wrapper design to perform the ATE based test for embedded cores [6]. Hence in our approach, it was decided to reuse the NoC to replace the traditional test-access busses. Two important boundary conditions have to be set for the use of a NoC as a TAM:

- The NoC itself should preferably be fault free / fault-tolerant, or the defective NoC segment should be known in advance
- The NoC should be able to setup the Core-Under-Test (CUT) in test mode and to provide the CUT with communication from the Test Pattern Generator (TPG) and to the Test Response Evaluator (TRE) to be discussed later.

Traditionally, a NoC consists of two parts: a router network and a series of network interfaces. The router network is responsible for routing the data across the IC, while a network interface act as a bridge between a router and a core. A fault-free NoC means the router network and the network interface should be fault-free (Figure 2).

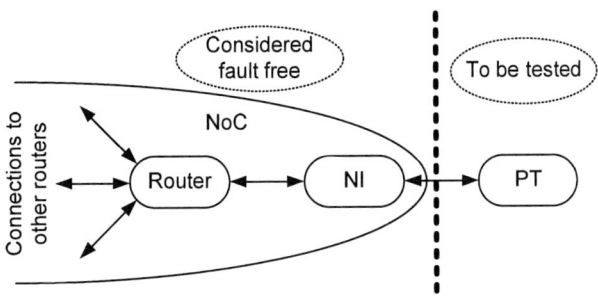

Figure 2. The PT and the NoC partitioning

With respect to the second aspect, similar wrapper design techniques as described in IEEE 1500 standard will be used, which is able to put the Processing Tile in test-mode if required. In the test-mode, the PT will be isolated from its normal input. Instead, the test-vectors from the Test-Pattern Generator will be fed to the CUT and its test response will be delivered to the Test Response Evaluator. When the PT is not in test mode, the wrapper should, from a functional point of view, be transparent.

To perform the on-chip self-diagnostics, a source to generate test patterns (Test Pattern Generator, TPG) and a sink to collect the test responses (TRE, Test Response Evaluator) will be attached to the NoC besides the Cores-Under-Test (CUT, namely the processing tile). As shown in Figure 3, several PTs receive the test patterns at the same time, thereby facilitating the evaluation of the response data. It is assumed that identical cores will always yield the same test response so a faulty response can always be discovered on the basis of a bit-by-bit comparison of the responses from several cores.

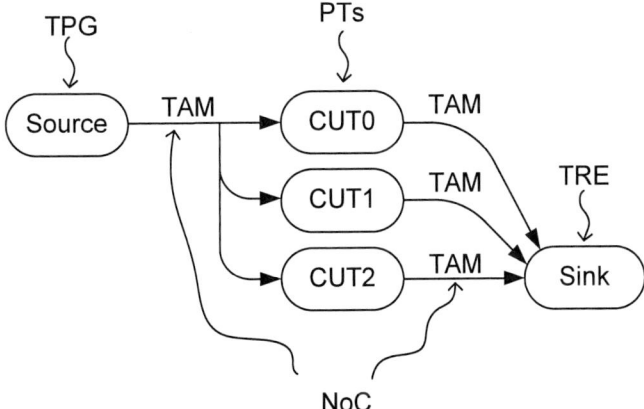

Figure 3. The output of the Test Pattern Generator (TPG) is fed to the different CUTs; the outputs of the CUTs are sent to the Test Response Evaluator (TRE)

As Figure 4 shows, the TRE has two connections to the router network, while the TPG only has one. All data coming from the TPG and going to the CUTs is always the same, as all CUTs receive the same test vectors. This means that the multicast function in the routers enables the

TPG to spread the patterns over the network, so that the single connection to the router network is considered to provide adequate bandwidth to enable a relatively short test time. However, all data from the different CUTs to the TRE is per definition unique. Therefore, the required bandwidth from the routing network to the TRE is larger than the bandwidth from the TPG to the routing network. Several data streams go into the TRE together at the same time, requiring more than one connection to the routing network.

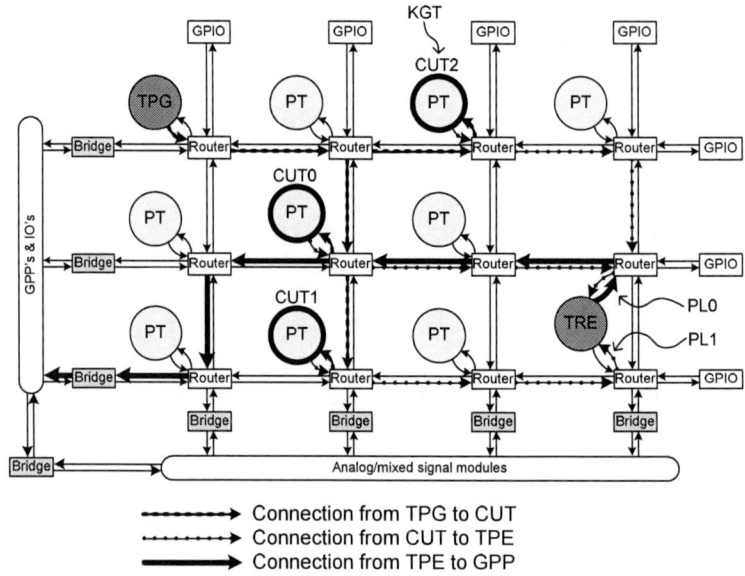

Figure 4. NoC-based Built-In Self-Diagnostics overview (the network interfaces are not shown)

An important issue is of course the relation between test-data volume and test times, actual application data traffic and the bandwidth of the NoC [7, 8].

As already indicated in section II, it requires 2,746 input vectors for testing a processing tile, each with a length of 23,145 elements. This results in more than 63 Mbit of test data volume, as well as more than 63 Mbit of response data volume; hence it is considered necessary to determine in an early stage whether the proposed test setup will lead to acceptable test times and availability of the system. Therefore, a theoretical analysis has been carried out. In this analysis, the following parameter settings have been used that would likely be implemented in the beam former application. The parameters are provided in Table 1:

Parameter	Value
Speed of the processing tile and the router network	100 MHz
Number of processing tiles to be simultaneously tested	3
Data width router network	16 bits
Availability of multicast function	yes
Number of connections from test pattern generator to router network	1
Number of connections from router network to evaluation unit	2

Table 1. Parameter settings used during test-time evaluation

Using the settings in Table 1, the test time is defined as a function of the available bandwidth on the path from TPG to CUTs (BW1) and of the available bandwidth from CUTs to TRE (BW2). The outcome of this analysis is graphically shown in Figure 5.

The analysis shows that a minimum diagnosis time of 400 ms can be achieved for the complete PT if the full bandwidth is assigned to the paths from TPG to CUTs and the paths from CUTs to TRE. When only e.g. 30% bandwidth is available, it is shown that the test time increases to roughly one second. It is stressed that this is the total time needed to test all PTs under test at that time (in this case this equals to three PTs). It is considered that the order of magnitude of these test times is acceptable. Since it is estimated that the application is always able to spare 30% NoC bandwidth for testing purposes, no problems are foreseen with regards to the test time. The diagnosis times for higher resolution [3, 4] are only fractions of the above.

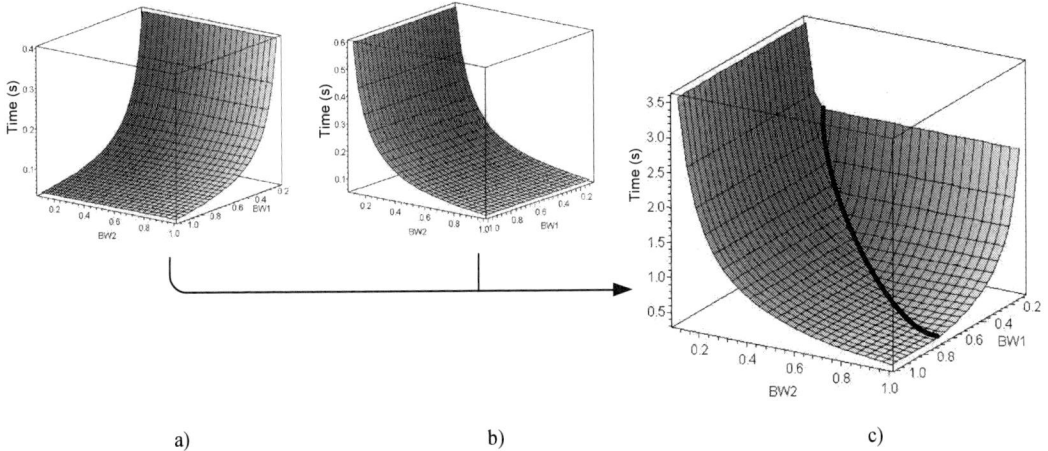

a)　　　　　　　　b)　　　　　　　　c)

Figure 5. a) The time required to deliver patterns from TPG to the CUTs, b) the time required for responses from the CUTs to TRE, c) the total required test time as function of BW1 and BW2.

The GPP in Figure 4 denotes the General Purpose Processor which is responsible for test control and scheduling tests and normal applications for the system. It is shown that data from the Test Pattern Generator is routed to different CUTs simultaneously. The CUTs route their response data to the Test Pattern Evaluator, which compares these responses. If all CUTs are fault free, all responses should be the same. In this case, the TRE signals the GPP that all CUTs are fault free. If one or more CUTs contain one or more errors, differences in response vectors will occur, which will be detected by the TRE. Should all CUTs present a different set of response data, the TRE will send an 'all faulty' command to the GPP. If two CUTs (or more in the case that there are more than three CUTs) have the same set of response vectors, they are considered to be correct. If one or more CUTs differ from these correct CUTs, the TRE will describe the GPP which CUTs differ from one another.

In case the GPP is informed that some PTs are faulty, it is the task of the GPP to re-map the software tasks to the correctly functioning PTs. Although this may reduce the available computing resources and thus lead to a lower Quality of Service (QoS), the chip as a whole can still be considered functionally correct, provided sufficient resource surplus is available.

4. Dependability issues, test response evaluator

The test response evaluator (TRE) has been constructed using a generic approach as much as possible. This enables the smooth fitting of the TRE to any routing network. The generics used are: the NoC data width is sixteen, the number of virtual channels (VCs) that a physical link can hold is four, and the maximum number of router hops from the TRE to the GPP is eight. The number of bits that are required to describe one hop over a router is two, the number of physical links that the TRE should have to the routing network is two, and the maximum number of CUTs that should be tested simultaneously is three. The width of the FIFOs that store the test responses is 16 and finally the depth of these FIFOs is eight. As the TRE disturbs the regularity of the system, it was also investigated to which degree a PT could perform this job; it turned out that this is not a favorable option. Figure 6 shows the setup of the TRE for the values of the generic example as provided above.

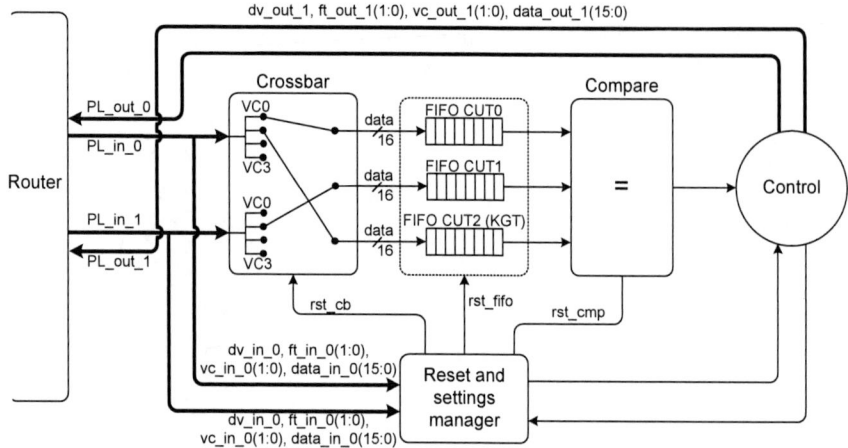

Figure 6. Example of the TRE in the case that there are two links to the router network and three possible CUTs

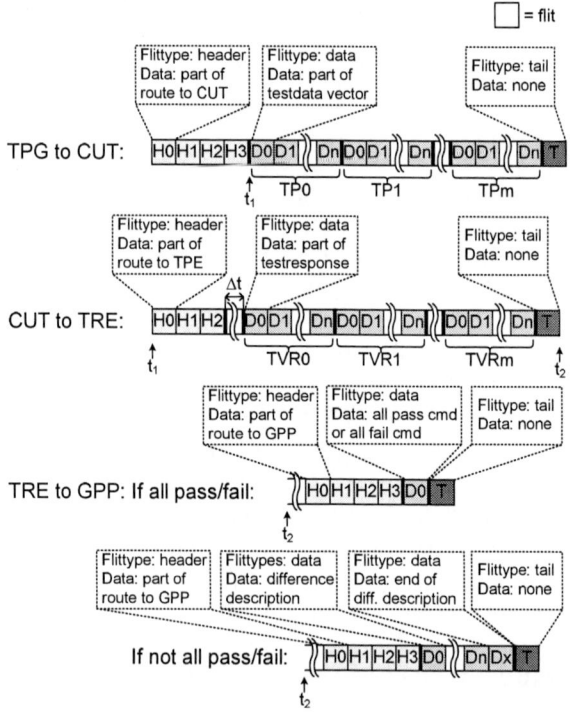

Figure 7. Packet description of the different data streams as depicted in Figure 4.

Flits that enter the TRE are routed to either the reset and settings manager, or to the crossbar. The reset and settings manager collects and stores virtually all settings in the design and resets the other blocks should the GPP request so. The crossbar connects any virtual channels from any physical link to any FIFO. As soon as at least one test response from every CUT is collected in every respective FIFO, the compare unit will examine whether these responses are the same. After this comparison, the FIFO storage space that was occupied by the responses can be freed, enabling the TRE to be implemented with relatively small FIFOs. Once all test data has been fed to the CUTs, the TPG generates a "test-done" signal. This signal is used to take the PTs out of testing mode. The "test-done" signal is also passed to the TRE. Once the TRE has received the "test-done" signals from all the CUTs, the control block sends the result of the test to the GPP, as is depicted in Figure 7, bottom part (TRE to GPP). The details concerning the link that the TRE needs to setup should be stored by the GPP into the reset and settings manager.

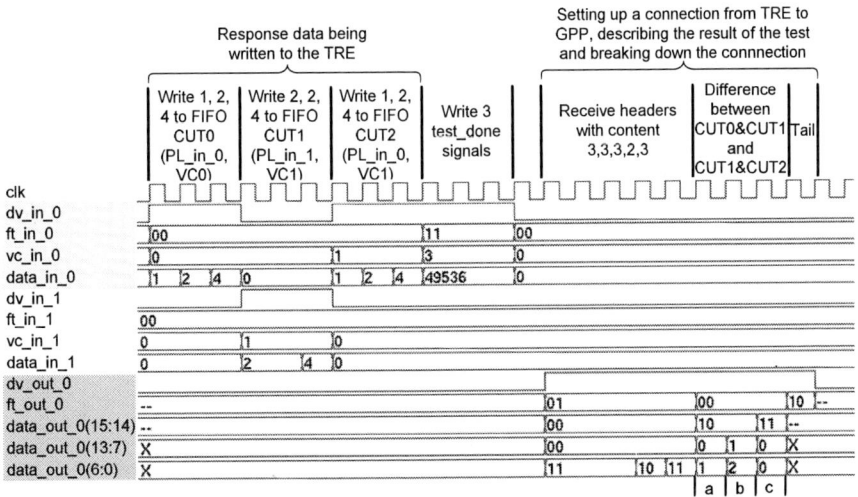

Figure 8. Response data is written to the TRE, the TRE does the comparison, after which the result of this comparison is send to the GPP.

Extensive simulations have been performed to verify the functional correctness of the design using various settings for the generics mentioned. For clarity, a basic simulation result is shown in Figure 8, showing only the signals required for understanding the TRE behavior.

Figure 8 shows how a configured TRE is used to compare the nine response vectors from the three CUTs (the configuration of the crossbar corresponds to the links shown in Figure 6). For clarity, in this simulation all response data from one CUT is send sequentially, while in a real-time situation it would be send in parallel. Also for clarity, a very small response set is send to the TRE: every CUT only sends 3 response vectors to the TRE before generating the "test-done" signal. The correct response should be 1, 2, 4, although the TRE is not aware of this on forehand. After the TRE receives the three "test–done" signals, it sets up a link to the GPP. This is accomplished by sending header flits (flit type "01") to one of the routers connected to the TRE. The content of the header flits describe the direction in which the link should be setup. In this case, as Figure 4 shows, the outcome of the comparison by the TRE should be routed to the GPP via PL0 of the TRE, after which the link is setup by taking three times a router hop to the left (this is shown in Figure 8 by the data_out bus being a decimal 3 for three clock cycles), one hop to the bottom (this is shown in Figure 8 by the data_out bus being a decimal 2 for one clock cycle), and again one hop to the left. Since the TRE did not receive all equal responses, nor did it receive all unique responses, it will send a packet to the GPP indicating that there was a difference in responses between CUT0 and CUT1 (timeslot 'a' in Figure 8, bottom), as well as between CUT1 and CUT2 (timeslot 'b' in Figure 8, bottom). This is coded on the data_out bus in the following way:

$$\{data\_out(15:14) = difference = "10"\} \& \{data\_out(13:7) = CUTx = 0\} \& \{data\_out(6:0) = CUTy = 1\}$$

Timeslot 'c' indicates that the TRE finished the difference description, after which a tail flit (flit type "10") breaks down the connection. Since the GPP now knows that the responses from CUT0 and CUT1 differ, as well as the responses from CUT1 and CUT2, it can conclude that, as apparently CUT 0 and CUT2 do not differ, CUT1 is faulty. This will cause a remapping of resources by the software so that the faulty core, CUT1, is no longer used.

Although the diagnostic outcome of this test is course (it only specifies at Montium tile level which device has (a) stuck-at fault(s)), the process has been extended after failure detection. In this case, the faulty Montium is again provided with additional (diagnostic) test vectors from the TPG, but now the routing is provided at sub-module level (ALU, Memory, interconnect segment). The routines (which modules and when) are pre-stored in the setting manager. Again a fault-free Montium is taken as reference. The power in this approach lies in pre-silicon test-vector generation and evaluation, and the highly reconfigurable TPG. It is noted that the PT cores have been enhanced with DfX hardware and test vectors, to enable diagnostics at the two other levels [3, 4].

5. Dependability issues, the test-pattern generator

Essential in terms of (in-field) dependability, and hence Built-In Self-Diagnostics, is the availability of an extremely flexible (reconfigurable) on-chip test-pattern generator. As the TPG disturbs the regularity of the system, it was also investigated to which degree a PT could perform this job; it turned out that this is not a viable option. The initial stage of our design contains elements of earlier references [9-14].

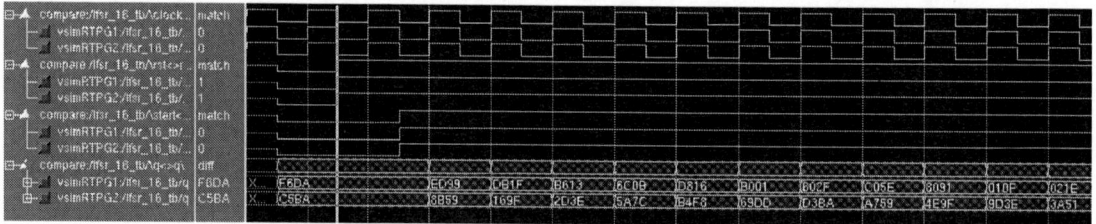

Figure 9. A simulation result of the highly reconfigurable Test-Pattern Generator (two seeds)

Among the reconfigurable parameters of the TPG are multi-seeds, multi-polynomials, and free-chosen lengths [9, 11]. Furthermore, bit-flipping/fixing techniques are being applied to approximate (diagnostic) deterministic tests with the fault coverage of Montium and sub-modules as parameter (DLBIST). In order to reduce hardware and software requirements, a new nested LFSR-based approach is being used, e.g. for generating multi-seeds. It has more flexibility than DBIST [13, 14], where deterministic patterns are encoded in the seeds only, although it is not an automatic design process. Both use TetraMAX as a basis.

Essential in the approach is the availability of all required test vectors, for the Montium as well as its (repeating) sub-modules for diagnostic purposes. This is a pre-Silicon generation and simulation effort. For this purpose a mix of TetraMAX and ModelSim simulations have been used to determine sufficient test coverage for the Montium and sub-modules diagnostics. The obtained key parameters from these simulations are stored on-chip. As an initial example, Figure 9 shows part of generated test vectors, shown here in hexadecimal values, of a required polynomial, with required length and seed. Subsequently, the clock, reset signal (RST), start signal (START) and finally output (q) is shown.

The speed of the TPG is expected to exceed the speed capability of the NoC and Montium tiles in 90nm CMOS technology (100 MHz). The TPG has to work in direct cooperation with the GPP and TRE. Potentially, by adaptation, also transition faults can be handled [15], but this option has not been used yet.

6. Conclusions

In this paper we have shown an effective approach towards increasing the dependability of a system with many identical highly reconfigurable processing tiles, interconnected by a NoC. Its application, a beam former, requires a high degree of dependability. Instead of using traditional (IEEE 1500) test-buses, we use the NoC as a test access mechanism and incorporate internal test-pattern generation, as well as a test-pattern evaluation infrastructural IPs. The diagnostic results at three hierarchical levels, based on the concept of multi-comparison, can be used to perform a re-mapping of the computing resources. Initial designs and simulations of the TPG and TRE have been carried out. The test times required for this diagnosis are small, assuming some part (30%) of the NoC bandwidth, thereby guaranteeing the required availability.

References

[1] B.D. van Veen and K.M. Buckley, "Beamforming: a versatile approach to spatial filtering," ASSP Magazine, IEEE, vol.5, no.2, pp.4-24, April 1988.

[2] P.M. Heysters, G.J.M. Smit, E. Molenkamp, and G.K. Rauwerda, "Flexibility of the Montium Coarse-Grained Reconfigurable Processing Tile", 4th PROGRESS Symposium on Embedded Systems, Nieuwegein, the Netherlands, 2003, pp. 102-108.

[3] H.G. Kerkhoff and J.J.M Huijts, "Testing of a Highly Reconfigurable Processor Core for Dependable Data Streaming Applications", IEEE DELTA Conference, Hong Kong, PRC, January 2008, pp. 38-44.

[4] H.G. Kerkhoff, O. Kuiken, X. Zhang, "Increasing SoC Dependability via Know Good Tile Testing", IEEE DSN Conference, Anchorage, USA, June 2008, pp. G6-G8.

[5] IEEE Computer Society, "IEEE Standard Testability Method for Embedded Core-based Integrated Circuits", 2005.

[6] A.M. Amory, K. Goossens, E.J. Marinissen, M. Lubaszewski and F. Moraes, "Wrapper design for the reuse of a bus, network-on-chip, or other functional interconnect as test access mechanism", IET Comput. Digit. Tech., January 2007, pp. 197–206.

[7] F.A. Hussin, T. Yoneda and H. Fujiwara, "Area Overhead and Test-Time Co-Operation through NoC Bandwidth Sharing", IEEE ATS, Fukuosa, Japan, 2007, pp. 459-462.

[8] E.J. Marinessen et al., "Bandwidth Analysis for Reusing Functional Interconnect as Test Access Mechanism", Proceedings of ETS, session 2B, Verbania, Italy, May 2008.

[9] G. Kiefer, H. Vranken, E.J. Marinessen and H-J. Wunderlich, "Application of Deterministic Logic BIST on Industrial Circuits", JETTA, vol. 17, issue 3/4, 2001, pp. 351 – 362.

[10] K. Chakrabarty, B. Murray and V. Iyengar, "Deterministic Built-In Test-Pattern Generation for High-Performance Circuits using Twisted-Ring Counters", IEEE Trans. on VLSI Systems, Vol. 8, no. 5, 2000, pp. 633-636.

[11] V. Gherman et al., "Efficient pattern mapping for deterministic logic BIST", ITC04, Paper 3.1, Charlotte (NC) USA, 2004, pp. 48-56.

[12] G. Papa, T. Garbolino, F. Novak, A. Hławiczka, "Deterministic Test Pattern Generator Design With Genetic Algorithm Approach", Journal of Electrical Engineering, Vol. 58, No. 3, 2007, pp. 121–127.

[13] M. Chandramouli, "How to Implement Deterministic Logic BIST", Synopsys Compiler Article, January 10th 2003.

[14] "DBIST Userguide", Synopsys, version Z-2007.03, March 2007, 193 pages.

[15] V. Gherman et al., "Deterministic logic BIST for transition fault testing", in Proc. ETS, Southampton, UK, 2006, pp. 225-231.

Network Fault Model for Dependability Assessment of Networked Embedded Systems *

F. Fummi, D. Quaglia, F. Stefanni

Dipartimento di Informatica, Università di Verona,

Strada le Grazie 15, I-37134, Verona, Italy

[franco.fummi|davide.quaglia|francesco.stefanni]@univr.it

Abstract

· ··· ····· ····· · ·· ···· ··· ····· ··· ····· ··· ····· ·· ···· ··· ···· ·· ··· ···· ······ ·· ····· ··· ··· ···· ··· ··· ···· ··· ····· · ···· ···· ·· · ··· ··· ···· ··· ···· ···· ·· ··· ··· ····· ··· ·· · ··· ·· ···· ··· ··· ··· ··· ····· ··· · ·· ··· ··· ···· ···· ····· · 1) ··· ····· ···· ···· ·· ···· ····· ····· · ··· ····· · ····· 2) ··· ····· ···· ········ ········ ····· 3) · ····· · ····· ··· ··· ···· ··· ·· ··· ···· ··· ···· ···· ···· · ···· ···· ··· ··· ···· ·· ····· ··· ···· ···· ···· ·· ··· ··· ··· ···· ···· ···· ·· ···· ··· ···· · ··· ····· ····· ···· ·· ··· ···· ·· ··· · ··· ·· ·· ···· ···· ··· ···· ····· · ····· ···· ··· http://sourceforge.net/projects/scnsl/·

1 Introduction

The high advances in network communications, both concerning protocols and actual HW/SW implementations, have increased the importance of networked embedded systems (NES) as building blocks of complex applications in many critical fields ranging from homeland security to healthcare and building automation. As depicted in Figure 1.a, the architecture of these distributed systems consists of an application plane laying over a network of interacting nodes; each node executes a piece of the whole application and exchanges packets with other nodes through the network [4].

A possible application-driven top-down design approach for this scenario consists in starting from the application to be provided and considering it as a kind of distributed application in which different HW and SW components in different network nodes interact through the network to provide the desired functionalities. In this case, the design space consists in the choice of different ·· ··· ······· –application code, middleware and operating system– ·· ··· ······· –CPU, memory, radio interface, and application-specific devices– and ····· ··· ····· ····· –protocols, number of nodes, their role, network topology. Such components may be either subject of the design process or already available on the market; furthermore they should be representable through models to allow the simulation of the whole architecture. An important step of the design flow of critical applications is the assessment of dependability which is the ability to deliver trusted service also in case of internal failures [2]. Fault simulation plays a decisive role in dependability assessment; if the use of traditional fault models is a challenging task for the ever more complex embedded systems, their use could be impossible for the validation of a distributed application made of hundreds of networked embedded systems. Therefore novel fault models must be found.

In this work we propose a novel fault model to represent the faults of a networked embedded system at a higher level of abstraction so that the dependability of the whole distributed application can be assessed efficiently. The paper describes ·· the proposed fault model in relation with existing

---

Research activity partially supported by the European project FP6-2005-IST-5-033506 ANGEL

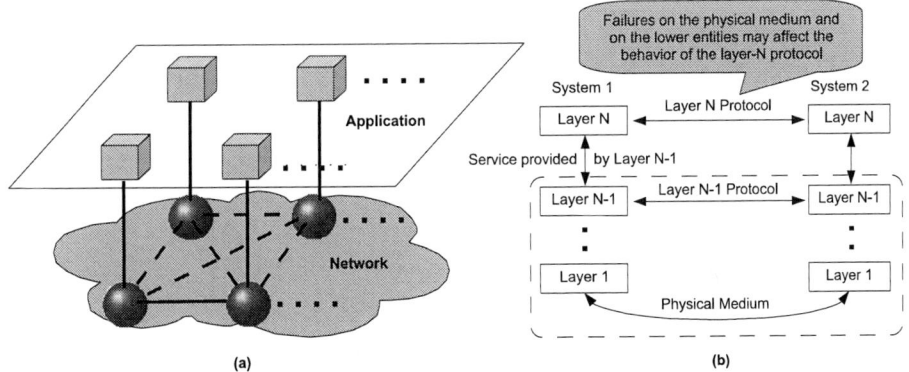

Figure 1. (a) Architecture of a distributed application build on networked embedded systems. (b) ISO/OSI reference model with entities, protocols and services.

ones, •• its possible application scenarios, and •• a SystemC tool for the simulation of both fault-free and faulty wireless sensor networks.

The paper is organized as follows. Section 2 reports some related works on the modelling of networked embedded systems. Section 3 describes the proposed Network Fault Model. Section 4 presents a SystemC library based on this fault model. Section 5 shows the validation of the fault model while Section 6 shows its application to a design problem. Finally, conclusions are drawn in Section 7.

2 Modelling of Networked Embedded Systems

When discussing about NES modelling a question arises regarding the boundary among HW, SW and network components since appropriate tools exist for each domain [9]. For example, in the scenario of two communicating nodes, network behavior depends both on protocol implementation (which in turn relies on HW and SW components) and on elements that are outside the transmitter and receiver such as the physical channel and some intermediate systems.

In this paper, we propose a possible way to abstract from these details by referring to the traditional ISO/OSI layered model reported in Figure 1.b. In this model, the interaction between different systems is represented with the so-called •• •• •• •• between •••• •• •••••• at the same level [16]. Since an entity at level N uses services provided by entities at level $N-1$ without knowing their implementation, then, ideally, the level-N protocol abstracts all the details of the interactions among lower level entities no matter such interactions depend either on HW/SW components of the involved systems or third-party elements like intermediate systems and physical links. For this reason, •• •••••• •• ••• •• •••••• are the focus of our fault model as explained in Section 3.

Different representation techniques are available in literature to model a protocol:

- • ••• •• •• ••• •• •••••••• •• ••••: the protocol is described in terms of processes containing actions and conditions [11].

- • ••• •• ••••: the protocol is graphically represented through a sequence of steps and conditions [1].

- • •• ••• •• •••• • •••• ••• •• • •: the protocol is described in terms of a finite state automaton in which the state of transmitter, receiver and channel are represented [1, 14].

- • •••• • •••: the protocol is described in terms of a Petri Net in which the state of transmitter, receiver and channel are represented [1].

The first and second approach are very similar while computational models and petri nets can be easily translated into FSA's; therefore, we used this last representation to derive our network fault model.

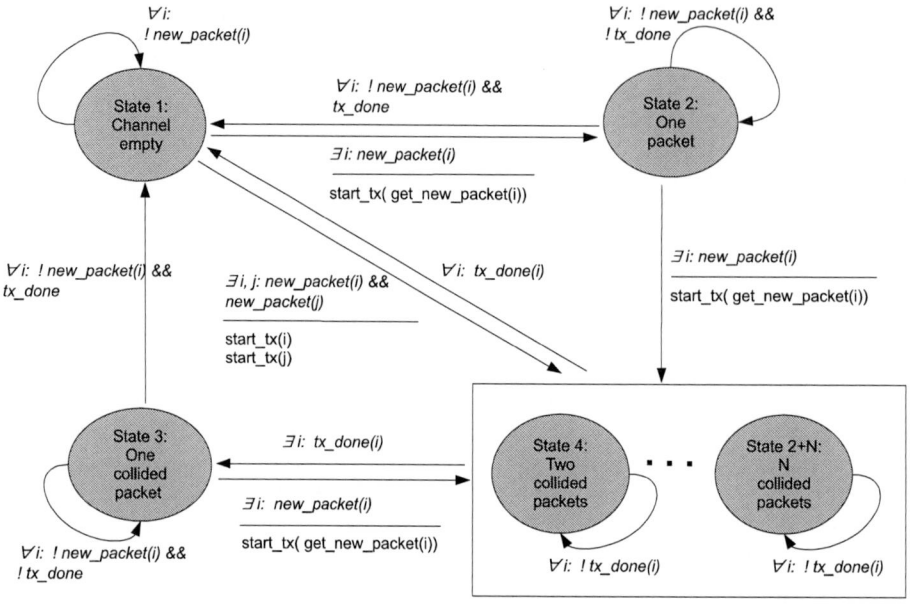

Figure 2. Finite state automaton of a transmission channel.

3 Network Fault Model

Our work focuses on distributed applications build over embedded systems which interact together through packet-based networks.

3.1 FSA of the channel

Figure 2 shows the FSA describing the status of a channel when up to N ISO/OSI level-2 entities interact together by transmitting packets on the physical medium. The FSA is based on the following assumptions:

- the channel status is sampled at a given spatial and temporal point;
- transmission and reception take place on a packet-by-packet basis even if the transmission/reception of a packet is not instantaneous;
- a •• •••••• takes place if two or more packets overlap partially or totally on the channel; in this case these packets are discarded by the receivers since their content has been corrupted;
- each entity is able to check if the channel is busy (carrier sense).

Four states represent channel status. In State 1 the channel is empty since no entity is transmitting. In State 2 only one packet is on the channel and its transmission never collided with another packet. In State 3 only one packet is on the channel but its content was corrupted by a previously sent packet which partially overlapped with it. In State 4 two packets overlap partially or totally on the channel. More states can be added after State 4 when more than two entities interact on the channel; in general, given N the number of entities, up to N packets may overlap on the channel. An objection may arise on this point since, from the channel perspective, N overlapped packets are not so different from two overlapped packets. However states have to be kept separated if we want to track the relationship between each collision and the responsible entities. Transitions between states are annotated with firing events (in italic) and actions. Possible events are the generation of a new packet from an entity and the completion of packet transmission. The number of states has a linear relationship with the number of entities while the number of transitions has a quadratic relationship with the number of entities.

LOST	$(1,2) \rightarrow (1,1)$, $(1,4) \rightarrow (1,1)$, $(1,4) \rightarrow (1,2)$, $(1,4) \rightarrow (1,3)$
	$(2,4) \rightarrow (2,1)$, $(2,4) \rightarrow (2,2)$, $(2,4) \rightarrow (2,3)$
	$(3,4) \rightarrow (3,1)$, $(3,4) \rightarrow (3,3)$
CORRUPT	$(1,1) \rightarrow (1,4)$, $(1,1) \rightarrow (1,3)$, $(1,2) \rightarrow (1,3)$, $(1,4) \rightarrow (1,3)$
	$(2,2) \rightarrow (2,3)$, $(2,4) \rightarrow (2,3)$
CUT	$(2,2) \rightarrow (2,1)$, $(2,4) \rightarrow (2,1)$
	$(3,4) \rightarrow (3,1)$, $(3,3) \rightarrow (3,1)$
	$(4,3) \rightarrow (4,1)$, $(4,4) \rightarrow (4,1)$, $(4,4) \rightarrow (4,3)$
DUPLICATE	$(1,1) \rightarrow (1,2)$, $(1,1) \rightarrow (1,4)$, $(1,1) \rightarrow (1,3)$, $(1,2) \rightarrow (1,4)$
	$(2,2) \rightarrow (2,4)$
	$(3,3) \rightarrow (3,4)$
	$(4,3) \rightarrow (4,4)$
CARRIER	$(1,2) \rightarrow (1,1)$, $(1,4) \rightarrow (1,1)$
	$(2,1) \rightarrow (2,2)$, $(2,1) \rightarrow (2,2)$, $(2,1) \rightarrow (2,3)$, $(2,2) \rightarrow (2,4)$
	$(3,1) \rightarrow (3,4)$, $(3,1) \rightarrow (3,3)$, $(3,3) \rightarrow (3,4)$
	$(4,1) \rightarrow (4,2)$, $(4,1) \rightarrow (4,4)$, $(4,1) \rightarrow (4,3)$
Meaningless	$(3,3) \rightarrow (3,2)$, $(3,1) \rightarrow (3,2)$, $(3,4) \rightarrow (3,2)$
	$(4,4) \rightarrow (4,2)$, $(4,3) \rightarrow (4,2)$

Table 1. Classification of single transition faults according to their effect on network behavior.

The representation of system behavior through finite state automata is a common approach and different fault models have been developed for this kind of representation. Among all, the so-called Single Transition Fault Model [3] assumes that one transition at a time changes its destination state. Table 1 reports the transition faults which can be obtained from the FSA of Figure 2 assuming $N = 2$. After having separated the meaningless transition faults, a classification has been applied to the meaningful faults according to the effect on the resulting packet sequence.

Five different effects can be highlighted as follows:

LOST : the packet disappeared from the channel.

CORRUPT : some bits of the transmitted packet were flipped.

CUT : some adjacent bits of the transmitted packet disappeared from the channel.

DUPLICATE : the transmitted packet was sent twice.

CARRIER : the test on the use of the channel is either positive or negative independently of the presence of concurrent transmissions.

We define this set the Network Fault Model (NFM). NFM can be used to model failures of the communication task among two or more entities. NFM can be applied to every packet-based protocol at every level of the ISO/OSI stack with one exception for the CARRIER fault which can be applied only at level 2 (datalink).

Compared to a traditional fault model (e.g., the reported Single Transition Fault Model), NFM provides more abstract faults which better describe wrong network behavior and allow a more efficient fault simulation. With the FSA representation and the Single Transition Fault Model, the number of possible faults is proportional to $E * T$ where E is the number of involved entities and T is the number of correct transitions; since T is proportional to E^2 then the number of possible faults is proportional to E^3 and, therefore, it may become very large when the distributed application involves a large number of nodes. Furthermore, some of these faults may be meaningless since do not correspond to any actual behavior (either correct or wrong). With NFM, the number of possible faults is reduced, being proportional to E and only faults corresponding to an actual network behavior are considered.

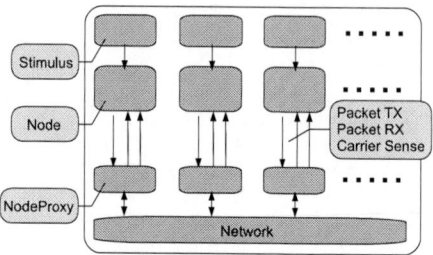

Figure 3. Architecture of a network scenario modeled with SCNSL.

Network Fault Model can be used to test the robustness of the application, i.e., its capacity to provide coherent services also in presence of failures. NFS can represent failures of the entities and of the channel between them and thus it can be viewed as a generalization of past works on network failure modelling which focused only on HW and SW failures [5, 6, 10, 15, 7, 12].

This fault model can also support the testing of the application to find design errors. Un-detected faults can reveal SW and HW components that are not used (dead components), communication steps which are useless and transmitted bits which come un-checked through the network.

When a testbench is used to test the application model, network faults can be used to support the automatic generation of exhaustive test patterns as in case of traditional fault simulation but with the advantage that the Network Fault Model is more abstract and therefore increases simulation performance.

Finally, the Network Fault Model can give support to formal verification to evaluate the accuracy of model checking as described in [8].

4 NFM-based SystemC simulation

In this Section a SystemC tool for network fault simulation is described; it consists of the SystemC Network Simulation Library which provides the primitives to build and simulate network scenarios and the Network Fault Simulation Library which implements the Network Fault Model and performs fault simulations.

4.1 SystemC Network Simulation Library

Figure 3 shows the architecture of a network scenario modeled with the SystemC Network Simulation Library. Four components are highlighted, i.e., the `Node`, the `Stimulus`, the `NodeProxy`, and the `Network`. Module `Node` models a network node; concerning the network it has two input ports and an output port to model network input, received signal energy and network output, respectively. The presence of a network port for each direction allows to model both wired and wireless interfaces; the input port reporting the received signal energy has been introduced since in some modern wireless interfaces it can be used both for carrier sense and localization techniques based on the evaluation of the received signal strength. Module `Node` has a set of properties which are used by the simulation framework to reproduce network behavior. Transmission •• •• represents the number of bits per unit of time which the interface can handle; it is used to compute the transmission delay and the network load. Transmission ••• •• is used to evaluate the transmission range and the signal-to-noise ratio. Module `Node` has also an input port to model application-specific features (e.g., a temperature sensor).

The data input port of each node can be bound to an instance of the module `Stimulus` which reproduces a generic environmental data source. It takes as input a clock signal as timing reference to synchronize the generation of data values. Different kinds of stimuli can be generated by sub-classes of this module.

Class `Network` is the core of the network simulator. It reproduces the behavior of the channel and manages the packet forwarding from the source node to destination nodes; transmission delay, path

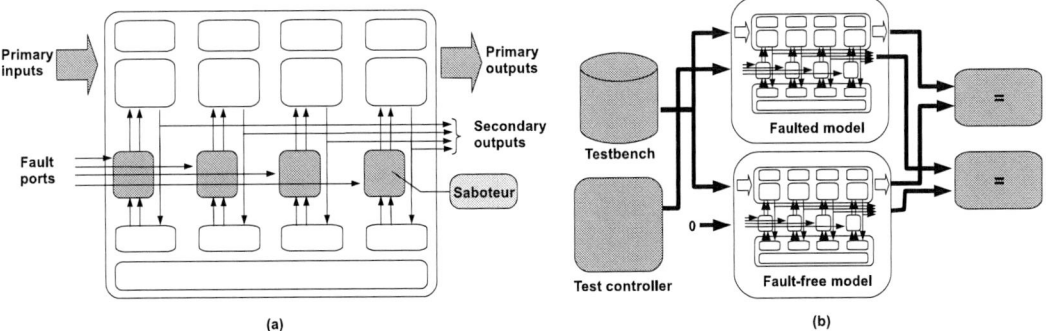

Figure 4. (a) Role of the saboteur in the Network Fault Simulation Library. (b) Architecture of the verification system based on the Network Fault Simulation Library.

loss, collisions, and the state of destination nodes are taken into account. To accomplish this task it maintains an object reference for each `NodeProxy` and keeps track of every on-going transmission.

Module `NodeProxy` is the interface between nodes and the network and each instance of `Node` must be bound to a different instance of `NodeProxy`. Each `Node` interacts with its own `NodeProxy` by using SystemC RTL ports only, while `NodeProxies` maintain an object reference of the `Network` and call its methods. By using `NodeProxy`, nodes can be designed as pure SystemC modules without object references to other non-SystemC classes; this approach enables the use of traditional hardware verification and synthesis tools. `NodeProxy` manages two node properties: node position and receiver sensitivity. Node •••••• is used to compute the path loss and to reproduce a mobile scenario. The •••••• ••• •••••• is the minimum signal power below which the packet cannot be received.

The SystemC Network Simulation Library reproduces the network behavior at packet-level. In the following text the behavior of the library is described with reference to a wireless scenario. When a node starts to transmit, its relative position to all other nodes in the same network is calculated, and the signal level in all those nodes is calculated according to a path loss rule. For each node, if the signal level is larger than the receiver sensitivity, then it can be detected and it may interfere with other on-going transmissions. If there are already on-going transmissions reaching the receiving node, then all those messages are marked as collided. Also, if there are other on-going transmissions which the currently sending node reaches with its transmission, then those messages are marked as collided as well. Since wireless nodes cannot detect collisions, a collided message is not interrupted and the channel remains busy.

4.2 Network Fault Simulation Library

Further modules are required to perform fault simulation. The first one is the so-called `Saboteur` which reproduces the network faults described in Section 3 according to a value on a specific fault port (Figure 4.a). For each node which will be subject to faults an instance of the `Saboteur` is created on the signal path from the `NodeProxy` to the `Node`.

As depicted in Figure 4.b, other components of the Network Fault Simulation Library are the `Test Controller` and the modules that compare the outputs of the faulted and fault-free models. Concerning the outputs of the model under test, it is worth to distinguish between the •••• ••• •••••• which represent the actual outputs of the application under test and the ••••• ••• •••••• which are extracted from the signals transmitting packets from nodes to the network. The observation of both primary and secondary outputs can give interesting information about the effect of faults on the node behavior as detailed in the experiment described in Section 5.

By using this implementation we made two different kinds of test. The first test (Section 5) aims at showing the effectiveness of the Network Fault Model to assess the dependability of a distributed system. The second test (Section 6) shows the use of the Fault Simulation Library to simulate an actual application based on a wireless sensor network; the tool supports the designer to determine the best trade-off between the number of nodes and application performance as a function of node and channel failure rate.

5 NFM validation

The test described in this Section aims at showing the effectiveness of the Network Fault Model to assess the dependability of a distributed system. The architecture of the verification setup is depicted in Figure 4.b. The application consists of five pairs of nodes implementing a datalink protocol to exchange packets. We tested six different protocols of increasing complexity and known robustness taken from literature [16]; in this way the proposed approach is validated against a well-known benchmark. The tested protocols have the following characteristics:

Protocol 1 : it is the simplest unidirectional transmission protocol since it does not care about channel reliability, buffer state, data availability.

Protocol 2 : it is a unidirectional stop-and-wait protocol; after sending a data packet, the transmitter waits indefinitely an acknowledge from the receiver.

Protocol 3 : it is a full unidirectional stop-and-wait protocol with a timed acknowledge mechanism.

Protocol 4 : it is a bidirectional one-bit sliding-window protocol in which data and acknowledge are sent in the same packet.

Protocol 5 : it uses a go-back-n strategy for incorrect packets.

Protocol 6 : it uses a selective repeat transmission for incorrect packets.

Table 2 reports the number of detected faults as a function of transmission protocol and fault type; each cell reports the number of detected faults by considering the primary and the secondary outputs, respectively. In columns regarding each fault type, the maximum number of detected faults is ten since five pairs of nodes are simulated. The last column reports the total number for all fault types. Results show that, for each fault type, the tested protocols behave in a different way, according to their features.

The analysis of the behavior of Protocol 1 as a function of the fault type shows that only half of the faults are detected since the receiver has no role in the data transmission (i.e., it behaves as a listener). This result demonstrates that the Network Fault Model can reveal dead components. With this very simple protocol, CARRIER is the only fault which may affect the sender's behavior, i.e., it represents a sender's failure while other fault types represent channel failures. For this reason the pair (Protocol 1, CARRIER) is the only one in which the fault is detected also by looking at the secondary outputs.

Except for the pair (Protocol 2, CARRIER) the number of faults detected by looking the secondary outputs is lower than by looking the primary outputs. The result for the pair (Protocol 2, CARRIER) is caused by the masking effect of collisions at application level; only by looking secondary outputs, i.e., channel status, it is possible to distinguish if a packet is dropped due to collision or if it has never been sent due to a node failure.

As expected, the least detected fault is CORRUPT since all protocols do not provide error detection on received data; this conjecture is confirmed by the result on the pair (Protocol 6, CORRUPT), the only one in which CORRUPT fault is detected even if on the secondary outputs; in fact, Protocol 6 is the only one in which protocol behavior depends on packet content (i.e., sequence number). This fact shows that the Network Fault Model can reveal weaknesses in case some packet fields are either useless or corrupted due to node or channel failures.

6 Case study

The case study aims at showing the use of the Network Fault Model and the Network Fault Simulation tool in a more practical context in which the designer of a wireless sensor network application needs to determine the minimum number of nodes required to monitor an environmental scalar value (e.g., the temperature) with a given precision and a given network/node reliability. A

Protocol type	Fault type					Total
	CARRIER	CORRUPT	CUT	DUPLICATE	LOST	
1	5–5	5–0	5–0	5–0	5–0	25–5
2	6–8	5–0	8–8	5–5	8–8	32–29
3	10–10	5–0	10–10	1–1	10–10	36–31
4	10–10	10–0	10–10	8–8	10–10	48–38
5	10–10	10–0	10–10	10–10	10–10	50–40
6	10–10	10–10	10–10	10–10	10–10	50–50

Table 2. Number of detected faults as a function of transmission protocol and fault type: each cell reports the number of detected faults by considering the the primary and the secondary outputs, respectively.

master node sends a data request to each sensor node deployed on the monitored area; the sensor node replies with the sensed data and then the master computes the average over all collected samples. The testbench reproduces the monitored area consisting of a scalar field with five sub-areas and different data values for each of them. Nodes are placed in the sub-areas with the aim of covering the whole area; nodes belonging to the same sub-area get the same sample. Network reliability is studied by using the Network Fault Model; in particular, the LOST fault type has been used. Let the •• • •• •• •• be the number of faulty nodes over the total number of nodes. Sample values do not change with time since in this work we considered only static transmission scenarios in which nodes position and failures do not change with time. In this scenario node communications reproduce a subset of the well-known IEEE 802.15.4 standard, i.e., peer un-slotted transmissions with acknowledge [13]. For all tested scenarios the simulation length was constrained by the fault-free simulation time.

Figure 5 plots the average value of collected data as a function of the number of sensor nodes for different values of the fault ratio. The true value of average is 1,657,509,420.6 (reported as a constant line on the plot). As expected, the accuracy of the result increases with the number of nodes except with 50% fault ratio because the time wasted for retransmission to faulty nodes is too high and the longer simulation time prevents the transmission of enough data.

The reported simulation results can be used to determine the minimum number of sensor nodes required to limit the error of the resulting sample average. For example, by allowing an error of 10% at least three sensor nodes are required assuming a near 0% fault ratio while a higher number is required in case of real-world failure-prone nodes (e.g., at least nine sensor nodes in case of a 33% fault ratio). This example shows the effectiveness of the Fault Simulation Library to support design-space exploration.

7 Conclusions

A novel fault model has been proposed to support the design of large distributed applications based on networked embedded systems. The Network Fault Model represents system and channel failures as a set of wrong behaviors in packet transmission. The approach has been compared with a traditional fault model based on finite state automata. A SystemC tool has been also developed to perform fault simulation according to this model. The Network Fault Model and the tool have been validated through the dependability assessment of well-known datalink protocols. Finally, the tool has been applied to a real-life sensor network application showing its valid support to design-space exploration.

References

[1] A. A. S. Danthine. Protocol representation with finite-state models. *IEEE Trans. on Communications*

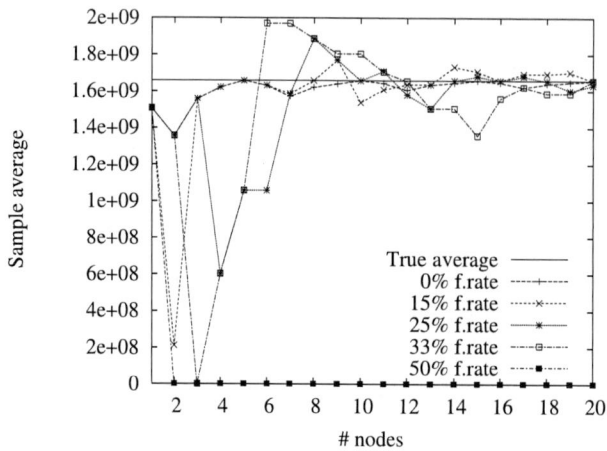

Figure 5. Average value of collected data as a function of the number of sensor nodes for different values of the fault ratio.

and Parallel Distrib. Syst., 28(4):632–643, Apr. 1980.

[2] A. Avizienis, J. Laprie, and B. Randell. *Fundamental Concepts of Dependability*. Research Report N01145, LAAS-CNRS, Apr. 2001.

[3] K.-T. Cheng and J.-Y. Jou. A single-state-transition fault model for sequential machines. *Computer-Aided Design*, pages 226–229, Nov. 1990.

[4] C.-Y. Chong and S. P. Kumar. Sensor networks: Evolution, opportunities, and challenges. *Proc. IEEE*, 91(8):1247–1256, Aug. 2003.

[5] M. El-Darieby and A. Bieszczad. Intelligent mobile agents towards network fault management automation. In *Proc. of the Sixth IFIP/IEEE International Symposium on Integrated Network Management*, pages 611–622, 1999.

[6] Denise W. Gurer et al. An artificial intelligence approach to network fault management.

[7] G. Jacques-Silva et al. A network-level distributed fault injector for experimental validation of dependable distributed systems. In *Proc. of International Computer Software and Applications Conference (COMPSAC)*, 2006.

[8] A. Fedeli, F. Fummi, and G. Pravadelli. Properties incompleteness evaluation by functional verification. *IEEE Trans. on Comp.*, 56(4):528–544, Apr. 2007.

[9] F. Fummi, D. Quaglia, F. Ricciato, and M. Turolla. Modeling and simulation of mobile gateways interacting with wireless sensor networks. In *Proc. IEEE Conf. on Design, Automation and Test in Europe (DATE)*, Mar. 2006.

[10] R. D. Gardner and D. A. Harle. Alarm correlation and network fault resolution using the kohonenself-organising map. In *Proc. of IEEE Global Communication Conference*, pages 1398–1402, 1997.

[11] Mohamed G. Gouda. Protocol verification made simple: a tutorial. *Computer Networks and ISDN Systems*, 25(9):969–980, 1993.

[12] W.-L. Kao and R. K. Iyer. DEFINE: a distributed fault injection and monitoring environment. In *Proc. of IEEE FTPDS*, pages 252–259, 1995.

[13] LAN/MAN Standards Committee of the IEEE Computer Society. IEEE Standard for Information technology - Part 15.4: Wireless Medium Access Control (MAC) and Physical Layer (PHY) Specifications for Low Rate Wireless Personal Area Networks (LR-WPANs). Sept. 2006.

[14] F. J. Lin, P. M. Chu, and M. T. Liu. Protocol verification using reachability analysis: the state space explosion problem and relief strategies. In *Proc. of the ACM workshop on Frontiers in Computer Communications Technology*, pages 126–135, New York, NY, USA, 1988. ACM.

[15] W. Najjar and J.-L. Gaudiot. Network resilience: A measure of network fault tolerance. *IEEE Trans. on Computers*, 39(2):174–181, Feb. 1990.

[16] A. S. Tanenbaum. *Computer Networks*. Prentice Hall, 2003.

Obtaining Microprocessor Vulnerability Factor Using Formal Methods

Syed Z. Shazli and Mehdi B. Tahoori
Northeastern University
Boston, MA
{sshazli,mtahoori}@ece.neu.edu

Abstract

· Microprocessor Vulnerability Factor ·

1 Introduction

A · · · · · · · · occurs when a radiation event causes enough of a charge disturbance to reverse or flip the data state of a logic gate, memory cell, register, latch, or flip-flop. The error is soft because the circuit/device itself is not permanently damaged by the radiation. The rate of soft errors in a system can be higher than that of all other errors combined, making soft errors one of the primary concerns for system reliability [5]. The vulnerability of VLSI systems to soft errors exponentially increases as an unwanted side effect of Moore's law.

To ensure reliability requirements, there is a target soft-error rate (SER) for each microprocessor design and significant pre-silicon analysis is performed to ensure a design adheres to this target. Therefore, accurately estimating the SER of a particular design early in the design stages is crucial to measuring a design's performance against its SER target.

This type of early-stage SER analysis has been greatly aided by the concept of · (AVF) and the introduction of · (ACE) Analysis as a method to estimate a structure's AVF [12][21]. The AVF of a processor structure is defined as the probability that a fault in that structure will result in a visible error in the final output of a program. This is a well-defined, measurable quantity that both yields insight into the behavior of a structure and allows a simple calculation to determine its failure rate. ACE Analysis estimates a structure's AVF by determining, during each cycle, whether a fault in a bit will propagate to the program's output. This allows a designer to estimate a structure's AVF in a single, fault-free, simulation run, significantly faster than prior techniques such as software fault-injection which require hundreds of runs to achieve statistical significance.

Although AVF analysis is developed to estimate the reliability parameters of a hardware structure, it also depends on vulnerability factors inherent to the program running on the microprocessors. In other words, the vulnerability parameters reported by ACE analysis are partly due to the processor structure and implementation (hardware) and partly related to the properties of the micro-code (software) running on the microprocessor and how different values at different cycles matter in correct execution (AVF of different programs running on the same microprocessor are different) [12].

1550-5774/08 $25.00 © 2008 IEEE
DOI 10.1109/DFT.2008.52

In this paper, we define a metric called • •••••••••••• • •• •••• •••• ••••• (MVF) which captures the reliability of a particular structure, independent of the vulnerabilities inherent to specific program running on the microprocessor. MVF is the probability that an error in any bit of the internal processor structure will result in an error in a program visible state. We also present a technique using Boolean Satisfiability (SAT) to obtain MVF of a high-level (behavioral or RTL) processor description. In our prior work, Boolean Satisfiability has been used to estimate the SER in early design stages [16]. However, only combinational circuits were experimented with. In this work, we transform the MVF computation problem into an equivalent Boolean satisfiability (SAT) problem and use state-of-the-art SAT solvers to obtain MVF. This approach allows us to obtain ••••• MVF values in contrast to fault simulation ••••• ••••• approaches. Moreover, the proposed technique is significantly faster than prior techniques such as software fault-injection which require hundreds of runs to achieve statistical significance [9]. We have developed an automated flow to convert high-level processor descriptions into SAT formulations for efficient MVF computation. We have implemented and applied this flow for commonly-used pipelined Reduced Instruction Set Computing (RISC) microprocessors.

The remainder of this paper is organized as follows: In Section 2, we present an overview of SER estimation. Description of MVF as well as our proposed MVF computation technique is presented in Section 3. We discuss experimental results for pipeline RISC processors in Section 4. Finally, Section 5 concludes the paper.

2 Background and Related Work

2.1 Soft Error Modeling and Estimation Techniques

Continuing technology scaling and the increase in frequency toward the GHz range have led to a number of consequences for the SER sensitivity of logic. First, critical charges are decreasing, due to smaller node capacitances and lower operating voltages. Furthermore, the propagation of a transient pulse (an SE glitch) has become much more efficient. Therefore, the SER sensitivity of logic is a substantial and growing concern [4].

Error Propagation Probability (EPP) of the gates has been used in [23] and [3] to measure the contribution of each gate to the overall soft error rate. A gate with a higher EPP means that a bit flip at the output of the gate is more likely to cause an error at the primary outputs of the circuit.

A study on the modeling of soft errors in combinatorial logic was performed at the VHDL level in [4]. Soft-Error Monte Carlo Model, or SEMM, was developed in [18]. The method does not require arbitrary fitting parameters and expensive high-energy beam testing. In [17], the effect of technology trends on electrical masking are studied.

The soft error sensitivity of various memory and logic devices used to create advanced commercial electronic systems as a function of technology scaling has been presented in [5].

The architecture-level approaches for soft error modeling include Fault Injection by Wang [20] and Classifying Architecturally Correct Execution bits (ACE bits) by Mukherjee [12]. The Architectural Vulnerability Factor of a processor structure is defined as the probability that a fault in that structure will result in a visible error in the final output of a program [12]. A bit in which a fault will result in incorrect execution is said to be necessary for ••• •••••••• ••• ••••••• •••••••• ; these bits are termed ACE bits. All other bits are un-ACE bits. An individual bit may may be ACE for a fraction of the overall execution cycles and un-ACE for the rest. Therefore, the AVF of a single bit can be defined as the fraction of cycles that the bit is ACE. The average AVF of an entire •••••••• can be computed as the weighted average of the AVFs of each structure for systems of reasonable size [10].

2.2 Boolean Satisfiability

The Boolean satisfiability (SAT) problem is the problem of deciding if there is a truth assignment for the symbols that appear in a Boolean function such that it assigns the value • ••• to the Boolean function. The boolean function is usually specified in product-of-sums or ••• •• ••••• •••• •• •••• or CNF. A CNF is a set of clauses. This form consists of the logical AND of this set of clauses. A clause is a set of literals and a literal is a variable or a negated variable. The CNF is satisfied under a given assignment for the variables if and only if (iff) all clauses are satisfied. A clause is satisfied iff at least one literal is satisfied. A literal is satisfied, iff the variable is not negated and has assigned the value 1 or the variable is negated and has assigned the value 0. If there is a satisfying assignment, the CNF is called satisfiable. Otherwise the CNF is called unsatisfiable [6].

Given a boolean expression for a logical network with n input variables, the SAT problem determines a combination of 1s and 0s for the n input variables such that the output of the boolean expression evaluates to 1 (SAT). If no such input combination exists such that the output is 1, the boolean expression is said to be unsatisfiable (UNSAT). A solver for boolean satisfiability must generate an input combination such that the output of the given boolean expression evaluates to 1 or reports that the expression is unsatisfiable if no such input combination exists.

Most modern SAT solvers are based on the DPLL algorithm which performs a branching search with backtracking. The algorithm is complete in the sense that it finds a solution if and only if the formula is satisfiable. Modern SAT solvers use a conflict analysis technique to analyze the reasons for a conflict. A good survey of modern SAT solvers is given in [13].

The success of SAT solvers has led to the increased popularity of a subtle derivative, the all-solution SAT solver. While in general, the SAT solvers seek to find a single solution to a SAT problem, the all-solution SAT solvers seek to find all possible solutions to a SAT problem. Typically, all-solution SAT solvers iteratively call a standard SAT solving procedure to find each solution to a problem. At each iteration, when the standard SAT solver returns a solution, a blocking clause is added to the problem to prevent it from discovering the same solution in future iterations. Since the number of solutions can be exponential to the problem size, 'compacting' the solutions at each iteration is critical for the efficiency of the solver [15].

Several RT-level test pattern generators based on boolean Satisfiability have been proposed in the literature [7] [22] [11]. Constraint propagation techniques between different domains have been explored to generate functional tests and high-level ATPG vectors on HDL descriptions in [7]. In [22], a linear programming based infrastructure is presented to solve the SAT problem for complex RTL designs. The approach handles both word-level arithmetic operators and bit-level Boolean logic. Recently, a SAT-based framework for automatically generating test programs that target gate level stuck-at-faults in microprocessors has been presented in [11]. The microarchitectural description of a processor is translated into a unified RTL circuit description. Test generation involved extracting justification/propagation paths from a module's I/O ports to primary I/O ports, abstraction of RTL modules in the justification/propagation paths and finally translation of these paths into Boolean clauses represented in CNF. The CNF clauses were sent to a SAT solver to find out satisfying input assignments.

3 Microprocessor Vulnerability Factor

We define Microprocessor Vulnerability Factor (MVF) as a metric to assess the reliability of the particular hardware implementation of an Instruction Set Architecture (ISA), independent of the applications running on this ISA. Since ISA is the interface between processor organization/structure (hardware) and instruction-level programming (software), MVF considers the reliability below ISA level. This is in contrast to AVF which is application-dependent and crosses ISA boundary. By separating application-independent vulnerabilities (captured in MVF) from application-related vulnerabilities (such as Program Vulnerability Factor) it is possible to optimize the overall reliability through hardware design and application/compiler optimization.

MVF is defined as the probability that an error in any internal bit (e.g. register file, pipeline stage registers, control finite state machine, internal registers) propagates to an architecturally visible state (e.g. ISA registers, memory interface). In other words, MVF is the probability of an internal error propagating to instruction boundaries. The error sites could be only sequential elements (flip-flops, latches, and registers) or can be extended to cover SEUs in combinational elements (logic gates). MVF for a processor structure is the average of per-bit MVF over all error sites.

Next, we present a technique for obtaining MVF of RTL processor descriptions based on Boolean Satisfiability (SAT).

3.1 MVF Computation Methodology

In order to obtain MVF, we compute ••••• ••••••••••• •••••••••• (EPP) which is the probability of error propagated from an error site to the primary outputs. EPP of an error site (bit in the processor description) is a ratio between the number of input combinations which resulted in an observable error and the total number of possible input combinations. In the proposed SAT-based approach, the processor description (RT-level) is modeled as a SAT instance. A faulty version and a fault-free version of the description are instantiated. In the faulty version, a fault is inserted by providing a statement in the HDL description in order to flip a bit on the error site. The outputs of fault-free design and the fault-inserted version are compared by XORing the corresponding outputs of these two version. The output of each XOR gate is 1 if and only if the error due to a bit-flip in the error site is propagated to that primary output. All the outputs are ORed together to generate a miter output, i.e. the error is propagated to the output if it appears on at least one primary output.

The overall description (including fault-free and fault-inserted versions, XORs, and OR gate) is then converted into a SAT instance. This SAT instance is only satisfiable, iff an input pattern is found that yields a wrong output value (1) in the presence of a fault.

The above approach needs to be slightly modified for pipeline structures. The difference lies in converting the circuit to its CNF equivalent. In pipelines, an error on one bit may take several cycles in order to be observable at the primary outputs because of sequential behavior of the circuits. As CNF form is purely combinational, multiple unrollings have to be performed to convert the circuit to a combinational one. In this case, the pipeline structure is converted into a combinational CNF by unrolling (i.e. copying) the combinational core of the sequential circuit n times, where n is equal to the number of pipeline stages in the processor. The bit flip is inserted in the particular combinational copy corresponding to the error site. The corresponding outputs of the fault-free and fault-inserted versions are XORed based on the pipeline structure of the processor. Specifically, the error is observable if it is propagated to architecturally visible states, which are ISA registers (register file, program counter) and the memory interface. Therefore, there is an error if the content of any ISA register is erroneous at the end of write-back stage (WB) or an incorrect value is written into the memory. The latter happens when an incorrect write-enable signal is asserted or incorrect value (address or data) is generated in the Memory Access stage (MEM). Based on the particular pipeline structure, the corresponding signals need to be XORed at particular combinational copy representing those pipeline stages.

This increases the size of the CNF and the SAT solvers take more time to find all possible satisfying input combinations. Also, the input space for sequential circuits will be much larger compared to combinational ones. The EPP due to a single bit i is computed as follows:

$$EPP_i = \frac{number\ of\ SAT\ instances}{2^{number\ of\ PIs \times loop\ unrollings}}$$

3.2 Automation of the Flow

In our implementation, unrolling based on number of pipeline stages is performed to obtain the (combinational) CNF. The (high-level) processor description in Verilog is first converted into the BLIF format. The unrolling, error insertion, and XORing are performed on the BLIF format and equivalent CNF was obtained. The CNF file thus obtained is sent to an all solutions SAT solver Relsat [14]. RELSAT finds the number of all satisfying input solutions to the given instance. The program uses the formula for finding EPP as given in the previous section to compute MVF. The algorithm used is shown in Figure 1.

4 Implementation and Results

We have implemented and applied SAT-based SER computation for combinational designs [16]. The results confirmed that this methodology outperforms fault simulation in terms of time taken to find SER of a circuit. Unlike fault simulation which is an ••••• ••••• method, the SAT-based technique is an ••••• method (100% accurate).

INPUT: Processor description in behavioral HDL
OUTPUT: MVF for the processor
ALGORITHM:
FIND_MVF (Processor P)

1. Make another copy of the module under test.
2. XOR the corresponding outputs of the two modules
   a.   ISA Registers after WB stage
   b.   Memory write values (data and address) after MEM (write) stage
   c.   Memory enable signal at all stages
3. OR all the outputs together to get a single output.
4. **For** each bit (error site) w in P **do**
   a.   XOR w with a 1 in the error-inserted module in the corresponding combinational copy
   b.   Unroll the description by the number of pipeline stages to obtain a pure combinational version
   c.   Convert this description into a CNF equivalent
        i.   Test for the property output =1
   d.   Send the circuit to all-solution SAT solver
   e.   Record the number of satisfying assignments (N_{SAT})
   f.   Compute EPP_w for wire w = $(N_{SAT})/2^{inputs}$
5. Find MVF_P using the formula

$$MVF_p = \frac{\sum\limits_{for.all.bits} EPP_w}{Total.number.of.bits}$$

Figure 1. SAT-based MVF Computation Algorithm

The tool was developed in C++, and experiments were carried out on pipeline RISC processor descriptions given in behavioral Verilog. As mentioned in the previous section, the processor description is converted from behavioral Verilog to CNF for sending to an all-solution SAT solver RELSAT [14]. The flow to convert to CNF is shown in Fig. 2. Synopsys was used to get a gate-level description in Verilog for the processor. Some utilities provided with the ABC Verification Suite [1] were used to convert the description to BLIF format. AIGER utilities [2] were used to convert from BLIF to CNF.

Figure 2. Flow from Behavioral Verilog to CNF

4.1   Five-stage Pipelined MIPS Microprocessor

MIPS is a widely used RISC processor as described in [8]. We have used a five-stage pipeline implementation of the 32-bit integer MIPS processor described in Verilog at RT-level. The pipeline is depicted in Fig. 3.

The microprocessor is interfaced with the memory. The primary inputs are 32 bits of instruction, memory address and data-in (each 32 bits), as well as control signals like reset. The outputs are memory write and data-out (each 32-bits) as well as read and write enable signals. We also put the architecturally visible registers ($0-$31) in the primary outputs. Since the combinational core is unrolled five times, the values of 32-bits of instruction inputs corresponding to cycle 2 to 5 are set to zero, representing no-op as the next instructions in the pipeline. This means that only one instruction is considered in the pipeline and all remaining are set to no-op. This allows us to analyze per-instruction MVF. We have also fixed the opcode for the instruction bits such that only valid MIPS instructions are used as the fault-free values (i.e. illegal instructions are avoided in fault-free inputs). After converting the description to the BLIF format, there are 1319 bits (flip-flops) which

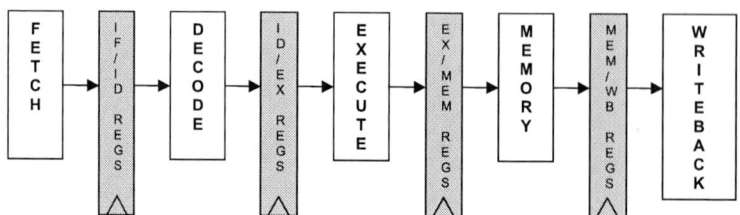

Figure 3. The MIPS 5-stage pipeline

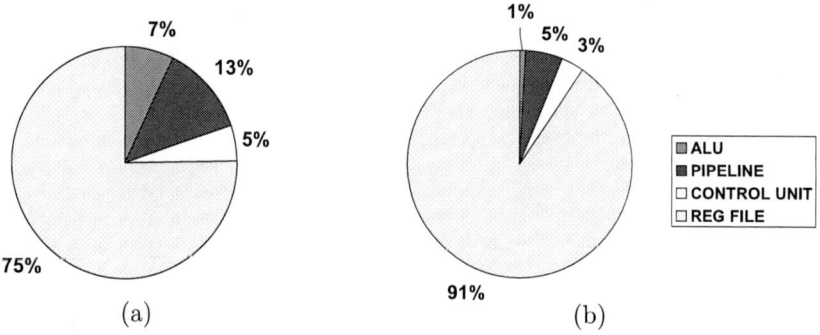

Figure 4. MIPS: (a) distribution of bits in different resources (b) MVF distribution for different resources

are used as the error sites. We obtained MVF for each bit in the processor structure for all 54 MIPS integer instructions. We grouped the internal bits based on major resources, namely registers on the boundaries of the ALU, pipeline registers, control FSM register, and register file. The distribution of bits in these resources is shown in Fig. 4(a). It can be seen that register file contains three quarters of all bits. The breakdown of overall MVF per resources is shown in Fig. 4(b). It shows that more than 90% of overall MVF is due to errors in the register file.

We have also looked at the bits with highest MVF. By sorting all 1319 bits based on their MVF in descending order and considering those in top 5 and 20 percentile, the distribution of most vulnerable bits in major resources is shown in Fig. 5. Top 5% consists of 66 most vulnerable FFs and top 20% contains 264 most vulnerable FFs. These results suggest that although control unit contributes to only 5% of total MVF, it contains 20%-26% of most vulnerable bits.

Figure 6 shows the distribution of flip-flops (bits) based on their MVFs. This shows that majority of bits have very large MVF. This is because register file constitute the majority of bits and since they are observable outputs, their MVFs are very high. We have also performed MVF analysis for

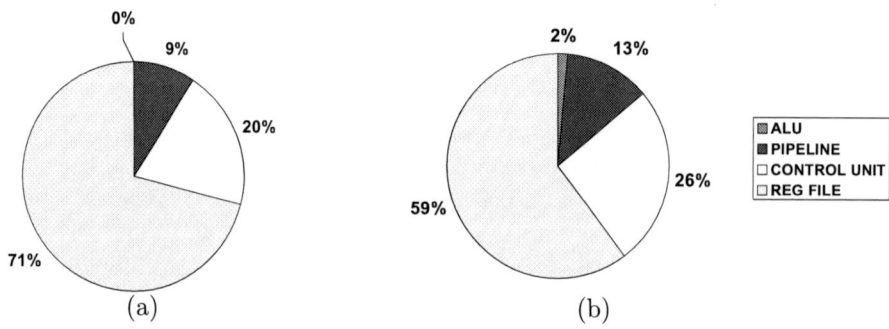

Figure 5. MIPS: (a) distribution of top 5% bits with highest MVF (b) top 20% bits

various instructions. We have grouped 54 integer instructions into four groups: arithmetic, logical, branch/jump, and store/load. The results are shown in Fig. 7. These results suggest that MVF variation for different instructions is almost negligible.

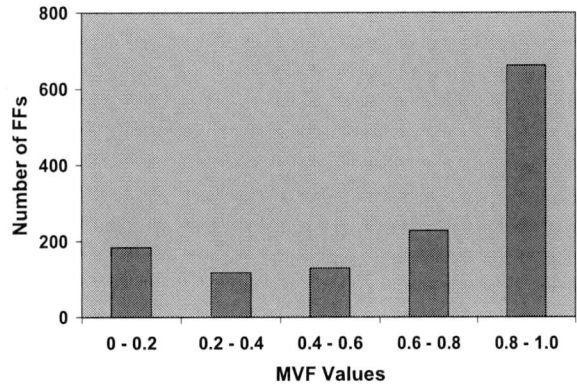

Figure 6. MIPS: Distribution of flip-flops (bits) based on MVF

4.2 Six-stage Pipelined UCore Microprocessor

UCore [19] is a 32-bit opencore RISC microprocessor which implements all the instructions of the MIPS32R2 Instruction Set. The processor has 6 pipeline stages: Instruction Fetch, Instruction Decode, Register Fetch, Execution, Memory Access and Write Back. It has a co-processor which implements some multimedia instructions. The primary inputs and outputs are similar to the MIPS processor described in Section 4.1. Since, the pipeline is six stages deep, the combinational core is unrolled six times. The value of the 32-bit instruction vector is set to 0 in cycles 2 to 6 representing a no-op. We applied the procedure given in the previous section and obtained a BLIF netlist consisting of 2289 flip-flops. These were marked as the error sites. The internal bits were grouped based on five resources namely ALU registers, pipeline registers, co-processor registers, control registers and the register file. The distribution of bits is shown in Fig 8 (a). Unlike five-stage MIPS, the number of bits in the pipeline registers is comparable to that of the register file for this processor. The overall MVF per resource is shown in Fig 8 (b). It shows that 61% of the overall MVF is due to the errors in the register file although the register file constitutes only 44% of the resources.

The distribution of most vulnerable bits in the five resources is shown in Fig 9. Top 5% consists of 114 most vulnerable FFs while top 20% contains 457 most vulnerable FFs. The results suggest that although pieline registers constitute 41% of the resources, they contribute to only 21% of the most vulnerable FFs. The register file on the other hand contributes to 60% of the most velnerable FFs.

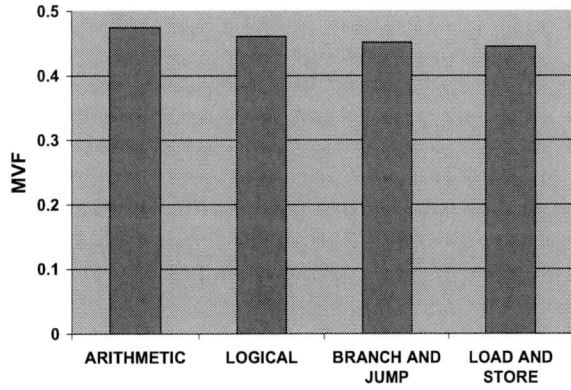

Figure 7. MIPS: MVF of different instruction classes

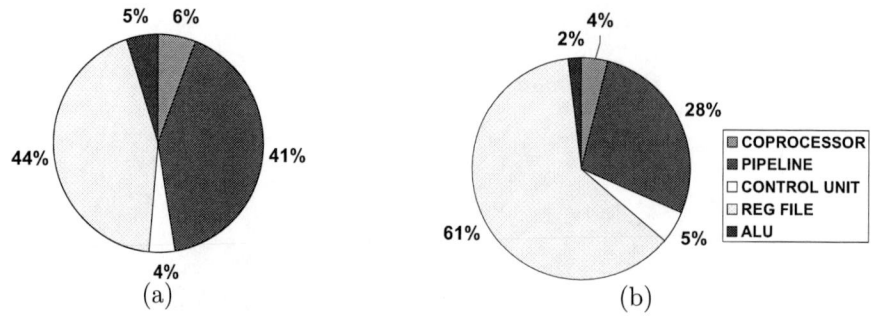

Figure 8. UCore: (a) distribution of bits in different resources (b) MVF distribution

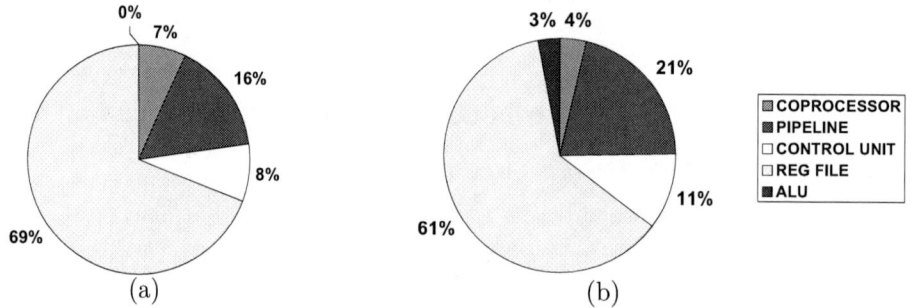

Figure 9. UCore: (a) distribution of top 5% bits with highest MVF (b) top 20% bits

Figure 10 shows a histogram of the number of bits and their corresponding MVFs. As compared to the results of Figure 6 it can be observed that more flip-flops have MVFs in the range $0.4 - 0.8$ in UCore compared to MIPS. This is because the number of bits in the pipeline registers is more in UCore compared to MIPS and they are not the most vulnerable ones.

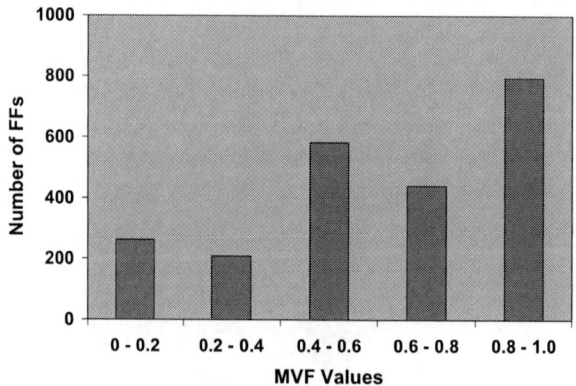

Figure 10. UCore: Distribution of flip-flops (bits) based on MVF

5   Conclusions

Reliability is increasingly becoming a major concern with the technology scaling at nanoscale. Radiation-induced soft errors are by far the major threat to the reliability of microprocessor. In this paper, we defined a metric called • •••••••••••• • •• •••••••••• •••••• (MVF) which captures the soft error rate of a particular implementation of the microprocessor. This metric represents hardware (implementation-dependent) vulnerabilities, separating it from vulnerability factors due to the micro-codes and programs running on the microprocessor (software). We also presented a flow based on Boolean Satisfiability (SAT) to accurately obtain MVF of processor descriptions at RT-level. Unlike fault simulation which is an ••••• ••••• method, the proposed technique is an ••••• method (100% accurate). Using this approach, we performed extensive soft error analysis of two pipeline RISC processors with respect to various resources and instructions.

Future work includes extension of this approach for more complex processor architectures, e.g. multi-cycle implementations of functional units and complex FSMs.

References

[1] ABC: A System for Sequential Synthesis and Verification, *www.eecs.berkeley.edu/ alanmi/abc/*.

[2] AIGER Libraries, *http://fmv.jku.at/aiger/*.

[3] H. Asadi and M. Tahoori, *Soft error hardening for logic level designs*, Proc. ISCAS, pp. 4139-4142, 2006.

[4] A.E. Baranski, L.W. Massengill, D.O. Van Nort, J. Meng, and B.L. Bhuva, *Single event faults in combinational logic modeling vulnerability during VHDL design*, Proc. SRC Top. Res. Conf. Rel., 2000.

[5] R. C. Baumann, *Radiation-Induced Soft Errors in Advanced Semiconductor Technologies*, IEEE Transactions on Device and materials reliability, **05**, no. 3, pp. 305-316, 2005.

[6] T. S Chaefer , *The complexity of satisfiability problems*, Proc. ACM Symp. Theory of Computing, pp. 216-226, 1978.

[7] F. Fallah, S. Devadas, and K. Keutzer *Functional Vector Generation for HDL models using Linear Programming and Boolean Satisfiability*, IEEE Transactions on CAD, **20**, no. 8, pp. 1003-1015, 2001.

[8] J.L. Hennessy and D.A. Patterson, *Computer Architecture: A Quantitative Approach*, 3rd Edition, 2003.

[9] S. Kim and A. K. Somani. *Soft error sensitivity characterization for microprocessor dependability enhancement strategy*, Proc. Int'l Conf. on Dependable Systems and Networks, pp. 416–428, 2002.

[10] X. Li, S. V. Adve, P. Bose, and J. A. Rivers, *Architecture-level soft error analysis: Examining the limits of common assumptions*, Proc. Int'l Conf. on Dependable Systems and Networks, pp. 266–275, 2007.

[11] L. Lingappan and N.K. Jha, *Satisfiability based Automatic Test Program Generation and Design for Testability for Microprocessors*, IEEE Transactions on VLSI, **15**, no. 5, pp. 518-530, 2007.

[12] S. S. Mukherjee, C. Weaver, J. Emer, S. K. Reinhardt, and T. Austin, "A Systematic Methodology to Compute the Architectural Vulnerability Factors for a High-Performance Microprocessor," Proc. IEEE/ACM MICRO, pp. 29-40, 2003.

[13] M. Prasad, A. Biere, and A. Gupta, *A Survey of Recent Advances in SAT-based Formal Verification*, Int'l Journal on Software Tools for Technology Transfer, **7**, no. 2. pp. 156-173, 2005.

[14] Relsat 2.1,*www.bayardo.org/resources.html*.

[15] S. Safarpour, A. Veneris and R. Drechsler *Integrating Observability Dont Cares in All-Solution SAT Solvers* Proc. ISCAS, pp. 1587-1590, 2006.

[16] S. Shazli and M.B. Tahoori, *A Framework based on Boolean Satisfiability for Soft Error Rate Computation in Early Design Stages*, Proc. Int'l Workshop on Resilience Assessment and Dependability Benchmarking, 2008.

[17] P. Shivakumar, M. Kistler, S. Keckler, D. Burger and L. Alvisi *Modeling the effect of technology trends on the soft error rate of combinational logic*, Proc. Int'l conference on dependable systems and networks, 2002.

[18] G.R. Srinivasan, *Modeling the cosmic-ray induced soft-error rate in integrated circuits: An overview*, IBM Journal of Research and Development, **40**, no. 1, pp. 77-89, 1996.

[19] UCore Processor Description, *http://www.opencores.org/projects.cgi/web/ucore/overview*.

[20] N. J. Wang, J. Quek, T. M. Rafacz, and S. J. Patel, *Characterizing the Effects of Transient Faults on a High-Performance Processor Pipeline*, Proc. Int'l Conf. on Dependable Systems and Networks, pp. 61-70, 2004.

[21] N. J. Wang, A. Mahesri and S. J. Patel, *Examining ACE Analysis Reliability Estimates using fault injection*, ACM SIGARCH Computer Architecture News, **35**, no. 2, pp. 460-469, 2007.

[22] Z. Zeng, P. Kalla and M. Ciesielski, *LPSAT: a unified approach to RTL satisfiability* Proc. DATE, pp. 398-402, 2001.

[23] Q. Zhou and K.Mohanram, *Transistor sizing for Radiation Hardening* Proc. IRPS, pp. 310-315, 2004.

IEEE International Symposium on Defect and Fault Tolerance of VLSI Systems

System Reliabilities when Using Triple Modular Redundancy in Quantum-Dot Cellular Automata

Timothy J. Dysart and Peter M. Kogge
Department of Computer Science and Engineering
University of Notre Dame
{tdysart, kogge}@cse.nd.edu

Abstract

Nanoelectronic systems are extremely likely to demonstrate high defect and fault rates. As a result, defect and/or fault tolerance may be necessary at several levels throughout the system. Methods for improving defect tolerance, in order to prevent faults, at the component level for QCA have been studied. However, methods and results considering fault tolerance in QCA have received less attention. In this paper, we present an analysis of how QCA system reliability may be impacted by using various triple modular redundancy schemes.

1. Introduction

In order to properly augment or even potentially replace CMOS circuitry, nanoelectronic systems face the challenge of providing reliable computation. These systems are likely to have a high percentage of manufacturing defects due to factors like their small size and utilizing some form of self-assembly in the manufacturing process. Some, but not all, of these defects may be problematic enough to cause errors in the system; as a result, some form of fault tolerance is a potential requirement for nanoelectronic systems.

Reconfiguration and redundancy have generally been utilized to provide fault tolerance in nanoelectronic systems. Systems using reconfiguration generally use crossbar arrays to implement logic and memory, and use the principles of the Teramac project[12] to build a functional system [10, 6, 24]. Redundant fault tolerant methods build on the work of von Neumann in [25]. One technique, multiplexing, utilizes large groups of NAND or majority gates to perform a computational stage; then randomly permutes and restores the result. Several articles, including [11, 23, 22, 2], have considered this approach. The other approach in [25] is N-modular redundancy which uses N, an odd number, copies of the same circuit to compute a result, and then a majority vote of the output selects the proper answer. Frequently, $N = 3$ to form systems with triple modular redundancy (TMR). Modular redundancy in nanoelectronic systems has been considered in [21, 1, 7].

This work considers quantum-dot cellular automata (QCA) because reconfigurable architectures [5] and custom logic circuits have been proposed. As a result, the reliability impact of each method needs to be studied to determine which method should be favored in QCA. Redundancy is of interest here since [7] surprisingly found that TMR negatively impacted the reliability of a simple electrostatic QCA circuit, and we need to further understand the impacts of using TMR in QCA systems with generic modules (circuits). Our results demonstrate that system reliability with TMR can be either positively or negatively affected depending on the choice of module, TMR style, QCA

1550-5774/08 $25.00 © 2008 IEEE
DOI 10.1109/DFT.2008.25

implementation, and component error rates. We also show that the components required to route signals between the modules and voting circuits impact system reliability more than the reliability of the generic module itself.

Section 2 provides background material. Sections 3 and 4 consider basic and cascaded TMR techniques. Section 5 determines which components have the greatest impact on system reliability when using TMR. Sec. 6 concludes.

2. Background Material [1]

2.1. Introduction to QCA

Quantum-dot cellular automata (QCA) has been implemented using charge-based, or electrostatic (ESQCA), and magnetic (MQCA) devices. In ESQCA, Coloumbic interactions are used to represent and move data[17]. In MQCA, a single, rectangularly shaped, nanomagnet will have its magnetization align along the long axis of the nanomagnet. Depending on the external forces, the magnetization of the nanomagnets will be either straight up or down and practical for computation [13]. Figure 1(a) shows sample cells of each type. In both implementations, the positioning and alignment of multiple cells determines which logical operation is performed.

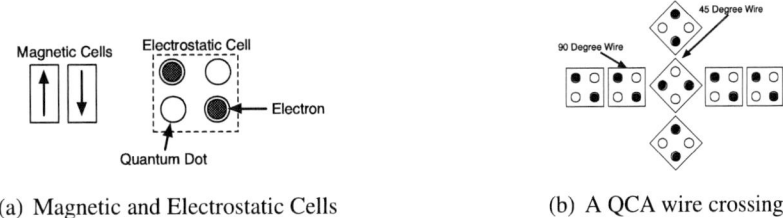

(a) Magnetic and Electrostatic Cells   (b) A QCA wire crossing

Figure 1. QCA components.

One limitation of an ESQCA implementation, especially if molecular cells are used, is that the wire crossing shown in Fig. 1(b) requires sub-Angstrom level precision in cell placement for the crossover to function properly [4]. However, a simple combinational circuit capable of crossing wires without using 45 degree wires can be formed from three XOR gates and implements the following two functions: A XOR B XOR A = B and A XOR B XOR B = A [3]. Specific implementations of this crossing circuit can be found in [8, 7]. In MQCA implementations, logical wire crossings may not be required since a wire crossing component may be possible [20].

2.2. PTM Framework

In this work, probabilistic transfer matrices (PTMs) [15, 16] are used to compute system reliabilities for systems made with error-prone components as outlined below. The PTM for a component/circuit/system with m inputs and n outputs is a $2^m \times 2^n$ matrix that, when error-free, is the truth table for the component (termed ideal transfer matrix, ITM). This matrix shows the relationship for all combinations of inputs and outputs for the component and with the specific relationship between input sequence i and output sequence j located at (i, j) in the matrix. If the component produces an incorrect output with probability p (error rate), regardless of input values, correct input/output combinations are represented by $1 - p$ and erroneous input/output combinations are represented by $p/(2^n - 1)$ in the PTM. To prevent a given input combination from having an output probability greater than one, the sum for the entries of a row in a PTM must be ≤ 1.

---

[1]Parts of this section are condensed from the material in [7, 8, 9].

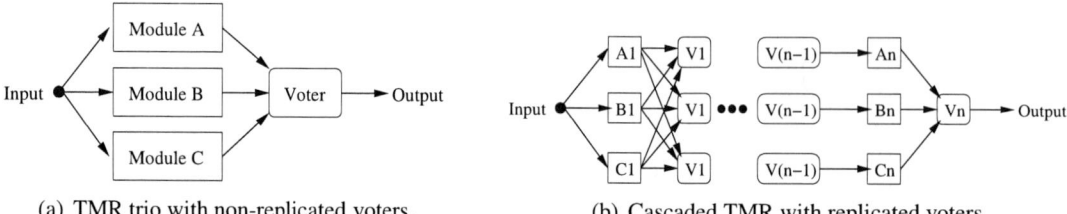

(a) TMR trio with non-replicated voters.　　　(b) Cascaded TMR with replicated voters.

Figure 2. Triple modular redundancy (TMR) configurations.

A circuit PTM is generated by first dividing the circuit into slices that are one component level each, i.e. no level has an interconnected series of components. Within a level the components are in parallel, and the Kronecker product is calculated for the level, thus giving each level a PTM. These level PTMs are then multiplied, from circuit inputs to outputs, using matrix-matrix multiplication to generate the circuit PTM. At each stage of this process, the size of the PTM corresponds to the number of inputs and outputs as discussed above.

Once the PTM for the circuit has been generated, it is element-wise multiplied by its ITM so that only desired input/output combinations result in the ETM (element-wise transfer matrix). The operation $v * ETM = v'$ is then performed, where v is a vector consisting of the probability of each input occurring. The resulting vector, v' is the probability of each output occurring. The exact reliability for the circuit is found by summing the elements of v'. [15, 7, 8, 9, 16] contain more detailed examples and discussion on this process.

Since QCA interconnect and logic share the same device, we must consider all circuit components faulty. As a result, we use the following component PTMs: AND, OR, and MAJ gates; inverters; wires; single bend wires; T-shaped fanouts; regular fanouts; wire crossing devices; and generic modules. The error rate for the AND and OR gates matches that of the majority gates because assuming that the fixed input to a majority gate (which would turn it into an AND or OR gate) is error-free, the overall complexity of the PTMs is reduced. Module ITMs are generated randomly based on the number of inputs and outputs the module contains. The error rate of the module is then used to create its PTM.

2.3. TMR Overview

Triple modular redundancy is a specific implementation of von Neumann's N-modular redundancy (NMR) where $N = 3$ [25]. With NMR, N instances of the same module perform the same computation and then a majority vote of the output(s) is taken. As long as $\lceil N/2 \rceil$ modules compute the output bit(s) properly, the system output is correct. This voting process enables a logical masking of potential errors in the modules.

[18] provides the background for the TMR models and reliability equations described here. The basic TMR model, also known as a trio, is shown in Fig. 2(a). The system reliability of this model, given a module reliability of R_M and a voting circuit reliability of R_v is

$$R = R_v * (3R_M^2 - 2R_M^3). \tag{1}$$

The reliability of a system can generally be improved if the modules above are only a part of the entire system. If we assume that a single module above is one of the m segments of a non-redundant system, then $R_M = R_0^{1/m}$ where R_0 is the reliability of the entire non-redundant system. This relationship assumes that each of the m modules has the same reliability. Dividing the system into m segments enables the use of cascaded TMR with non-replicated (not shown) or replicated voters as shown in Fig. 2(b).

For a cascaded TMR system with perfect voters, the reliability is found using Eq. 2 with R_T being the reliability for one trio. For a cascaded system with non-replicated faulty voters, the reliability is found using Eq. 3. If the voter circuits are replicated, then each module is in series with a voter circuit, thus $R_M = R_0^{1/m}$ is replaced with $R_M = R_v R_0^{1/m}$ and the resulting system reliability is Eq. 4.

$$R = R_T^m = (3R_0^{2/m} - 2R_0^{3/m})^m \tag{2}$$

$$R = (R_v * R_T)^m = (R_v * (3R_0^{2/m} - 2R_0^{3/m}))^m \tag{3}$$

$$R = (3R_v^2 R_0^{2/m} - 2R_v^3 R_0^{3/m})^m \tag{4}$$

3. Basic TMR in QCA

The first component of determining whether TMR may be an effective method for improving system reliability is to determine if the reliability of a trio can be improved. To do this, we have first considered generic modules with a given fixed error rate while varying the QCA component error rates with all components having uniform error rates. Doing this will identify component error rates where using TMR provides reliability gains.

The choice of error rates tested here has been driven by data in [14], which states that the error rate for CMOS transistors is approximately 10^{-10}, which would be an error rate of 10^{-9} for moderately sized gates (i.e. majority/XOR). Since the defect and fault rates for nanoelectronics is expected to be much higher, the error rates used here range from 10^{-8} to 10^{-1}. For this section, we have generated results for modules from one to four inputs and one to four outputs, but show only a subset of these results.

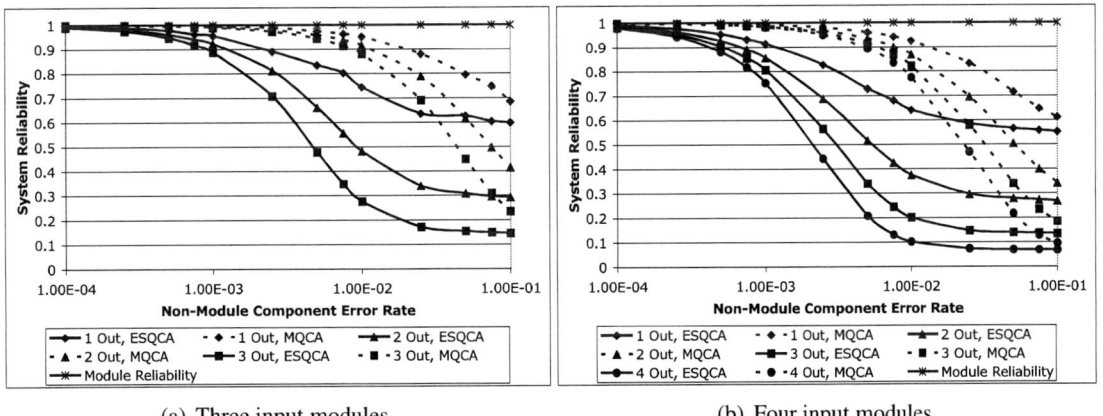

(a) Three input modules.   (b) Four input modules.

Figure 3. Basic TMR reliabilities for modules with $R_M = 1 - 1 \times 10^{-4}$.

We first consider how varying the number of outputs of a generic module influences the system reliability. This is done to show how routing these output signals negatively impacts system reliability. Results for three input modules are shown in Fig. 3(a) and for four input modules in Fig. 3(b). The four input cases show lower reliability since the input signal routing is more complex. For both cases, it is clear that the MQCA implementations, with their far simpler signal routing, have significantly higher system reliabilities than the ESQCA implementations.

In Fig. 4 we show the results when $R_M = 1 - 1 \times 10^{-4}$ for a multiplexor-like module (3 inputs, 1 output) and for an adder-like module (3 inputs, 2 outputs). The data lines labeled *Thy, FV* and

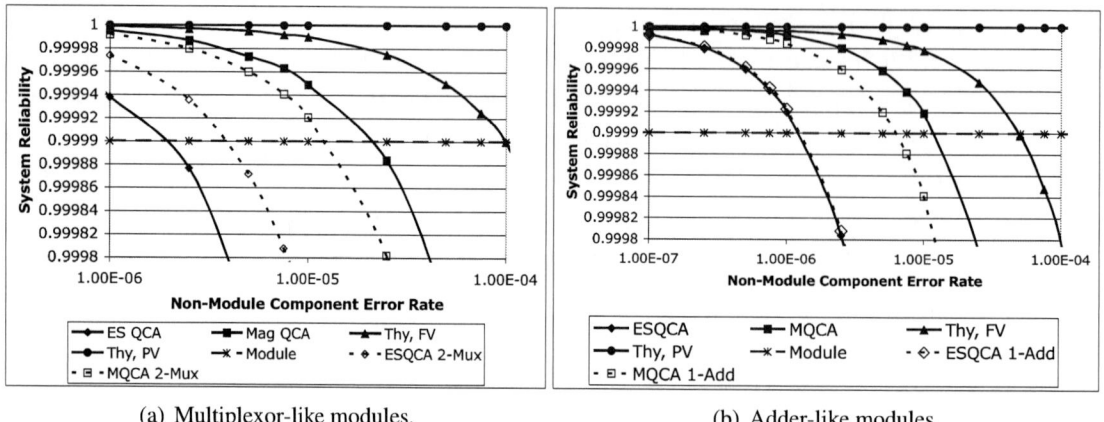

(a) Multiplexor-like modules.   (b) Adder-like modules.

Figure 4. Basic TMR reliabilities for modules with $R_M = 1 - 1 \times 10^{-4}$.

Thy, PV compute Eq. 1 (TMR theoretical reliability) with faulty voters (FV) using the non-module component error rate and perfect voters (PV). In both graphs we observe that the MQCA TMR implementations can improve reliability if the component error rate is near or below 1×10^{-5} and that the ESQCA TMR implementations can improve reliability if the component error rates are approximately an order of magnitude lower.

While we have identified component error rates where TMR does improve reliability, it is not clear if real circuits have reliabilities similar to that of the module at these component error rates. To provide this reference point, we have added the results from [9] to the graphs of Fig. 4. These results are labeled as *ES 2-Mux* and *Mag 2-Mux* in Fig. 4(a) and *ES 1-Add* and *Mag 1-Add* in Fig. 4(b). These results show that, for the component error rates shown in these graphs, certain modules may have a reliability similar to the $1 - 1 \times 10^{-4}$ considered here. These results also suggest that a basic form of TMR may aide system reliability for MQCA, but that TMR may negatively effect system reliability when using ESQCA.

4. Cascaded TMR

Since a basic form of TMR can sometimes improve system reliability in QCA systems, we consider if cascaded forms of TMR may improve system reliability. Here, we show results for one-input, one-output modules. Similar results are available for two-input, two-output modules, but are not shown due to page limits. Only modules with equal numbers of inputs and outputs are considered due to the matrix multiplication steps required by using PTMs. Larger modules are not considered since the memory requirements for these PTMs cannot be met with Matlab.

In the results presented in this section, each of the graphs have lines noted with *Thy* which considers an implementation with only faulty modules and voters. For the one-in, one-output module cases, these lines are equivalent to Eq. 3 in the non-replicated voters case and the reliability found by Eq. 4 multiplied by R_v in the replicated voters case. The final multiplication of R_v provides a final majority vote to ensure a single output bit rather than three copies of the same bit. One important thing to note about the *Thy* results is that while they are calculated via PTMs, they do not take full advantage of the PTMs power in that the multiple wrongs make a right residuals are eliminated. In other words, at each stage of the computation, all residuals that would exist in error states are eliminated and made zero. This has been done to match the algebraic forms used in TMR reliability calculations. These theory results are provided for reference only and are not directly compared to the QCA based implementations.

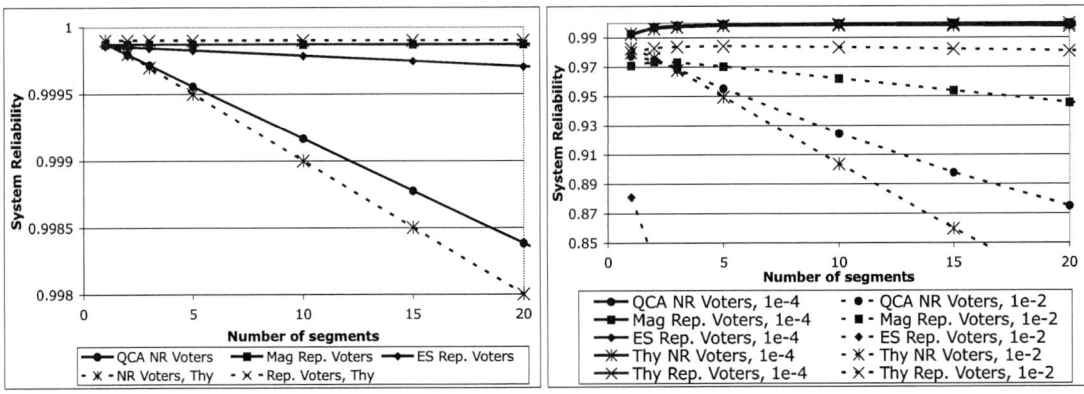

(a) $R_0 = 1 - 1 \times 10^{-4}$ and component error rate held constant at 1×10^{-4} (b) $R_0 = 1 - 5 \times 10^{-2}$ and component error rates of 1×10^{-4} and 1×10^{-2}

Figure 5. System reliabilities using cascaded TMR.

Figure 5 shows the system reliabilities that result when R_0 is held constant at $1 - 1 \times 10^{-4}$ and $1 - 5 \times 10^{-2}$. The component error rate is also held constant at 1×10^{-4} in Fig. 5(a) and additionally at 1×10^{-2} in Fig. 5(b). For these graphs, the lines marked *NR Voters* refers to using non-replicated voters while the other lines utilize replicated voters and are considered for both magnetic and electrostatic implementations. When the voters are not replicated and only one input, one output modules are considered, no wire crossings exist in the circuit, thus the electrostatic and magnetic reliabilities are equivalent.

Figure 5(a) shows that for a high R_0 and a low component error rate, cascaded TMR with replicated voters offer a higher reliability than cascaded TMR with non-replicated voters. This result is caused by the voting circuit putting an upper limit on the reliability that can be achieved by each segment when the voters are not replicated. This figure also demonstrates that for these error rates, cascaded TMR does not improve reliability (the lines are monotonically decreasing).

Figure 5(b) considers the case of a low R_0 and two values for the component error rate. For the higher component error rate (1×10^{-2}), the magnetic replicated voters case is more reliable than the other cases. Additionally, we observe that for this case, dividing the circuit into fewer than 5 stages can actually increase the reliability over the basic TMR configuration. For low component error rates (1×10^{-4}), the system reliability for all three cases is similar, and dividing the system into segments offers additional reliability benefits. For the non-replicated voters case, the reliability peaks when approximately 20 segments are used; for the ESQCA case, the peak occurs when approximately 25 segments are used; for the MQCA case, the reliability continues to increase to at least 50 segments (we do not increase the number of segments beyond 50).

In Figure 6 rather than show varying component error rates in the graphs, we show the impact of varying R_0 instead. Figure 6(a) holds the component error rate at 1×10^{-4} and considers R_0 values of $1 - 1 \times 10^{-4}$ and $1 - 1 \times 10^{-1}$. The solid lines in this graph are the same as those of Fig. 5(a). The dashed lines show the benefit of using cascaded TMR over the basic TMR when R_0 and the component error rate are both low. These dashed lines also show that replicated voters offer much greater reliability improvements than using a single voter in each segment.

In Fig. 6(b) a high component error rate of 5×10^{-2} is considered for a high value of R_0. In this scenario, the system reliability plummets as the value of m increases. While not shown, if R_0 is reduced to $1 - 1 \times 10^{-1}$, the initial system reliability is lower and a similar floor for each case is observed. These results suggest that the module error rate has a much smaller impact on the overall system reliability than the component error rate. This possibility is explored in the next section.

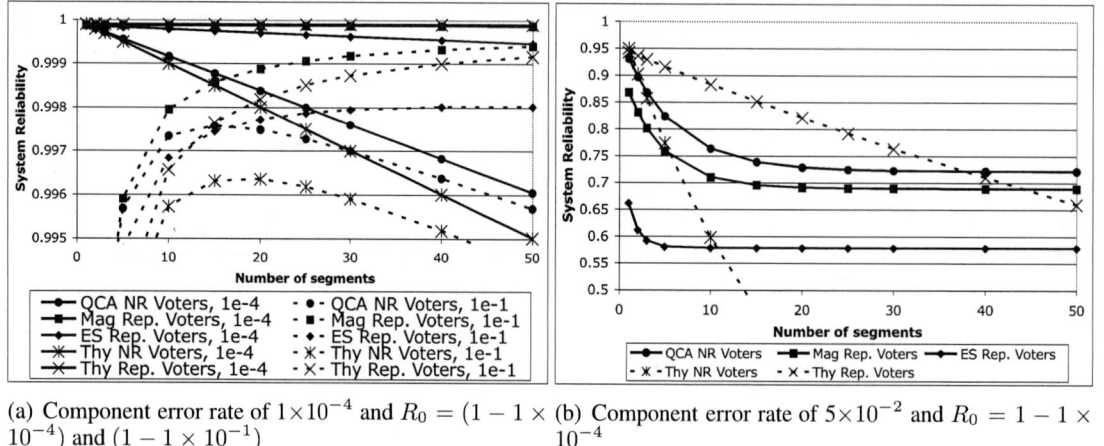

(a) Component error rate of 1×10^{-4} and $R_0 = (1 - 1 \times 10^{-4})$ and $(1 - 1 \times 10^{-1})$

(b) Component error rate of 5×10^{-2} and $R_0 = 1 - 1 \times 10^{-4}$

Figure 6. System reliabilities using cascaded TMR.

The main result from this section is that the use of cascaded TMR can be beneficial when the component error rate is much lower than the module reliability rate.

5. Linear Regression Results

In our previous work [7, 9], we have used regression analysis in an effort to determine which part of a circuit is most sensitive to changes in reliability. We have done similar work here for two and three input modules with one to three outputs. We are only considering a basic TMR implementation, not cascaded TMR implementations. Rather than test each component individually, we have grouped them together based on their purpose. For this section, the groupings are Logic (MAJ, AND, OR, NOT), Wiring (W, SB), Fanouts (FAN, TFAN, FAN3), Crossovers (for magnetic implementations), and Modules. The error rates for each of these groups has ranged from 1×10^{-8} to 1×10^{-2} by orders of magnitude (all combinations tested).

	Electrostatic			Magnetic		
	One Output	Two Output	Three Output	One Output	Two Output	Three Output
Logic	-4.9783	-11.5877	-22.4486	-0.9979	-1.9776	-2.9463
Wiring	-5.8052	-15.4439	-31.1184	-0.5619	-1.1027	-1.9121
Fanouts	-3.9613	-8.4430	-15.9296	-0.7723	-1.1651	-1.3698
Crossovers				-0.4119	-0.7010	-1.0386
Modules	-0.0445	-0.2165	-0.3115	-0.0438	-0.0583	-0.0885
Constant	0.9987	1.0011	1.0034	1	1.0001	1.0004

Table 1. Regression coefficients for two input modules.

	Electrostatic			Magnetic		
	One Output	Two Output	Three Output	One Output	Two Output	Three Output
Logic	-11.0164	-21.2017	-32.6620	-0.9897	-1.9614	-2.9209
Wiring	-13.8400	-28.0574	-43.0993	-1.5800	-2.7714	-4.0672
Fanouts	-8.8271	-16.5871	-25.0298	-1.2456	-1.8833	-2.2135
Crossovers				-1.2154	-1.9798	-2.6299
Modules	-0.1325	-0.1830	-0.2606	-0.0513	-0.0683	-0.0971
Constant	0.9944	0.9932	0.9908	1	1.0001	1.0004

Table 2. Regression coefficients for three input modules.

Tables 1 and 2 contain our regression results for both QCA implementations. In these tables each column must be considered individually; the coefficients cannot be compared between columns since each column represents a different circuit. The coefficients are negative because when the components have non-zero error rates, the reliability of the circuit is reduced. The larger the magnitude of the coefficient, the greater that group of components impacts reliability. The constant is a necessary regression term and is close to 1 since that would represent a perfect circuit.

For the two input module cases, the wiring and logic groupings have the largest impact on system reliability. Similar trends hold for the three input modules. These coefficients also suggest that module reliability has almost no real impact on system reliability. This trend was also observed in the cascaded TMR results. This strongly suggests that the reliability of the logic and wiring between the modules will determine the overall system reliability.

6. Conclusion

In this paper, we have focused on the reliability aspects of using TMR to improve system reliability. Other constraints, such as speed, power, and area are likely to play a role in deciding whether TMR is practical. We briefly comment on each of these constraints below

- *Area:* TMR requires at least three times the area just in the modules. In QCA, signal routing also adds area to the system since all signals are in the same plane. The number of of crossovers required when using TMR is dictated by the number of inputs and outputs to the modules. For input signal routing, $n^2 - n$ crossovers are required for the n bit input modules tested here. For output signal routing, the number of crossovers required is 1.5 times that of the input signal routing. Additionally, the largest number of crossovers that must be traversed by a signal is $2(n-1)$ where n is the number of input or output bits for the module.
  Since the crossover device proposed for MQCA is roughly the size of a majority gate, the increase in area for signal routing will probably be small compared to the amount of extra logic used in replicating the modules. However, in ESQCA, a crossover circuit requires nine majority gates and three inverters and the longest path through this circuit passes through four majority gates and two inverters. For modules with large numbers of inputs and outputs, this could cause a significant increase in area – provided that the module itself does not require a large number of crossovers.

- *Speed:* In QCA systems, the speed of the circuit is limited by the switching speed of the devices used to implement the system. Additionally, QCA systems utilize the principle of pipelining-in-wire to improve computational speed [19]. Since TMR will add extra logic, particularly if cascaded forms are used, the latency of the system will increase, but the throughput will remain the same.

- *Power:* The majority of the power dissipation in QCA systems will likely occur in the clocking wires rather than the devices themselves. As a result, power dissipation will increase when using TMR, but the exact amount of this increase will depend on the size and structure of the modules being replicated and the amount of logic that can be done in a clocking zone.

This paper provides another step forward in discerning how fault tolerance in QCA systems should be implemented by quantifying how QCA system reliability is effected by using TMR. Our results have shown that for ESQCA implementations, basic TMR implementations may lower system reliability, however, MQCA systems may show improved reliability. If cascaded forms of TMR are utilized, replicating the voting circuits in each segment is necessary. Additionally, cascaded TMR may only be beneficial for cases where the component error rate is significantly lower than $1 - R_0$ and the number of segments used is kept moderate. We have also shown that for any implementation of TMR, varying R_0 tends to have little to no impact; system reliability is impacted far more significantly by the wiring and logic between modules.

This work was supported by the U.S. Department of Defense and the National Science Foundation under grant CCF-0541324.

References

[1] D. Bhaduri and S. Shukla. Nanolab-a tool for evaluating reliability of defect-tolerant nanoarchitectures. *IEEE Trans. Nanotechnol.*, 4(4):381–394, July 2005.

[2] D. Bhaduri, S. Shukla, P. Graham, and M. Gokhale. Comparing reliability-redundancy tradeoffs for two von neumann multiplexing architectures. *IEEE Trans. Nanotechnol.*, 6(3):265–279, May 2007.

[3] A. Chaudhary, D. Z. Chen, X. S. Hu, M. T. Niemier, R. Ravichandran, and K. Whitton. Fabricatable interconnect and molecular qca circuits. *IEEE Trans. Computer-Aided Design*, 26(11):1978–1991, 2007.

[4] A. Chaudhary et al. Eliminating wire crossings for molecular quantum-dot cellular automata implementation. In *International Conference on Computer Aided Design*, pages 565–571, 2005.

[5] M. Crocker, X. S. Hu, and M. T. Niemier. Fault models and yield analysis for qca-based plas. In *Proceedings of the 17th International Conference on Field Programmable Logic and Applications*, pages 435–440, 2007.

[6] A. Dehon. Nanowire-based programmable architectures. *J. Emerg. Technol. Comput. Syst.*, 1(2):109–162, 2005.

[7] T. J. Dysart and P. M. Kogge. Probabilistic analysis of a molecular quantum-dot cellular automata adder. In *Proceedings of the 22nd IEEE International Symposium on Defect and Fault Tolerance in VLSI Systems*, 2007.

[8] T. J. Dysart and P. M. Kogge. Probabilistic analysis of a quantum-dot cellular automata multiplier implemented in different technologies. In *Proceedings of the 4th Workshop on Non-Silicon Computing*, 2007.

[9] T. J. Dysart and P. M. Kogge. Analyzing the inherent reliability of moderately sized magnetic and electrostatic qca circuits via probabilistic transfer matrices. *Submitted to IEEE Trans. on VLSI*, 2008.

[10] S. C. Goldstein and M. Budiu. NanoFabrics: Spatial Computing Using Molecular Electronics. In *International Symposium on Computer Architecture*, pages 178–189, 2001.

[11] J. Han and P. Jonker. A system architecture solution for unreliable nanoelectronic devices. *IEEE Trans. Nanotechnol.*, 1(4):201–208, Dec. 2002.

[12] J. R. Heath, P. J. Kuekes, G. S. Snider, and R. S. Williams. A defect-tolerant computer architecture: Opportunities for nanotechnology. *Science*, 280:1716–1721, June 1998.

[13] A. Imre. *Experimental study of nanomagnets for magnetic quantum-dot cellular automata (MQCA) logic applications*. PhD thesis, U. of Notre Dame, 2005.

[14] ITRS. International technology roadmap for semiconductors 2006 update edition. Technical report, ITRS, 2006.

[15] S. Krishnaswamy, G. F. Viamontes, I. L. Markov, and J. P. Hayes. Accurate reliability evaluation and enhancement via probabilistic transfer matrices. In *Proceedings of the Design, Automation, and Test in Europe Conference and Exhibition*, 2005.

[16] S. Krishnaswamy, G. F. Viamontes, I. L. Markov, and J. P. Hayes. Probabilistic transfer matrices in symbolic reliability analysis of logic circuits. *ACM Trans. Des. Autom. Electron. Syst.*, 13(1):1–35, 2008.

[17] C. Lent, P. Tougaw, W. Porod, and G. H. Bernstein. Quantum Cellular Automata. *Nanotechnology*, 4(1):49–57, 1993.

[18] R. E. Lyons and W. Vanderkulk. The use of triple-modular redundancy to improve computer reliability. *IBM Journal of Research and Development*, 6(2):200–209, Apr. 1962.

[19] M. T. Niemier and P. M. Kogge. Problems in designing with QCAs: Layout = timing. *Int. J. of Circ. Theory and App.*, 29:49–62, April 2001.

[20] M. T. Niemier, X. S. Hu, M. Alam, G. Bernstein, W. Porod, M. Putney, and J. DeAngelis. Clocking structures and power analysis for nanomagnet-based logic devices. In *Proceedings of the International Symposium on Low Power Electronics and Design*, pages 26–31, 2007.

[21] K. Nikolic, A. Sadek, and M. Forshaw. Fault-tolerant techniques for nanocomputers. *Nanotechnology*, 13(3):357–362, 2002.

[22] S. Roy and V. Beiu. Majority multiplexing - economical redundant fault-tolerant designs for nanoarchitectures. *IEEE Trans. Nanotechnol.*, 4(4):441–451, July 2005.

[23] A. Sadek, K. Nikolic, and M. Forshaw. Parallel information and computation with restitution for noi se-tolerant nanoscale logic networks. *Nanotechnology*, 15(1):192–210, Jan. 2004.

[24] D. B. Strukov and K. K. Likharev. Cmol fpga: a reconfigurable architecture for hybrid digital circuits with two-terminal nanodevices. *Nanotechnology*, 16(6):888–900, 2005.

[25] J. von Neumann. *Automata Studies*, chapter Probabilistic logics and the synthesis of reliable organisms from unreliable components, pages 43–98. Princeton University Press, 1956.

INVITED TALK

Error detection and tolerance for scaled electronic technologies

Kartik Mohanram, *Rice University*

This talk will summarize our design for reliability initiatives that anticipate the paradigm shift to error-aware and error-tolerant design of integrated circuits, both of which are required to address the problem of increasing hardware failures in future technology nodes. These concerns are only exacerbated as we look forward to emerging technology alternatives. Using graphene as an example, I will go on to describe the modeling, simulation, and design challenges that we believe are key to taming the complexity challenges associated with such scaled electronic technologies.

Speaker Bio − Kartik Mohanram received the B.Tech. degree in electrical engineering from IIT-Bombay in 1998, and the M.S. and Ph.D. degrees in computer engineering from University of Texas-Austin in 2000 and 2003 respectively. He is currently an assistant professor in the department of Electrical and Computer Engineering at Rice University. His primary research interests are in computer engineering and systems, with an emphasis on modeling, simulation, and computer-aided design of integrated circuits. He is a recipient of the NSF CAREER Award, the ACM/SIGDA Technical Leadership Award, and the A. Richard Newton Graduate Scholarship.

1550-5774/08 $25.00 © 2008 IEEE
DOI 10.1109/DFT.2008.65

SESSION 3
HOT TOPICS

IEEE International Symposium on Defect and Fault Tolerance of VLSI Systems

Hardware Trojan Detection and Isolation Using Current Integration and Localized Current Analysis

Xiaoxiao Wang, Hassan Salmani and Mohammad Tehranipoor
ECE Department
University of Connecticut

Jim Plusquellic
ECE Department
University of New Mexico

Abstract

This paper addresses a new threat to the security of integrated circuits (ICs). The migration of IC fabrication to untrusted foundries has made ICs vulnerable to malicious alterations, that could, under specific conditions, result in functional changes and/or catastrophic failure of the system in which they are embedded. Such malicious alternations and inclusions are referred to as Hardware Trojans. In this paper, we propose a current integration methodology to observe Trojan activity in the circuit and a localized current analysis approach to isolate the Trojan. Our simulation results considering process variations show that with a very small number of clock cycles the method can detect hardware Trojans as small as few gates without fully activating them. However, for very small Trojan circuits with less than few gates, process variations could negatively impact the detection and isolation process.

1. Introduction

Chip design and fabrication is becoming increasingly vulnerable to malicious activities and alternations with globalization. This has raised serious concerns regarding possible threats to military systems, financial infrastructures, transportation security and even household appliances. An adversary can introduce a Trojan designed to disable and/or destroy a system at some future time (we call it Time Bomb) or the Trojan may serve to leak confidential information covertly to the adversary. Trojans can be implemented as hardware modifications to application specific ICs (ASICs), commercial off the shelf (COTS) parts, microprocessors, or digital signal processors (DSPs), or as firmware modifications, e.g., to field programmable gate arrays (FPGA) bitstreams [1][2].

Unfortunately, the detection of such inclusions is difficult for several reasons: 1) Nanometer IC feature sizes and system complexity make detection through physical inspection and destructive reverse engineering difficult and costly. Moreover, destructive reverse engineering does not guarantee that ICs not destructively inspected are Trojan-free. Additionally, the adversary may insert the Trojan randomly in a large batch of fabricated chips. 2) Trojan circuits are by design activated under very specific conditions, which makes it difficult to fully activate them using random and functional stimuli and detect them using observation points (primary outputs and scan flip-flops). Moreover, existing automatic test pattern generation (ATPG) methods used in manufacturing test for detecting defects do so by operating on the netlist of the Trojan-free circuit specification. Therefore, existing ATPG algorithms cannot target Trojan activation directly.

Trojan detection methods can be applied immediately after the chip is returned to the customer, either as a die on a wafer or as a packaged chip, and/or they can be applied continuously during the lifetime of the system. For the latter case, board level support systems, such as trusted companions, are needed to carry out the monitoring. Although these types of approaches are of interest, the focus of this work is on 'time-zero' detection methods, i.e., methods applied before the chip is installed in the target system. This phase is referred to as *IC Authentication* that is done after manufacturing testing phase.

1.1 Prior Work

Security has become a new concern in the design and test of chips recently [3][4][5][6]. This trend has become more apparent with the advent of Cryptochips, which implement encryption and decryption algorithms in hardware [7]. Many researchers have been able to show that these chips are highly vulnerable to various power analysis [8][9], timing [10][11], and fault injection [12][13] attacks if not specially designed with countermeasures. If not considered carefully, strong encryption algorithms that would take years to crack by brute force can otherwise be defeated in a manner of weeks, days, or even hours through these side channel attacks. Recently, scan test has become a security risk to the intellectual property on the chip [14][15][16][17]. Such non-invasive attacks have also been used for extracting secret information such as keys used within ICs [3][18]. The Trojan detection is a new topic in hardware security area and there is a very limited prior work in this area. The authors in [19] propose the use of side-channel signals, e.g., transient power supply currents, to identify Trojans in chips. The method uses a global

1550-5774/08 $25.00 © 2008 IEEE
DOI 10.1109/DFT.2008.61

measurement of power supply transient signals to detect the Trojans which makes it difficult to target very small Trojans in presence of process variations.

1.2 Contribution and Paper Organization

Most Trojans inserted into a chip require power supply and ground to operate. The Trojans can be of different types and sizes and their impact on circuit power characteristics could be very large or very small. In this paper, we first develop a multiple supply transient current integration methodology to detect hardware Trojans in integrated circuits. We then develop a Trojan isolation method based on localized current analysis. We measure the current from various power ports or controlled collapse chip connections (C4s) on the die. Random patterns are applied to increase the switching in the circuit in a test-per-clock fashion [20].

The paper is organized as follows. Section 2 provides a taxonomy for Trojans. Section 3 presents the proposed multiple power supply transient current integration methodology. Section 4 presents the Trojan insertion procedure. The process variations' importance and effects when detecting Trojans will be discussed in Section 5. Section 6 presents the simulation results. Finally, Section 7 will conclude the paper.

2. Taxonomy

In order to develop methods designed to improve IC TRUST, it is essential to first define a taxonomy for Trojans. The Trojan classification scheme that we propose is derived from several fundamental characteristics of Trojans, including their physical, activation and action characteristics. Once a framework is established, we will be able to measure the effectiveness of the detection and isolation methods.

Malicious alternations to the structure and function of a chip can take many forms. We decompose the Trojan taxonomy into three principle categories as shown in Figure 1, i.e., according to their *physical*, *activation* and *action* characteristics. The physical characteristics of a Trojan are further partitioned into four categories; *type*, *size*, *distribution*, and *structure*. Our proposed taxonomy, therefore, describes Trojans using six attributes, including four physical, one activation and one action attribute. Although it is possible for Trojans to be hybrids of this classification, e.g., have more than one activation characteristic, we believe this taxonomy captures the elemental characteristics of Trojans and will be useful for defining the capabilities of various detection strategies.

Trojan Physical Characteristics: The *type* category partitions Trojans into *functional* and *parametric* classes. The functional class includes Trojans that are physically realized through the addition or deletion of transistors or gates, while parametric refers to Trojans that are realized through modifications of existing wires and logic. The thinning of a wire, the weakening of a transistor or any modification of a physical geometry designed to sabotage reliability or increase the likelihood of a functional or performance failure are examples of the latter.

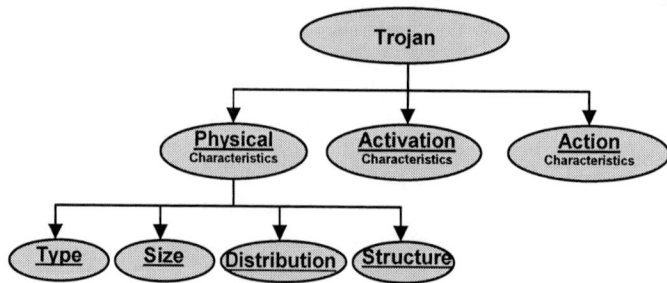

Figure 1. Taxonomy of Trojans.

The *size* category accounts for the number of components in the chip that have been added, deleted or compromised. Size of a Trojan can be an important factor during activation. A smaller Trojan has a higher probability for activation than a Trojan with larger number of inputs. The *distribution* category describes the location of the Trojan in the physical layout of the chip. For example, a *tight distribution* describes a Trojan whose components are topologically close in the layout while a *loose distribution* describes Trojans that are dispersed across the layout of the chip. Finally, the *structure* category describes the change in the layout structure. If the adversary is forced to regenerate the layout to be able to insert the Trojan in the circuitry, then the chip dimensions change. This change could result in different placement for some or all the design components.

Trojan Activation Characteristics: Activation characteristics refer to the criteria that causes the Trojan to become active and carry out its disruptive function. The adversary who inserted the Trojan will make it difficult for the user of the chip to activate it, in an effort to prevent 'accidental' activation and detection during the testing phase(s) of the chip and system. Therefore, activation of a Trojan can be considered a 'rare event' from a statistical perspective.

We use the term *stealthy activation* to describe the adversary's objective in this regard. We partition Trojan activation characteristics into two sub-categories, labeled *Externally-activated* and *Internally-activated*. In Externally-activated category, the Trojan can be activated externally by adversary in his/her time of choosing. This can be done by embedding a receiver or antenna on chip and controlling it through external signals. The Internally-activated category is divided into two subclasses, labeled *Always-on* and *Condition-based*. Always-on, as the name implies, indicates that the Trojan is always active and can disrupt the function of the chip at any time. The Condition-based subclass includes Trojans that are 'inactive' until a specific condition is met.

Trojan Action Characteristics: Action characteristics identify the types of disruptive behavior introduced by the Trojan. We partition Trojan actions into three categories; *Modify-function, Modify-specification,* and *Transmit-info.* As the name implies, the Modify-function class refers to Trojans that change the chip's function through additional logic or by removing or bypassing existing logic. The Modify-specification class refers to Trojans that focus their attack on changing the chip's parametric properties, such as delay. The latter class represents parametric Trojans that modify wire and transistor geometries. Lastly, the Transmit-info class refers to Trojans that transmit key information from design mission mode to an adversary.

3. Current Integration Method

3.1 Trojan Detection

A Trojan, when inserted into a chip, will most likely consume power. However, the Trojan's contribution to the total power consumption of the circuit depends heavily on its size and type. It also depends on its activation, that is, fully activated Trojan can consume more power than that of partially activated. A Trojan inserted in a chip will draw leakage current if it is powered on. Creating switching in the Trojan can further increase the amount of current drawn by the Trojan circuitry. We acknowledge that fully activation of a Trojan using structural and functional patterns would be extremely challenging and prohibitively expensive considering that the size and type of Trojan is unknown to us.

Partial activation of Trojans can be an effective way for Trojan detection and isolation using transient current-based side-channel analysis methods similar to our current integration method. A large number of transitions is generated when applying a pattern to the chip. Some of the Trojan inputs in the chip may also observe the transitions and in turn cause transitions in the Trojan circuitry as well. The switches in the Trojan will increase the local power consumption (i.e. current). The local power refers to the current drawn from the power port near the Trojan circuitry. The more the number of switching on the Trojan inputs and in the Trojan circuitry the larger the transient current. Since small Trojan sizes are expected to be inserted into chips by adversary to reduce the detection capability, the local current impact by Trojan could be more significant than the global current that can be measured only by power pins.

The amount of current a Trojan can draw could be so small that it can be submerged into envelop of noise and process variations effects, therefore, cannot be detected by measurement equipments. However, Trojan detection capability can be greatly improved when measuring currents locally and from multiple power ports/pads. Figure 2 shows our current (charge) integration methodology for detecting hardware Trojans. There are four power ports on the chips. The golden chip can be identified using an exhaustive test for a number of randomly selected chips. It can also be identified using the pattern set that will be used in our current integration method by comparing the results against each other for all the patterns. If the same results are obtained for all the selected chips, they can be identified as Trojan-free. We assume that adversary will insert the Trojans randomly in a selected number of chips. After identifying the golden chips, the worst-case charge will be obtained (dashed-line in the figure) in response to the pattern set. The worst-case charge is based on the worst-case process variations in one of the genuine ICs. Next, the pattern set will be applied to each chip and the current will be measured for each pattern locally via power ports or C4 bumps.

The figure shows the current waveform of n number of patterns applied to the chip. The figure also shows the charge variations with time for all the current waveforms obtained after applying the patterns. The charge corresponds to the area produced by each current waveform. $Q_n(t)$ denotes the accumulative charge after applying n patterns. Q_{thr} is the charge threshold to detect a Trojan which is in fact the resolution measurement defined by the instrumentation. When applying the patterns, the charge increases and is compared continuously against the worst-case charge calculated for golden chips. Once the difference between the two curves ΔQ is greater than Q_{thr} we consider a Trojan is detected. The number of patterns, n is expected to very small for large Trojans and large for very small Trojans.

By applying this integration method the small current difference between Trojan-inserted and Trojan-free circuits can be magnified through the charge integration process. By applying more number of patterns to the chip over time, larger current (or charge) difference will be created. In the figure, the curve with solid line shows the Trojan-inserted chip's accumulative charge.

Note that in this work, we assume that the IC authentication phase is done after manufacturing test. Therefore, the likelihood of encountering a defect during IC authentication would be very small. The existence of defect depends on the DPPM level for the manufactured chip in the foundry.

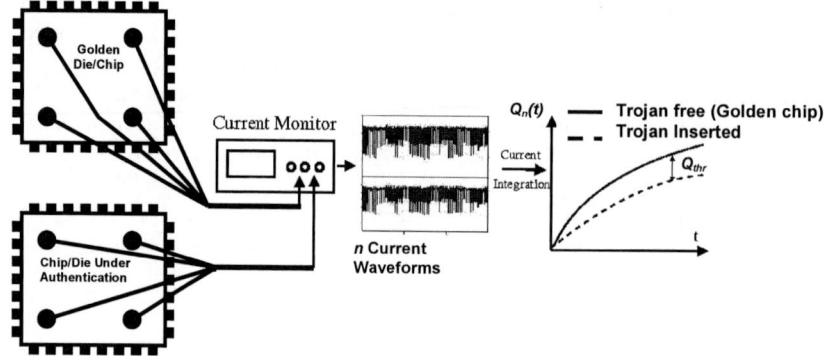

Figure 2. Current (Charge) Integration Method

If $I_{trojan\_free}(t)$ and $I_{trojan\_inserted}(t)$ denote the instantaneous supply current drawn by Trojan-free and Trojan-inserted circuit at time t, respectively, then the integrated current at time t for Trojan-free and Trojan-inserted circuit ($Q_{trojan-free}(t)$ and $Q_{trojan-inserted}(t)$) can be expressed by equations (1) and (2) (note that $dq = I \cdot dt$)

$$Q_{trojan-free}(t) = \int I_{trojan\_free}(t) \cdot dt \qquad (1)$$

$$Q_{trojan-inserted}(t) = \int I_{trojan\_inserted}(t) \cdot dt = \int (I_{trojan\_free}(t) + I_{trojan}(t)) \cdot dt \qquad (2)$$

where $I_{trojan}(t)$ denotes the current drawn by Trojan. Since same pattern set is applied to both golden chips and chip-under-authentication, the difference between $I_{trojan\_free}(t)$ and $I_{trojan\_inserted}(t)$ comes from (1) the additional current drawn by Trojan gates and (2) changes in the circuit current due to process variations. By integrating the charge along time axis for both chips, their cumulative difference at time t can increase as more number of patterns are applied.

Since we perform multiple power supply transient current analysis, the proposed integration method can be used for detecting both tight and loose distributed Trojans. During Trojan detection phase, the total supply current is integrated for chip-under-authentication and the golden chip separately. All Trojan gates located on chip will contribute to overall current consumption. Therefore, the total supply current's difference beyond the predefined threshold between the two chips will imply the existence of Trojan. We will show our simulation results in Section 6 for loosely distributed Trojans (e.g. Comparators).

3.2 Trojan Isolation

Trojan isolation process is done after detecting a Trojan in a chip. The Trojan isolation is essential in identifying the location of Trojan and possibly identifying the type of Trojan especially in terms of action characteristics. It is extremely valuable to find out what the adversary intended to carry out with the inserted Trojan.

Trojan isolation process is based on the fact that Trojan gates (similar to circuit gates) will draw more current from their nearest power port therefore more current difference occurs on the power ports near the Trojan gates. To further demonstrate the impacts on local power ports, we have inserted a Trojan (a 3-bit counter, tightly distributed) in ISCAS'89 s38417 benchmark which contains 8709 gates and 1636 FFs. In s38417 benchmark, we have considered 49 (7 × 7) power ports.

We have generated the layout of the circuit and inserted the Trojan circuit in a dead space in the layout and connected the clock to the Trojan. The Trojan was inserted close to power port 17 exactly at the coordinates of (950μm, 100μm) in the physical layout. Figure 3 shows the charge difference (ΔQ) of an array of power ports between Trojan-inserted and Trojan-free circuit obtained using post-layout simulation [21]. As seen, the maximum ΔQ happens on power port 17. The supply current difference falls drastically on neighboring power ports.

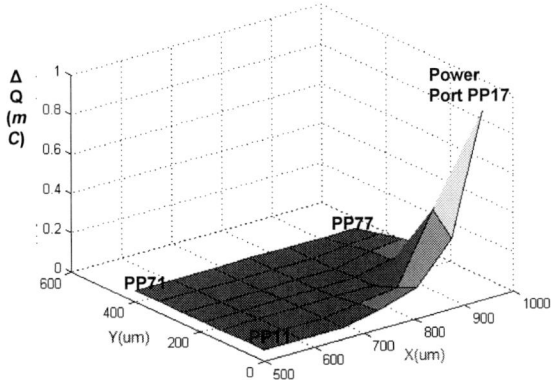

Figure 3. Current difference measured on power port 17

During Trojan isolation process the current of each power port is measured, integrated and compared with golden chip's current integration result separately for each power port. When there is a clear difference (depending on the pre-defined threshold) between the two chips, then the Trojan is assumed to be near the power port. However, if, for example, the Trojan is located right between power ports 13, 14, 23, and 24, then it will impact more than one power port when switching (see Figure 3). By comparing the currents drawn by each power port after applying the patterns, we will be able to identify the location of the Trojan between them. If the adversary distributes the Trojan in the entire circuit, the isolation process will be more difficult since the smaller portion of the Trojan will draw currents from different power ports.

3.3 Pattern Generation and Application

To detect and isolate a Trojan, a pattern set must be generated and applied to the chip during IC authentication step. Since the proposed method measures the current from various power ports, the patterns that generate localized switching would be most effective. However, generating such patterns will be computationally intensive. In this paper, we use random patterns that are effective in generating a large number of transitions in the circuit thus increasing the probability of partial activation of hardware Trojans.

The pattern generation for targeting a fault or defect is fundamentally different from targeting Trojans. For example, when detecting interconnect open defect, we generate patterns such that they target every node in the circuit for an open defect. However, this will not be the case for Trojan. A Trojan cannot be activated by activating a node. A transition on a node does not ensure a gate in the Trojan will be activated. To be able to activate a gate in the Trojan, we need to perform a procedure similar to multiple stuck-at faults [20] by generating switching on all combinations of two nodes assuming that there are two-input gates in the Trojan circuit. This procedure does not seem practical due to very high complexity. As mentioned earlier, the size, type, and location of a Trojan is unknown to us. To improve the probability of detection of a Trojan, it's best to increase number of switching in the circuit.

Another major difference between ATPG for defect/fault and Trojan is that a node is targeted with every pattern for defect/fault-oriented ATPG while a region is targeted using every pattern for Trojan-oriented ATPG. In both case, a large number of patterns must be applied. The total number of patterns to detect all fault depends on the number of nodes in the circuits but there is no limit on the number of patterns for Trojan detection. The number of patterns required to detect Trojans depends on the size and type of Trojans. A sequential Trojan that uses clock continuously consumes more power compared to monitor-like Trojans. Therefore the detection depends on the number of clock cycles needed to reach Q_{thr} charge over the worst-case charge. Note that in this work, there are no logic observation points. The patterns should generate switching in the circuit and the power ports are in fact the observation point for the side-channel signal, here current/charge. Also note that we use primary inputs and scan cells to apply patterns to the circuit. The pattern application is similar to *test-per-clock* [20] where a pattern is applied in every clock cycle. A random bit is shifted into the scan chain to generate a new random pattern in addition to applying a random pattern to primary inputs in every clock cycle. In a test-per-clock approach, the pattern application time will be quite short.

In this work, we apply random patterns to the chips. Some patterns may generate switching at the input of the Trojan and some may not. Those that do not generate a switching, will impact the charge based on the genuine gates switching and process variations. But, those patterns that activate part of the Trojan or its inputs, can have a extra current over the process variations and genuine gates switching current. Also, note that, since we apply the same patterns to Trojan-free and Trojan inserted circuit, the charge induced by genuine gates switching would be same and the difference is in process variation-induced current and Trojan gate switching.

3.4 Trojan Insertion

An adversary can exploit the dead spaces in the physical layout to insert small or large hardware Trojans. In this work, we generate the physical layout for ISCAS'89 s38417 benchmark. We then generate several copies of this benchmark to insert Trojans. In each layout, we use dead spaces to insert Trojan gates/circuitry. We insert two types of Trojans, *Counter* and *Comparator*, in a tightly and loosely distributed fashion, respectively. For distributed Trojan, we utilize small dead spaces to insert Trojan gates and connect them. When inserting Counter, we use those already existed Counters in standard cell library. We place them in available dead spaces in the physical layout. When designing a Comparator, however, we intentionally distribute the gates in difference locations on the layout. In all cases, we ensure not to change the s38417's original layout's form-factor.

We assume that the adversary has the knowledge of IC fabrication and testing, so he/she can design the Trojan circuit such that it will not to be detected during manufacturing test. The adversary is expected to ensure the layout of Trojan-inserted and Trojan-free circuits remain same by avoiding re-designing the physical layout. Any modifications to the layout of the circuit will change the position of cells and makes it easier to be detected, for instance, using circuit delay analysis.

4. Process Variations Impact on Trojan Detection

As technology scales to 45nm and below, the impact of process variations on current/power consumption is expected to be more significant than ever. Therefore, process variations should be considered during Trojan detection methods that rely on side-channel signals. Process variations can either help or harden the Trojan detection process. According to equation (4), which is current drawn by a single gate, decreasing voltage threshold V_{th}, channel length L, as well as oxide thickness T_{ox} will increase gate current. Conversely, increasing V_{th}, L and T_{ox} will decrease gate current consumption (I_D).

$$I_D = \frac{\mu C_{ox} W}{2L} (V_{GS} - V_{th})^2 (1 + \lambda V_{DS}) \qquad (4)$$

The following two scenarios will make the Trojan detection more difficult when considering process variations:
1. When process variations in Trojan-free circuit *increase* the transient current. This will make the current measured from a Trojan-free circuit closer to that of the Trojan-inserted circuit.
2. When the process variations in Trojan-inserted circuit *decrease* the current consumption. This will also make the current measured from the Trojan-inserted circuit closer to that of the Trojan-free circuit.

Similarly, the two scenarios that help make the Trojan detection process easier are:
1. When process variations in Trojan-free circuit decrease the current consumption.
2. When process variations in Trojan-inserted circuit increase the current consumption.

To generate the worst-case charge, we need to apply the worst-case process corners. Based on the above analysis, process corners that increase current consumption of a Trojan-free chip and decrease current for Trojan-inserted chip are the most difficult scenarios for Trojan detection. Although process variations may decrease the charge difference between the two charge curves (Trojan-inserted and Trojan-free), the current integration method would still be effective in detecting small Trojans since the integration effect can successfully increase the gap between the two curves with applying more patterns.

5. Simulation Results

The current integration method is applied for detecting Trojans inserted into s38417 benchmark. First, we generate 7 layouts for original s38417 benchmark using Synopsys physical design tools [21] in 180*nm* technology [22]. 1-bit, 3-bit, 7-bit, 9-bit Counter and 3-input, 5-input, 20-input Comparator Trojans are inserted into these seven layouts separately (i.e. only one Trojan in each layout). Table 1 shows the type, size, distribution and structure of the inserted Trojans in the layout. We also investigate the impact of process variations on Trojan detection. The wors-case process variations that we consider for a genuine IC during our simulation are shown in Table 2.

We first start with the results obtained from the circuits containing Counter Trojans. Figure 4 shows the simulation results obtained using Synopsys Nanosim [21] for s38417 containing a 1, 3, 7, and 9-bit Counters. The patterns are shifted into the scan chain with a frequency of 100MHz. As seen in the figure, for all four Trojans, after applying 15 clock cycles (equals 15 patterns), $\Delta Q \gg Q_{thr}$ where Q_{thr} is considered to be in the range of μC, which is easily detectable using measurement devices. The 1-bit Counter used here is a flip-flop with self loopback. The isolation method by measuring current on each power port would also be effective in detecting such Trojans. In general, detecting a Counter would be easier than a combinational Trojan since the Counter continuously receive the clock and consumes power. No process variations were considered for the results shown in Figure 4, although the process variations would not be significant enough to change the detection outcome for such Trojans.

Table 1. Trojan characterization

Trojan	Type	Size	Distribution	Structure
Counter	1-bit	0.04%	tight	no-change
	3-bit	0.10%	tight	no-change
	7-bit	0.31%	tight	no-change
	9-bit	0.42%	tight	no-change
Comparator	3-input	0.02%	loose	no-change
	5-input	0.04%	loose	no-change
	20-input	0.15%	loose	no-change

Table 2. Worst-case process variations applied to Trojan-free circuit during Trojan detection

	Inter-die	Intra-die
Threshold Voltage (V_{th})	5%	20%
Channel Length (L)	2%	8%
Oxide Thickness (T_{ox})	1%	4%

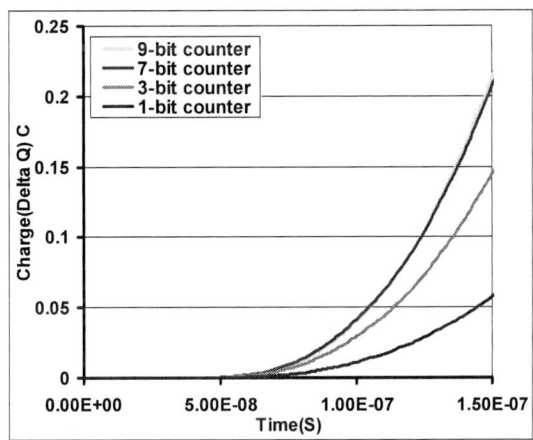

Figure 4. Charge measurement for s38417 with four Counter Trojans inserted into s38417

Figure 5 shows the simulation results for the circuit containing 3-bit Counter while considering the process variations for both Trojan-inserted and Trojan-free circuits. The process variations used in the Trojan-free circuit increases the current in the Trojan-free circuit while the variations used in the Trojan-inserted circuit reduced the total current. As seen, the Trojan-inserted circuit with process variations is still consuming more current when compared to the Trojan-free circuit with process variations. Our simulation for 5, 7, and 9-bit Counters have also shown similar results. The charge induced by worst-case process variations in presence of Trojans was still below the amount of charge contributed by the Trojans. Our simulation results showed that for the smaller Trojans, such as 2-bit and 1-bit Counters, the detection was not possible when considering worst-case process variations. However, we were able to detect 1-bit Counter when considering average process variations.

Figure 6 shows the simulation results when inserting a 20-input Comparator in s38417 circuit. The Comparator circuit is connected to 20 randomly select nodes in the circuit. This type of Trojan falls into the category of "loose distribution" and requires activation to generate transient current since it is a combinational circuit. Since fully activation is very time consuming and prohibitively expensive, we rely on partial activation of such hardware Trojan by applying random patterns. Figure 6 shows the results after applying random patterns to the circuit with and without Trojan considering process variations. As shown, the Trojan can be easily detected using the method in presence of worst-case process variations considered for Trojan-free circuit to identify the worst-case charge for genuine ICs. Lower variations were considered for Trojan-inserted circuit.

However, the results shown in Figure 7 shows the increase in charge difference between the two circuits, Trajan-free and Trojan-inserted. As seen, the current difference is significantly greater than Q_{thr} after applying only 8 random patterns. This shows that the patterns were able to partially activate the Trojan. We have also simulated the

Trojan-inserted circuit with the worst-case process variation but we were not able to clearly detect the Trojan. However, we acknowledge that using a lower process variation could potentially detect the Trojans. This was the case for 3-input Comparator as well. To further increase the probability of detection, more test patterns must be applied. The application time depends when the Trojan-inserted circuit results falls outside the Trojan-free circuit with worst-case process variations.

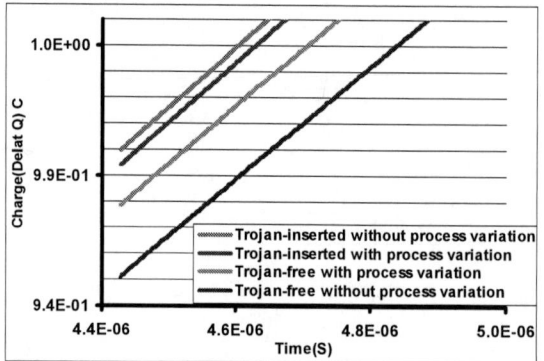

Figure 5. Charge measurement for s38417 with 3-bit Counter considering the worst-cased process variations for Trojan-free circuit

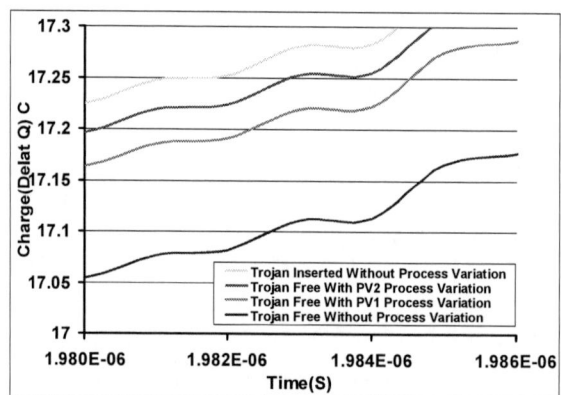

Figure 6 Charge measurement for s38417 with 20-input Comparator (here, PV2 represents the worst-case variations and PV1 represents lower variations)

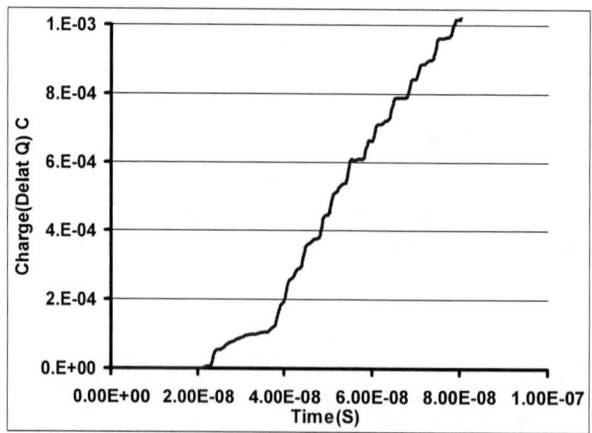

Figure 7. Charge measurement for s38417 with 5-input Comparator with no process variations

6. Conclusions and Future Work

We have presented a new current (charge) integration methodology for Trojan detection and isolation. The method measures the current locally from the on-die power ports. Comparing the results obtained for golden chips against the chip-under-authentication, the Trojan can be detected if the current integration results fall outside the golden chip results. We have shown that our method can easily detect Trojans as small as 0.1% the circuit area. We plan to improve the quality of test patterns using the layout-aware test pattern generation procedure we are developing. We also plan to deal with process variation impact in more accurate way by performing Monte Carlo simulation for all Trojan-free and Trojan-inserted circuits and measure the effectiveness of the proposed method.

5. AKNOWLEDGEMENT

The work of Xiaoxiao Wang, Hassan Salmani and Mohammad Tehranipoor was supported in part by NSF grant CNS-0716535. The work of Jim Plusquellic was supported in part by NSF grant CNS-0716559.

References

[1] http://www.acq.osd.mil/dsb/reports/2005-02-HPMS_Report_Final.pdf

[2] http://www.darpa.mil/mto/solicitations/baa07-24/index.html

[3] B. Yang, K. Wu, and R. Karri, "Scan Based Side Channel Attack on Dedicated Hardware Implementations of Data Encryption Standard," in *Proc. of the IEEE Int. Test Conf. (ITC)*, pp. 339.344, 2004.

[4] S. Ravi, A. Raghunathan, and S. Chakradhar, "Tamper Resistance Mechanisms for Secure Embedded Systems," in *Proc. of the 17th Intl. Conf. on VLSI Design*, pp. 605.611, 2004.

[5] P. Kocher, R. Lee, G. McGraw, A. Raghunathan, and S. Ravi, "Security as a New Dimension in Embedded System Design," in *Proc. of the 41st Annual Conference on Design Automation*, pp. 753.760, June 2004.

[6] K. Tiri and I. Verbauwhede, "A VLSI Design Flow for Secure Side-Channel Attack Resistant ICs," in *Proc. of Design, Automation and Test in Europe*, pp. 58.63, Mar. 2005.

[7] K. Hafner, H. C. Ritter, T. M. Schwair, S.Wallstab, M. Deppermann, J. Gessner, J. Koesters,W.-D. Moeller, and G. Sandweg, "Design and Test of an Integrated Cryptochip," *IEEE Design and Test of Computers*, pp. 6.17, Dec. 1991.

[8] P. Kocher, J. Jaffe, and B. Jun, "Differential Power Analysis," *Lecture Notes in Computer Science*, vol. 1666, pp. 388.397, 1999.

[9] G. B. Ratanpal, R. D. Williams, and T. N. Blalock, "An On-Chip Signal Suppression Countermeasure to Power Analysis Attacks," *IEEE Transactions on Dependable and Secure Computing*, vol. 1, no. 3, pp. 179.188, 2004.

[10] P. C. Kocher, "Timing Attacks on Implementations of Diffe-Hellman, RSA, DSS, and Other Systems," *Lecture Notes in Computer Science*, vol. 1109, pp. 104.113, 1996.

[11] J. Kelsey, B. Schneier, D.Wagner, and C. Hall, "Side Channel Cryptanalysis of Product Ciphers," in *Proc. of the European Symposium on Research in Computer Security*, pp. 97.110, Sep. 1998.

[12] D. Boneh, R. A. Demillo, and R. J. Lipton, "On the Importance of Checking Cryptographic Protocols for Faults," *Lecture Notes in Computer Science*, vol. 1233, pp. 37.51, 1997.

[13] E. Biham and A. Shamir, "Differential Fault Analysis of Secret Key Cryptosystems," *Lecture Notes in Computer Science*, vol. 1294, pp. 513.527, 1997.

[14] R. Goering, "Scan Design Called Portal for Hackers," Oct. 2004. [Online]. Available: http://www.eetimes.com/news/design/showArticle.jhtml?articleID=51200154

[15] S. Scheiber, "The Best-Laid Boards," Apr. 2005. [Online]. Available: http://www.reed-electronics.com/tmworld/article-/CA513261.html

[16] J. Lee, M. Tehranipoorand J. Plusquellic, "A Low-Cost Solution for Protecting IPs Against Side-Channel Scan-Based Attacks," in Proc.*VLSI Test Symposium (VTS'06)*, 2006.

[17] J. Lee, M. Tehranipoor, C. Patel and J. Plusquellic, "Securing Scan Design Using Lock & Key Technique," in *Proc. Int. Symp. on Defect and Fault Tolerance in VLSI Systems (DFT'05)*, 2005.

[18] D. H´ely, M.-L. Flottes, F. Bancel, B. Rouzeyre, N. B´erard, and M. Renovell, "Scan Design and Secure Chip," in *Proc. of the 10th IEEE Intl. On-Line Testing Symposium*, 2004.

[19] D. Agrawal, S. Baktir, D. Karakoyunlu, P. Rohatgi, B. Sunar, "Trojan Detection using IC Fingerprinting", Symposium on Security and Privacy, 2007, pp. 296 - 310.

[20] M. Bushnell, V. Agrawal, *Essentials of Electronics Testing*, Kluwer Publishers, 2000.

[21] Synopsys, "User Manual for SYNOPSYS Toolset Version 2005.09," Synopsys, Inc., 2005.

[22] http://crete.cadence.com, "0.18m standard cell GSCLib library version 2.0," Cadence, Inc., 2005.

IEEE International Symposium on Defect and Fault Tolerance of VLSI Systems

Built-In Proactive Tuning System for Circuit Aging Resilience

Nimay Shah[1], Rupak Samanta[1], Ming Zhang[2], Jiang Hu[1], Duncan Walker[3]

[1]Dept. of ECE, Texas A&M University, College Station

[2]SoC Enabling Group, Intel

[3]Dept. of Computer Science, Texas A&M University, College Station

E-mail: nimay_shah@tamu.edu, rupak9@tamu.edu, ming.y.zhang@intel.com,
jianghu@ece.tamu.edu, walker@cs.tamu.edu

Abstract

VLSI circuits in nanometer VLSI technology experience significant aging effects, which are embodied by performance degradation over operation time. Although this degradation can be compensated by over-design, it induces remarkable power overhead which is undesirable in tightly power-constrained designs. Dynamic voltage scaling (DVS) is a more power-efficient approach. However, its coarse granularity implies difficulty in handling fine-grained variations in the aging effects. We propose a Built-In Proactive Tuning (BIPT) system that allows each circuit block to autonomously tune its performance according to its own degree of aging. The BIPT system is validated through SPICE simulations on benchmark circuits with consideration of NBTI effect. The experimental results indicate that the proposed BIPT system leads to about 45% less power than the approach of over-design while maintaining the same performance. Compared to DVS, BIPT can achieve the same aging resilience with about 30% less power dissipation

1. Introduction

As VLSI technology scales to nanometer regime, circuit aging effects, such as NBTI (Negative Bias Temperature Instability) and HCI (Hot Carrier Injection) [10] become prominent. NBTI manifests itself by degradation of PMOS threshold voltage [6, 10] whereas HCI results in threshold voltage increase in mostly NMOS transistors. When technology scales from 180nm to 65nm, the MTTF (Mean Time To Failure) of processors due to aging effects is reduced by about 76% [10]. That is, if a chip would have previously lasted for 10 years, now it can perform well for about 2 years. Therefore, it becomes increasingly imperative to address the aging effect in chip design.

To handle the circuit aging problem, a common approach is to over-size transistors such that the aged performance can still meet specifications [6]. This approach is able to extend chip lifetime under the aging effect. However, it inevitably increases circuit power dissipation and therefore hits another wall of nanometer integrated circuit design – the increasingly tight power constraint. Over-sized transistors usually imply unnecessarily large timing slack and therefore wasteful power dissipation during the initial lifetime of a circuit. Alternatively, architectural approaches [9, 10] are suggested for mitigating the aging problem. One technique is architectural-level adaptation [9, 10] such as DVS (Dynamic Voltage Scaling). For instance, a chip can operate at relatively low supply voltage level when new and switch to higher supply voltage level when it gets aged. Such adaptation can avoid the wasteful power as compared to the oversized transistor approach. However, this is a coarse-grained technique – the supply voltage level is usually fixed for major partitions of the chip, if not across the entire chip. In general, the aging effects vary among different components of a circuit. In order to ensure the performance of an entire chip, the DVS must be performed according to the worst transistor aging. That is, strong aging of only 1% percent transistors may require the

1550-5774/08 $25.00 © 2008 IEEE
DOI 10.1109/DFT.2008.49

chip-level supply voltage increase although the other 99% transistors have very minor aging induced degradation.

In this paper, we propose a Built-In Proactive Tuning (BIPT) system to mitigate the aging problem in a power-efficient manner. This system includes a canary circuit which can generate predictive warning signals for performance degradation. According to the warning signal, circuit speed is tuned through body bias such that the performance degradation is compensated. The proactive tuning is performed offline, at power-on of the chip or periodically. Since aging is a slow change with time constant of weeks/months, periodic tuning of once in a few days is sufficient to capture the change. The offline tuning has the advantage of allowing relatively easy control on input vectors. When detecting performance degradation or circuit delay variation, one must consider the delay uncertainty due to different input vectors. Even if there is no warning signal for certain input vectors, there is still risk of delay errors under other input vectors. Therefore, we include a Test Pattern Generator (TPG) in the system in order to have large input vector coverage. TPG is usually a part of Built-in Self Test (BIST) hardware; thus, if a chip already has BIST circuit, TPG does not cause extra overhead. The proposed BIPT system has the following advantages:

- It can be applied at circuit block level instead of the chip level architectural approach [9, 10]. In other words, each block can be tuned according to its own degree of aging. Evidently, the finer granularity control allows improved power efficiency.
- Its performance degradation detection is obtained from the actual operating circuit as opposed to replica circuit in other adaptive design methods [12]. Since the detection is more direct, it is more reliable. Using TPG further improves the reliability of the detection.
- Its proactive nature can avoid the complex error correction schemes in retroactive systems [3, 4]. The retroactive systems rely on pipeline flush [4] or instruction replay [3] and therefore are restricted to processor designs. In contrast, our system can be applied to both processors and general sequential circuits.

Existing approaches have one or two of the above advantages, but none of them have all to the best of our knowledge. The work of [12] is a block level adaptive body bias technique. However, its delay variation detection is obtained from replica circuits which often have discrepancy from the actual operating circuits. The Razor based techniques [3, 4] use direct variation detection, but they rely on complex error correction method and are restricted to processor designs. Another retroactive method [7] is mainly targeted for fast variations such as voltage variations and hence complements our work. The canary circuit based predictive detection is proposed in [8]. However, it is applied with online tuning which suffers from delay uncertainty due to different input vectors. The recent work of [2] focuses on only the aging detection instead of an overall tuning system. Actually, the detection method in [2] can be easily adopted in our tuning system.

The BIPT system is validated through SPICE simulations on benchmark circuits with consideration of NBTI effect. Even with consideration of overhead due to TPG, canary and control circuit, the proposed BIPT system can lead to about 45% less power than the over-design approach while maintaining the same performance. Compared to DVS, BIPT can achieve the same aging resilience with about 30% less power dissipation.

2. Built-In Proactive Tuning System

2.1. Overview

The Built-In Proactive Tuning (BIPT) System consists of the existing main circuit augmented with a Test Pattern Generator (TPG), Body Bias Circuitry, Canary Circuit and Control circuit. Figure 1 shows these blocks and the corresponding interface signals.

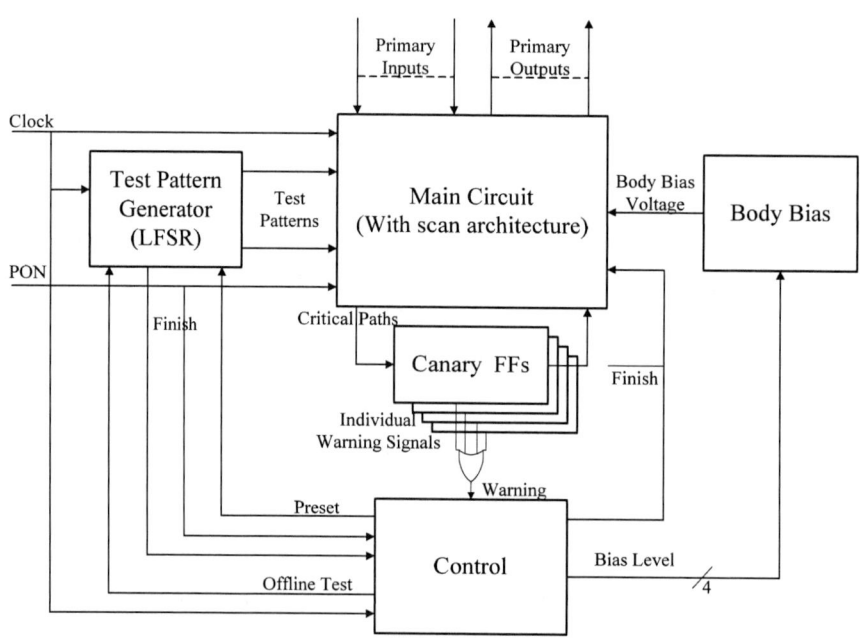

Figure 1. Overview of the proposed built-in proactive tuning system

At power-on or periodically, the BIPT system can launch test vectors from TPG and then tune circuit body voltage according to the observations from the canary circuit. Canary circuit plays the role of predicting aging-induced performance degradations (more details in Section 2.2). A *Warning* signal is generated by the canary flip-flops when the timing constraint is tight on one or more of the few critical paths where these are inserted. The top-level warning signal is the *OR* of all the individual canary flip-flop warning signals. The Linear Feedback Shift Register (LFSR) [1] is implemented as a pseudo-random test pattern generator which applies these random patterns when offline test is in progress. It is triggered by the *preset* signal from the control block. The control block monitors the status of all the blocks and issues control signals. *PON* is the power-on-reset signal which is an active high reset signal issued on start-up and basically triggers the offline test. *Offline test* is an active high signal indicating that offline test is in progress. The most critical activity performed by the control block is to monitor the warning signal from the canary circuit. Based on this signal, it appropriately sets the body bias to selective gates on the critical paths of the main circuit via the bias level signal passed to the body bias block. This interface and the body bias block are modeled as in [12]. The body bias is adaptive to the circuit state: the circuit automatically selects from 4 available options of forward body bias using a counter – decoder based scheme. This has been described in detail in section 2.3.

2.2 Canary Circuit

The canary circuit [8] is for detecting aging-induced performance degradation in a predictive manner. As shown if figure 2, a canary circuit consists of two flip-flops; a main FF and a canary FF. The main FF gets the direct input and the canary FF which serves as the checker part gets the input through a delay buffer. This delay in the input reaching the two flops serves as the guard band for error detection. The outputs from these flops are fed to an xor gate which functions as a comparator, outputting 1 when these are different and thereby predicting the occurrence of an error. Some advanced designs of canary circuits are proposed in [2, 13].

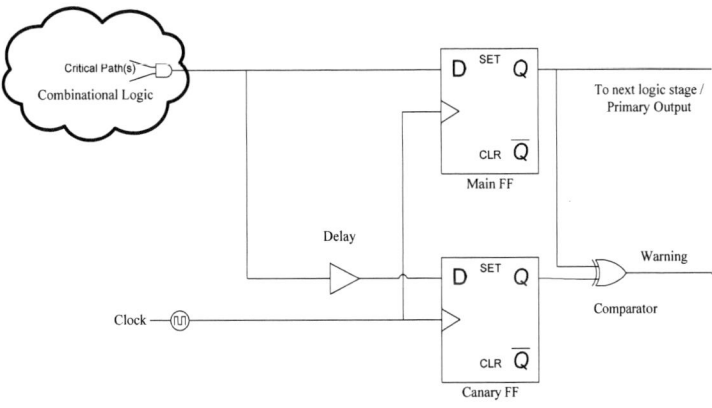

Figure 2. Canary circuit

Canary circuit is a typical case design alternative of Razor [4]. However, in contrast to Razor, which delivers a delayed system clock to the checker part (shadow FF), canary circuit delivers a delayed input signal to the checker part (canary FF). This simplifies the clock tree synthesis and routing as there is just one system clock now. Also, the delay buffer placed before the canary flop always has a positive delay, even if affected by aging, which makes the canary flip-flop recover from variation induced effects by itself. Canary circuit also predicts timing errors rather than detecting them afterwards. The predictive warning allows the user to take preventive measures before the timing violation actually occurs and thus the system does not run into any corrupt data states, except for errors that cannot be predicted such as single event upset (SEU) errors. However, aging induced timing violations can be predicted effectively by architectures such as canary circuit [2].

2.3. Test Pattern Generator and Control Circuit

Figure 3 shows the gate level implementation of the control circuit. *Finish* signal going high indicates the completion of offline testing, *PON* is the power-on-reset signal, *Warning* is the timing error prediction signal from the canary circuit and *Preset* is the active low signal to set the flip-flops in the LFSR to high state on power-on-reset. The preset generation circuit is shown in the dotted box in figure 3. The initial states of all the flip-flops in the LFSR on power-on-reset is '1', thus the starting seed for the LFSR is all 1's. The LFSR shown in figure 3 is a 12-bit LFSR; it implements a primitive polynomial to generate 4095 patterns (2^n-1; n=12) before returning back to the initial state of all 1's. The outputs of the flip-flops in the LFSR are fed to a scan chain through a mux-d connection. These connections are omitted in figure 3 for clarity.

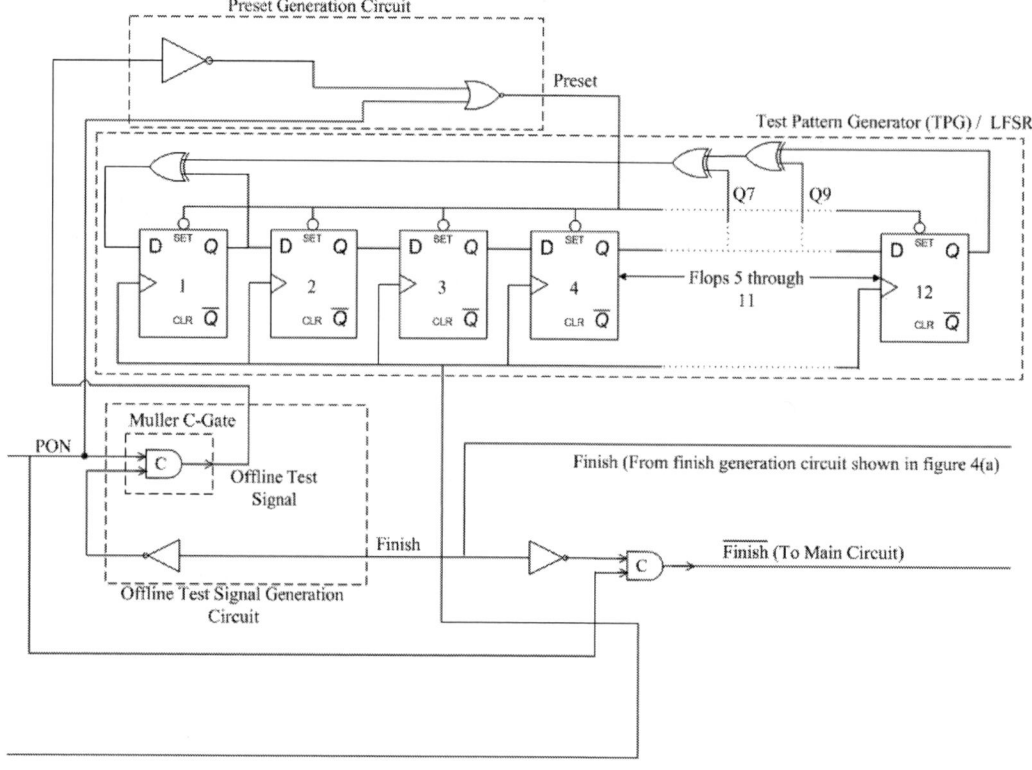

Figure 3. Test pattern generator and control circuit

The control circuit is triggered by the power-on-reset signal (PON), which remains high for one cycle on each power-on-reset of the chip. On each power-on-reset, the *offline test signal* triggers the offline testing. Generation of the offline test signal is shown in the box ifnfigure 3. The *offline test signal* is the input for preset generation circuit that presets the flip-flops in the LFSR to high, the initial seed for the test patterns in the LFSR. The *Finish* signal is generated by the circuit shown in figure 4(a). Its first stage consists of a 12-input *AND* gate and the second stage consists of a 2-input Muller-C gate. Muller-C gate is an AND gate for events i.e., it produces a high output when all the inputs are high and goes low only when all the inputs transition to low state. The description about Muller-C gates can be found in [16]. As shown in figure 4(a), the outputs of the flip-flops in the LFSR are connected to a 12-input *AND* gate. The output of this *AND* gate and $\overline{PON}$ feed to a 2-input Muller-C gate to produce the finish signal. On every power-on-reset, the flip-flops of the LFSR are preset to 1, thus the output of the *AND* gate rises high. Since *PON* is active high, $\overline{PON}$ is low at startup and thus finish stays at 0 initially. $\overline{PON}$ stays high for one clock cycle and then goes to low. When all the 4095 test patterns have been generated, the output of the *AND* gate goes high again and since $\overline{PON}$ is also high; finish goes high indicating the completion of offline testing. After finish goes to high, at the next clock edge, the output of the *AND* gate goes low due to a pattern other than all ones. However, the finish signal still stays high because of the property of the Muller-C gate to hold the previous value until both the inputs transition to the same value. In this case, although the output of the *AND* gate goes low, since $\overline{PON}$ is still high, finish stays high.

The possible timing violations in the critical paths i.e., the paths that are affected due to aging are predicted by the warning signal from the canary flip-flops. The critical paths that are affected by aging need to be corrected by application of suitable forward body bias. Since, the body bias generation circuit takes some time to apply correct bias to the devices on these critical paths, the LFSR needs to be stalled. In our approach, we stall the clock to the LFSR by using gated clock circuitry shown in figure 4(b). In figure 4(b), the circuit can stall the clock for one clock cycle, which is sufficient for us to change the body bias of the devices on the critical paths. However, if we need more time then the clock can be stalled for a longer period of time using cascaded Muller-C gates in figure 4(b).

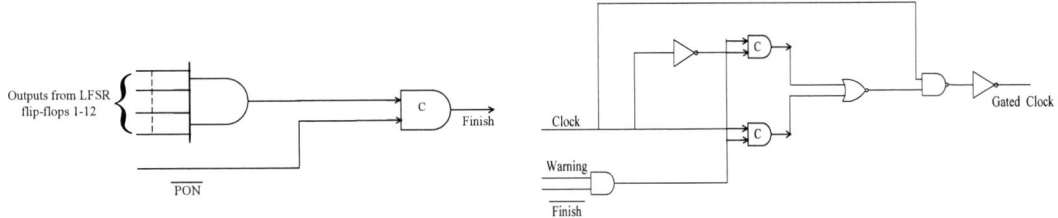

Figure 4(a). Finish signal generator   **Figure 4(b). Gated clock circuit**

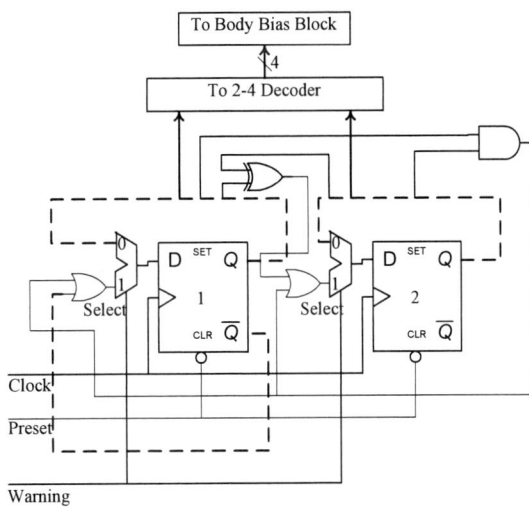

Figure 5. Generation of control signal to body bias block

Figure 5 shows the body-bias generation logic. As shown, it consists of a 2-bit up-counter, made up of two flip-flops. The four possible states of the counter translate into fours possible body bias levels to choose from. Level 0 is the no-bias condition; levels 1-3 are in the increasing order of the forward body biases. As stated earlier, we deal with aging degradation which monotonically degrades the circuit performance with time; thus, forward body bias or reduced reverse body bias is necessary to restore the circuit performance. The up-counter counts upward (increases forward bias / reduces reverse bias) when a warning signal is generated by the canary circuit. It counts upward till it reaches the highest forward body bias state / least reverse body bias (binary 11 in our case) and freezes in that state. We implement a four state counter as few forward body bias levels are sufficient for the circuits under consideration. However, larger number of forward bias levels can be generated by adding extra flip-flops in the body bias generation circuit. The outputs Q1 and Q2 of the counter go to a 2-to-4 decoder. The decoder outputs are inputs to the body bias circuitry which is implemented as in [12] and enable the appropriate body bias option.

101

3. Experiment Setup and Results

The experiments for offline testing are performed on ISCAS89 sequential benchmarks: s526 and s832. First, we augment these circuits with BIPT hardware. To do this, we determine the critical paths in these circuits by using a static timing analyzer written in C. The gate libraries needed for the static timer are characterized in HSPICE for 90nm model card from BPTM [http://www-device.eecs.Berkeley.edu/~ptm/]. The flip-flops at the output of the critical paths are replaced by canary circuits whose structure and operation is described in section 2.2 and in detail in [8]. Once the canary circuits are determined, we traverse the path from input of the canary FF in a breadth first manner till we reach either a flip-flop or a primary input. We replace the flip-flops by mux-d scan flops and add extra scan flip-flops for the primary inputs. The scan flip-flop has two inputs, one input is connected to the input of the original flip-flop and the other input is connected to the output of the LFSR. A scan-enable signal is used to select between the two inputs. *Finish* signal serves as the scan-enable for scan flip-flops in our design. It can as well be a user-defined input. The characteristics of the benchmarks pre- and post-BIPT processing are shown in Table 1. Column 3 shows the number of flip-flops originally in the design and column 7 shows the number of these flip-flops replaced by canary flops respectively. Column 6 shows the number of mux-d scan flops inserted in the design. To validate the BIPT system on these benchmarks, we consider the effect of NBTI induced PMOS threshold voltage (V_t) degradation in these circuits. Our simulations take into account the effect of both nominal V_t degradation and temporal variations in V_t degradation; using models as described in [5]. The other important task is to set the clock period for simulations and thus set the target performance for both the benchmarks. The clock period for the simulations is determined by applying a pre-defined V_{dd} to just the main circuit (without BIPT hardware). For this V_{dd}, we run simulations to find out the nominal clock period such that no error occurs during offline testing. We add a safety margin of 15% to this period and the resulting clock period becomes clock period for our simulations. For a V_{dd} of 1.15V, this final value is found to be 480ps (2.08 GHz) and 600ps (1.67 GHz) for s526 and s832 respectively. Since a circuit with BIPT hardware, doesn't need to operate with high safety margins, we set the clock period to be 480ps (600ps) and for this clock period we determine minimum V_{dd} such that no error occurs during offline testing. For both the benchmarks, V_{dd} is set to 0.925v for BIPT.

Table 1. Characteristics of ISCAS '89 benchmarks under consideration: pre-BIPT processing and post-BIPT processing

ISCAS '89 Benchmark	No of Gates	No of flip-flops (FF) (pre-BIPT processing)	No of Primary Inputs	No of Primary Outputs	No of mux-d scan-flops (post-BIPT processing)	No of FF replaced by canary FF (post-BIPT processing)
s526	193	21	3	6	12	4
s832	262	5	18	19	12	2

To evaluate the effectiveness of BIPT approach, we carry out two sets of simulations in HSPICE at 100° C: (a) Deterministic Simulations and (b) Statistical Simulations. For deterministic simulations, simulations are carried out for 0%, 5% and 10% of NBTI induced deterministic V_t degradation. We compare the total operating power consumed by BIPT scheme with the over-designed case as the baseline case. The power estimation of BIPT system includes power dissipation due to the TPG, canary circuit and control circuit. The over-design implemented here is a conservative scaling of V_{dd} level. In particular, the V_{dd} for the over-designed case is set such that it does not cause timing violations and meets the performance targets at 10% V_t degradation as well. This value is found to be 1.2V for both s526 and s832 for 2.08 GHz and 1.67 GHz respectively. On the other hand, BIPT scheme allows for typical case circuit design, and adapts to the degradation of the circuit during its lifetime. Thus, the operating voltage is kept at 0.925V for BIPT simulations. Figure 6 plots the power consumed for deterministic simulations for s526 and s832. From the simulation results, we can observe that, on an average, BIPT scheme leads to power savings of 45% compared to the over-designed case.

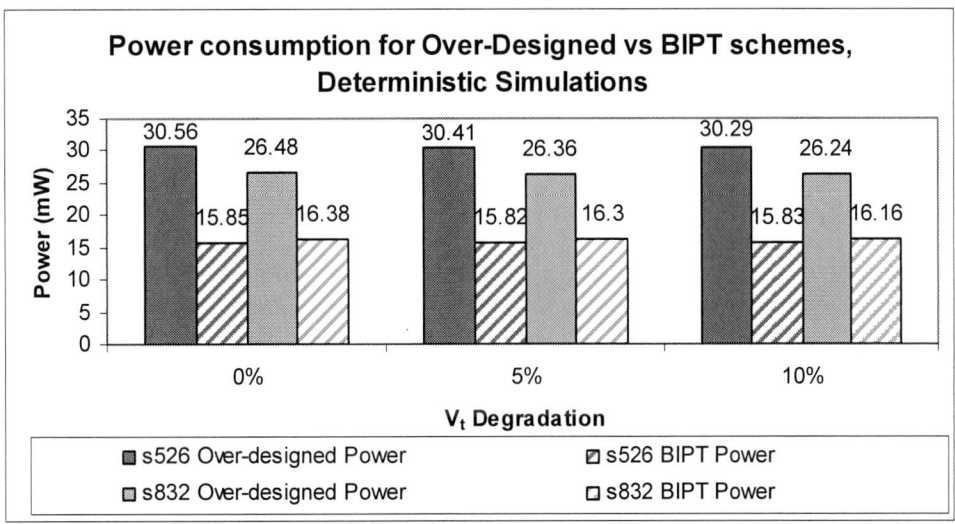

Figure 6. Power consumption for deterministic simulations

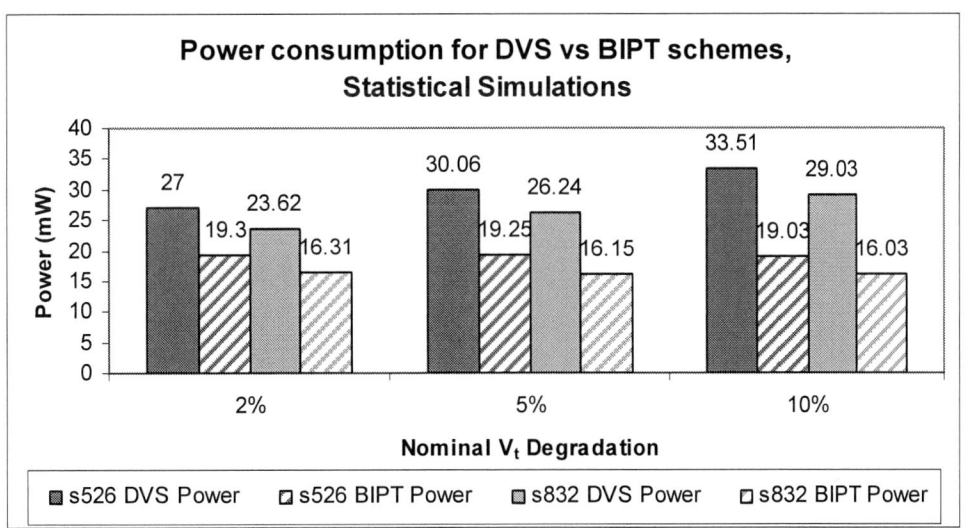

Figure 7. Power consumption for statistical simulations considering the temporal variations of NBTI effect

For statistical simulations, we take into account the statistical component of NBTI degradation over and above the nominal V_t degradation in lifetime V_t degradation, which takes into account the statistical variation in the underlying process causing V_t degradation [5]. We model the statistical V_t degradation as a Poisson random variable. We compare the power consumed by BIPT scheme with Dynamic Voltage Scaling (DVS) scheme. Thus, the dynamic voltage scheme serves as the baseline for statistical simulations. The simulations are carried out for statistical V_t variation over 2%, 5% and 10% of nominal value. The V_{dd} values for DVS are selected such that in each nominal case, the circuit is ensured to work for the worst statistical variation. Thus, for a transistor whose V_t is degraded by 5% (statistical variation component) over and above the 2% nominal degradation, V_{dd} is selected such that the circuit would still work without any timing violations if all transistors in the circuit were similarly affected. The operating voltages for the DVS schemes are found to be 1.15V, 1.2V and 1.25V for 2%, 5% and 10% degradations respectively. The operating voltage for BIPT case still remains at 0.925V. Figure 7 plots the power consumed for statistical simulations for s526 and s832. From the experimental results, we can observe that, on an average, BIPT scheme leads to power savings of 30% compared to the dynamic voltage scaling approach. Power saving here is less than the over-design case because dynamic voltage scaling scheme is an improvement over the over-designed approach. The power for DVS methodology increases as V_t degradation increases because of the fact that the voltage supply is varied keeping in mind the most degraded transistor.

4. Conclusions

In this paper, we propose a Built-In Proactive Tuning system that allows VLSI circuits to autonomously compensate aging-induced performance degradations. Due to its adaptive nature, BIPT is power-efficient and uses about 45% less power than over-design based aging compensation. Since it is a middle-grained approach, it can achieve 30% power reduction compared to the coarse-grained DVS method.

References

[1] M. Abramovici, M. A. Breuer and A. D. Friedman, "Digital Systems Testing and Testable Design", *IEEE Press*, New York, 1990.

[2] M. Agarwal, B. C. Paul, M. Zhang and S. Mitra, "Circuit Failure Prediction and Its Application to Transistor Aging", *IEEE VLSI Test Symposium*, 2007, pp. 277-286.

[3] K. A. Bowman, J. W. Tschanz, N. S. Kim, J. C. Lee, C. B. Wilkerson, S.-L. L. Lu, T. Karnik and V. K. De, "Energy-Efficient and Metastability-Immune Timing-Error Detection and Instruction-Replay-Based Recovery Circuits for Dynamic-Variation Tolerance", *IEEE ISSCC*, 2008, pp. 402-403.

[4] S. Das, D. Roberts, S. Lee, S. Pant, D. Blaauw, T. Austin, K. Flautner and T. Mudge, "A Self-Tuning DVS Processor Using Delay-Error Detection and Correction", *IEEE Journal of Solid-State Circuits*, Vol. 41, No. 4, April 2006, pp. 792-804.

[5] K. Kang, S. P. Park, K. Roy, and M. A. Alam, "Estimation of statistical variation in temporal NBTI degradation and its impact on lifetime circuit performance", *Proceedings of the 2007 IEEE/ACM ICCAD*, November 2007, pp 730-734.

[6] B. C. Paul, K. Kang, H. Kufluoglu, M. A. Alam and K. Roy, "Negative Bias Temperature Instability: Estimation and Design for Improved Reliability of Nanoscale Circuits", *IEEE Transactions on CAD of Integrated Circuits and Systems*, Vol. 26, No. 4, April 2007, pp. 743-751.

[7] R. Samanta, G. Venkataraman, N. Shah and J. Hu, "Elastic Timing Scheme for Energy-Efficient and Robust Performance", *IEEE ISQED*, 2008, pp. 537-542.

[8] T. Sato and Y. Kunitake, "A Simple Flip-flop Circuit for Typical-Case Designs for DFM", *IEEE ISQED*, 2007, pp. 539-544.

[9] J. Srinivasan, S. V. Adve, P. Bose and J. A. Rivers, "The Case for Lifetime Reliability-Aware Microprocessors", *ACM/IEEE ISCA* 2004, pp. 276-287.

[10] J. Srinivasan, S. V. Adve, P. Bose and J. A. Rivers, "Lifetime Reliability: Towards an Architectural Solution", *IEEE Micro*, Vol. 25, No. 3, May-June 2005, pp. 70-80.

[11] I. E. Sutherland, "Micropipelines", *Communications of the ACM*, Vol. 32, No. 6, June 1989, pp. 720-738.

[12] J. W. Tschanz, J. T. Kao, S. G. Narendra, R. Nair, D. A. Antoniadis, A. P. Chandrakasan and V. De, "Adaptive Body Bias for Reducing Impacts of Die-to-Die and Within-Die Parameter Variations on Microprocessor Frequency and Leakage", *IEEE Journal of Solid-State Circuits*, Vol. 37, No. 11, November 2002, pp. 1396-1402.

[13] M. Zhang, T. M. Mak, J. Tschanz, K. S. Kim, N. Seifert, and D. Lu, "Design for Resilience to Soft Errors and Variations", *Proceedings of the 13th IEEE IOLTS*, July 2007, pp. 23-28.

IEEE International Symposium on Defect and Fault Tolerance of VLSI Systems

Exploring Density-Reliability Tradeoffs on Nanoscale Substrates: When do smaller less reliable devices make sense?

Andrey Zykov and Gustavo de Veciana
Department of Electrical & Computer Engineering
The University of Texas at Austin

Abstract

It is widely recognized that device and interconnect fabrics at the nanoscale will be characterized by an increased susceptibility to transient faults. This appears to be intrinsic to nanoscale regimes and fundamentally limits the eventual benefits of the increased device density, i.e., the overheads associated with achieving fault-tolerance may counter the benefits of increased device density – density-reliability tradeoff. At the same time, as devices scale down one can expect a higher proportion of area to be associated with interconnection, i.e., area is wire dominated. This paper theoretically explores density-reliability tradeoffs in wire dominated integrated systems. We derive an area scaling model based on simple assumptions capturing the salient features of hierarchical design for high performance systems. We then evaluate overheads associated with using basic fault-tolerance techniques at different levels of the design hierarchy. This, albeit simplified model, allows us to tackle several interesting questions: When does it make sense to use smaller less reliable devices? At what scale of the design hierarchy should fault tolerance be applied in high performance integrated systems? Our analysis reveals two critical parameters, the technology and design scaling factors, which are key to predicting the reliability requirements for emerging technologies if traditional hierarchical design continues to be used.

1 Introduction

Future integrated systems will be implemented on increasingly dense substrates. These in turn are expected to be characterized by high densities of manufacturing defects and high rates of transient faults [1, 2, 3, 4]. In order to achieve a desired manufacturing yield and system reliability, engineers will have to apply defect- and fault-tolerance techniques. These techniques will necessarily incur overheads associated with additional circuitry and redundancy [5]. Such overheads take up area on the chip and thus increase power consumption. Thus, it is possible that the defect- and fault-tolerance overheads may grow faster than extra area afforded by the increased density of devices and wires the new technologies provide. This can happen if the reliability of the substrate drops quickly with increasing density. This paper develops a theoretical model to study when this is indeed the case.

Another key observation is that as technology scales down, the impact on area and power consumption of wires grows faster than that of devices (at least for high performance systems). This is supported empirically by Rent's Rule [6, 7]. It states that the number of external wires for a circuit sub-block is proportional to the size of this sub-block (e.g. in gates) to the power r where r is referred to as Rent's exponent and is typically about 0.6-0.7 for high performance systems. Rent's Rule can be viewed as arising from consistency across design hierarchy levels – i.e., reflects a design style based on interconnection of increasingly complex sub-blocks. The exponent may

1550-5774/08 $25.00 © 2008 IEEE
DOI 10.1109/DFT.2008.30

vary from system to a system (or sometimes even within a given system at different levels of hierarchy) reflecting not only design style, but also, the type of functionality being implemented. When Rent's exponent is greater than 0.5 for a 2D circuit, then the density of wires increases with system size growth provided that gates are tightly packed. This is reflected in the increase in metallization layers used for chips over the last decades – from 1 to 10 or more layers in today modern chips. It is technologically problematic to continue increasing the number of layers, so at some point it will become impossible to pack devices tightly due to the area required for wiring. At that point a chip's area will be wire dominated.

In such regimes (dynamic) power consumption may also be wire dominated. Indeed, static power consumption depends mostly on device physics, while dynamic power consumption is roughly proportional to load capacitance which already is dominated by wires in modern technologies. For example [8] consider a 3D design where devices were stacked vertically in 4 planes and show that the major effect is a reduction in wire length accompanied by a dramatic reduction in power consumption. Thus to understand the fundamental characteristics of future technologies, one must properly reflect wires' increasingly dominant influence on area, power and even performance.

In this paper we consider substrate technologies that are capable of delivering a high density of devices and wires but along with higher defect densities and transient fault rates. In Section 2 we propose and analyze an area scaling model for such technologies based on several natural assumptions. The model exhibits how area grows with complexity of the system (number of gates) in a wire dominated regime. Using this scaling model we consider the overheads associated with achieving fault-tolerance by applying spatial redundancy at different levels of system hierarchy. This simple model enables us to address two questions. When do smaller, but less reliable devices make sense, and at what level of the design hierarchy should fault tolerance be applied? The main contribution of this paper lies in combining a novel scaling model (capturing the wire dominated regimes of interest) with traditional reliability analysis to tackle these questions. This might be viewed in contrast to research initiated by [9] tackling computability with unreliable devices, but ignoring device and wiring overheads, see e.g., [10, 11]. By considering wire dominated regimes our work also differs from previous work considering reliability and overhead models based solely on gate count see e.g., [12]. Section 2 of this paper focuses only on gate reliability, while Section 3 motivates a general model both gates and wires may fail. This permits us to consider the manner in which device vs wire reliability impact the usefulness of a given technology. Section 4 offers some closing comments and perspective for this work.

2 Scaling Model and Basic Reliability Analysis

This section presents a novel area scaling model, capturing the wire dominated regime, which is then used to evaluate the density-reliability tradeoffs. We begin by carefully introducing several natural assumptions which underly our model.

2.1 Wire Dominated Area Scaling

The first assumption concerns interconnection across hierarchical levels. Traditional hierarchical design approaches to building increasingly complex systems are based on interconnecting subblocks. For example a pipelined CPU is realized based on blocks such as a fetch instruction stage, decode instruction stage, execute stage, registers file, external memory block, etc. The execute stage is itself built using different functional blocks (e.g., full word adders, multipliers, etc), where each block is built out of smaller blocks (e.g., one bit adders, etc). As a result when such systems are implemented on a substrate they lack structural regularity across hierarchical levels. By contrast, for intrinsically regular functions (e.g., memory arrays, FPGAs) one can adopt a more flat design style where the system is comprised of a large number of simple blocks. The implementation of such systems might eventually reflect regularity in placement and routing. In this paper we focus on hierarchical designs whose eventual implementations on a substrate would exhibit 'irregular' routing and placement across levels of the hierarchy.

Assumption 2.1. *(Hierarchical consistency) We consider systems designed in a hierarchical manner across multiple levels. Hierarchical consistency in interconnecting sub-blocks at different levels means that Rent's Rule should apply. Specifically,*

$$N_{ext}(M) = k_w M^r,$$

where $N_{ext}(M)$ is number of external wires for a block with M gates (or sub-blocks) and r is Rent's exponent (typically 0.6-0.7), k_w is a proportionality constant relating external wires to number of gates to the rth power.

Following [13] we refer to Rent's Rule as satisfying *hierarchical consistency*. Indeed, consider creating a block by composing P sub-blocks each comprised of M gates. By Rent's rule each sub-block has $N_{ext}(M)$ external wires and the number of external wires for the larger block should be $N_{ext}(M)P^r$. Yet the larger block has a total of MP gates, hence the total number of external wires should also be given by

$$N_{ext}(MP) = k_w(MP)^r = (k_w M^r)P^r = N_{ext}(M)P^r,$$

which exhibits the above mentioned hierarchical consistency. Note that Rent's Rule deals with logical wires, i.e., abstract connections among blocks [6, 7]. These logical wires may be implemented using one or more physical wires. So, for example, repeaters may be inserted along a logical wire subdividing it into several physical wires. The area cost of such a logical wire will be defined as its constituent physical wires and devices used to realize it. This leads us to our second assumption.

Assumption 2.2. *(Wire area) We assume block's area is the sum of its constituent gates and wires. The area of a wire is assumed to be proportional to its length, i.e.,*

$$A_w(l) = k_l l,$$

where $A_w(l)$ denotes the area of a wire of length l and k_l is a proportionality constant.

We measure length in linear minimal gate sizes, i.e., the linear minimal gate size l_g is 1. Similarly area is measured in minimal gate areas, so that minimal area of a gate is $a_g = l_g^2 = 1$. In these units the k_l reflects average area per unit length wire in units of minimal gate area. Note however that a chip may have several metal layers that would result in a smaller coefficient k_l, e.g., 10 metal layers at best gives 10 times the area to route wires, reducing the coefficient by a factor of 10.

In general we expect Assumption 2.2 to be reasonable. A wire's area is unlikely to grow sub-linearly in its length. In some cases it may grow super-linearly, e.g., if high performance is required, extra wide wires may be used to reduce resistance or extra repeaters to reduce latency. One can expect such wires to be only a small fraction which are on critical paths, and thus they would not significantly impact the overall scaling of area. As discussed in the introduction the dynamic power consumption of a wire is proportional to load capacitance, which in turn is roughly proportional to its area. Thus the total wire area can be used as a rough estimate dynamic power consumption.

The third assumption reflects our focus on hierarchically designed systems, which when mapped onto substrates exhibit irregular routing and placement.

Assumption 2.3. (*Irregular routing*) *The average length of wires used to interconnect sub-blocks having area A is proportional to their linear size, i.e.,*

$$L_w(A) = k_r \sqrt{A},$$

where $L_w(A)$ is the average length of wires interconnecting blocks having area A, and k_r is a proportionality constant reflecting the design's characteristics.

Note that interconnecting wires at a given scale, i.e., interconnecting blocks of a given size A, may have varying length, i.e., some may be short. Through Assumption 2.3 we posit that for systems which are hierarchically designed, resulting in irregular routing and placement, one should still expect the *average* length of such interconnections to be on the order of the linear size of the blocks they interconnect.

With these three assumptions in place one can show an area scaling law in system complexity (number of gates) capturing dominant role of wires on the area.

Theorem 2.4. (*Wire dominated area scaling*) *Under Assumptions 2.1-2.3 the growth in area A with system complexity M (in gates) satisfies the following differential equation:*

$$dA = \frac{A}{M}dM + k_l(k_r\sqrt{A})(k_w(1-r)M^{r-1}dM). \tag{1}$$

The solution to this equation for $r \neq 0.5$ is given by

$$A(M) = a_g(\sqrt{M} + tdM^r)^2 = a_g(M + 2tdM^{r+0.5} + (td)^2M^{2r}), \tag{2}$$

where $t = \frac{k_l k_w}{\sqrt{a_g}}$ is referred to as the technology scaling factor while $d = k_r\frac{1-r}{2r-1}$ is a design scaling factor.

We sketch the proof for Theorem 2.4 as follows. The differential growth in area represented by Eq. 1 has two terms on the right hand side. The first term can be interpreted as follows. Consider a block of area A with M gates, then the area per gate and internal wires for such a block is A/M thus if additional dM gates are added to create larger blocks, area should grow proportionally to A/M. The second term represents additional area associated with wires interconnecting blocks of size M. Consider a block of size M_2 consisting of M_2/M_1 sub-blocks of size M_1. By Rent's Rule the total number of external wires for all blocks of size M_1 is $\frac{M_2}{M_1}N_{ext}(M_1)$. This includes *some* of the internal and *all* of the external wires for the block of size M_2. However by Rent's Rule the number of external wires of the larger block is $N_{ext}(M_2)$, so the number of wires used to interconnect blocks of size M_1 within M_2 is $\frac{M_2}{M_1}N_{ext}(M_1) - N_{ext}(M_2)$. By hierarchical consistency and letting $M_2 = M + dM$ and $M_1 = M$ we obtain a differential number of interconnecting wires for blocks of size M in the form $\frac{M+dM}{M}N_{ext}(M) - N_{ext}(M + dM)$. This can be evaluated using Rent's formula. The second term also reflects our assumptions on the length and area of such wires, i.e., Assumptions 2.2 and 2.3. The solution Eq. 2 can be easily checked by substitution.

Note that the expected $A(M)$ growth includes a linear term in the number of gates, yet the other terms grow faster than linearly reflecting the dominant role of wires. Two key scaling parameters emerge. The first, called the *technology scaling factor*, depends on the average number of wires per gate k_w and wire length per linear gate length $k_l/\sqrt{a_g}$. The second, referred to as the *design scaling factor*, depends solely on characteristics of the design, i.e., on Rent's exponent r and k_r the parameter capturing the length of wires interconnecting blocks of similar size. The graph on the left in Fig. 1 exhibits the growth in area per gate, i.e., $A(M)/M$ for $d = 1$ for different technology scaling parameters; t=0.1, might be viewed as a baseline where $k_w = \sqrt{a_g}$, i.e., wire width is the same as minimal linear gate size and $k_l = 0.1$, e.g., 10 or so packed metallic layers for wiring.

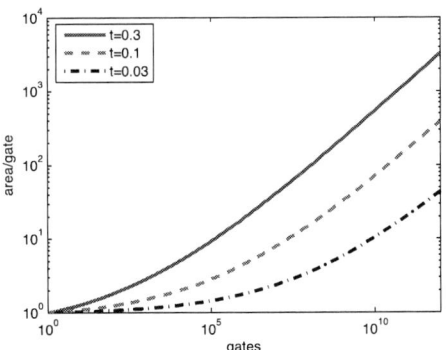 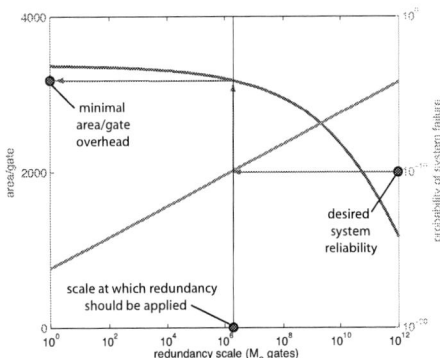

Figure 1. On the left the area $A(M)/M$ as a function of M for various technology factors and $d = 1$. On the right system overhead in area/gate and reliability as the hierarchical level at which redundancy is applied varies.

2.2 System Reliability and Fault-tolerance Overheads

To evaluate density-reliability tradeoffs, we need to characterize fault-tolerance overheads. The simplest way to achieve this is n-way spatial redundancy, i.e., replicate an unreliable sub-block n times and introduce bitwise majority voting to obtain reliable outputs. This approach achieves an exponential (in n) improvement in reliability with a linear (in n) area overhead. We recognize there are many alternatives to achieve fault-tolerance. For example, temporal redundancy requires much less overhead, but can only be applied at a sufficiently high architectural level (i.e., allowing "rollback") with blocks having sufficiently high reliability. Our motivation here is to consider high-performance, computation and/or control functions (e.g., those required to implement temporal redundancy) where spatial redundancy is a reasonable approach to achieve a significant boost reliability. We make the following assumption.

Assumption 2.5. *(Reliability and redundancy across hierarchical levels)*

a. *The probability of failure of a system is the sum of the failure probability of its constituent blocks. Thus a block of size M_0 gates has a probability of failure $p(M_0) = M_0 p_g$ where p_g is the probability of failure of a gate. Wires and voters are assumed to be reliable for now.*

b. *Spatial n-way redundancy is used to enhance a system's reliability. For a system of total size M_S gates, we assume redundancy can be applied at any of a continuum of hierarchical levels, indexed by the size of the blocks M_0, where M_0 can range from 1 to M_S gates.*

Assumption 2.5.a can be viewed as *consistency* assumption where failure probabilities are additive across constituent sub-blocks and scales. This corresponds to focusing on a regime where the failure probabilities are fairly low, and the probability of failure of a block of size M_0 gates is linear

$$p(M_0) = 1 - (1 - p_g)^{M_0} \approx M_0 p_g$$

if higher order terms can be ignored.

Assumption 2.5.b means that one may apply spatial redundancy to blocks of any size. In practice this would not be possible, but this idealization allows us to roughly investigate the granularity at which spatial redundancy should be applied. Specifically, if n-way redundancy is applied across blocks of size M_0 gates the system would have M_S/M_0 such blocks. Then for $n = 3, 5, 7 \ldots$ the probability of failure of an n-way redundant block of size M_0 is given by

$$\sum_{i=\frac{n+1}{2}}^{n} \binom{n}{i} (1 - M_0 p_g)^{n-i} (M_0 p_g)^i \approx \binom{n}{\frac{n+1}{2}} (M_0 p_g)^{\frac{n+1}{2}}.$$

Finally using Assumption 2.5.a the probability of failure for the overall system P_S composed of $\frac{M_S}{M_0}$ such blocks is $P_S = \frac{M_S}{M_0} \times \binom{n}{\frac{n+1}{2}} (M_0 p_g)^{\frac{n+1}{2}}$.

Ignoring voters and associated circuitry, and irrespective of the block granularity M_0 at which n-way spatial redundancy is applied the overall number of gates in the system increases by a factor of n. However if replication occurs at lower levels of the hierarchy, longer wires will be required at higher levels of the design hierarchy. Indeed these wires not only get replicated n times, but also become longer taking even more area. So the total area overhead of realizing n-way spatial redundancy will be higher if it is realized at a lower level of the design hierarchy.

To properly capture these overheads when n-way spatial redundancy is applied starting at a hierarchical level M_0 we modify Eq. 1 to reflect these redundancy overheads. For $M \leq M_0$ it remains the same which by Eq. 2 gives an area $A(M_0)$ for a block of size M_0. For $M > M_0$ this is modified as follows. The initial condition becomes $M = M_0$. The initial area with n-way redundancy at scale M_0 is $nA(M_0)$. The differential growth in area for a system with n-way redundancy and $M > M_0$ is now given by

$$dA = \frac{A}{M} dM + n k_l (k_r \sqrt{A})(k_w (1 - r) M^{r-1} dM. \tag{3}$$

This can be viewed as multiplying the design scaling factor by n to capture the additional overhead associated with redundant wires.

The graph on the right in Fig. 1 shows both the area per gate and the overall system reliability when 3-way redundancy is applied to blocks M_0 ranging from a single gate to the overall system size $M_S = 10^{12}$ for a fixed probability of gate failure $p_g = 10^{-14}$. As can be seen, if redundancy is applied at a higher level M_0 one sees a lower area overhead but also a lower reliability. Thus there is a highest scale M_0 at which one can apply redundancy to achieve a given overall system probability of failure P_S. This is exhibited graphically on the plot.

Using this model we can consider if it is worth moving to smaller less reliable gates. Consider a system of fixed complexity (number of gates without redundancy) $M_S = 10^{12}$ to be implemented on a fixed absolute area, with the fixed acceptable overall probability of failure $P_S = 10^{-16}$. Given we are using n-way redundancy, we can ask what is the maximum acceptable probability of gate failure p_g such that the overheads associated with reaching the desired P_S fit in the absolute area of interest. As we reduce gate size (area) a_g, i.e., increase the density of a technology, we expect to be able to afford higher overheads for fault-tolerance, allowing higher probabilities of gate failure. The left plot on the Fig. 2 exhibits curves for the maximum tolerable probability of failure for different gate sizes and different degrees of n-way redundancy. Horizontal axis on this graph is linear size of gates measured with respect to size of bigger gates of area a_{g0} that would give the same total system area if no redundancy is applied. Such big gates should have probability of failure p_g at most $P_S/M_S = 10^{-28}$ to provide target system reliability P_S. Note that exhibited curves have a finite domain representing what is possible when M_0 ranges from 1 to M_S.

These curves reflect limits on the reliability of gates, i.e., where the redundancy overheads to achieve overall system reliability consumes all extra area afforded by reduced gate size. All points below (better system reliability) and left (less area) from any point of these curves are acceptable. Points right or above all points of these curves are unacceptable, because system built using smaller

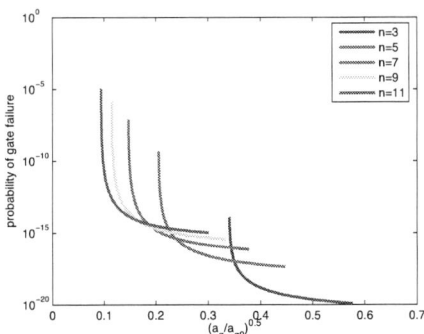 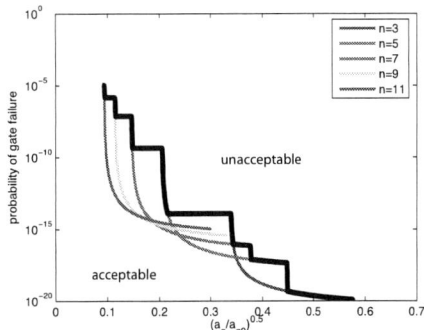

Figure 2. Minimum affordable gate reliability for reduced device size.

less reliable gates will occupy more area than non-redundant system built using bigger reliable gates. The right plot on the Fig. 2 shows the acceptable region for the various levels of redundancy.

3   Generalized Scaling Model

The results obtained in the previous section were predicated on both voters and wires being perfectly reliable. Let us reconsider these in turn. First we assumed voters can be assumed to be perfectly reliable at no cost. This is reasonable if redundancy is not applied at a very low hierarchical level. In this case each block's cost will be orders of magnitude higher than the cost of voters. Indeed, first the complexity of single bit voter is very small. Second the number of such voters is proportional to the number of external wires which by Rent's Rule scales as M^r ($r \approx$ $0.6 - 0.7$) which is small relative to M. Thus one could in principle use 'big' or more reliable devices to implement voters at a negligible area cost.

The second assumption that wires are reliable is harder to justify. On one hand it is likely that ionizing particles, as a source of soft faults, are more likely to impact active device areas than signals across wires. However it is not clear whether wires in emerging technologies might not also be vulnerable to ionizing particles. If this were the case, then a reasonable model would be a probability of wire failure which is proportional to its length (or equivalently its area). On the other hand it is widely recognized that other internal sources of transient errors are of critical concern. For example coupling among wires is data dependent and might be modeled as a probability of failure which is proportional to wire length. Also, delay variability, in some cases may be data dependent [1] and might again be modeled as a random event. Though in this case, it is not clear that the probability of failure is proportional to wire length, it is true that a longer wire would have a higher probability of failure. To better capture concerns with the reliability of wires and investigate their importance relative to gate reliability we shall revise Assumption 2.5 as follows.

Assumption 3.1. *(General reliability and redundancy across hierarchical levels)*

a. *The probability of failure of a system is the sum of the failure probability of its constituent blocks. A block of size M_0 gates and total wire length L_0 has a probability of failure $p(M_0) = M_0 p_g + L_0 p_w$ where p_g is the probability of failure of a gate and p_w is the probability of failure of a wire per unit length. Voters are assumed to be reliable.*

b. *Spatial n-way redundancy is used to enhance a system's reliability. For a system of total size M_S gates, we assume redundancy can be applied at any of a continuum of hierarchical levels, indexed by the size of the blocks M_0 where M_0 can range from 1 to M_S gates.*

---

[1]For example the critical path in an adder becomes important only for rare inputs resulting in carries having to be propagated across the entire word.

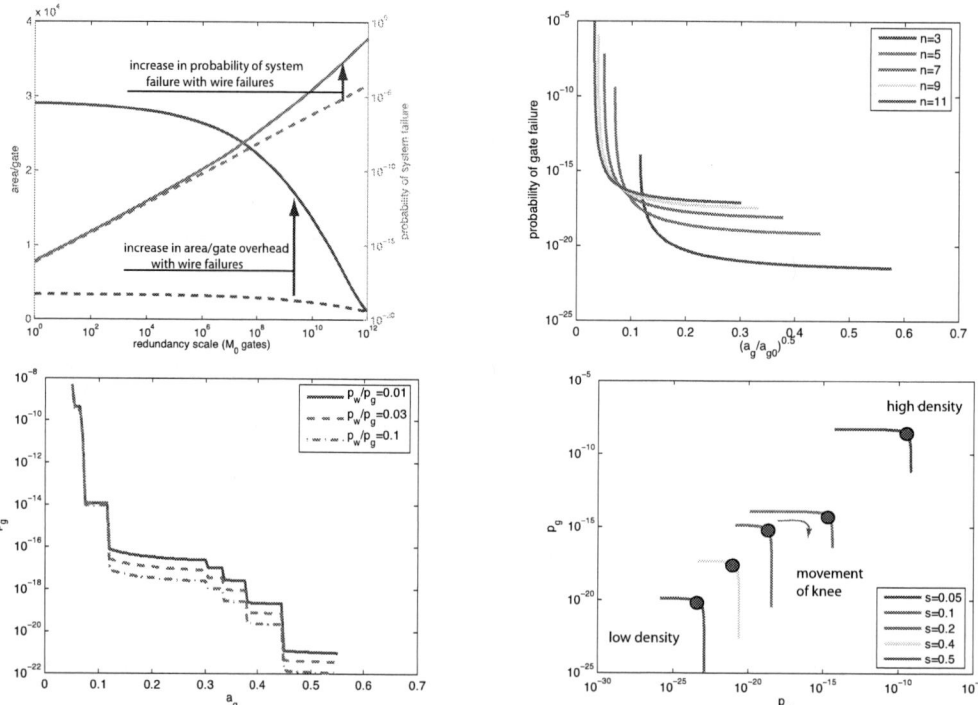

Figure 3. Top left, system overhead in area/gate and reliability as the hierarchical level which redundancy is applied varies. Top right and left bottom minimum affordable reliability for reduced device size. Bottom right, tradeoff between gate reliability (p_g) and wires reliability (p_w) at different reduced device sizes $s = (a_g/a_{g0})^{0.5}$.

c. *When redundancy is applied at level M_0 we assume that 'long' wires, i.e., interconnecting blocks of size M_0 or above are made reliable but have greater area per unit length by factor k_o.*

The key idea underlying this assumption is as follows. When spatial redundancy is applied at a certain scale, i.e., blocks of size M_0. 'Short' wires within the block are assumed to have a probability of failure which is linear in their length, and contribute to the block's failure. However 'long' wires that interconnect blocks of size M_0 and above, end up being too long and unreliable, i.e., reliability becomes wire dominated. Thus it makes sense to make long wires reliable. This can be achieved by dividing a wire into shorter sections and applying redundancy to these sections. Or possibly making long wires wider introduce a wider spacing among wires to reduce coupling. In either case making long wires reliable comes at some additional area overhead which in our assumption is modeled by the factor k_o. As in the previous section this increases design scaling factor d by a factor k_o in Eq 2. For example in the results below it is $k_0 = 3$ assuming 3-way redundancy is applied to sections of a long wire.

The top left graph in Fig. 3 exhibits the area/gate overhead and system reliability under our general model where redundancy is applied at different hierarchical levels M_0 and fixed $p_g = 10^{-14}$ and p_w ($p_w/p_g = 0.03$). For contrast, basic model (previously shown in Fig. 1) is included showing the dramatic impact of unreliable wires on overheads and system reliability. The top right graphs in the figure exhibit the new maximum possible probability of failure per gate that can be afforded as minimal device size gets smaller with respect to minimal size of reliable device. The bottom left graph shows only the bounding curves, but it does this for several rations p_w/p_g. These are akin to the results Fig. 2 for the basic model. Finally the graph on the bottom right shows maximum affordable probability of failure for gates (p_g) and wires (p_w) for various device scales. As can be seen in the figure that knee of curves moves to the right as we increase density (the total range of ratio p_w/p_g is fixed to $10^{-6} - 10^2$). That means that for higher density reliability of wires becomes more important than reliability of gates. This could be expected as for higher densities we are increasingly in a wire dominated regime, so their reliability is increasingly a concern.

4 Conclusion

In this paper we developed a new model area scaling for wire dominated systems to study density-reliability tradeoffs for future technologies. The motivation was to evaluate when smaller less reliable devices make sense and at what hierarchical levels (granularity) one should incorporate spatial redundancy. To our knowledge this is the first attempt to evaluate such tradeoffs. Perhaps the most interesting result emerging from our work is a study of the tension between reliability of devices vs wires vs density. Our results indicate how wire reliability becomes more critical as the technology density increases. Although area can be used as a crude proxy for power, it would be interesting to further enhance the model to capture the power density issues.

Acknowledgments: This work is supported by the Gigascale Systems Research Center (GSRC), under the 'Alternative' Theme.

References

[1] G. Bourianoff, "The future of nanocomputing," *Computer Magazine*, pp. 44–49, Aug. 2003.

[2] J. R. Heath, "A defect-tolerant computer architecture: Opportunities for nanotechnology," *Science*, vol. 280, pp. 1716–21, June 1998.

[3] M. Mishra and S. C. Goldstein, "Defect tolerance at the end of the roadmap," in *Proc. International Test Conference (ITC '03)*, 2003.

[4] "SEMATECH. International Technology Roadmap for Semiconductors - 2004 update on emerging research devices." http://www.itrs.net/Common/2004Update/2004Update.htm.

[5] D. P. Siewiorek and R. S. Swarz, *Reliable computer systems (3rd ed.): design and evaluation.* Natick, MA, USA: A. K. Peters, Ltd., 1998.

[6] B. Landman and R. Russo, "On a pin versus block relationship for partitions of logic graphs," *Computers, IEEE Transactions on*, vol. C-20, no. 12, pp. 1469–1479, Dec. 1971.

[7] P. Christie and D. Stroobandt, "The interpretation and application of Rent's rule," *IEEE Trans. Very Large Scale Integr. Syst.*, vol. 8, no. 6, pp. 639–648, 2000.

[8] Y. Xie, G. H. Loh, B. Black, and K. Bernstein, "Design space exploration for 3d architectures," *J. Emerg. Technol. Comput. Syst.*, vol. 2, no. 2, pp. 65–103, 2006.

[9] J. von Neumann, "Probabilistic logics and synthesis of reliable organisms from unreliable components," *Automata Studies*, pp. 43–98, 1956.

[10] B. E. Hajek and T. Weller, "On the maximum tolerable noise for reliable computation by formulas," *IEEE Transactions on Information Theory*, vol. 37, no. 2, pp. 388–391, 1991.

[11] D. Bhaduri and S. Shukla, "Reliability evaluation of von neumann multiplexing based defect-tolerant majority circuits," *Nanotechnology, 2004. 4th IEEE Conference on*, pp. 599–601, 16-19 Aug. 2004.

[12] K. Nikolic, A. Sadek, and M. Forshaw, "Fault-tolerant techniques for nanocomputers," *Nanotechnology*, vol. 13, no. 3, pp. 357–362, 2002.

[13] R. P. Feynman, *Feynman Lectures on Computation*, J. G. Hey and R. W. Allen, Eds. Boston, MA, USA: Addison-Wesley Longman Publishing Co., Inc., 1998.

IEEE International Symposium on Defect and Fault Tolerance of VLSI Systems

Impact of Technology and Voltage Scaling on the Soft Error Susceptibility in Nanoscale CMOS

Vikas Chandra
ARM R&D
vikas.chandra@arm.com

Robert Aitken
ARM R&D
rob.aitken@arm.com

Abstract

With each technology node shrink, a silicon chip becomes more susceptible to soft errors. The susceptibility further increases as the voltage is scaled down to save energy. Based on analysis on cells from commercial libraries, we have quantified the increase in the soft error probability across 65nm and 45nm technology nodes at different supply voltages using the Qcrit based simulation methodology. The Qcrit for both bit cells and latches decreases by ~30% as the designs are scaled from 65nm to 45nm. This decrease is expected to continue with further technology scaling as well. The results show that at nominal voltage, the Qcrit for a latch is just ~20% more than that of the bit cell in sub-65nm technology nodes. Further, as the voltage is scaled from 1V to 0.4V, Qcrit decreases by ~5X which substantially increases the probability of an upset if a particle strike happens. This work shows that in sub-65nm technology nodes with aggressive voltage scaling, it is equally critical to solve the soft error problems in logic (latches, flip-flops) as it is in SRAMs.

1 Introduction

Rapidly shrinking technology node and aggressive scaling of voltage have increased the probability of soft errors. Soft errors are radiation induced faults which happen due to a particle hit, either by an alpha particle from impurities in packaging material or a neutron from cosmic rays [1, 10, 16]. When particles strike the silicon substrate they create hole-electron pairs which are then collected by ▪▪ junctions via drift and diffusion mechanisms. This collected charge creates a transient current pulse and if it is large enough, it can flip the value stored in the state saving element (bit cell, latch etc.). These upsets are called Single Event Upsets (SEU). When particle strike happens in combinational circuit, the result is a glitch which can then propagate to a latch where it could be clocked in and incorrect data can be latched.

Embedded SRAMs are especially vulnerable to SEU due to small size of bit cell and its small node capacitances. However, soft error will just not impact SRAMs but latches/flip-flops and combinational logic as well. At the chip level, the contribution to the soft error rate (SER) from latches and flip-flops is growing [7, 13]. As technology node scales, the susceptibility of soft error increases due to a decrease in the node capacitances. The second facet to this problem is due to aggressive voltage scaling to save power. The dynamic power consumption is reduced quadratically with decreasing supply voltage and linearly with decreasing frequency. To decrease the dynamic power consumption, the supply voltage is scaled down at the cost of increased delay. Decreasing the supply voltage impacts soft error susceptibility as the charge needed to upset a node is a function of the voltage level.

There are many techniques which are currently used to protect SRAMs, namely parity and error correction codes (ECC). However, due to spatial distribution of latches/flip-flops on chip, it is expensive to apply similar techniques in logic. Also, it has always been assumed that latches/flip-flops are much more robust than SRAM bit cells due to their larger size devices and larger node capacitances. This might have been true in technology nodes greater than 90nm, but this certainly is not true in sub-65nm technology nodes [11]. In this work we have shown that with voltage scaling the latches become as critical as bit cells in terms of soft error susceptibility. We have analyzed the susceptibility of bit cells and latches in commercial 65nm and 45nm technology nodes.

The rest of the paper is organized as follows. Section 2 describes the background of soft errors, related work and the simulation methodology for measuring soft error susceptibility of a storage element. Section 3 analyzes the process and voltage scaling characteristics of SRAM bit cell and latch with respect to soft error susceptibility.

1550-5774/08 $25.00 © 2008 IEEE
DOI 10.1109/DFT.2008.50

Section 4 explains the design techniques used to reduce soft error probability. Section 5 summarizes the work and concludes.

2 Background

Figure 1 shows a scenario where an α particle strikes a PMOS transistor. The particle hit could be a glancing blow or a penetrating strike. When particles hit silicon, they generate hole-electron pairs which can then be swept to a diffusion junction if an electric field exists across the device. This has the effect of causing a short duration pulse of current. The effect of the strike on the diffusion can be modeled as a current source as shown in Figure 1. There are other mechanisms of soft errors besides α particles like neutrons, heavy-ions etc. Figure 1 is meant to be representative and not exhaustive.

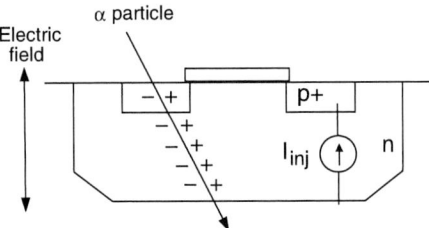

Figure 1: α particle hit on a PMOS transistor

A soft error will happen if the collected charge at a junction is equal to the critical charge ($Qcrit$). $Qcrit$ is defined as the minimum charge which is needed to flip the bit stored in the storage cell. To estimate a circuit's sensitivity to soft error, the $Qcrit$ values are calculated. The higher the value of $Qcrit$, the more difficult will be to flip the cell and hence it will be more robust. $Qcrit$ is proportional to node capacitance and supply voltage. With technology scaling, usually the node capacitance and supply voltage also scale down thus decreasing the value of $Qcrit$.

2.1 Simulation methodology

$Qcrit$ is an important parameter to characterize the sensitivity of a storage cell to a particle strike. A particle strike creates hole-electron pairs which when recombined forms a current spike. Hence, a particle strike at a node can be modeled as a current pulse. For a given current pulse waveform $i(t)$, $Qcrit$ is defined as the minimum time integral on $i(t)dt$ that results in the cell flip (Equation 1).

$$Qcrit = \int_0^t i(t)dt \tag{1}$$

There has been a lot of work on choosing the right shape of the current pulse. In [17], the authors use a triangular pulse and in [4], the authors conclude that the shape of the pulse will vary but it can be represented by a piecewise linear function with a peak corresponding to the funneling charge collection and a more slowly decaying tail for the diffusion charge collection. In [9, 15], the authors describe a double exponential current pulse, as described by Equation 2.

$$i(t) = \frac{Q_{total}}{\tau_f - \tau_r} \cdot \left(e^{-t/\tau_f} - e^{-t/\tau_r} \right) \tag{2}$$

In this equation, Q_{total} denotes the total amount of charge generated by the strike, τ_r and τ_f are the rise and fall time constants respectively. The value of τ_r is much smaller than τ_f and hence most of the charge collection happens right after the steep current rise. Another commonly used current pulse is given by Equation 3 [2, 14].

$$i(t) = \frac{Q_{total}}{\tau} \cdot \sqrt{\frac{t}{\tau}} \cdot e^{-t/\tau} \tag{3}$$

In Equation 3, Q_{total} is the amount of charge collected during the event and τ is is the time constant of the collection process. In reality, the peak value of the current pulse and the rise/decay time constant will depend on particle type, particle energy, angle of incidence etc. In order to avoid the complexities associated with choosing values for the models described by Equations 2 and 3, we elected to use a simple triangular model for the current pulse $i(t)$. We believe that this model is accurate for a relative comparison of $Qcrit$ across technology nodes and supply voltages. The rise time constant (τ_r) is chosen as 50fs and the decay time constant (τ_f) is 5ps. These values are representative of particle strike characteristics described in literature [8]. Based on these values, the current pulse is shown in Figure 2.

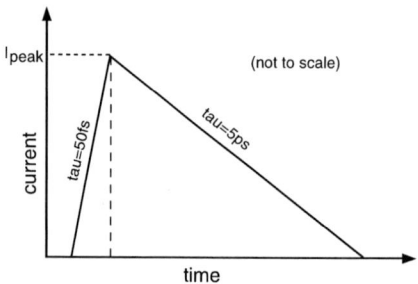

Figure 2: Current pulse shape showing rise and decay time constants

The area under the curve shown in Figure 2 denotes the charge dumped on (or taken away from) a node where the strike happens. For a more robust storage cell, the area under the curve will be larger. Since we fixed τ_r and τ_f, we change I_{peak} to vary the area under the curve. For a given set of operating conditions, I_{peak} is chosen so that the storage cell just flips. In other words, I_{peak} is proportional to the minimum amount of charge ($Qcrit$) needed to flip the cell.

3 Susceptibility analysis

As discussed in Section 2.1, $Qcrit$ is used as a metric to evaluate the susceptibility to soft errors. We discuss the susceptibility of two different classes of circuits:

- 6T bit cell

- Latch

These two classes of circuits represent the majority of storage cells in current VLSI implementations. A 6T bit cell is a dense, symmetrical storage cell consisting of two identical back-to-back inverters and is a basic building block of a high density SRAM. A latch is an asymmetrical storage cell which usually consists of an inverter and a clock (also called clk) controlled tri-state inverter. All the logic storage cells are built out of latches and flip-flops. A flip-flop consists of two latches which operate on different phases of clk. Sections 3.1 and 3.2 analyze these two circuits in terms of their soft error susceptibility. Both these circuits are from commercial libraries in 65nm and 45nm technology nodes. Also, the analysis in the following sections are based on detailed extracted netlists which include all the parasitic capacitances. This is important as parasitics change the value of $Qcrit$ substantially [6].

3.1 SRAM 6T bit cell analysis

A 6 transistor bit cell is shown in Figure 3. The storage inverters are connected to bit lines (bl and blb) through NMOS transistors. These NMOS transistors are driven by wordline (wl). During write operation, wl is asserted and one of the bit lines is pulled low to write a "1" or a "0". During read, the wordline is asserted but the bit lines are not driven. Depending on the value stored in the cell, one of the bit lines is pulled low and the voltage differential is sensed by a sense amplifier to read out the value. Usually a large number of bit cells connect to a bit line so the capacitance of the bit line is quite large. The factors which impact the critical charge of a bit cell are described in [6]. The two critical nodes in Figure 3 are shown as nodes $g1$ and $g2$. The criticality of the

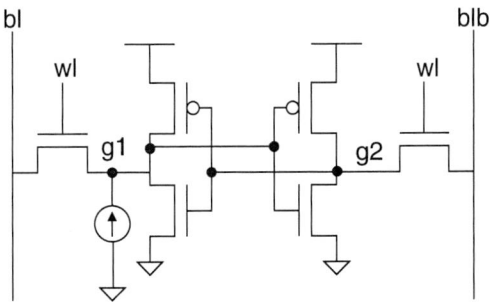

Figure 3: Particle strike modeling in a 6T bit cell

nodes results from the small size of the bit cells and their associated tiny capacitances (10^{-16} to 10^{-17} F). As described in Section 2.1, a particle strike is modeled as a current source. The current source shown in Figure 3 represents the charge generated at node $g1$ when it gets struck by a particle. Since the bit cell is symmetrical, the analysis of particle strike on node $g2$ will be identical to that of node $g1$.

Figures 4(a) and 4(b) show the impact of three different process corners on the $Qcrit$ of 65nm and 45nm bit cells respectively. Process corners denote the performance spread of transistors in silicon. The corners are given as fast (f), typical (t) and slow (s). The graph's data are individually normalized *wrt* the $Qcrit$ value at

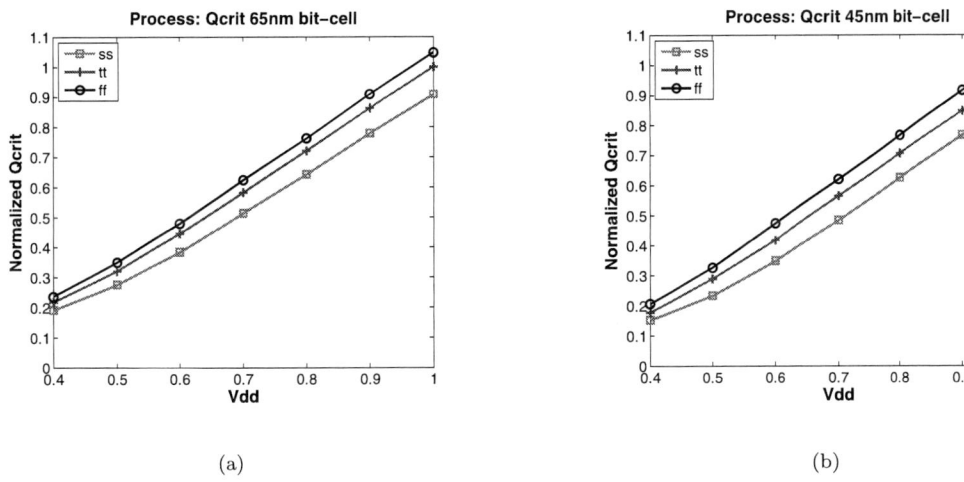

Figure 4: Impact of process corners on $Qcrit$ for a bit cell (a) 65nm process (b) 45nm process

the tt corner (PMOS typical, NMOS typical) and 1V supply. It is interesting to note that the normalized data look very similar for the two technology nodes. The $Qcrit$ decreases by a factor of 5X as the voltage is scaled from 1V to 0.4V. As expected, the ff corner is the most robust owing to faster transistors. In addition to the variation observed in $Qcrit$ within process generations, it is also interesting to compare the relative ratios of bit cell $Qcrit$ values at 45nm and 65nm nodes, as shown in Figure 5. At nominal voltage (1V), the value of $Qcrit$ decreases by a factor of 0.74 which is close to the •••••••• •••••• of the technology shrink. However, at lower voltage (0.4V) the factor is close to 0.6. So, voltage scaling increases the soft error upset probability by ~20%.

3.2 Latch analysis

A simple latch design is shown in Figure 6. When *clk* is "1", the latch is transparent and Q follows *Data*. When *clk* goes low, the input switch turns off and the latch closes with the value stored. During this close phase, the latch becomes vulnerable to soft error as a particle strike can flip the value stored in the latch. Unlike the bit cell described in Section 3.1, the latch is not symmetrical. In other words, nodes $n1$ and $n2$ behave differently

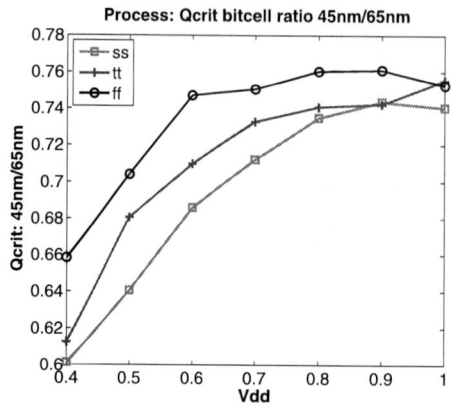

Figure 5: Ratio of bit cell $Qcrit_{45nm}/Qcrit_{65nm}$

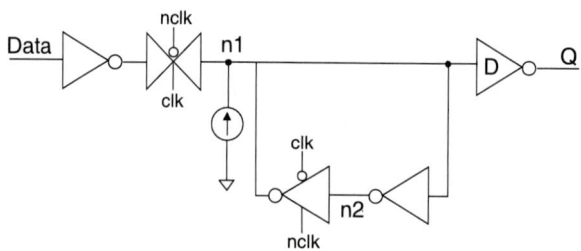

Figure 6: Particle strike modeling in a latch

when a particle strike happens. The two factors which cause the asymmetric behavior of nodes are driving gate strength and node capacitance. As can be inferred from Figure 6, both the factors will be different for nodes $n1$ and $n2$. For the latch we analyzed, node $n1$ has smaller $Qcrit$ than node $n2$. This is primarily due to weaker drive strength of the tri-state inverter (a factor of 2X) as compared to the drive strength of the inverter driving $n2$. Hence, to account for the worst case, we have done the particle strike analysis on the node $n1$. Since a flip-flop consists of two latches, the soft error analysis of a latch extends to a flip-flop as well.

Figures 7(a) and 7(b) show the impact of process variation on voltage scaling for the latch in 65nm and 45nm technology nodes respectively. The decrease in the value of $Qcrit$ is linear with respect to voltage. This can be attributed to the relation of charge to voltage ($Q = CV$). But as the voltage is reduced to the sub-threshold regime, the value of gate capacitance also begins to change. It is not shown in Figures 4 and 7, but the $Qcrit$ variation becomes non-linear wrt voltage near 0.2V or so. The data in each of the graphs in Figure 7 have been individually normalized wrt the $Qcrit$ value at the tt corner and 1V supply. If we compare the relative scaling of $Qcrit$ for bit cells and latches wrt voltage at 65nm and 45nm (Figures 4 and 7), we can see that the latches are more robust at lower voltage. The normalized $Qcrit$ for the bit cell is approximately 0.2 when the voltage is 0.4V, whereas for latches the relative $Qcrit$ only scales to 0.3 at 0.4V. Figure 8 shows the ratio of 45nm $Qcrit$ to 65nm $Qcrit$ for the latch. With voltage scaling this ratio decreases from 0.7 to 0.62. This behavior is very similar to that of the bit cell.

3.2.1 Latch drive strength variants

A standard cell library has typically many drive strength variants of a cell. We considered three drive strengths of the latch shown in Figure 6, namely A, B and C. For the library we analyzed, C is about 50% bigger than B which is twice the size of A. Drive strength of the latch shown in Figure 6 is controlled by the size of the inverter D. An increase in drive strength affects $Qcrit$ due to increase in the node capacitance of node $n1$. Figure 9 shows the increase of $Qcrit$ for drive strengths B and C • •• A. So, for a latch based design, increasing the drive strength will help to reduce soft error upset probability. For this work, we used a latch with a drive strength of A as it has a smaller $Qcrit$ than a latch with B or C drive strength.

118

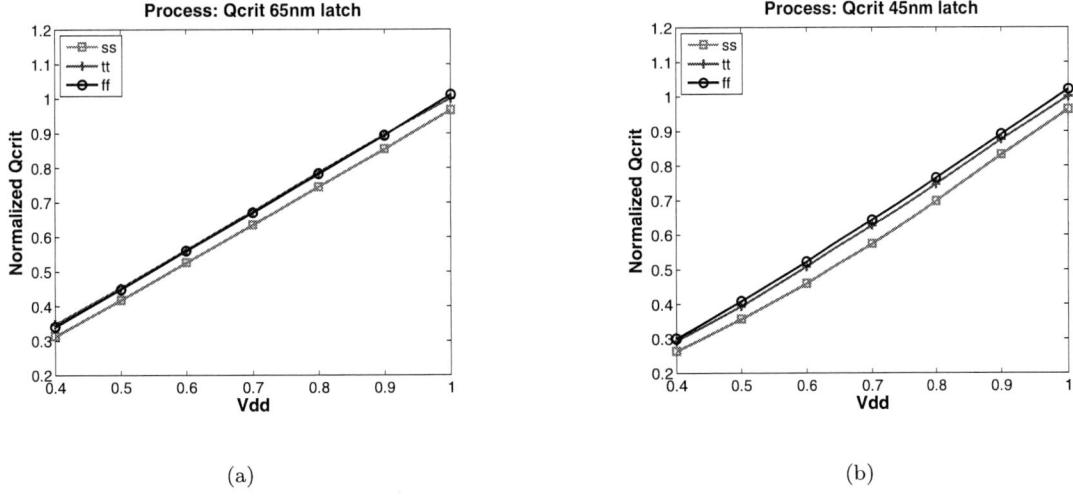

(a)   (b)

Figure 7: Impact of process corners on $Qcrit$ for a latch (a) 65nm process (b) 45nm process

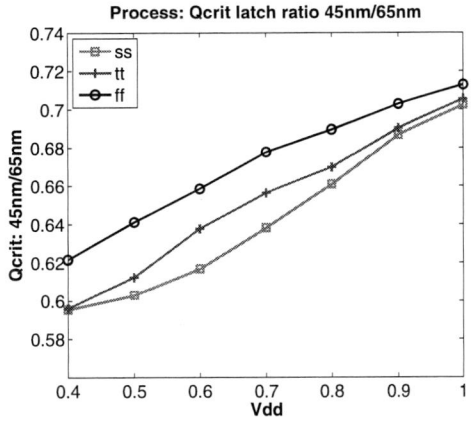

Figure 8: Ratio of latch $Qcrit_{45nm}/Qcrit_{65nm}$

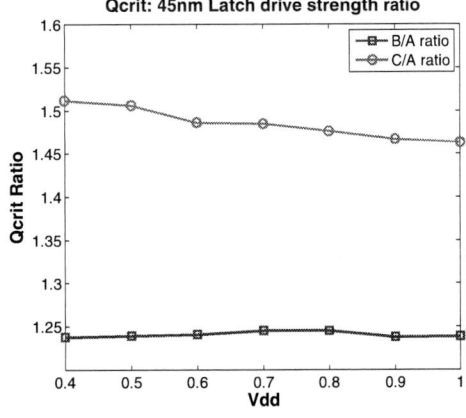

Figure 9: Effect of drive strength on $Qcrit$ in a latch

3.3 Bit cell and latch comparison

In a new technology node, a lot of time is spent on designing a bit cell. The important criteria being yield, density, leakage, performance, Vmin behavior and dynamic power. Typically a bit cell is designed to be as small as lithographically printable. On the other hand, latches (and other standard cells) do not use pushed lithography rules and are typically larger than bit cells. Hence there is a mismatch in the way these two circuits scale. Figures 10(a) and 10(b) show the ratio of latch to bit cell $Qcrit$ for 65nm and 45nm processes respectively. At 1V and tt corner, the latch is around 27% (ratio of 1.27) better than the bit cell for 65nm. For 45nm, the value drops down to 19% (ratio of 1.19). In other words, at nominal operating voltage the latch is marginally better than bit cell with regards to soft errors. As the voltage is scaled down, the latch starts to look better than the bit cell. For 65nm, at 0.4V, the best case ratio is 2.08 (ss corner) and the worst case ratio is 1.82 (ff corner). For 45nm, at 0.4V, the best case ratio is 2.05 and the worst case ratio is 1.72. Even in the low voltage regime, a latch is only twice as robust as a bit cell!

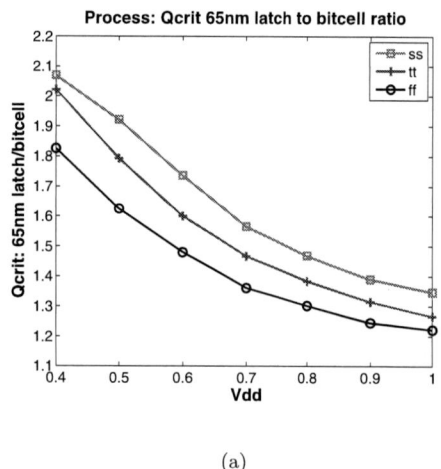

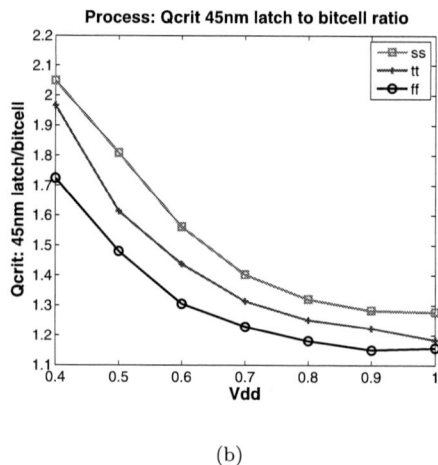

(a)   (b)

Figure 10: Ratio of latch to bit cell $Qcrit$ for (a) 65nm process (b) 45nm process

This observation tells us that soft errors in latches have already become as important as those in bit cells. With technology node scaling to 32nm and below, the soft error susceptibility of latches and bit cells is likely to become very similar. There are inexpensive error correcting schemes to correct soft errors in SRAMs. These schemes are not feasible for latches due to their spatially distributed nature. The techniques commonly employed to make the latch more robust are discussed in Section 4.

3.4 Temperature and $Qcrit$ scaling

Temperature also affects the susceptibility of a design to soft error. We analyzed the bit cells in 65nm and 45nm technology nodes for their sensitivity to temperature. Figures 11(a) and 11(b) show the effect of temperature on $Qcrit$ with voltage scaling for a bit cell. From the figures, it is clear that temperature does not change $Qcrit$ significantly (10% in 65nm, 15% in 45nm), although at higher voltage the $Qcrit$ difference caused by temperature is higher than at lower voltage. The spread at 1V in the 45nm node is higher than that in the 65nm node, although they behave more similarly at 0.4V.

In [9], the authors analyzed the impact of temperature on the SER of a 65nm SOI SRAM. It was observed that the $Qcrit$ decreases with increase in temperature. This trend was observed across all voltages. Our analysis shows a similar trend at higher voltages. However, we observed that at lower voltages the high temperature $Qcrit$ was higher than the low temperature $Qcrit$. At high voltage, 0C case has the best $Qcrit$ and 125C case has the worst $Qcrit$. At low voltage, 0C has the worst $Qcrit$ and 125C has the best $Qcrit$. This ••• •••••••• •••••••• effect happens for both 65nm and 45nm technology nodes (the inversion voltage is shown by the dotted lines in Figure 11). However, the voltage at which this inversion occurs reduces with technology scaling. For 65nm, the inversion voltage is 0.6V whereas for 45nm, the inversion voltage is 0.5V. The same effect was

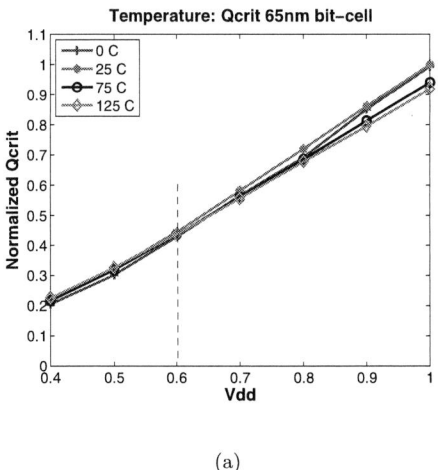
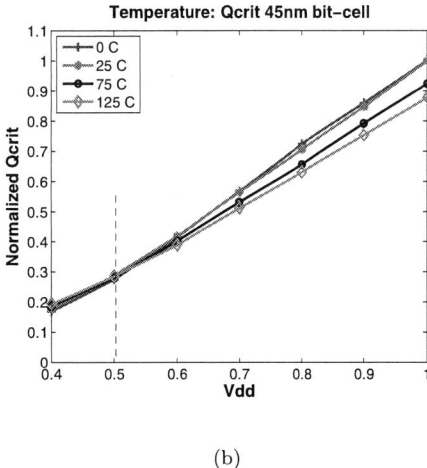

(a)   (b)

Figure 11: Impact of temperature on $Qcrit$ for bit cell in (a) 65nm process (b) 45nm process

observed for latches as well. In our opinion, this effect is related to the temperature inversion effect already being observed in nanometer bulk devices and which probably is not that prominent in SOI devices.

4 Soft error mitigation

Based on our analysis in Section 3, it is clear that as technology shrinks and voltage scales, soft error becomes an increasingly important reliability bottleneck. The techniques to reduce the probability of soft error can be employed at various levels of abstraction, including process, circuit and architecture.

- **Process:** Fully depleted SOI devices are much more robust against soft error due to lack of the bulk. Recently published data show that even partially depleted SOI based SRAMs are 5X more robust than bulk based SRAMs [3]. However, the SOI process is very different from the bulk process and volume manufacturing of fully depleted SOI devices is still challenging.

- **Circuit:** There are various techniques which can make a circuit more robust against soft errors (e.g. •••• •••• circuits). Common approaches include increasing the node capacitance by resizing [19] and changing the architecture of the storage cell [5, 12]. Another common technique is to add redundancy to latch/flip-flop which can detect and correct soft errors [18]. It is useful to note that redundancy is a kind of error correction, but not a very efficient one.

- **Architecture:** The architectural techniques used for SRAMs typically employ parity or ECC to detect errors. One common ECC implementation is SECDED (single error correct double error detect), which is used for processor caches.

All the techniques described above come at a cost. SOI has higher manufacturing cost than bulk, •••••••• cells usually have large area and power overhead and ECC techniques can have performance penalty during cache access. Nevertheless, these techniques, or others like them, will be needed to combat the substantial increase in soft error susceptibility due to technology and voltage scaling. These techniques must be an integral part of the design process, not an afterthought.

5 Conclusions

Technology scaling combined with voltage scaling exacerbates soft error upset susceptibility. The value of $Qcrit$ decreases by ~30% between the 65nm and 45nm technology nodes. This decrease will continue as the technology scales further down. With every technology shrink, the soft error susceptibility of a latch gets closer to that of a bit cell. For a 45nm process, the $Qcrit$ for the latch is just 19% higher than the bit cell at 1V. As the voltage is scaled down, the latch starts to look better but it is only ∼2x better at 0.4V. Further, with aggressive voltage scaling, the susceptibility to soft errors increases substantially. When voltage is scaled from 1V to 0.4V, the value of $Qcrit$ decreases by a factor of 5X. In summary, shrinking transistor geometries combined with voltage scaling make soft errors one of the most crucial reliability concerns. With the soft error susceptibility of latches approaching that of bit cells, increasing design and architecture effort is required in order to protect against soft errors.

References

[1] R. C. Baumann, "Soft errors in advanced semiconductor devices - Part I: The three radiation sources," *IEEE Transactions Device and Materials Reliability*, vol. 1, no. 1 pp. 17-22, Mar 2001.

[2] M. Baze and S. Buchner, "Attenuation of Single Event Induced Pulses in CMOS Combinational Logic," *IEEE Transactions on Nuclear Science*, Dec. 1997.

[3] E. H. Cannon *et al*, "SRAM SER in 90, 130 and 180 nm Bulk and SOI Technologies," *Intl. Reliability Physics Symposium*, pp. 300-304, 2004.

[4] C. Dai *et al*, "Alpha-SER Modeling and Simulation for Sub-0.25μm CMOS Technology," IEEE Symposium on VLSI Technology, pp. 81-82, June 1999.

[5] P. Hazucha *et al*, "Measurements and Analysis of SER-Tolerant Latch in a 90-nm Dual-V_T CMOS Process," *IEEE Journal of Solid-State Circuits*, Vol. 39, No. 9, Sep. 2004.

[6] T. Heijmen, "Factors that Impact the Critical Charge of Memory Elements," *IEEE Intl. On-Line Testing Symposium (IOLTS)*, 2006.

[7] T. Karnik, P. Hazucha and J. Patel, "Characterization of soft errors caused by single event upsets in CMOS process," *IEEE Transactions on Dependable and Secure Computing*, pp. 128-143, 2004.

[8] J. Keane *et al*, "Method for Qcrit Measurement in Bulk CMOS Using a Switched Capacitor Circuit," *NASA Symposium on VLSI Design*, June 2007.

[9] A. J. KleinOsowski *et al*, "Modeling Single-Event Upsets in 65-nm Silicon-on-Insulator Semiconductor Devices," *IEEE Transactions on Nuclear Science*, Vol. 53. No. 6, pp. 3321-3328, Dec. 2006.

[10] T. C. May and M. H. Woods, "Alpha-particle induced soft errors in dynamic memories," *IEEE Transactions on Electron Devices*, pp. 2-9, Jan 1979.

[11] S. Mitra, T. Karnik, N. Seifert and M. Zhang, "Logic Soft Errors in Sub-65nm Technologies: Design and CAD Challenges," *IEEE Design Automation Conference*, June 2005.

[12] R. Naseer and J. Draper, "DF-DICE: A Scalable Solution for Soft Error Tolerant Circuit Design," *ISCAS*, pp. 3890-3893, 2006.

[13] N. Seifert, D. Moyer, N. Leland and R. Hokinson, "Historical trend in alpha-particle induced soft error rates of the Alpha microprocessor," *IEEE International Reliability Physics Symposium (IRPS)*, pp. 259-265, 2001.

[14] P. Shivakumar, M. Kistler, S. Keckler, D. Burger and L. Alvisi, "Modeling the effect of technology trends on the soft error rate of combinational logic," *International Conference on Dependable Systems and Networks*, pp. 389-398, June 2002.

[15] G. R. Srinivasan *et al*, "Accurate, Predictive Modeling of Soft Error Rate Due to Cosmic Rays and Chip Alpha Radiation," *IEEE International Reliability Physics Symposium*, pp. 12-16, 1994.

[16] Y. Tosaka *et al*, "Impact of cosmic ray neutron induced soft errors on advanced CMOS circuits," *IEEE Symposium on VLSI Technology*, pp. 148-149, 1996.

[17] S. Walstra and C. Dai, "Circuit-Level Modeling of Soft Errors in Integrated Circuits," *IEEE Transactions on Device and Materials Reliability*, Vol. 5, No. 3, pp. 358-364, Sep. 2005.

[18] M. Zhang *et al*, "Sequential Element Design with Built-In Soft Error Resilience," *IEEE Transactions on VLSI*, Vol. 14, No. 12, pp. 1368-1378, Dec. 2006.

[19] Q. Zhou and K. Mohanram, "Transistor Sizing for Radiation Hardening," *Intl. Reliability Physics Symposium*, pp. 310-315, 2004.

SESSION 4
DESIGN FOR TESTABILITY

IEEE International Symposium on Defect and Fault Tolerance of VLSI Systems

Enhancing Silicon Debug via Periodic Monitoring

Joon-Sung Yang and Nur A. Touba

Computer Engineering Research Center
Department of Electrical and Computer Engineering
University of Texas at Austin, TX, 78712
E-mail : {jsyang, touba}@ece.utexas.edu

Abstract

Scan-based debug methods give high observability of internal signals, however, they require halting the system to scan out responses from the circuit-under-debug (CUD). This is time consuming as many scan dumps may be required. In this paper, conventional scan chains that have non-destructive scan out capability are configured to operate as multiple MISRs during system operation. Information from the multiple MISRs is monitored periodically to identify erroneous behavior. A procedure for constructing the MISRs to maximize debug capability is described. A three step process is used to zero in on the first clock cycle in which an error is present with a small number of scan dumps. Moreover, a method for bypassing errors is described to permit debug in the presence of multiple bugs.

1. Introduction

As technology advances, larger and denser devices are being manufactured with shorter time to market requirements. Identifying and resolving problems in integrated circuits (ICs) are the main focus of the pre-silicon and post-silicon debug process. As indicated in the International Technology Roadmap for Semiconductors (ITRS) [ITRS 05], post-silicon debug is a major time consuming challenge that has significant impact on the development cycle of a new chip.

Trace buffers are commonly used to capture data from some select signals to aid in the debug process [Abramovici 06], [Anis 07a, 07b], [Hopkins 06], [Yang 08]. They provide at-speed signal capture capability over a number of clock cycles which enhances the observability of the internal signals. More information for silicon debug can be achieved by using compression techniques which further improve the visibility for the debug process. However, trace buffer based debug techniques have limited capability due to limited on-chip storage space. The number of signals that are observed and the number of clock cycles over which the signal information is available is limited with trace buffers. The information provided by the trace buffer may not be sufficient to find both temporal (when) and spatial (where) information for failures in the silicon using only trace buffers.

Scan chains are used to support manufacturing testing and can be reused for post-silicon debug to increase debug capability [Carbine 97], [Hopkins 06], [Gu 06]. Scan dumps give high observability of internal signals and states after the occurrence of a triggering event. Scan dumps play a key role in binary search based debug [Yen 06] for observing the state of the circuit-under-debug (CUD). Binary search based debug involves iteratively dividing the search space in half until the first cycle that the error is activated and observed is found. In [Vermeulen 02], methods for using scan chains to further increase observability were introduced. Hardware debug modules integrated with scan chains are added to a chip and provide the capability to start, stop, reactivate, or single step execute the debug process with

1550-5774/08 $25.00 © 2008 IEEE
DOI 10.1109/DFT.2008.57

the scan chain values being delivered through the IEEE 1149.1 standard test access port (TAP). The drawback of binary search based debug is that it can require a large number of debug sessions to find the first failing cycle where each session requires halting the system to perform a scan dump.

Shadow flip-flops or latches are often used to provide a non-destructive scan out capability that preserves the existing system state. Many systems are fully scannable with non-destructive capability which is helpful for both test and debug [Carbine 97], [Vermeulen 02], [Kuppuswamy 04]. Note that while the system is running, shadow flip-flops or latches are not used for system operation. This fact is exploited in the work.

In this paper, we propose a new debug technique based on reusing non-destructive scan chains. The shadow flip-flops are configured to operate as multiple-input signature registers (MISRs) during system operation. The shadow flip-flops normally do not perform any function when the system operates, however, in the proposed method they are formed into multiple MISRs to enhance silicon debug capability. Compressed information from the multiple MISRs is monitored periodically with externally provided data to identify erroneous behavior. The MISRs are constructed based on structural information of the circuit to maximize their debug efficiency. A three step debug process is used to zero in on the first failing clock cycle trying to minimize the number of scan dumps. By providing high observability of the system state without the need for scan dumps, the proposed method can detect erroneous behavior far earlier than conventional debug methods can. In addition, we propose a debug method to bypass errors which can facilitate downstream debug. Because the presence of a bug may prohibit accurate downstream debug, a faulty response replacement or data masking method may be needed to assist in validating a system.

Sec. 2 describes how to configure multiple MISRs using non-destructive scan chains. Sec. 3 describes the features of the three step debug process, and Sec. 4 describes an error bypassing method, Sec. 5 shows the experimental results, and Sec. 6 concludes this paper.

2. Procedure for Configuring Multiple MISRs

Scan based debug gives greater observability than trace buffer based debug [Anis 07a, 07b], [Yang 08], however, observing the system state requires halting the system to perform the scan dump and hence is not suitable for at-speed debug [Hopkins 06]. It may take many cycles for an error in the system to propagate to a primary output where it can be observed, so there can be a long time gap between from when a bug is invoked and to when it is visible. Due to this latency, it can be time consuming to find the root cause of errors using only a scan dump based debug methodology.

As described in [Gu 02] and [Vermeulen 02], there have been several techniques proposed to reuse DFT logic for silicon debug. In this paper, conventional scan chains with non-destructive scan out capability are reused and configured to operate as a set of MISRs. A periodic checking scheme is proposed to monitor the states of a system without scan dumps. The MISRs configured from the shadow flip-flops in scan chains keep compacting the system state and linear compactors further compress the MISR signatures to greatly reduce the volume of debug data. By having only a very small amount of highly compressed data which represents the system state, it is possible to monitor this data and detect any misbehavior in the circuit much earlier than when it would normally become functionally observable at the chip pins. Structural information is used to configure the multiple MISRs in a way that helps to more rapidly diagnose the root cause of the erroneous behavior. By carefully configuring the MISR signatures, it is possible to extract spatial information about where the error is originating from.

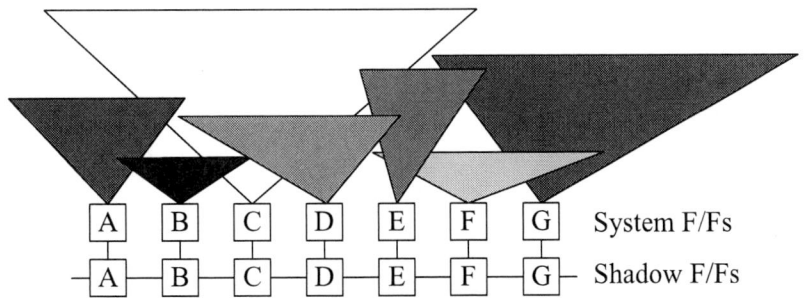

Figure 1. Scan Chain and Logic Cone

Fig. 1 shows an example of the logic cones driving the system flip-flops along with the shadow flip-flops present for non-destructive scan. The flip-flops are labeled as *A-G* for the illustrative purposes. By traversing the netlist, the degree of overlap between the logic cones that drive each flip-flop can be determined. The partitioning of flip-flops into multiple MISRs is based on this logic cone analysis. If only a single large MISR was constructed, it would require long wires to generate the feedbacks and may cause other issues related to the physical design. Partitioning the flip-flops into multiple MISRs addresses this problem and can also be used to spatially isolate the candidate error sites to speed up the debug process. The proposed approach is based on partitioning scan cells that are the most structurally related in the design together in the same MISR. This helps to minimize routing of the MISRs as well as maximize spatial diagnosis capability by reducing the probability that an error propagates to multiple MISRs.

In the proposed approach, the MISRs are configured using a graph that represents the degree of logic cone overlap between different flip-flops. In Fig. 2, each node corresponds to a flip-flop in Fig. 1. The weight on the edges corresponds to the amount of logic cone overlap between the logic cones of the corresponding nodes measured in terms of the number of gates that are shared between the two cones. For example, the logic cone driving flip-flop *A* overlaps with cones of *B* and *C,* and the degree of overlap is 20 and 10, respectively. MISRs are generated using an initial clustering procedure followed by an iterative merge & update procedure. The details of algorithm are described in the following subsections.

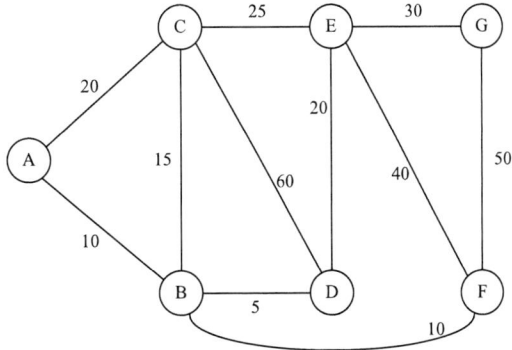

Figure 2. Graph Representation of Logic Cone Relation.

2.1. Initial Clustering Procedure

There are two inputs to the initial clustering procedure, one is the number of MISRs, n, and the other is the minimum size of a MISR, m, which determines the minimum aliasing probability. The initial clustering procedure select n clusters of the m most overlapped logic cones in each. The edge with the largest weight is selected as a starting point for the first cluster. Because the weight represents the logic cone overlap size, the largest weight has a higher probability of error propagation to both cones assuming that the functional and electrical bugs occur with Gaussian distribution. For the same reason, if there are equal size overlaps, the flip-flop driven by the largest logic cone and its neighbor flip-flop are selected. Therefore, the edge between node C and D is selected from Fig. 3(a), and node C and D are merged to begin constructing the first MISR (dashed circle in Fig. 3(a)).

The graph is updated after the nodes are merged. Nodes which are merged generate a composite node in a graph. From Fig. 3(b), since the node C and D are merged, they form a new node and the edge weights are updated. Additional nodes are added to the cluster in a greedy fashion by selecting the edge with the largest weight attached to the current cluster until the size of the cluster reaches m which ensures a certain minimum aliasing probability for the MISR. This process is repeated to create the n initial clusters.

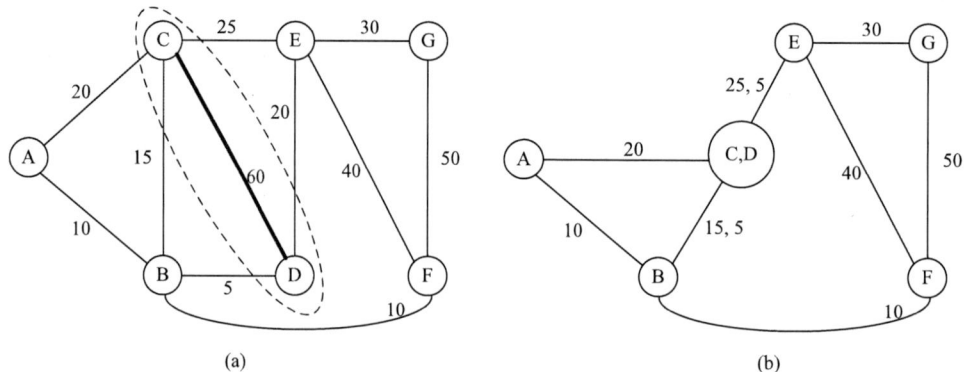

Figure 3. First Cluster Generation

2.2. Merge & Update Procedure

Once the initial clusters have been constructed, a merge & update procedure is iteratively performed to merge the remaining nodes together using the largest weight on an edge at each step. At the conclusion of the procedure, all the nodes will have been added to the initial clusters to form the n MISRs.

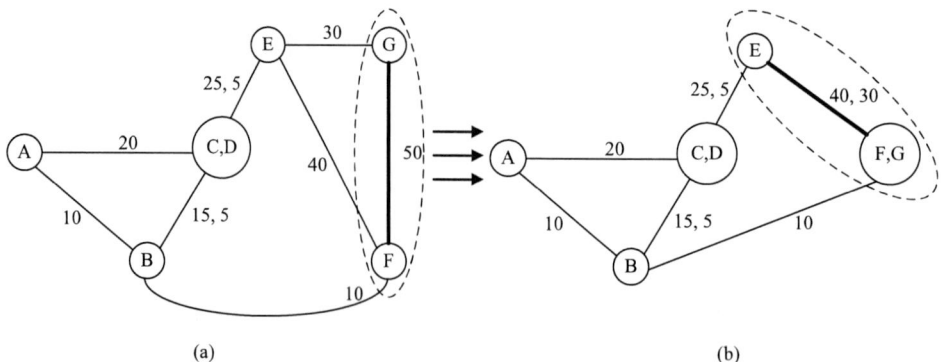

Figure 4. Merge & Update Process

In the merge & update process, since G and F have the largest weight edge, the second cluster is generated in Fig. 4(a). The iterative process merges node B into the second cluster.

128

In this small example, assume that the number of MISRs is 2, $n = 2$, and the minimum size of a MISR is 3, $m = 3$. The initial clustering procedure generates two clusters each with three nodes, A,C,D and E,F,G, as shown in Fig. 5(b). In Fig. 5(c), the final node B is merged with node A,C,D since it has more overlap with that cluster, and the two MISRs are finally generated.

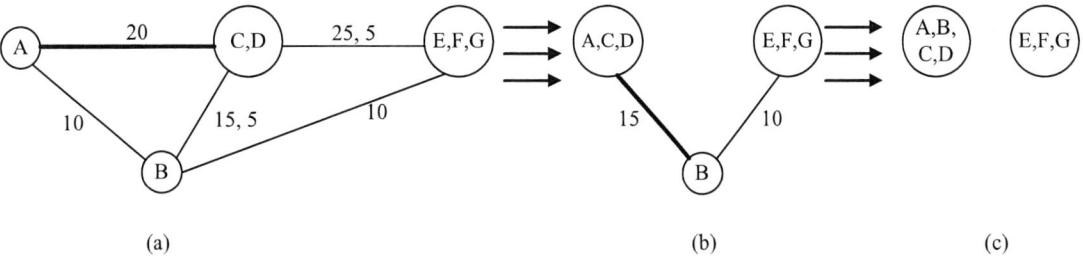

(a)   (b)   (c)

Figure 5. Merge & Update Example ($n = 2$ and $m = 3$)

3. Three Step Debug Process

Scan-based debug methods give high observability of internal signals by shifting out the internal state values. Conventional scan-based debug needs to stop the system to get the information from the CUD. Checking the internal states by scan dumps is a very time consuming task. If a large number of scan dumps is required, this may be an unattractive way of validating a chip. In Sec. 2, a procedure for constructing multiple MISRs is proposed. The MISRs keep compacting the internal state such that erroneous circuit responses will easily corrupt the MISR signatures. Therefore, if the erroneous input comes into the MISRs, although the internal states cannot be read out, the MISR signatures still provide a way to identify the error. In this section, we propose a technique to utilize the internal state information without scanning out the data via scan chains when they have non-destructive capability. A three step debug process is used to zero in on the first erroneous clock cycle. In the first step, single parity information is generated at every clock cycle for periodic monitoring. In the second step, more parity information is stored in a trace buffer to zoom in closer to the failing clock cycle. In the third step, MISR signatures are stored and checked so that the first erroneous cycle can be identified.

3.1. Step One : Checking Intermediate Parity

After configuring the MISR as described in Sec. 2, the MISRs are used to generate the signatures by compacting the outputs of the logic cones driving it. Linear compaction hardware is used to generate a single parity bit for each MISR signature. This can be done by XORing some subset of the flip-flops in the MISR. XORing the parities of the MISRs generates a single parity bit which represents the entire system state. Fig. 6. shows the logic cones with the MISRs and linear compaction circuits (which are simply XOR networks).

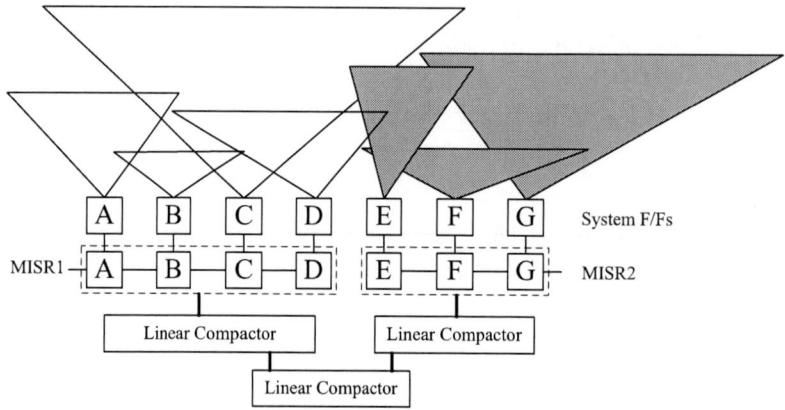

Figure 6. Configured MISRs, DFD and Logic Cones

The composite parity generated by combing the MISR parities is monitored periodically to identify erroneous behavior. It is compared with an externally provided golden parity from off-chip (either checked with an ATE or checked with data stored in some external memory). The periodic parity checking allows at-speed debug with far less volume than the conventional scan-based debug. Conventional scan-based debug requires to stopping the system to perform a scan dump and thus periodic monitoring is not feasible. The monitoring period is determined by the relationship between the system and the automatic test equipment (ATE) operating speed (or the speed of the external memory). Since the system typically runs internally at a higher frequency than the ATE does, the ATE can ideally check a golden parity bit at a period equal to the ratio of the internal chip clock rate to the ATE clock rate, i.e., *frequency of chip / frequency of ATE*. For example, if a chip operates at 3GHz and the ATE runs at 100MHz, the ideal monitoring period is 30 cycles.

Periodic real-time parity monitoring checks the highly compacted circuit response which is represented as a single parity bit and can detect the first erroneous clock cycle within some limited latency which depends on the frequency of the monitoring and the degree of aliasing. The debug session can be stopped when the first mismatching parity is found instead of waiting until the system fails in a functionally observable manner. Since only a single highly compacted bit is being monitored, aliasing is an issue. An even number of error bits will cause parity to alias and hence periodic monitoring would fail to detect a faulty state. However, because the MISRs continue to compact a corrupted signature after the first error, the cumulative aliasing probability exponentially decreases. The probability of 10 consecutive aliases when monitoring the parity is $1/2^{10}$ which is 1 in a thousand.

In addition to monitoring the parity, the proposed approach also involves storing the parity information in a trace buffer at every clock cycle. When the trace buffer gets full, the older data is overwritten so that when the period monitoring halts the debug session, the trace buffer contains the running history of the parity information for the most recent clock cycles. Even though there may be aliasing, the parity history is stored and can be used to find earlier erroneous cycles than the periodic monitoring did. If a 512 Byte trace buffer is used with a 500 cycle monitoring period (one check every 500 clock cycles), then the trace buffer has the history of last 8 monitoring periods. This can be used to more closely zero in on the first erroneous cycle and significantly reduces the debug search space.

3.2. Step Two : Storing Parity of MISRs

Periodic parity monitoring reduces the the range of cycles over which more careful debug is needed. Moreover, since the trace buffer contents provide cycle by cycle information, the first erroneous clock cycle can be approximated with even better clock resolution. This helps to find the neighborhood of the first erroneous clock cycle, however, it may not be able to zero in on the exact clock cycle since the parities stored in the trace buffer also have aliasing issues for even bit errors.

In step two, the parities of the MISRs are used to provide more specific information for identifying the first erroneous clock cycle as well as spatial information on where the error originates. By looking at the parity of each MISR signature and identifying which ones contain errors, information about which logic cones the errors are getting generated in can be deduced.

The periodic monitoring isolates the error to a small range and allows the debug process to be halted at that point. Periodic monitoring is not required in step two. The parities of the MISRs are stored in the trace buffer from a trigger point based on the information obtain in step one about the rough location of the first error. The triggering can be implemented similar to the internal breakpoint mechanism used in [Cabrine 97], [Anis 07a, 07b]. When the debug session stops, the trace buffer will contain each MISR's parity information for the last set of clock cycles. The number of cycles worth of data will be equal to *size of trace buffer/ number of MISRs*.

3.3. Step Three : Storing MISR signatures

After steps one and two, the search space for the first erroneous clock cycle is significantly reduced and some spatial information is available. However, the first erroneous clock cycle is not precisely known due to aliasing uncertainty. It is necessary to shift out MISR signatures and compare them with golden signatures to find the first corrupted signature. Scan dumps are only required in step three and only a small number of MISR signature scan outs are required. The volume of debug data can be reduced in comparison to full scan dumps. For example, if the first MISR signature is corrupted while the second MISR signature is clean in Fig. 6, the second MISR signature does not need to be investigated at the next scan out. A programmable counter can be used and only the MISR signatures of interest need to be compared to reduce the debug effort. In comparison, binary search based debug requires performing a full scan dump at every iteration.

When the first mismatching signature is found, it provides spatial diagnostic information as well because the MISRs are configured using structural information as described in Sec. 2. Multiple MISRs divide the logic cones into multiple regions and aids the debug process. Logic cones that are driving fault-free MISRs can be pruned out to reduce the space of possible root causes.

4. Error Bypassing

One difficult aspect of silicon debug is how to bypass errors in order to continue searching for additional bugs after the first bug is found. In a finite state machine (FSM), the state transition depends on the previous state and the circuit inputs. If a faulty response is found, the downstream state transitions are not guaranteed to follow the expected transitions. This makes downstream debug inefficient. A benefit of using scan chains is the accessibility to internal flip-flops. If the first bug in a design is detected and diagnosed, the downstream debug process needs to be able to continue to find additional bugs in a system. Using the proposed approach to precisely identify the first erroneous clock cycle, it is possible to determine what the correct state values should be from either simulation or emulation. By utilizing this information, golden values can be shifted in through the scan chains to return the system to the correct state and facilitate downstream debug. Because the system can be rerun with bypassing of the erroneous state, periodic monitoring can now be used to catch other bugs.

5. Experimental Results

The debug method proposed here tries to detect errors early on and thereby reduce the search space and number of debug sessions. Experiments were performed on the larger ISCAS-89 benchmark circuits [Brglez 89] and OpenRisc processor [OR1200]. Random faults were injected in the benchmark circuits and random input patterns were applied. MISRs were configured using the algorithm in Sec. 2 and the parity information was periodically monitored. The results obtained are shown in Table 1. The first and second columns show the circuit name and the number of scan elements. The third column indicates the first clock cycle in which an error becomes functionally observable at a primary output. The next three columns show the first clock cycle that the error is identified using the methods described in Secs. 3.1, 3.2, and 3.3, respectively. The last column shows the number of scan outs required to precisely identify the first erroneous clock cycle. The simulation results show that periodic monitoring detects the erroneous behavior quite close to the first erroneous clock cycle and that using the parities of the MISRs help give more precise information. Hence, very few scan outs are required.

Table 1. Erroneous Response Detection Clock Cycles

Circuit	Scan Size	Error at P.O.	Single Parity Monitor (Monitoring Period)		Parities of MISRs	First Erroneous Clock Cycle	Num. of Scan Out
s9234	211	493	100 200	(50, 100) (200)	45	43	3
s15850	534	868	100 200	(50, 100) (200)	7	7	1
s13207	638	828	600	(50, 100, 200)	597	597	1
s38584	1340	2362	1400 1800	(50, 100) (200)	1338	1338	1
s38417	1636	1680	1250 1300 1800	(50) (100) (200)	1221	1221	1
s35932	1728	210	100 200	(50, 100) (200)	6	5	2
OR1200	1989	1226	900 1000	(50, 100) (200)	834	834	1
		5712	4050 4200	(50) (100, 200)	3975	3975	1

Table 2 shows a comparison in terms of the number of debug sessions required using the proposed method and conventional binary search for the same designs and faults as in Table 1. The conventional binary search based debug is initiated when the errors are functionally observable at the primary outputs of the circuit, and it uses scan dumps to check the state of a system at the end of each debug session. However, with the proposed method, the three step process requires only 3 debug sessions and only a few number of scan dumps. MISR signatures can be stored in a trace buffer without additional debug sessions.

Table 2. Number of Debug Sessions

Circuit / Methodology	s9234	s15850	s13207	s38584	s38417	s35932	OR1200	O1200
Binary Search Debug	16	18	20	22	12	17	18	26
Proposed Debug	3	3	3	3	3	3	3	3

6. Conclusion

In this paper, a new debug technique using a three step process is proposed to zero in the first erroneous clock cycle using a small number of scan dumps. By using conventional scan chains with non-destructive scan out capability, multiple MISRs can be configured to provide high observability with periodic monitoring. Note that the proposed scheme can also be selectively applied to only part of a design, e.g., for newly implemented and unverified design blocks or parts of a design where bugs are more likely to originate from. The scan chains can also used to restore the system state with golden values to facilitate downstream debug.

Acknowledgements

This research was supported in part by the National Science Foundation under Grant No. CCR-0426608.

References

[Abramovici 06] Abramovici, M., P. Bradley, K. Dwarakanath, P. Levin, G. Memmi, and D. Miller, "A Reconfigurable Design-for-Debug Infrastructure for SoCs," *Proc. of Design Automation Conference*, pp. 7-12, 2006.

[Anis 07a] Anis, E., and N. Nicolici, "On Using Lossless Compression of Debug Data in Embedded Logic Analysis," *Proc. of Int. Test Conference*, Paper 18.3, 2007.

[Anis 07b] Anis, E., and N. Nicolici, "Low Cost Debug Architecture using Lossy Compression for Silicon Debug," *Proc. of Design, Automation, and Test in Europe*, pp. 1-6, 2007.

[Brglez 89] Brglez, F., D. Bryan, and K. Kozminski, "Combinational Profiles of Sequential Benchmark Circuits," *Proc. of International Symposium on Circuits and Systems*, pp. 1929-1934, 1989.

[Carbine 97] Carbine, A. and Feltham, D., "Pentium®Pro Processor Design for Test And Debug", *Proc. of Int. Test Conference*, pp. 294-303, 1997.

[Gu 06] Gu, X., Wang, W., Li, K., Kim, H. and Chung, S., "Re-Using DFT Logic for Functional and Silicon Debugging Test", *Proc. of Int. Test Conference*, pp. 648-656, 2006.

[Hopkins 06] Hopkins, A., and K. McDonald-Maier, "Debug Support for Complex Systems on-Chip: A Review," *IEEE Proc. on Computers and Digital Techniques*, Vol 153, No. 4, pp. 197-207, Jul. 2006.

[ITRS 05] "The International Technology Roadmap for Semiconductors," Semiconductor Industry Association, 2005.

[Kuppuswamy 04] Kuppuswamy, R., DesRosier, P., Fletham, D., Sheikh, R. and Thadikaran, P., "Full Hold-Scan Systems in Microprocessors : Cost/Benefit Analysis", *Intel Technology Journal*, Volume 08, Issue 01, 2004.

[OR1200] OPENCORES, *http://www.opencores.org*

[Vermeulen 02] Vermeulen, B., Waayers, T. and Goel, S.K., "Core-Based Scan Architecture for Silicon Debug", *Proc. of Int. Test Conference*, pp. 638-647, 2002.

[Yang 08] Yang, J.-S., and Touba, N. A., "Expanding Trace Buffer Observation Window for In-System Silicon Debug through Selective Capture", *Proc. of IEEE VLSI Test Symposium*, pp. 345-351, 2008.

[Yen 06] Yen, C-C., Lin, T., Lin, H., Yang, K., Liu, T. and Hsu, Y.-C., "Diagnosing Silicon Failures Based on Functional Test Patterns", *International Workshop on Microprocessor Test and Verification*, pp. 94-98. 2006.

IEEE International Symposium on Defect and Fault Tolerance of VLSI Systems

A Digital BIST for Phase-Locked Loops

Kevin Sliech, Martin Margala
Electrical and Computer Engineering Department
University of Massachusetts Lowell
1 University Avenue, Lowell, MA, 01854

ABSTRACT

This paper presents a conceptual implementation of a jitter measurement circuit with several BIST(Built-In Self Test) features for embedded phase-locked loops. We demonstrate a fully functional jitter measurement circuit capable of detecting cycle-to-cycle jitter. Proposed BIST logic provides additional information such as MAX jitter value, programmable threshold detection and most recent jitter result with low processing overhead.

1 INTRODUCTION

The phenomenal growth in today's information technology has increasing demand for multiple high speed integrated PLLs (Phase-Locked Loop). As the process technology scaling continues, signal reliability has emerged as a serious challenge. This demands new ideas in low-cost on-chip capabilities for signal quality monitoring. The idea behind the proposed low power solution is that it allows to detect and provide the system user with additional information such the lowest jitter in the circuit and the most recent jitter measurement. This improved design can be operated over a wide range of GHz frequencies.

There are two camps of research. First group, treating this challenge as a timing/delay measurement problem [9, 14]. The second group focuses on jitter test/measurement problem [2,8]. The solutions that have been presented in literature suffer from major problems. None of the solutions attempts to identify the optimum balance between the accuracy of the circuit and the area/power overhead. The solutions have either poor accuracy/sensitivity or large overhead.

This paper presents a low-cost simple implementation with enhanced BIST features. The solution presents an improved version of our previously proposed work [1] with a new jitter measurement circuit along with new digital BIST features. The previous work on jitter measurement circuit used an externally generated reference frequency for the comparator. Instead in this implementation, an embedded DAC(Digital-Analog Converter) is used to allow the user to select from different references. Moreover, it also uses an XOR design for phase difference detection. The previous work used also only a simple BIST function block. The BIST logic in this paper was designed to perform additional functions such maximum jitter measurement and user programmable threshold detection. Fig. 1 shows the basic building block of the improved version of jitter measurement circuit and BIST logic. The conceptual design was implemented in 0.18μm CMOS technology.

The rest of the paper is constructed as follows. Section 2 describes the overall architecture and the main building blocks of the design. Section 3 provides circuit specifications for several of the building blocks of the circuit and the initial the results. Section 4 concludes the paper.

2 PROPOSED ARCHITECTURE

This paper presents an embedded PLL jitter BIST circuit that is comprised of two parts: a modified version of the jitter estimate circuit presented in [1], and a block of digital BIST logic. The jitter estimate circuit interfaces directly to the PLL and presents a 6-bit value to the BIST logic that is proportional to the average amount of cycle-to-cycle jitter. The BIST logic outputs the most recent jitter measurement, the maximum measured jitter, and a flag that is asserted when the measured jitter crosses a user-defined threshold. A block diagram of the system architecture appears at the end of this paper.

Instead of simply presenting the most recent jitter measurement to the reset of the system, the BIST logic was included to perform two additional functions that serve to decrease system overhead. For example, a user may be interested in the maximum jitter measured over some period of time. This would

1550-5774/08 $25.00 © 2008 IEEE
DOI 10.1109/DFT.2008.62

require that the jitter measurement value be frequently polled and compared against a stored value. Instead of using CPU cycles to perform this function, we have implemented it in hardware with a small amount of additional logic. Flags are provided for both the maximum jitter detector and the user-defined threshold detector for use as interrupts when used in a CPU, FPGA, etc to eliminate the need for constant polling.

2.1   JITTER MEASUREMENT

The jitter measurement circuit implemented in this design is a modified version of the circuit found in [1]. Cycle-to-cycle jitter is measured by observing the 'up' and 'down' signals from the PLL's phase/frequency detector (PFD) with the circuit shown in Fig. 1. An XOR gate connected to these signals asserts its output high for the duration that the two signals are different, corresponding to a misalignment between the PLL clock and reference clock. This pulse is stretched out in time by an RC network before reaching an analog comparator. The longer the period of misalignment occurs, the more jitter there is, and thus the longer the pulse will be held up by the capacitor. The voltage across the capacitor is compared to a user controlled reference with an analog comparator. The output of this comparator is high when the capacitor voltage is below the reference, allowing a free-running counter to count. When the jitter is low enough that the voltage across the capacitor is below the reference, the comparator asserts its output high, resetting the counter back to 0.

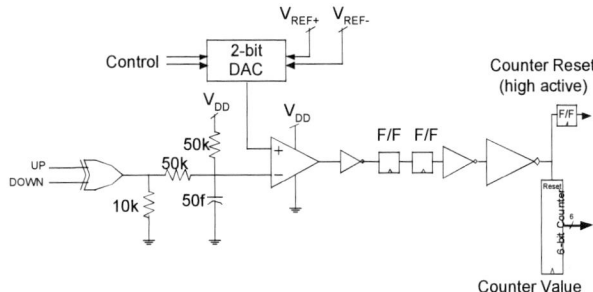

Fig. 1: Jitter measurement circuit.

The XOR gate was designed with minimum transistor sizes to minimize its intrinsic delay. The lower this delay is, the smaller the pulse the XOR can respond to, and the finer the resolution of the jitter measurement circuit is.

Four XOR topologies were investigated: Classic CMOS, pass transistor logic, transmission gate logic, and a novel 4T design. The minimum pulse at the inputs of the gate that would register a 1 on the output was found to be 50ps, 20ps, 80ps, and 40ps respectively. That said, the pass transistor circuit was selected for the design.

The charge pump and RC described in [1] for the jitter measurement circuit was implemented slightly differently. Instead of using a charge pump, the XOR directly drives the RC, as shown in Fig. 2. A 10kΩ resistor to ground is used at the output of the XOR to help discharge the output capacitance of the gate. A 50kΩ resistor to VDD was added to put a 600-700mV DC offset across the capacitor. The analog comparator will not switch if both inputs are close to ground, so this DC offset allows for faster, more reliable switching.

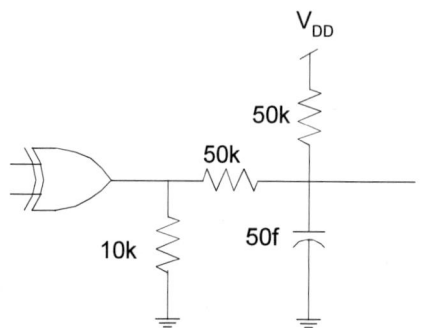

Fig. 2: XOR output conditioning circuitry.

The analog comparator is a simple CMOS differential amplifier whose output is fed into a common source amplifier for high gain, as shown in Fig. 4. The amplifier was designed with high open loop gain in order to switch with the smallest possible difference between the inputs. Some hysteresis was added by way of a feedback resistor to make the circuit more noise resistant when the comparator inputs are close in voltage.

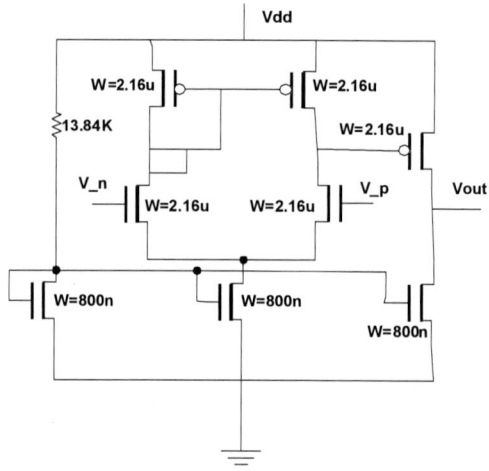

Fig. 3: Analog comparator topology.

The original jitter measurement described in [1] did not state how the reference voltage was generated. A static reference could be chosen and implemented with a simple voltage divider. With only a few additional transistors, a very simple DAC was created that allows the user to choose from different references.

A 2-bit digital to analog converter (DAC) is used to provide the reference voltage. This circuit allows the user to set the resolution of the jitter measurement circuit on-the-fly. By raising the threshold, the counter will spend less time counting, saving power at the expense of reduced resolution. Since the reference voltage generated by the DAC will have to source a negligible amount of current, the resistor ladder topology shown in Fig. 4 was used. A low-pass filter was added to the output to remove the high frequency switching noise that occurs when the ADC changes its output voltage.

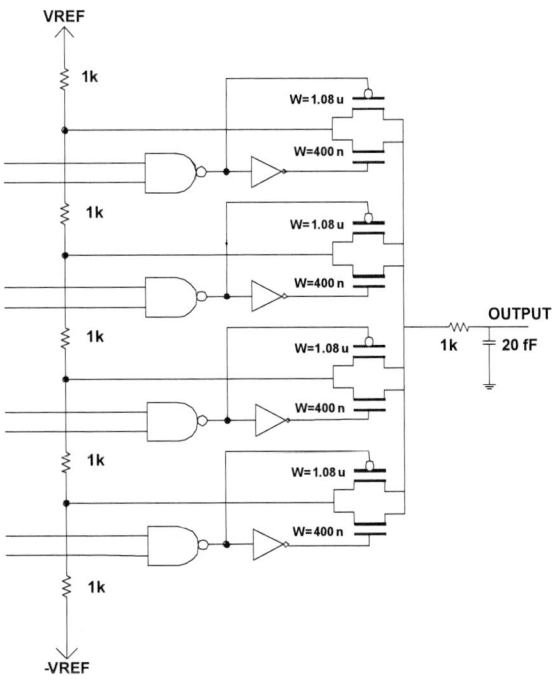

Fig. 4: DAC circuit topology.

The resolution of the 2-bit DAC is very coarse since there are only four possible output voltages. A ladder comprised of equal value resistors is used to provide the four voltages. Since the resistors are of equal value, the four possible output voltages are (Vref+-Vref-)/5 + Vref-, 2(Vref+-Vref-)/5 + Vref-, 3(Vref+-Vref-)/5 + Vref-, and 4(Vref+-Vref-)/5 + Vref-. Since both the top and bottom of the resistor ladder can be set to any voltages that the user desires, a fine resolution between two voltages close in magnitude can be achieved.

Under normal operating conditions, the counter is reset by the comparator depending on current jitter levels. An additional reset is included as a power saving feature by allowing the user to disable the counter when jitter measurements are not desired.

2.2 BIST LOGIC

The BIST logic block performs three functions for the user: a storage of the most recent jitter measurement, a storage of the maximum measured jitter since the last time the user reset this sub-component, and a flag assertion when the most recent jitter measurement exceeds a user-programmed threshold. The BIST logic block and its related I/O is shown in Fig. 5.

Before describing the main components of this block of logic, it is necessary to describe the flip-flop design used throughout these synchronous modules. A standard CMOS transmission gate D flip-flop with asynchronous low-active reset was implemented using transistors of minimum size. The architecture of this component is shown in Fig. 6.

Registers with parallel load capability were also required, and were implemented by adding a 2:1 multiplexor (MUX) to each flip-flop, as shown in Fig. 7.

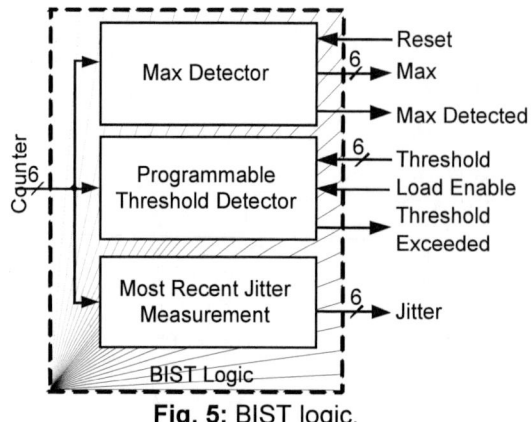

Fig. 5: BIST logic.

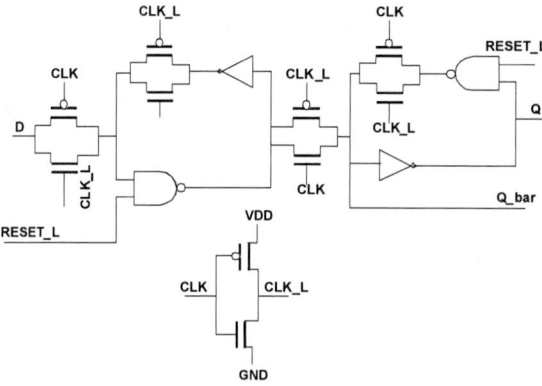

Fig. 6: D flip-flop with asynchronous reset.

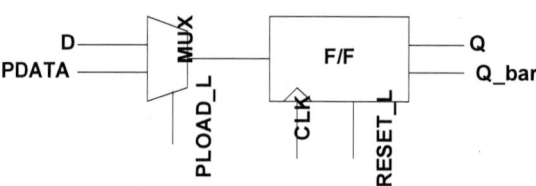

Fig. 7: Flip-flop with parallel load.

For addition, a CMOS full adder design was used. 1-bit full adders are strung together to form ripple-carry adders. This type of adder was chosen since adders of arbitrary size can be built quickly, unlike faster carry look-ahead types which could require integer multiples of 4 bits to be used. This allowed greater flexibility in the design should the number of bits for certain addition operations need to be modified as the design progressed.

The most recent jitter measurement is simply a register that is loaded with the jitter measurement circuit counter value when it is reset by the analog comparator. This is a critical function since the value of the counter as it is counting has no significance, it is only when it is reset that we are interested in the value. This load signal is registered and provided at the output of the BIST logic to signal when the register is updated.

Both the maximum measured jitter detection circuit and the user-defined threshold circuit use the same greater-than detection circuits. This function is performed by subtracting the locally stored value (user threshold or maximum detected jitter, depending on the circuit) with the most recent jitter measurement. If the result is positive, the most recent jitter measurement is larger than the stored value. Subtraction was performed by taking the two's complement of stored value and adding it with the most recent jitter measurement. This is implemented by summing the complement of the stored value with the most recent jitter measurement and setting the carry-in high. If the carry out of the add is a 0, the answer is positive,

138

meaning that the most recent jitter measurement is greater than the stored value.

The maximum measured jitter detection circuit compares the most recent jitter measurement to the contents of the maximum jitter register. When the most recent measurement is greater than the stored value, the stored value is updated with the most recent measurement. A reset is provided to allow the user to clear the stored value to all 0's and find a new maximum. When the maximum value is updated, a flag is set for one cycle. This circuit is shown in Fig. 8.

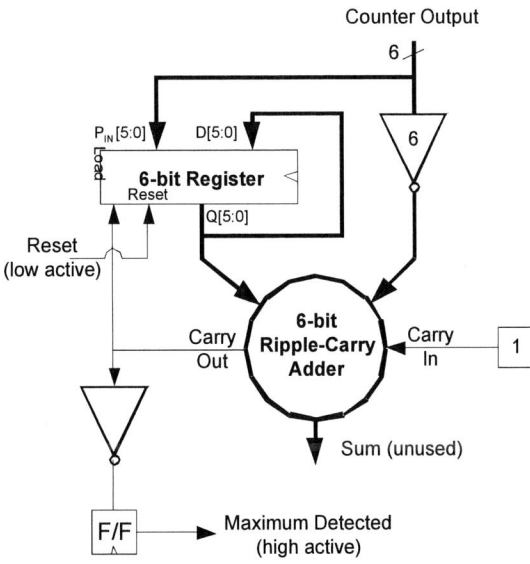

Fig. 8: Maximum jitter detector.

The user-defined threshold detection circuit is very similar to the maximum jitter detector. A register is provided that the user loads with a jitter threshold level. When the most recent jitter measurement exceeds this threshold, a flag is asserted. The value that exceeded the threshold is stored in the most recent jitter measurement register. This circuit is shown in Fig. 9.

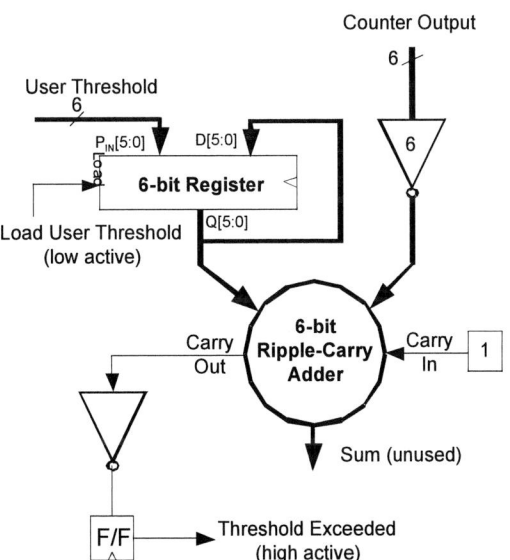

Fig. 9: Programmable threshold detector.

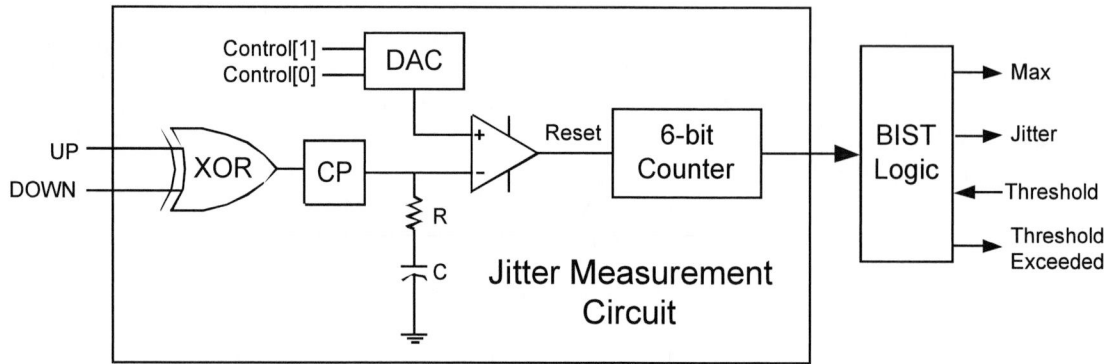

Fig. 10: Simplified architectural diagram of jitter measurement circuit with BIST logic.

The following Tab.s present a selection of simulated component parameters that directly impact the performance of the system described in this paper.

3   COMPONENT SPECIFICATIONS AND RESULTS

The following Tab.s present a selection of simulated component parameters that directly impact the performance of the system described in this paper.

Digital to Analog Converter		
Property	**Specification**	**Units**
Switching Time	740	ps (max)
00b: V_{out}	$(V_{ref+}-V_{ref-})/5 + V_n$	V (typ)
01b: V_{out}	$2(V_{ref+}-V_{ref-})/5 + V_n$	V (typ)
10b: V_{out}	$3(V_{ref+}-V_{ref-})/5 + V_n$	V (typ)
11b: V_{out}	$4(V_{ref+}-V_{ref-})/5 + V_n$	V (typ)

Analog Comparator		
Property	**Specification**	**Units**
t_{RISE}	650	ps (max)
t_{FALL}	860	ps (max)

Exclusive NOR Gate (XNOR)		
Property	**Specification**	**Units**
Minimum input hold time for output to go high	20	ps (max)

6-bit Counter		
Property	**Specification**	**Units**
P_{AVG}	148 (200MHz)	µw (typ)
f_{max}	500	MHz (min)

Tab. 1

Count Value for Simulated Misalignment Pulses	
Misalignment (ps)	Count

10	0
20	1
30	1
50	2
100	2
200	2
300	3
...	3
1000	3

During the investigation, it was discovered that the carry ripple adder was a performance limiting factor which constrained the maximum frequency of all synchronous elements. Therefore, the evaluation was performed at a system clock of 500MHz.

The jitter measurement circuit was simulated with various simulated PFD misalignment pulses ranging from 10ps to 1ns. The results appear in Tab. 1.

The count values are low for two reasons. First, the system clock is not nearly fast enough in comparison to the duration of the stretched misalignment pulse. Only a few clock cycles occur during even the longest pulses. Secondly, the count increases with the first few low jitter levels then stops increasing after 300ps because of the exponential nature of charging the capacitor. Small pulses have a large effect on the voltage across the capacitor, but once the pulses are sufficiently large, the waveform produced by the capacitor no longer changes significantly. This means that this circuit is most sensitive when measuring low levels of jitter.

To improve the jitter measurement circuit, the clock frequency needs to be increased. For this to be possible, the adders are being redesigned with speed as the primary consideration. Carry look-ahead adders are the first choice. Furthermore, the system voltage will be swept beyond the standard operating voltage of 1.4V to speed up the circuits. With the circuit running off of a faster clock, finer differences in clock misalignment could be measured.

The BIST circuitry performed as expected. The programmable threshold detector and maximum detector each showed correct operation when provided with simulated counter values over the full range of possible inputs. Additionally, all output signals were found to be properly timed with respect to each other.

Power measurements of each main circuit block were taken, as well as overall power consumption with a 100ps misalignment pulse and a 500 MHz clock. Tab. 2 summarizes these results.

Tab. 2

Power Measurements		
Circuit	Power (mW)	Type
Threshold Detector	1.23	Maximum
Maximum Detector	1.21	Maximum
Jitter Measurement	1.81	Typical
Entire Circuit	7.63	Typical

4   CONCLUSIONS

A low cost concept of a jitter measurement circuit with enhanced BIST functions described in this paper has been proved. Preliminary data showed good initial sensitivity and design limitations/improvements of the architecture. Further optimization analysis is on-going and updated results will be provided in a final paper.

REFERENCES

[1] S. Ali, M. Margala, "A 2.4-GHz Auto-calibration Frequency Synthesizer with on-chip Built-In-Self-Test Solution, in Proceedings of the IEEE International Symposium on Circuits and Systems (ISCAS), 21-24 May 2006.

[2] J.-C. Hsu, C. Su, "BIST for Measuring Clock Jitter of Charge-Pump Phase-Locked Loops", IEEE Transactions on Instrumentation and Measurement, vol. 57, no. 2, pp.276-285, February 2008.

[3] G. Yu, P. Li, "A Methodology for Systematic Built-in Self-Test of Phase-Locked Loops Targeting at Parametric Failures", in Proceedings of IEEE International Test Conference, pp.16.2.1-16.2.10, Santa Clara, October 2007.

[4] T. Xia, S. Wyatt, R. Ho, "Employing on-Chip Jitter Test Circuit for Phase Locked Loop Self-Calibration", in Proceedings of the 21st IEEE International Symposium on Defect and Fault-Tolerance in VLSI Systems (DFT'06), October 2006.

[5] J. M. Cazeaux, M. Omana, C. Metra, "Low-Area On-Chip Circuit for Jitter Measurement in a Phase-Locked Loop", in Proceedings of the 10th IEEE International On-Line Testing Symposium", (IOLTS), 2004.

[6] C. –C. Tsai and C.-L. Lee, "An On-Chip Jitter Measurement Circuit for the PLL", in Proceedings of the IEEE 12th Asian Test Symposium, 2003.

[7] T. Xia, J.-C., Lo, "Time-to-Voltage Converter for On-Chip Jitter Measurement", IEEE Transactions on Instrumentation and Measurement, vol. 52, no. 6, pp.1738-1748, December 2003.

[8] R. Voorakaranam, A. Chatterjee, "Low-Cost Jitter Measurement Technique for Phase-Locked Loops", in Proceedings of the 43rd IEEE Midwest Symposium on Circuits and Systems, pp. 956-959, Lansing MI, August 8-11, 2000.

[9] M.-C. Tsai, C.-H. Cheng, C.-M. Yang, "An All-Digital High-Precision Built-In Delay Time Measurement Circuit", in Proceedings of the 26th IEEE VLSI Test Symposium, pp.249-254, April 2008.

[10] X. Zhu, G. Sun, S. Yong, Z. Zhuang, "A Novel Method with Ps Accuracy for Time Interval Measurement", in Proceedings of the IEEE International Frequency Control Symposium, joint with the 21st European Frequency and Time Forum, pp.848-853, May 29 2007-June 1 2007.

[11] M.A. Abas, G. Russell, and D.J. Kinniment, "Built-in time measurement circuits – a comparative design study", IET Computers & Digital Techniques, pp. 87-97, Volume: 1, Issue: 2, March 2007.

[12] R. Datta, G. Carpenter, K. Nowka, J. A. Abraham, "A Scheme for On-Chip Timing Characterization", in Proceedings of the 24th IEEE VLSI Test Symposium (VTS), 30 April-4 May 2006.

[13] C.-K. Ong, D. Hong, K.-T. Cheng, L.-C. Wang, "A Scalable On-Chip Jitter Extraction Technique", in Proceedings of the 22nd IEEE VLSI Test Symposium (VTS), p. 267-272, 25-29 April 2004.

[14] M.A.Abas, G. Russell, D.J. Kinniment, "Design of Sub-10-Picoseconds On-Chip Time Measurement Circuit", in Proceedings of the IEEE Design, Automation and Test in Europe Conference and Exhibition, pp. 804 – 809, vol. 2, 16-20 Feb. 2004.

IEEE International Symposium on Defect and Fault Tolerance of VLSI Systems

On Optimizing Fault Coverage, Pattern Count, and ATPG Run Time Using A Hybrid Single-Capture Scheme for Testing Scan Designs

Shianling Wu[1,2], Laung-Terng Wang[1], Zhigang Jiang[1], Jiayong Song[1], Boryau Sheu[1], Xiaoqing Wen[2], Michael S. Hsiao[3], James C.-M. Li[4], Jiun-Lang Huang[4], and Ravi Apte[1]

[1] *SynTest Technologies, Inc., 505 S. Pastoria Ave., Suite 101, Sunnyvale, CA 94086, USA*
[2] *Dept. of Computer Science and Electronics, Kyushu Institute of Technology, Japan*
[3] *Dept. of Electrical and Computer Engineering, Virginia Tech, Blacksburg, VA, USA*
[4] *Dept. of Electrical Engineering, National Taiwan University, Taipei, Taiwan*

Abstract

This paper presents a hybrid automatic test pattern generation (ATPG) technique using the staggered single-capture scheme followed by the one-hot single-capture scheme for detecting structural faults, which are neither timing-dependent nor sequence-dependent in a scan design. Structural faults are also called combinational faults or DC faults, such as stuck-at faults and bridging faults. Typically, the one-hot scheme achieves near maximum fault coverage, takes shorter ATPG run time, but produces a large pattern count, whereas the staggered scheme produces smaller pattern count but needs long ATPG run time and may suffer from some fault coverage loss. The proposed hybrid technique is intended to optimize fault coverage with respect to the one-hot scheme by exploring trade-offs between pattern count and ATPG run time of multimillion-gate scan designs. Experimental results show that the proposed hybrid technique can achieve higher fault coverage and up to 4X smaller pattern count than the one-hot scheme.

1. Introduction

Scan design is a *design-for-testability* (DFT) technique in which the storage elements in a sequential circuit are converted into scan cells and these scan cells are then stitched together to form scan chains during scan testing [1-4]. By reconfiguring all storage elements into scan cells, the complexity of *automatic test pattern generation* (ATPG) for sequential circuits is transformed into manageable ATPG for combinational circuits. Since the late 1990s, scan design has become the most widely used DFT technique.

In recent years, with shrinking device geometry due to advances in design and manufacturing technology, circuits containing millions or tens of millions of logic gates and tens of clock domains are common. One key challenge in ATPG is how to generate a minimum set of test patterns that achieves the highest fault coverage with the shortest ATPG run time.

ATPG effectiveness is highly dependent upon the capture-clocking scheme used. Traditionally, **one-hot clocking** is used to sequentially test every clock domain one by one. This scheme has the luxury of testing each clock domain by deactivating test clocks to other clock domains; however, it often only saves ATPG run time but generates many more test patterns than optimally required resulting in higher test cost.

In this paper, we only consider **structural faults**, such as stuck-at faults, bridging faults, and I_{DDQ} faults. We first discuss two other single-capture clocking schemes, namely, **simultaneous clocking** and **staggered clocking** [5-7], which can be used to remedy the problems encountered in one-hot clocking. We show that using the simultaneous clocking scheme alone can result in the smallest pattern count but may lead to significant fault

1550-5774/08 $25.00 © 2008 IEEE
DOI 10.1109/DFT.2008.29

coverage loss. The staggered technique has shown to be effective in reducing pattern count [8-9] and achieving better N-detect fault coverage [10] but its ATPG time could be much longer. Lastly, we demonstrate that by using a hybrid scheme, which combines staggered clocking and one-hot clocking, one can optimize overall fault coverage, pattern count and ATPG time.

Most importantly, the proposed hybrid scheme (1) always achieves the same or higher fault coverage than the one-hot scheme alone, and (2) allows users to make a trade-off between pattern count and ATPG run time, without sacrificing the fault coverage of the design, in situations where staggered ATPG might take too long to finish. This hybrid scheme also addresses issues raised in [9] that it is uncertain whether the mixed clock-domain (one-hot clocking) and clock-concatenation (staggered clocking) approach could reach the one-hot fault coverage and whether it could complete the mixed ATPG run reaching the highest possible fault coverage within a specified run time for large designs.

This paper is organized as follows: Section 2 discusses various test timing control diagrams for detecting structural faults. Section 3 proposes the hybrid single-capture schemes. Section 4 shows results on two industrial designs, and Section 5 concludes.

2. Test Timing Control

An **intra-clock-domain structural fault** resides in one clock domain and gets detected within the same clock domain. An **inter-clock-domain structural fault** resides within a crossing clock domain and gets detected at the receiving clock domain.

Single-capture is a slow-speed test technique in which only one capture-clock pulse is applied to each clock domain. It is the simplest technique for testing all intra-clock-domain and inter-clock-domain structural faults. There are three capture-clocking approaches that can be used to implement the technique: (1) one-hot single-capture, (2) simultaneous single-capture, and (3) staggered single-capture. One major benefit of using these single-capture clocking schemes is that only a single, slow-speed *global scan enable* signal *GSE* is needed to drive each clock domain. This greatly simplifies physical implementation.

2.1. One-Hot Single-Capture

Using the **one-hot single-capture** approach, a capture pulse is applied to only one clock domain during each capture window, while all other test clocks are held inactive. An example timing diagram is shown in Figure 1. In the figure, since only one capture pulse (*C1* or *C2*) is applied during each capture window, this scheme can only test intra-clock-domain and inter-clock-domain structural faults. The main advantage of this approach is that the designer does not have to worry about clock skews between the clock domains since each clock domain is tested independently. The only requirement is that delays $d1$ and $d2$ be properly adjusted. Hence, this approach can be used for detecting structural faults. A major drawback is longer test time since all clock domains have to be tested one at a time.

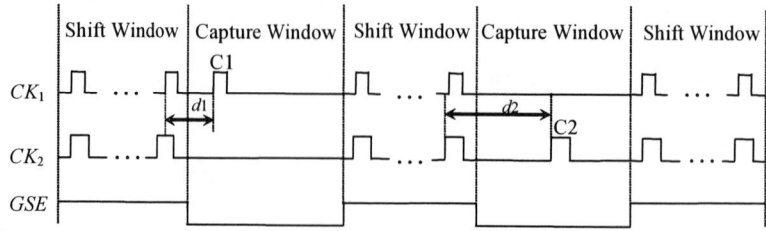

Figure 1. One-hot single-capture

2.2. Simultaneous Single-Capture

The long test time problem of one-hot single-capture can be resolved by using the simultaneous single-capture scheme illustrated in Figure 2. The **simultaneous single-capture** scheme allows testing to be performed on all clock domains in parallel. This scheme is quite helpful when signals in clock domains do not interact with each other. For clock domains where data may propagate from one clock domain to the other, the values of source scan cells in the originating clock domains will have to be forced to unknown values (X's) during ATPG in order to avoid pattern mismatches.

The major advantage of this approach is that all intra-clock-domain structural faults can be tested simultaneously thus yielding much shorter ATPG time and pattern count. However, it exposes the design to two major drawbacks which are not present in one-hot single-capture: (1) the forced X's on all source scan cells could cause significant fault coverage loss and (2) test compression may become very difficult with the need to mask off all X's captured when using simultaneous clocking.

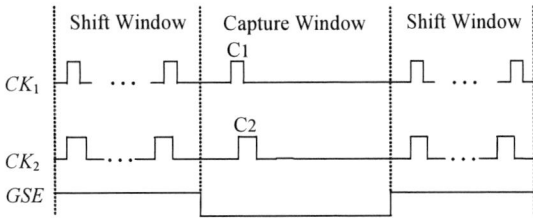

Figure 2. Simultaneous single-capture

2.3. Staggered Single-Capture

The fault coverage loss problem of simultaneous single-capture can be remedied using **staggered single-capture**. A test timing control example is shown in Figure 3. In this figure, capture pulses $C1$ and $C2$ are applied in a sequential or staggered order in the capture window to test all intra-clock-domain and inter-clock-domain structural faults in the two clock domains. If the two clock domains are synchronous to each other, then adjusting $d2$ will allow us to detect inter-clock-domain delay faults at-speed. In addition, since $d1$ and $d3$ can be adjusted to be as long as desired, a single, slow-speed global scan enable signal GSE can be used. This significantly simplifies physical implementation for designs with multiple clock domains. However, there may be some structural fault coverage loss among clock domains if an ordered sequence of capture clocks is used across all capture cycles. This fault coverage loss is mostly related to sequentially redundant faults that can only be detected when one-hot clocking is employed.

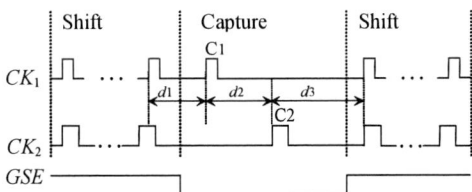

Figure 3. Staggered single-capture

3. Hybrid ATPG Techniques

This section describes two hybrid capture-clocking schemes that were patented in [11] and [12]. The proposed hybrid scheme combines two single-capture approaches, based on the result of clock grouping that identifies independent or non-interacting clock domains. The corresponding ATPG flow will also be described.

3.1. Clock Grouping

An effective technique for reducing ATPG time of a scan design is to first identify all clock domains that do not interact with each other. This is achieved by conducting **clock grouping** that analyzes all data paths in the circuit in order to identify all independent or non-interacting clocks. These clocks are then grouped together and applied simultaneously. If a data path originates at a clock domain and terminates at another clock domain, the two clocks controlling the data path need to be placed in different clock groups.

Independent clock groups identified by clock grouping can then be used for capture-clocking using the three basic single-capture schemes described in the previous section. Since all of the above-mentioned clocking schemes may result in fault coverage loss, large pattern count, and/or long ATPG time, we describe two hybrid schemes in the following for users to make a trade-off between pattern count and ATPG time while maintaining the same fault coverage achievable by the one-hot clocking scheme.

3.2 Simultaneous-Followed-By-One-Hot ATPG

One hybrid ATPG approach is to apply the simultaneous single-capture clocking scheme for all clock domains in the first phase and the one-hot single-capture clocking scheme in the second phase. In the first phase, all clocks are treated as a single clock. In practical applications, data can propagate from one clock domain to another clock domain in a predetermined clocking order. To avoid clock skew among interacting clock domains, one solution is to force all fanout branches of each originating flip-flop in an originating clock domain to unknown (X) values. Since all clocks are treated as a single clock, there is no need to perform clock grouping when the simultaneous scheme is employed. In the second phase, all remaining faults that are not detected in the first phase are targeted for one-hot ATPG. These remaining faults must include all faults from the first phase that are accidentally marked as untestable or undetected due to the existence of X's.

Since one-hot clocking is used in the second phase, this hybrid approach can achieve the same fault coverage as one-hot clocking alone but the hybrid approach often can generate fewer test patterns with shorter overall ATPG time when compared with the one-hot only approach. In addition, although clock grouping is not required in the first phase, performing one-hot ATPG based on clock groups rather than individual clocks, in general, can reap additional benefits of smaller pattern count and shorter ATPG time.

3.3. Staggered-Followed-By-One-Hot ATPG

Although above-mentioned hybrid approaches can lead to better results than using the one-hot approach alone, it may suffer from the same X-masking issue ATPG compression faces. If test response compaction logic in the circuit cannot effectively tolerate massive X's created by simultaneous clocking, a significant fault coverage loss may happen.

In order to solve this problem, another hybrid approach is to apply the staggered single-capture scheme in the first phase and the one-hot single-capture scheme in the second phase. In the first phase, all clock groups are specified in a predetermined, sequential or staggered order. ATPG is then conducted based on the staggered order. To reduce ATPG time, circuit model expansion based on the ordered sequences of clock groups is performed on the scan design during preprocessing. Since the staggered clocking scheme specifically deploys physically disjoint capture clock pulses from different clock domains (in our case, different clock groups), there is no need to insert X's at the fanout braches of each originating flip-flops in any originating clock domain. Therefore, the staggered approach will not create unnecessary X's that complicate test response compaction in a compression design. Since staggered clocking can cause the ATPG program to mark hard-detected faults as untestable or undetected due to the ordered sequence of clock groups, the second phase running one-hot ATPG is required to re-target those faults.

3.4 ATPG Flow

ATPG flow of the staggered-followed-by-one-hot scheme is shown in Figure 4 and summarized as follows: Starting with a scan-stitched netlist, clock grouping analysis is first conducted. With available clock grouping and capture order information, the ATPG program transforms the scan-stitched netlist into a combinational circuit ready for test pattern generation. Remodeling, learning, implication and other pre-processing steps are then conducted. During ATPG, switch criteria are closely monitored for the program to switch over from staggered ATPG to one-hot ATPG.

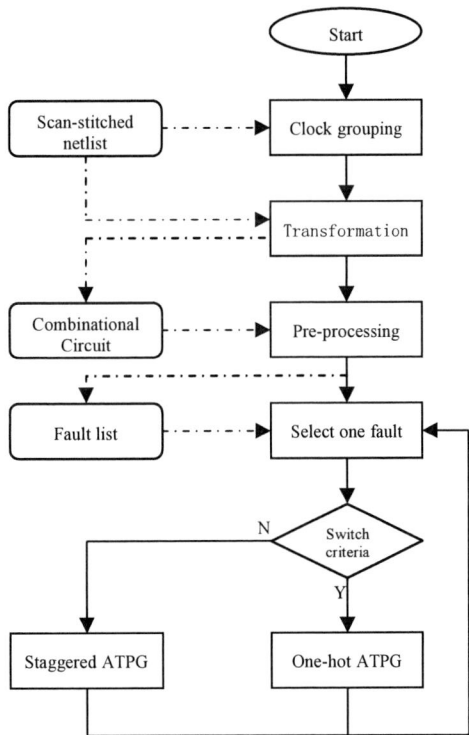

Figure 4. Hybrid ATPG flow chart

There are different ways to set the switch-over criteria such as certain increments in fault coverage over pattern count or a pre-determined amount of run time. For this paper, we define the switch-over criterion, called **switch ratio**, as the percentage of faults that have been processed during the entire ATPG process over total faults.

4. Experimental Results

The proposed hybrid staggered-followed-by-one-hot single-capture scheme has been applied to many industrial designs. We present the results on two multimillion-gate designs to illustrate the effectiveness of the proposed scheme.

4.1 Design Statistics

Table 1 summarizes the statistics of two industrial designs A and B. We developed a program to identify all independent clock groups. Because clock domains in the same clock group do not interact with one other, they can be enabled simultaneously during capture without adverse effect from clock skews.

Table 1. Design statistics

	Design A	Design B
# of Gates	1.2M	4.9M
# of Faults	3,954,048	11,395,890
# of Flip-Flops	105K	327K
# of Clock Domains	38	10
# of Clock Groups	7	8

4.2 Results

Stand-alone one-hot single-capture and staggered single-capture clocking schemes were first applied independently to both designs A and B. The platform used was a 2.5-GHz 64-bit PC running the Linux operating system. Table 2 summarizes the experimental results. Single stuck-at fault model was assumed in ATPG. The one-hot results were obtained by running one clock at a time, while the staggered results were obtained by running all clock groups in a given staggered order.

Table 2. Experimental Results on Designs A and B

	Design A One-Hot	Design A Staggered	Design B One-Hot	Design B Staggered
Hard-Detected Faults	3,650,163	3,639,949	10,432,447	10,442,183
Fault Coverage (%)	92.58%	92.32%	92.42%	92.51%
Pattern Count (one-hot / staggered)	6,339	1,464 (4.33X)	2,591	1,483 (1.75X)
ATPG Run Time	1:17:45	0:57:46	2:49:11	6:31:59
Memory Usage	1.6GB	2.0GB	6.0GB	7.2GB

The results show that one-hot clocking demands less memory than staggered clocking. At the cost of larger pattern count, one-hot clocking should have higher fault coverage and shorter ATPG run time than staggered clocking; however, higher fault coverage and shorter ATPG run were observed for staggered clocking for Designs B and A, respectively. The reason for the fault coverage increase may be because faults aborted in the one-hot clocking scheme (due to backtrack limit) got detected by patterns targeting other faults using the staggered clocking scheme. The reason for reduced run time may be due to staggered clocking proving many sequentially redundant faults, detectable only by one-hot clocking.

It is also interesting to note that, on average, ATPG time for stuck-at faults spent on Designs A and B using the staggered clocking scheme is quite linear, approximately one-hour per one million gates. This implies that ATPG on a 20-million gate design may take less than one day.

The results using the proposed hybrid clocking scheme on Designs A and B are listed in Tables 3 and 4, respectively. In our experiments, staggered clocking is switched to one-hot clocking after a certain percentage of faults have been processed. The switch ratio is shown in the first column of each table. In the other four columns, hard-detected faults, fault coverage, pattern count, and ATPG time are shown as $X + Y (Z)$. The first number, X, is the result using staggered clocking; the second number, Y, is from using one-hot clocking; and the third number, Z, is the sum of X and Y. The results on both tables show that as the switch ratio increases, the pattern count decreases and fault coverage increases. ATPG run time, generally speaking, increases except in the case of 95% switch ratio.

Table 3 further shows the effectiveness of the hybrid clocking scheme. First, the hybrid fault coverage is always higher than one-hot clocking alone. Second, the pattern count from using one-hot clocking is 3.70 (= 6339 / 1711) times the pattern count using the hybrid approach when the switch ratio is 100%. Lastly, as shown in the last row of the table, the staggered clocking can already achieve 83.99% fault coverage with merely 85 patterns; this is crucial when ATE memory size is a concern.

Table 3. Hybrid Clocking Results on Design A

Switch Ratio	Hard-Detected Faults	Fault Coverage (%)	Pattern Count	ATPG Time
100%	3,639,949 + 22,209 (3,662,158)	92.32% + 0.56% (92.88%)	1,464 + 247 (1,711) (3.70X)	0:57:46 + 0:12:42 (1:10:28)
99%	3,626,520 + 34,658 (3,661,178)	91.98% + 0.88% (92.86%)	848 + 1,028 (1,876) (3.38X)	0:36:46 + 0:19:45 (0:56:31)
95%	3,483,666 + 171,026 (3,654,692)	88.35% + 4.34% (92.69%)	201 + 2,939 (3,140) (2.02X)	0:08:38 + 0:36:55 (0:45:33)
90%	3,311,452 + 342,690 (3,654,142)	83.99% + 8.69% (92.68%)	85 + 3,533 (3,618) (1.75X)	0:05:45 + 0:41:08 (0:46:53)

Table 4 also shows the effectiveness of the hybrid clocking scheme on Design B. At each switch ratio, the hybrid scheme generates fewer test patterns and requires longer ATPG run time than the one-hot clocking scheme given in Table 2. Similar to Design A, the reduction in pattern count seems to fall within 20% when the switch ratio is 90% or lower. For

Design B, the fault coverage of the hybrid clocking scheme at each switch ratio is also found to be slightly higher than that of one-hot clocking scheme. These results are consistent with Design A.

Table 4. Hybrid Clocking Results on Design B

Switch Ratio	Hard-Detected Faults	Fault Coverage (%)	Pattern Count	ATPG Time
100%	10,442,183 + 651 (10,442,834)	92.51% + 0% (92.51%)	1,483 + 43 (1,526) (1.69X)	6:31:59 + 1:03:22 (7:35:21)
95%	10,119,541 + 318,037 (10,437,578)	89.65% + 2.82% (92.47%)	381 + 1,488 (1,869) (1.38X)	0:58:21 + 2:07:39 (3:06:00)
90%	9,572,488 + 863,072 (10,435,560)	84.80% + 7.65% (92.45%)	158 + 2,019 (2,177) (1.19X)	0:34:22 + 2:16:35 (2:50:57)

For Design A, Table 5 shows the difference between using one-hot and staggered clocking with 38 individual clocks *versus* 7 clock groups, respectively. The results indicate that performing ATPG based on clock grouping leads to much smaller pattern count and run time, but not necessarily higher fault coverage, in both one-hot and staggered cases.

Table 5. Experimental Results on Design A

	One-Hot 38 clocks	One-Hot 7 clock groups	Staggered 38 clocks	Staggered 7 clock groups
Hard-Detected Faults	3,650,163	3,653,526	3,641,227	3,639,949
Fault Coverage (%)	92.58%	92.66%	92.35%	92.32%
Pattern Count (38 clocks / 7 clocks)	6,339	3,982 (1.59X)	1,483	1,464 (1.01X)
ATPG Run Time	1:17:45	0:47:32	2:37:59	0:57:46

In summary, the experimental results on multimillion-gate designs show that (1) clock grouping effectively reduces pattern count and (2) hybrid clocking, on average, can result in 1.2X to 4X reduction in pattern count and slightly higher fault coverage than one-hot clocking alone. One-hot clocking, however, does have the benefit of shorter ATPG run time. Therefore, we recommend using one-hot clocking at an early development stage to predict the fault coverage and pattern count of a design. By the time the design is being taped out, one should apply hybrid clocking to reduce pattern count and improve the circuit's fault coverage to near maximum.

5. Conclusions

Scan designs containing multimillion logic gates are common today. When a scan design contains asynchronous clock domains, using the conventional one-hot and simultaneous single-capture schemes for capture clocking poses serious challenges. The drawback of the one-hot scheme is large pattern count, which leads to long test time and high test cost. The simultaneous scheme could reduce pattern count; however, the fault coverage loss and the negative impact on test compression may render the approach unacceptable.

This paper presents a hybrid *automatic test pattern generation* (ATPG) technique that combines the staggered single-capture and one-hot single-capture schemes to optimize fault coverage, pattern count, and ATPG run time for large scan designs. Experimental results on two industrial designs have demonstrated the effectiveness of the proposed scheme. When employed alone, the staggered scheme can drastically reduce pattern count at the expense of long ATPG time. In the hybrid scheme, ATPG switches from the staggered scheme to the one-hot scheme after a specified percentage of faults have been processed – this reduces ATPG run time at the cost of slight increase in pattern count. Since the one-hot scheme is applied after the staggered scheme, the hybrid scheme theoretically will achieve the same fault coverage as the one-hot scheme alone. In practice, however, we have observed higher fault coverage in all cases.

This paper only discusses capture-clocking schemes for detecting structural faults. While good results have been obtained, further investigation on how to automatically select the switch ratio remains to be explored. We plan to extend the hybrid scheme for detecting delay faults in scan designs. Through the experiments, we hope a near optimal solution among the three key ATPG parameters – fault coverage, pattern count and ATPG run time – could be empirically established.

6. References

[1]   M. L. Bushnell and V. D. Agrawal, *Essentials of Electronic Testing for Digital, Memory & Mixed-Signal VLSI Circuits*, Springer, Boston, 2000.

[2]   N. K. Jha and S. K. Gupta, *Testing of Digital Systems*, Cambridge University Press, London, 2003.

[3]   L.-T. Wang, C.-W. Wu, and X. Wen, Eds., *VLSI Test Principles and Architectures: Design for Testability*, Morgan Kaufmann, San Francisco, 2006.

[4]   L.-T. Wang, C. E. Stroud, and N. A. Touba, Eds., *System-on-Chip Test Architectures: Nanometer Design for Testability*, Morgan Kaufmann, San Francisco, 2007.

[5]   R. Apte, "Cutting SoC Test Costs with the Right Kind of Scan," *EDAVision Magazine*, February Issue, 2002 http://www.edavision.com/vision.php?article=200202/tool.html.

[6]   L.-T. Wang, P.-C. Hsu, and X. Wen, "Multiple-Capture DFT System for Detecting or Locating Crossing Clock-Domain Faults During Scan-Test," United States Patent No. 7,260,756, August 21, 2007.

[7]   SynTest Technologies, "ATPG User's Manuals – TurboScan[TM] and VirtualScan[TM]," SynTest Technologies, Inc., Sunnyvale, CA, 2008.

[8]   V. Jain and J. Waicukauski, "Scan Test Data Volume Reduction in Multi-Clocked Designs with Safe Capture Technique," in *Proc. IEEE Int. Test Conf.*, pp. 148-153, October 2002.

[9]   X. Lin and R. Thompson, "Test Generation for Designs with Multiple Clocks," in *Proc. ACM/IEEE Design Automation Conf.*, pp. 662-667, June 2003.

[10]   G. Bhargava, D. Meehl, and J. Sage, "Achieving Serendipitous N-detect Mark-Offs in Multi-Capture-Clock Scan Patterns," in *Proc. IEEE Int. Test Conf.*, Paper 30.2, October 2007.

[11]   L.-T. Wang, K. S. Abdel-Hafez, X. Wen, B. Sheu, and S.-M. Wang, "Smart Capture for ATPG (Automatic Test Pattern Generation) and Fault Simulation of Scan-Based Integrated Circuits," United States Patent No. 7,124,342, October 17, 2006.

[12]   K. S. Abdel-Hafez, L.-T. Wang, B. Sheu, Z. Wang, and Z. Jiang, "Method for Performing ATPG and Fault Simulation in a Scan-Based Integrated Circuit," United States Patent No. 7,210,082, April 24, 2007.

Analyzing the Impact of Fault-tolerant BIST for VLSI Design

W. Robert Daasch   Saurabh Jain   David Armbrust

Integrated Circuits Design and Test Laboratory
Department of Electrical and Computer Engineering
Portland State University

daasch@ece.pdx.edu   saurabhj@cecs.pdx.edu   lelanda@centurytel.net

Abstract

1. Introduction

This paper continues the exploration of robust DfT design through fault-tolerant methods [2]. Quadded gate inter-leaved node logic design has good fault tolerance properties but has not received much attention as the TMR-type alternatives. Quadded DfT design overheads are evaluated and tradeoffs for alternative fault tolerant synthesis options are discussed. Investigations of this kind are needed for the following reasons:

- Going forward validating device components (i.e. transistors) will be a mammoth task increasing test costs and test times. Fault tolerance is a plausible solution to control test costs.

- In the nano-scale environment it is harder to separate systematic, parametric and random fallout and DfT is an accepted norm for fast design turnaround time. When it works correctly, DfT offers distinct advantages for test, design debug and diagnosis. Increasing design complexity means DfT circuitry is also larger, more complex and more susceptible to faults rendering functionally correct die as a failure. Requiring still more area, fault-tolerant DfT amortizes test and debug circuitry by reducing DfT only fails.

- New materials and nano-scale feature sizes introduce different failure mechanisms which will have a considerable impact on the overall yield and reliability. Researchers

are proposing to mitigate these effects by using fault-tolerant designs [10, 2].

Historically, fault-tolerant and error masking techniques are used in fields such as aeronautical or bio-medical with mission critical components and strict reliability and fault-free system requirements. Pioneering work done by Von Neumann in 1950's provided the impetus for research in fault-tolerant circuit design [7]. His work suggested adding (redundant) logic to the circuitry. Assuming the logic failure rate is reasonable, a highly reliable design can be obtained [3].

Quadded logic, sometimes referred to as interwoven logic, is derived from Von Neumann's NAND multiplexing concept. Quadded logic is a gate level approach for designing circuits resilient to faults. J.G Tryon's initial work on quadded logic was in 1960's [12]. Jensen acknowledged any universal gate design can be quadded [5]. Although quadded • • • only design yields a highly fault-tolerant result it has a huge area overhead. Other fault-tolerant techniques replace interwoven logic with system redundancies, such as triple modular (TMR) [4] and N-tuple modular redundancy (NMR) [1].

Area and power increases have been studied for many fault-tolerant alternatives but not quadded logic designs. In this work, fault tolerance is limited to critical DfT subsystems. This choice was made because fully functional DfT subsystems are essential for yield learning and provide a well controlled system to learn about trade-offs involved in implementing quadded fault-tolerant design [6].

Fault tolerance is sound on its fundamentals yet answers to several key questions remain elusive. What are its costs compared to its benefits? At this point every fault-tolerant scheme needs careful study to ensure all viable alternatives are available to the design community. When is the right time for fault tolerance to be converted to practice? This paper takes a conservative step towards fault-tolerant design by synthesizing the critical DfT subsystem as fault-tolerant. Fault tolerant DfT offers a simple prototype system to gauge the consequences of incorporating fault tolerance on power, area, delay and analyzing the underlying trade-offs.

2. Quadding the design for fault tolerance

A quadded design is largely self-repairing for any single-stuck-at or single-stuck-open fault. A quadded equivalent of a 2-input logic gate requires quadrupling the gates and wires, and doubling the number of gate inputs. By interweaving duplicate signals, faults in quadded logic are masked within one or two logic levels. The quadded fault masking mechanism prevents erroneous signals from propagating to the design outputs. Quadding is possibly unique in its ability to decouple multiple-faults present in the design into independent single-stuck-at faults [5].

The obvious primary concern is the increase in area and interconnects. In the literature, the original design used a single universal gate and quadding it was extremely effective in achieving highly fault-tolerant design [5]. Single universal gate designs are known to be inefficient in gate count and quadding the design increases congestion in already congested routing channels. This paper introduces a quadding approach that improves upon the traditional approach. Instead of using a single universal gate throughout the design, the design is quadded using a larger gate library (e.g. buffers• • • • • • • • • • and • • • •) and sequential elements are modified to efficiently support quadding. Interconnect rules and flip-flop modifications are discussed next. The modified quad circuit has fault tolerance similar

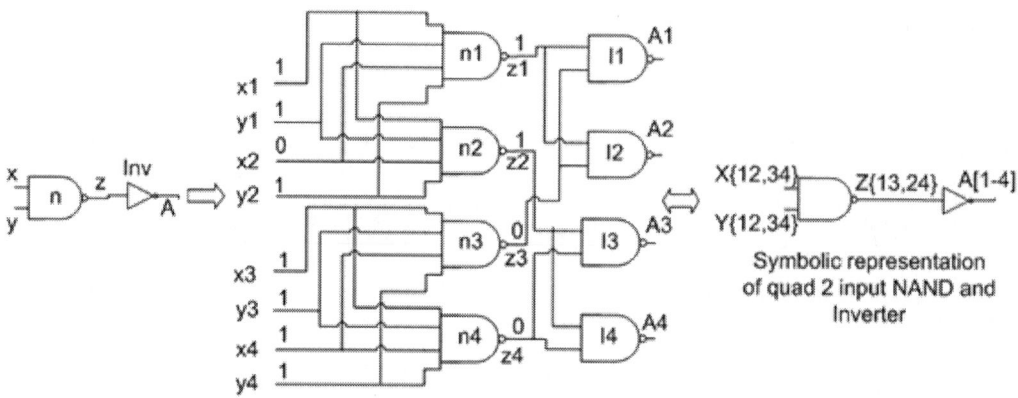

Figure 1. Quadded logic representation

to the universal gate approach with lower area (gate) overhead and reduced interconnect congestion. Other area and interconnection improvements may be possible by expanding quad specific cell libraries.

To implement a quadded structure each gate is cloned four times and each cloned gate has two copies of the original input signals. For example, Figure 1 shows a • • • • and an inverter combination and the details of its quadded equivalent. The gate • • has inputs • • • • • (separate but identical signals) and • • • • • (also separate/identical) and • • is one copy of the four outputs. Gates • • • • • and • • are similar with different inter-woven inputs. The far right shows a compact notation that summarizes the paired signal connections to the four logic gates. Figure 1 is an example of the input pattern •{• • • • •} and •{• • • • •}. Signals • • • • •) and • • • • • • are the inputs in the top two gates • • • • • hence the first part of the notation (e.g. •{• • • • •}) and • • • • • • • • • • • • are inputs for bottom two gates • • • • • (e.g. •{• • • • •}). The intermediate output, labeled • • • • •, etc., is shown with the different connectivity to gates • • • • • • • • • •, denoted as •{• • • • •}. The output • • • is a bus of four identical signals • • • • • • • • • • connecting to other downstream quadded gates in one of the three possible patterns (i.e. • {• • • • •}• • {• • • • •} • • • {• • • • •}).

To gain quadded logic fault repair, the basic units are the multiple gate copies (redundancies) and interweaving the signal inputs. Quadded logic error correcting capabilities is explained best by using the concept of controlling input logic values and their corresponding output values. In Figure 1, a single-stuck-at-1 • • • • • • at • • is masked at the output by the correct and controlling logic value • • • at the duplicate input • •. These • • • • • • • • • • • • • • • stuck-ats are corrected in a single logic stage and called • • • • • • • • • • • • • • • • • [1]. As a second example, the • • • • • on • • creates two incorrect logical • • • outputs at • • and • •. To fix these errors an additional logic stage is required. A fault that • • • • • • • • a gate's output is a • • • • • • • • • • • • • [1]. Error propagation rules determine connectivity (e.g. {• • • • •} • • {• • • • •}) and subsequent stage fault repair capability. For example, the • • • • • • errors from the • • • • gate in Figure 1 are fixed at • • • output by the two correct and controlling values on • •, and • •. Either • • • or • • • • with {• • • • •} connectivity will correct the error propagated from • • • • • • •. Shown in Figure 2, the error will propagate through • • • • • • • {• • • • •} connections anytime these gates follow the • • • •. When critical errors occur erroneous signals propagate on two out of the four signal channels and correct signals propagate on the remaining two channels. The effect of the fault propagates to a logic gate where the signal is not a non-controlling value. At this gate the propagated error

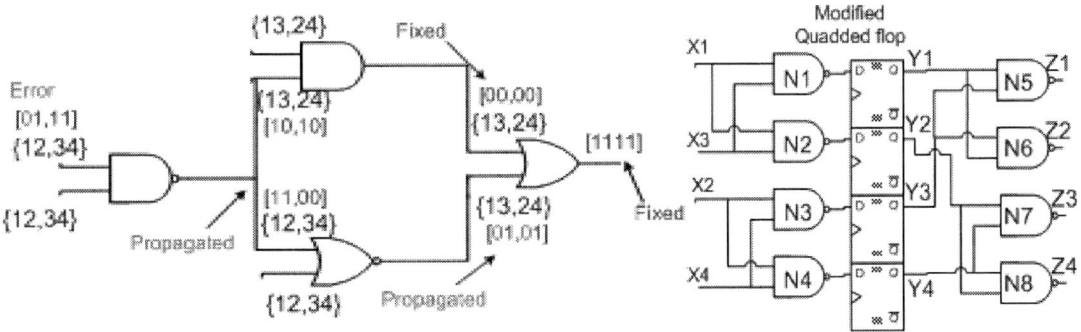

Figure 2. Error propagation and correction　　**Figure 3. Modified quadded flop**

Table 1. Connectivity and fault tolerance between 2-quad logic stages

Stage 1	OR	NOR	NAND	AND
NAND/OR	{12,34}/prop	{12,34}/prop	{13,24}/corr	{13,24}/corr
AND/NOR	{13,24}/corr	{13,24}/corr	{12,34}/prop	{12,34}/prop

is repaired by interweaving signals from a {•• •••} pattern to the {•• •••} pattern or vice versa. This ensures each pair of signal inputs has one correct and controlling value to repair the non-controlling propagated error.

In the traditional approach single-stuck-ats never propagate beyond two logic stages and multiple faults are generally not fatal for a quadded design [5]. Probabilistically additional critical-error propagation reduces the fault tolerance of the quadded logic. For every gate the fault propagates the critical-area increases for the appearance of a second fault. If the second fault occurs on the fault free signal paths the opportunity to correct the original fault is lost. It will be shown this critical-area to multiple faults is a design variable and adjustable at synthesis. Figure 1 and Figure 2 examples can be generalized to a set of simple synthesis rules which are summarized in Table 1. The left-most column represents the first gate and the remaining columns specify gates in second stage with their connectivity and tolerance to propagated faults.

Quadding a latch (or flip-flop) exploits its two consecutive • • • • gates. Buffer insertion into the flip-flop corrects any single fault propagating into a register cell. An example is shown in Figure 3. An internal fault in the flip-flop module is corrected in the flip-flop or latch fan-out. By registering all the output ports of the quadded design no single faults from within propagate to the BIST outputs. The only single faults not corrected by quadding are at the interface between the BIST and the inputs to the circuit under test. Other approaches to reduce area overhead at the expense of fault tolerance using flip-flops and inverters are possible. They have not been discussed for the sake of brevity.

3. Quadded design description

ISCAS-89 sequential benchmarks were used for the analysis of area, power and timing with and without quadded DfT. The ISCAS benchmarks are not large by current standards but their history provides a valuable basis for first comparison. Commercial tools were used throughout the course of this work. All physical designs use the standard cells from Virginia Tech, 240•• standard cell library based on the MOSIS scalable design rules [8]. There may be a concern about the usefulness of the results based on this library. The key is which results are vulnerable to the choice of library. Leakage estimates are the least reliable and are not included in this study. It is reasonable to assume area, dynamic power and delay results scale with technology [9]. Datasheets across four process nodes of industry cell libraries confirms simple scaling rules apply. Physical synthesis from Verilog netlists uses a LEF/DEF abstract model of the cell. Commercial tool placement and routing solutions would differ marginally with cells from 250•• or 45•• . Finally, published results from similar studies used a single scaling parameter to estimate technology node comparisons [11]. No more or less support for scaling results is contemplated in this study.

Two DfT components were added to each benchmark. First, all flops were stitched into multiple scan chains. Second, BIST (Built-In-Self-Test) was added to each design through the scan-chains. The BIST is a simplified STUMPS architecture (Self Test Using MISR Parallel Shift-register). The BIST design is a simple controller, a maximal length LFSR and the MISR (Multiple-Input-Signature-Register). The BIST did not use a phase shifter but a larger LFSR was used to generate test patterns with >••• fault coverage for all designs. The original BIST circuit synthesis used more than a single universal gate and two input gates were used throughout to limit fan-in effects on performance and dynamic power.

Quadded BIST (QBIST) circuits were created using a single pass synthesis tool based on lookup tables similar to Table 1. The following steps outline the generation of a quadded circuit using a gate mapped netlist:

1. Replace inverters by quadded • • • • equivalent design.

2. Insert single fault correcting buffers (similar to Figure 3) into flip-flops and latches.

3. Treat muxes as wires using connectivity rules governed by the logic preceding it. This approach saves area but as described earlier allows some faults to propagate.

4. Convert • • • 's to its equivalent • • • • form and apply • • • • quadding rules. • • • conversion is required because the symmetry of XOR logic lacks controlling and non-controlling values. Conversion of single • • • to • • • • equivalent means it shares fault correction properties of • • • • .

4. Results

For each ISCAS sequential benchmark, four designs were synthesized and simulated. For each design fault-coverage was maintained at 90% or more. This ensures that the BIST engine was not purposely made small at the expense of fault coverage. The baseline design represents the benchmarks original Verilog structural description. The baseline design used non-scan master-slave flip-flops in the registers. The next design (scan) is the scan insertion step with all flip-flops replaced with mux-scan flip-flops. This step provides a frame of reference for area increases from standard DfT insertion. The third design expanded the scan design by including the STUMPS-like BIST structure. This step measures DfT design impact from scan to BIST. The fourth design quadded the BIST block using the single-pass synthesis outlined earlier. This is the final step of the study and provides a basis for comparison for incorporating the quad design elements of gate redundancy and signal interleaving.

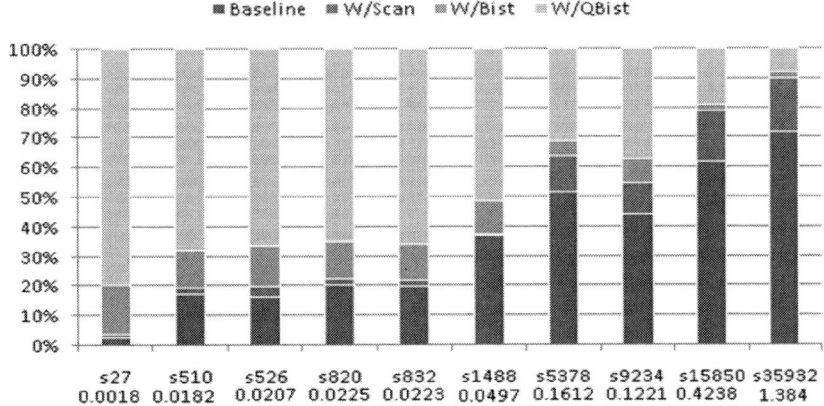

Figure 4. Area overhead of each design modification after QBIST insertion

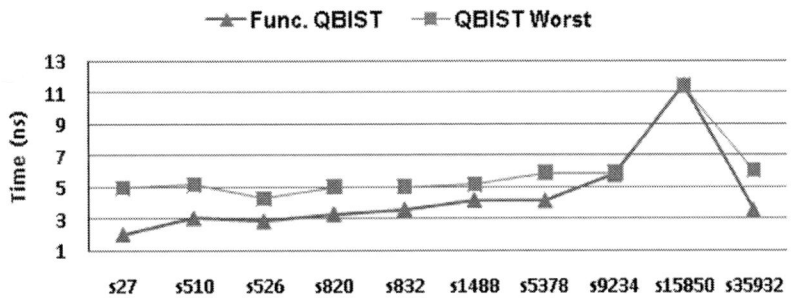

Figure 5. Critical path timing information for the circuits

4.1 Area comparison

The columns in Figure 4 display the baseline design and the area changes for each DfT step. Each column is normalized to the final QBIST area. Below each benchmark is the absolute area (in mm^2) of its baseline design. It is no surprise the QBIST area overhead for the debug benchmark (i.e. s27) is enormous, almost 80%. The larger benchmarks display the expected trend of approximately 10% for scan insertion and a decreasing fraction of area used by BIST and QBIST. For the largest post-scan insertion, s35932, the additional area for QBIST decreases to 10% of the total design area. Even with the large increases in gate and wire count, the QBIST area overhead trend is surprisingly similar to the overhead from scan insertion. The trend decreases with circuit size and complexity. Similar results were observed for total wire length in the four DfT designs. With technology scaling, the wire lengths scale linearly and the decreasing trend for overall wire delays and length is expected to continue [9].

4.2 Timing and power analysis

Power and timing performance was evaluated by running transistor level analog simulations and static timing analysis, respectively. Data from these experiments is shown in Figure 5 and Table 2. Scan insertion slowed the baseline designs by 15% and a small additional degradation was observed from BIST. The BIST slowing was caused by BIST structures loading some critical paths. In some cases, QBIST testing of design performance

Table 2. ISCAS benchmark power consumption baseline, scan, BIST, QBIST

ISCAS Benchmark	Power(mW)			
	Baseline	W/Scan	W/BIST	W/QBIST
s27	0.865	0.834	4.051	8.895
s510	3.077	3.495	6.751	12.643
s526	3.473	5.652	9.614	18.035
s820	2.243	3.328	6.280	12.671
s832	1.947	3.053	6.090	12.164
s1488	5.615	8.173	10.032	15.593

Table 3. Probability of multiple errors for different logic depths

Logic depth	Fault propagation area (mm^2)	Prob. of multiple faults	BIST area (mm^2)	Prob. of multi fail in QBIST	Prob. of multi fail in TMR
5	$2.07E-3$	$4.19E-12$	0.0035	$3.95E-10$	$7.20E-09$
9	$3.72E-3$	$1.36E-11$	4.00	$5.84E-08$	$9.32E-05$
12	$4.96E-3$	$2.41E-11$	7.00	$1.36E-07$	$2.83E-04$

was significantly slower than functional performance testing. The blue curve in Figure 5 shows post-QBIST functional path critical timing while red curve shows the best-case timing for the design while executing QBIST. While s15850 was an exception, the high fan-out within the QBIST logic results in critical path shifting from a functional logic path to a path in the quadded DfT. This suggests quadding can impact BIST at-speed testing. If quadding is restricted to DfT and is not used in a functional path the decreased performance is less of a concern. As in the area, QBIST performance losses tend to roll off as the design size grows. One explanation is the slower growth in depth of QBIST controller logic compared to overall controller complexity. QBIST driven performance changes can be expected to be manageable for the designs with larger flop-to-flop logic depth. Table 2 shows scan or BIST dynamic test power simulations for a sample of benchmark designs. The column labeled baseline is the baseline design with no DfT. The dynamic test power consumption grows with increasing area (Figure 4) because the dynamic power is a strong function of capacitive switching (i.e. area). Similar to area, the increases in power from a BIST design to its QBIST design slow with increasing design complexity.

4.3 Fault propagation and multiple faults

In the study the QBIST controller design faults propagate for maximum of 5 logic levels. The synthesis rules control the logic depth between critical-errors. Using ITRS 2007 defect densities ($1400/m^2$), Poisson statistics and an average logic cell area ($94\cdot{}^2$) basic calculations predict a negligible multiple fault probability (2.41E-11) for less than 12 logic stages. Many similar calculations are summarized in Table 3. The table clearly shows quad designs are statistically resilient to almost all single-stuck-at faults. Quad circuits are always susceptible to multiple faults.

In column 5 the QBIST multiple fault error levels are below 10 DPPM for a range of fault propagation logic depths and BIST sizes. QBIST design is assumed to be tolerant to all single-stuck-at faults. Column six of the table shows the multiple fault probability

Table 4. Impact of fault tolerant circuits on yield and die/wafer

Scan area (mm^2)	BIST area (mm^2)	QBIST area (mm^2)	Scan yield	BIST yield	Wafer radius (mm)	Useful area (mm^2)	# BIST die	QBIST lost no YL	QBIST lost with YL
1.74	1.77	1.93	0.998	0.998	80	18086	10193	845	841
140	142.8	151.2	0.822	0.818	150	63585	365	19	9
140	142.8	151.2	0.822	0.818	80	18086	104	5	2
200	204	216	0.756	0.752	150	63585	234	12	2
250	255	270	0.705	0.700	80	18086	50	2	0
300	306	324	0.657	0.652	150	63585	135	6	-2
400	408	432	0.571	0.565	150	63585	88	4	-3

for a TMR BIST design. For this calculation, the area overhead of the restoring organ of the TMR circuit is ignored. The probability of failure from multiple faults for QBIST (PQM) and TMR BIST (PTM) is computed using Poisson statistics. This column compares multiple fault probabilities of QBIST designs to TMR BIST designs. For all areas the QBIST design probability is smaller than TMR. This is attributed to local fault masking capability obtained from signal interweaving and the design level control of the susceptible area in a quadded design. In contrast, a TMR design to repair all faults (single and multiple) the fault must be propagated to the inputs of the restoring (voting) block.

4.4 Yield analysis

Redundancy necessarily increases design area and reduces the number of die per wafer. This reduction in chip count is offset somewhat by the fault-tolerant BIST yield improvements and the potential longer term advantages of yield learning (YL). Most memory designs (DRAM, Flash etc.) have long used redundant components to compensate for the lower yields of densely packed cells. An analogous use of redundancy and fault-tolerant design for random logic has not received the same level of attention. Table 4 shows the per wafer chip yield from adding fault-tolerant BIST. Each row in the table is a different sized die and wafer. All calculations use the ITRS 2007 defect density of 1400 defects/m^2. Column one displays die areas for the scan designs. Columns two and three display areas for BIST and QBIST DfT. The DfT areas are average results for ISCAS benchmark BIST and QBIST of 2% and 12%, respectively. Columns four and five are yield estimates for the scan and BIST designs. Column six sets the wafer radius for calculations. Column seven is the useful wafer area, assumed to be 90% of total wafer area. Column eight records the number of good parts per wafer (i.e. yield) for the BIST design. Column nine is the die lost per wafer due to QBIST additional area requirements.

As shown earlier, QBIST is resilient to almost all single-stuck-at faults and so the yield of QBIST block of the design is assumed to be 100%. This sets overall yield of the QBIST design equal to the baseline design with full scan insertion (W/Scan). The area penalty of fault-tolerant QBIST circuitry is amortized by no chip loss from DfT fallout. This has an indirect benefit of improving the rate of yield learning.

To gauge the benefit from fault-tolerant BIST to yield learning one more parameter is needed, the yield learning factor. For these calculations a conservative yield learning factor of 0.15 is assumed [6]. Even with the low learning rate quadding the design becomes a realistic alternative for designs with area larger than 250mm^2 (note rows where last column is ≤ 0). The die size for viable quadding decreases with increases in either the yield learning rate or the defect densities. Thus it can be seen that the designer can trade-off extra area in fault-tolerant QBIST for improved yield learning and its attendant benefits.

5. Conclusions

This research provides insight into yield and area tradeoffs for quadded logic and more generally fault-tolerant BIST. For more reliable designs or sub-modules the results show fault-tolerance with interleaved, quadded logic can be extended in a sensible manner into other functional blocks. The study can be used as a blueprint for evaluating fault-tolerant tradeoffs in other sub-systems. Defect tolerant DfT can improve yield learning and lower design turnaround time. Low DPPM requirements and faster time to market makes QBIST a plausible and worthwhile option to consider.

The results also show that the negative impact of QBIST on area, wire length, dynamic power and performance trends down with increasing design size and complexity. The overhead can be further reduced if quad customized standard cells are used. Designers can tradeoff the area and performance overheads by leveraging the advantage of having the DfT resistant to a defect level range. Without significantly affecting the final fault-tolerance, the generalized quadding approach reduces area overhead normally associated with a single universal gate design requirement for quad logic. Finally, the quadded design generator can be easily integrated into synthesis and design flows.

References

[1] A.E. Barbour and A.S. Wojcik. "A general constructive approach to fault-tolerant design using redundancy". *Computers, IEEE Transactions on*, 38(1):15–29, Jan 1989.

[2] F. Corno, P. Prinetto, and M. Sonza Reorda. "Self-checking and fault tolerant approaches can help BIST fault coverage: a case study". *European Design and Test Conference, 1996. ED&TC 96. Proceedings*, pages 610–, Mar 1996.

[3] J. Han. "Toward Hardware-Redundant, Fault-Tolerant Logic for Nanoelectronics". *IEEE Design & Test of Computers*, 22(4):328, 2005.

[4] Jie Han, Jianbo Gao, Yan Qi, Pieter Jonker, and Jose A. B. Fortes. "Toward Hardware-Redundant, Fault-Tolerant Logic for Nanoelectronics". *IEEE Des. & Test of Computers*, 22(4):328–339, 2005.

[5] P.A. Jensen. "Quadded Nor Logic. *IEEE Trans. Reliability*, 12(3):22–21, Sept 1963.

[6] Pranab K. Nag, Anne Gattiker, Sichao Wei, R.D. Blanton, and Wojciech Maly. "Modeling the Economics of Testing: A DFT Perspective". *IEEE Design and Test of Computers*, 19(1):29–41, 2002.

[7] J. Von Neumann. "Probabilistic logics and synthesis of reliable organisms from unreliable components". *Automata Studies*, pages 43–98, 1956.

[8] Jos. B. Sulistyo and Dong S. Ha. "A New Characterization Method for Delay and Power Dissipation of Standard Library Cells". *VLSI Design*, 15(3):667–668, Jan 1989.

[9] D. Sylvester and K. Keutzer. "Getting to the bottom of deep submicron". *Computer-Aided Design, 1998. ICCAD 98. Digest of Technical Papers. 1998 IEEE/ACM International Conference on*, pages 203–211, 8-12 Nov 1998.

[10] M. Tehranipoor, Rad, and R.M.P. "Built-In Self-Test and Recovery Procedures for Molecular Electronics-Based Nanofabrics". *Computer-Aided Design of Integrated Circuits and Systems, IEEE Transactions*, 26(5):943–958, May 2007.

[11] N.A. Touba and E.J. McCluskey. "Logic synthesis of multilevel circuits with concurrent error detection". *Computer-Aided Design of Integrated Circuits and Systems, IEEE Trans. on*, 16(7):783–789, Jul 1997.

[12] J.G Tryon. "Redundancy Techniques for Computing Systems". chapter Quadded Logic, pages 205–228. Spartan, Washington,DC, 1962.

INVITED TALK

Targeting "Zero DPPM" – Can we ever get there?

Nilanjan Mukherjee, *Mentor Graphics*

Certain mission critical applications such as automotive, medical, aerospace, etc. have been voicing the need for "Zero DPPM" for shipped parts. With rapid scaling of semiconductor devices along with technological innovations that include material and process changes, some of the current DFT techniques as well as the conventional fault models are inadequate to achieve such an ambitious goal. This talk will focus on discussing the latest fault models that are being used in these market segments to improve the overall test quality. There is an increasing trend to generate "defect-aware" test vectors that maximize the chances of detecting potential manufacturing defects. Digital tests are also being designed to play an important role in detecting systematic defects rather than just random defects, thereby making it possible to identify quickly process related issues. The talk will highlight some of the above techniques to improve the overall test quality, and conclude by briefly presenting ways to manage exponential increase in test costs while guaranteeing the highest level of test quality.

Speaker Bio – Nilanjan Mukherjee received a Ph.D. degree from McGill University, Montreal, Canada. He currently leads a technical group in the Design to Silicon division at Mentor Graphics Corporation. At Mentor Graphics, he was a co-inventor of the EDT technology and was a lead developer for TestKompress®. His research focuses on developing next generation test methodologies for DSM designs, test data compression, test synthesis, memory testing, and fault diagnosis. Prior to joining Mentor Graphics, he worked at Lucent Bell Laboratories in New Jersey.

Dr. Mukherjee has published more than 40 technical articles in various IEEE journals and conferences and is a co-inventor of 14 US patents. Dr. Mukherjee was the co-recipient of the Best Paper Award at the 1995 IEEE VLSI Test Symposium, the best student paper award at the Asian Test Symposium in November 2001, and recently, the prestigious 2006 IEEE Circuits and Systems Society Donald O. Pederson Outstanding Paper Award recognizing the paper on embedded deterministic test published in the *IEEE Transactions on Computer-Aided Design of Integrated Circuits and Systems*. Dr. Mukherjee has presented tutorials at several conferences including ITC, DAC, VLSI Design, and have offered DFT seminars on behalf of Mentor Graphics in the US and India.

SESSION 5
POSTERS

IEEE International Symposium on Defect and Fault Tolerance of VLSI Systems

A BIST Technique for Crosstalk Noise Detection in FPGAs

Waleed K. Al-Assadi, senior Member, IEEE and Sindhu Kakarla
Department of Electrical and Computer Engineering
Missouri University of Science and Technology, Rolla, MO 65409 USA
{sk9qd, waleed}@mst.edu

Abstract

As Integrated Circuits are migrated to more advanced technologies, it has become clear that crosstalk noise is an important phenomenon that must be taken into account. Also, crosstalk noise has emerged as a serious problem in recent years, because more and more devices and wires have been packed on electronic chips. Despite being more immune to crosstalk noise than their ASIC (Application Specific Integrated Circuit) counterparts, the dense interconnected structures of FPGAs (Field Programmable Gate Arrays) invite more vulnerabilities to crosstalk noise. Due to the lack of electrical detail concerning FPGA devices it is quite difficult to test the faults caused by crosstalk noise. This paper proposes a new approach for detecting effects such as glitches and delays in transition due to crosstalk noise in FPGAs. This approach is similar to the BIST (Built-in Self Test) technique in that it incorporates the test pattern generator to generate the test vectors and the analyzer to analyze the crosstalk faults without any extra overhead for testing.

1. Introduction

Crosstalk noise occurs when a change in voltage on one trace causes a corresponding change in voltage on a nearby trace. As Integrated Circuits continue to migrate to more advanced technologies, crosstalk noise due to inter-wire capacitance of FPGA has become a critical concern for electronic designers due to the following factors: (1) interconnect scaling in one dimension, (2) continuous reduction in a device feature size, and (3) increase in the domination of interconnect capacitance to the total interconnect capacitance [1]. If the affected trace (victim) is a global signal such as a clock or reset, the induced pulse might cause the circuit to incorrectly change the state, which in effect would manifest itself as a functional error or change the switching time of the victim. A long switching time may lead to critical timing failures.

A typical FPGA structure is composed of Configurable Logic Blocks (CLBs), I/O blocks, and programmable interconnects. Each CLB consists of one or more basic logic elements (BLEs) each of which contains a look-up table (LUT), a flip-flop, and a multiplexer. The interconnect programmability is implemented by switches inside switch boxes and connection boxes [1]. Connection boxes allow CLB pins to connect to tracks and switch boxes are used to build connections of appropriate lengths from prefabricated wire segments along the tracks. FPGAs have become popular design fabrics because of their faster time-to-market, re-programmability, low non-recurring engineering costs, and easy debugging.

Process variations and manufacturing defects worsen the problem of noise and delay effects leading to unexpected unpredictability and increase in coupling capacitances and mutual inductances between interconnects [3]. Due to the present dense interconnect structure in FPGAs; coupling capacitance dominates which results in crosstalk noise in FPGAs. This might cause neighboring switching nets (aggressors) to introduce noise pulses to their quiet neighbors (victims), resulting in a functional failure if the noise-induced incorrect

1550-5774/08 $25.00 © 2008 IEEE
DOI 10.1109/DFT.2008.14

value is latched. As crosstalk noise in FPGAs has become a serious problem, this paper proposes a new approach similar to the BIST technique of detecting crosstalk faults among interconnects in FPGAs.

This paper is organized as follows: Section 2 reviews the previous work that has been done on techniques used to detect crosstalk noise in FPGAs and crosstalk reduction techniques, Section 3 proposes a new FPGA crosstalk noise test architecture, and section 4 presents a new version of interconnect routing cost function. Finally, Section 5 concludes the paper.

1.1. Nature of capacitive crosstalk

The effects of crosstalk can be modeled by an increase or decrease in the wiring capacitance of the victim net, which has a significant effect on the delay of a net [4]. Coupling between a pair of interconnects can result in two different crosstalk effects: a glitch or a delayed transition, depending on the nature of the signal transitions in interconnects as shown in Figure 1(a) and 1(b), respectively [5,6]. The glitches due to crosstalk noise may have either positive or negative on the quiet neighbors and the delay transition may be either a rise time or a fall time delay transition. In addition to these effects, damped oscillations may be imposed on top of a glitch. However, if the damping is large enough to alter the state of the circuit, then the oscillations can be approximated as either a glitch or a delayed transition [5]. As far as FPGAs are concerned the effect of crosstalk noise among interconnects is more observable as a delay variation (i.e., significant effect on delay of the net) or spurious transition [4].

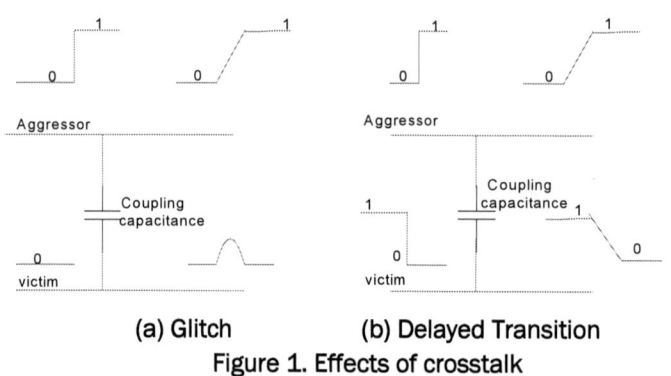

(a) Glitch   (b) Delayed Transition

Figure 1. Effects of crosstalk

This paper proposes a novel FPGA crosstalk test architecture for detecting crosstalk effects such as positive and negative glitches, as well as the rise and fall time delays among the interconnects in FPGAs.

2. Previous work

Crosstalk in FPGAs is mainly caused by coupling capacitance coming from the physical adjacency between nets, possibly introducing noise pulses and delay variations in the quiet neighbors (victims). The major difficulties in testing the FPGA system for faults induced due to crosstalk noise are mainly a result of the almost complete absence of electrical detailing concerning an FPGA device. Crosstalk can become a real problem in two conditions: (1) if an aggressor switches at a specific time window with respect to the transition of a victim so as to have a significant effect on a victim's transition, called temporal correlation; (2) if an aggressor switches in an opposite or the same direction with respect to a victim's transition [1]. Most of the previous work on the crosstalk noise in FPGAs is focused on the crosstalk induced delay, developing routing algorithms for crosstalk reduction for the applications that are eventually downloaded on to the FPGAs. Among the techniques currently used for minimizing crosstalk effects are efforts to explore ways to reduce capacitive coupling.

Coupling capacitance can be dramatically reduced if the spacing between the adjacent wires is increased. A reduction in the total loading capacitance can also be obtained by the wire spacing technique wherein the circuit's timing and power characteristics can be improved as well. The reduction of coupling capacitance will be directly converted into a reduction in delay variation due to crosstalk. The effectiveness of the wire spacing technique depends on the available routing space in an FPGA. If the total area of an FPGA structure is decided by the transistors, and if the extra spacing between wires does not cause any area penalty, then the wire spacing proves effective. However, for the architectures in which wires decide the total area, the wire spacing technique is ineffective. The other commonly used technique is power or ground shielding [1]. In this technique the VDD or GND track between two signal wires is placed to eliminate the effects of neighbors switching on the victim explicitly. Shielding is not as effective in improving the timing as is wire spacing with the same area overhead. The advantage of shielding is that it provides a good delay prediction, which introduces more confidence to noise control. This technique is mostly used for long lines crossing the entire FPGA structure and also for some dedicated interconnects such as clock and reset. The authors in [1] stated that crosstalk-induced delay can be easily accumulated on long parallel wires and should be avoided. They proposed a new switch box design in which long parallel wires are reduced. In this paper's new approach to detecting crosstalk effects in FPGA, preference is given to testing the interconnect wires before they hit a switch box. In [4], a crosstalk-aware router is proposed, which is the enhancement of the Versatile Place and Route [VPR] timing router, by assuming a simple model in which crosstalk between two traces is modeled by a change in the effective capacitance seen by both traces. The VPR timing router is enhanced by modifying the cost function which specifies the complexity of routing process and timing model optimized for the delay that may be induced due to crosstalk [7]. The work in [8] proposes a new approach for crosstalk reduction through area routing using multiple possible connections between two points that are to be connected. In this approach the crosstalk effects between interconnects are avoided by routing the segments in various layers either as L-shaped or as Z-shaped, where the L-shape or Z-shape indicates the direction of the routes. In [9], an online testing of non-logical faults such as crosstalk faults is proposed using an N-bit two rail code checker and functional block based on a finite state machine. This, however, limits the detection of crosstalk faults to the faults in CLBs. The functional block consists of N-basic cells, each connected to a monitored interconnect line and able to concurrently detect an undesired transition of the monitored line [9, 10]. The work in [11] presents an iterative logic array (ILA) method of finding delay faults among interconnects wherein the FPGA under test is configured to have one or more independent iterative logic arrays, each of which propagates a signal transition. An ILA is a series of logic blocks and the associated interconnects under test. Considering the crosstalk noise in FPGA devices, the detection of interconnects that may be affected by the crosstalk noise should be done at the manufacturing level. In this model, the newly proposed test architecture can be connected in a scan chain fashion.

3. Crosstalk noise detection in FPGAs

This paper proposes a novel test architecture for the detection of crosstalk-affected interconnects at the manufacturing level, considering all the factors resulting in an increase in coupling capacitance among interconnects in an FPGA,. The proposed approach concentrates mainly on MOTP (Manufacturing oriented testing procedure) of FPGA, where the FPGA device is tested for the crosstalk noise irrespective of the application that is being downloaded onto the device based on the Maximum Aggressor Fault model, wherein among the bunch of interconnects one interconnect is a victim and all other interconnects are aggressors.

The Maximum Aggressor Fault Model is a high- level representation of all physical defects and process variations that lead to crosstalk errors, including positive glitches, negative glitches, and delays in transitions. The maximum aggressor model defines one victim line in the circuit and the signal on this victim line is disturbed by signals on the surrounding lines, called aggressors. For the aggressor lines it is assumed that all signals have identical polarity and switch, at same moment of time. Due to this assumption, the voltage level on all the aggressors will be identical, preventing them from affecting each other [5]. This study's main objective is to develop a technique similar to BIST for detecting the crosstalk faults among interconnects under maximum aggressor fault model in an FPGA device by connecting the CLBs of an FPGA in a scan chain fashion based on the boundary scan architecture without using any extra hardware [3].

This study concentrates on detecting effects such as glitches, or delayed transition among the interconnect structure in FPGA devices. An interconnect can act either as a victim or an aggressor, so the complete testing procedure is done for all the interconnects, assuming them to be victim or aggressor each time. Efficiency of detecting the crosstalk noise depends on the test vectors being applied to the interconnects. In FPGAs a regular counter cannot be used for detecting the effects due to crosstalk as it leads to a computationally expensive approach. Figure 2 shows an example of an interconnect system in which the transitions can detect the glitches, rise time, and fall time delay transitions due to the crosstalk noise in an interconnect system.

As shown in Figure 2, the second interconnect line is assumed to be a victim and other lines act as aggressors. This is shown as an example of how the test pattern generator in our design generates all the patterns assuming that each interconnect can act as a both victim and aggressor for other interconnects.

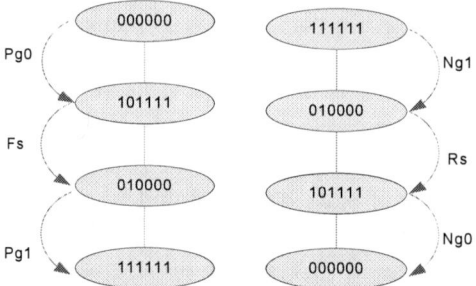

Figure 2. Test vectors generated by Test Pattern Generator

The victim interconnects may be either at static high or static low logic values, which may be affected by the transition of the aggressors. Pg0, Pg1 and Ng0, Ng1 are positive and negative glitches, respectively. Rs and Fs are rising and falling skews in the victim line. Pg0 is the glitch that occurs on the victim line at static low values due to the transition in the aggressors from '0' to '1'. Pg1 is a glitch that may be manifested in the victim line, which is static high due to transition in the aggressors from '0' to '1'. Similarly Ng0, Ng1 correspond to the glitches on the victim line with the aggressors transitioning from '1' to '0'. Based on this model of different crosstalk effects, new test architecture is proposed to detect the crosstalk noise among FPGA interconnects.

Careful study of Figure 2 shows that the victim line undergoes transition for every two clock cycles, whereas the aggressors undergo transition for every clock cycle. Based on this, a test pattern generator is obtained by configuring each CLB at the source side of the wire under test (WUT). This CLB is configured with the test architecture as shown in Figure 3. To evaluate the presence or absence of crosstalk effects as stimulated by the test pattern generator on the source side of the WUT, a new block called an "analyzer" is presented at the destination side of the WUT. The CLBs on the destination side of the concerned WUT are configured with the architecture of the analyzer as shown in Figure 4.

3.1. Architecture of Test Pattern Generator

The test architecture for the test pattern generator shown in Figure 4 is configured in the CLBs at interconnects input sides.

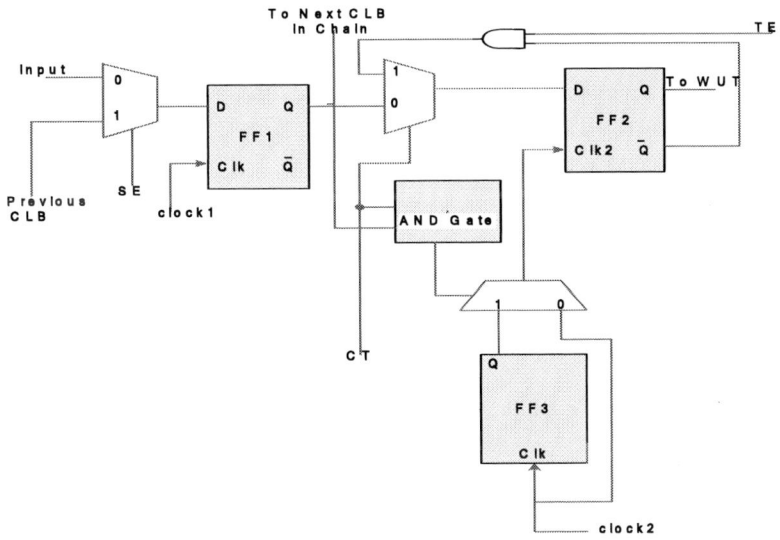

Figure 3. Test Pattern Generator of one cell (CLB) in scan chain

The test pattern generator is controlled by external primary inputs, SE (Shift Enable), and CT (Crosstalk), Test-enable (TE). For the input SE at logic'1', the initial vectors for the test pattern generator are scanned in and shifted. Flip-flop FF1 stores the victim-select data indicating the line that is assumed to be a victim in a bunch of interconnects. The AND gate controls the transitions on the victim interconnects and aggressor interconnects based on the victim select data and other input pin CT. Flip-flop FF2 stores the initial vectors that are scanned in before the actual testing procedure begins. Flip-flop FF3 is used to delay the transitions in interconnects that are assumed to be victims to two clock cycles based on the observations from Figure 2. The multiplexer at the output of FF3 is used to set the frequency of the clock in flip-flop FF2 depending on the output from the AND gate (i.e., whether interconnect line is assumed to be a victim or as an aggressor). The output of flip-flop FF2, Q, is connected to the interconnect wire under test (WUT) and the other output of flip-flop FF2 is fed back to one of the inputs of AND gate in the feedback. TE controls the other input of the AND gate in the feedback and the output of the AND gate is given to the multiplexer followed by the FF1.

171

3.2. Operation of Test Pattern generator

The primary inputs SE, CT and TE control the overall operation of the test pattern generator. The following are the steps the CLBs configured Figure 3 undergoes.

(a) Initial vectors "000000" is scanned into flip-flop FF1 with SE='1', CT='0' and TE='0'.

(b) In the next clock cycle with SE='1', CT='0' and TE='0' initial vector "000000" stored in FF1 is passed to FF2 and so, flip-flop FF2 contains initial vector "000000".

(c) The testing procedure starts with the enabling of the crosstalk noise pin CT. With SE='1', CT='1' and TE='0' victim select data (representing which line is a victim line) is scanned in and stored in FF1. Since in this example second interconnects line is assumed to be the victim line, victim-select datum here would be "010000".

(d) In the next clock cycle with SE='1', CT='1' and TE='1',

   (i) For the first-interconnect line output of flip-flop FF1 is '0' (which is the victim-select data scanned-in at step2), which makes the output of the AND gate '0'. The output of flip-flop FF2 undergoes transition from '0' to '1'for every clock cycle due to the feed-back path in Figure 3.

   (ii) For the second-interconnect line, output of flip-flop FF1 is '1' (victim-select data scanned-in at step2), which makes the output of AND gate '1' (given an output of flip-flop FF1='1', CT='1') and the clock for flip-flop FF2 is delayed. The output of flip-flop FF2 undergoes a transition from '0' to '1' for the two clock cycle shown in Figure 2 because the clock at FF2 is delayed.

   (iii) Similarly, the third to sixth interconnect lines undergoes transition from '0' to '1' at output of FF2 at every clock cycle as victim-select datum (scanned-in at step 2) is '0' for these interconnects.

   Through these steps vector "101111" is generated which is the next test vector in the test sequence as shown in Figure 2.

(e) The procedure repeats steps c to d and different test vectors as in the test sequence are generated automatically

Assuming a six-interconnects system and considering the second interconnect as a victim net appropriate test patterns are generated. After the generation of test patterns for the second victim line, the line rotates with one-hot encoded data obtained by scanning in a '0'. For example, if the second line is a victim, the victim select datum is "010000"; for the third interconnect to be a victim, '0' is scanned in to shift the victim select data one position to the right as "001000". In this manner all the interconnect lines in a bunch are assumed to be a victim at least once.

For the generation of test patterns, only the initial vectors and victim select data are to be scanned in and the rest of the patterns are generated by the pattern generator itself. This structure is configured in a scan chain for all the CLBs that are connected to the WUTs and the scan procedure is enabled by the input pin SE. The main advantage of this pattern generator is that only initial vectors "000000" and "111111" need to be scanned in. The remaining test vectors will be generated by the testing architecture. After the generation of patterns and detection of the faults assuming that one of interconnects in a bunch is a victim, a '0' is scanned into FF1 to change the victim and the procedure repeats.

3.3. Analyzer

As explained, the CLBs at the destination side of the WUTs are configured with the test architecture called the "analyzer", as shown in Figure 4. The analyzer consists mainly of a simple XOR gate for which one of the inputs is from the interconnects under test (WUT) and the other input is from the local test generator, which generates the same test patterns as that of the original pattern generator. However, the local test generator is delayed by placing the basic inverters "inv". The number of inverters that should be placed depends on the specifications of the wire delay for the type of interconnects that are under test. The delay is used in order to synchronize test patterns generated locally and patterns from the wire under test (WUT). The output of the XOR gate is given to flip-flop FF. An FF output '1' indicates that WUT is affected by crosstalk. The output of the flip-flop from the analyzer at the destination side of WUTs is connected as a scan chain.

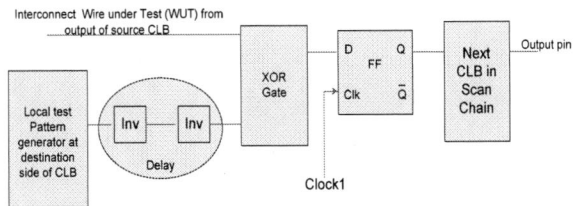

Figure 4. Block diagram of Analyzer

3.4. Overall test architecture

The overall testing procedure is conducted for horizontal parallel wires and vertical parallel wires in an FPGA. As an example, the overall test architecture for a set of six horizontal interconnects is shown in Figure 5.

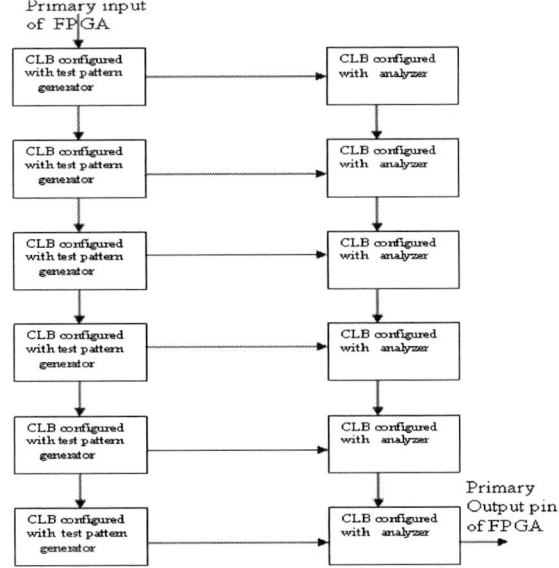

Figure 5. Overall Test architecture for set of interconnects

The pattern generators in Figure 5 at the source side of WUTs are the CLBs configured with architecture as shown in Figure 3 and the analyzers at the destination side of the WUTs are the CLBs configured with architecture as shown in Figure 4. At the destination side of WUTs, the output of the flip-flops from the CLBs configured with the architecture in Figure 4

173

is connected in a scan-chain fashion and the output of the scan-chain is directly fed to the primary output pin of the FPGA.

After the testing procedure is conducted, data from the output pins of the FPGA should be stored in a parameter file and the routing procedure should use the parameter file for any application that is to be routed in the FPGA. If a particular interconnect is affected by crosstalk, then using the parameter file that interconnects is avoided for routing the given application by modifying the routing cost function. Conversely, shielding (inserting VDD or GND between those two signal wires) may be conducted for those interconnect.

4. Enhanced cost function

In the routing scenario, fitness of any segment is evaluated from the cost-function, and the maze-routing algorithm is used for routing the nets in an FPGA [7]. The cost function is used to obtain the most appropriate path between the CLBs to be connected among all the possible paths.

Cost function parameters are defined as follows:

Cost (i) = Cost for routing a particular segment i

Crit = Criticality of the currently routed net

Delay (i) = Elmore delay of the segment i

b (i) = Base cost of using segment i

h (i) = Historical congestion cost of using segment i

p (i) = Present congestion cost of using segment i

The cost function is defined as

$$Cost(i)=Crit*delay(i)+(1-Crit)*b(i)*h(i)*p(i) \qquad (1)$$

The criticality of a net is close to 1 if the net is close to the critical path of the circuit [4]. The base cost is the basic cost of using a resource. The historical congestion cost prevents the nets from being routed to tracks that have led to bad routing solutions in the past. The present congestion cost indicates the congestion cost of using the segment for a given application. In the routing procedure, if the parameter file indicates that a particular interconnect line is affected due to crosstalk noise, then a penalty function (which indicates an increase in the cost of routing a segment) is added to the original cost function. Using the new approach that is proposed for detection of crosstalk noise the cost function for routing the nets in an FPGA can be modified as follows:

$$Cost(i)=Crit*delay(i)+(1-Crit)*b(i)*h(i)*p(i)+Penalty \qquad (2)$$

Where Penalty = 0 if $i \neq j$

$$Penalty=\sum_{j=0}^{n} (1- Crit)*b(j)*h(i)*p(j) \text{ if } i = j \qquad (3)$$

In these equations, 'j' is the track number or segment from a parameter file obtained after a testing procedure and 'n' is the number of times that particular segment appears in a parameter file.

This study draws from equation 3 the important conclusion that, for interconnects affected by the crosstalk noise, the base cost and the present congestion cost of the segment will be increased with the penalty function. This increases the cost of routing a particular segment. Using this vital information, the route which carries more probability of being affected by crosstalk may be avoided for application by the routing algorithm (maze routing algorithm), which essentially depends on the cost function.

5. Conclusion

This paper proposed a new framework for detecting crosstalk noise among interconnects in FPGAs where the testing is done irrespective of the intended application. The proposed framework presents test architectures for the CLBs to be configured at the source and destination side of WUTs to detect the effects of crosstalk noise. The main advantage of this approach is that the testing architecture includes crosstalk effects such as positive and negative glitches on the victim interconnects, as well as delays in transition. An improved version of the existing cost function for routing in FPGAs which simultaneously incorporates the crosstalk noise effects such as glitches and delays in the same expression is also presented. The data from a parameter file (described in section V) are used to modify the cost function for routing any application on an FPGA such that it will be unaffected by crosstalk noise. The information obtained from the parameter file can be used by the application configuring process so that the routing can be done with minimal crosstalk effects. The experimental results for the proposed approach detecting crosstalk noise effects in FPGAs are yet to be observed. In future work this testing procedure will be correlated to the BIST technique usually used to find logical or interconnect faults in FPGAs in a such way that the same technique can be used to detect stuck-at faults along with crosstalk faults.

6. References

[1] Yajun R., Marek–Sadowska, "Crosstalk Noise in FPGAs", In Proc. of Design Automation Conference (DAC'03), pp.944-949, 2003.

[2] Nourani M, Attarha A., "Built-In Self-Test for Signal Integrity" In Proc. of Design Automation Conference (DAC'01), pp: 792-797, 2001.

[3] Ahmed N., Tehranipour M., Nourani M., "Extending JTAG for Testing Signal Integrity in SOCs", In Proc. of the Design Automation and Test in Europe Conference and Exhibition (DATE'03), 1530- 1591/03, 2003.

[4] Wilton S., "A Crosstalk- Aware Timing- Driven Router for FPGAs", FPGA '01, pp.21- 28, 2001.

[5] Bai X., Dey S., Rajski J., "Self – Test Methodology for At-Speed Test of Crosstalk in Chip Interconnects", In Proc. of Design Automation Conference (DAC,'01), pp: 619-624, 2000.

[6] Chen W., Gupta S. K., Breuer M. A., "Test Generation in VLSI Circuits for Crosstalk Noise", In Proc. IEEE International Test Conference (ITC) pp: 641-650, 1998.

[7] Hur W. S., Jaganathan, Lillis J., "Timing Driven maze routing", In proc. Int. Symp. on Physical Design, Pages 208- 213, 1999.

[8] Smey, M. R., Swartz B., Madden H.P, "Crosstalk Reduction in Area Routing", In Proc. of the Design Automation and Test in Europe Conference and Exhibition (DATE'03), 2003.

[9] Metra C., Pagano A., Ricco B., "On-line Testing of Transient and Crosstalk Faults Affecting Interconnections of FPGA Implemented Systems", In Proc. IEEE International Test Conference (ITC), pp:939-947, 2001.

[10] Abramovici M. and Stroud C.,"BIST-Based Detection and Diagnosis of Multiple faults in FPGAs", in Proc. of IEEE International Test Conference (ITC), pp.785- 794,2000.

[11] Chmelar E., "FPGA Interconnect Delay Fault Testing" in Proc. Of IEEE International Test Conference (ITC), pp: 1239-1247, 2003.

IEEE International Symposium on Defect and Fault Tolerance of VLSI Systems

A Fault Tolerance Aware Synthesis Methodology for Threshold Logic Gate Networks

Manoj Kumar Goparaju   Ashok Kumar Palaniswamy   Spyros Tragoudas

Department of Electrical and Computer Engineering

Southern Illinois University Carbondale

Carbondale, IL 62901.

{goparaju,ashok,spyros}@engr.siu.edu

Abstract

Threshold Logic technology is conceived as the crucial alternate emerging technology to CMOS implementation in nanoelectronic era. The gate that is implemented with threshold logic is called a Threshold Logic Gate (TLG). Threshold gates are very fast and implement complex functionalities thus reducing the logic levels in the circuit implementation. Extensive research has been done in the development of suitable synthesis methodologies in the past, predominantly greedy. In this work, a synthesis methodology is proposed for increased fault tolerance. Experimental results demonstrate the effectiveness of the proposed method both in terms of resulting TLG count in the network implementation and reliability.

1: Introduction

A gate that implements threshold logic is called a Threshold Logic Gate (TLG). Gates implemented with threshold logic offer the capability of realizing complex Boolean functions using a smaller number of logic gates or fewer logic stages. A threshold logic gate is implemented by constituting weights w_i for each input i and a weight w_0 which is called the threshold value of the gate. The logic value of the gate is determined by comparing the weighted sum of input weights against the threshold value of the gate. Namely, the logic output of the TLG is determined to be 1 if $\sum_{i=1}^{n} w_i x_i \geq w_0$ and 0 otherwise [18].

Consider the following example from [2].

Example 1: Function $f = x_1 x_2 + x_1 \bar{x}_3 + x_2 \bar{x}_3$ is implemented using threshold logic with weight configuration $\{w_1, w_2, w_3: w_0\}$. For each input pattern an inequality that involves w_i must be satisfied. The weight configuration $\{w_0, w_1, w_2, w_3\}$ of threshold logic gate for f is the solution set satisfying the inequalities of f. This can be obtained with a Integer Linear Program (ILP) problem formulation. The input-output relationship of the f is given in Table-1. The table also lists corresponding inequalities associated with each input pattern.

The popular nano-technological implementation of threshold logic is using the MOBILE element [20, 19]. A MOBILE is a pseudo-dynamic clocked logic circuit consisting of a FET that is monolithically integrated with a resonant tunneling diode (RTD) [19]. TLG gates are also designed with CMOS technology. An extensive study of CMOS implementations can be found in [1].

Figure 1 depicts the MOBILE[19] implementation of f. The MOBILE implementation has been considered for illustration purpose only. It should be noted that proposed methodologies are applicable to the design of TLG in general irrespective of technology of implementation.

1550-5774/08 $25.00 © 2008 IEEE
DOI 10.1109/DFT.2008.44

The functionality of a TLG may be affected with change in the weights of the inputs and also due to change in the threshold [6, 10]. In a MOBILE, the weights in the TLG are determined by the areas of the RTDs. Poor overgrowth in fabrication, temperature changes [11, 13], changes in current density [4], mismatch between photomask layout and design layout [9] are some of the factors influencing the change in the weight values [6, 10]. An ATPG methodology [2] has been devised to inspect such weight defects in threshold logic gate networks.

Example 2 : Consider function f in Example 1, and let an ILP program generate the solution $\{w_1, w_2, w_3: w_0\}$ as $\{2, 2, -2: 1\}$. Deviations d_i at individual weights $w_i, 0 \leq i \leq 5$ can be above or below the ideal value. Assume that all d_i are 0.5. Consider the inequality corresponding to pattern 010. d_i value is added such that worst case would apply: $w_2 - 0.5 \geq w_0 + 0.5 \Leftrightarrow 2 - 0.5 \geq 1 + 0.5 \Leftrightarrow 1.5 \geq 1.5$.

The inequality corresponding to 010 is still valid even in the presence of deviations. Similarly it can be verified that all the remaining inequalities are also valid in the presence of $d_i = 0.5$. But if d_i is 0.6 instead of 0.5 then some of inequalities including 010 would result in faulty output. Hence it can be observed that different weight configurations tolerate different deviations.

The TLG implementing f will be able to tolerate deviations up to 1.0 if we choose weight configuration as $\{4, 4, -5: 2\}$. This shows that selection of weight configuration plays a major in achieving good fault tolerance.

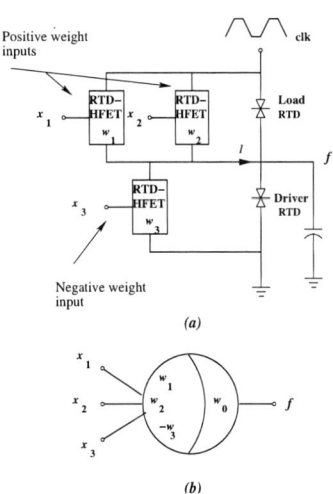

Figure 1. MOBILE implementation of f

Example 2 assumes that all weights have the same deviation d_i. However, when the designer specifies weights w_i and w_j, different deviations d_i and d_j may result after the manufacturing process and functionality may still hold for such d_i values.

Table 1 Input-Output relations of f

x_1	x_2	x_3	f	Inequality
0	0	0	0	$0 < w_0$
0	0	1	0	$w_3 < w_0$
0	1	0	1	$w_2 \geq w_0$
0	1	1	0	$w_2 + w_3 < w_0$
1	0	0	1	$w_1 \geq w_0$
1	0	1	0	$w_1 + w_3 < w_0$
1	1	0	1	$w_1 + w_2 \geq w_0$
1	1	1	1	$w_1 + w_2 + w_3 \geq w_0$

In a recent work [6], a threshold logic design methodology is developed to design fault tolerant threshold logic functionalities by an appropriate weight assignment to each input and the threshold value such that $d = min\{d_i\}$ is maximized. This will result to a designed function with maximum fault tolerance because after manufacturing, defects can be tolerated to the maximum possible extent. The factor d is termed as the $fault\text{-}tolerance$ (FT) of the function f[6].

The list of FT values for all possible 2-input functions is given in Table 2 as listed in [6].

For a given functionality, different weight configuration offer different FT values. The weight configuration that offers highest possible FT is chosen. Also, it can be observed from Table 2 that different functionalities pose different FT values. This observation lays foundation for this work. The traditional synthesis process should be modified to have such high fault tolerant cells in the circuit.

Table 2 The FT of 2-input functions [6]

Function f_i	Optimum weight assignment set $\{w_1, w_2, w_0\}$	fault-tolerance FT
f_1	{-25, -25, -9}	7.99
f_2, f_4	{-25, 16, 6}	4.99
f_7	{-18, -18, -25}	3.5
f_8	{18, 18, 25}	3.49
f_{11}, f_{13}	{-16, 25, -6}	4.99
f_{14}	{25, 25, 9}	7.99

In this paper, an effective heuristic synthesis process is proposed that employs such fault tolerant design process to generate circuits with a small number of threshold logic clusters offering high fault tolerance. Threshold logic functions have ability to implement complex functions with a single gate. Owing to this intrinsic characteristic of threshold logic functionalities, the area is usually optimized. Many synthesis processes have been developed [5, 7, 8, 12] exploiting this property. However, to the best of authors' knowledge, none of the previous work in threshold logic concentrated on *reliability* aspect of the synthesis process. A robust synthesis process is presented that guarantees high fault tolerance while reducing the number of threshold logic clusters. The logic synthesis of TLG networks in this paper is implemented with technology mapping principles employing suitable threshold logic functional cells.

The paper is organized as follows. Section 1 gives preliminaries and briefly describes the the fault tolerant design methodology for TLG. The approach is based on a novel ILP formulation which guarantees optimality for any function. Section 2 presents proposed fault tolerance aware synthesis methodologies. Section 3 presents experimental results and section 4 concludes.

2: Fault tolerant synthesis process

All the methodologies proposed so far follow predominately a greedy approach. A greedy methodology does not always guarantee good fault tolerance characteristic for the synthesized circuit. Consider a synthesis process as in [7]. The network is synthesized employing a clustering process of Boolean nodes in a brute force method. The clustering process involves grouping different Boolean gates such that the functionality of the resulting node cluster is a viable threshold logic function. In particular, any function that to be implemented with threshold gate should be unate w.r.t each and every variable, x_i. A function, $f(x_1, x_2, ..., x_n)$, is said to be positive (negative) in variable x_i if there exists a disjunctive or conjunctive expression of f in which x_i appears in uncomplemented (complemented) form only [18].

The clustering process is initiated from output node and the Boolean nodes in the network are assigned to some cluster in a greedy fashion. The selection of Boolean nodes for clustering employs graph traversal schemes like depth first search or breadth first search starting from the node under consideration. The clustering process for a node terminates once the fanin of the cluster reaches the fanin bound. The process is repeated recursively until all nodes in the original circuit are included in one of threshold logic clusters in the circuit.

The clustering process is illustrated with the following example. Consider the following Boolean network implementation in Figure 2 that needs to be synthesized with threshold logic gates. Assume that the fanin for the threshold function is limited to 4 and breadth first search(BFS) is employed for clustering process. The clustering process is initiated from the output node, $G7$. A BFS is initiated to select the nodes to be combined together in a cluster as they are encountered. Nodes $G7, G6, G5$ are chosen to be included in the threshold logic cluster for $G7$. The fanin for resulting cluster is 4 and hence the clustering process stops here.

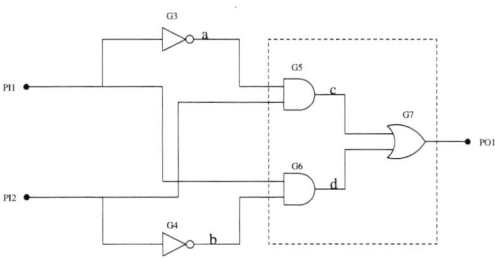

Figure 2. Boolean network under consideration

The FT of the resulting cluster $G7 - G5 - G6$ can be determined using the ILP method developed in [6] and is equal to 0. This implies that even a small deviation in the weight implemented in the cluster leads to malfunction of the whole circuit and hence results in poor reliability of the circuit. However, if the gates $G5, G3$ were chosen for the cluster with $G7$, the FT of resulting would have been 2.66. Thus, the above greedy process may not achieve optimization aspect, *reliability*, of the synthesis process. Even though it is successful in achieving the area optimization it might fall short of securing agreeable FT characteristic for the resulting synthesized circuit.

A modified greedy approach is proposed to overcome this shortcoming. It effectively makes use of various graph traversal schemes to identify all possible threshold logic functional clusters for a given node and the functional combination that would offer highest possible FT is selected. The process is implemented repeatedly until all nodes have been included in one of threshold logic functional clusters. That way, while determining the structural threshold logic blocks the synthesis process optimizes the two important aspects of quality of circuit performance - *area* and *reliability*.

Owing to the intrinsic characteristic feature of threshold logic of being able to implement complex functionalities, the area is reduced immensely. As the threshold logic gates at each stage are chosen with possible high FT value, the reliability of circuit will be improved by many folds. The proposed fault tolerant synthesis process is termed as **FTS** and is explained with following example.

Consider the Boolean network in Figure 2 that need to be synthesized with threshold logic functionalities. First, the clustering process starts from the output node, $G7$. The possible number of threshold logic functionalities that can be constituted with $G7$ with fanin within the prescribed bound are determined. Various graph traversal algorithms can be employed in determining potential candidate list for grouping. Only the functionalities that would resolve to threshold logic are considered. In our approach, nodes with fanout branches are omitted for clustering and the fanin bound is set to 4. All such possible 4 input clusters of $G7$ are listed in Figure 3(b).

The ILP methodology is then employed to determine the FT value for each potential cluster in the list. The FT values for all possible clusters with $G7$ are shown in the Figure 3(b). The functional cluster with highest FT is chosen for clustering to enhance the *reliability* characteristic of the resulting circuit. It can be seen from Figure 3(b), such fault tolerant combinations for node $G7$ are $[G7 - G5 - G3]$ and $[G7 - G6 - G4]$ offering the highest $FT = 2.66$. The combination $[G7 - G5 - G3]$ is chosen. The corresponding cluster will be replaced with respective TLG cell,

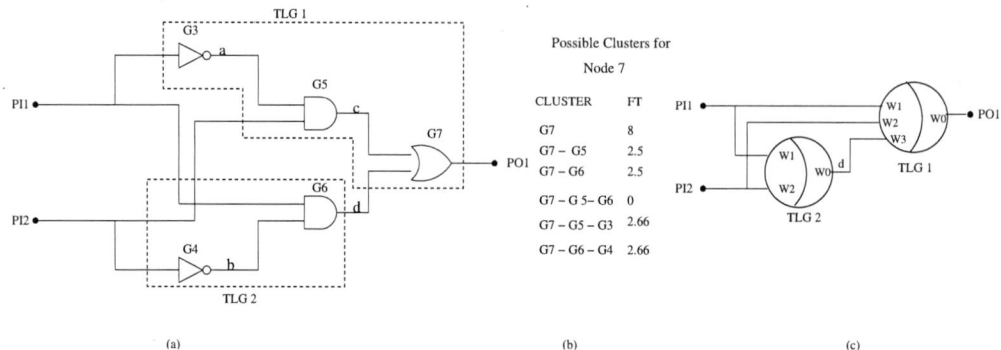

Figure 3. FTS process on Boolean network in Figure 2

$TLG1$. It should be noted that though the cluster with just $G7$ offers the highest FT of 8, it is not selected as it would result in cluster with one node. This leaves nodes $G6$ and $G4$ left in the network to be clustered. The synthesis process is then implemented on these nodes resulting in cluster $[G6 - G4]$. This cluster is shown as $TLG2$ in Figure 3(a). The complete synthesized TLG network is shown in Figure 3(c).

It can be observed from the Figure 3(c) that the the number of logic implementations reduced to great extent with implementation in threshold logic. The traditional CMOS implementation in Figure 3(a) demands 4 logic gates while the threshold logic implementation in Figure 3(c) requires only 2 logic gates. This has the effect of reducing the power consumption and minimizing the circuit area resulting in faster circuits. The proposed method guarantees the fault tolerance of the circuit. With the proposed method FTS, the FT is improved to 2.66 against 0 with a greedy process.

The FTS focuses on increasing the FT while choosing clusters for the nodes in the network. In the process of choosing fault tolerant cluster, for a certain node, a cluster with less number of nodes may be favored to the cluster with more number of nodes. For this reason, FTS methodology will result in circuit with high fault tolerant characteristic but might have some area overhead. However, it will be shown in experimental results that this area overhead is minimal and can safely be disregarded.

To investigate the possible existence of accord between *area* and *reliability*, a heuristic synthesis process is developed for area optimization (**FTSA**). FTSA is a variation of proposed FTS methodology. Here in this method while choosing the cluster for a node, the cluster with highest number of nodes is chosen. If there exists more than one such node, then the cluster that offer higher FT is chosen. Even though the FTSA methodology does not choose cluster with high FT, it is still better than normal greedy synthesis process. It does a search in the proximity of node under consideration and chooses the cluster with better FT which is lacking in traditional methods. That way, clusters that would result in low FT value as low as 0 can be omitted while reducing the cluster count.

3: Experimental results

In this section experimental results are presented to demonstrate the need for proposed fault tolerance aware synthesis process for threshold logic gate networks. The methodology has been implemented on $ISCAS$'85 benchmarks. The proposed ILP method was implemented using the *lp_solve* tool [16]. The experiments with synthesis algorithm were conducted using C language on

SunBlade 1000 workstation with a 2 Gigabyte RAM.

In our experiments, it was assumed that the range of weights (values) that can be manufactured is below 5. Also, the binary search on the FT, d, is implemented by restricting the precision to two decimal digits in the calculation of d. If the manufacturing process allows for higher precision then the weight assignment may be different and the FT value slightly higher. The fanin bound was set to 4.

The lowest FT value of a threshold cluster in the synthesized circuit indicates the high vulnerability of the synthesized circuit to weight discrepancies. Thus, the lowest FT (FT_{lowest}) value possible in the circuit indicates reliability of the circuit and hence considered to be FT of the whole circuit ($FT_{circuit}$). The number of clusters with lowest FT_{lowest} indicates the probability of occurrence of circuit malfunction.

The number of TLG clusters indicate the *area* of the synthesized network and FT_{lowest} indicate the *reliability* characteristic of the network.

Table 3 Comparative analysis of Synthesis Algorithms

Benchmark circuit	Brute force[7]			FTS			FTSA		
	Total Clusters	FT_{lowest}	#Clusters with FT_{lowest}	Total Clusters	FT_{lowest}	#Clusters with FT_{lowest}	Total Clusters	FT_{lowest}	#Clusters with FT_{lowest}
C17	4	2	0	4	2	2	4	2	2
C880	210	0	40	279	1.33	25	246	0.75	1
C1355	298	1.33	0	298	1.33	8	298	1.33	8
C1908	714	0	105	645	1.33	62	623	0.75	2
C2670	832	0	117	846	0.75	1	797	0.67	4
C3540	1122	0	79	1127	1.33	113	1061	0.75	11
C5315	1717	0	222	1663	1.33	348	1651	0.67	4
C6288	1456	2.5	0	1456	2.5	464	1456	2.5	464
C7552	2489	0	479	2374	1.33	220	2356	0.67	1

Table 3 list the results. Column2 refers to brute force synthesis methodology similar to the one proposed in [7]. Column 3 corresponds to the proposed enhanced fault tolerant synthesis process, FTS. Column 4 depicts the experimental results corresponding to FTSA method. Subcolumn 2-1 refers to the total number of TLG clusters in the network due to greedy process. Similarly subcolumn 3-1 corresponds to the total number of clusters in the network due to FTS method and subcolumn 4-1 corresponds to total number of TLG clusters corresponding to FTSA method. Subcolumn 2-2, subcolumn 3-2 and subcolumn 4-2 corresponds to the FT_{lowest} value respectively for Brute force method, FTS method and FTSA method. The number of clusters with FT_{lowest} with implementation of each method respectively listed in columns 2-4, 3-4 and 4-4.

It can be seen from the column 2-1 that traditional synthesis method results in FT values as low as 0. That means, even the slightest change in the weight values in the respective clusters the circuit cease to function properly. The adverse impact of the process can be seen predominantly in the circuits C2670, C5315 and C7552 where the number of clusters with $FT = 0$ is high implying the high probability of malfunction. The proposed methods FTS and FTSA overcome this problem by not choosing the cells with lower FT values thus optimizing *reliability* criteria of the synthesis process.

It can be observed from the *'Total Clusters'* columns of respective methods that even if the number of clusters in the circuit with FTS and FTSA process is greater than the number of clusters resulted with traditional method but the overhead is minimal. Thus the proposed fault tolerant synthesis process fairs well even when optimizing area.

The impact of proposed methodologies can be realized from the *'#Clusters with FT_{lowest}'* columns. The number of clusters with lowest FT value is large in case of Brute force method implying that the potential fault sites in the circuit are large in number. This is improved to a great extent with the proposed methodologies. The FT_{lowest} is improved compared to brute force method and the number of clusters with the lowest FT is decreased. Hence, the probability of fault is reduced.

The FTS process results in highest reliable clusters compared to FTSA process. The area is better optimized with FTSA process. The FTS process is recommended for the manufacturing process which is prone to defects. The FTSA process is desirable for the manufacturing process which is less susceptible to discrepancies and area can be optimized.

4: Conclusion

An ILP methodology has been effectively utilized to determine the fault tolerant characteristic of any threshold logic function. Optimum weight assignments for threshold logic gates targeting high fault tolerance can be determined. Cell libraries can be generated using high fault tolerance gates. Its been demonstrated that if the traditional synthesis process is modified to include these cell libraries, high fault tolerance from manufacturing process can be guaranteed. Enhanced synthesis processes have been presented that enable the effective optimization of area and reliability.

References

[1] V. Beiu, J. M. Quintana, and M. J. Avedillo, "VLSI implementations of threshold logic - A comprehensive survey", *IEEE Trans. Neural Networks*, vol. 14, pp. 1217-1243, September 2003.

[2] M.Goparaju and S.Tragoudas,"A Novel ATPG Framework to Detect Weight Related Defects in Threshold Logic Gates",*VLSI Test Symp.*,pp.323-328,April 2008

[3] A. Luck, S. Jung, R. Brederlow, R. Thewes, K. Goser, and W. Weber, "On the design robustness of threshold logic gates using multi-input floating gate MOS transistors", *IEEE Trans. Electron. Devices*, vol. 47, pp. 1231-1240, June 2000.

[4] W. Prost et al., "Manufacturability and robust design of nanoelectronic logic circuits based on resonant tunneling diodes", *Int. J. Circ. Theory Appl.*, vol. 28, no. 6, pp. 537-552, Nov. 2000.

[5] T.Gowda and S Vrudhula,"A decomposition based approach for synthesis of multi-level threshold logic circuits",*Proc. of ASP-DAC*,pp.125-130,January,2008.

[6] M.Goparaju and S.Tragoudas,"Fault Tolerant Design Methodology for Threshold Logic Gates and Its Optimizations",*Proc. of the 8th Intl. Symp. on Quality Electronic Design*,pp.420-425,March 2007.

[7] R. Zhang, P. Gupta, L. Zhong and N.K. Jha, "Threshold Network Synthesis and Optimization and Its Applications to Nano Technologies", *IEEE Transactions on Computer-Aided Design Of Integrated Circuits And Systems*, vol. 24, pp. 107-118, January 2005.

[8] M.J. Avedillo and J.M. Quintana,"A Threshold Logic Synthesis Tool for RTD Circuits",*Proc. of the Digital System Design, EUROMICRO Systems*, pp. 624-627,2004.

[9] J.M. Jerke, "Semiconductor measuresment technology: Optical and dimensional-measurement problems with photomasking in microelectronics", *Special Publication National Bureau of Standards, Washington, DC. Optical Physics Div*, vol. 76, October 1975.

[10] S. Dechu, M.Goparaju and S. Tragoudas, "A metric of tolerance for the manufacturing defects of Threshold logic gates", *The 21st IEEE International Symposium on Defect and Fault Tolerance in VLSI Systems*, pp. 318-326, October 2006.

[11] W. Prost, U. Auer, C. Pacha, A. Brennemann, K. Goser, and F.J. Tegude, "Novel MOBILE gates with low-power, relaxed parameter sensitivity and increased driving capability", *Proc. 11th Int. Conf. Indium Phosphide and Related Materials*, pp. 411-414, May 1999.

[12] A. L. Oliveira and A. Sangiovanni-Vincentelli, "LSAT - An algorithm for the synthesis of two level threshold gate networks", *Proc. Int. Conf. Computer-Aided Design*, pp. 130-133, November 1991.

[13] X.L.E. Wernersson, N. Carlsson, B. Gustafson, A Litwin and L Samuelson, "Lateral current-constriction in vertical devices using openings in buried lattices of metallic discs", *Applied Physics Letters* vol. 71, no. 19, pp. 2803-2805, 1997.

[14] H. zdemir, A. Kepkep, B. Pamir, Y. Leblebici and U. ilingiroglu, "A capacitive threshold-logic gate", *IEEE J. Solid-State Circuits*, vol. 31, pp. 1141-1150, August 1996.

[15] H.-Y. Kwon, K. Kotani, T. Shibata, and T. Ohmi, "Low power neuron-MOS technology for high-functionality logic gate synthesis", *IEICE Trans. Electron.*, vol. E80-C, pp. 924-929, July 1997.

[16] M.Berkelaar, *lp_solve*, version 4.0, www.cs.sunysb.edu/algorithm/implement /lp_solve/.

[17] G.De Micheli, "Synthesis and optimization of digital circuits", McGraw-Hill Inc., NJ, 1994.

[18] S. Muroga, "Threshold Logic and its Applications", John Wiley, New York, NY, 1971.

[19] K. J. Chen, T. Akeyoshi, and K. Meazawa, "Monolithic integration of resonant tunneling diodes and FET's for monostablebistable transition logic elements (mobile's)", IEEE Electron Device Lett.,vol.16, no.1, pp. 70-73, Feb 1995.

[20] T. Akeyoshi, K. Meazawa, and T. Mitzutani, "Weighted sum threshold logic operation of MOBILE (Monostable-Bistable Transition Logic Element) using resonant-tunneling transistors", *IEEE Electron Device Lett.*, vol.14, no. 10, pp. 475-477, October 1993.

IEEE International Symposium on Defect and Fault Tolerance of VLSI Systems

A Framework to Evaluate the Trade-off among AVF, Performance and Area of Soft Error Tolerant Microprocessors[1]

GONG Rui, DAI Kui, WANG Zhiying

School of Computer, National University of Defense Technology, Changsha, China
{rgong, daikui, zywang}@nudt.edu.cn

Abstract

Because of the increasing susceptibility of the integrated circuits to soft errors, many techniques have been proposed in all the design levels to reduce the AVF (Architecturally Vulnerable Factor) of microprocessors with extra performance and area overheads. These overheads have a negative impact on the reliability. Conventional reliability evaluation frameworks do not take both performance and area overheads into account. A new metric, mMWTF (modified Mean Work To Failure), is proposed in this paper to capture the trade-off among AVF, performance and area. A quantitative approach to evaluate mMWTF is also presented, in which fault injection is used to estimate the AVF. To modify the conventional fault injection methods which inject only SEU (Single Event Upset), a new method is proposed to injects both SEU and MBU (Multi Bits Upset), the latter of which happens more frequently with the shrinking feature size. Because of the new metric and the new fault injection method, the framework presented in this paper is more accurate than conventional ones. As a case study, two control flow checking techniques are proposed and evaluated in this paper. The evaluation results demonstrate that the techniques with better balance among AVF, performance and area can better improve the reliability of microprocessors.

1. Introduction

One of the most critical challenges in microprocessor design is the soft errors caused by high-energy particles from cosmic ray or packaging material. Transistor's vulnerable cross-section collects these high-energy particles, which may finally cause undesirable results. A device's soft error rate depends on the dose of radiation, the amount of stored charge and the area of vulnerable cross-sections [1]. Some researches demonstrate that the soft error rate of a single device remains roughly constant or slightly decreasing for several technology generations in the next decade [2]. But the soft error rate of a whole microprocessor keeps increasing exponentially with the exponential increase of the devices in a single chip. Consequently, microprocessors, especially those for critical applications, may adopt various hardening techniques to tolerate soft errors. These techniques can reduce the AVF (Architecturally Vulnerable Factor) [3] of microprocessors by masking, detecting or recovering from soft errors. Meanwhile, these techniques bring some performance and/or area overheads. The probability of the particle striking during a given task increases with the execution time. And the striking probability is also proportional to the chip area exposed to radiation. So the performance and area overheads induced by soft error tolerant techniques have a negative impact on reliability.

Quantitative reliability evaluation is an important stage in the design process. With the evaluation results, designers can select the most appropriate techniques in various hardening approaches to satisfy the given reliability specifications. Conventional evaluation framework

---

[1] This work is supported by the National High Technology Research and Development Program of China under grant No. 2007AA01Z101 and the National Natural Science Foundation of China under grant No. 60773024.

1550-5774/08 $25.00 © 2008 IEEE
DOI 10.1109/DFT.2008.9

can not measure the trade-off among AVF, performance and area, thus it is neither a complete nor an accurate approach.

A framework to quantitatively evaluate the reliability with the consideration of both performance and area overheads is presented in this paper. A new metric, *mMWTF* (modified Mean Work To Failure), is proposed to capture the trade-off among AVF, performance and area. A quantitative approach to evaluate *mMWTF* is also presented. The performance can be evaluated by running a suite of benchmarks and the area can be obtained by synthesis using commercial EDA tools. The fault injection on the RTL (Register Transfer Level) model is used to estimate the AVF in this paper. Conventional fault injection methods inject SEU (Single Event Upset) only. A new fault injection method is also proposed in this paper to inject MBU (Multi Bits Upsets) by adding SRAM output pin into the fault injection locations. Two control flow checking techniques based on 8051 architecture, CFCCH (Control Flow Checking by Compiler signatures and Hardware checking) and CFCCH-R (CFCCH-Recovery), are proposed in this paper. Compared to traditional technique CFCSS (Control Flow Checking by Software Signatures) [4], these two techniques have less performance overheads. And CFCCH-R can automatically recover from control flow errors. As a case study, all the aforementioned three techniques are evaluated in our framework. The evaluation results demonstrate that CFCCH achieves better trade-off among AVF, performance and area, thus is more preferable than CFCSS and CFCCH-R.

2. Related Work

One of most important metric in reliability evaluation is Mean Time To Failure (*MTTF*), which is the expected time to failure. The Failure In Time (*FIT*) is the number of failures in one billion hours, which can be derived from *MTTF* as $FIT=10^9/MTTF$. Another metric, Mean Time Between Failure (*MTBF*) is related to MTTF as $MTBF=MTTF+MTTR$, where *MTTR* is the Mean Time To Repair. Because *MTTR* of a microprocessor is usually orders of magnitude smaller than *MTTF*, *MTTF* and *MTBF* can be used synonymously [5]. As MTTR-related information typically resides outside the microprocessors, *MTTF* is a more appropriate term for designers who do not have such information. *MTTF* can be computed as *MTTF=1/raw error rate*, where *raw error rate* is the number of faults in a given unit of time.

According to [3], these raw faults may not result in execution errors because of the masking capability of the architecture itself. For example, data corruptions in branch predictor or operand code of a NOP (No Operation) instruction would not cause incorrect execution. Mukherjee, et al., [3] defined AVF as the probability that a raw fault will cause an actual error. The hardening techniques can reduce the AVF. And *MTTF* can be expressed more accurately as *MTTF=1/(raw error rate×AVF)*. But *MTTF* considers only the positive impact of reducing AVF, and does not measure the negative effects of performance and area.

Weaver, et al., [5] proposed a novel metric, Mean Instruction To Failure (*MITF*), as *MITF=IPC×Frequency×MTTF=(Frequency×IPC)/(raw error rate×AVF)*, where *IPC* is Instructions Per Clock. *MITF* is the average number of instructions that can be executed between two failures. Compared to *MTTF*, *MITF* captures the trade-off between AVF and performance. But *MITF* is not appropriate in some cases. For example, inserting large amount of NOP instructions in a program will reduce the AVF, and keep the same frequency and *IPC*, causing the increase in *MITF*. But the number of useful instructions executed between two failures does not increase.

To solve such problems of *MITF*, Reis, et al., [6] proposed Mean Work To Failure (*MWTF*) as the average works that can be executed between two failures. *MWTF* is defined as

$MWTF=1/(raw\ error\ rate \times AVF \times execution\ time)$, where execution time is the time to execute a given unit of work. $MWTF$ is an appropriate metric to measure the trade-off between AVF and performance, but still does not take the area overhead into account.

The most important step in the evaluations with these aforementioned metrics is to estimate AVF accurately. Faults are injected into the RTL models of picoJave II and Alpha 21164 in [7] and [8] respectively to compute the AVF. The fault model of such injection experiments is SEU only. Mukherjee, et al. [3], estimate the upper bound of AVF by systematic level analysis. This method can estimate AVF at an early stage of the design process without RTL model, but it is not so accurate as fault injection experiment. And its fault model is still SEU only.

The metric proposed in this paper, $mMWTF$, modified the definition of $MWTF$ to take both the performance and area overheads into account. Thus $mMWTF$ is a more appropriate metric to capture the trade-off among AVF, performance and area. In order to estimate AVF more accurately, fault injection in RTL model is used in our framework. As the probability of MBU in SRAM increases sharply with the shrinking feature size, a new method is proposed in this paper to inject both SEU and MBU to simulate the raw faults in the deep sub micro era.

To verify the validation of the evaluation framework proposed in this paper, three Control Flow Checking (CFC) techniques have been evaluated using the metric of $mMWTF$. CFC is a commonly used method to detect wrong execution trace. CFCSS is a software implemented CFC technique [4], which has a relatively high performance loss. A checking IP is proposed in [9] to accelerate the checking process. Goloubeva, et al, [10] uses the additional executable assertions to check the control flow. Two CFC techiniques, CFCCH and CFCCH-R, are proposed in this paper. CFCCH uses software signature and hardware checking, thus it can reduce the performance loss. CFCCH-R implements fault recovery based on CFCCH.

3. Metric

In order to distinguish from the *raw error rate* in the last section which is not related to the die area, we define modified raw error rate ($mRER$) as the number of raw faults during a given unit of time in a given unit of die area. The unit of $mRER$ is $s^{-1}mm^{-2}$, different from s^{-1} of *raw error rate*.

A new metric, modified Mean Work To Failure ($mMWTF$) is defined as

$$mMWTF = \frac{1}{mRER \times A \times AVF \times execution\ time}$$

, where A is the die area. So any technique is worthwhile only if it increases $mMWTF$, i.e., if it reduces AVF to a greater degree than the product of area and performance it increases.

The probability of soft error is proportional to the die area and exposure time. So the term of $mRER$ used to define $mMWTF$ is identical for microprocessors with the same manufacture technology in the same radiation environment. The relative reliability of different microprocessors with the same technology and radiation dose can be expressed by relative $mMWTF$ as

$$\frac{mMWTF_1}{mMWTF_2} = \frac{A_2 \times AVF_2 \times execution\ time_2}{A_1 \times AVF_1 \times execution\ time_1}.$$

Compared to conventional reliability metric, $mMWTF$ considers both performance and area overheads. So it is a more accurate metric to measure the trade-off among AVF, performance and area, and can conduct the designers to choose appropriate technique from various hardening methods more accurately.

4. Methodology for Evaluation

4.1. Execution Time

Execution time is defined as the time to execute a given unit of work. And *mMWTF* is the average amount of such work that can be executed between two failures. A given unit of a work is an abstraction of concept whose specific definition depends on the applications. For server applications, the work can be defined as a transaction. In other applications, it can be a suite of benchmarks. The execution time can be derived by running benchmarks on RTL or cycle-accurate performance model.

4.2. Architecturally Vulnerable Factor

AVF can be estimated by systematic analysis [3] or fault injection [7][8]. Systematic analysis is faster but less accurate than fault injection. So we use fault injection method to obtain more accurate AVF. Besides, we modified the conventional method to inject both SEU and MBU faults to evaluate the behaviors of microprocessors in deep sub micro era.

The framework of fault injection is shown in Fig. 1. During the fault injection experiment, two RTL models, golden and injected microprocessors, run the same benchmarks. SEU and MBU are injected into injected RTL model by fault injector. And the golden model serves as a golden run for comparisons by the controller.

The Fault Injection Locations (FILs) of our method includes the inputs of flip-flops and the data outputs of SRAM. If flip-flops are chosen by the fault injector at Fault Injection Point (FIP), the input signal of chosen flip-flop is flipped for one cycle, so that the corrupted date can be latched to simulate an SEU. Compared to flip-flops, SRAM is more susceptible to soft errors. But the data in the internal SRAM array can not be modified without interrupting the normal execution of benchmarks. If SRAM is chosen as the FIL, one or more adjacent output data bits are upset for one cycle during the first output enable period after the FIP to simulate an SEU or MBU in SRAM. So the fault model of our injection experiment includes both SEU and MBU.

A large number of faults are injected to obtain statistically significant AVF as the probability of an injected fault caused a wrong execution. It should be noted that the particle-induced faults include SET, SEU and MBU. The SET pulse may not be latched by storage cells. But in our injection method, every fault is effective to cause data corruptions. Furthermore, the SRAM data corrupted by SEU or MBU may not be used. But our method makes every fault in SRAM to be read. So the injection method used in this paper is pessimistic, resulting in the upper-bound estimation of AVF.

4.3. Area

The accurate die area of microprocessor can only be obtained after layout. For the microprocessors using standard cells and semi-custom design flow, the whole die area is composed of cell area and routing area. In commercial EDA tools, utility is defined as the proportion of cell area to the whole die area. The utility can be the same for the microprocessors with the same technology, if layout by the designer with the same experience. So the cell area derived from synthesis is enough to evaluate the relative *mMWTF*.

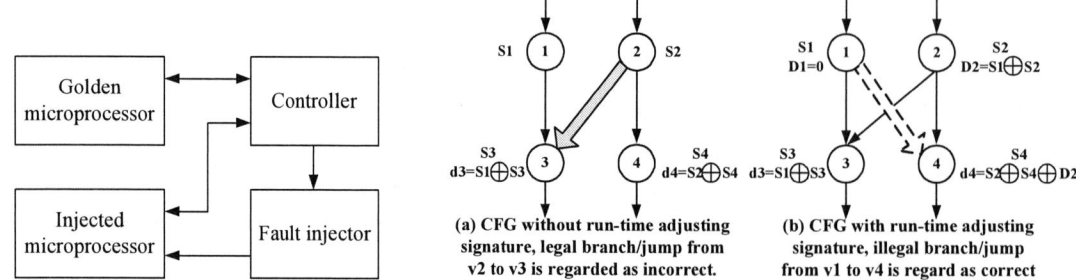

Figure 1. Framework of fault injection

(a) CFG without run-time adjusting signature, legal branch/jump from v2 to v3 is regarded as incorrect.

(b) CFG with run-time adjusting signature, illegal branch/jump from v1 to v4 is regard as correct

Figure 2. Control flow graph

5. Control Flow Checking

Control flow error means that the program is executed in an incorrect trace and can not finish correctly. Many Control Flow Checking (CFC) techniques have been proposed to detect control flow errors. The basic idea of CFC is that the control flow graph of program is determined when compiling. So the wrong control flow can be detected by comparing run-time trace with compiler determined graph. A novel software implemented CFC technique, Control Flow Checking by Software Signatures (CFCSS), is proposed by Oh, et al. [4]. It can be implemented on any microprocessors without any hardware modification. But if implemented on 8-bit 8051 architecture, there would be huge performance loss. Two CFC techniques on 8051 architecture, CFCCH and CFCCH-R, are presented in this section to address performance loss and recovery mechanism, respectively.

5.1. CFCCH

CFCSS defines the Control Flow Graph (CFG) as a directed graph, CFG=(V, E), where V={v | v is a basic block}, E={<vi, vj>|existing a branch or jump from vi to vj}. Assign a unique signature Si to every vertex vi. If $\exists$ <vi, vj>$\in$ E, then the XOR difference from vi to vj is dj=Si$\oplus$Sj. When the branch/jump from vi to vj is taken, compute the run-time signature sj=Si$\oplus$dj. If there is no control flow error, sj=Si$\oplus$dj=Si$\oplus$(Si$\oplus$Sj)=Sj. If the run-time signature sj is not identical with the compiler determined signature Sj, a control flow error is detected.

Because there are multi-fan-in vertices in CFG, only signature Sj and signature difference dj are not enough. As is shown in Fig. 2(a), the signature difference in v3 is determined by v1, thus d3=S1$\oplus$S3. Then the correct branch/jump from v2 to v3 would be detected as illegal. To sovle such problem, assign a run-time adjusting signature Di into every basic block. When a branch/jump is taken, calculate the run-time signature sj=Si$\oplus$Di$\oplus$dj. As is shown in Fig. 2(b), run-time adjusting signature of v1 is D1=0. Determine the signature difference of v3 by v1, then d3=S1$\oplus$S3. Assign the signature S2 to v2, and guarantee S2$\neq$S1. Because S3=S2$\oplus$D2$\oplus$d3=S2$\oplus$D2$\oplus$(S1$\oplus$S3), so D2=S1$\oplus$S2. Run-time adjusting signature solves the problem of multi-fan-in vertices, but it also induces aliasing. Signature difference of v4 is determined by v2, and is computed as d4=S2$\oplus$S4$\oplus$D2=S1$\oplus$S4. So the illegal control flow transfer for v1 to v4 would be considered as correct branch/jump. To solve such aliasing, CFCSS computes ss=Si$\oplus$ds when entering a single-fan-in vertice vs while computes sm=Si$\oplus$Di$\oplus$dm when entering multi-fan-in veritce vm.

The checking operation of CFCSS can be accomplished via common instructions. But if CFCSS is implemented on 8-bit 8051 architecture, each basic block would be inserted 11-byte signature instructions with 13-cycle performance loss. To solve the performance problem of CFCSS, Control Flow Checking by Compiler signatures and Hardware checking (CFCCH) is proposed in this paper. Instead of 11-byte signature instructions of CFCSS, CFCCH inserts only 3-byte signature data for every basic block, i.e. signature difference di, signature Si and run-time adjusting signature Di. The compiler signature algorithm of CFCCH is shown below.

```
Assign a unique signature S_i to every vertex v_i
Insert S_i in the head of v_i
for every vertex v_j in CFG do
  if v_j is NOT a multi-fan-in node then
    Select {v_i}=pred(v_j)
    Calculate signature difference of v_j, d_j=S_i⊕S_j
    Insert d_j before S_j in the head of v_j
  else
    Select v_i□pred(v_j), calculate d_j=S_i⊕S_j
    Insert d_j before S_j in the head of v_j
    for every vertex v_k□pred(v_j) do
      Calculate run-time adjusting signature, D_k=S_k⊕S_i
      Inset D_k after S_k in the head of v_k
    endfor
  endif
endfor
```

For hardware checking, two Special Function Registers, Sreg and Dreg are added to latch the signature Si and run-time adjusting signature Di of current basic block. The hardware checking is triggered by branch/jump instruction. The hardware checking spends only 3 extra cycles, which is far less than CFCSS. The hardware operations are shown in Table 1.

Table 1. Hardware checking operations

Cycle	Operation
1	Read di from program memory, update Sreg= Sreg⊕Dreg⊕di
2	Read Si from program memory, compare with Sreg. If not identical, set error flag
3	Read Di from program memory, update Dreg=Di

Note the aliasing in CFCCH. When entering a basic block, the hardware does not know whether it is a multi-fan-in vertex or not. So hardware compute the run-time signature as Sreg=Sreg⊕Dreg⊕di, which would do cause an aliasing. But the aliasing only happens when the destination address of branch/jump in v1 is exactly changed to the head of v4 as shown in Fig. 2(b).

The checking in CFCCH is executed by hardware, thus its performance overheads are much less than CFCSS. And in CFCSS, if branching into an incorrect basic block, the control flow error would not be detected until entering the next basic block. In this period, the microprocessor may produce further wrong results. The checking operation in CFCCH is triggered by every branch/jump instruction, making the control flow error to be detected immediately it happens.

5.2. CFCCH-R

Soft errors can be classified into Silent Data Corruption (SDC) and Detected Unrecoverable Error (DUE) based on whether they have been detected by error detection mechanisms. CFCSS and CFCCH do not have recovery mechanism. They reduce SDCs but at the same time increase DUEs. In order to reduce DUEs, CFCCH-R is proposed to provide recovery mechanism based on CFCCH.

The state of microprocessor can be recovered only if the correct context has been saved. The context is the state related data, which reside in special function registers (SFRs) and internal RAM in the 8051 architecture. In CFCCH-R, SFRs and internal RAM are saved and recovered by different mechanisms. Redundancy scheme is used for saving and recovering SFRs, which are distributed in different modules. Every SFR is backed up to a backup register at the saving point. Once detecting a control flow error, SFR is recovered according to its corresponding backup register. Both the SFR saving and recovering needs only one cycle. Internal RAM in 8051 architecture, like register file in RISC architecture, occupies large die area. There will be huge area overheads to save and recover internal RAM by the same redundancy scheme as SFRs. So a different approach, write buffer, is adopted. Data are only stored in write buffer unless the execution trace is proved to be correct. Once detecting a control flow error, the context can be recovered by invalidating all the data in the write buffer in one cycle. At the context saving point without control flow errors, all the write buffer data are stored into the internal RAM. The context saving cycle varies according to the amount of data in the write buffer.

It must be guaranteed that the microprocessor is in the correct execution trace at the context saving point. There are two types of saving point in our scheme. One is after control flow checking in the head of every basic block if not detecting any error. The other saving point is the write buffer overflow. If there are multi stores to internal RAM in the same basic block, the write buffer may overflow. At this point, all the write buffer data must be stored into internal RAM. All the SFRs are also backed up at the same time. Once detecting illegal execution trace, the microprocessor can roll back to the head of predecessor basic block or after some store instruction in the predecessor basic block. CFCCH-R adds recovery mechanism to CFCCH, so it can reduce DUEs. But compared to CFCCH, context saving causes more performance and area overheads.

6. Evaluation and Discussion

The aforementioned three CFC techniques, CFCSS, CFCCH and CFCCH-R, have their own advantages and disadvantages. For the area overhead, CFCSS is implemented purely in software level, thus has no area overhead; CFCCH adds some components for hardware checking; CFCCH-R has the largest area overheads due to its hardware context saving mechanism. For the performance overhead, CFCSS needs extra 13 cycles for every basic block; while CFCCH only spends 3 cycles on every checking; CFCCH-R has a higher performance overheads than CFCCH because of the context saving. For the AVF, if the trace switches into the middle of an illegal basic block in CFCSS, there will be a period of time before this error to be detected at the head of next basic block, during which further errors may be produced; the control flow errors can be detected in CFCCH once they happen because hardware checking is triggered by the transferring of control flow, but there may be aliasing; CFCCH-R is helpful to reduce DUEs for its recovery mechanism, but it still has aliasing. So it is difficult for designers to choose among these three techniques to achieve better reliability without quantitative evaluation. As a case study, the three CFC techniques are evaluated in the framework proposed in this paper to measure the trade-off among AVF, performance and area.

A suite of benchmarks runs at the RTL model to obtain the performance parameters. The execution cycles is shown in Fig. 3, in which the data is normalized to a non fault tolerant baseline microprocessor, NOFT. CFCSS has the highest performance loss due to its 13-cycle checking operation. While in CFCCH, only 3 cycles are added to every basic block, resulting in 9%~76% overheads. The performance overhead of CFCCH-R is higher than CFCCH, due to the extra context saving point. But it is still less than that of CFCSS.

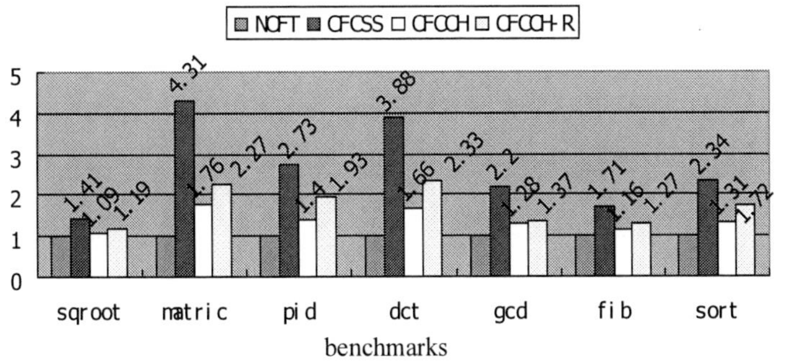

Figure 3. Normalized execution cycles

Table 2. AVF

	AVF_{SDC}	AVF_{DUE}	AVF
NOFT	72.38%	0%	72.38%
CFCSS	1.08%	68.56%	69.64%
CFCCH	1.15%	71.98%	73.13%
CFCCH-R	1.24%	32.1%	33.34%

Table 3. Area (mm^2)

	cell	memory	total
NOFT	0.244	0.099	0.343
CFCSS	0.244	0.099	0.343
CFCCH	0.248	0.099	0.347
CFCCH-R	0.318	0.114	0.432

Fault injection is used to estimate AVF. The fault model includes both SEU and MBU. And the faults are injected into internal RAM and all the SFRs. Faults are also injected into write buffer in CFCCH-R to simulate SEU and MBU in it. To obtain statistically significant results, 10000 faults are injected into each microprocessor. The upper bound of AVF is shown in Table 2, where the AVF_{SDC} is the ratio of SDCs to the total injected faults and AVF_{DUE} is that of DUEs to total faults. There is no fault detection mechanism in NOFT, so the AVF_{DUE} of NOFT is 0. And 27.62% faults injected into NOFT are masked by the architecture itself. CFCSS and CFCCH detect many control flow errors, resulting in high AVF_{DUE} of 68.56% and 71.98% respectively. But there are still some SDCs because CFCSS and CFCCH have no mechanism to detect the computational errors. CFCCH-R reduces DUEs via its recovery mechanism, but there are still 32.1% faults resulting in DUEs because of data corruptions in backup SFRs or write buffer. It is indicated from Table 2 that in the 8051 architecture, almost all the faults will finally cause control flow errors that would be detected by CFC techniques. Around 1% faults will cause computational errors that are shown as SDCs. Furthermore, the AVF_{SDC} of the three CFC techniques are nearly the same. So these techniques have nearly the same ability of control flow error detection, which is determined by the same signature algorithm. Because SDC is more critical than DUE, AVF_{SDC} is used to evaluate *mMWTF*. So *mMWTF* is the average work that can be executed between two SDCs. The percentage of SEU and MBU caused errors are shown in Fig. 4. It is indicated from Fig. 4 that nearly 50%~60% errors are caused by MBUs. So it is very important to use the fault model including both SEU and MBU.

The four microprocessors are synthesized in SMIC 90nm CMOS process under the timing constraint of 100MHz. The area is shown in Table 3. There is no hardware modification in CFCSS, thus no area overhead. The cell area of CFCCH is a little bigger than NOFT because of its hardware checking components. The total area of CFCCH-R is 25.7% larger than that of NOFT due to its backup registers and write buffer.

The three CFC techniques are evaluated using *MTTF*, *MWTF* and *mMWTF* as metrics respectively, and the normalized reliability to NOFT is shown in Fig. 5. Evaluating with

MTTF does not consider performance and area overheads. So relative *MTTF* only depends on AVF$_{SDC}$. CFCSS, CFCCH and CFCCH-R increase *MTTF* compared with NOFT by 76×, 63× and 58× respectively. When evaluated in *MWTF*, CFCSS has the lowest reliability due to its huge performance loss. CFCCH has the highest *mMWTF*, nearly the same as its *MWTF*, because it has only 1.02% area overhead. So CFCCH is regarded to be the best trade-off among AVF, performance and area. Note the comparison between CFCSS and CFCCH-R. When using *MWTF* as the metric, CFCCH-R is much better than CFCSS, which has the biggest performance loss. But the area overhead of CFCCH-R is 25.7%. So the *mMWTF* of CFCCH-R is nearly the same as CFCSS. Thus these two techniques can be regarded as achieving the same reliability. It is indicated that taking area into account is very important for the techniques with large area overheads, such as CFCCH-R.

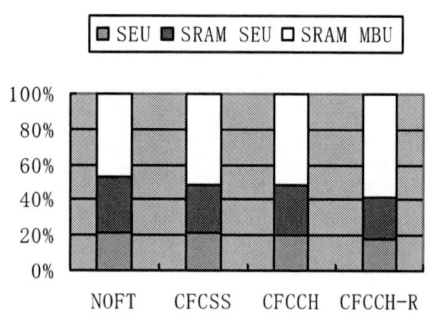

Figure 4. SEU and MBU caused error   **Figure 5. Normalized Reliability**

7. Conclusion

A new framework to evaluate the trade-off among AVF, performance and area is proposed in this paper. Compared to conventional reliability metric, the new metric proposed in this paper, *mMWTF*, is more accurate and appropriate for reliability evaluation. Two control flow checking techniques based on 8051 architecture, CFCCH and CFCCH-R, are proposed and evaluated in our framework. The evaluation results indicate that compared to CFCCH-R and traditional CFCSS, CFCCH achieves better trade-off among AVF, performance and area.

References

[1] P. Shivakumar, et al., "Modeling the effect of technology trends on the soft error rate of combinational logic", In Proceedings of International Conference on Dependable Systems Networks (DSN'02), 2002, pp. 389-398.

[2] T. Karnik, et al., "Scaling trends of cosmic rays induced soft errors in static latches beyond 0.18 μ ", Symposium on VLSI Circuits, Digest of Technical Papers, 2001, pp. 61-62.

[3] S. S. Mukherjee, et al., "A systematic methodology to compute the architectural vulnerability factors for a high-performance microprocessor", In Proceedings of IEEE/ACM International Symposium on Microarchitecture (Micro'03), 2003, pp. 29-40.

[4] N. Oh, et al., "Control flow checking by software signatures", IEEE Transactions on Reliability, vol. 51(2), 2002, pp. 111-122.

[5] C. Weaver, et al., "Techniques to reduce the soft error rate of a high-performance microprocessor", In Proceedings of International Symposium on Computer Architecture (ISCA'04), 2004, pp. 264-275.

[6] G. A. Reis, et al., "Design and evaluation of hybrid fault-detection systems", In Proceedings of International Symposium on Computer Architecture (ISCA'05), 2005, pp. 148-159.

[7] S. Kim, et al., "Soft error sensitivity characterization for microprocessor dependability enhancement strategy", In Proceedings of International Conference on Dependable Systems Networks (DSN'02), 2002, pp. 416-515.

[8] N. J. Wang, et al., "Characterizing the effects of transient faults on a high-performance processor pipeline", In Proceedings of International Conference on Dependable Systems Networks (DSN'04), 2004, pp. 61-70.

[9] C. A. L. Lisboa, et al., "Online hardening of programs against SEUs and SETs", In Proceedings of IEEE International Symposium on Defect and Fault-Tolerance in VLSI Systems (DFT'06), 2006, pp. 280-290.

[10] O. Goloubeva, et al., "Soft-error detection using control flow assertions", In Proceedings of IEEE International Symposium on Defect and Fault Tolerance in VLSI Systems (DFT'03), 2003, pp. 581-588.

IEEE International Symposium on Defect and Fault Tolerance of VLSI Systems

A Power Efficient Masking Technique for Design of Robust Embedded Systems against SEUs and SETs

M. Fazeli[1], S.G. Miremadi[2]

Department of Computer Engineering, Sharif University of Technology,
Tehran, Iran
[1]*m_fazeli@ce.sharif.edu,* [2]*miremadi@sharif.edu*

Abstract

In this paper, an SET and SEU tolerant latch suitable for use in embedded systems called SETUR (Single Event Transient and Upset Robust latch) is presented and evaluated. The SETUR is based on the use of a redundant feedback line and a CMOS delay element to tolerate the effect of the SETs occurring in the input line of the latch as well as SEUs occurring inside the latch. The experimental results show that the probability of an SET resulting in a soft error can be reduced up to 90% by choosing a proper delay value. The soft error rate of the SETUR due to SEUs occurring inside the latch is reduced by 95% while having lower area, power and performance overhead than the previously proposed latches.

1. Introduction

Single Event Upsets (SEUs) caused by energetic particles i.e., alpha and neutron particles have long been the major source of soft errors in memory elements [1-5]. But due to smaller feature sizes, lower voltage levels, higher operating frequencies, and reduced logic depth of today's digital circuits; the soft error rate due to particle strikes in the combinational parts i.e. Single Event Transient are severely increasing[6-10]. A study shows that the soft error rate per chip of logic circuits increases nine orders of magnitude from 1992 to 2011 [9]. Based on these facts, the utilization of masking techniques which consider both SEUs and SETs to increase circuit reliability are of decisive importance in deep submicron circuits.

One effective way to overcome the SEU effects at circuit level is to triplicate each latch of the system i.e. TMR-latch. Although the TMR-latch is highly reliable and widely used, it suffers from high area and power consumption overheads which are not acceptable for applications where area and power consumption are the primary concerns. In contrast, some other techniques try to reduce these overheads by employing redundant components inside the latch [11-15]. In these techniques, some extra transistors are used to either suppress or recover from the SEU effect. Although these techniques have lower area and power consumption overheads than TMR-latch, they are still vulnerable to soft errors caused by relatively high energetic particles and cannot be employed in highly reliable systems. In addition, these techniques are designed just for SEU tolerance while the SETs are becoming a serious concern in DSM technologies.

Using a delay element and a voting circuitry is a viable solution for masking the effect of SETs occurring in the combinational part [16][23-28]. In such techniques, the outputs of the combinational parts are replicated in a way that two or three redundant versions of the output line are delayed with different amounts of delay and the delayed version of the output line are compared using a voting element which can be a majority voter or a filtering circuit. However, due to large amount of redundancy, these techniques suffer from high area and power consumption overheads. Some other techniques such as [29][30] use the transistor resizing technique to mitigate the SETs effect. In these techniques, the transistors located in the SET vulnerable paths are selected and their sizes are increased in way that the SETs occurring in the combinational part are attenuated when crossing the resized transistors. These techniques consume less power and occupy less area overhead but they just reduce the probability of soft error occurrence and there is still probability that an SET caused by high energy particle causes a soft error.

1550-5774/08 $25.00 © 2008 IEEE
DOI 10.1109/DFT.2008.33

193

In this paper, an SET/SEU tolerant latch based on the joint use of transistor resizing and delayed replication is proposed. A redundant feedback line along with a filtering circuit is used to protect the SETUR from SEUs caused by particle strikes in the internal nodes of the latch. In addition, a CMOS delay element is utilized in the design of the SETUR to mask the SETs occurring in the combinational parts. The SEU and SET injection experiments have been carried out to investigate the reliability of the SETUR latch. The power consumption, area, and performance overheads of the SETUR latch are also measured and compared with previously proposed SET/SEU tolerant latches.

The rest of the paper is organized as follows. Section 2 presents the delay based SET/SEU-tolerant latch designs. In section 3, the design of proposed latch is presented. The soft error rate estimation method used in this paper is explained in section 4. The experimental results are reported in section 5. Finally section 6 concludes the paper.

2. The Delay Based SET/SEU-Tolerant Latch Designs

Using a delay element along with a voting circuit at the output parts of a combinational circuit [16][23-28] is a common technique to filter out the SETs and various efficient delay elements have been designed for this purpose [31]. The voter can be a majority voter or a filtering circuit called C-element.

Figure 2, shows an SET/SEU tolerant latch based on the use of a delay element and a C-element. To explain about how the latch can mask the SETs and SEUs, we will first describe the C-element circuit.

The C-element is a state holding element, and it has the basic property that inverts its inputs only if both of its inputs are of identical logic value. If the two inputs of this circuit have different values, the previous output value will be retained. Figure 1.a shows the transistor schematic of the C-element and Figure 1.b shows the output of the C-element for some sample input values.

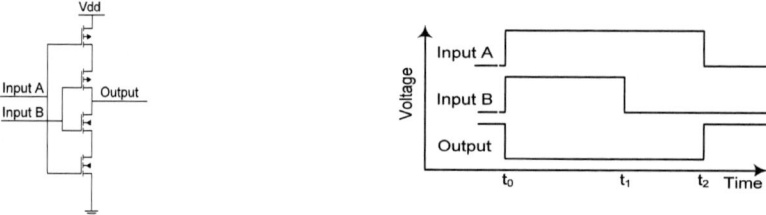

Figure 1. The C-element (a), Sample input/output of the C-element (b)

To show how the latch shown in Figure 2 can mask the SETs occurring in the output of the combinational part, suppose that an SET has occurred in the output line of the circuit shown in Figure 2 (see Figure 3.a). The initial value of the output in this example is '0' and in the time interval (T0, T1) it becomes '1' due to a particle strike so that the delayed output is '1' during time interval (T2, T3). Based on the behavior of the C-element, the value of the output does not change during the time intervals (T0, T1) and (T2, T3). If the delay time is less than the possible transient pulse width, the C-element may pass the transient to its output. Figure 3.b illustrates a case that the amount of the delay is less than the SET pulse width. As shown in Figure 3.b, the output has overlapped with the delayed output during the time interval (T1, T2). Consequently, both inputs of the C-element have the same value in this time interval so that the C-element acts as an inverter during the time interval (T1, T2). During the time interval (T2, T3), the out signal and its delayed version have different value and the C-element holds its previous output value. As it can been seen in Figure 3.b, the occurred SETs in the out signal at time T0 has passed the C-element i.e. the output of the C-element has been affected by the SET.

To mask the effect of particle strike inside the latch (Figure 2), a redundant latch is exploited and also a C-element is used to combine the outputs of the main and redundant latch. Based on the fact that the probability that two latches are simultaneously affected by particle strikes is so low; the C-element used in the latch can effectively mask the SEU occurred in one of the latches. In the case of having an SEU in one of the latches, the output of the C-element becomes float i.e. it is not connected to a driving source so that a keeper is needed for the latch correct operation.

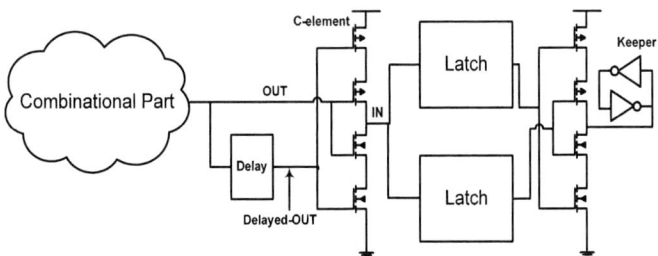

Figure 2. A technique to tolerate SETs using C-element and delay element

The other possible technique to overcome the SEUs and SETs based on the delay element is to use of two delayed version of combinational output along with a majority voter for SET tolerance purpose (Figure 4). In this technique, the second delay element delays the output value twice than the first one. If the amount of delay is more than the SET pulse width, it will be guaranteed that in each time interval at least two of the three inputs of the voter have identical values. In order to overcome the SEUs, the latch is triplicated and again a voter is used to filter out the effect of SEUs occurring in one the latches. Both of the mentioned techniques have two drawbacks. First, Although using delay element and a C-element or a voter filter out the effect of SETs occurring in the output of the combinational part, both techniques are still vulnerable to SETs occurring in the C-element or voter output line i.e. if an SET occurs in the output line of the C-element or voter, it will captured by the latches, and no filtering mechanism exists to tolerate the SET. Second, both of them especially one based on triplication and voter have high area and power consumption overhead especially static power.

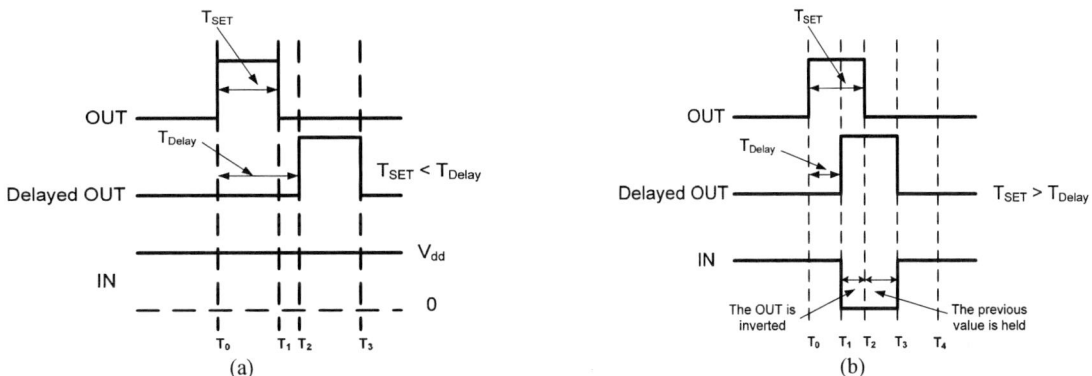

Figure 3. An example of an SET occurring on the output line in the SET-tolerant Circuit based on C-element:
(a) $T_{SET} < T_{Delay}$, (b) $T_{SET} > T_{Delay}$

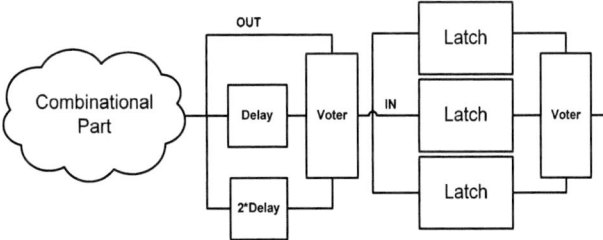

Figure 4. A technique to tolerate SETs using voter and delay elements

3. The SETUR Latch

The joint use of a delay element and a C-element is exploited in the design of the SETUR latch in a way that the SETUR latch can simultaneously tolerate the SETs and SEUs. The main advantage of the SEUTR latch as compared to those introduced in previous section is that the SETUR has lower area and power overhead. Therefore it is suitable for the use in application where power is a main concern.

In Figures 5 and 6, the conventional latch and the SETUR latch are depicted. In the design of the SETUR (Figure 6), there are two redundant feedback lines called original feedback and redundant feedback. These two feedback lines are combined by using a C-Element. Having two feedback lines guarantees that there are always two copies of the stored value inside the latch. Since a C-element is already used in the SETUR to tolerate the SEUs occurring inside the latch, contrary to the technique shown in Figure 2, it is not required to use additional C-element for masking the SETs occurring in the input line. In fact, in the design of the SETUR, the C-element is used simultaneously for masking the SEUs and SETs. The only element that should be used to make the SETUR robust against SETs is a delay element. By using a delay element and a C-element along with a redundant feedback, the SETUR can effectively mask the SEUs and SETs. The only thing that should be considered is the amount of delay applied to the input line. Several studies shows that the SET pulse width can be up to even few nanoseconds depending on the energy of the particle or the amount of deposited charge [20][21][22] in DSM technologies. But it should be noted that the probability of a particle strike decrease exponentially as the amount of deposited charge or the amount of particle energy increases. Based on this observation the delay element can be adjusted by the amount of the required reliability.

As mentioned the SETUR latch can mask both SETs and SEUs. In the case of SETs occurring in the input line of the latch which is the output of a combinational part, the use of delay element with a proper amount makes the two inputs of the latch to have different values. In this case the C-element filters out the effect of the SET. The main problem is choosing the proper amount of delay such that required SET filtering is gained.

The SETUR latch is also robust against particle striking an internal node of the latch. The SEU susceptible nodes of the SETUR latch are shown in the Figures 6 in the bold format. It should be noted that the nodes which are directly connected to supply voltage and ground are not susceptible to the particle strikes. In the other words, the diffusion area is SEU susceptible. Based on this fact, the nodes A, B and C in the original feedback line and the nodes A', B' and C' in the redundant feedback line and the output nodes are the SEU susceptible nodes of the SETUR latch. The SETUR latch is robust against the particle strike in the nodes A, B, A', B'. That is because a C-element is used to isolate the two feedback lines. It means that, if a voltage transient due to a particle strike occurs in the mentioned nodes, the transient do not affect the corresponding node in the other feedback line. Therefore only one input line of the C-element is affected and the other one holds the correct value. In this case the C-element filters out the transient and after a while the affected node will be overwritten by the correct value. The important thing that should be noted is: the robustness of the nodes A, A', B and B' are independent of the amount of deposited charge by the particle or the particle energy. In the other words, regardless of the particle energy, the mentioned nodes are completely robust against particle strike.

The main weakness of the proposed latch is that the nodes C, C' and output are vulnerable to particle strikes. If a transient occurs in the output line, since it simultaneously propagates to two feedback lines, the two input of the C-element are affected by the particle strike at the same time and the transient will result in soft error inside the latch. Although these nodes are still SEU susceptible, the Q_{crit} of these three nodes can be significantly increase by properly sizing the transistor T3, T4, T5, T6 without degrading the latch performance. Moreover, even with a minimum sized T3, T4, T5, T6, the SETUR has lower SEU susceptibility than that of the internal nodes of the conventional latch. It should be also noted that the nodes C and C' are less SEU susceptible than the output node. In fact, the node C is vulnerable to a particle strike when the latch output is 1 or the both PMOS transistors T3 and T4 are conducting. Moreover, if this condition occurs, the striking particle should be a negative one, i.e. the particle should discharge the load capacitance of the node. Since in our experiment, it is assumed that the probability of having logic value of 0 and 1 is equal and also the negative and positive particles occurrence probabilities are equal, the SEU susceptibility of nodes C and C' are 50% of the output node. Although increasing the size of transistors T3, T4, T5 and T6 can increase the Qcrit which in turn make the output node more robust against particle strikes; it affects the total power consumption and occupied area of the latch which will be discussed more in the next section.

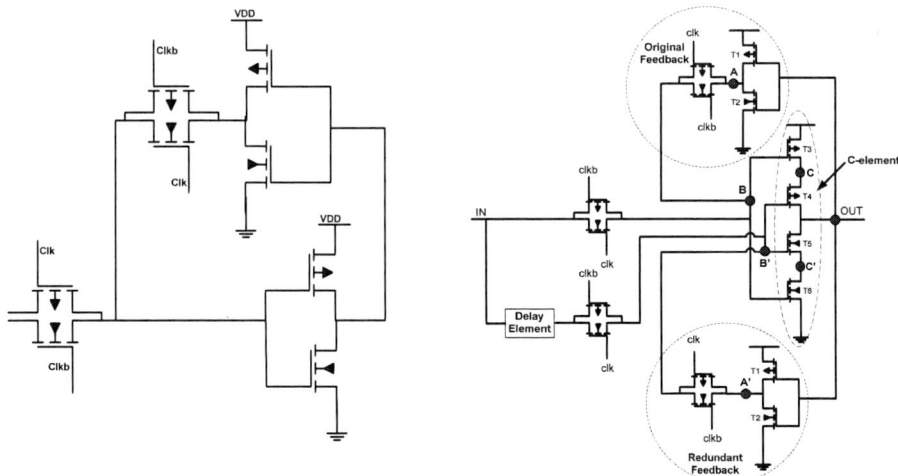

Figure 5. The conventional latch   Figure 6. The SUTR Latch

4. The Soft Error Rate Estimation Approach

The impact of particles hit on a circuit node in digital circuits is modeled by an independent current source at a drain node (Eq. 1) [17].

$$I_{Particle} = \frac{Q}{T\sqrt{\pi}} \sqrt{\frac{t}{T}} e^{\frac{-t}{T}} \qquad (1)$$

where, T is a time constant and technology dependent parameter. This parameter is measured for different technology size in [17]. The parameter t represents the time in which a particle is starting to produce the current.

When a particle strikes a sensitive node of a latch or more generally a memory element, an SEU occurs if collected charge Q exceeds critical charge Q_{CRIT} of a circuit node. In other words, Q_{CRIT} can be defined as the minimum charge collected due to a particle strike that can cause an SEU [17]. The Q_{CRIT} is measured by the equation 2:

$$Q_{CRIT} = \int_0^{t_f} I_d \qquad (2)$$

where, Id is the drain current caused by the particle strike. Tf is the time duration from when a particle strikes a sensitive node of the latch until the latch starts to continue the flipping process and called flipping time [18]. In order to measure the Q_{CRIT} of a sensitive node, the independent current source is applied to the node and the tf is extracted and the Q_{CRIT} of the node is measured by solving the equation 2.

After the Q_{CRIT} of all susceptible nodes of the latch is extracted, the probability of an SEU occurrence in the latch when a particle strikes a sensitive node can be measured by the equation 3:

$$P_{SEUi} = \frac{WoV_i}{T_{Clk}} \cdot \frac{A_i}{A_{Total}} \cdot (P_0 \cdot P_{Positive} \cdot \int_{Q_{Crit}}^{\infty} Prob(Q) \cdot d(Q) + P_1 \cdot P_{Negative} \cdot \int_{Q_{Crit}}^{\infty} Prob(Q) \cdot d(Q)) \qquad (3)$$

where the P_{SEUi} is the probability of soft error occurrence of the node i, WoV_i is the windows of vulnerability [15] which is defined as the time interval in which the latch is vulnerable to SEUs. The WoV of the latch is the summation of the latch setup time and the time duration in which the latch is in its keeping state. The $Area_i$ is the occupation area of the sensitive node, in fact the Ai is the drain area of the sensitive node and A_{Total} is the total occupied area of the latch. The Prob(Q) is the probability distribution of the deposited charge.

The integral in equation 3, shows that the probability that a particle deposits more charge than the critical charge of the node. P_0 (P_1) is the probability of the node i having the value of 0 (1). $P_{Positive}$ ($P_{Negative}$) is the probability of a particle deposits a positive charge (negative charge) i.e. if a node is storing the value of 1 and a particle strike tries to charge the already charged equivalent capacitance of the node, it does not cause an SEU and vice versa. In our experiments we assume that the probability of having value of 0 and 1 in a specific node are equal so that the $P_0=P_1=0.5$. It is also assumed that the probability of a positive and negative particle strike are equal so that $P_{Positive}=P_{Negative}=0.5$.

The Prob(Q) is obtained from hazucha's empirical model [17] and is shown in equation 4:

$$\text{Prob}(Q) = {k}/{Q_S} \cdot e^{(-q/Q_s)} \tag{4}$$

where the Q_S is the charge collection slope and K is a constant and both of them depends strongly on supply voltage and doping profile, i.e. they are technology dependent parameters. The value of Q_S and K are also obtained from [17] for the used technologies. The total SER of a latch can be measured by the summation of all susceptible nodes PSEUi according to equation 5:

$$P_{SEU} = \sum_{i=1}^{n} P_{SEUi} \tag{5}$$

5. The Experimental Results

As it was mentioned in section 4, the SETUR latch is vulnerable to particle strikes occur in the nodes C, C' and output (Figure 6). To overcome this problem, the resizing of the C-element transistors is used to increase the critical charge of these nodes. Although the resizing results in higher critical charge which in turn leads to lower total soft error rate of the latch, but it also increases the both static and dynamic power consumption, input to output delay and also the occupied area. To show the effect of resizing on the mentioned parameters, the SETUR latch is simulated by HSPICE tool for two DSM technologies including 45nm and 90nm using predictive transistor model [19]. The length and widths of the transistors T3, T4, T5, T6 in the SETUR are sized for different (W/L) ratios namely 1, 2, …,9 and the SER improvement of the output node is measured and depicted in Figure 7.

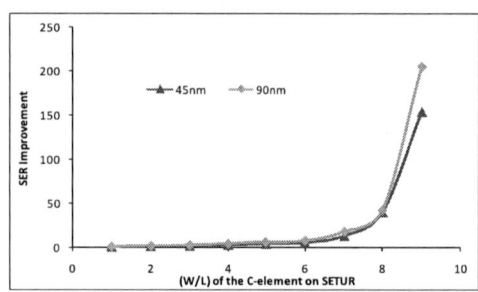
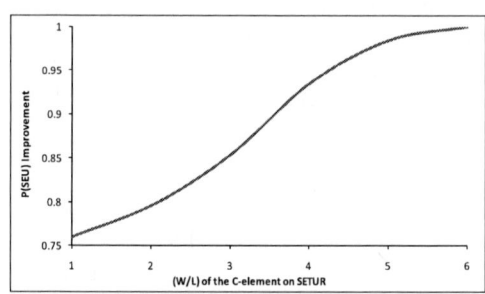

| Figure 7. SER improvement of the output node as a function of (W/L) | Figure 8. Total SEU tolerance improvement of the SETUR latch as function of (W/L) |

In Figure 7, the horizontal axis shows the different values of (W/L) and the vertical axis shows the SER improvement of the node out with respect to the minimum size transistor i.e. (W/L)=1. As (W/L) increase the SER improvement increases even for (W/L) ratios more than 7 the amount of improvement is significant.

The Figure 8, shows the total PSEU improvement of the SERUT as a function of (W/L) with respect to a conventional latch (Figure 5), i.e. a latch with no SEU protection mechanism. As shown in Figure 8, even with the minimum sized transistors the total PSEU improvement of the SERUT is more than 75% and it reaches to about 95% when the (W/L) ratio is 4.

In Figures 9 and 10, the increase in dynamic and static power consumption of the SETUR latch is reported as a function of SER improvement with respect to the SETUR latch with minimum sized transistors. It can be inferred from the Figure 9 and 10 that for 90nm and 45nm technologies, the dynamic and static power consumption do not increase significantly as the SER improvement of the output node increases. It should be noted that as it is shown in Figures 7 and 8, by choosing the (W/L)=4, we gain about 3 times improvement in SER and about 95% improvement in the total P(SEU) of the SETUR as compared to the conventional latch. Based on these two observations, for having about 95% reliability, it is enough to choose the (W/L)=4 (Figure 8) resulting in 3 times improvement in SER of the output node (Figure 7) and just about 20% increase in the dynamic power in 45nm technology. This claim is also true for 90nm technology due to have same behavior with the 45nm especially in SER improvements less than 50 times. The important gained result is the very low increase in static power as the SER improvement increases. As shown in Figure 10, the static power increases at most 18% when the SER improvement varies from 1 to 200 times. Since the static power is becoming a serious concern in DSM technologies, the SETUR can be an effective design for DSM and ultra DSM technologies.

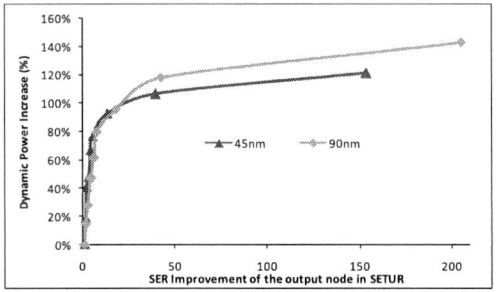

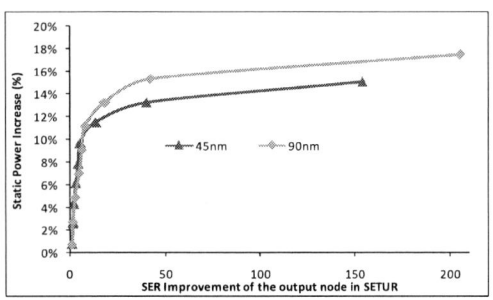

Figure 9. Dynamic power increase of the SETUR as a function of output SER improvement

Figure 10. Static power increase of the SETUR as a function of output SER improvement

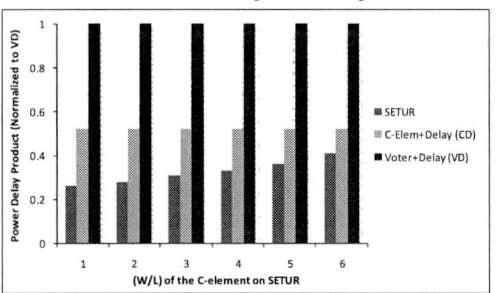

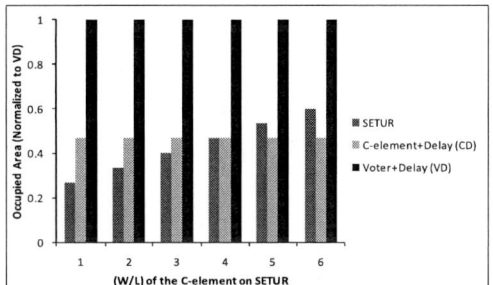

Figure 11. Power Delay Product of the SETUR and other technique as a function of (W/L)

Figure 12. Occupied area of the SETUR and other technique as a function of (W/L)

The transistor resizing technique also results in latch input to output delay increase. To measure the input to output delay increase of the SETUR the Power Delay Product (PDP) metric is used. The PDP is interpreted as energy per result and is the most widely used metric to compare low power design techniques due to incorporation of both performance and power overheads in the comparison [15]. In order to investigate the PDP efficiency of the SETUR latch, some simulations have been carried out for extraction of the PDP for the SETUR and two techniques introduced in the section 2 and the simulation results are shown in Figure 11. As shown in Figure 11, the PDP of the SETUR latch is extracted for different (W/L) ratios. It should be noted that resizing is just applied to the SETUR, since the two other techniques do not use resizing technique. In other words, theses two technique have the reliability improvement of 100% as compared to conventional latch. Based on what can be observed from Figure 11, the PDP of the SETUR is almost less than the two other techniques when the (W/L) changes from 1 to 6. In the case of (W/L)=4, where the reliability of three design are almost the same, the PDP improvement of the SETUR is about 20% as compared to C-element+Delay technique and about 70% improvement as compared to Voter+Delay technique (the results are normalized to the Voter+Delay design). Same as our previous simulation, the occupied area of the SETUR latch is a function of (W/L) ratio of the C-element transistors. The Figure 12, shows the occupied area of the SETUR latch as compared to the two other designs. The results are normalized to the occupied area of the Voter+Delay design. As shown in Figure 12, the occupied area of the SETUR latch is less than the two others for (W/L) ratios less than 4 and it slightly increases as the (W/L) increases.

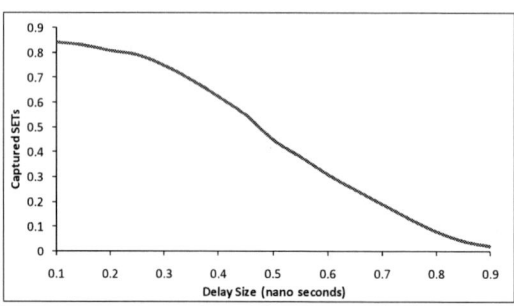

Figure 13. Probability of SET being captured as a function of the amount of delay

As it was mentioned in the section 3, the SET tolerance capability of the SERUT latch likes the other introduced techniques depends on the amount of the applied delay to the input line of the latch which in turn depends on the SET pulse widths. In order to show the effect of the amount of delay in SET tolerance capability of the SETUR, the independent current source with different values for particle deposited charge Q for 45nm technology size is applied to the input line of the latch when the latch is transparent. The results of the SET injections are shown in Figure 13. Based on the results, as the amount of delay increases, the probability of an SET being captured by the SETUR latch decreases such that for a delay value equal to 0.9ns second, less than 10% of injected SETs are captured by the latch. Although according to some studies [20][21][22] the SET pulse widths may be even more than 2 nanoseconds, but it should be noted that the SET pulse widths depends on the amount of the deposited charge. In addition, the probability of a particle strike with specific amount of charge decreases exponentially as the amount of deposited charge increases.

6. Conclusions

In this paper, a novel SEU/SET-tolerant latch called SETUR which is based on the joint used of delayed replication and transistor resizing has been proposed and evaluated. The proposed latch not only has the ability to tolerate the particle strikes occurring inside the latch but also prevents the SETs that occur in the combinational part from resulting in SEUs in the latch. This is why this latch is called SEU/SET-hardened. To precisely evaluate the reliability of the proposed latch, the circuit level SEU/SET injection experiments have been carried out using HSPICE tool. The SEU/SET-injection results showed that, the SETUR latch can effectively mask the SETs occurring in its input line if the amount of delay is properly chosen. In addition, it was shown that, by properly transistor resizing, the SETUR latch consumes lower power and occupies lower area than the two other widely used designs while having almost the same SEU tolerance capability.

7. References

[1] R. C. Baumann, "Radiation-induced soft errors in advanced semiconductor technologies," Device and Materials Reliability, IEEE Transactions on, vol. 5, no. 3, pp. 305-316, 2005.

[2] T. Heijmen, D. Giot, and P. Roche, "Factors that impact the critical charge of memory elements," Proceedings of the 12th IEEE International Symposium on On-Line Testing, pp. 57-62, 2006.

[3] T. Heijmen, P. Roche, G. Gasiot, K. R. Forbes, and D. Giot, "A Comprehensive Study on the Soft-Error Rate of Flip-Flops From 90-nm Production Libraries," Device and Materials Reliability, IEEE Transactions on, vol. 7, no. 1, pp. 84-96, 2007.

[4] J. Patel, "Characterization of Soft Errors Caused by Single Event Upsets in CMOS Processes," IEEE Transactions on Dependable and Secure Computing, vol. 1, no. 2, pp. 128-143, 2004.

[5] F. X. Ruckerbauer and G. Georgakos, "Soft Error Rates in 65nm SRAMs--Analysis of new Phenomena," Proceedings of the 13th IEEE International On-Line Testing Symposium, pp. 203-204, 2007.

[6] J. Benedetto, P. Eaton, K. Avery, D. Mavis, M. Gadlage, T. Turflinger, P. E. Dodd, and G. Vizkelethyd, "Heavy ion-induced digital single-event transients in deep submicron Processes," Nuclear Science, IEEE Transactions on, vol. 51, no. 6, pp. 3480-3485, 2004.

[7] K. J. Hass and J. W. Ambles, "Single event transients in deep submicron CMOS," Circuits and Systems, 1999. 42nd Midwest Symposium on, vol. 1 1999.

[8] R. Rajaraman, J. S. Kim, N. Vijaykrishnan, Y. Xie, and M. J. Irwin, "SEAT-LA: A soft error analysis tool for combinational logic," Intl. Conf. on VLSI Design (VLSID), pp. 499-502, 2006.

[9] P. Shivakumar, M. Kistler, S. W. Keckler, D. Burger, and L. Alvisi, "Modeling the effect of technology trends on the soft error rate of combinational logic," Dependable Systems and Networks, 2002. Proceedings. International Conference on, pp. 389-398, 2002.

[10] C. Zhao, X. Bai, and S. Dey, "Evaluating Transient Error Effects in Digital Nanometer Circuits," Reliability, IEEE Transactions on, vol. 56, no. 3, pp. 381-391, 2007.

[11] S. R. Whitaker, J. Canaris, and K. Liu, "SEU Hardened Memory cells For CCSDS Reed Solomon Encoder," IEEE Trans. on Nuclear Science, NS-38(6):1471-1477, Dec. of 1991.

[12] Y. Zhao, S. Dey, "Separate Dual-Transisto Registers –A Circuit Solution for On-Line Testing of Transient Error in UDSM-IC", in Proc. of 9th IEEE Int. On-Line Testing Symp. (IOLTS'03), pp. 7-11, 2003.

[13] T. Monnier, F. M. Roche, G. Cathebras, "Flip-flop Hardening for Space Applications", Proc. IEEE Workshop on Memory Technology, Design and Testing, pp. 104-107, 1998.

[14] Liang Wang, Suge Yue, and Yuanfu Zhao, Senior Member. "Low-Overhead SEU-Tolerant Latches". in the Proceedings of International Conference on Microwave and Millimeter Wave Technology, pp. 1-4, China, April 2007.

[15] M. Omana, D. Rossi, C. Metra, "Latch Susceptibility to Transient Faults and New Hardening Approach", IEEE Trans. On Computers, Vol. 56, n. 9, Sept. 2007, pp. 1225-1268.

[16] S. Mitra, M. Zhang, S. Waqas, N. Seifert, B. Gill, K. S. Kim, Combinational Logic Soft Error Correction," In Proc. of the IEEE International Test Conference, November, 2006, Santa Clara, California, USA.

[17] P. Hazucha, C. Svensson: "Impact of CMOS technology scaling on the atmospheric neutron soft error rate", IEEE Trans. On Nuclear Sc. Vol. 47, n. 6, Dec. 2000, pp. 2286-2294.

[18] A. Ejlali, B.M. Al-Hashimi, M.T. Schmitz, P. Rosinger, S.G. Miremadi, "Combined time and information redundancy for SEU-tolerance in energy-efficient real-time systems", IEEE Trans. on Very Large Scale Integration Systems, Vol. 14, April 2006, Issue: 4, pp. 323- 335.

[19] W. Zhao, Y. Cao, "New generation of Predictive Technology Model for sub-45nm design exploration," ISQED, San Jose, CA, March 2006, pp. 585-590.

[20] J. M. Benedetto, P. H. Eaton, D. G. Mavis, M. Gadlage, ,T. Turflinger, "Digital Single Event Transient Trends With Technology Node Scaling", IEEE TRANS. ON Nuclear Science, Vol. 53, No. 6, pp. 3462-3465, December 2006.

[21] M. J. Gadlage, R. D. Schrimpf, J. M. Benedetto, P. H. Eaton, D. G. Mavis, M. Sibley, K. Avery, T. L. Turflinger, "Single Event Transient Pulse widths in digital Microcircuits", IEEE Trans. on Nuclear Science, Vol. 51, No. 6, pp. 3285-3290, December 2004.

[22] J. M. Benedetto, P. H. Eaton, D. G. Mavis, M. Gadlage, T. Turflinger, "Variation of Digital SET Pulse Widths and the Implications for Single Event Hardening of Advanced CMOS Processes", IEEE Trans. on Nuclear Science, Vol. 52, No. 6, pp. 2114-2119, December 2005.

[23] M. Zhang and N. Shanbhag, "A transient-tolerant high-performance circuit style," IEEE Workshop on System Effects of Logic Soft Errors, April 2006, Urbana Champaign, Illinois.

[24] S. Mitra, M. Zhang, N. Seifert, T.M. Mak, S. K. Kee, "Soft Error Resilient System Design through Error Correction", IFIP International Conference on Very Large Scale Integration (ISVLSI'06), October 2006, pp. 332 – 337.

[25] W. Wang, H. Gong, "Edge Triggered Pulse Latch Design With Delayed Latching Edge for Radiation Hardened Application", IEEE Trans. On Nuclear Science, Vol. 51, No. 6, December 2004.

[26] M. Nicolaidis, "Design for Soft Error Mitigation", IEEE Trans. on Device and Materials Rreliability, VOL. 5, NO. 3, SEPTEMBER 2005.

[27] L. Anghel, D. Alexandrescu, M. Nicolaidis, "Evaluation of a Soft Error Tolerance Technique Based on Time and/or Space Redundancy", Proceedings of the 13th Symposium on Integrated Circuits and Systems Design, Manaus, Brazil, September 2000.

[28] S. Krishnamohan, N. R. Mahapatra, "A Highly-Efficient Technique for Reducing Soft Errors in Static CMOS Circuits", Proceedings of the IEEE International Conference on Computer Design, pp. 126-131, San Jose, California, October 2004.

[29] Q. Zhou, K. Mohanram, "Gate sizing to radiation harden combinational logic " IEEE Transactions on Computer-Aided Design of Integrated Circuits and Systems, Volume 22, Issue 1, Jan. 2006, pp. 155 – 166.

[30] Q. Zhou and K. Mohanram, "Transistor sizing for radiation hardening," in Proc. Int. Reliability Physics Symp., Phoenix, AZ, 2004, pp. 310–315.

[31] K. Gyudong, K. Min-Kyu, C. Byoung-Soo, K. Wonchan, "A low-voltage, low-power CMOS delay element", IEEE Journal of Solid-State Circuits, Volume 31, Issue 7, pp. 966 – 971, Jul 1996.

IEEE International Symposium on Defect and Fault Tolerance of VLSI Systems

Can knowledge regarding the presence of countermeasures against fault attacks simplify power attacks on cryptographic devices?

Francesco Regazzoni[1], Thomas Eisenbarth[2], Luca Breveglieri[3], Paolo Ienne[4], and Israel Koren[5]

[1] ALaRI - University of Lugano, Lugano, Switzerland. Email: regazzoni@alari.ch

[2] Horst Görtz Institute for IT Security, RUB, Bochum, Germany. Email: {eisenbarth,cpaar}@crypto.rub.de

[3] DEI - Politecnico di Milano, Milano, Italy. Email: Luca.Breveglieri@elet.polimi.it

[4] I & C - EPFL, Lausanne, Switzerland. Email: Paolo.Ienne@epfl.ch

[5] University of Massachusetts, Amherst, MA, USA. Email: koren@ecs.umass.edu

Abstract

Side-channel attacks are nowadays a serious concern when implementing cryptographic algorithms. Powerful ways for gaining information about the secret key as well as various countermeasures against such attacks have been recently developed. Although it is well known that such attacks can exploit information leaked from different sources, most prior works have only addressed the problem of protecting a cryptographic device against a single type of attack. Consequently, there is very little knowledge on how a scheme for protecting a device against one type of side-channel attack may affect its vulnerability to other types of side-channel attacks. In this paper we focus on devices that include protection against fault injection attacks (using different error detection schemes) and explore whether the presence of such fault detection circuits affects the resistance against attacks based on power analysis. Using the AES S-Box as an example, we performed attacks on the unprotected implementation as well as modified implementations with parity check circuits or residue check circuits (mod3 and mod7). In particular, we focus on the question whether the knowledge of the presence of error detection circuitry in the cryptographic device can help an attacker who attempts to mount a power attack on the device. Our results show that the presence of error detection circuitry helps the attacker even if he is unaware of this circuitry, and that the benefit to the attacker increases with the number of check bits used for the purpose of error detection.

1. Introduction

Security plays a fundamental role in today's world: the rapid growth of embedded devices executing security-sensitive applications, and the global interest in doing on-line business pose new concerns for system designers. Unfortunately, as has become evident in recent years, the use of strong cryptographic algorithms can not guarantee a sufficient level of privacy and security. In fact, increasingly simpler and cheaper attacks on cryptographic algorithms are being developed. Unlike mathematical approaches, the so called *side channel attacks* exploit the weaknesses of the hardware and/or software platform on which the algorithm is implemented in order to acquire sensitive information, rather than attempting a direct attack on the algorithm. One of the most successful examples of such attacks is that of power analysis that exploits the correlation between the power consumed by a device and the data being processed. The effectiveness of such attacks is very high

1550-5774/08 $25.00 © 2008 IEEE

DOI 10.1109/DFT.2008.53

because they do not require any particular knowledge about the implementation of the device. Besides power attacks, other side-channel attacks have been developed and shown to be very effective; for example, an attacker can get access to sensitive information by maliciously injecting faults and analyzing the faulty behavior of the system.

These unconventional forms of attack have been studied in the past and some solutions to counteract them have been proposed [2, 9, 16, 17, 18, 22, 23]. Still, there is currently no perfect protection against power attacks. However it is, possible to make the task of the attacker more difficult and more time consuming by applying several countermeasures at different levels. Similarly, fault injection attacks can be protected against using, for example, robust error detection schemes.

The focus of most previous work has been on a single type of attack, and as a result, it is not clear whether and how a countermeasure that defeats one particular attack affects the robustness against a different attack.

In this paper we concentrate on devices that are protected against fault injection attacks using different error detection schemes, and our goal is twofold: investigate whether one of the circuits is easier to attack than the others, and find out whether knowledge about the presence of an error detection circuit can be exploited by the attacker. For our study we use the AES algorithm as an example and we consider hardware implementations of the non linear transformation (Sbox) within AES. We have added error detection circuits based on parity checks as well as residue checks modulo 3 and 7, to the original implementation, and we attacked them using the *Correlation Power Attack* [6]. To compare the implementations we analyzed simulated data as well as current consumption traces obtained from transistor level simulation. The simulations, used with different attack hypothesis, allowed us to evaluate the impact of the known presence of error detection circuitry on the effectiveness of such power attacks.

The rest of the paper is organized as follows. Section 2 summarizes previous research efforts involving fault injection and power analysis attacks. Section 3 introduces the cryptographic algorithm we used as a case study, the AES, and describes our circuits for error detection. The simulation environment as well as the results are presented in Section 4. Section 5 concludes the paper.

2. Related work

Since the introduction of side-channel attacks by *Kocher et al.* [12], a large number of publications have addressed this problem. This is because such attacks – that target the device that executes the algorithm rather than the mathematical structure of the algorithm – are very powerful and often reasonably cheap. Since the attacker usually needs physical access to the device, the security threats are most severe for secure embedded system designs, in particular, for smart cards. Among the so called *side-channels* attacks, time, power, electromagnetic emanation and the deliberate injection of faults are of particular relevance [1, 11, 12].

The problem was addressed by a large number of previous works, where the common approach for defeating power analysis attacks has been to remove as much as possible the correlation between the power consumption and the data being processed, by using a combination of *hiding* and *masking* [15]. The proposed countermeasures act at different levels of the design process, ranging from algorithmic techniques [9, 23], through architectural approaches [16, 17] down to hardware-level methods [18, 22]. Still, despite the substantial amount of research, a perfect protection against such attacks is not yet available.

Defeating fault injection attacks [3, 4] is, in comparison, a simpler task since robust and efficient protection schemes can be developed based on error detection codes which have been traditionally used in data transmission, especially for dealing with noisy channels. Parity check is an example of a classical code that was adapted to the needs of cryptographic devices. New solutions tailored to the specific needs of cryptography have also been developed. In this case, the error protection is mainly based on *Concurrent Error Detection (CED)* techniques. Typically, every time an error is detected, the detection circuit stops the normal execution of the algorithm to prevent the generation of the wrong output. As a result, the attacker is unable to view and analyze the faulty output.

Clearly, the correctness of the output can be verified by duplication of the computation either in area or in the time domain. However, both of these methods are expensive since they either double the execution time or the area requirements.

In addition to the above mentioned general approaches, some publications have focused on particular cryptographic algorithms or on particular classes of algorithms. Wolter et. al. [24] presented an implementation of the IDEA algorithm in which data are first encrypted and then, as a check, decrypted with the result compared to the original plaintext. Public key algorithms are analyzed in [8], where the authors propose an approach for providing error detection and correction by means of redundant arithmetic based on finite rings. Although comprehensive, the proposed implementation is complex and results in a higher area overhead compared to other methods.

In their work [10], Karri et al. propose a CED that is tailored to substitution-permutation network ciphers and compares the modified parity of the input with the parity of the output.

A residue-based error detection scheme for RSA was proposed in [5]. Though there remains a small possibility of undetected errors, the area overhead of the proposed scheme is very small.

The CED scheme proposed for AES in [2] uses one parity bit for every internal state byte of the AES. This scheme, which requires a limited amount of area for its implementation, detects all odd errors and in many cases even errors as well.

The main limitation of all the previous research efforts is that they target only one specific attack. Instead, while designing a scheme for protection against a given attack, it is crucial to also take into account how the implemented countermeasure would affect other possible attacks. This problem was addressed only in very few previous publications. Maingot *et al.* [13] have analyzed the impact of four different differential fault analysis countermeasures on the power analysis resistance. Their study, that was carried on using gate level simulation, shows that the power analysis vulnerability depends on the particular error detection code used.

In [14], the authors compare different error detection codes in order to provide hints about the best code selection for secure chips. The authors show that a complementary parity scheme that can improve the circuit robustness, induces higher overhead.

Transistor level simulation was performed in [20], where the authors analyze an error detection code based on parity check and discuss how this protection could affect the resistance against power based attacks and the role played by measurement noise. This paper extends the work reported in [20] by analyzing the impact of other error detection schemes (mod 3 and mod 7 residue checks) to find out which scheme is less vulnerable to power attacks. Furthermore, in this paper we study the benefits that attackers may enjoy if they are aware of the presence of error detection circuits in the cryptographic device.

3. AES and error detection circuits overview

The AES (Rijndael) [7] algorithm implements a block cipher for symmetric key cryptography. The block size is 128 bits, while the key size is 128, 192 or 256 bits. During the encryption process, four different transformations are iterated a number of times depending on the key size.

The four basic transformations are: *ShiftRows*, *SubBytes* (using SBoxes), *MixColumns*, and finally *AddRoundKey*. The added key is different in each round and these round keys are generated by a *key schedule* routine that takes the secret key and executes an expansion as specified in the standard. The same basic transformations are used during decryption, but they are applied in reverse order.

For the AES S-box, we implemented four versions of the non-linear function. The first circuit implements the non linear transformation as described in the standard, while in all the other three we added logic to provide error detection. We considered three types of error detection circuits: *parity based* and *residue code modulo 3* and *modulo 7*.

The parity check we used is the one proposed in [2]: a single even parity bit is added to every byte. The number of additional bits required for error detection based on residue code depends on the particular modulus used.

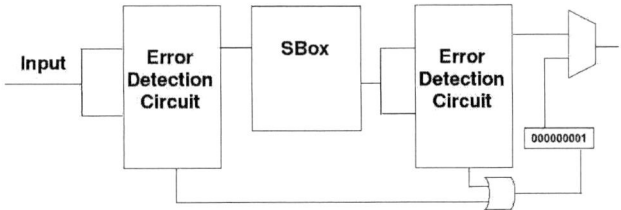

Figure 1. Block diagram of the Sbox with an attached generic error detection circuit.

Figure 1 shows an AES S-box with an added error detection circuit. The error detection circuit checks the correctness of the input and the output of the S-box. When new data enters the S-box, the check bits are separated from the data bits and an error detection is performed. If no error is detected, the 8 data bits enter the S-box circuit. The S-box produces then the result of the non-linear transformation plus the corresponding check bits. At this point the second check is performed, again as described before. If no error is detected in both checks, the output of the S-box can be forwarded to the next round transformation, otherwise, a faulty output composed of all zeros except the right most bit is generated to signal the error.

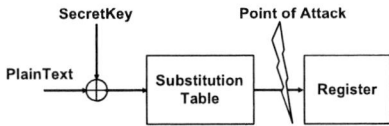

Figure 2. Our attack point: the output of the SBox.

4. Results of power attacks using simulated data

In this section we describe the circuits considered in this study and discuss the results we obtained when mounting attacks on simulated data.

We have analyzed three possible circuits for adding error detection capabilities to the AES S-Box: parity code, residue code modulo 3 and residue code modulo 7. As previously described, the goal of this paper is twofold: investigate whether one of the circuits is easier to attack than the others, and find out whether knowledge about the presence of an error detection circuit can be exploited by the attacker.

To perform such an evaluation, we consider the situation depicted in Figure 2: the input plaintext has a size of 8 bits, and only one Sbox is used following the secret key addition. In this case, our attacks target the full output of the non linear transformation. The circuit used for attacking the implementation with error detection circuits is similar to the one depicted in Figure 2, but the output of the Sbox has a number of check bits that depends on the specific scheme used: one extra bit for parity, two extra bits for residue modulo 3 and three extra bits for residue modulo 7.

To evaluate the resistance against power analysis attacks, we performed a Correlation Power Analysis (CPA) that evaluates all the key guesses using statistical correlation. In particular, the correct key guess is the one that shows the highest value for the correlation between the power consumption and the hypothesized Hamming weight.

We divided our attack into two steps. During the first one, faster but less precise, we mounted our CPA on simulated data. In the second step, we used power consumption traces collected using transistor level simulation, which are much closer to a real world attack situation. The use of simulated data has a major advantage: it is available at an early stage of the design flow and thus is of a particular interest for determining the correct point of attack and for estimating the minimum number of measurements needed to distinguish the correct key from the others.

During the first step, we obtained simulated data using the approach described in the work of Örs et al. [19] to have a first estimate of how knowledge about the error detection circuitry could help the attacker. As previously described, the target of our attack is the output of the non linear transformation that in many proposed AES architectures is stored in a register. Thus, we have developed a simulator that, using a fixed key and N random plaintexts, writes the Hamming weight of the Sbox output at each encryption cycle. We have performed this step for the normal Sbox as well as for the Sboxes with added error detection circuits. Then, using the same plaintext of the previous step, we have calculated the Hamming weight of the 8 least significant bits of the Sbox output. We have then calculated the correlation between the traces generated during step two and the ones produced during step one, increasing the number of considered plaintexts from 1 to N. In this case, the used attack hypothesis is always of 8 bit, thus this represents the situation in which the attacker is unaware of the presence of the error detection. Finally, we have calculated the Hamming weight of the Sbox output including also the check bits generated by the error detection circuit when present, and as before, we have calculated the correlation between the traces generated during step one and those produced during this step, increasing at each run the number of considered plaintexts.

The attacks we performed show that in the case of a single Sbox, the correlation for the case where the presence of the error detection circuit is known, is equal to 1, while for the case where the presence of the error detection is unknown to the attacker the correlation is lower, but the value never goes below 0.95. Based on this we can conclude that being unaware of the presence of the error detection circuits will not adversely impact the effectiveness of the power attack. We should keep in mind that in this case the values for the correlation are high because the simulations have been carried out in a noise free environment.

To have a more realistic analysis and to be able to compare different error detection circuits, we developed *VHDL* codes for all of them and performed transistor level power simulation using NANOSIM, as described in the work of Regazzoni et al. [21]. The technology library we used is the UMC 0.18μm and the options of the tool were set to provide the highest possible resolution both for time and current. As in the previous series of attacks, we randomly generated N plaintexts and using them we simulated the circuit keeping the key constant. We then added a white Gaussian noise to each trace: this noise is normally distributed with mean equal to 0 and a given variance, and mimics the typical noise generated by the measurement instruments. In particular, to simulate different noise conditions, we generated several sets of traces, obtained by adding noise with a different variance to the same base traces generated at transistor level. We finally performed CPA calculating the correlation between the Hamming weight of the Sbox output and the power traces, increasing at each run the number of considered plaintexts.

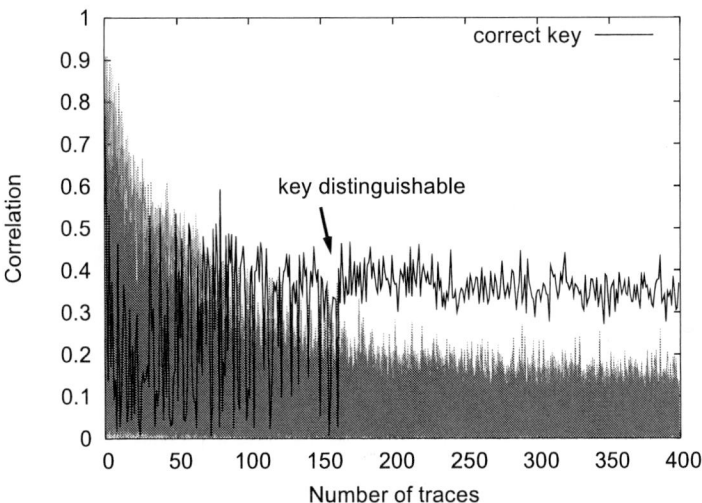

Figure 3. Attack on the AES Sbox.

Our results based on these experiments show that the presence of an error detection circuit clearly helps the attacker. These results confirm the recent observation of [20] that added parity bits have a negative impact on the Power Analysis resistance, and extend the same observation to more complex error detection schemes that have not been analyzed in the past. Furthermore, we were able to observe that there is a strong relation between the success of a power analysis attack and the number of redundancy bits used for error detection. As can be seen from Figure 3, in the case of the Sbox without error detection circuits, the correct key starts to be clearly distinguishable after 160 traces. The required number of traces is smaller when there is a parity bit (see Figure 4 where the correct key is distinguishable after 130 traces) or residue modulo 3 check bits (see Figure 5 where the correct key starts to be distinguishable after 100 traces). This indicates that the more check bits the code has, the quicker the CPA attack gets the secret key. We want to emphasize that the noise added to the traces is normally distributed, thus while considering how the error detection code helps the attacker we are not interested in checking whether we can guess correctly all the keys starting from a specific number of traces, but we are more interested in checking the trend of the time instant when the correct key becomes visible.

Additionally, to completely evaluate the impact of known presence of error detection on the success attack rate, we mounted a CPA including in the attack hypotheses the knowledge about the error detection circuit as well as without including it. Figures 5 and 6 show attacks on a Sbox that implements a residue code modulo 3, with the first figure showing the results for the case in which the presence of the detection code was known to the attacker, while the second figure presents the case when the attacker is unaware of the presence of error detection circuits. As can be seen from these figures, in both cases the correct key starts to be distinguishable soon after 100 traces and in any case it is distinguishable using fewer traces than in Figure 3, where there are no error detection circuits. Being aware of the presence of the residue code provides a small benefit to the attacker, but there is not statistically relevant evidence that the knowledge regarding the presence of a code to protect the SBOX, makes the attack significantly easier. This means, in other word, that the

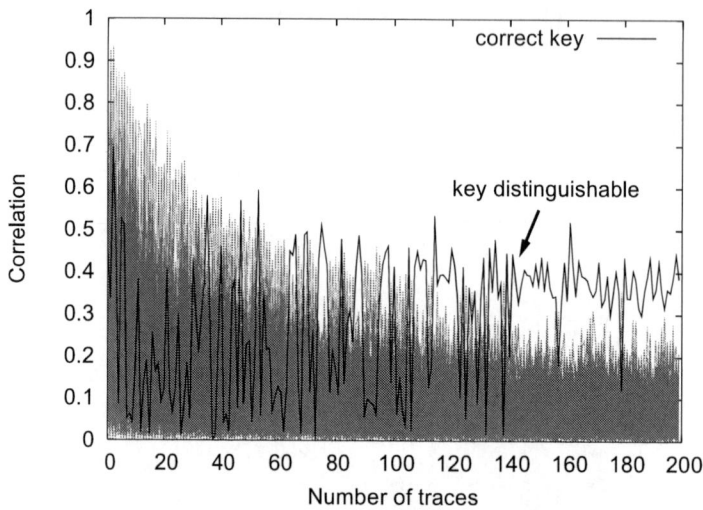

Figure 4. Attack on AES Sbox with parity based error detection.

presence of the redundancy helps the attacker even if he is not aware of it. This is due to the fact that the redundancy bits, even if not explicitly included in the hypothesis, are not random (and thus comparable to noise), but they depend on the bits targeted by the attack.

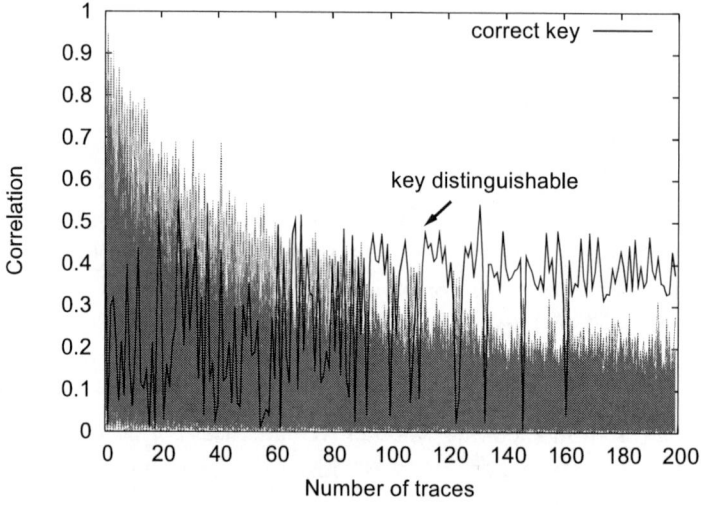

Figure 5. Attack on AES Sbox with residue code modulo 3 based error detection.

5. Conclusions

In this paper, we have evaluated the effect that different error detection circuits may have on the resistance to power analysis attacks. We focused in particular on the non linear transformation of the AES. We discussed how the attacker's knowledge of the presence of error detection circuits could affect the effectiveness of side channel attacks based on power consumption. Our results show that in a noisy situation, the check bits could help the attacker in any case, even if the presence of error

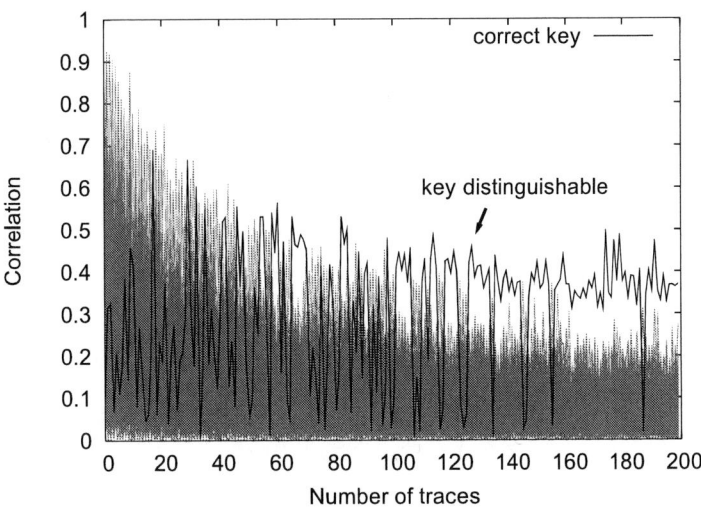

Figure 6. Attack on AES Sbox with residue code modulo 3 based error detection – the presence of the detection circuit is unknown to the attacker.

detection is unknown to the attacker. Furthermore, we showed that the higher the number of check bits is, the easier the attack is.

References

[1] F. Bao, R. H. Deng, Y. Han, A. B. Jeng, A. D. Narasimhalu, and T.-H. Ngair. Breaking public key cryptosystems on tamper resistant devices in the presence of transient faults. In *Security Protocols, 5th International Workshop*, vol. 1361 of *Lecture Notes in Computer Science*, pp. 115–124. Springer Verlag, 1998.

[2] G. Bertoni, L. Breveglieri, I. Koren, P. Maistri, and V. Piuri. Error analysis and detection procedures for a hardware implementation of the advanced encryption standard. In *IEEE Transactions on Computers*, vol.52, pp. 492–505, 2003.

[3] E. Biham and A. Shamir. Differential Fault Analysis of Secret Key Cryptosystems. *Advances in Cryptology - CRYPTO*, p. 513, 1997.

[4] D. Boneh. On the Importance of Eliminating Errors in Cryptographic Computations. In *Journal of Cryptology*, vol.14, pp. 101–119. Springer, 2001.

[5] L. Breveglieri, I. Koren, P. Maistri, and M. Ravasio. Incorporating Error Detection in an RSA Architecture.

[6] E. Brier, C. Clavier, and F. Olivier. Correlation power analysis with a leakage model. In *Cryptographic Hardware and Embedded Systems — CHES 2004*, vol. 3156 of *Lecture Notes in Computer Science*, pp. 16–29. Springer, 2004.

[7] J. Daemen and V. Rijmen. AES Proposal: Rijndael. *http://csrc.nist.gov/CryptoToolkit/aes/rijndael/Rijndael.pdf*, 1999.

[8] G. Gaubatz and B. Sunar. Robust Finite Field Arithmetic for Fault-Tolerant Public-Key Cryptography. In *2nd Workshop on Fault Diagnosis and Tolerance in Cryptography - FDTC'05*, 2005.

[9] T. Izu, B. Möller, and T. Takagi. Improved elliptic curve multiplication methods resistant against side channel attacks. In *Progress in Cryptology — INDOCRYPT 2002*, vol. 2551 of *Lecture Notes in Computer Science*, pp. 296–313. Springer Verlag, 2002.

[10] R. Karri, G. Kuznetsov, and M. Gössel. Parity-based concurrent error detection of substitution-permutation network block ciphers. In *Cryptographic Hardware and Embedded Systems — CHES 2003*, vol. 2779 of *Lecture Notes in Computer Science*, pp. 113–124. Springer, 2003.

[11] P. C. Kocher. Timing attacks on implementations of Diffie-Hellman, RSA, DSS, and other systems. In *Advances in Cryptology — CRYPTO '96*, vol. 1109 of *Lecture Notes in Computer Science*, pp. 104–113. Springer Verlag, 1996.

[12] P. C. Kocher, J. Jaffe, and B. Jun. Differential power analysis. In *Advances in Cryptology — CRYPTO '99*, vol. 1666 of *Lecture Notes in Computer Science*, pp. 388–397. Springer Verlag, 1999.

[13] V. Maingot and R. Leveugle. Error Detection Code Efficiency for Secure Chips. Electronics, Circuits and Systems, ICECS '06, Dec. 2006.

[14] V. Maingot and R. Leveugle. On the use of error correcting and detecting codes in secured circuits. *Microelectronics and Electronics Conference, 2007. RME. Ph.D. Research in*, pp. 245–248, July 2007.

[15] S. Mangard, E. Oswald, and T. Popp. *Power Analysis Attacks*. Springer, 2007.

[16] D. May, H. L. Muller, and N. P. Smart. Non-deterministic processors. In *Information Security and Privacy — ACISP 2001*, vol. 2119 of *Lecture Notes in Computer Science*, pp. 115–129. Springer Verlag, 2001.

[17] D. May, H. L. Muller, and N. P. Smart. Random register renaming to foil DPA. In *Cryptographic Hardware and Embedded Systems — CHES 2001*, vol. 2162 of *Lecture Notes in Computer Science*, pp. 28–38. Springer Verlag, 2001.

[18] S. W. Moore, R. J. Anderson, P. Cunningham, R. Mullins, and G. Taylor. Improving smart card security using self-timed circuits. In *Proceedings of the 8th International Symposium on Asynchronous Circuits and Systems (ASYNC 2002)*, pp. 193–200. IEEE Computer Society Press, Apr. 2002.

[19] S. B. Örs, F. K. Gürkaynak, E. Oswald, and B. Preneel. Power-analysis attack on an asic AES implementation. In *ITCC (2)*, pp. 546–552. IEEE Computer Society, 2004.

[20] F. Regazzoni, T. Eisenbarth, J. Großschädl, L. Breveglieri, P. Ienne, I. Koren, and C. Paar. Power Attacks Resistance of Cryptographic S-boxes with added Error Detection Circuits. In *Proceedings of the 21st IEEE International Symposium on Defect and Fault-Tolerance in VLSI Systems (DFT'07)*, 2007.

[21] F. Regazzoni, S. Badel, T. Eisenbarth, J. Großschädl, A. Poschmann, Z. Toprak, M. Macchetti, L. Pozzi, C. Paar, Y. Leblebici, and P. Ienne. A Simulation-Based Methodology for Evaluating the DPA-Resistance of Cryptographic Functional Units with Application to CMOS and MCML Technologies. In *International Symposium on Systems, Architectures, Modeling and Simulation (SAMOS VII)*, 2007.

[22] K. Tiri, M. Akmal, and I. M. Verbauwhede. A dynamic and differential CMOS logic with signal independent power consumption to withstand differential power analysis on smart cards. In *Proceedings of the 28th European Solid-State Circuits Conference (ESSCIRC 2002)*, pp. 403–406, Sept. 2002.

[23] C. D. Walter. MIST: An efficient, randomized exponentiation algorithm for resisting power analysis. In *Topics in Cryptology — CT-RSA 2002*, vol. 2271 of *Lecture Notes in Computer Science*, pp. 53–66. Springer Verlag, 2002.

[24] S. Wolter, H. Matz, A. Schubert, and R. Laur. On the VLSI implementation of the international data encryption algorithm IDEA. In *IEEE International Symposium on Circuits and Systems, ISCAS'95*, vol. Ĩ, pp. 397–400, 1995.

IEEE International Symposium on Defect and Fault Tolerance of VLSI Systems

Modeling and Evaluation of Threshold Defect Tolerance *

Z. Patitz and N. Park
Department of Computer Science
Oklahoma State University, Stillwater, OK 74078-1053
npark@a.cs.okstate.edu

Abstract

This paper presents a theoretical study on the threshold defect tolerance level. A new defect level model is proposed and used as a basis for the study in order to facilitate a theoretical modeling and evaluation of defect tolerance under a circumstance in which conventional testing processes are not practically viable. A comprehensive and thorough defect classification and characterization is conducted under specific design constraint when traditional testing is not practically deployable. An approach to an extensive theoretical and parametric compilation and optimization will be introduced in order to reveal a theoretical threshold defect tolerance level at which each choice defect tolerance method and strategy can be justified. The simulation results reveal there exists a theoretical threshold of defect tolerance level that will help guide the test and defect tolerance engineers to identify the optimal and viable testing and defect tolerance level.

1 Introduction

To meet the challenges of future technologies for circuits we need a robust approach for the realization of defect level with high confidence of a circuit [1, 2, 3, 4, 5, 6, 7]. The needs of emerging technologies are centered around defect tolerance, both in the manufacturing stage and during operation. To properly define the effects of defects on emerging technology we must stray from current theory and practice on defect level and compose a more complete model that can encompass the wide array of defects that may occur on a diverse set of devices including gates, wires, interconnects, etc. The nature of the emerging technologies is probabilistic and therefore is ideally suited for a statistical model [9][12].

Due to the complexity and device density it may not be economically or physically possible to fully test (or test at all) circuits at the sub-micron scale or beyond. This raises several important questions: What kind of yield can we achieve? What level of redundancy or other defect tolerance is optimal?; Is it worth the overhead in testing to reduce the defect level?; What is the balance between testing and defect tolerance with these technologies to make them economically feasible, and, therefore, practical for production and use?

Some of the current issues in testing also motivate the investigation of limited- or non-tested circuit defect tolerance. The International Technology Roadmap for Semiconductors [8] discusses these issues in-depth. Some of the problems facing a circuit design and production cycle that incorporates testing include: IDDQ implementation efficacy challenges; Overtesting; Damage caused by testing; Erratic, non-deterministic device behavior; and Defects occurring in test-only circuitry [8, 13].

N. Park the contact author.

1550-5774/08 $25.00 © 2008 IEEE
DOI 10.1109/DFT.2008.56

Furthermore, deep submicron technologies are inherently fragile due to the small, molecular scale of the design. For such technologies testing and defect discovery becomes increasingly difficult [10]. Also, while transistors and geometries continue to reduce in size, the cost of testing is not reducing [13].

The aim of the work is ultimately to establish a theoretical foundation for economical decision making on whether to test; how far to test; whether testing is feasible; and, otherwise, whether there is an alternative to defect tolerance without testing.

In this work, a new defect tolerance for the circuits and systems under the circumstances where little or no testing is allowed or feasible, is to be investigated. New methodologies for the design for defect tolerance will be presented and theoretically validated.

This paper is organized as follows: in Sections 2 and 3, defects and defect level classification, characterization and theoretically possible optimization method will be presented; then in section 4, the proposed defect tolerance tolerance level will be presented with simulation results; followed by conclusions in the last Section.

2 Defect Classification

To properly and fully classify defects we must encompass all possible sets of defects. Therefore, we use the following four divisions to create our ten defect classes.

Defects Defined: Our first division is defined defects. A defect is either defined or not defined. A defined defect is one that is contained within the defect model and is known to occur. Defined defects may be tested for and/or tolerated. Defects that are not defined are defects that may happen but are not known, or cannot be defined well enough to include in a defect model. An undefined defect may be tolerated, but cannot be tested for. Undefined defects that do not occur are not explicitly included in the classification scheme, though their values precisely oppose the occurring undefined defects.

Defect Present: Each defect has a probability of occurring. A defect is either present or not by virtue of that probability. The probabilities of defined and undefined defects are different values (variables) inherited from the architecture under examination.

Defect Found: For each tested defect, the defect is either found or not. The quality of the test dictates if a defect is present and not found by a test. Tests can also detect defects that are not present, which is a false positive and also undesirable behavior. By definition, undefined defects do not have a probability of being found.

Defect Tolerated: Each defect class has a probability of being tolerated or not. The tolerance level for defined and undefined defects may differ in many circumstances which is discussed in further detail later.

Below is a table displaying the full coverage of possible defects and their effect on defect level and cost. Each classification has two divisions. Only those defect classes that are possibly predictable and effect the defect and/or quality level are included in the table. Each defect class has listed a probability with respect to a variable set and associated v_{ijkl} listed.

Defined_i	Yes (v_1)				No (v_0)
$\text{Defect}_j(p)$	Yes (v_{11})		No (v_{10})		Yes (v_{01})
$\text{Defect Found}_k(q)$	Yes (v_{111})	No (v_{110})	Yes (v_{101})	No (v_{100})	N/A
Tolerated_1 (t)	v_{1111} tpq	v_{1101} $tp(1-q)$	v_{1011} $t(1-p)\gamma q$	v_{1001} $t(1-p)(1-\gamma q)$	v_{011} $t'u$
Not Tolerated_0	v_{1110} $(1-t)pq$	v_{1100} $(1-t)p(1-q)$	v_{1010} $(1-t)(1-p)\gamma q$	v_{1000} $(1-t)(1-p)(1-\gamma q)$	v_{010} $(1-t')u$

We can now define each v_{ijkl} as a probability. Let p be the probability that a defect occurs and let q be the quality of the test used. The defect tolerance level of a defect is defined by t. For undefined defects we use the probability t' to represent the defect tolerance. In many cases t' will be inherently less than t, but for simplicity of the model, we will assume $t' = t$ for the time being. This assumes that the defect tolerance, say redundancy, is equally implemented across defined and undefined defect possibilities. Further assumptions can be made with respect to t' when analyzing specific architectures and/or technologies. The constraint of $t' \leq t$ will hold though.

The quality of the test used is the probability that a test will find a defect. For the test quality we have two variables. The first variable, q, is the probability that the test catches a present defect. The second variable, γ, is the probability that the test finds a defect that does not exist, a false-positive. We assume that $0 \leq \gamma \leq 1$, where if $\gamma = 0$ the test only finds true defects and if $\gamma = 1$ every defect the test set finds is a false-positive. That is, γ is the probability that a found defect is not actually a defect. This assumption makes the false-positive detection probability the same or less than the found defect probability. With this is mind, the actual test quality becomes $q - \gamma q$. We can view γ as the unreliability of the test set with respect to false-positives.

As can be seen in the table, each defect has a corresponding v_{ijkl}, where i is whether or not it is defined, j is whether the defect actually occurred or not, k is whether the test found the defect or not, and l is whether the defect is tolerated or not. For undefined defects, k is omitted and instead correspond to a v_{ijl} value. A more in-depth description of each v_{ijkl} fo llows:

- v_1: (**1**) The set of all defined defects.
- v_0: (**u**) The set of all undefined defects.
- v_{11}: (**p**) The set of all defined defects that occur.
- v_{10}: (**1 − p**) The set of all defined defects that do not occur.
- v_{111}: (**pq**) The set of all defined defects that occur and are detect ed by testing.
- v_{110}: (**p − pq**) The set of all defined defects that occur and are not detected by testing (missed).
- v_{101}: (**$\gamma q - p\gamma q$**) The set of all defined defects that do not occur but are detected by testing (false-positive).
- v_{100}: (**$1 - \gamma q - p + p\gamma q$**) The set of all defined defects t hat do not occur and are not detect by testing.
- v_{1111}: (**tpq**) A present defect that has been detected by testing and is tolerated.
- v_{1110}: (**$pq - tpq$**) A present defect that has been detected by testing and is not tolerated.
- v_{1101}: (**$tp - tpq$**) A present defect that has not been detected by test ing and is tolerated.
- v_{1100}: (**$p - pq - tp + tpq$**) A present defect that has not been detected by testing and is not tolerated.
- v_{1011}: (**$t\gamma q - tp\gamma q$**) A non-present defect that has been d etected by testing and is tolerated.
- v_{1010}: (**$\gamma q - p\gamma q - t\gamma q + tp\gamma q$**) A non-presen t defect that has been detected by testing and is not tolerated.
- v_{1001}: (**$t - t\gamma q - tp + tp\gamma q$**) A non-present defect that ha s not been detected by testing and is tolerated.
- v_{1000}: (**$1 - \gamma q - p + p\gamma q - t + t\gamma q + tp - tp\gamma q$**) A non-present defect that has not been detected by testing and is not tolerated.
- v_{011}: (**$t'u = tu$**) An undefined defect (present) that is tolerated.

- v_{010}: ($u - t'u = u - tu$) An undefined defect (present) that is not tolera ted.

In the next section we will formally define the defect level of tested and untes ted circuits as well as formally define other important variables which we will use in our defect level analysis.

3   Defect Level Characterization

Each defect is defined by the probabilities of each v_{ijkl}. A defect D_i^v is defined as follows:

$$D_i^v = \{v_{1111}^i, v_{1110}^i, v_{1101}^i, v_{1100}^i, v_{1011}^i, v_{1010}^i, v_{1001}^i, v_{1000}^i, v_{0101}^i, v_{0100}^i\} \tag{1}$$

If a defect is not defined all v_{ijkl} have a zero value except $v_{0101} = t'u$ and $v_{0100} = (1 - t')u$, where u is the probability that the undefined error will occur. We denote the set of all defects by $D = \{D_1, ..., D_n\}$. Alternatel y we can define a defect by its p, q, t, γ, and u values:

$$D_i = \{p_i, q_i, t_i, t_i', \gamma_i, u_i\} \tag{2}$$

We informally define the *defect level* of a circuit as the probability that a defect occurs (or is perceived to have occurred) and is not tolerated. The defect level is, in essence, the probability that a circuit is defective or is falsely detected as defective. The *tested defect level* is then defined as

$$
\begin{aligned}
DL_T &= \prod_{i=1}^{n}(v_{1110}^i + v_{1100}^i + v_{1010}^i + v_{010}^i) \\
&= \prod_{i=1}^{n} p_i - t_i p_i + \gamma_i q_i - p_i \gamma_i q_i - t_i \gamma_i q_i + t_i p_i \gamma_i q_i + u_i - t_i u_i
\end{aligned}
$$

where n is the number of defects, t is the *defect tolerance level*, and v_{ijkl}^i, t_i, p_i, and γ_i are the v_{ijkl}, t, p, and γ of the ith defect. Note that included in the tested defect level is the group of falsely-positive detected defects that are not tolerated. The more simplified and generalized equation becomes

$$DL_T = p - tp + \gamma q - p\gamma q - t\gamma q + tp\gamma q + u - tu \tag{3}$$

for all defects having equal probabilities associated with them. We alternately assume that defect types can be grouped where they have effectively equal probabilities and calculate DL_T for each defect type, reducing the calculations involved and increasing the accuracy of the measurement. This is dependent on the architecture and the defect tolerance scheme. DL is intended as a generalized equation for later expansion according to the architecture under examination.

We define the *untested defect level* as

$$
\begin{aligned}
DL_U &= \prod_{i=1}^{n}(v_{1110}^i + v_{1100}^i + v_{010}^i) \\
&= \prod_{i=1}^{n}(p_i - t_i p_i + u_i - t_i u_i) \\
&= p - tp + u - tu \tag{4}
\end{aligned}
$$

DL_U is simply DL_T where $q_i = 0$ for all $1 < i \leq n$, since not testing is the same using tests with zero quality. Also, note that if $\gamma_i = 0$ for all $1 \leq i \leq n$ in the DL_T equation, then $DL_U = DL_T$.

The next definition of importance that must be made is that of the *test coverage*. The test coverage is ratio of detected defects to defined defects and is defined below as d.

$$
\begin{aligned}
d &= \frac{|v_1|(v_{111})}{|v_1|} \\
&= pq
\end{aligned}
\tag{5}
$$

including only those defects caught that actually occur, simply the probability of a fault occurring and being detected. The value $|v_1|$ is the size of the set of defined defects, i.e. the number of known possible defects. Alternately, if we include the false-positive detection of faults we get

$$
\begin{aligned}
d_T &= \frac{|v_1|(v_{111} + v_{101})}{|v_1|} \\
&= pq + \gamma q - p\gamma q \\
&= (p + \gamma(1 - p))q
\end{aligned}
\tag{6}
$$

As can be seen, including false-positive detection, we get an increase in the test coverage (proportional to $\gamma(1 - p)$), albeit an undesirable increase.

Now we must define the *defect tolerance level* of the system. The defect tolerance level of the system is the ratio of tolerated defects to total defects. Without including false-positive defect detection we get

$$
\begin{aligned}
\alpha &= \frac{v_{1111} + v_{1101} + v_{011}}{v_{11} + v_{01}} \\
&= \frac{tp + tu}{p + u} \\
&= t
\end{aligned}
$$

which supports the claim that t is the defect tolerance level. Yet, we need to include false-positive defect detection.

The tested defect tolerance level is defined by

$$
\begin{aligned}
\alpha_T &= \frac{v_{1111} + v_{1101} + v_{1011} + v_{011}}{v_{11} + v_{101} + v_{01}} \\
&= \frac{tp + t\gamma q - tp\gamma q + tu}{p + \gamma q - p\gamma q + u} \\
&= t
\end{aligned}
$$

which includes the false-positive detection of defects in the form of v_{1011} in the numerator and v_{101} in the denominator. The addition of the false positive detection does not effect the calculated defect tolerance, as expected.

The untested defect tolerance level is is defined by

$$
\alpha_U = \frac{v_{1101} + v_{011}}{v_{11} + v_{01}}
$$

$$= \frac{tp - tpq + tu}{p + u}$$

$$= \frac{tp - tu}{p + u}$$

$$= t \qquad (7)$$

since $q = 0$ for untested circuits (third step reduction), as expected.

The previous three equations establish that for tested and untested systems, the defect tolerance level remains t, even with the inclusion of false-positive defect detection.

With this formalization, we may move onto the analysis of the defect level of tested and untested circuits in the next section.

4 Threshold Defect Tolerance Level

The previous section defined the components needed to produce an in-depth analysis of tested versus untested defect level, namely DL_T, DL_U, d_T, α_U and α_T.

We will now define the equation for quality level as

$$QL = \alpha(1 - Y^{1-d}) + Y^{1-d} \qquad (8)$$

where Y is yield. From this we can derive the equation for quality level with testing as

$$QL_T = \alpha_T(1 - Y^{1-d_T}) + Y^{1-d_T}$$
$$= t(1 - Y^{1-(pq+\gamma q - p\gamma q)}) + Y^{1-(pq+\gamma q - p\gamma q)} \qquad (9)$$
$$(10)$$

and quality level without testing, QL_U, as

$$QL_U = \alpha_U(1 - Y) + Y \qquad (11)$$
$$= t(1 - Y) + Y$$

since, by definition, the test coverage $d = 0$ without testing.

Our primary goal is to determine what untested defect tolerance level we must achieve in order to have $QL_U \geq QL_T$. We define α'_U to be the threshold defect tolerance level α_U where $QL_T \geq QL_U$. We now derive an equation for α_U for $QL_T = QL_U$,

$$QL_U = QL_T$$
$$\alpha_U(1 - Y) + Y = \alpha_T(1 - Y^{1-d_T}) + Y^{1-d_T}$$
$$\alpha_U = \frac{\alpha_T(1 - Y^{1-d_T}) + Y^{1-d_T} - Y}{(1 - Y)}$$

which results in an α'_U value of

$$\alpha'_U \geq \frac{\alpha_T(1 - Y^{1-d_T}) + Y^{1-d_T} - Y}{(1 - Y)}$$
$$\geq \frac{\alpha_T(1 - Y^{1-(pq+\gamma q - p\gamma q)}) + Y^{1-(pq+\gamma q - p\gamma q)} - Y}{(1 - Y)} \qquad (12)$$

which is the untested defect tolerance level needed to achieve for the untested quality level to exceed the tested quality level.

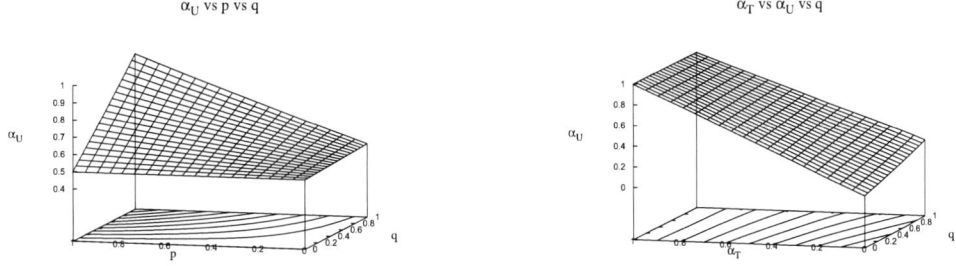

Figure 1. Illustration of α'_U in surface plots with contours. The left plot shows α_U with respect to p and q with $\alpha_T = 0.50$, $\gamma = 0.05$, and $Y = 0.90$. The right plot shows α_U with respect to α_T and q with $\gamma = 0.05$, $p = 0.20$, and $Y = 0.90$.

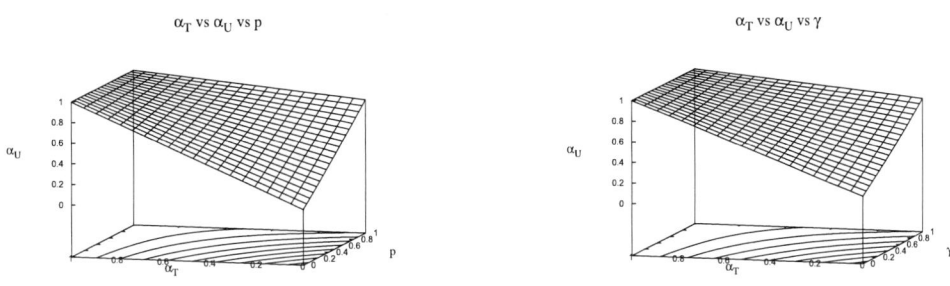

Figure 2. Illustration of α'_U in surface plots with contours. The left plot shows α_U with respect to p and α_T with $q = 0.80$, $\gamma = 0.05$, and $Y = 0.90$. The right plot shows α_U with respect to α_T and γ with $q = 0.80$, $p = 0.20$, and $Y = 0.90$.

Figures 1 and 2 show minimum α'_U values with respect to different variables. Figure 1 (left) shows α_U with respect to p and q. As can be seen, as the quality of the test (q) and the probability of defects (p) increase, α_U must be increased to achieve $\alpha_U = \alpha_T$.

Figures 1 (right) and 2 (left and right) show minimum α'_U with respect to α_T and q, p, and γ respectively. All three plots show similar behavior with the q plot showing the least slope and the p plot showing the most. The γ plot shows how much false-positives can effect the tested defect tolerance level and, thus effect the untested defect tolerance level needed to achieve $\alpha_T = \alpha_U$.

All four plots of in Figures 1 and 2 use synthetic values to accurately illustrate the behavior of equation 12.

5 Conclusions

This paper has presented a comprehensive and thorough defect classification and characterization to study and identify a threshold defect tolerance level. The proposed methods have been investigated under a specific design constraint when traditional testing is not practically deployable. An approach to an extensive theoretical and parametric compilation and optimization has been introduced; and it has revealed there exists a theoretical threshold test coverage and defect tolerance level at which each choice of testing and defect tolerance method and strategy can be justified. In Figure 1, we showed that there exists a point or points at which decreasing defect coverage will allow for increases in defect tolerance levels to negate the positive impacts of testing.

References

[1] T. W. Williams, "Test Length in a Self-Testing Environment", IEEE Design and Test of Computers, Vol. 2, No. 2, Apr. 1985

[2] R. W. wadsack, "VLSI: How Much Fault Coverage is Enough?", Proc. IEEE Int. Test Conference, Oct. 1985

[3] F. J. Meyer and N. Park, "Predicting the Yield Efficacy of a Defect-Tolerant Embedded Core", IEEE Trans. on Computers, Vol.52, No.11, Nov. 2003

[4] N. Park and F. Lombardi, "Analysis of Stratified Testing for Multichip Module Systems", IEEE Trans. on Reliability, Vol.51, No.1, Mar. 2002

[5] M. Choi, N. Park, F. Lombardi and V. Piuri, "Quality Enhancement of Reconfigurable Multichip Module Systems by Redundancy Utilization" IEEE Transactions on Instrumentation and Measurement Vol. 51, Number 4, Aug. 2002

[6] N. Park, F. Meyer and F. Lombardi, "Quality-Effective Repair of Multichip Module Systems", Journal of Systems Architecture 47, Euromicro, Elsevier, pp. 883-900, 2002

[7] N.-J. Park, K.M. George, N. Park, M. Choi, Y.B. Kim, F. Lombardi, "Environmental-Based Characterization of SoC-Based Instrumentation Systems for Stratified Testing", IEEE Transactions on Instrumentation and Measurement, Vol.54, No.3, June 2005

[8] International Technology Roadmap for Semiconductors, "International Technology Roadmap for Semiconductors (ITRS) 2006," *http://public.itrs.net*, 2006.

[9] R. Iris Bahar, Dan Hammerstrom, Justin Harlow, William H. Joyner Jr., Clifford L au, Diana Marculescu, Alex Orailoglu, Massoud Pedram "Architectures for Silicon Nanoelectronics and Beyond," Computer, vol. 40, no. 1, pp. 25-33, Jan., 2007.

[10] Mehdi B. Tahoori "Defects, Yield, and Design in Sublithographic Nano-electronics," Proceedings of the 2005 20th IEEE International Symposium on Defect an d Fault Tolerance in VLSI Systems, pp. 3-11, Oct., 2005.

[11] Mahim Mishra and Seth C. Goldstein "Defect Tolerance at the End of the Roadmap," Nano, quantum and molecular computing: implications to high level desi gn and validation, pp. 73-108, 2004.

[12] T. W. Williams "The New Frontier for Testing: Nana Meter Technologies," Test Symposium, 1998. ATS '98. Proceedings. Seventh Asian, pp. 2-6, Dec. 1998.

[13] Kwang-Ting Cheng, Sujit Dey, Mike Rodgers, and Kaushik Roy "Test Challenges for Deep Sub-Micron Technologies," Proceeding of the 37th Design Automation Conference, 142-149, June, 2000.

[14] Ramyanshu Datta, Antony Sebastine, Ravi Gupta, Whitney J. Townsend, and Jacob A. Abraham "Test and Debug in Deep-Submicron Technologies," 5th IBM Austin Center for Ad-

vanced Studies Conference, 142-149, Feb, 2004.

Defect Tolerance for a Capacitance Based Nanoscale Biosensor

Glenn H. Chapman
School of Engineering Science,
Simon Fraser University,
Burnaby, B.C., Canada, V5A 1S6
glennc@ensc.sfu.ca

Vijay K. Jain
University of South Florida
Tampa, Florida, USA 33620
jain@eng.usf.edu

Abstract

A capacitance based nanoscale biosensor and its defect tolerance are explored. The sensor consists of a microchamber that can be filled with the fluid under test. In a two step procedure the capacitance is measured between the upper plate and the lower plate first for (a) benign fluid, and then for (b) the test fluid, potentially containing the antigens. In each of these tests, an on-chip oscillator provides a test signal of selectable frequency. As shown in the paper the output signal can be processed by on-chip digital modules to estimate the capacitance values. A decision is then made not only on the presence/absence of the antigens, but also on the level of concentration in the medium. We describe the fabrication steps for the sensor plane, the 3-D architecture and the detection methodology, efficient circuits on the analog plane, and the J-platform on the digital plane. However, due the confluence of diverse technologies involved, the probability of defects is higher than that encountered in the usual 2-D device . Therefore, we propose defect tolerance for the planes of this 3-D biosensor. For example, on the sensor plane multiple chambers for the test fluid of defect resistant design are provided, rather than just one, and their measurements used for reliable estimates.

1. Introduction

Small portable biosensors, which are really detection labs on a chip, are of considerable interest to the biomedical community. These devices detect nanoscale biomolecules, for example antigens, in a fluid being investigated. The proposed sensor consists of a microchamber that can be filled with the fluid under test. In a two step procedure the capacitance is measured between the upper plate and the lower plate first for (a) a benign fluid, and then for (b) the fluid under test, potentially containing the antigens. The fundamental basis of the nanoscale biosensor presented here is AC capacitance measurement where these biomaterials form as part of the dielectric in a parallel plate capacitor. Then changes in the amount of biological material, say when a virus is bound to an antigen receptor, result in a change in the capacitance of the system. While capacitance measurement, including that at micro scales, has been adopted by several authors [1]-[6], bulky and expensive external equipment (i.e., signal source, measuring equipment, and PC processing, etc.) is still needed. What differentiates the concept presented here is the integration of all sensing and processing on to a single 3-D chip. The potential advantages are the elimination of not only the expensive external equipment, but also the need for an expert to configure all of the pieces together for the ultimate measurements. Also, the sensitive capacitance measurements are not overshadowed by the parasitic effect of the interconnections to external equipment. Furthermore, if manufactured on a mass scale, such devices could become disposable, thus offering the benefit of one-time use. We also remark that while optical measurements have also been suggested by some authors [7], the problem of external equipment has not been eliminated by them.

The idea in 3-D chip is to fabricate separate wafers for each plane (sensors, analog VLSI circuits, digital processing, etc), mechanically polish them to a thin structure, add inter-plane vias [8], and separate into chip blocks, thus creating a stacked 3-D structure. We employ the reconfigurable *J*-platform [9], which uses *coarse-grain* VLSI cells with high functionality, performance, and reconfigurability. These include a Universal Nonlinear (UNL) cell [10], an extended multiply accumulate (MA_PLUS) cell [11], and a Data-Fabric (DF) cell [11]. The coarse-grain approach has the benefits of reduced external interconnect, much reduced design time, and manageable testability. Thus sensing and processing are all accomplishes in a single chip. A further advantage is the scalability. That is, the device could be targeted to biological particles that are of a size 5 to 10 microns in diameter down to viral particles that are of a size 30 to 100 nm in diameter, for example the HIV virus.

However, due to the confluence of diverse technologies involved, the probability of defects in such sensors is higher than that encountered in the usual 2-D devices. Therefore, we propose defect tolerance for the sensor plane of this 3-D biosensor. Specifically, on the sensor plane four chambers for the test fluid are provided, rather than just one, and their measurements used for obtaining reliable estimates. An example is presented in the paper.

2. 3-D Capacitive Nano Biosensor

The proposed 3-D SoC, shown in Fig. 1, has multiple planes. The top plane houses the capacitive biosensor. The key player is the micro- nano- chamber in which the biofluid under test resides. Its floor and ceiling are conductive to provide the capacitor plates, and its size is tailored to the size of the particles of interest. The particle size could range from multi-micrometer diameter down to viral particles that are of 30 nm to 100 nm diameter. For specificity, we will assume the former case and take the bio-particle size to be $d=5$ μm. In the literature the height of such chambers is taken to be two to three times the anticipated largest diameter. Thus, here the chamber height is taken to be 10 μm. While the height is small, the sides of the chamber are provided to be on the order of multi-mm. This choice enables adequate amount of fluid in the chamber to minimize statistical variations in the measurements. For specificity, we will indeed take the chamber to be square on its top, with $L = 2$ mm. Hereafter, we will refer to the chamber as a microchamber. In the interest of defect tolerance for the planes of this 3-D biosensor, redundancy is provided at all planes. In particular, four microchambers for the test fluid are provided on the sensor plane, rather than just one, and their measurements compared for reliable estimates.

The next plane houses the analog circuitry including the ADCs. Finally the third – the digital plane, contains the intelligence of the device. We present the architecture and methodology for the detection of diseased cells using the *J*-platform [9] on the digital plane, which employs coarse-grain VLSI cells with high functionality, performance, and reconfigurability. These VLSI cells include a Universal Nonlinear (UNL) cell [10], an extended multiply accumulate (MA_PLUS) cell [11], and optionally a Data-Fabric (DF) cell. The advantages of such an approach are high performance, small area and low power compared to FPGAs, and greater flexibility over ASICs. We also discuss the analog plane wherein the necessary signals are generated for the excitation of an electrical circuit that is employed to capacitively sense the biological cells in the sample fluid.

It is useful to make a few remarks. First, the testing of the device after fabrication does not necessarily involve the use of bio-particles. Often, polystyrene beads are used for preliminary tests before bio-particle testing. Second, after successful fabrication and test of this device, a more ambitious goal in the future will be to adapt the concepts to the more challenging problem of sensing viral particles on an integrated basis in a 3-D chip.

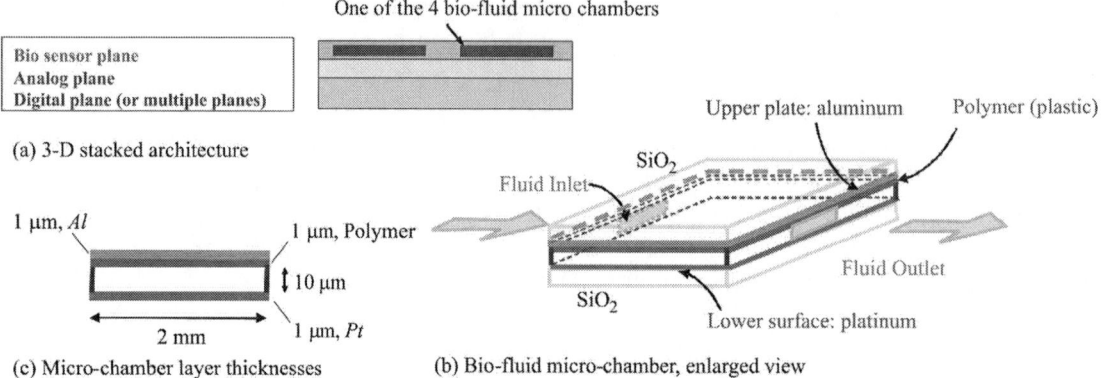

Fig. 1: Integrated 3D system on a chip biosensor

(a) 3-D stacked architecture

(c) Micro-chamber layer thicknesses

(b) Bio-fluid micro-chamber, enlarged view

3. Fabrication of Fault Tolerant Biofluid Microchambers

The key to the creation of Biofluid microchambers for the capacitive sensors is the fabrication of a layered sensor structure starting with a platinum (conductive) lower plate, sealed microchannels for the fluid flow, and a capacitive upper plate covering these microchannels. To fabricate such a chamber we have developed [12] a technique using SU-8, a high contrast negative photoresist that has been used for creating thick-film 3D structures in MEMS for close to 10 years. After development SU-8's biocompatibility and processing simplicity have led many researchers to use it for defining micro-channels for micro-fluidic applications. An important characteristic for the current processes is that SU-8 has a very high optical absorption coefficient for deep-UV light but is weakly absorbing at wavelengths >350 nm, which allows thick layers up to 250 µm thick to be patterned for MEMS structures.

For the creation of encapsulated micro-channels in SU-8, the toughest problem is the fabrication of the top sealed cover layer. One common technique has been the use of sacrificial materials to support the top layer during its fabrication, which must then be subsequently removed [13]. Using sacrificial materials to provide support for the roof of the channel during the encapsulation requires additional processing steps in order to add and pattern the spacing layer, the add and pattern the top coat, and finally remove the sacrificial film. Moreover it is difficult to remove this sacrificial material from deep within large chambers as would be required for this application, as it is hard for any etching fluid to penetrate deeply and remove a material solid enough to support the roof layer.

The microchambers must span several millimeters in size. To accomplish this we will create a series of support pillars throughout the chambers, with a top covering that seals the chamber. We have developed a single mask process that makes use of the change in the optical absorption properties of this resist. For UV light > 365 nm exposures SU-8 is only weakly absorptive and so can create solid structures where exposed to the UV from a mask to the depths of several hundred microns. However for shorter wavelengths, 254 nm, .the SU-8 is highly absorptive and makes structures only to depths of about a micron. Fig. 2, illustrates the dual exposure steps required for the fabrication of micro-channels. Starting first with a SU-8 coated silicon wafer as shown in Figure 2(a), the creation of micro-channels using the dual exposure technique requires a mask with alternating transparent and opaque structures. As shown in Fig. 2 (b), the opaque areas protects the SU-8 from the 365 nm UV exposure in a regular mask aligner and defines regions which will eventually form the encapsulated channels. Conversely, the transparent lines allow the SU-8 to be exposed and define the channel's support posts or sidewalls. To create the short wavelength exposure for the encapsulation layer, a Stratalinker 2400 UV cross-linker at 254 nm is used to flood-expose in Fig. 3(c), with the top layer thickness being set by the duration of the exposure. See reference [12] for the full process details.

Once both exposures are complete, the wafer is baked during which time the crosslinking of the SU-8 occurs. After the post-exposure bake, the wafer is cooled and placed in SU-8 developer with gentle agitation from a magnetic stir rod. The unexposed channels are rapidly attacked by the developer. By removing the unexposed   SU-8, as in Fig. 2(d), the developer creates the micro-channels starting from the open ends. After development, the SU-8 is left to dry naturally. In this system the posts need to be about 75 microns apart to support the encapsulation top layer.

Using this dual exposure method the microchambers would be created as follows. First the bottom conductor would be deposited on the surface. Platinum would be used for this due to its biocompatibility and non-reactive characteristics. Then the microchannels would be created using the SU-8 dual exposure process. For the capacitive design the height of the channels should be between 2-3 times that of the reactive biosensor material. In the targeted system materials of 3-10 μm are expected, so this creates a chamber height of 8-30 microns depending on the materials used, well within the capabilities of the SU-8 dual exposure system which can create channels over 250 μm in height. The thickness of the top, or encapsulation, layer will be set by the flood exposure duration to be near 1 μm so that it does not dominate the capacitance of the system. Finally an aluminum film will be deposited on top of the structure forming the upper conductive layer of the capacitive system. Because the SU-8 encapsulation layer is in direct contact with the microfluid chamber it is not necessary to use a biocompatible film like platinum for this upper layer. The upper conductor will provide additional strength to the roof of the chamber. As the encapsulation layer is in direct contact with the microfluid chamber it is not necessary to use a biocompatible film like platinum for this upper layer. The upper conductor will provide additional strength to the roof of the chamber

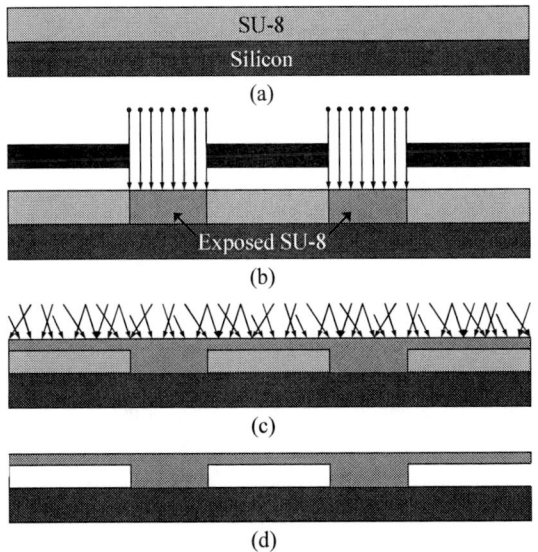

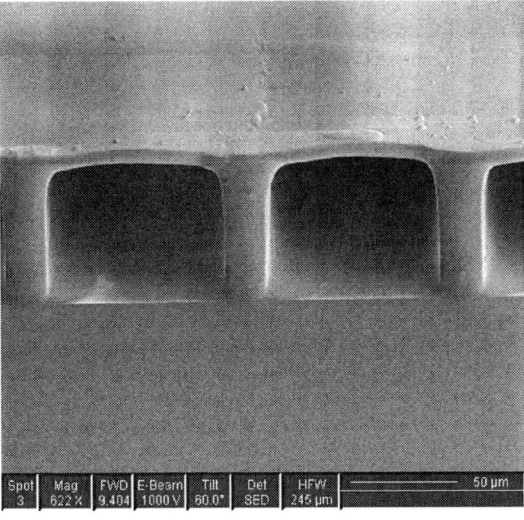

Fig. 2 Process for creating micro-channels using 2 exposure λ. (a) SU-8 spun onto silicon wafer. (b) Exposure mask with 365 nm defining channel walls. (c) Flood-exposure SU-8 with uncollimated 254 nm. (d) Post-bake and development of the SU-8 channels.

Fig. 3 SEM image of SU-8 micro-channels showing the support posts and the top encapsulation layer.

It is important to note that microfluidic channels have a significant fabrication failure rate in general. The issue is that to create the chamber the SU-8 developer must penetrate the entire length of the chamber. With SU-8 the top cover cannot span a very wide a distance (about 100 μm) before the ceiling becomes unstable, so often several parallel channels are used. However by using a large chamber with pillars, what can be called a Cathedral chamber, rather than narrow channel system, creating effectively a wide chamber, the flow of the fluid, and the fabrication developer, is enhanced.

In addition the design will actually consist of 4 separate chambers, with the bottom and top conductors being patterned to create 4 separate capacitive sensors, and the SU-8 pattern having thicker vertical channel walls separating the separate chambers. The electronic sensing systems for each chamber will be separate but feed by multiple connections to top/bottom plates, creating 4 devices each with its own signal output.

Experiments have shown the most likely failure mechanism in the microchambers is blockage of the channels at the SU-8 development stage because the developer has not penetrated far enough, and not a breaking of the encapsulation layer. Indeed for the microchamber length (about 2 mm) and large width due to the creation of support pillars rather than narrow channels, even this even is unlikely. Finally to prevent the blockage of inlet or outline apertures causing a chambers to die there will be a mixing chamber the beginning and end, which feeds into the 4 separate capacitive chambers.

If one of the chambers becomes blocked, either at fabrication time or by particles during operation, the result will be a significant change in that chamber's capacitance. This will be detected in the capacitance and fault tolerance sections 5 and 6 of this paper.

4. Detection Approach

A possible approach, as developed here, is a two-frequency test. For the R-C circuit, *formed by a resistor in series with the bio-fluid microchamber*, the *ratios of the amplitudes of the voltage outputs* (at the two frequencies), for the two cases of (a) benign fluid (B), and (b) test fluid potentially with antigens (T), are easily shown to be

$$\rho_0 = \left.\frac{V(f_2)}{V(f_1)}\right|_{Benign\ fluid} = \sqrt{\frac{RC_0(2\pi f_1)^2 + 1}{RC_0(2\pi f_2)^2 + 1}}$$

$$\rho_1 = \left.\frac{V(f_2)}{V(f_1)}\right|_{\substack{Test\ fluid, \\ potentially \\ with\ antigens}} = \sqrt{\frac{RC_1(2\pi f_1)^2 + 1}{RC_1(2\pi f_2)^2 + 1}}$$

In our simulations, the parameter ρ appears to lead to a reliable discriminant statistic. Note that ρ_0 is the measured ratio that results with benign fluid (B), while ρ_1 is the measured ratio that results with test fluid (T). Use of this ratio statistic eliminates the need for costly calibration for the unpredictable overall circuit gain. A differential oscillator, not shown here, is specifically designed for the test signals.

Its voltage output, used for exciting the R-C circuit under test, is a sinusoid whose frequency is selected, from f_1 or f_2, by the binary input F_{SEL}. Note that the load resistance equals R_1 when F_{SEL} is LOW and $R_1 \| R_2$ when F_{SEL} is HIGH. Correspondingly,

$$f_1 = \frac{1}{2\pi(R_{D1}C_L)} = \frac{1}{2\pi(R_1 C_L)}; \qquad f_2 = \frac{1}{2\pi(R_{D2}C_L)} = \frac{1}{2\pi(R_1 \| R_2)C_L}$$

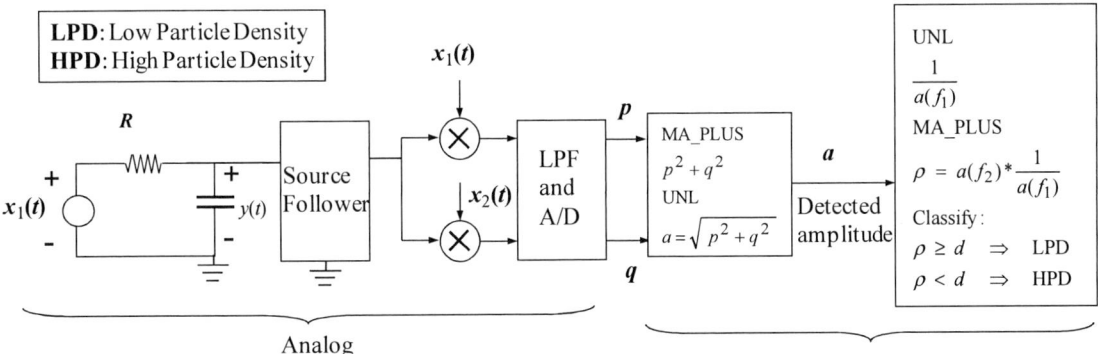

Fig. 4 Microchamber capacitance and resistor circuit, its excitation, processing of response and classification

The in-phase signal $x_1(t)$ is used to excite the R-C circuit which simply consists of a resistor in series with the capacitance formed by the top and bottom conductive plates of the microchamber. The R-C output $y(t)$ is measured across the capacitance, and then processed. The signal processing and The in-phase signal $x_1(t)$ is used to excite the R-C circuit which simply consists of a resistor in series with the microchamber capacitance. The R-C output $y(t)$ is measured across the capacitance, and then processed. The signal processing and decision making entails both analog circuitry and digital cells as shown in Fig. 4. The mixer (multiplier) circuit will be presented in the final paper. Note that the analog test is repeated for both test frequencies f_1 and f_2, and then processed on the digital plane. In the digital part, the coarse-grain cells of the *J*-platform [11] are used, namely MA_PLUS and the UNL. Both of these are highly efficient single-cycle cells, intended to be used in systolic architectures.

5. Differential Capacitance Estimates

We assume that the test fluid is concentrated with the potential antigen by use of a suitable Centrifugal Filter Device. As a result, we will assume a volume concentration ranging from 0.001 to 0.01 on a volumetric basis.

For analysis purposes, we make the following assumptions:

- The capacitance resulting from two layers formed by (a) the plastic on top, (b) fluid without a particle is given by the formula

$$c_0 = \left[\frac{h_{plastic}}{\varepsilon_{plastic}} + \frac{h_{chamber}}{\varepsilon_{fluid}} \right]^{-1} \quad \text{per unit area}$$

- The capacitance resulting from four layers formed by (a) the plastic on top, (b) fluid on top of a particle, (c) the particle, and (d) fluid below the particle is given by the formula

$$c_1 = \left[\frac{h_{plastic}}{\varepsilon_{plastic}} + \frac{h_{top\_fluid}}{\varepsilon_{fluid}} + \frac{d_{particle}}{\varepsilon_{particle}} + \frac{h_{bottom\_fluid}}{\varepsilon_{fluid}} \right]^{-1} \quad \text{per unit area}$$

- The particles contained in the fluid are singly aligned, that is no two particles appear in a vertical column.

On an absolute basis this is apparently an unrealistic assumption, but is reasonable on a statistical basis.

In addition to the microchamber specifications and the particle size given in Section 2, the following dielectric constants are used. Adaptation to other numbers is straightforward.

$$\varepsilon_0 \quad = 8.845 \times 10^{-14} \ \text{F/cm}$$

$$\varepsilon_{plastic,rel} = 4 \quad \Rightarrow \quad \varepsilon_{plastic} = 35 \times 10^{-14} \text{ F/cm}$$

$$\varepsilon_{fluid,rel} = 20 \quad \Rightarrow \quad \varepsilon_{fluid} = 177 \times 10^{-14} \text{ F/cm}$$

$$\varepsilon_{particle,rel} = 40 \quad \Rightarrow \quad \varepsilon_{particle} = 354 \times 10^{-14} \text{ F/cm}$$

Denoting the volume concentration of the particles as F and the top area of the chamber as A, it can be shown that the area occupied by the antigens as G, it can be shown that

$$G = \frac{F A}{4} \qquad \text{(assuming that the chamber height} = 2 * \text{particle diameter)}$$

Then, the differential capacitance, defined as $\Delta C = C\big|_{with\ particles} - C\big|_{benign\ fluid}$, can be calculated. It is plotted in Fig. 5 for various values of volume concentration F. In particular, for $F = 0.001$, $\Delta C = 2.4$ fF, and for $F = 0.1$, $\Delta C = 236$ fF. Recall that we have assumed that the test fluid has been concentrated with the potential antigen by use of a suitable centrifugal filter device.

Important Remark: Although we have indicated a two-step procedure, it is entirely possible to use two of the microchambers simultaneously (one with (a) benign fluid, and the other with (b) the test fluid potentially containing the antigens). A differential circuit could then be used to discern the differences between the two chamber capacitances.

Preliminary results are shown in Table 1, showing the clear separability of Low or no Particle Density (LPD) and High Particle Density (HPD). The densities in the two cases were taken to be 0.001 and 0.1.

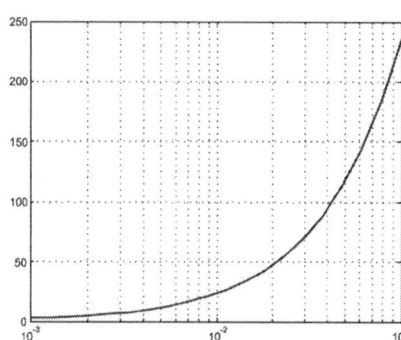

Fig. 5 Plot of differential capacitance ΔC vs. volume concentration F

Table 1 Results of two frequency tests on the microchambers (applied voltage=20 mV)

	Test 1		Test 2		Ratio
	f_1 (MHz)	V_1 (mV)	f_2 (MHz)	V_2 (mV)	$\rho = V_2/V_1$
Low Particle Density (LPD)	10	20.0	100	16.1	0.805
High Particle Density (HPD)	10	1.4	100	0.4	0.286

6. Defect Tolerance Through Redundancy

6a Modeling Assumptions
1. The analog and digital processing plane is shared by all microchambers.
2. The analog and digital processing plane (Plane-2) is tested before bonding with the sensor plane (Plane-1). No gross redundancy is provided on plane, albeit cell level redundancy may be provided as discussed in the authors' previous papers.

6b Basic Functionality
In view of the above assumptions we need only consider the yield as follows:

$$Y_1 = Y_M * Y_B$$

where Y_1 denotes the effective yield of a microchamber

Y_M denotes the yield of a microchamber on the sensor plane

Y_B denotes the yield of the alignment and bonding process

The discussion below focuses on Y_M, the yield of the microchamber.

Let $f_1(x)$ denote the probability distribution of defects at an inlet, based either on the diameter or the area. Further, let X_1 denote the maximum permissible value of x, i.e., if x equals or exceeds X_1 then the inlet is considered inoperable. In the literature, the pdf for defect diameter is found to resemble the exponential pdf. Then, the yield of an inlet is

$$H_1 = \int_0^{X_1} f_1(x) \ dx$$

Similarly, H_2 is defined for the outlet. Finally, let H_3 denote the yield of the chamber part of the microchamber. Then the yield of a single microchamber is

$$Y_M = H_1 H_2 H_3$$

In turn, the yield for a single microchamber Y_1 becomes known using a reasonable estimate for the bonding yield . We call this the basic functionality yield.

Note that the probability of complete failure is then $F = (1-Y_1)^4$,

and the probability of a functional device is $Y = 1 - F = 1 - (1-Y_1)^4$.

6c Measurement Variability

We model the measurement variability by a Gaussian random variable. Then, the capacitance measured by a single successful microchamber is a random variable

$$C_1 = \mathcal{N}(\overline{C}, \sigma_1{}^2)$$

where $\overline{C}$ denotes the mean and σ_1^2 the variance. We take the overall measured capacitance estimate to be the mean from all successful microchambers. Then the variance of the capacitance estimate is given by

$$\hat{\sigma}^2 = \sum_{m=1}^{4} \left\{ \binom{4}{i} Y_1{}^m (1-Y_1)^{(4-m)} \right\} \left(\frac{\sigma_1{}^2}{m} \right) / Y$$

In the extreme case when $Y_1 = 1$, $\hat{\sigma}^2 = \sigma_1{}^2 / 4$, so that the measurement variance is reduced by a factor of 4.

6d An Example:

Consider that $H_1 = 0.94$, $H_2 = 0.94$, $H_3 = 0.96$, and $Y_B = 0.95$, so that

$$Y_M = H_1 H_2 H_3 = 0.848, \text{ and } Y_1 = Y_M Y_B = 0.848*0.95 = 0.806$$

and, the probability of a functional device is

$$Y = 1 - (1-Y_1)^4 = 1 - (1-0.806)^4 = 0.998$$

Also, let $\overline{C} = 25$ fF, $\sigma_1^2 = 16$ fF^2, then for the capacitance estimate the overall mean is $\mu_c = 25$ fF, and the effective variance is 5.4 which has been reduced to one third of the value obtained from just one microchamber.

6e Benefit of Defect Tolerance

It is apparent that the four fold redundancy in microchambers not only improves the basic functionality, but it also leads to improved measurements.

7. Conclusions

We have presented a 3-D capacitance based nanoscale biosensor and its defect tolerance. The potential advantages are the elimination of not only the expensive external equipment, but also the need for an expert to configure all of the pieces together for the ultimate measurements.

Also, the sensitive capacitance measurements are not overshadowed by the parasitic effect of the interconnections to external equipment. Furthermore, if manufactured on a mass scale, such devices could become disposable, thus offering the benefit of one-time use.

However, due the confluence of diverse technologies involved, the probability of defects is higher than that encountered in the usual 2-D devices. Therefore, we have proposed defect tolerance through redundancy for the sensor plane of this 3-D biosensor. Specifically, on the sensor plane four microchambers of a defect resistant design for the test fluid are provided, rather than just one, and their measurements are averaged for reliable estimates. Thereby, not only the basic functionality is improved, but it also leads to enhanced estimates of capacitance.

8. References

[1] D. M. Hanna, B. A. Gross, S. Kandlikar, E. Lempicki, B. A. Oakley, and G. A. Stryker, "Detection of Vesicular Stomatitis Virus using a Capacitive Immunosensor," *Proc. IEEE Engineering in Medicine and Biology*, 27th Annual Conference, pp. 534-537, 2005.

[2] X.S Guo, X.S.Fang, X.L Yang, Y.Q Chen, L.R Wang, "Capacitive monitoring of the antigen-antibody reactions enhanced by nanogold," *Proc. IEEE Engineering in Medicine and Biology,* 27th Annual Conference, pp. 1260-1263, 2005.

[3] D. Jiang, J. Tang, B. Liu, P. Yang, X. Shen, J. Kong, "Covalently coupling the antibody on an amine-self-assembled gold surface to probe hyaluronan-binding protein with capacitance measurement," *Biosensors and Bioelectronics (Elsevier)*, pp. 1183-1191, 2003.

[4] L. L. Sohn, O. A. Saleh, G. R. Facer, A. J. Beavis, R. S. Allan, and D. A. Notterman, "Capacitance cytometry: Measuring biological cells one by one," *Proc. National Academy of Sciences*, pp. 10687–106890, Sept. 2000.

[5] A. Romani, N. Manaresi, L. Marzocchi, G. Medoro, A. Leonardi, L. Altomare, M. Tartagni, Guerrieri, "Capacitive sensor array for localization of bioparticles in CMOS lab-on-a-chip," *Proc. Int. Solid States Circuits Conf.*, paper #12.4, 2004.

[6] D. M. Hanna, B. A. Oakley, and G. A. Stryker, "Using a system-on-a-chip implantable device to filter circulating infected cells in blood or lymph," *IEEE Trans. on Nanobioscience*, pp. 6-13, March 2003.

[7] N. Manaresi, A. Romani, G. Medoro, L. Altomare, A. Leonardi, M. Tartagni, R. Guerrieri, "A CMOS chip for individual cell manipulation and detection," *Proc. Int. Solid States Circuits Conf.*, paper #11.1, 2003.

[8] G. H. Chapman, V. K. Jain, and S. Bhansali, "Inter-plane via defect detection using the sensor plane in 3-D heterogeneous sensor systems," *Proc. IEEE Int. Symposium on Defect and Fault Tolerance in VLSI Systems*, pp. 158-168, 2005.

[9] V. K. Jain, S. Bhanja, G. H. Chapman, and L. Doddannagari, "A highly reconfigurable computing array: DSP plane of a 3-D heterogeneous SoC," *Proc. IEEE Int. System on a Chip Conf.*, pp. 243- 246, Sept. 2005.

[10] V. K. Jain, and E. E. Swartzlander, Jr., "32 bit single cycle nonlinear VLSI cell f*or the ICA Algorithm,"* Proc. *Int. Conf. on Acoustics Speech and Signal Processing*, March 2008.

[11] V. K. Jain, and S. Shrivastava, "Rapid system prototyping for high performance reconfigurable computing," *Design Automation for Embedded Systems J.,* pp. 339-350, August 2000.

[12] J.M. Dykes, D.K. Poon, J. Wang, D. Sameoto, J.T.K. Tsui, C. Choo, G.H. Chapman, A.M. Parameswaren, and B.L. Gray, "Creation of embedded structures in SU-8", Proc. SPIE Photonics West Microfluidics, BioMEMS, and Medical Microsystems V, v 6465, pp 64650N1-N12, San Jose, Jan 2007.

[13] L. J. Guerin, M. Bossel, M. Demierre, S. Calmes, and P. Renaud, "Simple and low cost fabrication of embedded micro-channels by using a new thick-film photoplastic," presented at Solid State Sensors and Actuators, 1997. TRANSDUCERS '97 Chicago., 1997 International Conference on, 1997.

[14] B. Razavi, *Design of Analog CMOS Integrated Circuits.* McGraw-Hill Publishers, 2001.

IEEE International Symposium on Defect and Fault Tolerance of VLSI Systems

Fault Detection of Bloom Filters for Defect Maps

Jae-Young Choi, Yoon-Hwa Choi

Department of Computer Engineering, Hongik University, Seoul, Korea
yhchoi@cs.hongik.ac.kr

Abstract

Bloom filters can be used as a data structure for defect maps in nanoscale memory. Unlike most other applications of Bloom filters, both false positive and false negative induced by a fault cause a fatal error in the memory system. In this paper, we present a technique for detecting faults in Bloom filters for defect maps. Spare hashing units and a simple coding technique for bit vectors are employed to detect faults during normal operation. Parallel write/read is also proposed to detect faults with high probability even without spare hashing units.

1. Introduction

Bloom filters have been used for many network applications, such as packet filtering, routing lookups, web cache sharing, and string matching. A Bloom filter is a simple space-efficient randomized data structure representing a set to support membership queries. It can provide constant lookup times at the cost of small false positives independent of the number of strings in the database, as long as the memory used by the structure scales linearly with the number of strings stored in it [1-3].

A defect-tolerant memory architecture using Bloom filters for defect maps has been presented in [5]. An improved space efficiency has been achieved in realizing defect maps. A redundant nanoscale memory, consisting of multiple nano-modules sharing the same address space, was also addressed for enhanced memory configurability. More recently, a two-level redundancy scheme, both in module-level and row-column-level, has been proposed to reduce crossbar-area overhead in designing nano-scale memory [6]. Although Bloom filters can considerably reduce the space required for defect maps, they have to be reliable to be practically useful. In most applications of Bloom filters, false positives can be tolerated as long as the their probability is sufficiently small. The same is true for Bloom filters realizing defect maps since the defect-free locations incorrectly identified as defective due to false positives can simply be treated as unusable. False positives and false negatives induced by a fault in Bloom filters, however, may cause a fatal error in memory access, unless an efficient error detection mechanism is provided.

In this paper, we investigate the impact of a fault in Bloom filters for defect maps and present a simple technique for detecting faults in the Bloom filters during normal operation. Both spare hashing units and a bit-vector for parity are used to monitor the hash functions and to detect errors in the bit vectors of Bloom filters. Parallel write/read is also addressed to enhance fault detection and to detect faults with high probability even without spare hashing units.

The rest of the paper is organized as follows. In Section 2, a Bloom filter is briefly introduced. Section 3 presents a simple technique for detecting faults in Bloom filters for defect maps in a

1550-5774/08 $25.00 © 2008 IEEE
DOI 10.1109/DFT.2008.41

redundant memory system. Parallel write/read is discussed in Section 4 to improve fault detection capability and to detect faults even without spare hashing units. Conclusions are made in Section 5.

2. Bloom Filters

A Bloom filter is a data structure that stores a given set of signatures in a bit vector M of m bits, initially all set to 0, with k independent hash functions h_1, h_2,...,h_k that map each element of a set $S=\{x_1,x_2,....,x_n\}$ to the set $\{1,2,...,m\}$. A typical Bloom filter is shown in Fig. 1. For each string $x \in S$, the Bloom filter computes k hash functions on it, producing k hash values ranging from 1 to m. The filter then sets k bits in an m-bit vector M at the addresses corresponding to the k hash values. To see if a given string y belongs to S, we apply the k independent hash functions to y, resulting in a set of locations. If all the bits (in M) at the locations corresponding to the k hash values are set to 1, the filter accepts y with high probability as a member of S. On the other hand, if any of the mapped locations is zero, y is determined to be a non-member. Thus a Bloom filter could result in false positives. However, it can never generate false negatives, where an item is rejected while it actually belongs to the set. In most applications, false positives may be acceptable as long as they occur with a small probability.

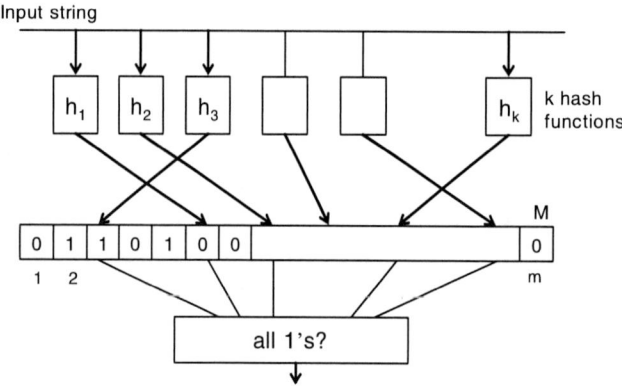

Figure 1. A Bloom filter with k hash functions

3. Bloom Filters for Defect Maps in Nanoscale Memory

Multiple nano-modules that share the same address space can be used to tolerate defects in a nanoscale memory system as shown in Fig. 2, where r nano-modules of 2^m cells each are used to construct a memory module of 2^m cells. The redundant memory architecture consists of three major functional blocks: r nano-modules of 2^m cells, r Bloom filters, and a selection logic. Each nano-module can be constructed as a crossbar-based molecular memory [4]. Bloom filters and a selection logic are expected to be implemented with CMOS technology. The nano/CMOS hybrid technology [7][8][9] can be used to integrate them. Each nano-module has its own defect map, implemented in its associated Bloom filter. In the figure, BF_i stores the defect map of the nano-module N_i. In other words, the locations of defective devices in N_i are programmed in the Bloom filter BF_i. To access the memory with a given address a, the Bloom filters of the r nano-modules are first queried to see if the corresponding locations are defective. If at least one of them shows a 0, the defects can be tolerated and one of the nano-modules N_i with $BF_i(a)=0$ will be selected by

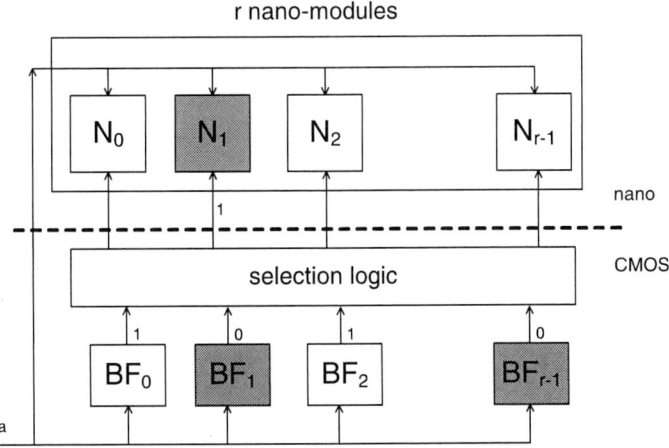

Figure 2. A memory module consisting of r nano-modules and defect maps

the selection logic and the location a of the selected nano-module will be accessed. If the first 0 (from left) is to be selected, N_1 will be the nano-module to be accessed in Fig. 2.

Although Bloom filters may reduce the space required for defect maps, some of the defect-free memory cells cannot be used due to the false positives. If the design parameters are properly chosen, the false positive probability can be made negligibly small, and thus the memory utilization will be improved. Faults in a Bloom filter, however, may cause false negatives and fault-induced false positives to occur, resulting in incorrect read/write accesses. As an illustration, consider the following memory module shown in Fig. 3, where four nano-modules $N_{0\sim3}$ sharing the same address space are used to tolerate defects in the system. The Bloom filter associated with each nano-module is assumed to be programmed and a memory access with an address a is in progress. N_1 and N_3 at location a are defect-free and $BF_1(a)=BF_3(a)=0$ (i.e., no false positive) for the input a. That is, both memory cells can be used as part of the functioning memory. Since N_0 and N_2 are defective at location a, $\mathrm{BF}_0(a)$ and $\mathrm{BF}_2(a)$ must be 1 as marked in the figure.

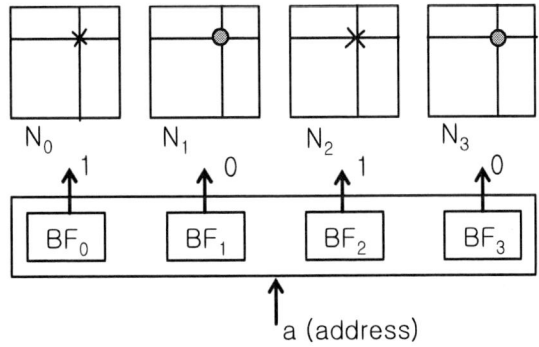

Figure 3. Faults in Bloom filters for defect maps

Due to a fault in the Bloom filters, several different situations may occur. Among others we illustrate the following two representative cases to point out the importance of fault detection in Bloom filters for defect maps.

- BF_1 is faulty such that $BF_1(a)=1$ instead of 0. Then the selection logic will select N_3 instead of N_1, resulting in an incorrect memory access unless parallel write/read is performed.

- BF_0 is faulty such that $BF_0(a)=0$ instead of 1. Then the selection logic will select N_0 instead of N_1, resulting in an incorrect memory access.

Although there are some other cases to be discussed depending on the location of a faulty Bloom filter, it is clear that false positive or false negative induced by a fault causes an incorrect memory access. Without fault detection, correct memory function cannot be guaranteed.

The memory architecture we present in this paper is shown in Fig. 4, where it consists of multiple modules ($G_0,G_1,...,G_{b-1}$) and each of them is composed of r nano-modules ($N_0,N_1,...,N_{r-1}$) sharing the same address space. For a given address, its module number is decoded to select one of the modules $G_0, G_1,....,G_{b-1}$ and one of the r nano-modules of the selected module will be accessed with the offset value. The k hashing units, shared by all the b modules (also shared by r nano-modules in each module), will provide k hash values for the address (offset) and the r bit-vectors $M_0, M_1,..., M_{r-1}$ of the selected module will be looked up for the selection logic to select the nano-module.

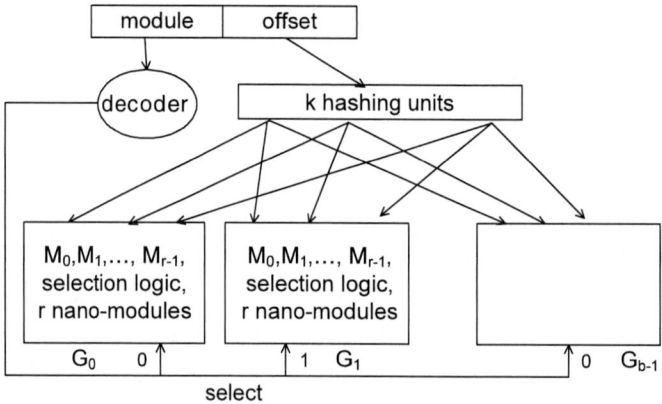

Figure 4. Sharing k hashing units in Bloom filters for defect maps.

In our fault detection of Bloom filters for defect maps, we assume that there is a single fault in the hashing units and bit-vectors. Hence a faulty Bloom filter might generate either an incorrect hash value or incorrect output in the bit vector (M_i) lookup. An incorrect hash value results in a lookup with an incorrect address. In order to detect a fault in Bloom filters for defect maps, we add w spare hashing units for duplication as shown in Fig. 6 and a parity vector for checking errors in table lookups as shown in Fig. 5. The bit-vector for parity is determined by the even or odd parity scheme across the r bit-vectors. Since all the r vectors are looked up simultaneously along with the bit-vector for parity, any single error in the lookups can readily be detected.

Detecting a fault in hashing units is done by comparisons with spare hashing units. Since hashing units will be shared by all the nano-modules in the memory system, the required 100% overhead (i.e., $w=k$) for a complete checking might be acceptable. Reduction in the number of spare hashing units can be made by performing partial checking with w spare hashing units, where k is a multiple of w (i.e., $k = c \cdot w$). In the case of $c = 2$, each spare hashing unit monitors two hashing units, one at a time, alternately. That is, each hashing unit is tested every other cycle for fault detection. The two hashing units to be compared must receive the same predetermined random numbers although they are not shown in Fig. 6. In a class of hash functions called H_3 [10], for example, random numbers to the hashing unit for h_1 also need to be given to hs_1 for comparing the resulting hash values.

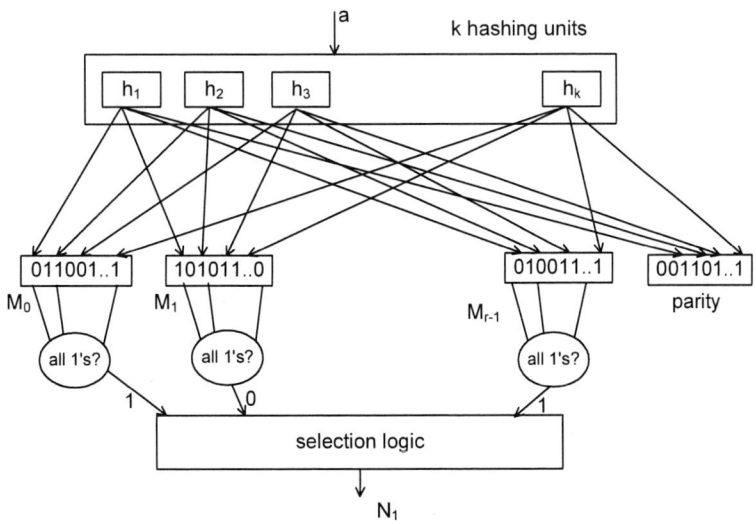

Figure 5. Parity vector for detecting errors in bit-vector lookups

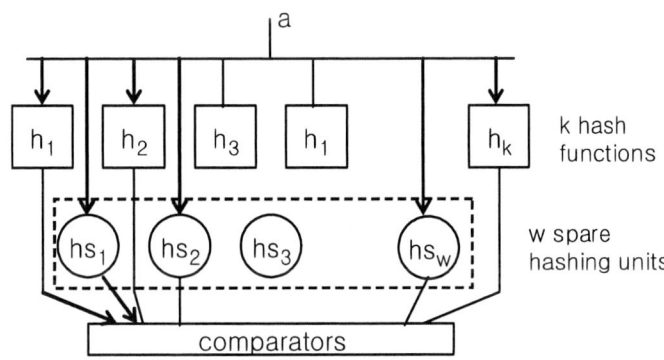

Figure 6. Spare hashing units to monitor hash functions of the Bloom filters

In the case where a spare hashing unit is faulty, the error it generates will be detected and subsequent recovery actions need to be taken, although the error does not cause any harm on the memory operation. When $c \geq 2$, an erroneous hash value generated by a faulty hashing unit will not be detected unless the faulty unit is under comparison at the time. This will cause some delay in fault detection depending on the value of c and the structure of hashing units. A delay of approximately $c - 1$ cycles will be needed in the worst case under the assumption that a faulty hashing unit always generates an incorrect hash value.

Although duplicating hashing units and comparing the resulting hash values will immediately detect errors, it might be desirable to reduce the hardware overhead. Parallel write/read may be attractive in reducing or eliminating the spare hashing units and detecting most of the faults in Bloom filters during normal operation. In the next section, we will present the usefulness of parallel write/read in fault detection of Bloom filters for a nanoscale redundant memory system.

4. Parallel Write/Read for Fault Detection

In accessing nano-modules for a memory read or write operation, parallel write/read may greatly enhance fault detection in Bloom filters for defect maps. For a Bloom filter with a set S of n elements to support membership queries and k hash functions of range m, the probability that a particular bit of the bit vector M of length m is still 0 is given by $P_0 = (1 - \frac{1}{m})^{kn} \approx e^{-\frac{kn}{m}}$. It is well known that false-positive probability can be minimized when k is equal to $ln\,2 \cdot (\frac{m}{n})$ [3]. If the condition is satisfied, P_0 becomes 0.5.

To justify the usefulness of parallel write/read, we first assume that the Bloom filter design parameters, k and m, are chosen in such a way that approximately half of the bit vector M are filled with 0. This assumption is not necessary but we use the condition for simplicity of presentation.

To see what happens for faults in a hashing unit, we consider the following two cases. Suppose that a hashing unit h_i of a Bloom filter is faulty such that $\tilde{h}_i$ is actually performed instead of h_i. We also assume that for a given input x the faulty hashing unit generates any number $\tilde{h}_i(x)$ between 1 and m. In other words, the table (M_i) lookup with $\tilde{h}_i(x)$ will be 1 with a probability of 0.5. A nano-module N_i is currently under consideration.

(1) The cell at location a in N_i is defective. In this case, $BF_i(a)$ is expected to be 1. Apparently $M(h_i(a))$ must be 1. Due to the fault in the hashing unit $M(\tilde{h}_i(a))$ can be 0 with a probability of 0.5. Since $M(h_j(a))$=1 for $j \neq i$, the cell will be incorrectly determined as defect-free with a probability of 0.5.

(2) The cell at location a in N_i is defect-free. In this case, $BF_i(a)$ is expected to be 0. $M(h_i(a))$ can be either 0 or 1. Due to the fault in the hashing unit, $M(\tilde{h}_i(a))$ can be 1 with a probability of 0.5. The cell will be determined to be defective only when $M(h_i(a))$=0, $M(\tilde{h}_i(a))$=1, and $M(h_j(a))$=1 for $j \neq i$. Hence the false alarm probability is $\frac{1}{2^{k+1}}$. If k=10, for example, the incorrect decision will be made with a probability of 0.0005.

In summary, due to a faulty hashing unit a defective cell will be incorrectly identified as defect-free with a probability of 0.5, while a defect-free cell will be determined to be defective with an extremely low probability of $\frac{1}{2^{k+1}}$. Hence parallel write/read can greatly enhance fault detection during normal operation. In Fig. 3, for example, $BF_0(a)$, $BF_1(a)$, $BF_2(a)$ and $BF_3(a)$ are expected to be 1, 0, 1, and 0, respectively. If $BF_0(a)$=0 instead of 1 due to a faulty hashing unit, $BF_2(a)$ will also be 0 with the same probability of 0.5. On the other hand each of $BF_1(a)$ and $BF_3(a)$ will remain at 0 with a probability of 0.9995 (for k=10). The selection logic will then select the first one (i.e., N_0), resulting in an incorrect memory access. If parallel writes/reads are performed, however, all the fault-free cells will store correct data on writes. On read, both defective cell(s)(incorrectly identified as defect-free) and defect-free cells (correctly identified) provide the stored data. Although the defective cell could provide an incorrect data, it can be detected by a simple comparison. If all the data read in parallel match, the fault is masked. Otherwise, there is a fault in the Bloom filters. In the case where parity checking is in place to detect an error in table lookups, the location of a fault could be narrowed down to hashing units.

In general, the probability that at least one defect-free cell will be involved in a memory write/read access with an address a depends on the number of nano-modules that have defect-free cell at address a. If there are q defect-free cells, the probability that at least one will be involved in the memory operation when there is a faulty hashing unit is $1 - (\frac{1}{2^{k+1}})^q$. If q=2 and k=10, for example, at least one defect-free cell will be involved in memory read/write with a probability $1 - \frac{1}{2^{22}}$. Hence we can claim that parallel write/read can detect faults in Bloom filters with an extremely high probability.

For a given defect rate p a functioning memory module can be constructed with multiple nano-modules sharing the same address space. If sufficient redundancy is involved, most of the addresses will be supported by at least two nano-modules. For p=0.05, a fully functional module of 2^{12} can almost surely be constructed with 7 nano-modules of the same size. An address in that case will be supported by only a single nano-module with a probability less than 10^{-7}.

5. Conclusions

In this paper, we have presented a simple technique for detecting faults in Bloom filters for defect maps in a nanoscale memory. Faults in two major functional blocks of a Bloom filter, hashing units and bit-vectors, are detected during normal operation using spare hashing units and parity checking. Parallel writes/reads, when accessing memory, are shown to improve fault detection capability and to detect faults in Bloom filters even without spare hashing units.

References

[1] B. Bloom, "Space/time tradeoffs in hash coding with allowable errors," Communications of the ACM 13:7, 1970, pp. 422-426.

[2] S. Dharmapurikar, P. Krishnamurthy, T.S. Sproull, J.W. Lockwood, "Deep packet inspection using parallel Bloom filters," IEEE Micro, Jan-Feb. 2004, pp.52-61.

[3] A. Broder and M. Mitzenmacher, "Network applications of Bloom filters: A survey," Internet Mathematics, Vol.1, No. 4, 485-509.

[4] Y. Chen el al., "Nanoscale molecular-switch crossbar circuits", *Nanotechnology*, 14(2003), pp. 462-468.

[5] G. Wang, W. Gong and R. Kastner, "On the use of Bloom filters for defect maps in nanocomputing," IEEE ICCAD, Nov. 2006, pp.743-746.

[6] Y.-H. Choi and M.-H. Lee, "A defect-tolerant molecular-based memory architecture," IEEE DFTS, Sept. 2007, pp. 143-151.

[7] M.M. Ziegler and M.R. Stan, "The CMOS/nano interface from a circuit perspective," Int. Symp. Circuits and Systems, May 2003, pp. 904-907.

[8] M.M. Ziegler and M.R. Stan, "CMOS/Nano codesign for crossbar-based molecular electronic systems," IEEE Trans. Nanotechnology, Vol. 2, No.4, Dec. 2003, pp. 217-230.

[9] X. Ma, D.B. Strukov, J.H. Lee, K.K. Likharev, "Afterlife for silicon: CMOS circuit architectures," IEEE Nano 2005, pp. 175-178.

[10] M.V. Ramakrishna, E. Fu, and E. Bahcekapilli, "Efficient hardware hashing functions for high performance computers," IEEE Trans. Computers, Vol. 46, No. 12, Dec. 1997, pp. 1378-1381

IEEE International Symposium on Defect and Fault Tolerance of VLSI Systems

Fault Tolerant Schemes for QCA Systems

Xiaojun Ma and Fabrizio Lombardi
Dept of Electrical and Computer Engineering
Northeastern University
Boston, MA 02115
{xma,lombardi}@ece.neu.edu

Abstract

Quantum-dot Cellular Automata (QCA) is a promising nanotechnology that offers significant improvements over CMOS. QCA is limited by the high fault rate in manufacturing. New fault tolerant schemes are required to reliably assemble a system with fault-prone QCA technology. This paper analyzes and compares several fault tolerant schemes for QCA systems. It is shown that majority multiplexing (Maj-MUX) is a high capacity fault tolerant scheme specially suitable for QCA. The fault tolerant capability of QCA using Maj-MUX is investigated in detail in this work. The signal restoration speed of the Maj-MUX scheme is reported. Compared with NAND-multiplexing, the Maj-MUX can tolerate a higher fault rate and restore signals at a lower overhead.

Index terms: QCA, reliable computing, Fault tolerance.

1 Introduction

Quantum-dot Cellular Automata (QCA) is a promising emerging technology that relies on novel design concepts. In QCA, information is transferred and transformed by Columbic interactions among basic elements (referred to as cells) rather then electrical currents as in CMOS-based VLSI; hence, energy dissipation of QCA circuits is small. Different QCA circuits have been designed [1][2] and different implementations of QCA cells have been proposed [3][4]. Modeling of QCA operation have been addressed with respect to the effects of clocking [5] and energy under innovative computational paradigms, such as reversible computing [6].

As in other nano-scale technologies, the manufacturing process of QCA suffers from a high fault rate; fault characterization of QCA has been pursued in previous works [7][8]. To assemble a reliable computing system using QCA, fault tolerance must be utilized. Unfortunately, fault tolerance schemes proposed for VLSI are not fully adequate to handle the expected fault rates of QCA. A novel fault tolerant scheme referred to as *majority multiplexing* has been proposed in [9]. It combines the NAND-multiplexing scheme (originally proposed in [10]) with a 3-input MV (majority voter) to provide high fault tolerant capabilities. However, [9] did not provide the lower bound of the tolerable fault rate of the computing modules in the Maj-MUX system. In addition, restoration speed was not considered in [9]. Since MV is the basic logic unit, fault tolerance by majority voting is a promising scheme for QCA systems.

In this paper, different fault tolerant schemes, such as TMR (Triple Modular Redundancy), NAND-multiplexing and Maj-MUX, are compared in terms of fault tolerant capacity and signal restoration speed. As MV is the basic device construct in QCA, Maj-MUX for QCA implementation is investigated in detail. It is shown that for QCA the Maj-MUX can tolerate a higher fault rate and restore signals at a lower overhead compared to NAND-multiplexing.

This paper is organized as follows: Section 2 gives the background on QCA technology. A brief review on available hardware redundancy techniques is given in Section 3. For QCA, the performance of the majority multiplexing technique is presented and compared with NAND-multiplexing in

1550-5774/08 $25.00 © 2008 IEEE
DOI 10.1109/DFT.2008.12

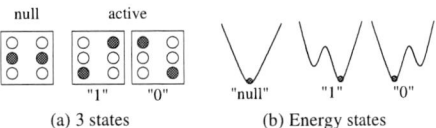

(a) 3 states   (b) Energy states

Figure 1. Tri-state model for clocked molecular QCA

Section 4. Conclusion and discussion on the majority multiplexing technique for QCA systems are provided in Section 5.

2   Review of QCA

Quantum-dot Cellular Automata (QCA) relies on the Coulombic interaction between device cells to implement unique paradigms. In its simplest form, a QCA cell can be viewed as a set of four charge containers or "dots" (two dipoles), positioned at the corners of a square [1]. The cell contains two extra mobile electrons that can tunnel between dots within the cell. Recent developments in QCA manufacturing show a promising future for molecular assembly of QCA devices.

A tri-state cell [5] has been proposed for clocked molecular QCA, as shown in Figure 1. This cell has six "dots" and two extra mobile electrons. An induced electric field mechanism operates as a clocking scheme to modulate the inter-dot tunneling barrier of QCA cells. When the electric field is applied to pull the electrons into the two middle dots, the cell is in the "NULL" state. Otherwise, the cell is in an active state and the electrons are forced to the four corner dots by Coulombic repulsion. The two possible states are defined as logic "0" and logic "1". A *four-phase clocking scheme* has been proposed in [11].

The four phases are *Relax, Switch, Lock* and *Release*. In the relax phase, a QCA cell is in "NULL" state; it is in active state for lock phase. The switch phase is the transition from relax to lock, and the release phase is the transition from lock to relax. A QCA circuit is partitioned into clocking zones and arranged in this periodic fashion such that zones in the lock phase are followed by zones in the switch, release and then relax.

Figure 2 shows the basic logic units for QCA, including logic gates (inverter and MV) and other interconnect devices (binary wire and inverter chain) [1]. Other circuits such as memory cells and adders, have also been proposed [2]. A QCA MV is implemented with only 5 QCA cells and is used as the basic functional element in the general design of QCA circuits.

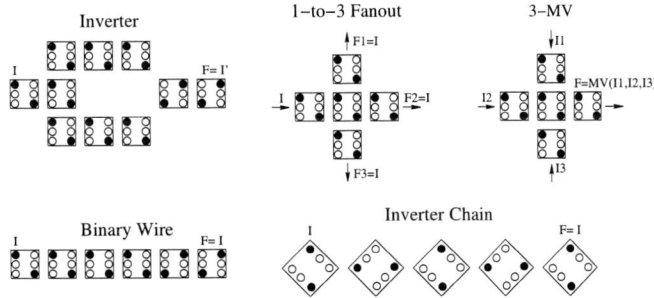

Figure 2. Basic circuit units of molecular QCA

3   Hardware Redundancy Techniques

As briefly reviewed in this section, different types of redundant systems have been proposed and used for VLSI. They are also applicable to emerging technologies such as QCA.

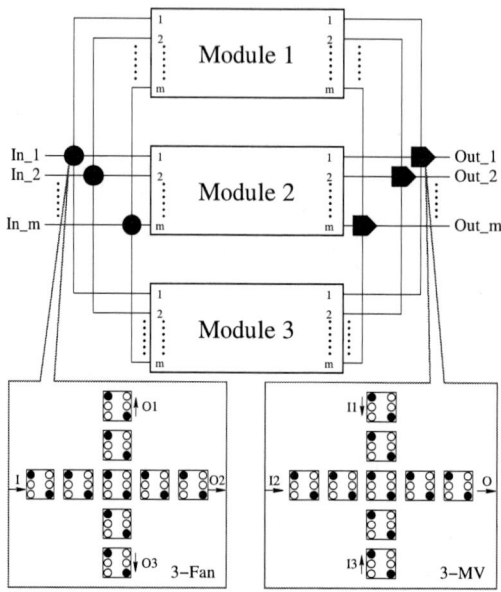

Figure 3. A TMR system in QCA

TMR (Triple Module Redundant) is a widely used fault tolerant technique. In QCA, a TMR system (Figure 3) generates a correct result at the output when at most one module is faulty. TMRs can be cascaded to further improve the system's reliability. Signal *reliability* is defined as the probability of the signal to be fault free or correct. If the MV is assumed to be fault free, then every stage in the cascaded TMR system can improve the signal reliability as $R_{out} = (R_{in})^3 + 3(R_{in})^2(1 - R_{in})$, where R_{in} is the reliability of the input signal. The reliability of the outputs of the TMR stage (R_{out}) is higher than the reliability of the inputs (R_{in}) when $R_{in} > 50\%$. If this is extended to the *NMR* (N-Module-Redundant) system, then the reliability is improved as $R_{out} = \sum_{i=0}^{(N-1)/2} \binom{N}{i}(R_{in})^{N-i}(1 - R_{in})^i$, when $R_{in} > 50\%$. The TMR system is advantageous for QCA because the 3-input MV is also the basic QCA device.

The reliability of a TMR system with a non-perfect MV is $R_{sys} = R_{MV} \times [(R_m)^3 + 3(R_m)^2(1 - R_m)]$, where R_{sys} is the system reliability, R_m is the reliability of a module and R_{MV} is the reliability of the non-perfect MV. For an improvement in system reliability due to TMR, it is required that:

$$R_{m,a} < R_m < R_{m,b} \tag{1}$$

$$\text{where} \quad R_{m,a} = \frac{3R_{MV} - \sqrt{9R_{MV}^2 - 8R_{MV}}}{4R_{MV}}; \quad R_{m,b} = \frac{3R_{MV} + \sqrt{9R_{MV}^2 - 8R_{MV}}}{4R_{MV}}$$

So, R_{MV} must be $> \frac{8}{9} \approx 0.8889$ to improve over a module reliability. With $R_{MV} > 0.8889$, a module with $R_m \in (R_{m,a}, R_{m,b})$ can utilize the fault tolerant capabilities of TMR.

If the reliability of a module is too low, a concatenated TMR system (Figure 4) can be employed. Let R_{m_i} denote the reliability of stage i, which is the reliability of the output signal of a module at stage i. All three modules at stage i have the same output reliability R_{m_i}. By dividing a large module into n serially connected stages, the reliability of each stage (R_{m_i}) is suitable for a TMR scheme. The reliability of a concatenated TMR system with n stages is

$$R_{sys1} = \prod_{i=1}^{n} R_{MV} \times [R_{m\_i}^3 + 3R_{m\_i}^2(1 - R_{m\_i})]$$

This reliability is limited by the reliability of the MVs (i.e. R_{MV}). To avoid this bottleneck, a TMR system can be modified as shown in Figure 5. The reliability of this n-stage system (concatenated

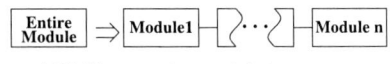

a) Divide executive module into stages

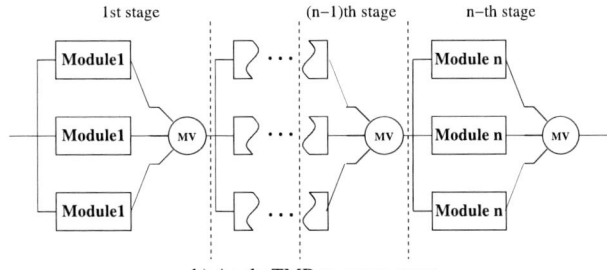

b) Apply TMR to every stage

Figure 4. Concatenated TMR System

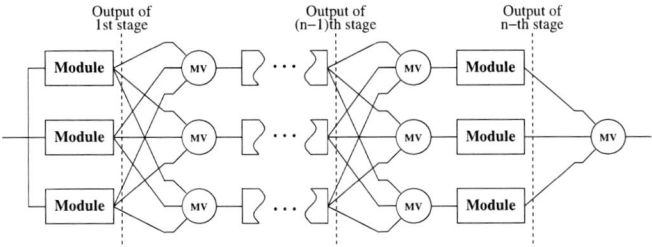

Figure 5. A Concatenated TMR system with MV redundancy

TMR system with MV redundancy) is

$$
\begin{aligned}
R_{sys2} &= [R_{m\_1}^3 + 3R_{m\_1}^2(1 - R_{m\_1})] \times \\
&\prod_{i=2}^{n}[(R_{m\_i} \times R_{MV})^3 + 3(R_{m\_i} \times R_{MV})^2(1 - R_{m\_i} \times R_{MV})] \\
&\times R_{MV}
\end{aligned}
$$

Note that $R_{m\_i} \times R_{MV}$ denotes the reliability of an MV concatenated with a module at stage i (as shown in Figure 5).

MV redundancy can improve the reliability of a concatenated TMR system (i.e. $R_{sys2} > R_{sys1}$) when

$$
R_{m\_i} > \frac{3}{2(1 + R_{MV})} \tag{2}
$$

Dynamic redundancy is commonly used for a system with a high failure rate; it consists of a TMR core with a bank of spares. A dynamically redundant system can tolerate more faulty modules than an NMR system. For example, a dynamically redundant system with five redundant modules can tolerate up to 3 faulty modules, while a NMR system with 5 modules can tolerant at most 2 faulty modules. However, this technique requires a more complex circuitry than other techniques; thus, it has a higher hardware cost and probability of failure in the fault tolerant circuit.

NAND multiplexing [10] uses NAND gates and random permutation multiplexing to restore a bundle of faulty copies of the same signal. As shown in Figure 6, there are N_{bundle} redundant copies of the computing module in a NAND Multiplexing scheme. The multiplexing unit U randomly permutates the signals and the NAND gates are used to restore the signals. Although quantitative analysis of NAND multiplexing is difficult, a probabilistic analysis has shown that this technique provides better fault tolerant performance under a high fault rate than NMR [9], albeit a high redundancy is needed.

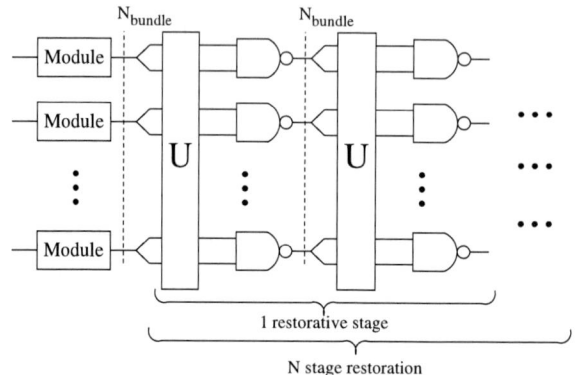

Figure 6. A NAND multiplexing system

A NAND multiplexing system with NAND gates as computing moduleshas been analyzed in [10] and it has been shown that with an extremely large N_{bundle}, the tolerable fault probability of a NAND gate must be at least 0.0107. A more accurate estimation of the tolerable fault probability has been pursued by a probabilistic analysis in [12] and it has been proved that NAND gates with a fault rate ϵ smaller than $\epsilon_0 = \frac{(3-\sqrt{7})}{4} \approx 0.08856$ can restore faulty signals from an computing module to a distinguishable level. With multiple levels of restorative stages and a large amount of redundancy, the restored signal fault probability is a function of ϵ only.

4   Majority multiplexing in QCA

For a QCA system with a high failure rate, a possible approach for establishing the most suitable fault tolerant technique must consider tolerating both high manufacturing (permanent) and operational (transient) fault rates. Due to limitations in current fabrication technology, a QCA system is likely to be unreliable when manufactured (at time 0) and the treatment of transient faults during time $[0, t]$ has not yet been addressed. In this paper, the fault probability is analyzed by considering only manufacturing faults.

None of the traditional fault tolerant techniques by themselves can provide a satisfactory solution for QCA. They are either unable to deal with the high fault rate of QCA devices, or have unacceptable redundancy levels. For QCA, due to the compact implementation of a majority voter, a cascaded voting scheme provides a good basis for a fault tolerant solution. Due to its good capability in restoring signals, the use of a MV in place of a NAND gate in a NAND multiplexing technique is intuitively appropriate (Figure 7). This arrangement was originally proposed in [9] and is generally referred to as *majority multiplexing* (Maj-MUX). In this section, the fault tolerant capabilities as well as signal restoration speed have been reported.

4.1   Fault tolerance

The fault tolerant capability of a Maj-MUX scheme is analyzed next under two scenarios (perfect and non-perfect) for the multiplexing unit.

First, assume a perfect multiplexing unit. Using the method in [10], [9] has shown that the tolerable MV fault rate of Maj-MUX scheme must be at least 0.0197. In this paper, the method of [12] is employed to pursue a more accurate estimate of the lower bound of MV fault rate that is required by Maj-MUX scheme to improve system reliability.

Assume the inputs of the MVs have an equal fault probability given by x and the fault rate of the MVs is ϵ. Then, the probability x_1 of the MV outputs being faulty, is:

$$x_1 = 1 - (1 - \epsilon)[(1 - x)^3 + 3x(1 - x)^2] \tag{3}$$

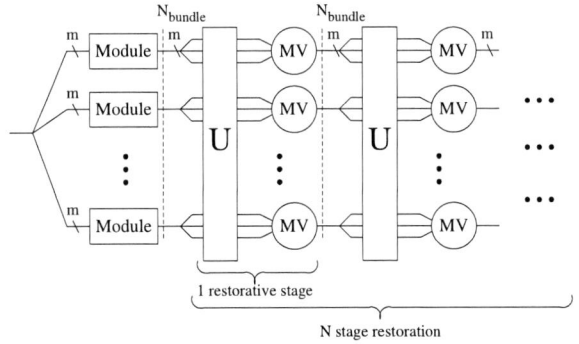

Figure 7. A majority multiplexing system

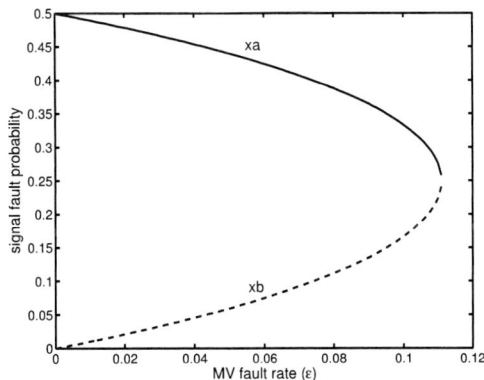

Figure 8. Range of fault probability improvement for Maj-MUX

The worst case scenario is analyzed, so fault compensation in masking is not considered. The reliability is improved when $x_1 < x$. Thus, $(2\epsilon - 2)x^3 + (3 - 3\epsilon)x^2 + \epsilon > x$. Since $x \le 1$, then the solution of this inequality is

$$x_b < x < x_a \tag{4}$$
$$\text{where} \quad x_a = \frac{(1 - \epsilon) + \sqrt{(9\epsilon - 1)(\epsilon - 1)}}{4(1 - \epsilon)}; \quad x_b = \frac{(1 - \epsilon) - \sqrt{(9\epsilon - 1)(\epsilon - 1)}}{4(1 - \epsilon)}$$

If $\frac{1}{9} < \epsilon < 1$, then (4) cannot be satisfied. Only when $\epsilon \in [0, \frac{1}{9}]$, the signals with $x \in [x_b, x_a]$ can be restored to a fault probability equal to x_b. Figure 8 plots the signal fault probability versus the MV fault rate.

Next, consider a faulty multiplexing unit. For a Maj-MUX system implemented in QCA, the interconnections in the random multiplexing unit (U in Figure 9) are important sources of error. Let the probability that faults in a multiplexing unit results in a signal error, be denoted by μ. The signal fault probability after restoration is:

$$x_1 = 1 - (1 - \mu)(1 - \epsilon)[(1 - x_0)^3 + 3x_0(1 - x_0)^2] \tag{5}$$

To improve system reliability, it is required that $x_1 < x_0$. By substituting $(1 - \mu)(1 - \epsilon)$ with $(1 - \beta)$, (5) has the same form as (3). So the solution is also in the same form:

$$x_d < x < x_c \tag{6}$$
$$\text{where} \quad x_c = \frac{(1 - \beta) + \sqrt{(9\beta - 1)(\beta - 1)}}{4(1 - \beta)}; \quad x_d = \frac{(1 - \beta) - \sqrt{(9\beta - 1)(\beta - 1)}}{4(1 - \beta)}$$

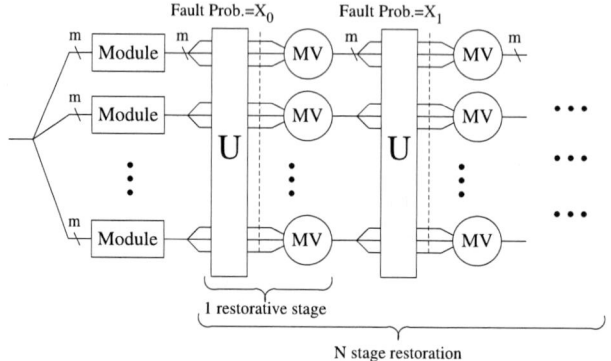

Figure 9. A Maj-MUX system considering faults in the multiplexing unit

If $\frac{1}{9} < \beta = (\mu + \epsilon - \mu\epsilon) < 1$, $x_1 < x_0$ cannot be satisfied. Only when $\beta \in [0, \frac{1}{9}]$, the output signals from the computing modules with fault probability $x \in [\frac{x_b - \mu}{1 - \mu}, \frac{x_a - \mu}{1 - \mu}]$ can be restored to a fault probability $= \frac{x_b - \mu}{1 - \mu}$, after the Maj-MUX.

4.2 Restoration speed of multiplexing

As in previous works, restoration speed is defined in this paper as the fault probability improvement that can be achieved with one restorative stage; it is commonly used as a figure of merit to establish the number of restorative stages that are needed to assemble a reliable system.

For a NAND multiplexing system, the reliability (i.e. the probability of being fault free or correct) of a signal after one restorative stage is given as follows (where x is the probability of the signal being faulty prior to restoration):

if input=1:

$$P[\text{FF after 1 nand}] = (1 - \epsilon)(1 - x)^2 \tag{7}$$

$$P[\text{FF after 1 stage}] = (1 - \epsilon)(2 - P[\text{FF after 1 nand}]) \times P[\text{FF after 1 nand}] \tag{8}$$

if input=0:

$$P[\text{FF after 1 nand}] = (1 - \epsilon)(1 - x^2) \tag{9}$$

$$P[\text{FF after 1 stage}] = (1 - \epsilon)P[\text{FF after 1 nand}]^2 \tag{10}$$

For a Maj-MUX system, the faulty probability after one restorative stage is given by

$$P[\text{FF after 1 stage}] = (1 - \epsilon)[(1 - x)^3 + 3x(1 - x)^2] \tag{11}$$

Figure 10 shows the signal reliability after different numbers of restorative stages. The NAND multiplexing and Maj-MUX systems can be compared under different values of ϵ. The Maj-MUX scheme has a faster signal restoration speed than the NAND-MUX scheme. For example, with the error rate of MV and NAND both at $\epsilon = 0.03$ and a signal reliability before restoration of 0.8, Maj-MUX needs 4 restorative stages to recover the signal to achieve full fault tolerance, while NAND-MUX needs 6 stages.

5 Conclusion and Discussion

In this paper, a number of available fault tolerant techniques have been considered for QCA implementation. Due to the compact implementation of a majority voter (MV) in QCA, it has

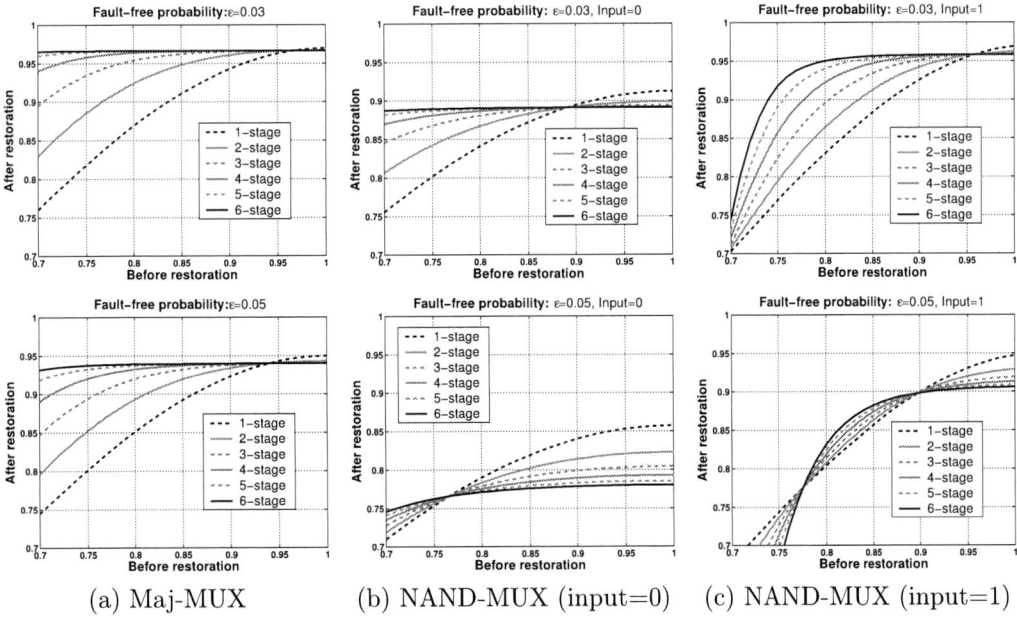

(a) Maj-MUX   (b) NAND-MUX (input=0)   (c) NAND-MUX (input=1)

Figure 10. Comparison of restoration speed for Maj-MUX and NAND-MUX

been shown that majority multiplexing (Maj-MUX) is most promising for QCA. The fault tolerant capacity and signal recovery speed of Maj-MUX have been investigated in detail.

It has been shown that the Maj-MUX scheme is attractive for QCA due to the following features:
(1) A Maj-MUX scheme requires a reliability bound for the majority voter (0.8889) that is easier to meet than NAND multiplexing for a NAND gate (0.91144). In QCA, a MV has a every compact implementation. As a MV requires only 5 QCA cells, it is possible to reach the gate reliability requirement of the Maj-MUX scheme. This advantage makes the Maj-MUX scheme suitable for QCA implementation.

(2) Given a sufficient number of restorative stages and redundancy, the tolerable fault rate of a computing module is very high (for example, if the fault rate of the MV is 0.1, signals can be recovered as long as the module fault rate is ≤ 0.333). This fault tolerant bound and fault probability of the restored signals are mostly determined by the reliability of the restorative stages. The restored signal reliability is $\frac{(1-\epsilon)-\sqrt{(9\epsilon-1)(\epsilon-1)}}{4(1-\epsilon)}$. Using Taylor's expansion, the reliability is $\epsilon + 3 \times \epsilon^2 + O(\epsilon^3)$. For $\epsilon < 0.1$, the restored signal reliability is approximately ϵ (i.e. same as the reliability of MV).

(3) As shown in Figure 10, the Maj-MUX scheme has a better restoration speed than NAND-MUX.

However, the following disadvantages are incurred using the Maj-MUX.

(1) The redundancy level considered in this work is still rather extensive. Ultimately, the fault tolerant capability of this scheme will be limited by the redundancy rate that the overall system can afford.

(2) An implementation of Maj-MUX will require a large amount of wire crossing devices in QCA. The reliable operation of the wire crossing device is therefore crucial for assessing the applicability of Maj-MUX.

(3) A multiplexing scheme (using a MV or a NAND gate) can preserve a high reliability of a system, however the outputs of a signal must be generated in bundles. So, for a traditional output signal, a threshold (or voted) logic must be provided to reduce the bundle-signal to a bit-signal. The reliability of these "final" output gates may affect the overall system reliability.

References

[1] P. D. Tougaw and C. S. Lent, "Logical devices implemented using quantum cellular automata," *Journal of Applied Physics*, vol. 75, no. 3, pp. 1818–1825, 1994.

[2] K. Walus, A. Vetteth, G. A. Jullien, and V. S. Dimitrov, "Ram design using quantum-dot cellular automata," in *NanoTechnology Conference*, vol. 2, 2003, pp. 160–163.

[3] G. Toth, "Correlation and coherence in quantum-dot cellular automata," Ph.D. dissertation, University of Notre Dame, 2000.

[4] W. Hu, K. Sarveswaran, M. Lieberman, and G. H. Bernstein, "High-resolution electron beam lithography and dna nano-atterning for molecular QCA," *IEEE Trans. on Nanotechnology*, vol. 4, no. Issue 3, pp. 312–316, May 2005.

[5] C. S. Lent, M. Liu, and Y. Lu, "Bennett clocking of quantum-dot cellular automata and the limits to binary logic scaling," *Nanotechnology*, vol. 17, pp. 4240–4251, 2006.

[6] X. Ma, J. Huang, and F. Lombardi, "A model for computing and energy dissipation of molecular QCA devices and circuits," 2008, to appear in ACM Journal on Emerging Technologies in Computing Systems(JETC).

[7] J. Huang, M. Momenzadeh, M. Tahoori, and F. Lombardi, "On the evaluation of scaling of QCA devices in the presence of defects," *IEEE Trans. on Nanotechnology*, vol. 4, no. 6, pp. 740–743, 2005.

[8] M. Ottavi, M. Momenzadeh, and F. Lombardi, "Modeling QCA defects at molecular level in combinational circuits," in *Proc. IEEE Intl. Symposium on Defect and Fault Tolerance in VLSI Systems (DFT)*, Monterey, Oct 2005, pp. 208–216.

[9] S. Roy and V. Beiu, "Majority multiplexing - economical redundant fault-tolerant design for nano architectures," *IEEE Trans. on Nanotechnology*, vol. 4, no. 4, 2005.

[10] J. von Neumann, "Probabilistic logics and synthesis of reliable organisms from unreliable components," *Automata Studies*, pp. 43–98, 1956.

[11] K. Hennessy and C. S. Lent, "Clocking of molecular quantum-dot cellular automata," *Journal of Vaccum Science and Technology*, vol. 19, no. 5, pp. 1752–1755, 2001.

[12] W. Evans and N. Pippenger, "On the maximum tolerable noise for reliable computation by formulas," *IEEE tran. Information Theory*, vol. 44, no. 3, pp. 1299–1305, May 1998.

On Reducing Circuit Malfunctions Caused by Soft Errors[*]

Ilia Polian[1], Sudhakar M. Reddy[2], Irith Pomeranz[3],
Xun Tang[2], Bernd Becker[1]

[1]*Institute for Computer Science*
Albert-Ludwigs-University
Freiburg, Germany
{polian\becker}informatik.uni-freiburg.de
[2]*ECE Department*
University of Iowa
Iowa City, Iowa, USA
{reddy\xutang@engineering.uiowa.edu}
[3]*School of ECE*
Purdue University
West Lafayette, Indiana, USA
{pomeranz@ecn.purdue.edu}

Abstract

Soft errors due to radiation are expected to increase in nano-electronic circuits. Methods to reduce system failures due to soft errors include use of redundancy and making circuit elements robust such that soft errors do not upset signal values. Recent works have noted that electronic circuits have partial intrinsic immunity to soft errors since single event upsets on a large percentage of signal lines do not cause errors on circuit outputs. Using ISCAS-89 benchmark circuits we present experimental evidence that the partial immunity to single event upsets is in most cases due to redundancy in the circuits and thus immunity to soft errors may not be available in irredundant circuits. Thus goals on immunity to soft errors may not be achievable in highly optimized circuits without adding circuit redundancy and/or relaxing the requirements on system failures due to soft errors.

---

[*] This work was supported in part by the DFG project RealTest (BE 1176/15-1), Alexander von Humboldt Foundation and by SRC Grants 2007-TJ-1642 and 2007-TJ-1643.

1550-5774/08 $25.00 © 2008 IEEE
DOI 10.1109/DFT.2008.20

1. Introduction

Soft errors are a major factor limiting dependability of micro- and nanoelectronic circuits [Shivakumar 02]. While it is possible to protect circuits against soft errors, hardening solutions impose prohibitive costs in terms of area and power consumption. Recent publications suggest that a significant fraction of soft errors do not have any critical effect on [Wang 04, Li 07]. Moreover, some authors were even able to locate spots in the circuit such that a soft error on these spots did not result in critical errors on system outputs.

In [Polian 06], over 70% of the logic signals in an MPEG motion estimator were shown to be non-critical in the sense that soft errors on these signals led to errors on circuit outputs which were guaranteed to disappear after a few clock cycles. Thus single event upsets on a large majority of circuit lines in the MPEG motion detector effect circuit outputs for only a very limited time and specifically the errors caused by single event upsets disappear after only a few clock cycles. This concept was generalized in a probabilitic way in [Hayes 07]. In [Seshia 07], the properties which are used to formally verify a telecommunication chip are proven to hold even under errors in some of the flip-flops. In [Nowroth 08], errors on over 54% of the flip-flops in a JPEG compressor were shown not to result in perceptible reduction of image quality.

The fact that a significant fraction of spots in the system is non-critical, i.e., the system continues to operate meaningfully even in the presence of errors on these spots, allows the application of area- and power-efficient selective hardening techniques where only selected parts of a circuit are hardened against soft errors [Mohanram 03, Hayes 07]. On the other hand, it is counter-intuitive that errors on some parts of the system seemingly do not matter. In this paper, we study the possible reasons for the large number of non-critical spots. We evaluate the following three hypotheses.

First, the non-critical spots may be redundant. We call a spot redundant if its logical value has no influence on the circuit's operation. Permanent defects and soft errors on a redundant spot never produce effects on the circuit outputs. Since redundant spots can be eliminated from the design without changing its functionality, most circuits contain no or very few redundant spots. Hence, we assume that only a small fraction of non-critical spots are redundant.

The second hypothesis is that the non-critical spots are single-cycle redundant, i.e., a soft error on that spot with a duration of one cycle is not propagated to the circuit outputs irrespective of the circuit's state or input values. This is equivalent to the error effect being eliminated by a mechanism called logical masking [Shivakumar 02]. There are two other masking mechanisms based on the observation that realistic soft errors are pulses of width significantly less than one clock cycle. Hence, single-cycle redundancy is a stronger requirement than immunity to the majority of the realistic soft errors, though it is a weaker requirement than redundancy. In general, single-cycle redundancy is not a sufficient condition to eliminate logic associated with a spot from the circuit.

The third hypothesis is that the non-critical spots are neither redundant nor single-cycle redundant. In this case, errors on non-critical spots do not violate the essential properties of the system either because no system errors occur under normal operating conditions or because the errors caused were within acceptable limits.

In this paper, we focus on the relationship between the first and the second class of the non-critical spots, i.e., the connection between redundancy and single-cycle redundancy. Intuitively, single-cycle redundancy is a weaker requirement than redundancy. Hence, one may be tempted to believe that there are more single-cycle-redundant spots than redundant spots. We generated empirical data to quantify the numbers of redundant and single-cycle redundant spots in ISCAS circuits to resolve this question.

The remainder of the paper is organized as follows. The method to identify redundant and single-cycle redundant spots is presented and the empirical data is reported in Section 2. The results are discussed in Section 3. Section 4 concludes the paper.

2. Identification of Redundant and Single-Cycle Redundant Spots

We determined the numbers of single-cycle-redundant and redundant spots in ISCAS-89 sequential benchmark circuits based on the observation that a (single-cycle-) redundant spot corresponds to a (single-cycle-) untestable fault on this spot. We used a modified version of a tool that identifies redundant stuck-at faults in sequential circuits [Reddy 99]. This tool identified almost all redundant spots in ISCAS-89 circuits. A spot is redundant if the fault-free circuit and the faulty circuits are synchronizable and there is no test to detect the fault [Pomeranz 96]. The method uses indirect implications derived through sequential static learning to determine that no tests exist under normal circuit operation using an Iterative Logic Array (ILA) of finite length ($m+n$) to model the sequential circuit [Reddy 99]. The ILA model used is shown in Figure 1 given below. It should be noted that the fault being considered is injected only in the last m right most cells and the n left most cells cells are fault-free. If the effect of the fault under consideration cannot be propagated to a primary output or to a the pseudo-primary outputs (the horizontal outputs of the right most cell of the ILA) then the fault is untestable [Reddy 99] and also redundant if the faulty circuit is synchronizable [Pomeranz 96]. To establish that a single-cycle fault is redundant we use the ILA of Figure 2 in which the target fault is injected only in the (m+1)th cell and the requirement that the faulty circuit is synchronizable is removed since single cycle faults do not effect synchronizability of a circuit.

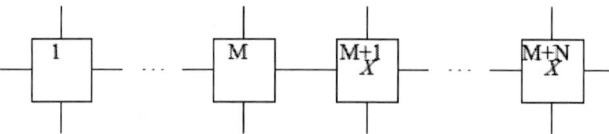

Figure 1: ILA model used to determine redundant spots

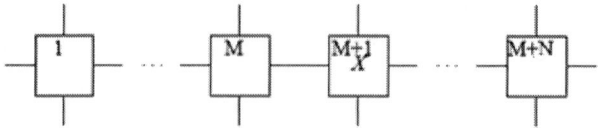

Figure 2: ILA model used to determine single-cycle redundant spots

3. Experimental Results

The results on redundant spots and single cycle redundant spots are given in Tables 1 and 2. We used the set of uncollapsed single stuck-at faults in the combinational logic of the synchronizable ISCA-89 benchmark sequential circuits. In Table 1, after the circuit name we give the total number of stuck-at faults followed by the number of (permanent) redundant faults and single-cycle-redundant faults. In reporting the numbers of faults we consider stuck-at-0 and stuck-at-1 as separate faults. It can be seen that the numbers of stuck-at faults that are redundant and the number of single cycle redundant faults are the same in many of the circuits. In thirteen circuits (shown by entries in bold) where the numbers of single cycle redundant faults are higher than the number of redundant stuck-at faults the difference is small with few exceptions. For all the circuits in our experiment the set of single-cycle-redundant faults contains the set of redundant stuck-at-faults. However, we have not been able to prove that this observation will hold in general.

In Table 2 we give the number of flip-flops at which the output stuck-at faults are redundant or the single-cycle redundant faults at the output of the flip-flops are redundant. After the circuit name we give the total number of delay flip-flops (DFFs) in the circuit followed by the numbers of stuck-at or single-cycle redundant faults at the outputs of flip-flops. In the last two columns we give the numbers of DFFs at the outputs of which both stuck-at and stuck-at faults are redundant and both the single-cycle faults are redundant. These are the flip-flops that need not be augmented to be fault-tolerant to transient faults as suggested in [Seshia 07]. From Table 2 it can be seen that number of circuits in which flip-flop outputs that have redundant stuck-at and/or single-cycle redundant faults are small. Furthermore the numbers of redundant and single cycle redundant faults at the outputs of flip-flops are identical.

Table 1: Redundant faults in combinational logic

circuit	total number of faults	permanent redundant	single cycle redundant
S208	**416**	**121**	**124**
S298	596	60	60
S344	670	18	18
S349	680	22	22
S382	764	8	8
S386	**772**	**76**	**81**
S400	800	20	20
S420	**840**	**438**	**447**
S444	888	30	30
S499	954	163	163
S526	1052	56	56
S641	**1278**	**140**	**152**
S713	**1426**	**218**	**230**
S820	**1640**	**60**	**71**
S832	**1664**	**79**	**96**
S838	**1676**	**1070**	**1107**
S953	**1906**	**10**	**14**
S967	**1928**	**21**	**24**
S1196	**2392**	**3**	**5**
S1238	2476	83	83
S1269	**2522**	**3**	**4**
S1423	2846	40	40
S1488	2976	68	68
S1494	2988	94	94
S3384	**6750**	**1**	**17**
S4863	9598	210	210
S5378	10590	2112	2112
S35932	71224	7344	7344

Table 2: Redundant faults in D-flip-flops

circuit	total number of DFFs	redundant faults	DFF	double redundant DFFs	
		permanent	single cycle	permanent	single cycle
S208	8	0	0	0	0
S298	14	0	0	0	0
S344	15	0	0	0	0
S349	15	0	0	0	0
S382	21	0	0	0	0
S386	6	0	0	0	0
S400	21	0	0	0	0
S420	16	11	11	0	0
S444	21	0	0	0	0
S499	22	0	0	0	0
S526	21	0	0	0	0
S641	19	0	0	0	0
S713	19	0	0	0	0
S820	5	0	0	0	0
S832	5	0	0	0	0
S838	32	27	27	0	0
S953	29	0	0	0	0
S967	29	0	0	0	0
S1196	18	0	0	0	0
S1238	18	0	0	0	0
S1269	37	0	0	0	0
S1423	74	0	0	0	0
S1488	6	0	0	0	0
S1494	6	0	0	0	0
S3384	183	0	0	0	0
S4863	104	0	0	0	0
S5378	179	79	79	23	23
S35932	1728	0	0	0	0

3. Discussion

In the benchmark circuits we studied the sizes of the sets of single-cycle-redundant and redundant spots are very similar for combinational logic and always identical for flip-flops. Thus, soft error immune signal lines of the circuits are essentially the same as the redundant logic lines of the circuits. This suggests that there is very little difference between redundancy and single-cycle redundancy in practice. If a spot is single-cycle redundant, it is most probably also redundant and can be eliminated. On the other hand, given that a circuit is optimized and thus contains very little redundant spots, it also contains very little single-cycle redundant spots. Hence, hypothesis two is as unlikely to explain the presence of a large number of non-critical spots as hypothesis one. Consequently, hypothesis three must be valid: most non-critical spots are neither redundant nor single-cycle redundant.

The observations above seem to suggest that the circuits analyzed were unintentionally over-specified and, consequently, over-designed. In our opinion, the key to the design of dependable systems is the understanding of critical versus non-critical errors with respect to the specification. For instance, soft errors in performance-enhancing modules of a microprocessor, such as branch predictors, are typically non-critical, except in application with hard real-time constraints. Tolerating occasional pixel deviations in imaging application such as video processing may be preferred to spending additional hardware and power consumption for massive redundancy needed to prevent the errors from happening. A range of applications, e.g., in communication, can handle certain classes of errors at the system level. Additional fields in which this concept appears to be applicable include recognition, mining, synthesis, tracking and control.

In our opinion, design and implementation of cost-effective nanoelectronic systems will have to incorporate dependability aspects with respect to soft error resilience as part of specification. We need novel methods of *specification-aware synthesis*, which satisfy the dependability requirements by a carefully chosen mix of techniques such as hardware redundancy, commit-rollback recovery, error-resilient information coding and algorithmic fault tolerance in software. This is in contrast to today's approach where a circuit is designed first and then its dependability is enhanced without modifying the circuit's Boolean function. The future specification formalisms must distinguish between functionality considered as essential, and functions that are allowed to fail occasionally due to soft errors.

4. Conclusions

We demonstrated that the concepts of redundancy and single-cycle redundancy essentially fall together in practice. As a consequence, most non-critical spots are neither redundant nor single-cycle redundant. This suggests large headroom for design of resilient systems which does not rely exclusively on hardware fault tolerance. Without loosening the system specification, employing massive redundancy for full protection of the circuits cannot be avoided.

References

[Hayes 07] J. Hayes, I. Polian, and B. Becker, An analysis framework for transient-error tolerance, VLSI Test Symp., 2007.

[Li 07] X. Li and D. Yeung. Application-level correctness and its impact on fault tolerance. Int'l Symp. on High Perf. Comp. Arch., 2007.

[Mohanram 03] N. Mohanram and N. Touba. Partial error masking to reduce soft error failure rate in logic circuits. Int'l Symp. on DFT, 2003.

[Nowroth 08] D. Nowroth, I. Polian and B. Becker. A study of cognitive resilience in a JPEG compressor. Int'l Conf. on Dependable Systems and Networks, 2008

[Polian 06] I. Polian et al., Low-cost hardening of image processing applications against soft errors, Int'l Symp. on DFT, 2006.

[Pomeranz 96] I. Pomeranz and S.M. Reddy, On removing redundancies from synchronous sequential circuits with synchronizing sequences, IEEE Trans. on Computers, pp. 20-32, Jan. 1996.

[Reddy 99] S.M. Reddy, I. Pomeranz, X. Lin and N.Z. Basturkmen, New procedures for identifying undetectable and redundant faults in synchronous sequential circuits, VLSI Test Symp., 1999.

[Seshia 07] S.A. Seshia, W. Li, and S. Mitra. Verification-guided soft error resilience. DATE Conf., 2007.

[Shivakumar 02] P. Shivakumar et al., Modeling the effect of technology trends on the soft error rate of combinational logic, Int'l Conf. on Dependable Systems and Networks, 2002.

[Wang 04] N.J. Wang, et al.. Characterizing the effects of transient faults on a high-performace processor pipeline. Int'l Conf. on Dependable Systems and Networks, 2004.

IEEE International Symposium on Defect and Fault Tolerance of VLSI Systems

Realization of L2 Cache Defect Tolerance Using Multi-bit ECC

Hongbin Sun, Nanning Zheng
Xi'an Jiaotong University, China
{sunsir,nnzheng}@mail.xjtu.edu.cn

Tong Zhang
Rensselaer Polytechnic Institute, USA
tzhang@ecse.rpi.edu

Abstract

This paper presents a design solution that enables the use of powerful multi-bit error-correcting code (ECC) to realize L2 cache defect tolerance at minimal latency and silicon area cost. This work is motivated by the observation that the continuous CMOS Technology scaling may result in an increasing level defect density and make conventional cache memory defect tolerance strategies inadequate. The basic idea is to complement conventional L2 cache core with two separate fully associative caches, one stores multi-bit ECC check bits for realizing area-efficient selective multi-bit ECC protection and another one stores the most recently decoded multi-bit ECC codeword to minimize the impact of explicit multi-bit ECC decoding on L2 cache access latency. Its effectiveness has been demonstrated using SimpleScalar and Cacti tools. At the defect density of 0.5%, this design approach can maintain almost the same instruction per cycle (IPC) performance over a wide spectrum of benchmarks compared with ideal L2 cache without defects, while only incurring less than 2.5% of silicon area overhead.

1 Introduction

As CMOS technology approaches its end-of-roadmap physical limit, there has been a dramatic increase of process variations and reliability degradation, posing significant challenges to robust circuit design [2]. Since SRAM cells usually have the smallest feature size and the highest sensitivity to transistor mismatches, the design of SRAM cells is much more vulnerable to variations than that of logic circuits. This makes SRAM increasingly subject to defects [4, 14], leading to a renewed interest in SRAM defect tolerance with the focus on SRAM-based cache in processors [3, 8, 9, 15].

This work is interested in improving SRAM-based L2 cache defect tolerance capability by using error-correcting code (ECC). In conventional design practice, memory defects are compensated by using spare (or redundant) rows, columns, and/or words to repair (i.e., replace) the defective ones, while soft errors are compensated by ECC such as *signle-error-correcting and double-error-detecting* (SEC-DED) codes that are widely used in L2 cache of modern microprocessors [7]. However, as the SRAM defect density increases (e.g., up to $\sim 10^{-3}$ as suggested in [4]), traditional redundancy repair methods will probably be inadequate to handle such a large amount of random defects. This naturally motivated recent research efforts to explore the potential of extending the role of ECC for defect tolerance [8, 9, 12, 13]. Intuitively, as memory defect densities increase, the use of multi-bit ECC such as *double-error-correcting and triple-error-detecting* (DEC-TED) may be preferred or even inevitable, especially for L2 (and L3) cache. Nevertheless,

1550-5774/08 $25.00 © 2008 IEEE
DOI 10.1109/DFT.2008.16

compared with conventional SEC-DED, multi-bit ECC tends to incur much higher coding redundancy, decoding latency and energy overhead. Hence, a straightforward use of multi-bit ECC in cache memory, i.e., uniformly protecting all the cache lines with the same multi-bit ECC, will incur prohibitive cost in terms of cache access latency and silicon area.

In this paper, we develop a design approach to enable the use of multi-bit ECC to realize L2 cache defect tolerance at minimal latency and silicon area cost. To minimize the silicon area cost, we may simply apply multi-bit ECC only to those cache blocks whenever necessary by using a separate fully associative cache to store the multi-bit ECC check bits. To minimize the impact of multi-bit ECC decoding on overall L2 cache access latency, we use another fully associative cache to store the most recently accessed cache blocks that are protected by multi-bit ECC, which may obviate most explicit execution of multi-bit ECC due to the cache access locality characteristics. The effectiveness of this defect tolerant L2 cache architecture has been successfully demonstrated using SimpleScalar and Cacti tools. Results show that, even at the defect density of 0.5%, this design approach can maintain almost the same instruction per cycle (IPC) performance over a wide spectrum of benchmarks compared with ideal L2 cache without defects, while only incurring less than 2.5% of silicon area overhead.

2 Background

In modern microprocessors, each L2 cache block is typically partitioned into several equalized sub-blocks and each sub-block is protected with either ECC for soft error tolerance in write-back cache or error detection code for soft error detection in write-through cache. In this work, we categorize cache sub-blocks based on the number of defective cells they contain as follows.

- A sub-block that does not have any defective cells is called a good sub-block, which is denoted as *g-sub-block*.

- A sub-block that has one defective cell is called a single-defect sub-block, which is denoted as *s-sub-block*.

- A sub-block that has two or more defective cells is called a multi-defect sub-block, which is denoted as *m-sub-block*.

Furthermore, a cache block that only contains g-sub-blocks is called a *g-block*, a cache block that contains both g-sub-blocks and s-sub-blocks is called an *s-block*, and a cache block that contains one or more m-sub-blocks is called an *m-block*. Given the defect density λ and assuming all the cell defects are random and uniformly distributed, we can calculate the probabilities of the occurrence of different types of cache sub-block and blocks. For example, assuming each cache sub-block stores 64 bits and each cache block contains 8 sub-blocks (i.e., the L2 cache configuration in AMD Opteron processor [7]), we can obtain the corresponding probabilities as shown in Fig. 1. As defect densities increase, the percentages of m-sub-blocks and m-blocks will quickly become non-negligible, e.g., under the defect rate of 0.005, more than 4% of cache sub-blocks will contain multiple defective cells and over 28% percent of cache blocks are m-blocks. Clearly, even for write-through cache, the conventional single-error-correcting codes will be inadequate under certain defect densities and multi-bit ECC appears to be inevitable.

However, a straightforward use of multi-bit ECC in L2 cache will incur too much speed and area penalty and hence is infeasible in practice. For example, compared with the simple SEC-DED codes, the BCH-based DEC-TED codes that can correct up to two errors and detect the occurrence of three errors demand more check bits as listed in Table 1. More importantly, the decoding latency of a DEC-TED code can be much longer than that of a SEC-DED code. The decoding

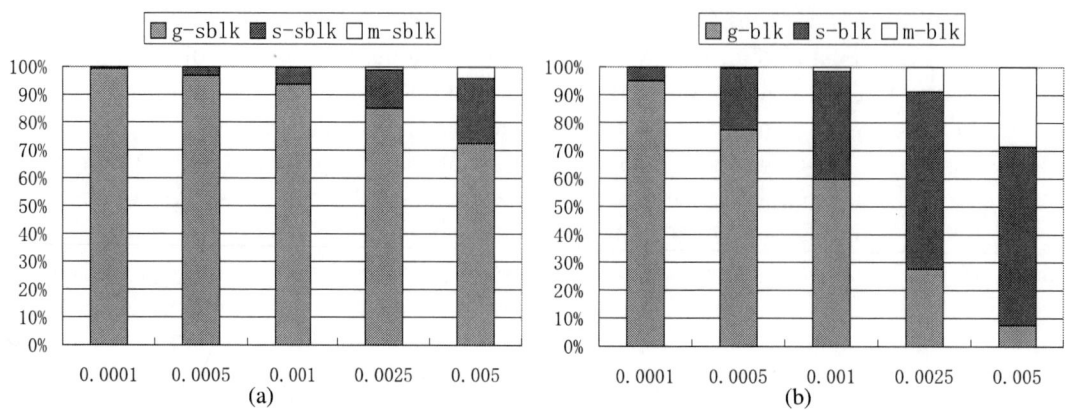

Figure 1: Probabilities of the occurrence of different types of (a) sub-blocks and (b) blocks under different defect densities.

Table 1: Number of check bits in SEC-DED and DEC-TED codes.

Data bits	32	64	128
SEC-DED check bits	7	8	9
DEC-TED check bits	12	14	16

of SEC-DED code only involves simple XOR operations and can be efficiently parallelized while the decoding of BCH-based DEC-TED involves more complex Galois Filed arithmetics and even its fully parallel realization, at the cost of a significant silicon area overhead, still suffers from relatively long latency [10]. Although there exist other types of DEC-TED codes with modestly improved decoding latency [6], they will incur much more check bits than BCH-based DEC-TED codes.

3 Proposed L2 Cache Defect Tolerant Design Strategy using Multi-bit ECC

In this work, we develop an L2 cache architecture that can use multi-bit ECC to realize defect tolerance at minimal latency and silicon area cost. The two *underlying themes* are (i) avoid the execution of multi-bit ECC decoding on the access critical path as much as possible to minimize the impact on cache latency, and (ii) try to use just enough multi-bit ECC coding redundancy for those defective cache blocks and sub-blocks, instead of uniformly protecting the entire L2 cache using multi-bit ECC, to minimize the silicon cost. As illustrated in Fig. 2, the developed L2 cache architecture includes a conventional L2 cache core, a built-in self-test (BIST) unit, a fully-associative multi-bit ECC cache (denoted as M-ECC cache), and a fully-associative pre-decoding buffer. The BIST unit is used to identify the location of all the cache sub-blocks with more than one defective cells (i.e., all the m-sub-blocks). Facilitated with the BIST unit, the location information can also be periodically updated to handle the defects that develop in the field. The locations of m-sub-blocks stored in the tag portion of the fully associative M-ECC cache that stores the multi-bit ECC check bits. The fully associative pre-decoding buffer stores the most recently accessed m-blocks, which aims to avoid the execution of multi-bit ECC decoding on the cache access critical path. The architecture and operation of this developed L2 cache design are further explained as follows.

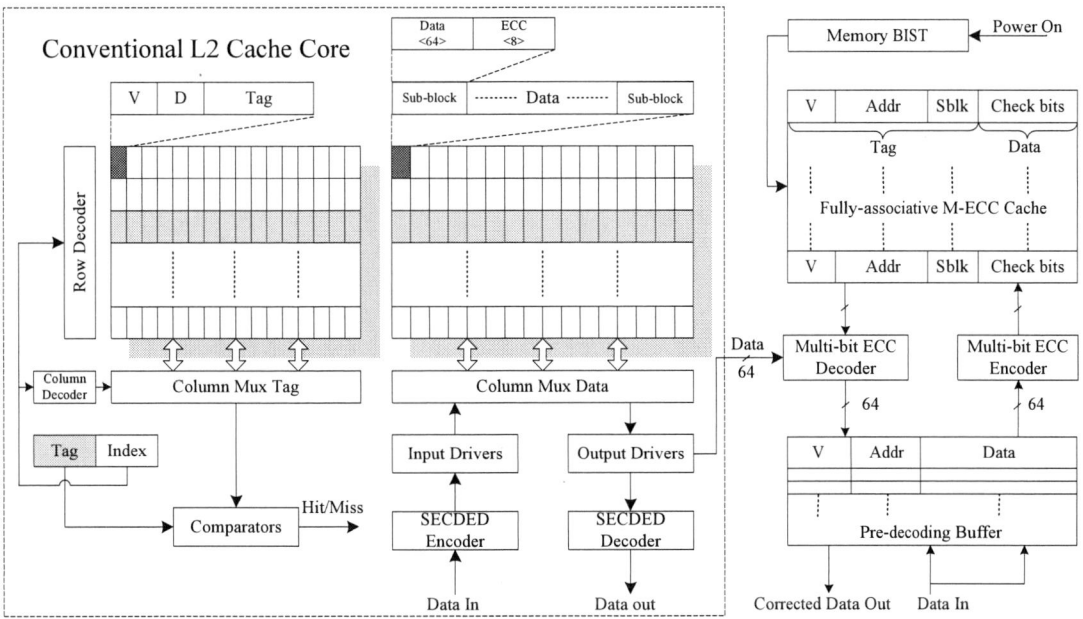

Figure 2: Architecture of the proposed fault tolerant L2 cache using Multi-bit ECC.

Given the available defect information provided by the BIST, it is obvious that we should only use multi-bit ECC whenever necessary. Therefore, it is straightforward that we may only use the conventional SEC-DED code to the entire L2 cache core, complemented with a fully associative cache to store the multi-bit ECC check bits only for m-sub-blocks. The tag of the M-ECC cache stores the location of the m-sub-blcoks identified by the BIST. One valid bit is associated with each entry in the M-ECC cache to indicate whether it indeed contains the check bits for one m-sub-block. Combined with the power-on BIST, this architecture can readily realize the tolerance to defects that develop in the field. In a straightforward manner, whenever the L2 cache is being accessed, the M-ECC cache will be searched. If a matching address is found in M-ECC cache, which means the cache block being accessed contains m-sub-block(s), then we have: (i) in case of write operation, the data of the m-sub-blocks will be encoded by the multi-bit ECC and the corresponding check bits will be stored in the M-ECC cache; (ii) in case of read operation, the corresponding multi-bit ECC check bits will be fetched and used in the multi-bit ECC decoding to recover the correct data. Since the M-ECC cache is fully associative and its size is relatively small compared with the L2 cache core, the M-ECC cache access tends to be insignificant. However, the multi-bit ECC decoding may take many clock cycles and is on the L2 cache read critical path in case of an m-block is being hit. This will greatly degrade the L2 cache performance as the number of m-blocks increases.

To address the above issue, we propose to further add a fully associative pre-decoding buffer to hold the data of those most recently m-blocks, as illustrated Fig. 2. This essentially leverages the well-known fact that cache access has a large degree of spatial and temporal localities. This fully associative pre-decoding buffer employs the least recently used (LRU) policy for replacement when it is full. As a result, the explicit multi-bit ECC decoding can be avoided when a matching is found in this buffer, as shown in Fig. 3. Each L2 cache access to an m-block will update this pre-decoding buffer. In case of an L2 cache read hit, this pre-decoding buffer is searched first. If a hit occurs, the cache block data stored in this buffer will be sent out without incurring M-ECC

257

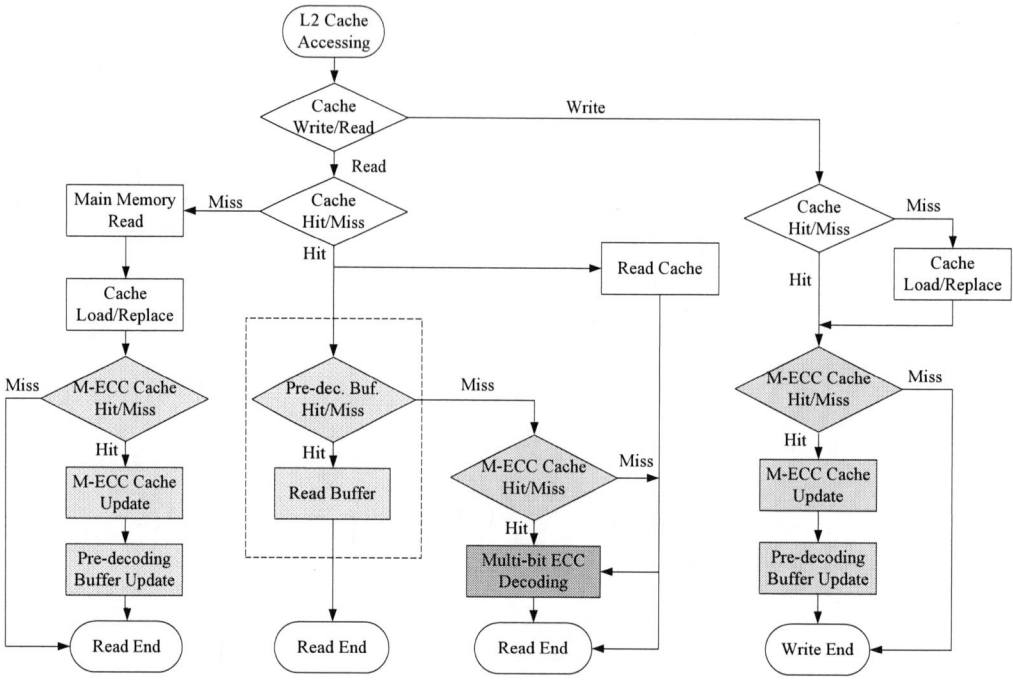

Figure 3: Flow chart of operation in the fault tolerant L2 cache.

cache access and multi-bit ECC decoding. By combining the M-ECC cache and the pre-decoding buffer, this new L2 cache can effectively enable the use of strong multi-bit ECC to realize defect tolerance at minimal access latency and silicon cost.

4   Performance Evaluation

To evaluate the performance of the above presented L2 cache architecture, we carry out simulations using the popular *SimpleScalar 3.0* simulator [5]. Table 2 lists the simulator configuration parameters. The cache memory defect density is set as 0.005, and this work assumes the use of BCH-based DEC-TED code that adds 14-bit check bits for each 64-bit sub-block, and cache sub-blocks contain more than two error are repaired by redundancy. We set the size of M-ECC cache and pre-decoding buffer as 16KB and 4KB, respectively. We use cache modeling tool Cacti 6.0 [11] to estimate the characteristics of the main L2 cache core, M-ECC cache and pre-decoding buffer at 45nm node as listed in Table 3. We set the execution of BCH DEC-TED decoding takes 82 cycles in total.

Table 4 shows the 12 benchmarks used in our evaluation, including seven SPEC2000 integer benchmarks and five SPEC2000 floating-point benchmarks [1]. For each benchmark, we simulated the sequence of instructions which have relatively high L2 access rate. Table 4 lists the number of instructions skipped to reach the phase start, i.e. the FFWD (fast-forward) column, and the number of instructions simulated, i.e. the RUN column. Table 4 also shows the average number of L1 cache misses and L2 cache hits per million instructions and the corresponding average instruction per cycle (IPC). For the purpose of comparison, we considered the following three different scenarios in simulations:

- Base : The L2 cache does not have any defects and has no M-ECC cache and performs as the conventional cache.

Table 2: Simulator configuration parameters.

Configuration Parameters	Value
Processor	
Functional Units	4 integer ALUs, 4 FP ALUs
	1 integer and 1 FP multiplier/divider
LSQ / RUU Size	8 / 16 Instructions
Fetch/Decode/Issue/Commit Width	4 Instructions/cycle
Fetch Queue Size	4 Instructions
Cycle Time	1 ns
Cache and Memory Hierarchy	
Instruction L1 Cache	64KB, 2-way, 64-byte blocks, 1 cycle latency
Data L1 Cache	64KB, 2-way, 64-byte blocks, 1 cycle latency
United L2 Cache	1 MB, 8-way, 64-byte blocks, 13 cycle latency
Main Memory	300 cycle latency
Branch Logic	
Predictor	combined, bimodal 2KB table
	two-level 1KB table, 8 bit history
BTB	512 entry, 4-way
Miss-prediction Penalty	3 cycles

Table 3: Cacti report for L2 cache core, M-ECC cache and Pre-decoding buffer.

Memory	Size	Associativity	Block size (byte)	Access time (ns)	Cycle time (ns)	Area (mm^2)
L2 cache core	1MB	8	64	1.60	0.18	4.3483
M-ECC cache	16KB	full	2	0.39	0.05	0.0840
Pre-dec. buffer	4KB	full	64	0.31	0.05	0.0225

Table 4: Benchmarks used for performance evaluation

SPEC2000	Phase FFWD	Phase RUN	L1 misses / Million ins.	L2 hits / Million ins.	IPC
164.gzip	1.0B	400M	2875	4468	1.7520
175.vpr	100M	400M	1343	1469	1.6048
176.gcc	1.0B	400M	48618	75468	1.3526
177.mesa	500M	400M	6329	6585	1.5300
179.art	1.2B	300M	35055	24021	0.1872
181.mcf	200M	300M	9461	7684	0.4504
183.equake	100M	400M	3816	3795	1.7033
188.ammp	100M	300M	16825	983	0.1830
197.parser	100M	400M	1962	2652	1.2910
255.vortex	1.0B	400M	19409	20115	1.1866
256.bzip2	200M	400M	4585	5326	1.2925

- M-ECC : The L2 cache is equipped with M-ECC cache, but has no Pre-decoding buffer. The scheme has to search the M-ECC cache with each access, and suffer the decoding delay with each M-ECC hit.

- M-ECC-buf : The L2 cache is equipped with M-ECC cache, and has Pre-decoding buffer to mitigate the delay penalty due to Multi-bit ECC decoding.

Fig. 4 shows the normalized IPC comparison. Most of the benchmarks have severe performance degradation for the M-ECC scheme, especially for those benchmarks with high L2 cache hit rate. This is because the multi-cycle decoding delay significantly increases the average accessing time of L2 cache. Clearly, the performance degradation is reduced dramatically when using the M-ECC-buf scheme, where the IPC performance can be almost as good as the Base scheme. This is because most of decoding processes have been eliminated due to the use of the pre-decoding buffer.

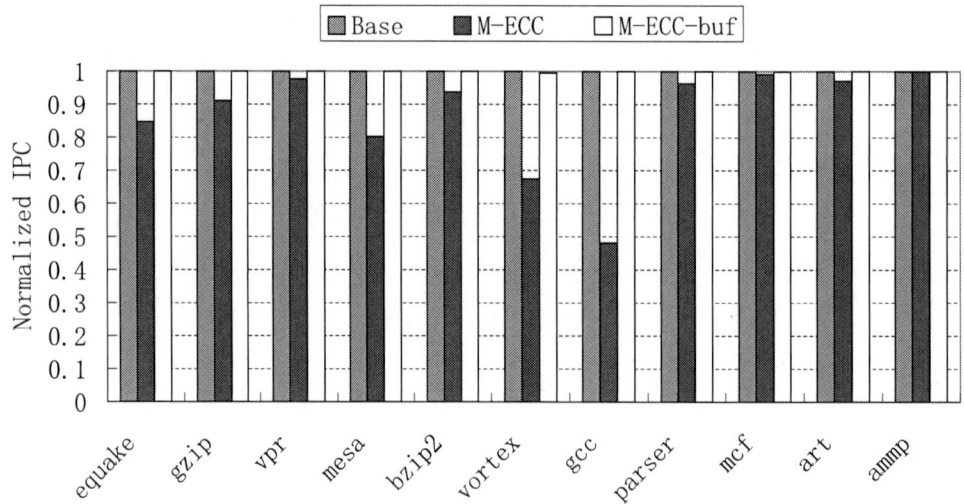

Figure 4: Normalized IPC comparison among Base, M-ECC and M-ECC-buf schemes.

Since the size of the pre-decoding buffer directly affects the average L2 cache access latency, we further evaluated its size vs. hit rate. First, we vary its size from 16 blocks to 128 blocks (the L2 cache is fixed to be 1MB, 8-way associative as shown in Table 2). The results of the pre-decoding hit rate are shown in Fig. 5. Clearly, the pre-decoding hit rate increases with the increase of the pre-decoding buffer size. For pre-decoding buffers with the size of 64 blocks and above, the pre-decoding hit rates are larger than 98.4%, and hence a large fraction of multi-bit ECC decoding can be avoided. Furthermore, fixing the pre-decoding buffer as 64 blocks, we conduct experiments to examine the pre-decoding buffer hit rate by varying the number of associativity of the L2 cache. Fig. 6 shows the pre-decoding buffer hit rates with the number of L2 cache associativity varying from 4-way to 16-way. With the increase of set associativity, the pre-decoding buffer hit rate decreases. Nevertheless, even for the 16-way associative L2 cache, a 64-block pre-decoding buffer can still achieve a hit rate as high as 98%. The above results further demonstrates that 64-block pre-decoding buffer appears to be an appropriate choice in this experiment setup.

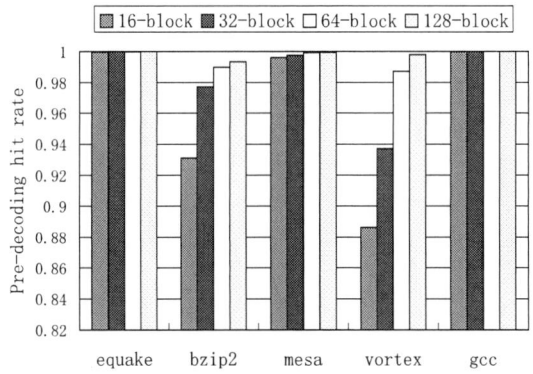

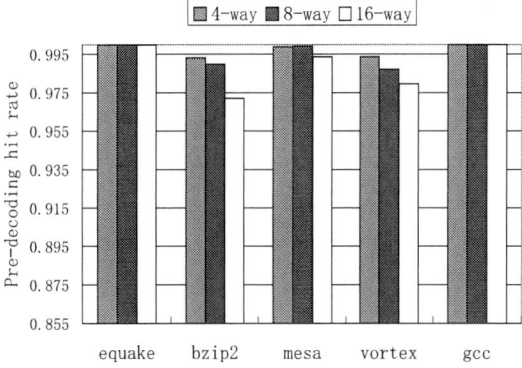

Figure 5: Pre-decoding hit rates for different Pre-decoding buffer size.

Figure 6: Pre-decoding hit rates for different L2 cache set associativity.

5 Conclusion

This paper develops a solution to improve the reliability of L2 cache by using advanced multi-bit ECC at minimal cost of latency and silicon area. The developed new defect-tolerant L2 cache architecture uses a fully associative cache to store the multi-bit ECC check bits and a fully associative buffer to minimize the impact of multi-bit ECC decoding on L2 cache performance. This L2 cache architecture has been thoroughly evaluated by using processor simulation and memory modeling tools and the effectiveness has been successfully demonstrated at the defect density of 0.5%.

References

[1] Standard performance evaluation corporation. *http://www.spec.org*, 2000.

[2] International technology roadmap for semicondutor. *http://public.itrs.net*, 2007.

[3] A. Agarwal, B. C. Paul, H. Mahmoodi, A. Datta, and K. Roy. A process-tolerant cache architecture for improved yield in nanoscale technologies. *IEEE Transactions on Very Large Scale Integration (VLSI) Systems*, 13:27–38, Jan. 2005.

[4] A. Bhavnagarwala, S. Kosonocky, C. Radens, K. Stawiasz, R. mann, Q. Ye, and K. Chin. Fluctuation limits & scaling opporturnies for CMOS SRAM cells. In *Proc. International Electron Devices Meeting*, pages 1551–1558. IEEE, 2005.

[5] D. Burger and T. Austin. The simplescalar tool set, version 2.0. In *University of Wisconsin-Madison Computer Sciences Department Tech. Report*, 1997.

[6] K. Chakraborty and P. Mazumder. *Fault-Tolerance and Reliability Techniques for High-Density Random-Access Memories*. Prentice Hall, 2002.

[7] J. L. Hennessy and D. A. Patterson. *Computer Architecture A Quantitative Approach*. Morgan Kaufmann, 2006.

[8] L. Hung, H. Irie, M. Goshima, and S. Sakai. Utilization of SECDED for soft error and variation-induced defect tolerance in caches. In *Proc. Design, Automation & Test in Europe Conference & Exhibition (DATE'07)*, pages 1–6. IEEE, 2007.

[9] J. Kim, N. Hardavellas, K. Mai, B. Falsafi, and J. Hoe. Multi-bit error tolerant caches using two-dimensional error coding. In *Proc. the 40th Annual ACM/IEEE Int. Sym. on Micro. (Micro-40)*, pages 197–209. ACM/IEEE, 2007.

[10] S. Lin and D. Costello. *Error Control Coding: Fundamentals and Applications*. Prentice Hall, 1983.

[11] N. Muralimanohar, R. Balasubramonian, and N. Jouppi. Cacti 6.0: A tool to understand large caches. In *HP Research Report*, 2007.

[12] M. Nicolaidis, N. Achouri, and L. Anghel. A diversified memory built-in self-repair approach for nanotechnologies. In *Proc. of IEEE VLSI Test Symposium*, pages 313–318, April 2004.

[13] M. Spica and T. M. Mak. Do we need anything more than single bit error correction (ECC)? In *Proc. of International Workshop on Memory Technology, Design and Testing*, pages 111–116, Aug. 2004.

[14] J. Srinivasan, S. Adve, P. Bose, and J. Rivers. The impact of technology scaling on life-time reliability. In *Proc. International Conference on Dependable Systems and Networks*, pages 177–186. IEEE/IFIP, 2004.

[15] W. Zhang. Replication cache: A small fully associative cache to improve data cache reliablility. *IEEE Trans. on Computers*, 54:1547–1555, Dec. 2005.

Selective Hardening of NanoPLA Circuits

Ilia Polian

Computer Architecture Group
Institute for Computer Science
Albert-Ludwigs-University
Georges-Köhler-Allee 51
D-79110 Freiburg i. Br., Germany
polian@informatik.uni-freiburg.de

Wenjing Rao

ECE Department
University of Illinois at Chicago
851 S. Morgan St., MC 154,
1020 SEO building
Chicago, IL, USA
wenjing@ece.uic.edu

Abstract

Nanoelectronic components are expected to suffer from very high error rates, implying the need for hardening techniques. We propose a fine-grained approach to harden a promising class of nanoelectronic circuits, called NanoPLAs, against errors. An analytical procedure and simulations are both incorporated into the algorithm to identify the most critical error locations. By targeting errors with the largest impact for a given circuit, the method can provide significant reliability boost at low cost. Furthermore, the method yields a plethora of alternative designs, trading off hardening costs against circuit robustness. In many cases, solutions found achieve both lower cost and higher robustness compared with the duplication-based hardening strategy introduced before.

Keywords: NanoPLA circuits, robust design, selective hardening, fault tolerance

1: Introduction

Next-generation electronic circuits based on nanotechnology are commonly expected to outperform today's CMOS microelectronics with respect to integration density (area), power consumption, and computation speed [1, 2]. The flip side of the coin is the increased susceptibility of these circuits to errors. Systems employing nano components will presumably have to deal with non-negligible error rates [2, 3, 4, 5]. The error effects can be corrected using techniques such as hardware / time redundancy, error-correcting information encoding, software-based fault tolerance, or combinations thereof [6, 7, 8, 9, 10, 11, 12].

The optimal error correction strategy strongly depends on the expected error rates. For instance, commit-rollback recovery is only effective when error rates do not exceed a certain threshold. On the other hand, hardware redundancy imposes considerable costs. This is particularly true for nano components in which the integration of a large number of basic elements into a working system often becomes a challenge due to strict constraint of localized interconnections. Moreover, even massively redundant design techniques such as triple-modular redundancy fail to provide the flexibility to deal with the high and variable fault rates. Under such a severe reliability challenge, employing even more redundancy aiming at the worst case scenario will increase the cost to a level at which the advantage over CMOS becomes questionable. As a consequence, the optimal strategy should probably combine some degree of hardware-level hardening to tune the error rate of the component to a desired value and a system-level error correction mechanism which works sufficiently well for that error rate.

In this paper, we propose a strategy to selectively harden nanoelectronic circuits belonging to the class of NanoPLAs. Extensive research work has been carried out for NanoPLAs

in [13, 14, 15, 16, 17]. A NanoPLA logic is built on a crossbar architecture, consisting of two sets of perpendicular nanowires. Self-assembly based fabrication can be used to place a nanoelectronic device at each crosspoint, with two distinct states of connecting and disconnecting the two wires. The reconfigurability in nanoelectronic devices enables a nano crossbar to implement arbitrary functions in a two-level PLA logic form.

Reliability concern regarding CMOS PLAs, including online fault detection and diagnosis, has been existing in the literature for a long time [18, 19, 20, 21, 22, 23]. Focusing on the fault tolerance of NanoPLAs, hardening techniques have been proposed in [24, 12]. These techniques are associated with rather high cost and target at all the possible errors unanimously. The proposed strategy hardens a NanoPLA circuit at a per-defect based fine-grain level. The strategy pinpointedly determines the elements in the circuit with the largest contribution to the error rate and hardens them with higher priorities. This allows to achieve the desired robustness at minimized cost.

The algorithm to select the elements to harden optimizes a cost function which combines the two criteria: robustness improvement and hardening cost. A limited number of candidate locations are shortlisted. The robustness improvement and the area cost of hardening are determined for shortlisted candidates, and the candidates with the best overall outcome is selected. Experiment on NanoPLAs constructed from benchmark circuits confirm that focusing on most critical error locations leads to a spectrum of designs, which allow the designer to select the cost-minimal alternative which satisfies the robustness requirement given by the system-level considerations.

2: Hardening Techniques for NanoPLA Circuits

2.1: NanoPLA circuits

NanoPLA circuits have structure similar to conventional Programmable Logic Arrays (PLA). An (unhardened) NanoPLA circuit has n *input wires* $i_1, \ldots, i_n$, l *output wires* $o_1, \ldots, o_l$, and m *row wires* $r_1, \ldots, r_m$ used for product terms. The input wires are connected with row wires by the *AND plane* formally described by an $m \times n$ matrix $\mathcal{A}$ with $\mathcal{A}_{jk} = 1$ if input wire i_k is connected with row wire j and $\mathcal{A}_{jk} = 0$ otherwise. The output wires are connected with row wires by the *OR plane* described by an $m \times l$ matrix $\mathcal{O}$ with $\mathcal{O}_{jk} = 1$ if output wire o_k is connected with row wire j and $\mathcal{O}_{jk} = 0$ otherwise.

When a Boolean vector $(v_1, \ldots, v_n)$ is applied to the input wires, the row wires and the output wires assume Boolean values given by the mapping *val*. The value $val(r_j)$ of row wire r_j is the conjunction of values on the input wires connected with row wire r_j, while the value $val(o_k)$ of output wire o_k is the disjunction of values on the row wires connected with output wire:

$$val(r_j) = \bigwedge_{\mathcal{A}_{jk}=1} v_k \qquad val(o_k) = \bigvee_{\mathcal{O}_{jk}=1} val(r_j). \qquad (1)$$

Obviously, NanoPLAs implement two-level sum-of-product functions. In principle, any arbitrary Boolean function can be transformed into the two-level sum-of-product form.

While PLAs implemented in CMOS are easily integrated into a larger CMOS circuit, embedding NanoPLAs into CMOS circuitry is expected to require special nano-CMOS interface modules [15, 16, 17]. We assume that such a module is present on each input and each output of the (unhardened) NanoPLA. In this way, input values for the NanoPLA are reliably generated, and the output values are reliably read out, using surrounding CMOS electronics.

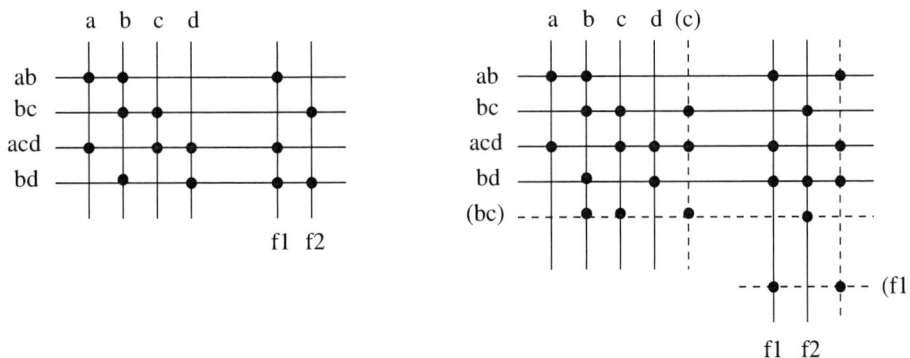

Figure 1. An example of hardening NanoPLAs

2.2: Error model

We focus on the main types of errors due to the fabrication mechanism of NanoPLAs. The behavior of devices sandwiched between the two sets of nanowires will be heavily impacted by the process variation in a self-assembly based fabrication. In addition, the "on" and "off" states of each device can be influenced dynamically by the environmental effects and background noise. At the logic level, existing connections between two wires can disappear, and absent connections can appear. Consequently, based on the error direction and location, four error modes can be distinguished: appearance error in the AND plane, disappearance error in the AND plane, appearance error in the OR plane, and disappearance error in the OR plane. We denote an error leading to appearance of a new connection at the j's row and k's column in the AND plane as $\mathcal{A}_{jk}^{\uparrow}$. An error leading to disappearance of an existing connection of such a place is denoted as $\mathcal{A}_{jk}^{\downarrow}$. Similarly, appearance / disappearance errors at the j's row and k's column in the OR plane are written as $\mathcal{O}_{jk}^{\uparrow}$ and $\mathcal{O}_{jk}^{\downarrow}$, respectively. Obviously, an appearance (disappearance) error is only defined for connected (unconnected) pairs of wires.

When considering errors in conventional technologies such as CMOS, the error rate is often assumed to be so low that maximally one error can be present in the circuit. This assumption is overly restrictive for NanoPLA circuits for which high error rates are expected. In this paper, we assume errors on individual locations (pairs of wires) to be independent stochastic processes. We define four probabilities: $p(\mathcal{A}^{\uparrow})$, $p(\mathcal{A}^{\downarrow})$, $p(\mathcal{O}^{\uparrow})$ and $p(\mathcal{O}^{\downarrow})$. If wires i_k and r_j are connected (i.e., the device at the position (j, k) is turned on: $\mathcal{A}_{jk} = 1$), then an error makes such a connection disappear (i.e., error $\mathcal{A}_{jk}^{\downarrow}$ may occur) with probability $p(\mathcal{A}^{\downarrow})$. Probability $p(\mathcal{A}^{\uparrow})$ describes the likelihood of unconnected wires in the AND plane to erroneously become connected. $p(\mathcal{O}^{\downarrow})$ and $p(\mathcal{O}^{\uparrow})$ are the respective probabilities for the OR plane.

When simulating an input vector under the employed error model, errors are injected at all locations simultaneously with the probabilities mentioned above. This means that no errors, one error or several errors may be injected. This methodology is essential for accurate simulation of selectively hardened NanoPLAs. As described below, hardening a NanoPLA against an error introduces new hardware which is also error-prone. We assume that errors occur on the redundant hardware with the same probability as on the original hardware.

2.3: Hardening techniques

We next introduce techniques to pinpointedly harden a NanoPLA against an error based on the error model used: $\mathcal{A}_{jk}^{\uparrow}$, $\mathcal{A}_{jk}^{\downarrow}$, $\mathcal{O}_{jk}^{\uparrow}$, and $\mathcal{O}_{jk}^{\downarrow}$. Different from the traditional fault tolerance schemes such as TMR, the hardening technique is based on Boolean tautology. Applying such a strategy eliminates the voting stage in TMR which imposes significant performance and area overhead in a two-level logic structure.

A hardening process deals with a particular error at a time. When the NanoPLA is hardened against error $\mathcal{A}_{jk}^{\uparrow}$, the occurrence of such an error will not generate any erroneous effect at the output for all the inputs. Comparing to the original area cost of the unhardened NanoPLA as $m \cdot (n + l)$, we also derive the area cost of hardening using the proposed techniques.

Hardening the NanoPLA against a **disappearance error in the AND plane** $\mathcal{A}_{jk}^{\downarrow}$ is done by introducing a new input wire $i_{k'}$, connected with the same input signal and the same row wires as i_k: $\mathcal{A}_{jk'} := \mathcal{A}_{jk}$. This hardening step also immunizes the NanoPLA against all disappearance errors on input wire i_k.

Hardening the NanoPLA against an **appearance error in the AND plane** $\mathcal{A}_{jk}^{\uparrow}$ is done by adding a new row wire $r_{j'}$ connected with the same input and output wires as row wire r_j: $\mathcal{A}_{j'k} := \mathcal{A}_{jk} \ \forall k \in \{1, \ldots, n\}$, $\mathcal{O}_{j'k} := \mathcal{O}_{jk} \ \forall k \in \{1, \ldots, l\}$. The hardened NanoPLA is also immune against all other appearance errors on row wires r_j and $r_{j'}$ in the AND plane. Moreover, it is immune against **disappearance errors on that rows in the OR plane**.

Finally, to harden the NanoPLA against an **appearance error in the OR plane**, $\mathcal{O}_{js}^{\uparrow}$, a new output wire $o_{s'}$ duplicating all the connections on wire o_s is introduced, with $\mathcal{O}_{js'} = \mathcal{O}_{js}$. The output wires o_s and $o_{s'}$ are ANDed together, as the result of output s. There are two methods to implement this AND operation. First, a new AND plane could be introduced and the AND operation could be implemented using the same technology as in the first AND plane of the NanoPLA. The disadvantage of this solution is the susceptibility of the new AND plane to appearance errors (disappearance errors are protected by the logic in the OR plane). The second solution is to use nano-CMOS interfacing on output wires o_s and $o_{s'}$ and adding a CMOS AND gate to connect that wires. In this paper, we choose the second solution and assume that the CMOS and gate is not susceptible to errors.

The *area cost* of hardening the NanoPLA against an error is calculated as the increase in its area due to hardening. For hardening the NanoPLA against an $\mathcal{A}_{jk}^{\uparrow}$ or $\mathcal{O}_{jk}^{\downarrow}$ error, the area overhead is the length of the added wire, or $n + l$. For hardening against an $\mathcal{A}_{jk}^{\downarrow}$ error, the new input wire adds the area cost of m. To harden a NanoPLA against an $\mathcal{O}_{jk}^{\uparrow}$, one new output wire of length m must be complemented by a CMOS gate. Assuming that one CMOS transistor occupies ten times an area of a NanoPLA switch, a two-input CMOS gate amounts to an extra area cost of 40. Clearly, the NanoPLA sizes n, m and l are increased by hardening which results in a higher area cost of subsequent hardening steps. It is possible to harden the NanoPLA multiple times against the same error, reducing its susceptibility to multiple errors.

Applying the hardening operations for every input wire, row wire and output wire results in a NanoPLA *completely protected against all modeled single errors*. This is similar to the AOA architecture from [24, 12], except that the AND operation with newly introduced output wires was implemented as an AND plane and not in CMOS. The area cost of the NanoPLA thus hardened is $4 \cdot m \cdot (n + l) + 40 \cdot l$. Figure 1 shows an example of partially hardening a NanoPLA circuit. The left side shows the original circuit, while the right side shows the hardened circuit with duplicated variable c, product term bc and output $f1$.

Procedure calculate_OER
Input: NanoPLA C, number of vectors S,
convergence bound ε
Output: Observed error rate OER
int iterations = 0, no_errors = 0;
double OER = $-\infty$, OER_old;
input_vector B;
(1) **do**
(2)   no_iterations := no_iterations + 1;
(3)   OER_old := OER;
(4)   **for** i := 1 **to** S **do**
(5)   B := random_input_vector();
(6)   **if** (good_sim(C, B) $\neq$ error_sim(C, B))
(7)   **then** no_errors := no_errors + 1;
(8)   **end for**
(9)   OER := no_errors / ($S\cdot$ no_iterations);
(10) **while** ($|$OER $-$ OER_old$| > \varepsilon$);
(11) **return** OER;
end calculate_OER;

Procedure harden
Input: NanoPLA C, area bound ref_area
Output: Hardened NanoPLA $\hat{C}$
error_array error_list, short_list;
NanoPLA $\hat{C}$, $\tilde{C}$;
(1) $\hat{C} := C$;
(2) Create error_list; mark all errors unprotected;
(3) Calculate det. probabilities of all errors;
(4) Sort error_list according to det. probability;
(5) **while** (area($\hat{C}$) < ref_area) **do begin**
(6)   short_list := 5 unhardened errors with
             largest det. prob. + 5 random errors
(7)   **for each** error e from short_list **do**
(8) $\tilde{C} := C$ hardened against error e;
(9)   cost(e) := area($\tilde{C}$) $\cdot$ calculate_OER($\tilde{C}$);
(10)   **if** (e marked protected)
                 then cost(e) := cost(e) $\cdot$1.5;
(11)   **end for**
(12) e_{opt} := e with minimal cost;
(13) $\hat{C} := \hat{C}$ hardened against error e_{opt};
(14)   Output area($\hat{C}$), calculate_OER($\hat{C}$);
(15)   Mark protected errors in error_list;
(16) **end while**
(17) **return** $\hat{C}$;
end harden;

(a) (b)

Figure 2. Algorithm to estimate observed error rate (a) and to harden a NanoPLA (b)

3: Selective Hardening Algorithm

3.1: Estimation of Observed Error Rate (OER)

Accurate estimation of the error rate under error model from Section 2.2 is essential for guiding the hardening process of the partially hardened NanoPLA. When S input vectors are applied to a NanoPLA with errors injected according to probabilities $p(\mathcal{A}^{\uparrow})$, $p(\mathcal{A}^{\downarrow})$, $p(\mathcal{O}^{\uparrow})$ and $p(\mathcal{O}^{\downarrow})$, the *Observed Error Rate* (OER) is defined as the number of input vectors producing erroneous output divided by S. OER is therefore a metric for a NanoPLA's robustness.

It is possible to determine OER by simply simulating S random input vectors, thus value of OER depending on both the random vectors and the errors injected. In general, the accuracy of estimation increases with S. To bound the inaccuracy in OER estimation without resorting to computationally prohibitive values of S, we employ the adaptive procedure outlined in Figure 2 (a). The number of vectors simulated is iteratively increased by S until the difference between two last values of OER falls below ε. Larger values of S and smaller values of ε yield more accurate result with longer simulation time, and vice versa.

3.2: Hardening

Selective hardening of a NanoPLA is done using the procedure from Figure 2 (b). All modeled errors are organized in an error list. We use the *Detection Probability* (DP) of an error as a quick estimate of its contribution to circuit-wide OER. The calculation of this number is based on *expected signal probabilities* on row wires. Assuming equiprobable input patterns, the expected signal probability on row wire r_j is

$$ESP(r_j) = 1/2^{|\{t \in \{1,\ldots,m\} \mid A_{jt}=1\}|}. \tag{2}$$

For an error in the AND plane ($\mathcal{A}_{jk}^{\uparrow}$ and $\mathcal{A}_{jk}^{\downarrow}$) to be stimulated, all input wires connected to row wire r_j except input wire i_k must be set to logic-1. Furthermore, for such an error to manifest at an one output wire connected with r_j, all other row wires connected with it should carry logic-0. The detection probability thus is

$$DP(\mathcal{A}_{jk}^{\uparrow}) = DP(\mathcal{A}_{jk}^{\downarrow}) = \frac{1 - \left(1 - \prod\limits_{\mathcal{O}_{jv}=1} \left(\prod\limits_{\mathcal{O}_{wv}=1, w \neq j}(1 - ESP(r_w))\right)\right)}{2^{|\{t \in \{1,\ldots,k-1,k+1,\ldots,m\} \mid A_{jt}=1\}|}}. \tag{3}$$

The denominator accounts for error propagation to the row wire. The numerator is the probability that the propagation is blocked in the OR plane for every connected output wire, subtracted from one.

For errors $\mathcal{O}_{jk}^{\uparrow}$ and $\mathcal{O}_{jk}^{\downarrow}$, the detection probabilities are

$$DP(\mathcal{O}_{jk}^{\uparrow}) = DP(\mathcal{O}_{jk}^{\downarrow}) = \prod\limits_{\mathcal{O}_{wk}=1, w \neq j}(1 - ESP(r_w)). \tag{4}$$

The error list is sorted by the detection probability. Furthermore, each error is assigned a *protection status* which indicates if it has been protected by any of the hardening operations done so far. The initial protection status is 0 for all errors.

The NanoPLA is hardened iteratively: one error per iteration is selected (lines 6–12) and the NanoPLA is hardened against this error (lines 13–15). This process is terminated when the area cost of the NanoPLA exceeds a user-defined bound (Line 5), although other criteria are possible. The selection is done in two steps: first, a *shortlist* of errors is generated by taking 5 unprotected errors with the largest single-error detection probability from the error list accompanied by 5 random errors, protected or not (Line 6).

For each of the errors in the shortlist, we calculate the area cost which a NanoPLA hardened against that error would have. We also calculate the OER of that NanoPLA using Procedure calculate_OER (Lines 8–9). We compute the *cost function* as the product of these two numbers and finally select error e_{opt} with the lowest cost function (Line 12). This compensates for potential inaccuracy of single-error detection probability as a metric. In addition, we bias the algorithm towards selecting errors not yet protected (Line 10).

The NanoPLA is then hardened against error e_{opt} and the protection status of newly protected errors is updated and the area cost and the OER of the solution found so far are reported. It is important to point out that we repeat the measurement of the OER in Line 14 rather than simply using the value computed in Line 9. This is done to handle the situations in which the OER estimated by procedure calculate_OER happens to be much lower than the actual OER due to fortunate selection of input vectors and error locations. These situations cannot be ruled out if procedure calculate_OER is called frequently. By repeating the calculation, the likelihood of reporting overly optimistic numbers is greatly reduced.

4: Experimental Results

We derived benchmark NanoPLAs from LGSynth93 by setting all input and output don't cares in these circuits to 0 and keeping the structure of the circuit unchanged. Based on this setup, we ran the proposed algorithm (procedure harden from Figure 2 (b)). For reference,

NanoPLA	n	l	m	Unhardened		Complete		First lower OER			First larger cost		
				cost	OER	cost	OER	iter	cost	OER	iter	cost	OER
9sym	9	1	87	870	0.002186	3520	0.000000	–	–	–	71	3575	0.001230
alu4	14	8	1028	22616	0.005095	90784	0.000000	–	–	–	493	90842	0.001250
b12	15	9	431	10344	0.010636	41736	0.000625	12	14020	0.000620	457	42567	0.000000
bw	5	28	87	2871	0.047290	12604	0.001664	99	10764	0.001244	120	12659	0.001025
clip	9	5	167	2338	0.013896	9552	0.000300	–	–	–	108	9584	0.003176
con1	7	2	9	81	0.040658	404	0.005633	9	202	0.004845	20	420	0.001850
duke2	22	29	87	4437	0.045856	18908	0.000500	129	17650	0.000500	141	18910	0.000656
e64	65	65	65	8450	0.113185	36400	0.002200	194	34970	0.001844	202	36492	0.001700
ex5p	8	63	256	18176	0.028365	75224	0.000280	79	33568	0.000243	329	75395	0.000275
inc	7	9	34	544	0.048000	2536	0.001883	44	2113	0.001662	53	2566	0.001442
misex1	8	7	32	480	0.042828	2200	0.002167	37	1679	0.001575	52	2288	0.001255
misex2	25	18	29	1247	0.119100	5708	0.004200	57	4448	0.004055	75	5752	0.002989
misex3c	14	14	305	8540	0.020464	34720	0.000520	–	–	–	226	35148	0.000964
rd53	5	3	32	256	0.027029	1144	0.001200	34	988	0.000400	39	1178	0.000683
rd73	7	3	141	1410	0.004800	5760	0.000100	92	5722	0.000080	94	5774	0.000025
rd84	8	4	256	3072	0.010172	12448	0.000300	–	–	–	102	12501	0.001224
sao2	10	4	58	812	0.014083	3408	0.000667	9	1174	0.000471	75	3522	0.000471
sqrt8	8	4	40	480	0.022337	2080	0.001100	37	1778	0.000977	44	2103	0.000807
squar5	5	8	32	416	0.078462	1984	0.004381	39	1668	0.004014	49	2048	0.002746
vg2	25	8	110	3630	0.015758	14840	0.000300	81	10060	0.000257	143	14893	0.000300
xor5	5	1	16	96	0.010620	424	0.001500	14	288	0.001489	24	429	0.001450

Table 1. Cost and observed error rate of unhardened, completely hardened, and selectively hardened NanoPLAs

we estimated the observed error rate (using Procedure calculate_OER from Figure 2 (b)) for the original NanoPLA and the NanoPLA completely hardened against single errors as discussed in the end of Section 2.3. We used the cost of the completely hardened NanoPLA as the cost limit for Procedure harden.

We set error probabilities $p(\mathcal{A}^\uparrow)$, $p(\mathcal{A}^\downarrow)$, $p(\mathcal{O}^\uparrow)$ and $p(\mathcal{O}^\downarrow)$ to identical value of $1/(m \cdot (n + l))$. This implies that the expected number of errors injected per input vector is one. We employed this very aggressive error injection rate to study the performance of our algorithm for a technology which might be classified as totally unreliable.

We set parameters S and ε of Procedure calculate_OER to 10^4 and 10^{-5}, respectively, when evaluating the actual OER of the found (intermediate) solution (Line 14 of procedure harden) and estimating the OER of unhardened and completely hardened NanoPLA. When grading the OER of shortlisted hardening candidates (Line 9 of Procedure harden), we use $S = 10^3$ and $\varepsilon = 10^{-4}$. This yields less accurate results but saves computational effort.

Table 1 summarizes the results. Columns 1 through 4 contain the NanoPLA name, its number of input wires, row wires and output wires. Columns 5 through 8 quote the cost and the OER of the unhardened NanoPLA and the NanoPLA completely hardened against all single errors. Note that the latter OER is typically not 0 because multiple errors occur.

While Procedure harden yields a plethora of solutions (NanoPLAs with different degrees of hardening), we report two of the data points. The first solution with an OER below that of the completely hardened NanoPLA is given in Columns 9 through 11 (circuits for which no lower OER was achieved are marked by '–'). The first solution which exceeded the cost of the completely hardened NanoPLA is quoted in Columns 12 through 14. For both solutions, the iteration in which they have been obtained, their cost and their OER are reported. The complete solution space produced by Procedure harden is shown in Figure 3 in graph form. The cost and the OER of the NanoPLA obtained after a number of iterations are plotted, cost and OER of the completely hardened NanoPLA are shown as horizontal lines for reference.

Both the cost and the robustness are monotonic in the number of iterations of Procedure harden. While monotonic increase can be observed for cost in Figure 3, the OER (which should be tracking the robustness) is not decreasing monotonically in all cases. This is because the OER is measured using random input vectors and randomly injected errors. If the NanoPLA has been protected against an error, this error may never be injected during

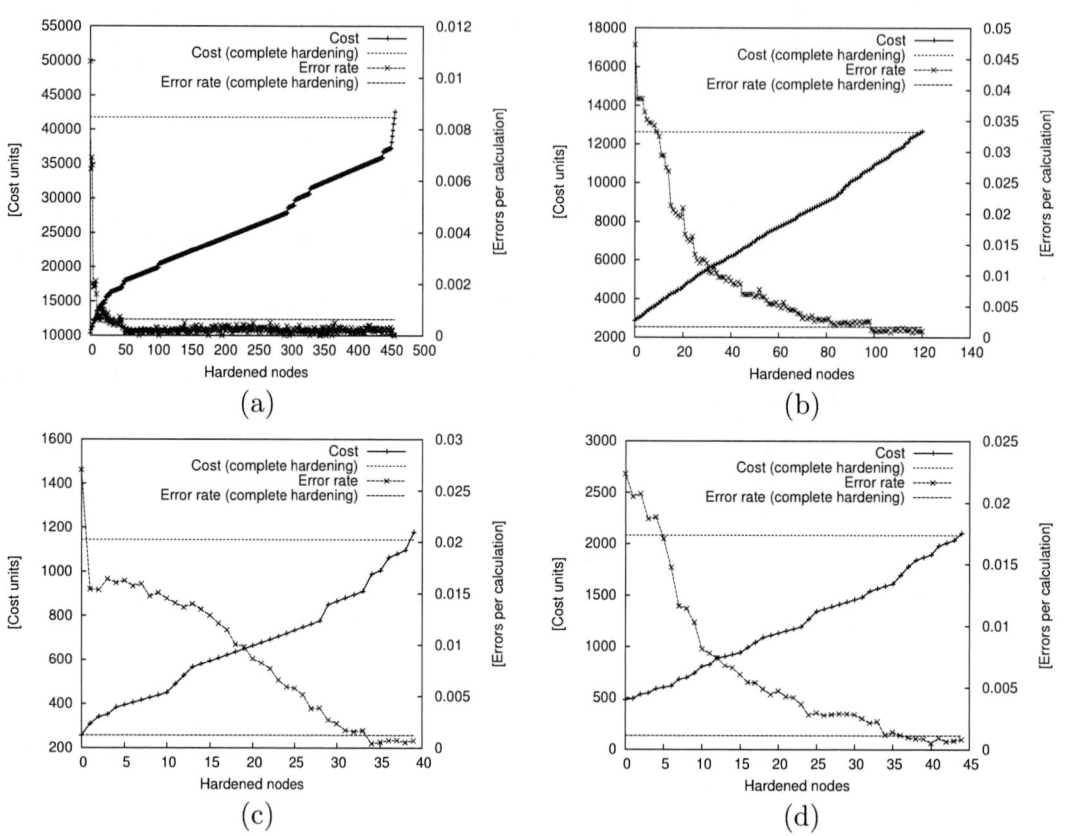

Figure 3. All (intermediate) solutions found by Procedure harden for NanoPLAs b12 (a), bw (b), rd53 (c) and sqrt8 (d)

the next invocation of Procedure calculate_OER, while errors against which the NanoPLA is unprotected are injected instead. As a consequence, the OER may actually increase.

While it appears advisable to set up some sort of a "reference error injection experiment" using the same input vectors and error injection sites to avoid the counter-intuitive OER increase, it is not possible to perform such an experiment in a meaningful way. The error injection sites cannot stay constant because new hardware is added to the NanoPLA and errors must be injected on locations not present in previous iterations. A fixed set of input vectors cannot be used because Procedure calculate_OER terminates on convergence which happens after a different number of vectors for different simulations. Last but not least, having reference parameters would bias the experiment towards hardening the NanoPLA against the errors most critical for these parameters rather than on average. Hence, we are forced to accept the OER increase due to measurement inaccuracy.

From Table 1, the proposed procedure quite often produces solutions with both cost and OER below the completely hardened NanoPLA. This must not be over-emphasized due to the above-mentioned measurement inaccuracy, yet it demonstrates that it is possible to outperform the complete hardening by directed search. The solution with equivalent cost typically has lower OER than the completely hardened NanoPLA. However, the main feature of the method is the large range of alternative NanoPLA it provides. This allows the system integrator to pick the implementation with maximal error rate the application can tolerate (or resort to a CMOS implementation if such an error rate cannot be achieved). Given the measurement inaccuracy, it appears advisable to validate the found solution by a more thorough simulation or to use a safety margin.

5: Conclusions

We demonstrated the ability of selective hardening to perform fine-grained control of robustness in NanoPLA circuit minimizing the area cost. The proposed method is based on a fully-automated sequence of systematic design transformations. The approach is transparent, i.e., it does not change the circuit's functionality and does not require extra control functionality such as commit-rollback architectures. Hence, the process does not require any specific information on the design and can be done by the system integrator with no need to interact with the designer.

Our future research will focus on systems in which certain failing patterns are acceptable. We will identify error behavior which is non-critical with respect to the application and consider only critical error behavior as optimization target. Another promising research direction is the application of our approach to other classes of nanoarchitectures.

6 References

[1] European Commission. *Technology Roadmap for Nanoelectronics*, 2001.

[2] ITRS. *International Technology Roadmap for Semiconductors Emerging Research Devices*, 2006.

[3] P. Beckett and A. Jennings. Towards nanocomputer architecture. In *Asia-Pacific Computer System Architecture Conference*, pages 141–150, 2002.

[4] M. Forshaw, R. Stadler, D. Crawley, and K. Nikolic. A short review of nanoelectronic architectures. In *Nanotechnology*, volume 15, pages 220–223, 2004.

[5] R. I. Bahar, D. Hammerstrom, J. Harlow, W. H. Joyner, C. Lau, D. Marculescu, A. Orailoglu, and M. Pedram. Architectures for silicon nanoelectronics and beyond. *IEEE Computer*, 40(1):25–33, January 2007.

[6] J. H. Patel and L. Y. Fung. Concurrent error detection in alus by recomputing with shifted operands. *IEEE Transactions on Computers*, 31:589–592, December 1982.

[7] K. Nikolic, A. Sadek, and M. Forshaw. Fault-tolerant techniques for nanocomputers. In *Nanotechnology*, pages 357–362, 2002.

[8] W. Rao, R. Karri, and A. Orailoglu. Fault tolerant arithmetic with applications in nanotechnology based systems. In *ITC*, pages 472–478, 2004.

[9] Y. Qi, J. Gao, and J. A. B. Fortes. Markov chains and probabilistic computation - a general framework for multiplexed nanoelectronic systems. *IEEE Transactions on Nanotechnology*, 4(2):194–205, March 2005.

[10] J. Han, J. Gao, Y. Qi, P. Jonker, and J. A. B. Fortes. Toward hardware-redundant, fault-tolerant logic for nanoelectronics. *IEEE Design and Test of Computers*, 22(4):328–339, July-August 2005.

[11] W. Rao, A. Orailoglu, and R. Karri. Towards nanoelectronics processor architectures. *JETTA Special Issue on Test, Defect Tolerance, and Reliability of Nanoscale Devices*, 23:235–254, 2007.

[12] W. Rao, A. Orailoglu, and R. Karri. Fault tolerance approaches to nanoelectronic programmable logic arrays. In *IEEE/IFIP International Conference on Dependable Systems and Networks (DSN)*, pages 216–224, 2007.

[13] P. J. Kuekes, D. R. Stewart, and R. S. Williams. The crossbar latch: Logic value storage, restoration, and inversion in crossbar circuits. *Journal of Applied Physics*, 97(3):034301, July 2005.

[14] G. Snider, P. J. Kuekes, and R. S. Williams. Cmos-like logic in defective, nanoscale crossbars. *Nanotechnology*, 15:881–891, Aug 2004.

[15] A. DeHon and M. J. Wilson. Nanowire-based sublithographic programmable logic arrays. In *FPGA*, pages 123–132, 2004.

[16] A. DeHon. Array-based architecture for fet-based, nanoscale electronics. *IEEE Transactions on Nanotechnology*, 2(1):23–32, 2003.

[17] D. B. Strukov and K. K. Likharev. Cmol fpga: A reconfigurable architecture for hybrid digital circuits with two-terminal nanodevices. *Nanotechnology*, 16:888–900, Apr 2005.

[18] V. K. Agarwal. Multiple fault detection in programmable logic arrays. *IEEE Transactions on Computers*, 29:518–522, June 1980.

[19] J. Khakbaz and E. J. McCluskey. Concurrent error detection and testing for large pla's. *IEEE Journal of Solid-State Circuits*, 17(2):386–394, April 1982.

[20] W. K. Fuchs, C. R. Chen, and J. A. Abraham. Concurrent error detection in highly structured logic arrays. *IEEE Journal of Solid-State Circuits*, 22(4):583–594, August 1987.

[21] T-Y. Chang and C-L. Wey. Design of fault diagnosable and repairable pla's. *IEEE Journal of Solid-State Circuits*, 24(5):1451–1454, October 1989.

[22] P. K. Lala and D. L. Tao. On fault-tolerant pla design. In *IEEE Southeastcon*, pages 945–947, 1990.

[23] M. Demjanenko and S. J. Upadhyaya. Yield enhancement of field programmable logic arrays by inherent component redundancy. *TCAD*, 9(8):876–884, 1990.

[24] W. Rao, A. Orailoglu, and R. Karri. Logic level fault tolerance approaches targeting nanoelectronic plas. In *IEEE Design, Automation, and Test in Europe (DATE)*, pages 865–869, 2007.

IEEE International Symposium on Defect and Fault Tolerance of VLSI Systems

Soft Error Hardened FF Capable of Detecting Wide Error Pulse

Shuangyu RUAN[†]   Kazuteru NAMBA[‡]   and   Hideo ITO[‡]

† ‡ Graduate School of Advanced Integration Science, Chiba University

1-33 Yayoi-cho, Inage-ku, Chiba-shi, Chiba, 263-8522 Japan

E-mail: † ruanshuangyu@graduate.chiba-u.jp,   ‡ {namba, h.ito}@faculty.chiba-u.jp

Abstract

In the recent high-density and low-power VLSIs, occurrence of soft errors becomes significant problems. Recently, soft errors frequently occur on not only memory system but also circuits. Based on this standpoint, constructions of soft error tolerant FFs have been proposed. The FFs consist of some master and slave latches and C-elements. In the FFs, soft error pulses occurring on combinational parts of logic circuits are corrected as long as the width of the pulses is narrow, that is within a specified width. However, soft error pulses or other error pulses having wide width are neither detected nor corrected in the FFs. This paper presents a construction of another soft error tolerant FFs being added some latches and delay elements into the conventional soft error tolerant FFs. The proposed FFs have capability detecting error pulses having wide width as well as capability correcting those having narrow width. The proposed FFs are also capable of detecting hard errors. This paper also presents scan FFs facilitating delay fault testing and soft error tolerant FFs for two-rail logic circuit based on the proposed FFs. The evaluation shows that the area of the proposed FF is up to 66% larger than that of the conventional soft error tolerant FFs.

1   Introduction

In recent high-density and low-power VLSIs, soft errors frequently occur. Soft errors are radiation-induced transient error pulses caused by neutrons from cosmic rays and alpha particles from packaging material [1, 2]. Traditionally, only soft errors occurring on memory system seriously affect the operation of VLSI systems. However, in recent VLSI systems, soft errors frequently occur on latches and combinational parts of logic circuits. So, occurrence of soft errors on logic circuits becomes significant problems [1, 2].

From this standpoint, many soft error tolerant methods [3-6] and soft error analysis methods [7] for soft errors occurring on combinational parts of logic circuits were proposed. In [3], time redundancy method is proposed. In [4], Delay-Assignment-Variation (DAV) based optimization method is proposed. In [5], the techniques using Schmitt trigger circuit is proposed by the authors' group. In [6], the techniques using C-element circuit is proposed. Moreover, constructions of soft error tolerant FFs capable of correcting soft errors occurring on latches as well as those on combinational parts has been proposed [8]. The FFs consist of some master and slave latches and C-elements. In the FFs, soft error pulses occurring on combinational parts of logic circuits are corrected as long as the width of the pulses is narrow within a specified one. However, soft error pulses having wide width outside the specified one are neither detected nor corrected in the FFs. In recent VLSI, such error pulses with wide width sometimes occur due to several reasons such as power-supply noise, and thus they are not negligible. Therefore, it is necessary to detect or correct such error pulses occurring in a system.

The recent high-density VLSIs have also led to the increasing of delay faults caused by manufacturing defects. Therefore, there is a need for exploiting a manufacturing testing method capable of detecting such delay faults [9]. Two-rail logic circuit design [10, 11] is known as a dependable design capable of detecting single fault like dual module redundancy system. Two-rail

1550-5774/08 $25.00 © 2008 IEEE
DOI 10.1109/DFT.2008.22

272

logic circuit design has higher delay fault testability than dual module redundancy system, and thus will be widely used [11].

This paper presents a construction of soft error tolerant FFs capable of detecting error pulses with wide width as well as correcting error pulses with narrow width and ones occurring on the combinational parts or the FFs. The proposed FFs are also capable of detecting hard errors. The proposed FFs are constructed by being added some latches and delay elements into the existing soft error tolerant FFs. This paper also presents scan FFs facilitating delay fault testing and soft error tolerant FFs for two-rail logic circuits based on the proposed FFs.

The rest of this paper is organized as follows: Section 2 is preliminary. Section 3 explains the proposed FFs, and evaluation results of the proposed methods are given in section 4. Section 5 explains scan FFs facilitating delay fault testing and soft error tolerant FFs for two-rail logic circuit based on the proposed FFs. Finally, section 6 concludes the paper.

2 Previous work
2.1 Error correction using duplication

Figure 2.1 shows a construction of soft error tolerant FFs using duplication [8]. The FFs are capable of correcting soft errors occurring in combinational logic circuits and latches. The FF is based on a master slave FF and comprises four latches; Latch1, Latch2, Latch3 and Latch4. Latch1 and Latch2 function as master latches. Meanwhile, Latch3 and Latch4 function as slave latches. The FF also comprises two C-elements; C1 and C2. The structure of the C-element is illustrated in Figure 2.2. Table 2.1 shows the truth table of the C-elements. Every C-element has two inputs, and it works as an inverter if the values of both inputs are the same. If they differ from each other, the output of the C-element becomes in high impedance state, and thus it retains the previous logic value kept by weak keepers (loops of two inverters) placed at the output of the C-elements. In circuits using the FF, the clock cycle has to be set to be longer than $T + \tau$, where T is the maximum propagation delay time of the combinational logic circuit and τ is a time longer than the expected maximal time of (correctable) soft error pulse width.

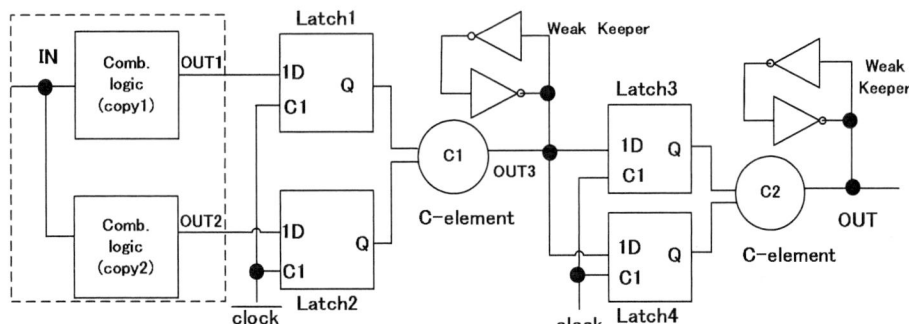

Figure 2.1 Construction of soft error tolerant FF using duplication [8].

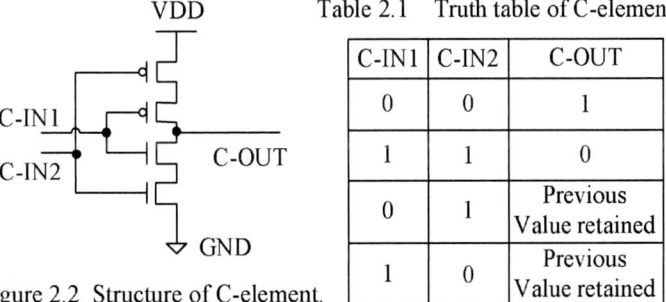

Figure 2.2 Structure of C-element.

Table 2.1 Truth table of C-element.

C-IN1	C-IN2	C-OUT
0	0	1
1	1	0
0	1	Previous Value retained
1	0	Previous Value retained

Suppose a soft error pulse whose width is narrower than τ occurring in a combinational logic circuit. Figure 2.3 (a) shows an example of a waveform chart in such case. The Figure illustrates waveforms of inverted clock signals, output signals of the combinational logic circuit (i.e. the input signal of the master latches Latch1 and Latch2), and the output of the C-element C1 connected back to master latches. In the example, the soft error occurs, after the output values of the combinational logic circuit are decided, and before the clock signal changes from 1 to 0 closing the master latches. There is a period that the input values of both C-elements correspond to the correct value (logic 0, in the example) before the master latches are closed. In general, there is always such period as long as the width of every soft error pulse is narrower than τ; because there is always a period longer than τ after the output values of the combinational logic circuit are decided and before the master latches are closed. It is true even if the value in one of the latches is incorrect after the master latches are closed. Therefore, even if soft error pulses with width narrower than τ occur, the C-element C1 outputs correct value (logic 1, in the example). Needless to say, even if soft error pulses occur before the output values of the combinational logic circuit are decided or after the master latches are closed, the pulses do not affect operation of the circuit because the output value of the combinational logic circuit in such period are not used in the circuit.

Suppose a soft error pulse occurring in a master (slave) latch of the FF when the latch is closed. The FF comprises two master (slave) latches. The latches always have the same value if no error occurs. So, just before the error occurs, the latches have the same correct value and thus the output value of the C-element C1 (C2) is correct. After the error occurs, the output value of the latch that error occurs becomes incorrect. However, that of the other latch is still correct. So, the input values of the C-element C1 (C2) differ from each other and the output values of the C-element retains the previous logic values, i.e. the correct values. If a soft error pulse occurs in a latch when the latch is opened, the error is corrected as errors occurring on combinational logic circuit.

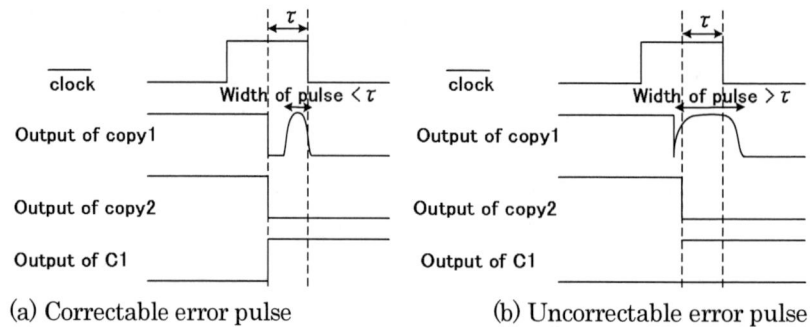

(a) Correctable error pulse   (b) Uncorrectable error pulse

Figure 2.3 Correctable and uncorrectable error pulse on the FFs using duplication.

Suppose a soft error pulse occurs in the C-element C1 or the corresponding weak keeper. If the slave latches are closed, the output values of the C-elements are not used in the FF. So, the error pulse does not affect value of the FF. Meanwhile, if the slave latches are opened, the pulse is output from the FF while it is not stored in the FF. The error is equivalent to errors occurring on the combinational logic circuit connected back to the FF, and is corrected in the next stage's FF connected from the combinational logic circuit even if the pulse is not masked and not attenuated. Needless to say, pulses masked or attenuated do not affect operation of the circuit. Soft error pulses occurring in the C-element C2 or the corresponding weak keeper is also equivalent to the errors occurring on the combinational logic circuit connected back to the FF and is corrected in the same way.

In sum, even if soft errors occur on circuit with the FF, only if the width of the pulses is narrower than τ, the errors are corrected and do not affect the operation of the circuit.

Figure 2.3 (b) illustrates an example of a waveform chart in the case that a soft error pulse with width wider than τ occurs at the input of the FF. The error pulse starts before the output values of the combinational logic circuit are decided and ends after the clock signal changes from 1 to 0 closing the master latches. There is not a period that the input values of both C-elements correspond to the correct value after the output values of the combinational logic circuit are decided and before the master latches are closed. Then, the output values of C1 and the FF end up with incorrect value.

2.2 Error correction using time-shifted output

Figure 2.4 shows a construction of the soft error tolerant FF using time-shifted outputs [8]. The construction is similar to FF shown in Figure 2.1. Unlike Figure 2.1, the combinational logic circuit is not used duplicated and a delay element of delay τ is inserted at the input of one of master latches (Latch 2, in Figure 2.4). In circuits using the FF, the clock cycle is set to be longer than $T+2\tau$.

The FF is capable of correcting soft errors with width narrower than τ occurring on combinational logic circuits and latches, like the FF in Figure 2.1 as discussed below.

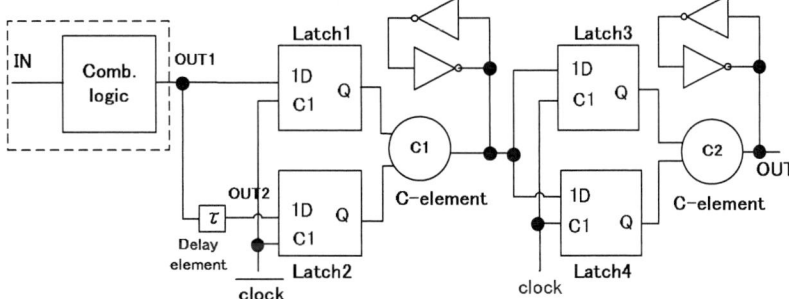

Figure 2.4 Construction of soft error tolerant FF using time-shifted output.

Suppose a soft error pulse whose width is narrower than τ occurring in the combinational logic circuit. Figure 2.5 (a) shows an example of a waveform chart in such case. Since the input values of Latch2 have delay time τ, the error pulses do not appear at the inputs of both latches simultaneously. Since, there is always a period longer than 2τ after the input value of a latch is decided and before the latch is closed, and the input values of Latch2 is delayed time τ, there is a period of time τ after the input values of both latches are decided and before either of latches (Latch 1) is closed. From those, there is always a period that the input values of both C-element C1 correspond to the correct value before the master latches are closed, even if soft error pulses with width narrower than τ occurs. Therefore, the C-element C1 outputs correct value.

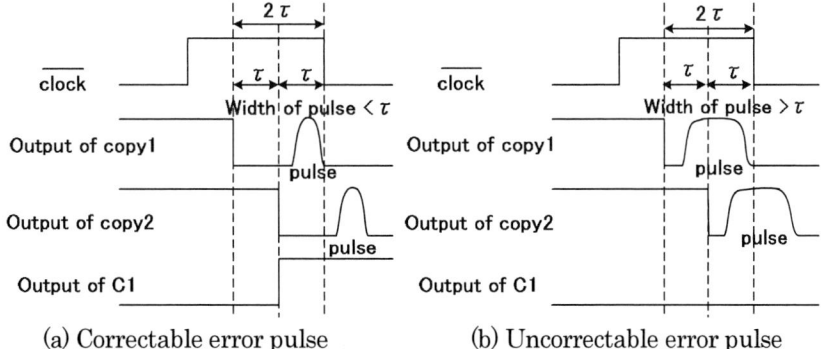

(a) Correctable error pulse   (b) Uncorrectable error pulse

Figure 2.5 Correctable and uncorrectable error pulse on the FFs using time-shifted outputs.

Figure 2.5 (b) illustrates an example of a waveform chart in the case that a soft error pulse with width wider than τ occurs in the combinational logic circuit. There is a period that error pulses appear at the inputs of both latches simultaneously, and thus there is not a period that the input values of both C-elements correspond to the correct value after the output values of the combinational logic circuit are decided and before the master latches are closed. So, the output values of C1 and the FF end up with incorrect value. This case should be improved one by the proposed design.

3 The proposed technique

Figure 3.1 shows a structure of the proposed soft error hardened FF capable of detecting wide error pulse with width wider than τ. It is based on a master slave FF and comprises six latches; ML1, DL1, ML2, DL2, SL1 and SL2. The latches ML1, DL1, ML2 and DL2 function as a master latch while the latches SL1 and SL2 function as a slave latch. The FF also comprises three C-elements; C1, C2 and C3. Signal ckb is expressed as the inversion of clock signal ck. While the slave latch part of the proposed FF is constructed in the same manner as the conventional FF shown in Figure 2.4, the master latch part of proposed FF is constructed by duplicating the master latch part of the conventional FF in Figure 2.4. Concretely speaking, ML1, DL1 and C1 form the same construction of the master latch part of the conventional FF. In addition, ML2, DL2 and C2 also form the same construction. The outputs of the C-elements C1 and C2 are connected to the inputs of a two-input XOR gate. The XOR gate detects uncorrectable wide error pulses. The output value becomes 1 only when uncorrectable errors occur. An XOR gate can be constructed as shown in Figure 3.2. In the XOR, the inverted values of those at inputs of IN1 and IN2, i.e. BIN1 and BIN2, are required. The outputs of the C-elements C1 and C2 are connected to the weak keeper, and the reversal value of the signals IN1 and IN2 are obtained by the inverter in the weak keeper can be used as the inputs of the XOR gate BIN1 and BIN2.

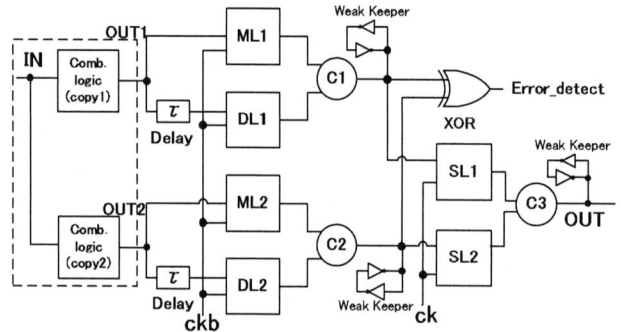

Figure 3.1 Proposed soft error hardened FF.   Figure 3.2 XOR gate.

Here, how to correct and/or detect soft errors are explained.

(1) Soft error pulses with width narrower than τ :

The master latch part of the proposed FF is composition by duplicating the master latch part of the conventional FF shown in 2.2. Needless to say, each of them works just like the master latch part of the conventional FF. So, even if soft error pulses with width narrower than τ appear at the input of the FF, the output values of the master latch part, i.e. the output values of the C-elements C1 and C2, become correct. So, the input and output values of SL1, SL2 and the C-element C3 also become correct. Moreover, both input values of the XOR become the same correct values and no errors are detected. In sum, the FF works correctly.

(2) Soft error pulses occurring in FF:

The pair of latches in the FFs, (ML1, DL1), (ML2, DL2), and (SL1, SL2), certainly takes the same value when no errors occur. Therefore, the soft error pulses can be corrected like FF discussed in 2.1 and 2.2.

(3) Soft error pulses with width wider than τ / Hard errors:

Suppose an error pulse occurs on the combinational logic circuit copy1, without lose of generality. Since ML1, DL1, and the C-element C1 form the same construction as the master latch part of the conventional FF shown in 2.2, from the discussion in 2.2, the output value of the C-element C1 may become incorrect. Needless to say, the output value of the C-element C2 is correct because it

276

does not depend on the output of copy1. If the output value of the C-element C1 is correct, the FF works correctly like the case (1). If it is incorrect, both input values of the XOR gate differ from each other and the error is detected.

4 Evaluation

In the evaluation, the proposed FF shown in Figure 3.1 is designed. For comparison, the conventional FFs shown in Figure 2.1 and Figure 2.4 are also designed. Each layout of FFs is designed in a $0.18\,\mu$ m technology with Virtuoso Layout Editor. Layout of transistors affects several characteristic of the latches such as soft error tolerant capability and area overhead. In the evaluation, every layout is designed to minimize the area overhead. The areas of the proposed FF and the conventional FFs are shown in Table 4.1. Those of the delay elements in the FF shown in Figure 2.4 and proposed FF are not included. The area of the proposed FF is up to 66% larger than that of the conventional soft error tolerant FFs.

Table 4.1 Area of FFs.

FF	Area Ratio
Conventional FF of Figure 2.1 [8]	1.00
Conventional FF of Figure 2.4 [8]	1.00
Proposed FF of Figure 3.1	1.66

Figures 4.1 (a), (b) and (c) show examples of waveform chart of the following signals in the proposed FF in cases that soft error pulses with widths of 0.4ns, 0.5ns and 0.8ns, respectively: the inverted clock signals, the output signals of the combinational logic circuits (i.e. the input signals of the master latches Latch1 and Latch2), the output of the C-element C1 connected to master latches, and the output of the XOR connected back to the C-elements C1 and C2. Figure 4.1 (d) shows an example of waveform that a hard error occurs. The horizontal axis illustrates time (ns), and the vertical axis illustrates voltage (V). The simulation was made with HSPICE in the $0.18\,\mu$ m technology. Supply voltage is 1.8V and a clock cycle is 4ns. Each delay element consists of ten inverters and is with a delay time of 0.47ns. The driving capability of the C-element is strongly depend on the size of the inverter whose output is the output of the weak keeper. In the simulation, the channel length of NMOS and PMOS in the inverter is set to $0.68\,\mu$ m. The channel widths of NMOS and PMOS in the inverter weakly are set $0.55\,\mu$ m and $0.22\,\mu$ m, respectively. Note that the channel widths of PMOS and NMOS have to be larger than $0.5\,\mu$ m and $0.2\,\mu$ m in the technology, respectively. The soft error pulses occur at the output of the combinational logic circuit copy1. It means that the error pulses affect the output of the C-element C1 and do not affect that of the C-element C2. The pulse occurs at the time about 7-8ns. In the time, the correct value is zero and the pulses change the value from zero to one.

In the example shown in Figure 4.1 (a), a narrow soft error pulse with the width of 0.4ns occurs. The waveform at the output of the C-element C1 is pretty similar to that of the C-element C2. It means that the soft error pulse is corrected. Similar results are obtained for soft error pulses with the width of 0.10-0.47ns.

In the example shown in Figure 4.1 (b), a soft error pulse with the width of 0.5ns occurs. Like this, in case that soft error pulses with the width of 0.48-0.52ns, for just a moment, the output voltages of the C-element C1 and the XOR gate change to about 1.0V and 0.78-1.08V, respectively. However, they return to the correct voltages, soon. The change does not affect the output value of the FF and the circuits connected to the XOR gates. So, we can regard the soft error pulses are corrected.

In the example shown in Figure 4.1 (c) and (d), a wide error pulse with the width of 0.8ns and a hard error occur, respectively. The output value of the C-element C1 becomes incorrect and that of the XOR gate turn on. It means that the error pulse and the hard error are detected. Similar results are obtained for soft error pulses with the width of 0.53ns or wider and for hard errors.

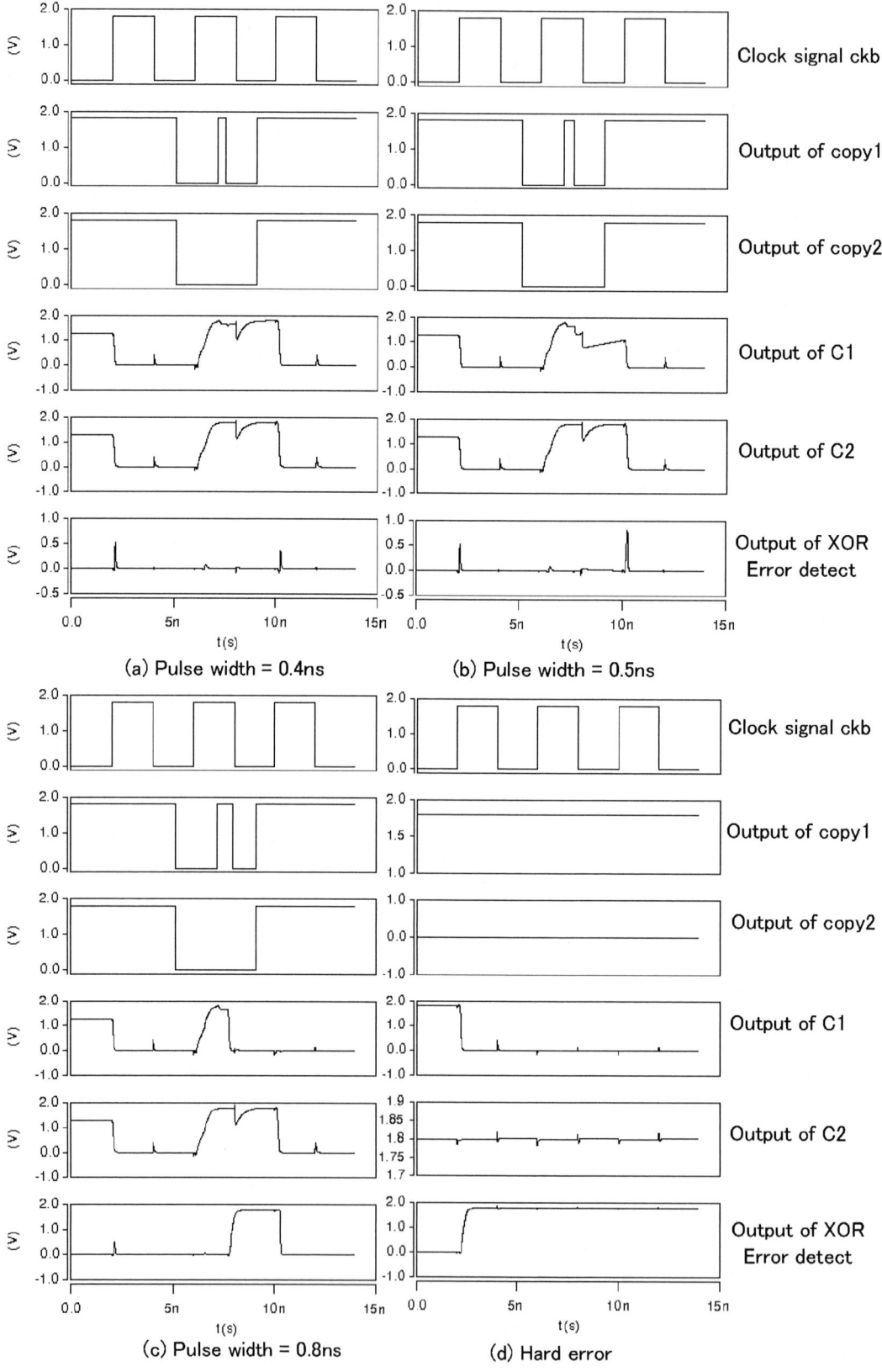

Figure 4.1 Results of error pulses.

5　Discussion

5.1　Proposed scan FFs with delay fault testability

We can obtain scan FFs with delay fault testability as well as soft error tolerant capability by improving the proposed FFs shown in Figure 3.1. Figure 5.1 shows a construction of the scan FFs. Unlike the structure of C-element shown in Figure 2.2, the C-elements C1 and C2 comprised in the proposed scan FFs are with enable signal EN.

In normal operations, Scan and EN are set to logical value 0. The outputs of the combinational logic circuits are selected as the input of master latches. In addition, transistors whose gates are connected from EN or $\overline{EN}$ turn off and the C-elements C1 and C2 work as ordinary C-elements like Figure 2.2. The clock signal ck1 and inverted one are supplied to the master latches ML1, DL1, ML2, DL2; and the clock signal ck is supplied to the slave latches SL1, SL2. The scan FF works just like the proposed FF shown in Figure 3.1.

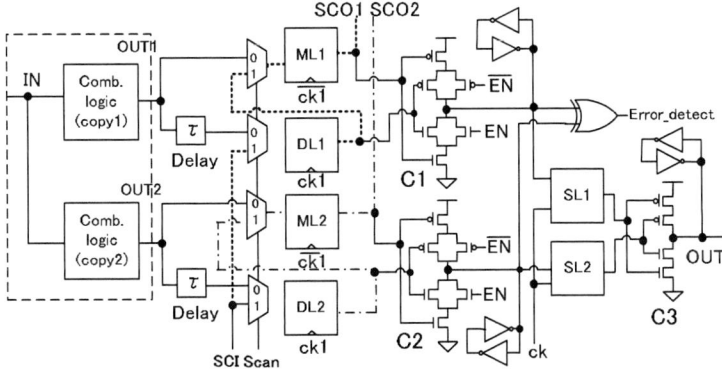

Figure 5.1 Structure of proposed scan FFs.

In scan test operation, Scan and EN are set to logical value 1. The scan input SCI is selected as the inputs of the master latches DL1 and DL2. The outputs of the master latches DL1 and DL2 are selected as the input of the master latches ML1 and ML2. In addition, a clock signal ck1 is supplied to the DL1 and DL2; and the inverted clock signal $\overline{ck1}$ is supplied to the ML1 and ML2. A pair of the master latches ML1, DL1 works as a master slave FF. Another pair of ML2, DL2 works as a FF, too. In the scan FF, two scan paths are constructed. Since EN=1, each C-element works as an inverter whose input is the output of ML1 or ML2.

Delay fault testing is made as follows. First, an initial pattern is shifted to DL1, ML1, DL2, and ML2 through two scan chains. After the scan shift operation, the values in DL1, ML1, DL2, and ML2 become the same. Next, logical value 1 is applied to ck (clock of slave latches SL1, SL2) and EN. The initial pattern is copied from ML1 to SL1 through the inverter C1 and from ML2 to SL2 through the inverter C2. Next, ck is set to 0 and a second pattern is shifted to DL1, ML1, DL2, and ML2. During the scan shift operation, the initial pattern is kept in SL1 and SL2. By setting ck to 1 again, we change the output of the scan FFs from the initial pattern to the second pattern. Next, the test response is captured into DL1, ML1, DL2, and ML2. If the output of either ML1 or ML2 is incorrect, the output value of the XOR gate turns on and the fault is detected. The value in DL1 and DL2 are scan out from SCO1 and SCO2 and observed.

5.2　Proposed FFs for two-rail logic circuit

A code which contains two code words (0, 1) and (1, 0) corresponding to binary information bits '0' and '1' is called a two-rail code. A logic circuit constructed by encoding each value of every line in an ordinary logic circuit with the two-rail code is called a two-rail logic circuit.

Figure 5.2 illustrates the structure of the proposed FF for two-rail logic circuit. If no errors occur, the values at the corresponding outputs of two-rail logic circuits are encoded, and differ from each other. The structure of the FF is almost the same as that of the FF shown in Figure 3.1 except the following points. An inverter is placed on an input of the FF. By inverting the output of the two-rail logic circuit X0 with an inverter, we make the correct input value of two inputs of the FF the same just like dual module redundancy system. In addition, a C-element C4 in addition to the C-element C3 is placed at the output of the FF. While the inputs of the C-element C3 are connected back to the Q outputs of the slave latch SL1 and SL2, those of the C-element C4 are connected back to the

$\bar{Q}$ outputs of the SL1 and SL2. The output values of the C-elements C3 and C4 differ from each

other as long as the value are correct.

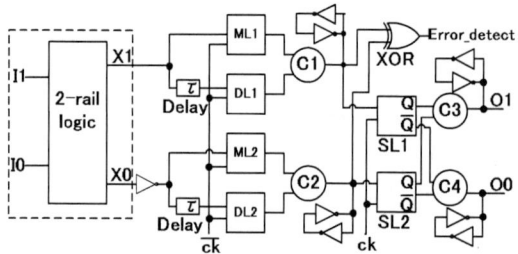

Figure 5.2 Structure of proposed FF for two-rail logic circuit.

6   Conclusion

This paper has presented a construction of soft error tolerant FFs. The proposed FFs have capability detecting soft error pulses having wide width outside a specified one as well as capability correcting those having narrow width within the specified one. The proposed FFs are also capable of detecting hard errors. This paper also presents scan FFs facilitating delay fault testing and soft error tolerant FFs for dual-rail logic circuit based on the proposed FFs. The evaluation shows that the area of the proposed FF is up to 66% larger than that of the conventional soft error tolerant FFs.

Acknowledgment

This work is supported by VLSI Design and Education Center (VDEC), the University of Tokyo in collaboration with Rohm Corporation, Toppan Printing Corporation, Synopsys, Inc. and Cadence Design Systems, Inc.

This research was partially supported by the Grant-in-Aid for Scientific Research (C) No.19560335.

References

[1] S. Mitra, N. Seifert, M. Zhang, Q. Sbi and K.S. Kim, "Robust system design with built-in soft-error resilience," IEEE Des. & Test Comput., pp.43-52, Feb. 2005.

[2] T. Karnik, P. Hazucha and J.Patel, "Characterization of soft errors caused by single event upsets in CMOS processes," IEEE Trans., Dependable & Secure Comput., vol.1, No.2, pp.128-143, 2004.

[3] M. Nicolaidis, "Time redundancy-based soft-error tolerance to rescue nanometer technologies," Proc. IEEE VLSI Test Symp., pp. 86-94, 1999.

[4] Y. S. Dhillon, A. U. Diril, A. Chatterjee, and C. Metra, "Load and logic co-optimization for design of soft-error resistant nanometer CMOS circuits," Proc. 11th IEEE International On-Line Testing Symposium, pp. 35-40, 2005.

[5] Y. Sasaki, K. Namba and H. Ito, "Soft error masking circuit and latch using Schmitt trigger circuit," Proc. 21st IEEE Int'l Symp. Defect Fault Tolerance VLSI Syst., pp.327-335, 2006.

[6] M. Fazeli, A. Patooghy, S.G. Miremadi, A. Ejlali, "Feedback redundancy: a power efficient SEU-tolerant latch design for deep sub-micron technologies", 37th Annual IEEE/IFIP International Conference on Dependable Systems and Networks, pp.276-285, 2007.

[7] J. M. Cazeaux, D. Rossi, M. Omana, C. Metra, and A. Chatterjee, "On transistor level gate sizing for increased robustness to transient faults," Proc. 11th IEEE International On-Line Testing Symposium, pp.23-28, 2005.

[8] S. Mitra, M. Zhang, S. Waqas, N. Seifert, B. Gill, K. S. Kim,   "Combinational logic soft error correction," 2006 IEEE International Test Conference, pp.824-832, 2006.

[9] A. Krstic and K.-T. Cheng, Delay fault testing for VLSI circuits, Kluwer Academic Publishers, 1998.

[10] F.F. Sellers, M.-Y. Hsiao, and L.W. Bearnson, Error detecting logic for digital computers, McGraw-Hill, 1968.

[11] K. Namba, H. Ito, "Delay fault testing for two-rail logic circuits," IEICE Tech. Rep., FIIS2007-217, 2007. (in Japanese)

IEEE International Symposium on Defect and Fault Tolerance of VLSI Systems

XOR-based Low Cost Checkers for Combinational Logic

C. A. L. Lisboa, L. Carro
Universidade Federal do Rio Grande do Sul
Instituto de Informática, Programa de Pós-Graduação em Computação
Porto Alegre, RS, Brasil
{calisboa, carro}@inf.ufrgs.br

Abstract

Radiation induced transient faults, formerly a concern mainly for memory devices, became one important element contributing to the increase of SER of combinational logic too. Conventional mitigation techniques based on time or space redundancy, either will no longer cope with the long duration transient faults predicted for future technologies, or impose heavy penalties in terms of area, power, and/or performance. In such scenario, the development of new low cost techniques to detect transient faults in combinational logic is a mandatory issue. This paper proposes one alternative for the implementation of XOR-based low cost checkers for combinational circuits, able to detect errors with much less overhead than conventional techniques.

1. Introduction

The effects of Single Event Transients (SETs) on combinational logic, formerly almost neglected [1], are becoming a significant matter of concern among the fault tolerance community. This is due to the ever shrinking dimensions of devices manufactured with new CMOS technologies, which bring along lower operating voltages and smaller critical charges, making these devices more susceptible to radiation induced transients.

This paper introduces an innovative approach for the design of low cost checkers able to detect single event upsets affecting combinational logic. The proposed checkers use parity verification to detect errors, and reduced area XOR gates in order to reduce the overhead of the parity verification circuits.

Parity has been largely used in communication channels as a means to detect erroneous messages. By computing the parity of a message both in the sender and the receiver sides one could detect an eventual bit change that occurred during the transmission. As the overall effect of a SET in combinational logic is effectively a bit-flip, one could try to use the much consolidated field of fault detection in messages to protect combinational circuits. The primary difficulty with this approach is that the size of the input string is different from the size of the output string. Moreover, the whole idea of a logic function is to transform the input string, rather than preserve it, as in message transmission. In this research, we include an additional output that changes the parity according to the parity of the inputs, allowing the use of traditional parity checking as a strategy for error detection.

The goal of this research is to define a set of functions for which the proposed approach requires lower area overheads than the classic duplication with comparison technique (DWC) [2]. In order to evaluate the area overhead, the number of transistors required to implement each function has been determined using the SIS tool [3], and a version of the 44-2.genlib cell library modified to supply the area in terms of transistors. To calculate the area required for the parity checking circuits, the 4-transistors XOR and XNOR gates proposed in [4] have been used.

1550-5774/08 $25.00 © 2008 IEEE
DOI 10.1109/DFT.2008.35

281

The paper is structured as follows: Section 2 discusses related work, Section 3 introduces the proposed approach using a well known 3-input circuit as an example, Sections 4 and 5 discuss the generalization of the technique for circuits with more than 3 inputs and shows the experimental results obtained for functions with 3 to 7 inputs. Section 6 summarizes the conclusions and points to future work.

2. Related work

The mitigation of transient errors affecting combinational logic has been traditionally done through the use of redundancy, either in time or space.

Techniques based on time redundancy, such as those proposed in [5, 6, 7], while able to detect short duration transients, may become impracticable for use with future technologies, due to the prediction of long duration transients that would impose an unbearable performance overhead on them [8, 9].

Space redundancy is another alternative, and techniques based on duplication with comparison or triple modular redundancy (TMR) [10] are the most commonly used. Duplication with comparison allows the detection of errors affecting one of the duplicated modules, but requires the recomputation of the results in order to correct the error. Triple modular redundancy, in turn, allows the selection of the correct result by comparing the results generated by three modules and, in case of discrepancy, choosing the output generated by the majority (two) of the modules. Both techniques rely on the single fault model, being useless in the presence of multiple faults affecting different modules. Besides that, they impose overheads from more than 100 to more than 200 percent, respectively, making them not feasible for use in systems where area and power consumption are a major concern, such as in the embedded systems arena.

This work investigates the alternative of using low cost checkers to cope with single faults affecting combinational logic circuits, focusing on low area overhead alternatives. With this purpose, the proposed technique uses parity verification circuits to detect the errors, and builds those circuits using chains of XOR gates with reduced area, which require only 4 transistors for each XOR gate of the chain, except for the last one, that requires 6 transistors.

3. The proposed technique

Given an n-input combinational circuit, the area overhead imposed by the use of duplication and comparison is higher than 100 percent, due to the need for a comparator, besides the two copies (modules) of the circuit to be protected. An example of this approach for a 3-input function is shown in Figure 1.

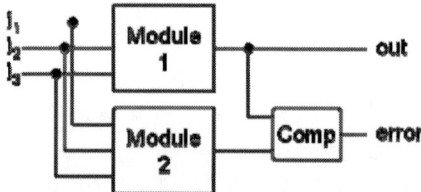

Figure 1. Using duplication and comparison with a 3-input circuit

For single output circuits, the comparator shown in Figure 1 can be easily implemented with a single XOR gate, which requires 4 transistors.

In this work, our goal is to define innovative checker circuits that can be used to detect single errors affecting combinational logic with lower area overhead than that imposed by duplication with comparison. For a given n-input combinational circuit, the proposed checkers contain the additional logic required to generate an extra output, hereinafter named *checker function*, in such a way that the parity of the output bits generated by the circuit (usual output bit plus checker bit), for each of the 2^n possible combinations of the inputs, is equal to (or different from) the parity of the n input bits. Besides the logic of the checker function, circuits to check the parity of inputs and outputs must be added, and a comparator to activate an error signal when the calculated parity bits are different is also required. The schematic diagram of the proposed solution, also for a 3-input function, is shown in Figure 2.

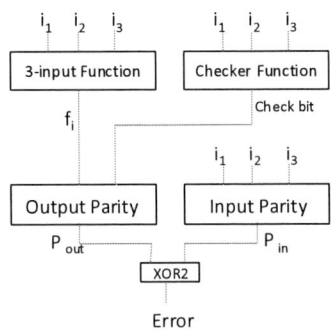

Figure 2. Schematic diagram of the proposed checker for a 3-input circuit

Considering that an *n*-input circuit can generate $2^{(2^n)}$ different functions, and that for each possible function we need to verify the feasibility of the proposed approach when the parity used for the checker function is equal to or different from that of the inputs for a given minterm, the exhaustive exploration of the design space for functions with more than 4 inputs would require the generation of a huge number of checker functions, making this task too much time consuming. Therefore, in this work we present the results of the exhaustive design space exploration for 2, 3, and 4-input functions, and for a small fraction of the possible 5-input functions, which is enough to show that the proposed approach is able to provide checker circuits with area overheads lower than that of DWC for a significant number of functions. Within this section, only 3-input functions will be used to present and discuss our proposed technique.

In order to define the nomenclature that will be used in this work, Table 1 shows the 8 minterms of a 3-input function, which will be named m_0 .... m_7, and the possible outputs, according to the 256 different functions, named f_0 ... f_{255}. As one can see, the number of a given function is defined by the set of output bits generated by that function for all minterms, with the bit corresponding to m_0 being the most significant one.

Table 1. The 256 functions that can be generated by a 3-input circuit

	i_1	i_2	i_3	f_0	f_1	f_2	f_3	f_4	...	f_{126}	f_{127}	f_{128}	f_{129}	...	f_{251}	f_{252}	f_{253}	f_{254}	f_{255}
m_0	0	0	0	0	0	0	0	0	...	0	0	1	1	...	1	1	1	1	1
m_1	0	0	1	0	0	0	0	0	...	1	1	0	0	...	1	1	1	1	1
m_2	0	1	0	0	0	0	0	0	...	1	1	0	0	...	1	1	1	1	1
m_3	0	1	1	0	0	0	0	0	...	1	1	0	0	...	1	1	1	1	1
m_4	1	0	0	0	0	0	0	0	...	1	1	0	0	...	1	1	1	1	1
m_5	1	0	1	0	0	0	0	1	...	1	1	0	0	...	0	1	1	1	1
m_6	1	1	0	0	0	1	1	0	...	1	1	0	0	...	1	0	0	1	1
m_7	1	1	1	0	1	0	1	0	...	0	1	0	1	...	1	0	1	0	1

3.1. Definition and Generation of the Parity Function

To illustrate how the parity function is generated, we show in this subsection how the proposed scheme would be used to separately check the outputs of two possible 3-input functions, which in this case are the outputs of a full adder, one of the most common 3-input circuits used in functional units.

3.1.1. Using Parity of Outputs Equal to the Parity of Inputs

In this first example, we will generate parity functions that provide for the outputs the same parity of the inputs for a given minterm. In the next subsection we will explore the complementary alternative.

Note that two separate parity functions are being used for the outputs of the full adder. This is necessary to avoid the generation of circuits that share components, when using the SIM tool to calculate the areas, since this would introduce the possibility of one error affecting a shared component being propagated to more than one output bit, which would preclude the use of parity. This single error assumption is consistent with the assumptions used for DWC and TMR based designs.

In Table 2, Co is the carry out bit (which would correspond to f_{23} in Table 1), S is the sum bit (f_{105}), and P_{Co} and P_S are the checker functions (f_{126} and f_0) generated according to the above mentioned rule. As defined by our technique, for each minterm the parity of the inputs (i_1, i_2, and i_3) is the same as that of the outputs (Co and P_{Co} for the carry, and S and P_s for the sum).

Table 2. Independent checker functions for the carry out and sum bits of a full adder – using parity of outputs equal to that of the inputs

	i_1	i_2	i_3	Co	P_{Co}		i_1	i_2	i_3	S	P_S
m_0	0	0	0	0	0	m_0	0	0	0	0	0
m_1	0	0	1	0	1	m_1	0	0	1	1	0
m_2	0	1	0	0	1	m_2	0	1	0	1	0
m_3	0	1	1	1	1	m_3	0	1	1	0	0
m_4	1	0	0	0	1	m_4	1	0	0	1	0
m_5	1	0	1	1	1	m_5	1	0	1	0	0
m_6	1	1	0	1	1	m_6	1	1	0	0	0
m_7	1	1	1	1	0	m_7	1	1	1	1	0

As one can see in Table 2, for a 3-input function the parity of the 8 minterms is always the same: even parity for minterms m_0, m_3, m_5, and m_6, and odd parity for minterms m_1, m_2, m_4, and m_7. Therefore, the parity functions P_{Co} and P_S are calculated to provide to the 2 outputs generated for a given minterm the same parity of the inputs for that minterm. Once the parity function for a given circuit is defined, the next step is to create a circuit able to generate that function, by applying the usual design and synthesis techniques. As mentioned in the introduction, in this work the SIS tool has been used to synthesize the circuits required for each function and the corresponding area has been calculated using a modified version of the 44-2.genlib cell library, which supplies the area as the number of transistors required to implement the function. The full_simplify SIS command [3] has been used for minimization, and the resulting boolean expressions for each circuit are presented below.

The checker function circuits for both functions can be easily derived from the truth tables shown in Table 2, and can be represented by the following expressions:

$$P_S = 0 \qquad\qquad (1)$$
$$P_{Co} = i1.i3' + i1'.i2 + i2'.i3 \qquad\qquad (2)$$

The implementation of (2) requires 18 transistors, while (1) requires only a wire connection to GND, i.e., no transistors. The boolean expressions for calculation of Sum and Cout, are:

$$\text{Sum} = i1.i2.i3 + i1.i2'.i3' + i1'.i2.i3' + i1'.i2'.i3 \quad (3)$$
$$\text{Cout} = i1.i2 + i1.i3 + i2.i3 \qquad\qquad (4)$$

The comparison of the circuits required to implement expressions (1) and (2) with those required for expressions (3) and (4), shows that some checker functions can be more expensive, in terms of area, than the corresponding main functions, while others are not. In this example, the generation of P_{Co} (area = 18) is more expensive than the generation of Cout (area = 16), while the opposite is true for the function Sum (area = 24) and its counterpart P_S (area = 0).

3.1.2. Using Parity of Outputs Different from the Parity of Inputs

In this subsection, we further explore our design space, and the checker functions P_{Co} and P_S are calculated in order to provide to the 2 outputs generated for a given minterm the complementary parity of the inputs for that minterm:

$$P_S = 1 \qquad\qquad (5)$$
$$P_{Co} = i1.i2.i3 + i1'.i2'.i3' \qquad\qquad (6)$$

This modification does not imply any additional cost to the parity verification circuit, nor it affects the implementation of the main functions, which in this case are the same described by expressions (3) and (4) above. However, depending on the function being protected, it may lead to a cheaper implementation of the parity generation circuit. Table 3 shows the output tables of the resulting checker function circuits for P_S and P_{Co} in this case.

Table 3. Independent checker functions for the carry out and sum bits of a full adder – using parity of outputs different from that of the inputs

	i_1	i_2	i_3	Co	P_{Co}		i_1	i_2	i_3	S	P_S
m_0	0	0	0	0	**1**	m_0	0	0	0	0	**1**
m_1	0	0	1	0	**0**	m_1	0	0	1	1	**1**
m_2	0	1	0	0	**0**	m_2	0	1	0	1	**1**
m_3	0	1	1	1	**0**	m_3	0	1	1	0	**1**
m_4	1	0	0	0	**0**	m_4	1	0	0	1	**1**
m_5	1	0	1	1	**0**	m_5	1	0	1	0	**1**
m_6	1	1	0	1	**0**	m_6	1	1	0	0	**1**
m_7	1	1	1	1	**1**	m_7	1	1	1	1	**1**

The area for implementation of (5) and (6) is also 0 and 18, respectively, which means that, in this specific case, the use of the same type of parity of the inputs or its complement has no influence on the area of the checker function circuits.

However, as will be seen later, our experiments have shown that the complexity of the parity generator circuit may vary, for the same main function, depending on the choice between equal or different parities for inputs and outputs.

Before we proceed with the generalization of the analysis, it is important to remind that one additional cost must be taken into account before we claim that a given checker requires less area than duplication with comparison: the cost of verification of the parity, which may be significant for simple circuits, as discussed in the following subsection.

3.2. Checking the Parity of Inputs and Outputs

Assuming that the inputs of the circuit to be protected are driven from an storage element that is protected using an error detection and correction (EDAC) scheme, or from another combinational circuit that detects errors using checkers such as the ones proposed in this work, we can assure that the values of the inputs are always correct, and so it is not necessary to check the input values.

As opposed, whenever a transient fault affects the circuit to be protected or the checker function generator, causing an error in one output, given the single fault hypothesis that is used in this work the parity of the outputs of the main circuit and the checker will change, and will be different from that adopted at design time (parity of outputs equal to the parity of the inputs or not).

Considering the first alternative, i.e., the parity of outputs must be equal to that of the outputs, for a given function f_i, $i = 0...255$, in order to raise one error line when the parity of the inputs and outputs is not the same we need only a circuit that implements the following boolean function:

$$ERROR = (i_1 \oplus i_2 \oplus i_3) \oplus (f_i \oplus P_{fi}) \tag{7}$$

where P_{fi} is the output of the chosen checker function for f_i. This circuit does not depend on the function to be protected itself, only on the number of inputs and outputs. The first sub-expression between parentheses calculates the even parity of the inputs (hereinafter denoted by P_{in}), while the second one calculates the even parity of the output plus the generated checker bit (parity of Co and P_{Co} in our example, hereinafter denoted by P_{out}). The ERROR signal is raised whenever P_{in} is different from P_{out}.

Using the reduced area XOR gates, the total area required for the implementation of this parity checker is equivalent to 16 transistors.

If the parity of the outputs had been chosen to be different from that of the inputs, the only difference in the parity verification circuit would be the use of 4-transistor XNOR gates or to complement the logic of the error signal, i.e.:

$$!ERROR = (i_1 \oplus i_2 \oplus i_3) \oplus (f_i \oplus P_{fi}) \tag{8}$$

As the number of inputs of the circuit increases, the verification circuit requires additional XOR gates in the chain, i.e., more 4 transistors for each additional input. For a circuit with I inputs, I-1 XOR gates are necessary to calculate the parity of the inputs, one XOR gate to calculate the parity of the outputs, and one XOR gate to compare the results. Therefore, the total number of XOR gates required to check the parity is given by:

$$\text{Number of XOR gates} = (I - 1) + 1 + 1 = I + 1 \tag{9}$$

and the number of transistors by:

$$\text{Number of transistors} = 4 \times I + 4 \tag{10}$$

3.3. Total Cost of the Proposed Checker

As seen in the preceding subsections, the proposed checkers are composed by a checker bit generator circuit and a parity verification circuit (to generate the parity of inputs, parity of outputs, and compare them), which must be added to the circuit that generates the main function, as shown in Figure 2.

Table 4 summarizes the area information previously discussed for both options. In the third and fourth columns, the two different values shown are for use of parity equal to/different from that of the inputs.

Table 4. Area required for implementation (check bit with equal/different parity)

Function	Main circuit	Check bit generator	Parity verification	Total checker	Checker overhead
Sum (f_{105})	24	0 / 18	16	16 / 34	67% / 142%
Cout (f_{23})	16	18 / 0	16	34 / 16	213% / 100%

For the above examples (Sum and Co), we have seen that there is no difference, in terms of area, between the use of the parity of the outputs equal or different from that of the inputs, because this does not change the size of the parity generation circuit.

Not surprisingly, for small circuits like the ones used in this example, the parity generation and verification network imposes a significant area overhead when compared to the size of the main function. Due to the same reason, the proposed approach is not suitable for circuits with only 2 inputs. As a rule of thumb, one could say that the use of such checkers would be feasible only if the area required by the checker (check bit generation + parity calculation + parity comparison) is equal to or less than that of the circuit to be protected. Otherwise, the duplication and comparison technique would be a better alternative.

However, for the specific circuits used in our example, the overhead imposed by duplication and comparison is also higher than 100%, as shown in Table 5. Therefore, for the Sum function, the lower area overhead would be obtained when using our proposed approach, with parity of outputs equal to that of the inputs (67% overhead), while for the Cout function the use of our approach with parity of outputs different from that of the inputs (100% overhead) provides an overhead lower than that of DWC.

Table 5. Area required for implementation of duplication with comparison for the same functions

Function	Main circuit	Duplicated circuit	Comparator	Total Checker	Checker Overhead
Sum (f_{105})	24	24	4	28	117%
Co (f_{23})	16	16	4	20	125%

While the purpose of the examples presented in this section was to explain the basic concepts behind our proposed technique, they also correspond to functions in which the overhead imposed by this technique is smaller than that of DWC. As will be shown in the following section, this is not the case for all possible functions.

4. Generalization of the approach

As seen in the previous section, there are $2^{(2^n)}$ different functions that can be generated using an n-input circuit. For each of them, we can decide to generate the parity of the outputs equal to that of the inputs, or its complement. Therefore, the total number of possible combinations of functions and parity generators is $2 \times 2^{(2^n)}$, and grows very fast with the number of inputs, as shown in Table 6. This fact makes the exhaustive analysis for all functions very time consuming.

Table 6. Number of minterms, functions, and combinations for n-input circuits

n	3	4	5	6
minterms: 2^n	8	16	32	64
functions: $2^{(2^n)}$	256	65,536	4,294,967,296	2^{64}
combinations: $2 \times 2^{(2^n)}$	512	131,072	8,589,934,592	2^{65}

In this study the exhaustive analysis has been developed for all possible 3-input and 4-input functions, and among the 2^{32} possible 5-input functions a subset has also been analyzed. A summary of the obtained results for all 3- and 4-input functions is shown in Table 7, where one can see that the percent of cases in which the approach proposed in this work has overhead (in number of transistors used to synthesize the function using the SIS tool) less than or equal to that of the duplication with comparison approach is almost the same.

Table 7. Functions for which the proposed approach has overhead ≤ DWC

Number of inputs	3	4
Total number of functions	256	65,536
Our approach uses less area than DWC in	15	7,481
% of functions in which our approach requires less area	5.9%	11.4%
Average area reduction w. r. t. DWC	9.6%	8.6%
Our approach uses the same area as DWC in	9.0%	2.6%
Our approach uses the same or less area than DWC in	14.9%	14.0%

For the 5-input functions, a relatively small (when compared with the available design space) subset of functions has been selected for analysis: 16,384 consecutive functions, in the range 1,771,468,394 – 1,771,484,777. This range has been selected for the experiments because its center is the even parity of the inputs function (1,771,476,585). Each function and the corresponding check bit generation functions, for parity of outputs equal and different from the inputs, have been synthesized and the area of the resulting 32,768 combinations compared to that of DWC. For this specific group of functions, our proposed approach has proven to be a better solution, in terms of area overhead, for 16,324 functions, in some cases using parity equal to that of the inputs, and in other cases different. The average area reduction, when compared to the area required by DWC, was 22.48%.

Through this analysis, we confirmed that there are many functions for which the use of the proposed approach provides an area overhead lower than that of duplication with comparison, and therefore our technique could be a better solution for protection of the corresponding circuits against single event upsets.

5. Sample results of the proposed technique for 6- and 7-input functions

We have manually analyzed a few functions in this group, and the results for those in which our approach is more suitable are shown in Table 8. The numbers of the functions have been replaced by letters (A, B, C and D) in the table, due to the large number of digits required to represent the function numbers at this level, which would be irrelevant information in the current context. Those results are included here to emphasize that there are cases in which our approach offers a dramatic overhead reduction when compared to DWC (see functions A, C and D).

As already mentioned for the 5-input functions, the exhaustive analysis of 6- and 7-input functions can no longer be done with the available tools. The best solution is the development of a specific tool that, given the definition of a function, automatically checks if our proposed solution is the best alternative in terms of area overhead. Also, for multiple output functions, this tool could tell the designer for which outputs our approach is the most suitable one, and for which outputs duplication with comparison is better, thereby allowing a faster decision making process during the early design stages.

Table 8. Comparison between the proposed approach and DWC for selected 6- and 7- input functions (areas in number of transistors)

Function	A		B		C		D	
Number of inputs	6		6		7		7	
Parity type	P_{eq}	P_{ne}	P_{eq}	P_{ne}	P_{eq}	P_{ne}	P_{eq}	P_{ne}
Function area	146	146	50	50	258	258	276	276
Check bit gen. area	22	28	6	4	44	48	48	44
Verification area	28	28	28	28	32	32	32	32
Total area of our approach	196	202	84	82	334	338	356	352
Area overhead of our approach	34%	38%	68%	68%	30%	31%	29%	28%
DWC area	296		104		520		556	
DWC area overhead	103%		108%		102%		101%	

P_{eq} : check bit generated for parity of outputs equal to that of inputs P_{ne} : not equal

6. Conclusions and Future Work

This paper has described a proposed methodology for the design of low cost checkers able to detect single faults affecting combinational circuits using less area overhead than the classic duplication with comparison approach. One of the area reduction factors is the use of low cost XOR gates in the parity verification circuits.

The methodology has been explained using a well known 3-input, 2-output circuit, and then the results obtained with an exhaustive design space exploration for 3-input and 4-input functions have been presented. Also, selected 5-, 6-, and 7-input functions have been analyzed, and the results show that the proposed approach becomes more interesting as the number of inputs of the circuit to be protected increases.

The next step in the study of this technique will be the development of a tool to automatically synthesize the check bit generation functions according to the type of parity (equal to or different from that of the inputs) that has lower area overhead considering the circuit to be protected. This tool will also be able to detect when our approach can be used in substitution to DWC.

6. References

[1] Heijmen, T. "Radiation Induced Soft Errors in Digital Circuits: A Literature Survey". Philips Electronics Natl. Lab., Netherlands, Report 2002/828, August, 2002.

[2] Wakerly, J. F. Error detecting codes, self-checking circuits and applications. New York: North-Holland, 1978.

[3] Sentovich, E. M., et all, "SIS: A System for Sequential Circuit Synthesis", Electronics Research Laboratory, U. C. Berkeley, 1992. Available at: http://www.eecs.berkeley.edu/Pubs/TechRpts/1992/ERL-92-41.pdf, last accessed March 11, 2008.

[4] Wang, J-M., Fang, S-C., and Feng, W-S., "New Efficient Design for XOR and XNOR Functions on the Transistor Level", IEEE Journal of Solid-State Circuits, Vol. 29, No. 7, July 1994, pp 780-786.

[5] Austin, T.; Blaauw, D.; Mudge, T. and Flautner, K. "Making Typical Silicon Matter with Razor". *IEEE Computer*, vol. 37, nr. 3, pages 57-65, Los Alamitos, CA, March, 2004.

[6] S. Mitra, N. Seifert, M. Zhang, Q. Shi, and K. S. Kim, "Robust system design with built-in soft-error resilience", Computer, Vol. 38, No 2, pp. 43-52, 2005.

[7] Anghel, L., Lazzari, C. and Nicolaidis, M., "Multiple Defects Tolerant Devices for Unreliable Future Technologies", in Proceedings of the 7th IEEE Latin-American Test Workshop (LATW 2006), IEEE Comp. Soc, March 2006, pp. 186-191.

[8] Lisboa, C. A., and Carro, L. "System Level Approaches for Mitigation of Long Duration Transient Faults in Future Technologies", *Proc. of 12th European Test Symposium – ETS 2007*, 2007.

[9] Lisbôa, C. A. L., Argyrides, C., Pradhan, D. K., and Carro, L., "Algorithm Level Fault Tolerance: a Technique to Cope with Long Duration Transient Faults in Matrix Multiplication Algorithms", to appear in *Proceedings of the 26th IEEE VLSI Test Symposium (VTS 2008)*, San Diego, CA, USA, April/May 2008.

[10] Johnson, B. W., "Design and Analysis of Fault Tolerant Digitals Systems: Solutions Manual. Reading", MA: Addison – Wesley publishing Company, October 1994.

IEEE International Symposium on Defect and Fault Tolerance of VLSI Systems

Minimization of CTS of k-CNOT Circuits for SSF and MSF Model

Muhammad Ibrahim, Ahsan Raja Chowdhury and Hafiz Md. Hasan Babu
Department of Computer Science and Engineering
University of Dhaka, Dhaka-1000, Bangladesh

E-mail: ibrahimmuhammad313@yahoo.com, farhan717@univdhaka.edu, hafizbabu@hotmail.com

Abstract

In this paper, we consider the problem of testing reversible circuits for a particular fault model: Stuck-at Fault Model. We propose a design-for-test construction technique for k-CNOT circuit having $k \geq 1$ and only 2 test vector (i.e. minimal) suffice as their complete test set. We have also shown the way to exploit our method for the case of 0-CNOT circuits. Finally we provide some experimental results for the proposed method and compare it with existing method to show how the proposed one outperforms the existing one both in terms of number of test vectors of complete test set and number of gates need to be replaced in the design-for-test.

1. Introduction

The more logic elements are packed in conventional computer, the more heat is dissipated. When a computational system erases a bit of information, it must dissipate ln 2 x kT energy, where k is Boltzmann's constant and T is the temperature. In room temperature, this is about 2.9 x 10^{-21} joules [1, 2]. Reversible circuits dissipate very low power and essential components of quantum logic [3].

Testing is one of the final phases of the production of a circuit and is one of the paramount issues we have to address in the production cycle as we must furnish flawless (at least to an acceptable level) working products to customers. A test set is a set of test vectors that are applied to the circuit in question to detect faults, if any. Generating efficient test sets is hard for conventional irreversible circuits though, it is significantly simpler for the case of reversible circuits. Agrawal [4] showed that fault detection probability is greatest when the information output of a circuit is maximized. This suggests that it may be easier to detect fault in reversible circuits which are information lossless.

Research reported so far in the field of reversible circuit testing regarding the topic in question is not bountiful. Some existing works are hinted below.

Patel et al. [5] formulated the problem of constructing minimal test set as an Integer Linear Program (ILP) with binary variables. They also used an algorithm that simplifies the aforementioned method for a large circuit by breaking it into smaller sub-circuits [6]. Chakraborty [7] showed that all stuck-at faults in an n-wire k-CNOT circuit $C(n)$ with $k \geq 2$ can be detected using at most n test vectors (TV), namely, n weight-1 vectors. $C(n)$ has a complete test set (CTS) of size n for multiple stuck-at faults (MSF). He proved this using *fixing* property. He has also shown that if an n-wire k-CNOT circuit ($k \geq 0$) is modified by introducing an extra wire that feeds a new control input added to every k-CNOT gate, the resulting circuit has CTS of size 3 for MSFs. Ito et al. [8] showed that it is NP-hard to

1550-5774/08 $25.00 © 2008 IEEE
DOI 10.1109/DFT.2008.38

generate a minimum complete test set for stuck-at faults on a set of wires of a reversible circuit.

We propose methods which exploit *fixing property* of digital circuit, like [7], to produce CTS for a given circuit. We apply a test vector at input of the circuit and observe if the output is same as the input. If it does, then we proceed to the next test vector. If it does not, then some wire(s) of the circuit must be faulty. Our target here is to reduce the size of the complete test set to 2 i.e. minimum using some design-for-test (DFT) of the circuit if needed. We do not consider fault diagnosis in this paper.

2. Assumption

We follow the following assumptions throughout the whole paper if otherwise not stated:

(a) An n-wire reversible circuit consisting of m number of k-CNOT gates is given ($k \geq 2$ and $m \geq 0$). If we have 0-CNOT and/or 1-CNOT gates in the original circuit, then a design-for-test (DFT) of the circuit can be constructed easily by adding at most 2 extra wires (so that 1-CNOT and 0-CNOT gates becomes 2-CNOT gates having necessary extra control input(s) from the extra wire(s)), so number of wires becomes $n + 2$ and all the methods are applicable on the DFT. Thus, without losing any generality, we assume that we have only k-CNOT gates with $k \geq 2$. When we used any k-CNOT gates with $k = 0$ or 1, we specify explicitly.

(b) Let any test vector (TV) be $t_1 t_2 \ldots t_n$ where t_i is i^{th} bit.

(c) Let g_i be the set of positions i.e. wire indices where gate g_i has control inputs ($1 \leq i \leq m$). For example, $g_i = \{g_{i1}, g_{i2}, \ldots, g_{ip}\}$ ($1 \leq p \leq n - 1$) means gate g_i has control inputs on wires having indices $g_{i1}, g_{i2}, \ldots, g_{ip}$.

(d) 0^k (1^k) means k consecutive zeros (ones).

(e) In spite of altering any bit in any level of the circuit except the last, if the final output of the circuit is equal to the input, then the input vector is *fixed* by definition. But this concept leads to some extra complexity. So throughout the whole paper, when we say a vector is *fixed* by a circuit, we mean that no bit of the input vector is altered by any gate across the whole circuit.

(f) The whole paper is *fixing property* oriented unless otherwise stated. For example, if we say a test vector is invalid for a circuit, it tacitly implies that this test vector is not *fixed* by that circuit and vice versa.

3. Proposed Method

Before introducing our method, we provide a definition used throughout the paper.

3.1. Description of a Gate and Description Table of a *k*-CNOT Circuit

Let a gate g has wire indices $g_1, g_2, \ldots, g_p$ as control inputs. Then we call the sequence $g_1, g_2, \ldots, g_p$ the description of the gate g. If g is 0-CNOT gate, then it has no description as we are concerned here with only the control inputs. If we construct a table where the rows are descriptions of the gates of a circuit, then we call this data structure "**Description Table**" of the circuit.

Example 1. Figure 1(b) shows the description table of the circuit shown in Figure 1(a).

3.2. Generating Minimum Number of Test Vectors (i.e., 2)

The following theorem shows that getting minimum |CTS| i.e. 2 for a circuit is not rarity – there are more than 2^n - 1 circuit configurations having |CTS| = 2 for an n-wire k-CNOT ($k \geq$ 2) circuit.

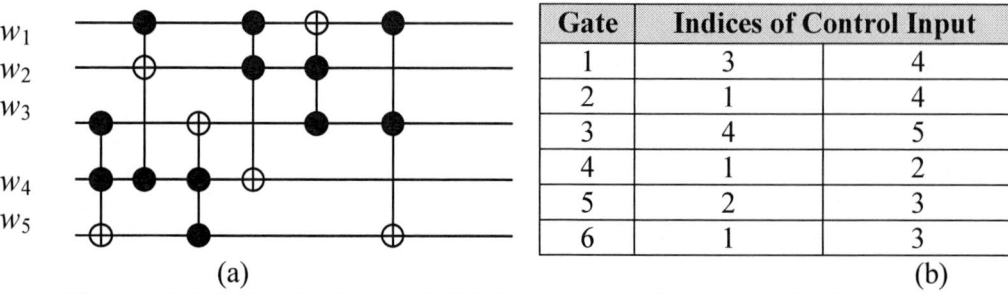

Figure 1. (a) Original circuit (b) Corresponding *description table*

Theorem 1: Let an n-wire circuit consisting of m number of k-CNOT ($k \geq 2$) gates is given. Then for single stuck-at fault (SSF): **(a)** Minimum size of CTS is 2. (This is also true for $k \geq 0$), **(b)** If $|CTS| = 2$, then $TV_2 = inverse$ (TV_1) and vice versa (*inverse* means t_i is altered for all i) (This is also true for $k \geq 0$), **(c)** There may be more than $2^n - 1$ circuit configurations where $|CTS| = 2$. ∎

Proof: (a) If $m = 0$ i.e. there is no gate in the circuit, then 2 TVs – 0^n and 1^n suffice as CTS because they set all wires of all levels to both 0 and 1. Otherwise, we prove it by contradiction. Let $|CTS| = 1$. This means t_i in TV_1 is either 0 or 1, but not both. So we cannot test i^{th} wire for stuck-at-t_i fault (SA-t_i fault) (Lemma 1) with mere TV_1. Hence $|CTS|$ cannot be 1. So to test i^{th} wire for stuck-at t_i (SA-t_i) fault, we need at least another TV such that t_i in TV_2 or in some other subsequent TV will be $\overline{t_i}$ of TV_1, thus completes the proof.

(b) We prove it by contradiction. Let $|CTS| = 2$ and t_i in TV_2 is same as it was in TV_1. This means i^{th} wire is not tested for SA-t_i fault, so at least another TV is needed, which is a contradiction of the assumption.

(c) If $m = 0$, then $TV_1 = 0^n$ and $TV_2 = 1^n$ (or $TV_1 = 1^n$ and $TV_2 = 0^n$) suffice as CTS. Let S be the set of the rest $2^n - 2$ input vectors mentioned in statement. Now if we can show that for each input vector $inp\_vctr \in S$, there exist at least one circuit where $TV_1 = inp\_vctr$; $TV_2 = inverse$ (inp_vctr) and $|CTS| = 2$, then the proof completes.

Let $S_1 = \{i_1, i_2, \ldots, i_p\}$ and $S_2 = \{j_1, j_2, \ldots, j_q\}$ ($1 \leq p \leq n-1$, $1 \leq q \leq n-1$, $p + q = n$ and $S_1 \cap S_2 = \Phi$) where S_1 be the set of positions in TV_1 which are 1 and S_2 be the set of positions in TV_1 which are 0. Now if the circuit in question has $|CTS| = 2$, then, according to part (b), the second and last TV, $TV_2 = inverse$ (TV_1) and S_1 be the set of positions in TV_2 which are 0 and S_2 be the set of positions in TV_2 which are 1. Now we will show that there exists at least one circuit where all the above statements are true. We can easily find a circuit where any arbitrary input vector, $inp\_vctr \in S$ can be TV_1 because what we need to make sure is that there is no gate g_i in the circuit such that description of g_i is not in S_1 i.e. $g_i \subseteq S_1$. Likewise, for TV_2 to be valid and to be the last TV of CTS, there can be no g_i such that $g_i \subseteq S_2$ for all i ($1 \leq i \leq m$). (If it be so, then the TV will not be *fixed* by the gate g_i). The reader can easily verify that to find such a circuit is always possible because we have all the choices at our disposal to select gates to construct a desirable circuit. Moreover, it is also obvious that we can have more than one circuit that has $|CTS| = 2$ with same TV_1 and TV_2 – what we need is to just add as many gates as we wish without violating the rule that every gate has 0 in at least one of its control inputs. □

Theorem 2: Let an n-wire circuit consisting of m number of k-CNOT ($k \geq 2$) gates is given. Let $S = \{\Phi\}$. Now let S is built as follows: if $s_j \notin g_i$, then $S \leftarrow S \cup \{g_{i1}\}$ for all i and j ($1 \leq j$

$\leq |S|,\ 1 \leq i \leq m)$. If there exists no gate g_i such that $g_i \subseteq S$, then this circuit has 2 test vectors as CTS for SSF model. ∎

Proof: According to Theorem 1, minimum $|CTS|$ is 2 and in that case, $TV_2 = inverse\ (TV_1)$ and vice versa. Actually S (mentioned in the statement) is the set of positions of the 1st TV where t_i should be 0 in order to *fix* the TV through the circuit. This means i^{th} wire is tested for SA-1 fault, so they are needed to be tested for SA-0 fault, so these s_i must be 1 in the 2nd TV to have $|CTS| = 2$ for the circuit. So for the 2nd TV to be valid (and to be the last TV), we must make sure that every gates fixes 2nd TV i.e. each gate contains at least one control input which is 0 i.e. there should be no gate such that $g_i \subseteq S$ i.e. at least one control input of g_i must have 0 value. □

3.3. Basic Algorithm

Now we present the method in form of algorithms that show the way to modify the n-wire k-CNOT circuit with $k \geq 1$ so that there exist just 2 TVs i.e. minimum, as CTS for the modified circuit i.e. the DFT. Our algorithm also produces these 2 TVs. We dub this algorithm "Basic algorithm" because it is later used in an extended version.

Algorithm 1: **Get_S**

Input: *Description table* of a circuit.
Output: A set S, possibly empty, of integers those are the bits of TV_1 that will be 0 – implying the bits of TV_2 that will be 1.

Step 1: Compute S, initially empty.
Step 2: for each gate g **do**
    if no element of the row of *description table* corresponding to g is in S,
      then add first element of that row in S.
return S

Algorithm 2: **Get_G**

Input: A set S of integers, possibly empty.
Output: A set G containing the set of gate indices in the *description table* that 'create problem' for the circuit to have 2 TVs as CTS.

Step 1: Compute G, initially empty.
Step 2: for each gate g **do**
    if all the elements of the row of *description table* corresponding to gate g are in S,
      then add g to G.
return G

Algorithm 3: **Get_TV_from_S**

Input: A set S of integers.
Output: A TV that will be fixed through the circuit at hand.
Step 1: Declare a *TV*
Step 2: for all i in $\{1, 2, \ldots, n\}$, **do**
    if i is in S,
      then set t_i of *TV* to 0,
    else set t_i to 1
return *TV*

Algorithm 4: **Process_description_table**

Input: *Description table* of a circuit.
Output: *Description table* of the circuit with some sorting applied according to section 3.4.
Step 1: Sort each row of the description table in increasing order of their elements.
Step 2: Treat a row of a description table as a single element and then sort all rows in decreasing order of their first element.
return *Description table*

Algorithm 5: **Minimum_Complete_Test_Set_Generator (MCTSG)**

Input: The configuration of a k-CNOT ($k \geq 1$) circuit is in *description table* as input. Input pattern is as follows: each gate occupies 1 row in the *description table*. The elements of each row are the wire indices of a particular gate where this gate has control inputs. Inversion input wire index is not entered. Each row is terminated with a -1 value which works as sentinel.

Output: Two test vectors – TV_1 and TV_2 as CTS of the circuit. The original circuit may be modified to a DFT.

Step 1: Call *process_description_table* procedure.

Step 2: Call *get_S* procedure.

Step 3: Call *get_G* procedure.

Step 4: If G is empty, then no need to construct DFT for this circuit to get 2 TVs as CTS, because the desired 2 TVs are 'readily available' for this circuit. So go to step 9 then.

Step 5: Construct a DFT as follows: add an extra wire to the circuit and assign the last (i.e. maximum) index to this extra wire. Replace each k-CNOT ($k \geq 1$) gate of G by ($k + 1$) gate where the extra control input comes from the newly added extra wire.

Step 6: Regenerate *description_table* of the DFT by modifying *description_table* of the original circuit because some gates are replaced while constructing the DFT.

Step 7: Call *Process_description_table* procedure. Note that this time *description_*table contains the description of the DFT, not of the original circuit.

Step 8: Call *Get_S* procedure on the DFT.

Step 9: Call *Get_TV_from_S* procedure and print the TV as TV_1.

Step 10: Compute $\overline{S} = \{1, 2, \dots , n\} - S$.

Step 11: Call *Get_TV_from_S* procedure on $\overline{S}$ and print the TV as TV_2 – the final TV.

return TV$_1$ and TV$_2$

The following example demonstrates how MCTSG works.

Example 2: Consider the circuit of Figure 2(a). Now we describe the important steps of the main procedure of MCTSG on this circuit. The *description table* of this circuit after sorting is shown in Figure 2 (c) (Step 1). Then S for the original circuit = $\{1, 2, 4\}$ (Step 2). Then $G = \{4, 6\}$ (Step 3). So 4th and 6th gates are replaced to construct a DFT shown in Figure 2(b) (Step 5). Now the *description table* of this DFT after sorting is shown in Figure 2 (d) (Step 6 and 7). Then $S = \{1, 2, 4\}$ (Step 8). So 1st TV is: 0 0 1 0 1 1 (Step 9). Then $\overline{S} = \{3, 5, 6\}$ (Step 10). So 2nd and last TV is: 1 1 0 1 0 0 (Step 11).

3.4. Sorting the *Description Table*

Now we elucidate the sorting performed in the *process description table* algorithm. First we present a lemma.

Lemma 1: If the index of newly added wire in the DFT is the first element of any row of the *description table* of the DFT, then MCTSG may not work correctly. ∎

Proof: We prove it by giving a counter example. Let *description* of a circuit is $\{(5), (5, 4)\}$. Hence $S = \{5\}$ at step 2, $G = \{1\}$ at step 3. Step 7 finds the *description* of the DFT as $\{(5, 6), (5, 4)\}$. $S = \{5\}$ at step 8 and 2 TVs are 1 1 1 1 0 1 and 0 0 0 0 1 0. But if $\{(6, 5), (5, 4)\}$, S = $\{6, 5\}$ at step 8. So TV1 = 1 1 1 1 0 0, TV2 = 0 0 0 0 1 1; reader can verify that which are wrong.

But in some cases, in spite of being the index of newly added wire index in the DFT the first element of a row of the *description table* of the DFT, MCTSG may work correctly. Consider a circuit having *description table* $\{(1, 2), (3, 1)\}$. Now the DFT constructed by MCTSG has *description table* $\{1, 2), (4, 3, 1)\}$. The reader can verify that there is nothing wrong here to produce the CTS in spite of the first position of the newly added wire (index 4) in the DFT in second row. □

294

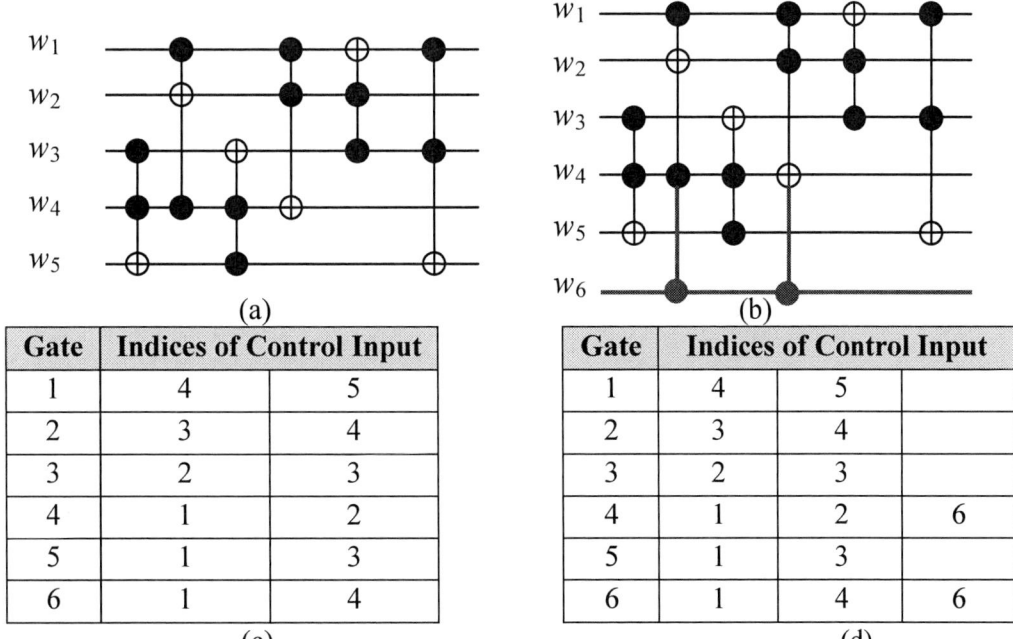

Gate	Indices of Control Input	
1	4	5
2	3	4
3	2	3
4	1	2
5	1	3
6	1	4

(c)

Gate	Indices of Control Input		
1	4	5	
2	3	4	
3	2	3	
4	1	2	6
5	1	3	
6	1	4	6

(d)

Figure 2. (a) Original circuit, (b) corresponding DFT of Figure 2(a), (c) *description table* of the circuit of Figure 2 (a) (d) *description table* of the DFT of Figure 2 (b). (Both the tables are sorted according to the proposed style)

If no sorting is performed in the procedure *process_description_table*, the correctness of the proposed method will not be hampered as long as we are concerned with the precaution mentioned in Lemma 1. To comprehend the reason behind applying the sorting, let us consider the following arguments.

In the first place, we sort each row of *description_table* in increasing order of its elements. Now we will investigate the effect if the rows of *description_table* are sorted by *description_table[i][1]* i.e. the first element. Consider the *description table* of Figure 3(b) and Figure 3(c) where the rows of *description_table* are sorted by increasing and decreasing order of the *description_table[i][1]* respectively ($1 \leq i \leq m$). Figure 3(a) shows the original *description table*.

Gate	Indices of Control Input	
1	3	4
2	4	1
3	4	5
4	1	2
5	3	2
6	3	1

(a)

Gate	Indices of Control Input Gate	
1	1	2
2	1	3
3	1	4
4	2	3
5	3	4
6	4	5

(b)

Gate	Indices of Control Input Gate	
1	4	5
2	3	4
3	2	3
4	1	2
5	1	4
6	1	3

(c)

Figure 3. (a) Original *description_table* for the circuit in Figure 1(a), (b) Rows sorted in increasing order of the first elements and each row is also sorted in increasing order (c) Rows sorted in decreasing order of the first elements each row is also sorted in increasing order.

Table 1 Comparison of different sorting styles of *description tables* shown in Figure 3

Description table	CTS	Number of gates need to be replaced	Indices of the gates in the corresponding *description table* need to be replaced
Figure 3 (a)	((0 1 0 0 1), (1 0 1 1 0))	3	1, 2, 6
Figure 3 (b)	((0 0 0 0 1), (1 1 1 1 0))	5	1, 2, 3, 4, 5
Figure 3 (c)	((0 0 1 0 1), (1 1 0 1 0))	1	4

Table 1 shows that the sorting style adopted in Figure 3 (c) yields most efficient result. The reason may be like this: in both cases we choose the first control input k of a gate (i.e. first element of a row in *description_table* i.e. *description_table[i][1]*) to include in S. In the first case of our example given above, the abovementioned first control input k may appear only in the rows where *description_table[i][1]* is the first element – possibly few because it is not natural to have all the gates connected with wire 1. But in the second case, the first control input k may appear throughout all subsequent rows in the *description_table* because the rows are sorted in increasing order, therefore possibly reducing the size of S. Now the reader can verify that reduction of the size of S increases the possibility of yielding efficient result.

We can conclude from the above discussion that MCTSG with sorting according to the discussed style *may* outperform MCTSG without sorting. We say 'may' because we are not providing any rigorous proof of our conjecture. For this same reason we are not ruling out the possibility of getting worse result after sorting according to the abovementioned style for a particular circuit.

3.5. Algorithm for k-CNOT ($k \geq 0$) Circuits

If 0-CNOT gates exist in a circuit, then MCTSG does not work because MCTSG works on 1-CNOT circuit. Also, we did not include the information of 0-CNOT gates in the *description table* of the circuit. So to deal with k-CNOT circuit with $k > 0$ we need to do something more than MCTSG. We here adopt a simple approach: we convert 0-CNOT circuit into 1-CNOT circuit and then use MCTSG.

Algorithm 6: **Extended_MCTSG**

Input: The configuration of a k-CNOT ($k \geq 0$) circuit is in *description table* as input. Input pattern is as follows: each gate occupies 1 row in the *description table*. The elements of each row are the wire indices of a particular gate where this gate has control inputs. Inversion input wire index is not entered. Each row is terminated with a -1 value which works as sentinel.
Output: Two test vectors – TV_1 and TV_2 as CTS of the circuit. The original circuit may be modified to a DFT.

Step 1: If some 0-CNOT gate exist(s) in the circuit, then list the 0-CNOT gates of the circuit. Else go to step 3.
Step 2: Construct a DFT as follows: add an extra wire to the circuit and assign the last (i.e. maximum) index to this extra wire. Replace each 0-CNOT gate of the circuit by 1-CNOT gate where the control input of this new gate comes from the newly added wire.
Step 3: Call MCTSG on the DFT constructed from step 2, if any. Note that configuration of this DFT will be taken as input of MCTSG and this yields no error for MCTSG because the DFT is necessarily a 1-CNOT circuit. If no DFT is constructed in step 1-2 (i.e. no 0-CNOT gate exists in the circuit), then call MCTSG on the original circuit.

return *TV₁* and *TV₂*

Example 3: Consider the circuit of Figure 4(a). Figure 4(b) shows the DFT constructed according to Algorithm 6. After applying Extended MCTSG on this DFT, we get 2 TVs – 0 1 0 1 1 0 1 and 1 0 1 0 0 1 0 as its CTS.

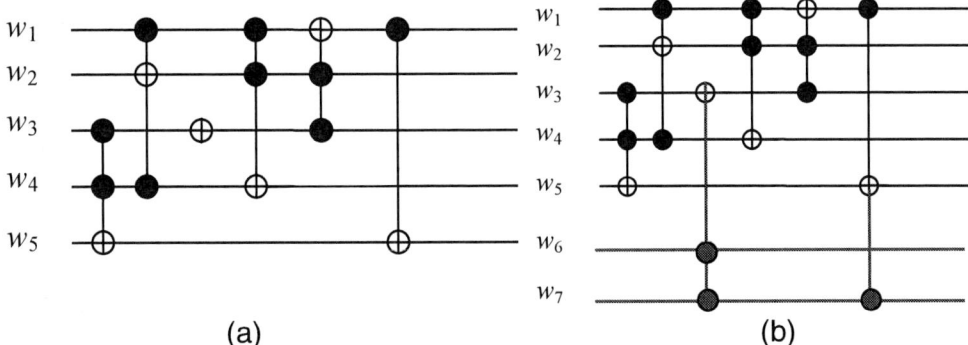

(a)   (b)

Figure 4. (a) Original circuit, (b) Corresponding DFT of Figure 4(a)

4. Experimental Results

We implemented MCTSG in an Intel Pentium 4 Desktop, CPU 1.7 GHz, 256 MB RAM and Microsoft Windows Xp using C programming language and ran it on a number of benchmark circuits. The result is depicted in Table 2. The designs of the benchmark circuits are taken from [9]. It is important to note that the machine-readable versions we used, posted on [9], have a characteristic: all the elements (discarding the *inversion* input of a gate) of a row are sorted in increasing order. So, the result shown in Table 2 produced by MCTSG is based on this structure, but actually MCTSG not necessarily needs the rows sorted.

We used *fixing property* of reversible circuits as the key technique. Here lies the similarity between [7] and us. As we said earlier that [7] showed that the size of CTS can be reduced to 3 for any circuit consisting of k-CNOT gates ($k \geq 0$) but **always** needs modification of the circuit by adding one extra wire and replacing **every** k-CNOT gate by $(k+1)$-CNOT gate. As there may be hundreds of thousands of gates in a real-world circuit, it may be suitable to opt for our algorithm in this regard, which **may** need to construct DFT by adding maximum two extra wires (in some cases it does not need to construct DFT even) and the number of gates replaced by it is may be far less than the total number of gates. Obviously, the less the number of control inputs in k-CNOT gate, the cheaper the gate is. Moreover, in our case |CTS| may be only 2, but minimum |CTS| of [7] is 3. The number of TVs generated by another method of [7], which generates exactly n TVs as CTS for an n-wire k-CNOT circuit ($k \geq 2$), is necessarily worse than ours in terms of |CTS| in all cases. The number of gates replaced by MCTSG is same as [7] in some cases and less in rest of the cases but never more than [7].

5. Conclusion

Based on fixing property, we presented two methods to construct DFT for k-CNOT circuits with $k \geq 1$ and $k \geq 0$ so that size of complete test set for SSF and MSF model is reduced to only 2 i.e. minimal. We also provided some experimental results and comparison with the existing method to show that how the proposed method outperforms the existing one.

Table 2 Experimental Results

Bench-mark Circuits	Model	No. of gates	No of Bits	Bits of DFT		Number of TVs		Number of gates to be replaced	
				*	**	*	**	*	**
2of5	GT	18	6	8	7	2	3	11	18
4_49	GT	16	4	6	5	2	3	16	16
3_17	NCT	6	3	5	4	2	3	6	6
5mod5	GT	17	6	7	7	2	3	17	17
6sym	NCT	20	10	11	11	2	3	17	20
9sym	NCT	28	12	13	13	2	3	21	28
Ham 3	NCT	5	3	4	4	2	3	5	5
Ham7	GT	23	7	8	8	2	3	17	23
Ham 15	GT	132	15	16	16	2	3	132	132
Hwb4	GT	17	4	5	5	2	3	17	17
Hwb5	GT	55	5	6	6	2	3	55	55
Hwb6	GT	126	6	7	7	2	3	126	126
Hwb7	GT	289	7	8	8	2	3	289	289
Hwb8	GT	637	8	10	9	2	3	637	637
Hwb9	GT	1544	9	10	10	2	3	1544	1544
Hwb10	GT	3631	10	11	11	2	3	3631	3631
Hwb11	GT	9314	11	12	12	2	3	9314	9314
Mod5	NCT	8	5	6	6	2	3	8	8
Rd32	NCT	4	4	5	5	2	3	3	4
Rd53	GT	30	7	9	8	2	3	18	30
Rd73	NCT	20	10	11	11	2	3	12	20
Rd84	NCT	28	15	16	16	2	3	16	28
Xor5	NCT	4	5	6	6	2	3	4	4

* According to the PROPOSED Design
** According to the Existing Design of [7]

6. Reference

[1] C. Bennett, "Logical reversibility of computation", IBM Journal of Research and Development 17 (1973) 525–532.

[2] R. Landauer, "Irreversibility and heat generation in the computational process", IBM Journal of Research Development 5 (1961) 183–191.

[3] http://www.zyvex.com/nanotech/reversible.html

[4] V.D. Agrawal, "An information theoretic approach to digital fault testing", IEEE Trans. Con Comp. (1981) 582-587.

[5] K.N. Patel, J.P. Hayes, I. L. Markov, "Fault testing for reversible circuits", VLSI Test Symposium, 21 (2003) 410– 416.

[6] J.S. Allen, Jacob D. Biamonte, Marek A. Perkowski, "ATPG for Reversible Circuits using Technology-Related Fault Models", International Symposium on Representations and Methodologies for Emergent Computing Technologies, Tokyo, Japan, September (2005).

[7] A. Chakraborty, "Synthesis of Reversible Circuits for Testing with Universal Test Set and C-Testability of Reversible Iterative Logic Arrays", 18th International Conference on VLSI Design (2005) 249-254.

[8] I. Shigeru, I. Yusuke, T. Satoshi, U. Shuichi, "On the Complexity of Fault Testing for Reversible Circuits", IEIC Technical Report (Institute of Electronics, Information and Communication Engineers) 105 (2005) 13-16.

[9] D. Maslov, G. Dueck, and N. Scott, Reversible Logic Synthesis Benchmarks Page. http://webhome.cs.uvic.ca/~dmaslov/

KEYNOTE TALK

Architectural Vulnerability Factor (or, does a soft error matter?)
Shubu Mukherjee, *Intel*

With each technology generation, we are experiencing an increased rate of cosmically-induced soft errors in our chips. We are starting to see a dark side to Moore's Law in which the increased functionality we get with our exponentially increasing number of transistors is being countered with a exponentially increasing soft error rate. This will take increasing effort and cost to cope with. Architectural solutions for this problem are inherently expensive and often not cost-effective for the commodity processor market.

A key aspect of estimating a processor's soft error rate is to compute the architectural vulnerability factor (AVF) of its constituent structures. A structure's (AVF) as the probability that a fault in that particular structure will result in an error in the final output of a program. A structure's error rate is the product of its raw error rate, as determined by process and circuit technology, and the AVF. Processor designers can use these AVF estimates to determine which structures need protection (e.g., structures with high AVF are likely to be protected).

Computing AVFs of complex structures, such as the instruction queue or data translation buffer tag, can be quite involved. To guide such complex AVF calculation, we identify numerous cases, such as prefetches, dynamically dead code, and wrong-path instructions, in which a fault will not affect correct execution. Our simulations suggest that AVFs of different structures can vary widely and provides insight into which structures are potential candidates for protection.

In this talk, I will outline how the soft error problem is affecting microprocessor design, how to compute the AVFs for various processor structures, and how to estimates AVFs at run-time using a new concept called Quantized AVFs.

Speaker Bio – Shubu Mukherjee is a Principal Engineer and Director of Intel's SPEARS Group (Simulation and Pathfinding of Efficient and Reliable Systems). The SPEARS Group is responsible for spearheading architectural change and innovation in the delivery of enterprise processors and chipsets by building and supporting simulation and analytical models of performance, power, and reliability. Dr. Mukherjee is widely recognized both within and outside Intel as one of the experts on architecture design for soft errors. He has made pioneering contributions towards the design of Redundant Multithreading (RMT) techniques, architectural vulnerability modeling for soft errors, creation of performance modeling infrastructure called Asim (jointly with Dr. Joel Emer), design of the Alpha 21364 interconnection network, and the creation of the first shared memory prediction scheme.

Prior to joining Intel, Dr. Mukherjee worked at Compaq for 3 years and Digital Equipment Corporation for 10 days. Dr. Mukherjee received his B.Tech. from the Indian Institute of Technology, Kanpur and M.S. and PhD from the University of Wisconsin-Madison. He was the General Chair of ASPLOS (Architectural Support for Programming Languages and Operating Systems), 2004. He has co-authored over 40 external papers. He holds 16 patents and has filed over 25 more in Intel. Dr. Mukherjee's book titled, *Architecture Design for Soft Errors* appeared in the market in February 2008. Dr. Mukherjee serves on the Editorial Board of IEEE Computer Architecture Letters (CAL), as an Associate Editor of *IEEE Transactions of Secure and Dependable Computing (TDSC)*, on National Science Foundation (NSF) panels, on numerous technical program committees, on Intel Corporation's patent committee, on the Advisory Board of Green Street Studios, and on the Board of Trustees of Merrimack Repertory Theatre.

SESSION 6

RELIABILITY AND FAULT TOLERANCE

Automatic Detection of In-field Defect Growth in Image Sensors

Jenny Leung, Glenn H. Chapman

School of Engineering Science
Simon Fraser University
Burnaby, B.C., Canada, V5A 1S6
jla98@sfu.ca, glennc@ensc.sfu.ca

Israel Koren, Zahava Koren

Dept. of Electrical and Computer Engineering
University of Massachusetts
Amherst, MA, 01003
koren,zkoren@ecs.umass.edu

Abstract

Characterization of in-field defect growth with time in digital image sensors is important for measuring the quality of sensors as they age. While more defects were found in cameras exposed to high cosmic ray radiation environments, comparing the collective growth rate of different sensor types has shown that CCD imagers develop twice as many defects as APS imagers, indicating that CCD imagers may be more sensitive to radiation. The defect growth of individual imagers can be estimated by analyzing historical image sets captured by individual cameras. This paper presents a defect tracing algorithm, which determines the presence or absence of defects by accumulating Bayesian statistics collected over a sequence of images. Recognizing the complexity of image scenes, camera settings, and local clustering of defects in color images (due to demosaicing), refinements of the algorithm have been explored and the resulting detection accuracy has increased significantly. In-field test results from 3 imagers with a total of 26 defects have shown that 96% of the defects' dates were identified with less than 10 days difference compared to visual inspection. In addition to our continuous study of in-field defects in high-end digital SLRs, this paper presents a preliminary study of 10 cellphone cameras. Our test results address the comparison of defects types, distribution and growth found in low-end and high-end cameras with significantly different pixel sizes.

1. Introduction

Over the last decade, digital imagers have become increasingly popular in many products. Unfortunately, like all microelectronic devices, digital imagers are prone to develop defects over their lifetime. Furthermore, unlike digital circuits, imager pixels are analog devices, so defects that would not affect digital devices will manifest themselves in these pixels. Advancements in image processing techniques have significantly improved the quality of digital images; however, very little has been done to address in-field defects in digital imagers. Moreover, ignoring the presence of defects during the processing of images causes faulty pixels to smear into neighboring pixels and significantly degrade the image quality. Our previous study has shown that defects in digital imagers are permanent and increase in number continuously over time and consequently, such defects will manifest themselves in all captured images [1]. Although defects can be hidden by sending the camera back for factory calibration, this is in most cases very expensive and time consuming or simply impossible. Thus, with the integration of image sensors in many portable devices, exploration of in-field defect correction techniques is needed.

In our on-going study [1], we have characterized in-field defects by analyzing their spatial distribution, and tracing the growth of defects in a set of semi-professional cameras. Our study has shown that defects are not likely related to material degradation. In fact, the initial analysis of

results in another study [2], had suggested that the defect rate would be higher for sensors that have been through more transatlantic/pacific flights, which we have also seen in our tested cameras. However, to verify the driving force of high defect rates during long high-altitude air flights, we must collect defect data from a wider range of cameras. Our recent work [3] has focused on developing a set of software tools that allow us to collect in-field defect samples from a wider set of imagers. In particular, we have proposed a defect-tracing algorithm that can identify the first appearance of defects through analysis of a sequence of images. The development of this algorithm has not only allowed us to better observe the quality of an aging image sensor; it can potentially become an embedded solution to correct any identified defects as they develop. In this paper, we extend this work by applying the algorithm to image data sets from our tested cameras as well as explore possible enhancement to improve the accuracy of the detection. To further extend our study of high-end DSLRs, we started experiments on cellphone cameras, which have smaller pixels and different characteristics compared to DSLRs.

2. Defects Characteristics

Expanding our on-going study [4] we are currently identifying in-field defects from 12 semi-professional cameras that include sensor technologies from both Charge Couple Device (CCD) and Active Pixel Sensors (APS). While our previous study had concluded that defects only impact single isolated pixels, in any pictures, a faulty pixel will appear as a cluster in color images due to their spreading during internal image processing steps such as noise reduction, color interpolation (demosaicing), and image compressing. Defect clusters are more noticeable than a single faulty pixel; therefore, defects in image sensors are highly undesirable. Thus, a detailed study of in-field defects can provide us with better understanding of the defect source mechanism.

2.1 Defect Identification

Our recent laboratory calibration result on 12 cameras in Table 1 again shows that hot pixels are the dominant type of defects in all tested cameras. In-field defects are identified by performing a dark-frame calibration where a set of images is captured in the absence of light with increasing exposure settings. With the new imagers added to our collection of cameras and additional defects appearing in the other cameras, our defect number had increased from 98 [1] to 136 allowing us to increase the statistical relevance of our defect analysis.

Based on our laboratory calibration, we have identified two types of hot pixels: standard hot pixel and partial stuck hot pixel. The dark-frame response of the hot pixels is shown in Figure 1. A standard hot pixel is characterized by an illumination independent component that increases with exposure time, as shown in curve (a). On the other hand, a partially stuck hot pixel has an additional offset that can be observed even in the absence of light, as shown in curve (b). The presence of a dark current will reduce the dynamic range of the pixel in addition to the offset. The phenomenon of these two types of hot pixels can be summarized by

$$f_{Hot-Pixel}(I_{photo}, I_{Dark}, T_{Integration}) = m \cdot (I_{photo}T_{Integration} + I_{Dark}T_{Integration}) + b , \qquad (1)$$

I_{photo} is the incident illumination on the pixel, $T_{intergration}$ is the exposure duration, I_{dark} is a unique dark current at each pixel site, and b is the additional offset found in partially stuck hot pixels.

Table 1. Summary of identified hot pixels from all tested cameras.

Camera	Number of defects found			Total
	Hot			
	No offset	W/ offset	*Total*	**Total**
A	0	11	11	11
B	17	0	17	17
C	6	5	11	11
D	0	0	0	0
E	26	0	26	26
F	0	5	5	5
G	0	2	2	2
H	3	0	3	3
I	2	17	19	19
J	4	27	31	31
K	9	1	10	10
L	0	2	2	2
			Cumulative total	**137**

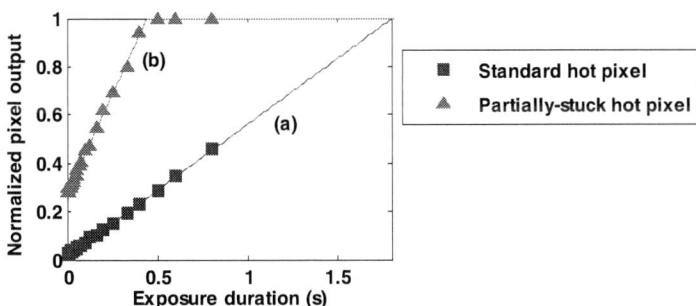

Figure 1. Dark Response of (a) standard hot pixel and (b) partially stuck hot pixel

2.2 Temporal and spatial growth of defects

In our previous study [1] we have shown that defects develop continuously throughout the sensor lifetime and these in-field defects are irrevocable. To better understand the defect source mechanism we have analyzed the spatial distribution of defects using the most recent collection of defects as shown in Figure 2a, see [1] for the analysis procedure. When there is local clustering of defects we should have observed multiple local peaks around long and short distances. However, with a broad distance distribution and an average distance of about 10mm, there is clearly no indication of defect clusters. While the minimum distance of 17μm between faulty pixels, where pixels size is ~(6-7 μm), this result is again consistent with our claim that defects are caused by a random process and not by material degradation. The most likely cause is cosmic ray induced defect damage which is a purely time random process [1].

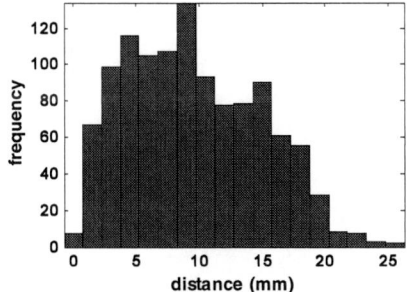
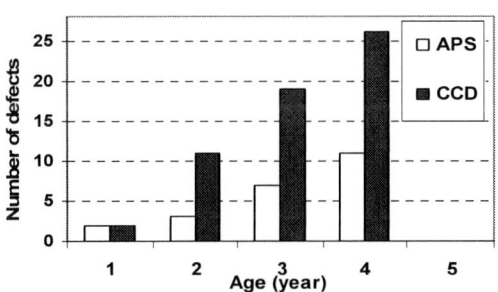

(a) Spatial distribution of defects   **(b) Temporal growth of defects**
Figure 2. Defects growth for APS and CCD cameras.

In Figure 2b we compare the number of defects collected from the 5 APS and 7 CCD cameras, using the average number of defects collected at various ages of each sensor type. As this plot indicates, the development of defects is a continuous process in both sensors; however, there is a significant difference in the defect growth rate between the two types of sensors. The average defect growth rate of APS imagers is ~2.2 defects/year while for CCD imagers it is ~5.2 defects/year; thus we are seeing twice more defects in CCD than in APS imagers at the same age, yet both detectors have approximately the same image area and pixel area. We observed high defect numbers in cameras that have been through high altitude and long flights due to the ~100 times higher radiation levels during those flights [2]. However, this is more noticeable in CCD imagers, where we observed as many as 20 defects within one year from a camera that has been on four international flights. For this reason imagers with many air flights were not considered in this growth analysis so it not skewed by the external conditions. Although there is no clear explanation to the different defect rate, our preliminary results suggest that CCD may be more sensitive to cosmic ray radiation.

The temporal growth rate of defects in individual sensors can be obtained by analyzing the full historical image collection from these imagers. Defects within pictures can often be identified by visual inspection; however, this manual technique is time consuming and due to privacy concerns we cannot gain access to a wider range of datasets. In our most recent study [3] we had proposed a defect-tracing algorithm that can automatically detect the first appearance of a defect by analyzing a sequence of color images. Simulation experiments have shown the effectiveness of this algorithm. In this paper we extend this algorithm and use it to experimentally determine the time development of defects from our set of tested cameras. We determine the accuracy of the detection algorithm based on comparison with the detect dates found from visual inspection. Once this algorithm is proven to operate with sufficient accuracy, we can provide end-users with a software tool that will collect defect growth data by analyzing their image dataset.

3. Defect tracing algorithm

Automatic defect tracing from color images enables a more quantitative analysis of how in-field defects develop over the lifetime of digital imagers. More importantly, with the ability to detect the appearance of defects, we can develop an in-field scheme for the correction of in-field defects. As described in Section 2, hot pixels are the dominant defects found in all tested cameras, thus the main interest of our tracing algorithm is to estimate the first appearance date of these hot pixels by analyzing the image datasets from individual imagers.

Most camera users capture images in RGB color mode, where each color is composed of red, green and blue (RGB) values. One main challenge in detecting defects in color images is the irreversible post-processing algorithm applied to the captured images prior to the observed output. In any digital camera, the initial image captured is called a raw image where each pixel will only record one of the three color channels (red, green or blue) as shown in Figure 3. To produce a color image, demosaicing, an internal color interpolation algorithm is used to interpolate the two missing color channels at each pixel site. As one can expect, if a pixel is defective, interpolating with the inaccurate response of the pixel will spread the error to neighboring pixels; thus we will observe a cluster of defective pixels in color images as shown in Figure 4 [3]. While digital raw photos permit us to extract the pixels before this spread, most photos are not saved in that format.

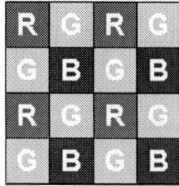

Figure 3. Raw image pattern.

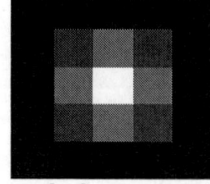

Figure 4. Demosaic image output with defect

3.1 Pixel value estimation

In order to estimate the first defective date for any faulty pixel, our algorithm needs to detect the presence or absence of the known hot pixels, obtained from simple calibration tests with digital raw images at the identified locations. In our algorithm, each sensor is represented by an array of $W \times H$ pixels, and we denote the output of each pixel by $y_{i,j}$ where (i,j) is the location of the pixel. To properly identify defects in images, we model the behavior of each pixel with Equation (1), where I_{dark} is zero for a good pixel, and the locations and unique dark current values of each hot pixel are mapped out using a dark frame calibration.

The expected value of a pixel, denoted by $z_{i,j}$, can be estimated by interpolation with neighboring pixels. Given a good interpolation scheme, the interpolation error $e_{i,j} = y_{i,j} - z_{i,j}$, of a good pixel is approximately zero; however, in the case of a hot pixel, the dark current and the additional offset will result in a larger interpolation error. By collecting the image-wide interpolation error we can derive the interpolation error Probability Density Function (PDF), $p_E(e)$ the probability of a good pixel given its value y_k and the interpolation z_k, and the Cumulative Density Function (CDF), $P_E(e)$, which provide a statistical measure of the state of the defective pixel, are given in Equations (2) and (3) (see [3] for the detailed derivation of these equations)..

$$Prob(y_k \mid Good) = p_E(y_k - z_k) \tag{2}$$

$$Prob(y \mid Hot) = \frac{1}{255 - \Delta_{min} + 1} \cdot \left[P_E(y - z - \Delta_{min}) - P_E(y - z - 255) \right] \tag{3}$$

Because most picture sets only have access to jpg pictures which contain the demosaic spread defects we need to compensated for that in our search. The key to estimating the presence of a faulty pixel is related to the accuracy of the interpolation scheme from the neighboring pixels. With a simple 3×3 averaging, we would expect to have sufficient accuracy; however, as indicated by Figure 4, due to the application of the demosaicing process to color images, the nearest neighbors are the most distorted by the defects. Because single defects appear as local clusters in color images, the expected output of a defective pixel would be more accurate if we employ a larger interpolation region that are further away, such as 5×5 and 7×7 and eliminate the nearest neighbor region (because those will be affected by the defect presence) with a ring averaging mask shown in Figure 5. With the ring averaging, we can estimate pixel output and avoid the problem of the local defect cluster.

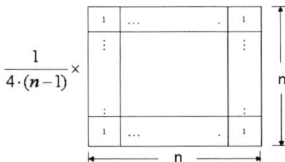

Figure 5. Ring averaging coefficient mask.

3.2 Bayesian accumulation detection

Pixel estimation can provide a measure of the likelihood of a pixel being good or defective at one single image. However, the decision on the status of a pixel cannot rely solely on analysis of one image. External factors such as the complexity of an image scene, the capturing mode such as ISO, and the exposure setting will affect the visibility of defects in different images. To accumulate statistics over a sequence of images, we use Bayesian statistics as shown in Equations (4) and (5) which evaluate the likelihood of having a good/hot pixel at the k-th image.

$$Prob(Good \mid y_k) = \frac{Prob(y_k \mid Good) \cdot Prob(Good \mid y_{k-1})}{Prob(y_k \mid Good) \cdot Prob(Good \mid y_{k-1}) + Prob(y_k \mid Hot) \cdot Prob(Hot \mid y_{k-1})} \tag{4}$$

$$Prob(Hot \mid y_k) = 1 - Prob(Good \mid y_k).\qquad(5)$$

By accumulating statistics on the status of pixels using Equation (4), detection of a good pixel is indicated by $Prob(Good|y_k) \sim 1$, and when $Prob(Good|y_k)$ is ~ 0, this will indicate that the pixel has become defective as shown in Figure 6. However, when a large number of images is accumulated, saturation of the accumulated probability may become a problem. In addition, because most regular images are captured in short exposure range, hot pixels with low dark current magnitude are not easily observed, and the change in the accumulated probability will not be reflected. Thus, to better detect the instantaneous change of the pixel status, a sliding window approach is used as shown in Figure 7. The sliding window approach will determine the status of the pixel by accumulating statistics from the n most recently analyzed images. Previous simulation results [3] have shown that a window length of 5 or 7 tends to attenuate the detection date whereas with a length of 3, more emphasis is put on the recent images and older imagers are ignored. Thus we will focus on using a window length of 3 only.

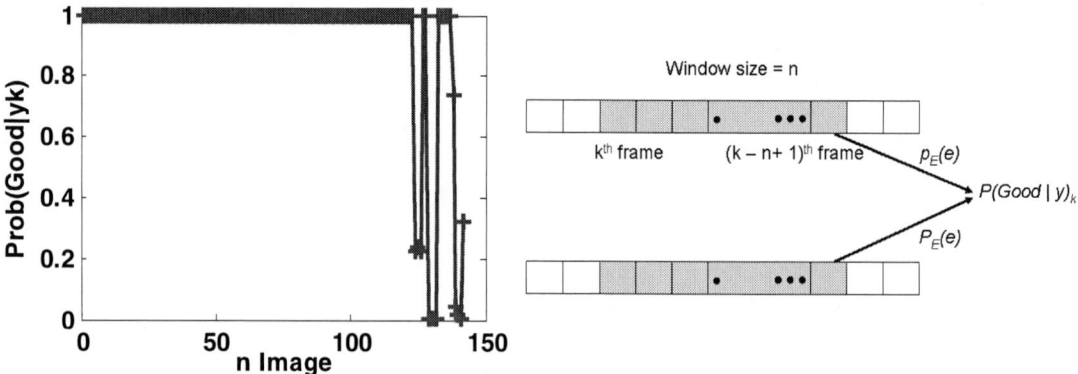

Figure 6. Probability accumulated over sequence of images

Figure 7. Sliding window

3.3 Local region analysis

Analyzing pictures is a complicated procedure; external factors such as the complexity of image scenes, ISO settings, exposure setting, and dark current magnitude will all affect the performance of our detection procedure. Local regions with edge or fine details tend to have more color variation, thus large estimation errors are unavoidable. Images taken at high ISO setting are grainier; therefore, a region with similar colors can still result in large variations. To improve the accuracy of the algorithm, these external factors must be taken into consideration. Because fine details are usually localized in a region; ignoring images with lots of details will potentially flush away other useful information. Instead of discarding images, we applied a post-procedure where we attempted to correct the detected defect date by incorporating some knowledge of local region around the defect. The complexity of any local region can be measured by evaluating the mean and variance of each color channel separately. Given these two measurements, we can simply set a threshold on these parameters and correct detection caused by inaccurate estimation of a pixel output. In our experiments, we performed detection with and without local statistics and analyzed the trade-off with the additional correction procedure.

4. Automated defect growth detection

Using the developed software, we accessed the historical images of three of our test cameras. The three cameras are 2 to 5 years old, with resolution varying from 6M to 10M and a total of 26 hot pixels. The development dates of defects were detected both by the automatic defect-tracing algorithm and by visual inspection. To evaluate the performance of our detection algorithm, we calculate the detection error that is the difference between the algorithm-detected defect date and the observed defect date. Due to the spreading of defects we have experimented with 3×3, 5×5 and 7×7 ring averaging. In addition, to better examine the improvement of the local analysis correction, we compared the detected dates with and without local analysis correction. Figure 6 summarizes our results by plotting the frequencies of the detection errors.

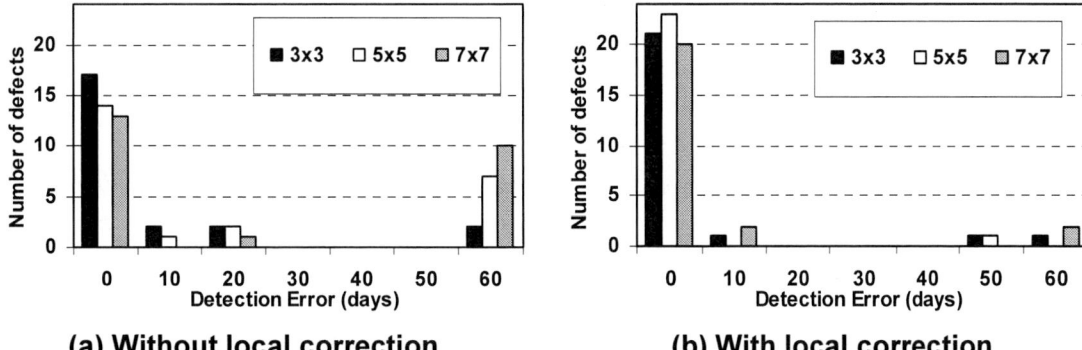

(a) Without local correction   (b) With local correction
Figure 8. Detection error (a) without local correction, (b) with local correction.

As shown in Figure 8a, the detection error results without correction with a 3×3 averaging achieved the highest accuracy among the three interpolation schemes where 71% of the defects were identified within 10 days of the observed date. But errors of up to 60 days were found. With 5×5 and 7×7 ring averaging we observed significantly more outliers with error >60 days. The majority of detection errors are caused by false identification of defects when the faulty pixel is located in a fine detailed color region. Thus, to correct these errors, a localized analysis around the defective region can help determine if the detection is simply an interpolation error or the first appearance of a defect. By recognizing details from the local region analysis, false detection can be corrected and the results in Figure 8b show a significant improvement in the accuracy of the detection. In fact, the 5×5 ring averaging now achieves the highest accuracy with 96% of defects identified with less than 10 days error. Although the accuracy with 3×3 averaging had increased to 88%; it is clear that this averaging scheme still has some large outliers of 50 – 60 days. As discussed in Section 3, the spreading of defects caused by the demosaicing algorithm is most significant in the nearest neighbors, thus 3×3 averaging cannot provide a good approximation to the expected pixel value. Defects can become undetectable when the neighboring pixels are highly distorted by the defects, thus these outliers simply indicate the limitation of the 3×3 interpolation scheme. On the other hand, with 7×7 ring averaging, 83% of defects were detected with error <10 days. However, the large interpolation region required by this averaging scheme induces more estimation errors; thus this interpolation does not provide a very robust detection result. In all cases it must be remembered that the detection error is highly dependent on the dates on which the images were taken. Because most camera users do not capture images at a steady rate, inherently, part of the detection error can be caused by large gaps between two sequential image. This error is simply inevitable where when the detection is difference by one image can be interpreted as anything from one day to 50 days as some picture data bases show.

Knowing the defect development date of each defect, we can estimate the defect growth rate for all three tested cameras. Figure 9 shows a comparison of the defect growth rates based on visual inspection and on the detection algorithm with the 5×5 interpolation scheme. In particular, a linear fit function is used to estimate the defect growth rate for each test camera.

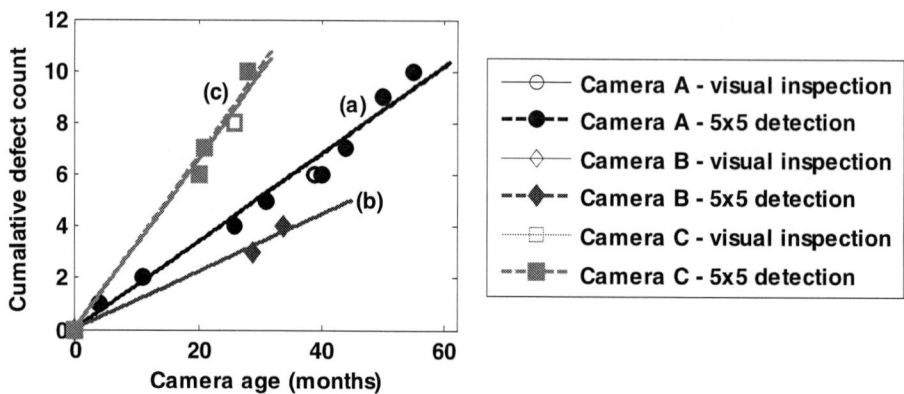

Figure 9. Temporal defect growth of (a) camera A, (b) camera B, (c) camera C.

The observed and detected defect growth rates (found by the detection algorithm) with different interpolation schemes are summarized in Table 2. Again, by comparing the detect growth rate to that estimated from visual inspection, the results with 5×5 averaging achieve the best approximation of the growth rate where camera A, and Bare both APS imagers and camera C is a CCD imager.

Table 2. Defect growth rate results from detection algorithm with local correction.

	Observed defect growth rate (defects / year)	Detected defect growth rate (defects / year)		
		3x3 Averaging	5x5 Averaging	7x7 Averaging
Camera A	2.04	1.99	2.03	2.08
Camera B	1.34	1.32	1.34	1.34
Camera C	3.94	4.04	4.04	3.74

5. Defects characterization in cellphone cameras

The popularity of integrated image sensors in devices such as cellphone has increased over the past few years. To extend our analysis, we expanded our data collection to include a set of high-end cellphone cameras. In this study we collect in-field defects from a set of 10 mobile phone cameras of the same model. Cellphone cameras do not have the advanced capability of capturing raw images, have no explicit control on exposure setting, and standard laboratory dark calibration cannot be performed on these cameras. To estimate the number of defects in these cameras, we performed dark frame calibration by capturing dark images in color mode, and the results are summarized in Table 3.

Table 3. Summary of identified defects cellphone cameras.

Camera	Hot pixels		Total
	No offset	w/offset	
A	6	3	9
B	11	2	13
C	6	2	8
D	4	2	6
E	6	6	12
F	12	2	14
G	10	4	14
H	6	4	10
I	9	5	14
J	7	10	17
		Cumulative total	117

With shrinkage in the pixel dimensions to 2.2 μm, defect clusters in these sensors are more likely to occur and the impact of hot pixels can be more significant. Moreover these cameras only have 5.7x4.3 mm sensors, only 7% the size of the 24x15 mm DSLR sensors. Because internal processing algorithms such as demosaicing and image compression will distort the defects in color images, we cannot at this point conclude if defects in these sensors are cluster-free nor can we provide an estimate of the dark current magnitude. However, a first approximation on the number of defects in each camera does indicate that with the smaller pixel size we are seeing more defective pixels per unit area than in regular cameras. As opposed to regular digital cameras, only simple procedures are taken to correct manufacture time defects; thus the defects identified in these cellphone cameras include both in-field and manufacture time defects, and will require a totally different analysis approach.

6. Conclusion

Defect developing is inevitable in any aging digital images. Our continuous analysis of spatial distribution and temporal growth of defects with an expanding number of defects (137 defects so far) has again shown no indication of material source related defects. Moreover, the higher number of defects observed in imagers that were exposed to high radiation environments indicates that cosmic ray radiation is the probable defect source. First approximation on defect growth rates based on a collection of APS and CCD imagers showed that the average growth rate of APS imagers is ~2.2 defects/year while for CCD imagers it is ~5.2 defects/year. While CCD developing defects at twice the rate of APS sensors, the suggestion is that CCD sensors may be more sensitive to cosmic ray radiation. With the development of an automated defect-tracing algorithm, we are able to detect the defect development date and estimate the defect growth rate of individual imagers by analyzing historical images from individual cameras, permitting this to eventually be expanded to many cameras. The accuracy of the detection algorithm is limited by false detections caused by scene complexity; however, by incorporating knowledge of the local region around the defect, we are able to correct some false defect detections. Our recent in-field tests on three imagers 2-5 years old with a total of 26 defects has shown that 96% of the defect dates were identified within 10 days of the visually identified dates. To expand our in-field defect analysis, we have extended our study to 10 cellphone cameras with a much smaller pixel size. Preliminary results showed that these cameras had ~117 defects prior to shipment by the manufacturer, because only simple techniques were used to map out manufacture time defects. Thus, defects identified in these cameras will include both manufacture-time and in-field defects and will require a new set of analysis tools.

7. References

[1] J. Leung, J. Dudas, G. H. Chapman, I. Koren, Z. Koren, "Quantitative Analysis of In-Field Defects in Image Sensor Arrays," *Proc. of the 2007 Intern. Symposium on Defect and Fault Tolerance in VLSI*, pp.526-534, Rome, Italy, Sept 2007.

[2] A. J.P. Theuwissen, "Influence of Terrestrial Cosmic Rays on the Reliability of CCD Image Sensors," IEDM 2005, San Francisco, CA, 2005.

[3] J. Leung, J. Dudas, G.H. Chapman, Z. Koren, and I. Koren, "Characterization of defect development During Digital Imager Lifetime," Proc. Electronic Imaging, Sensors, Cameras, and Systems for Industrial/Scientific Application IX, San Jose, Jan. 2008.

[4] J. Dudas, L.M. Wu, C. Jung, G.H. Chapman, Z. Koren, and I. Koren, "Identification of in-field defect development in digital image sensors," Proc. Electronic Imaging, Digital Photography III, v6502, 65020Y1-0Y12, San Jose, Jan. 2007.

IEEE International Symposium on Defect and Fault Tolerance of VLSI Systems

Material Fatigue and Reliability of MEMS Accelerometers

Xingguo Xiong, Yu-Liang Wu*, and Wen-Ben Jone**
Department of Electrical and Computer Engineering,
University of Bridgeport, Bridgeport, CT 06604, USA
Department of Computer Science and Engineering*,
The Chinese University of Hong Kong, Shattin, Hong Kong
Department of ECECS, University of Cincinnati, Cincinnati, OH 45221, USA**
Email: xxiong@bridgeport.edu, ylw@cse.cuhk.edu.hk*, wjone@ececs.uc.edu**

Abstract

MEMS (Microelectromechanical System) reliability has been a very important issue, especially for safety-critical applications. Due to the diversity and multiple energy domains involved, MEMS devices are vulnerable to various failure mechanisms. MEMS reliability under different failure mechanisms should be analyzed separately. Since most of MEMS devices contain movable parts, material fatigue and aging under long-term repeated cycling load may lead to potential device failure, which in turn degrades the device reliability. In this paper, the reliability of poly-silicon MEMS comb accelerometers under material fatigue failure mechanism is analyzed. Based on ANSYS stress simulation, the mean-time-to-failure (MTTF) lifetimes and failure rates for both BISR (built-in self-repairable) and non-BISR poly-silicon MEMS comb accelerometers are derived. Simulation results show that the fatigue lifetime of MEMS accelerometers made by poly-silicon material can be good enough for general purpose applications. However, for some "weak" devices with certain structure defects, the material fatigue and aging may become potential threats. Compared to non-BISR design, BISR MEMS accelerometer demonstrates effective reliability improvement due to redundancy repair. MEMS reliability under material fatigue for other MEMS materials will be further studied in the future.

Keywords: material fatigue, reliability, failure analysis, MEMS accelerometer, redundancy repair.

1. Introduction

MEMS reliability has been a very important issue as more and more MEMS (Microelectromechanical Systems) technologies have been commercialized. MEMS reliability is especially important for those safety-critical applications, such as automobile, biomedicine and aerospace. Due to the diversity and multiple energy domains involved in MEMS working principles, MEMS devices are vulnerable to many more failure mechanisms during their fabrication as well as in-field usage. Some possible failure mechanisms [1][2] for MEMS include material fatigue, mechanical fracture, stiction, wear, delamination, residual stress, etc. Further, there are also some environmentally induced failure mechanisms, such as shock, vibration, humidity, particle contamination, electrostatic discharge. Most of them are MEMS-specific which are very different from those of VLSI chips.

MEMS reliability under shock environment has been discussed in [3]. MEMS reliability under other failure mechanisms (e.g. vibration, stiction) have been reported [4][5]. Previously we developed MEMS reliability models for both BISR (built-in self-repairable) and non-BISR accelerometers and analyzed device reliability for both designs under Z-axis shock environment [6]. Most MEMS devices, including comb accelerometers, contain movable components. A MEMS device relies on the movable components to perform its specific function. However, the long-term cycling movement will lead to the material fatigue and material aging. Fatigue of a material is the process of damage and failure due to cycling loading, so that the structure will crack even at stress well below the material's ultimate strength. The fatigue behavior of a ductile material such as metal has been well researched, and is generally modeled with a stress versus life (number of cycles to failure) curve (also known as a S-N curve) as shown in Figure 1[7].

From Figure 1, we can see that the cycling load leads to the decrease of the material fracture stress. Eventually, the material will crack at a stress lower than its original material strength. If the material is loaded at a higher stress level, the number of cycles to failure will be smaller. Once the amplitude of the cycling load is known, one can predict the fatigue life (in number of cycles to failure) from the S-N curve. Fatigue has been estimated to account for up to 80% to 90% of mechanical failures in engineering structures [7]. Thus, fatigue is directly related to the mechanical long-term reliability. With the commercialization of MEMS devices, researchers have also begun looking into the fatigue behavior in a MEMS material. Most MEMS devices involve movable components, and this makes the material fatigue a major reliability concern. Whether the re-

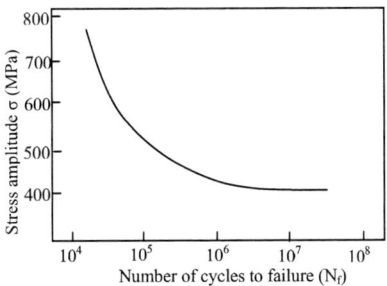

Figure 1. A typical S-N curve of a ductile material[7]

peated movement of MEMS components will induce fatigue in its structural material and how this will affect the reliability of MEMS remain very interesting research topics.

Besides the material fatigue under cyclic loading, it also has been observed that MEMS material properties, such as Young's modulus, etc, will gradually shift after long-term repeated cycling load, which is often in the order of billions of cycles [8]. This material properties shift due to long-term cyclic loading is sometimes referred to as material aging. The shift in Young's modulus leads to the change of the device resonant frequency as well as the sensitivity of an accelerometer (for example). It turns out this will degrade the sensor output signal. Other properties, such as the dampening coefficient and resistance of a material, may also change due to long-term cyclic loading. The mechanical and electrical parameters degradation of a p+ silicon cantilever beam due to fatigue is discussed in [8]. According to their results, after about 1.9×10^9 cycles of loading, the Young's modulus of p+ silicon shifts from 129MPa to 136MPa, an increase of about 5.4%. This leads to a corresponding sensitivity decrease of 5.4%. For some applications, this may be totally unacceptable and it will be treated as a device failure. Thus material fatigue and aging are really important concerns for MEMS.

In this paper, we analyzed MEMS reliability for both BISR and non-BISR poly-silicon comb accelerometers under material fatigue failure mechanism. Based on ANSYS stress simulation, the mean-time-to-failure (MTTF) lifetimes and failure rates for both BISR and non-BISR poly-silicon MEMS comb accelerometers are derived. Simulation results show that the fatigue lifetime of MEMS accelerometers made by poly-silicon material can be good enough for general purpose applications. However, for some "weak" devices with certain structure defects, the material fatigue may become potential threats. Material aging may also degrade the long-term device reliability. Compared to non-BISR design, BISR MEMS accelerometer demonstrates effective reliability improvement due to redundancy repair.

2. Reliability of MEMS accelerometers in Material Fatigue

2.1. Fatigue Analysis and Cycles to Failure

The fatigue behavior of metal materials in MEMS, such as aluminum used in digital micromirror device (DMD), can use the already existing theories and models. Silicon has been widely used in MEMS for its superior electrical and mechanical properties [9]. As a brittle material, silicon was believed not to be susceptible to dynamic fatigue. However, in 1992, the fatigue of silicon was demonstrated in [10] using a specimen with dimension in the range of microns. Ever since, many researchers have made tremendous efforts to explore the fatigue mechanism and behavior of both single crystal silicon [11] and poly-silicon [12][13].

Currently, the fatigue mechanism and model for MEMS materials such as silicon and poly-silicon are still far from being mature. In [14], $3.5\mu m$ thick and $50\mu m$ wide poly-silicon tensile specimens under cycling loading were used to investigate the long-term mechanical fatigue behavior of poly-silicon material for MEMS applications. Based on the experimental data, the S-N curve of poly-silicon under cycling loading is plotted. They observed that the tensile strength $\sigma_c = 1.1GPa$ of virgin samples is reduced by about 35% to a fatigue strength of $\sigma_f = 0.70GPa$ after 10^9 cycles. They varied the test frequency between 20 and 6000Hz, and no influence of the test frequency on the fatigue behavior was observed in this range. That is, the number of cycles N_f to failure does not depend on the frequency f of the cyclic loading.

The mean-time-to-failure (MTTF) of a device can be calculated by [14]

$$MTTF = N_f \cdot T = N_f/f,$$

where $f = 1/T$, T is the period of cyclic loading. Hence, the MTTF does depend on the frequency of the cycling load. The samples which are experienced cycling load with higher frequency will fail after a shorter time. Further, no endurance limit (stress below which failure will never occur) was observed in their experiments. That is, even very small stress in cycling load will also induce material fatigue of poly-silicon. Hence, in some MEMS devices, even if the displacement (and thus the induced stress) in normal operation is very small, fatigue still exists in the poly-silicon material. The only difference is the fatigue life will be much longer compared to the case of large stress. Based on the experiment data, the S-N curve for poly-silicon material can be plotted. The experimental data can be fitted with a power law as shown in Figure 2[14]. According to [14], the number of cycles (N_f) to failure of the poly-silicon sample in cycling load can be predicted using the following experimental formula

$$N_f = (\frac{\sigma_f}{\sigma_c})^{1/m}.$$

where σ_c is the mean tensile strength of poly-silicon, $\sigma_c = 1.10 GPa$ in their research. Further, σ_f is the applied maximum stress during cycling (fatigue strength), and m is a constant. Using the least square fit analysis on the experimental data, they found $m = -0.02$. This experimental equation is very useful to estimate the lifetime (number of cycles to failure) of the material under cycling load, or the maximum allowed peak stress for a required lifetime if the initial strength σ_c is known. In reality, the mean tensile strength (σ_c) of brittle materials (such as poly-silicon) shows a large scattering. The above results are also derived using sinusoidal wave loading, while other loading waveforms may or may not change the behavior. However, the above equation is very helpful for us to have an order-of-magnitude evaluation on the device reliability.

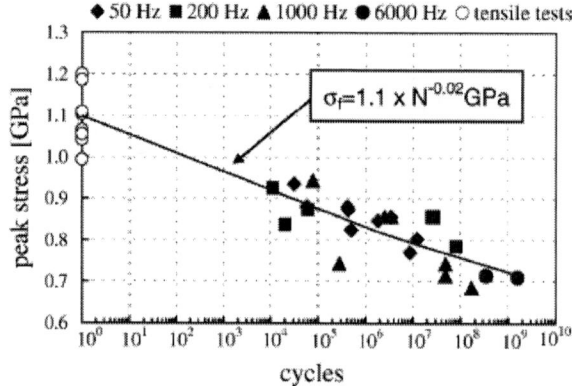

Figure 2. The S-N curve of poly-silicon tensile specimens with cyclic loading [14].

2.2. Reliability Analysis by Cycles to Failure

In our previous work [15][6], we introduced a BISR (built-in self-repairable) MEMS comb accelerometer design. The device consists of six identical modules (for example), and each module has its own beams, mass and fixed/movable finger structures. Among them, four modules are connected together as the main device, while the remaining two modules serve as redundancy. The movable parts of each module are physically connected to those of adjacent modules through the common anchors, and signals sensed by all movable fingers in the device are connected to the sensing circuit directly. However, the fixed fingers of each module are connected to the modulation signal circuit through switches made of analog MUXes. By turning on or off these switches, we can determine whether a module works as part of the main device or the redundant device. If a module is tested as faulty, the control circuit will permanently exclude the module from the main device and replace it with a good redundant module (if there is any). Thus, after repairing, the main device can still be ensured to work properly. Because each module has its own independent beam and mass structure, a faulty module does not affect the function of other modules. For example, even if the movable part of one module is broken or stuck to substrate, the movable parts of other modules can still move freely and work jointly to ensure the function of the main device. Due to the redundancy repair, the yield and reliability of the BISR design can be effectively improved compared to non-BISR design. The BISR design introduces area overhead as well as sensitivity loss due to modularized design. The sensitivity loss can be compensated by beam width or electrostatic compensation [15]. The MEMS comb accelerometer is fabricated with poly-silicon surface-micromachining technique. In this paper, we evaluate the reliabilities of example designs for both non-BISR and BISR devices with poly-silicon material fatigue. The non-BISR device has a beam width of $W_{bnsr} = 3.2 \mu m$. For the BISR device, we consider two cases: one is the case with beam width compensation (which is called BWC device, $W_{bbwc} = 2.0 \mu m = 0.63 W_{bnsr}$), and the other is the case with electrostatic force compensation (which is called EFC device, $W_{befc} = 3.2 \mu m = W_{bnsr}$). The design parameters of both the non-BISR and BISR devices are shown in Table 1 [6].

316

Table 1. Non-BISR/BISR accelerometers design parameters for reliability evaluation.

Design parameters	non-BISR device	BWC BISR module	EFC BISR module
beam width W_b	$3.2\mu m$	$2.0\mu m$	$3.2\mu m$
beam length L_b	$310\mu m$	$310\mu m$	$310\mu m$
mass width W_m	$100\mu m$	$100\mu m$	$100\mu m$
mass length L_m	$534\mu m$	$132\mu m$	$132\mu m$
movable finger width W_f	$4\mu m$	$4\mu m$	$4\mu m$
movable finger length L_f	$160\mu m$	$160\mu m$	$160\mu m$
number of finger groups N	48	12	12
device thickness t	$2.0\mu m$	$2.0\mu m$	$2.0\mu m$

The structural diagram of the non-BISR MEMS accelerometer for this analysis is shown in Figure 3. In our analysis, we first use ANSYS to extract the stress distribution of the comb accelerometer in response to a given cyclic input stimulus (acceleration) with frequency f_0. With this result, we can find the maximum stress in the device. The maximum stress generally occurs in the beams of the accelerometer, since they have maximum deformations in the device. Thus, we only need to consider the MTTF of the beams for device reliability. With this maximum stress value, we can derive the number of cycles N_{fbnsr} to failure for one beam using the above equation by considering the poly-silicon material fatigue. The lifetime or the MTTF of one beam under cyclic loading can be expressed as

$$MTTF_{bnsr} = \frac{N_{fbnsr}}{f_0}.$$

Since there are four beams in the non-BISR device and they can be described with series reliability model, the total mean-time-to-failure $MTTF_{nsr}$ of the non-BISR device is

$$MTTF_{nsr} = \frac{1}{4}MTTF_{bnsr} = \frac{N_{fbnsr}}{4f_0}.$$

Thus, the failure rate of the device can be calculated as

$$\lambda_{nsr} = \frac{1}{MTTF_{nsr}} = \frac{4f_0}{N_{fbnsr}}.$$

Finally, the reliability function $R_{nsr}(t)$ of the MEMS device can be expressed as

$$R_{nsr}(t) = e^{-\lambda_{nsr}t}.$$

This is our strategy in evaluating the reliability of the non-BISR MEMS device for material fatigue due to cyclic loading.

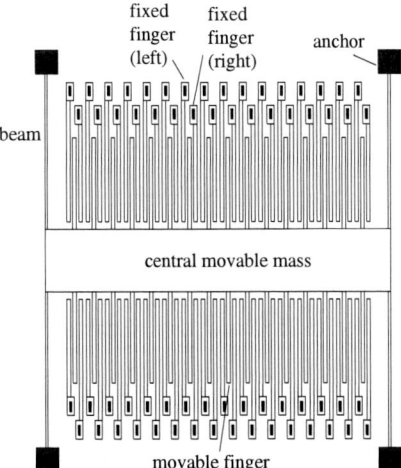

Figure 3. The structural diagram of a non-BISR MEMS device.

The structural diagram of the BISR comb accelerometer is shown in Figure 4. Similarly, we use ANSYS to extract the maximum stress in each BISR module in response to the same cyclic loading as that of the non-BISR device. With this maximum stress value, we can derive the number of cycles N_{fbmod} to failure of one beam in a BISR module. The lifetime or the MTTF of one beam under cyclic loading can be expressed as

$$MTTF_{bmod} = \frac{N_{fbmod}}{f_0}.$$

Since there are four beams in each BISR module, the total mean-time-to-failure $MTTF_{mod}$ of one BISR module is

$$MTTF_{mod} = \frac{1}{4}MTTF_{bmod} = \frac{N_{fbmod}}{4f_0}.$$

Thus, the failure rate of one BISR moduel can be calculated as

$$\lambda_{mod} = \frac{1}{MTTF_{mod}} = \frac{4f_0}{N_{fbmod}}.$$

Finally, the reliability function $R_{mod}(t)$ of one BISR module can be expressed as

$$R_{mod}(t) = e^{-\lambda_{mod}t}.$$

The reliability function of the whole BISR device based on 4-of-6 redundancy is

$$
\begin{aligned}
R_{sr} &= R_{mod}^6 + 6R_{mod}^5 Q_{mod} + 15R_{mod}^4 Q_{mod}^2 \\
&= (e^{-\lambda_{mod}t})^6 + 6(e^{-\lambda_{mod}t})^5(1 - e^{-\lambda_{mod}t}) + 15(e^{-\lambda_{mod}t})^4(1 - e^{-\lambda_{mod}t})^2.
\end{aligned}
$$

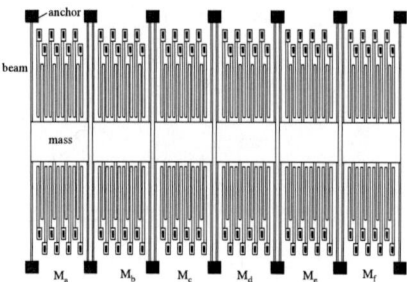

Figure 4. The structural diagram of the BISR MEMS accelerometer.

By comparing the reliability functions of both non-BISR and BISR devices, we can quantitatively evaluate the reliability enhancement of the BISR design.

3. Results and Discussions

3.1. Material Fatigue and MEMS Reliability

In our simulation tests, we applied sinusoidal wave acceleration along the X direction (the sensitive direction of the accelerometer) as the cyclic load. The acceleration frequency is $1kHz$, and the acceleration amplitude is $500g$. This mimics the cyclic loading of the comb accelerometer in accelerated testing. The maximum vibration amplitude $A(f)$ of the accelerometer in sinusoidal actuation can be expressed as [13]

$$A(f) = \frac{F_0/m}{[(f_0^2 - f^2) + (\frac{f-f_0}{Q}^2)]^{1/2}}.$$

where F_0 and f are the amplitude and frequency of the cyclic loading separately, f_0 and m are the resonant frequency and mass of the accelerometer respectively, and Q is the quality factor of the accelerometer. Since the frequency of cyclic loading (acceleration) is lower than the resonant frequencies of non-BISR/BISR devices, and the maximum vibration amplitude of the accelerometer is limited by the capacitance gap, the resonant vibration effect is ignored in our analysis.

An example ANSYS stress simulation of a BWC BISR module is shown in Figure 5. According to the stress distribution contour plots of ANSYS simulation, it is clearly shown that the maximum stress occurs at the end of each beam (either the anchor end or the end connected to the movable mass) in response to the X axis acceleration input. This indicates that these locations on the beams are most vulnerable to the fatigue failure due to cyclic loading. This also demonstrates that the beams have the lowest reliability compared to other components (mass, fingers, etc) in material fatigue. Thus, their lifetimes determine the lifetime of the entire comb accelerometer. The lifetimes of other components (mass, movable finger, anchor, left/right fixed fingers, etc.) under material fatigue can be treated as infinity, and their failure rates can be treated as zero. Assume an accelerated fatigue testing with cyclic loading of $500g$ along X direction (device sensitive direction) with frequency of $1kHz$, the ANSYS stress simulation results and estimated failure rates for one beam in both non-BISR and BISR devices are shown in Table 2.

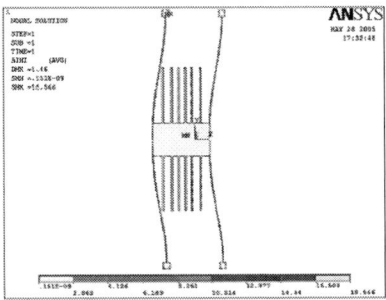

Figure 5. ANSYS stress analysis of one BWC BISR module.

Table 2. The ANSYS stress analysis result for one beam of both BISR and non-BISR devices.

Items	non-BISR device	BWC BISR module	powered EFC module	unpowered EFC module
Displacement	$1.3\mu m$	$1.3\mu m$	$1.3\ \mu m$	$0.3\ \mu m$
Maximum stress σ_f	22.67 MPa	13.30MPa	22.7 MPa	5.53MPa
No. of cycles to failure N_f	1.98×10^{84}	7.53×10^{95}	1.98×10^{84}	8.58×10^{114}
Lifetime (MTTF)	$1.98 \times 10^{81} sec$	$7.53 \times 10^{92} sec$	1.98×10^{81}	$8.58 \times 10^{111} sec$
Failure rate λ (sec^{-1})	5.05×10^{-82}	1.33×10^{-93}	5.05×10^{-82}	1.17×10^{-112}

Based on the expected lifetime data derived above, we can evaluate the reliability functions of both non-BISR and BISR devices using the MTTF of a beam n the non-BISR device. The failure rate λ_{bnsr} of one beam can be determined by

$$\lambda_{bnsr} = \frac{1}{MTTF_{bnsr}} = 5.05 \times 10^{-82} sec^{-1}.$$

The reliability function $R_{nsr}(t)$ of the non-BISR device is thus

$$R_{nsr}(t) = e^{-4\lambda_{bnsr}t} = e^{-2.02 \times 10^{-81} \cdot t}.$$

Similarly, the failure rate λ_{bmod} of one beam in BWC BISR module is

$$\lambda_{bmod} = \frac{1}{MTTF_{bmod}} = 1.33 \times 10^{-93} sec^{-1}.$$

The failure rate λ_{mod} of each BISR module is

$$\lambda_{mod} = 4\lambda_{bmod} = 5.32 \times 10^{-93} sec^{-1}.$$

The reliability function R_{mod} of one BISR module is therefore

$$R_{mod} = e^{-\lambda_{mod}t} = e^{-5.32 \times 10^{-93} \cdot t}.$$

Finally, the reliability function $R_{sr}(t)$ of the BWC BISR device can be derived using 4-out-of-6 reliability model [15]

$$
\begin{aligned}
R_{sr}(t) &= R_{mod}^6 + 6R_{mod}^5 Q_{mod} + 15R_{mod}^4 Q_{mod}^2 \\
&= (e^{-\lambda_{mod}t})^6 + 6(e^{-\lambda_{mod}t})^5(1 - e^{-\lambda_{mod}t}) + 15(e^{-\lambda_{mod}t})^4(1 - e^{-\lambda_{mod}t})^2 \\
&= (e^{-5.32 \times 10^{-93} \cdot t})^5(1 - e^{-5.32 \times 10^{-93} \cdot t}) + 15(e^{-5.32 \times 10^{-93} \cdot t})^4(1 - e^{-5.32 \times 10^{-93} \cdot t})^2.
\end{aligned}
$$

The reliability function of the EFC BISR device can be calculated in a similar way. Following the above analysis, the reliability function curves for both non-BISR and BWC/EFC BISR devices are shown in Figure 6. From the curves, we can see that the MTTF of the non-BISR device is about 10^{80} sec. The MTTF of the BWC device is increased to about 10^{92} sec. This is because the beam width of BWC BISR device is shrunk, and hence for the same displacement the maximum stress in the beam of the BWC BISR device is smaller than that of the non-BISR device. This leads to longer fatigue life compared to the non-BISR device. For the EFC BISR device, we analyzed both the unpowered and powered cases. In the powered case, the electrostatic force compensates the device sensitivity to the same value as the non-BISR device. The beam width of the EFC device is the same as that of the non-BISR device. Thus, the maximum stress in the EFC device beam is almost the same as that in non-BISR device. This leads to almost the same fatigue life (MTTF) for the powered EFC device as that of the non-BISR device. However, for the unpowered EFC device, the displacement for the same acceleration input is almost $\frac{1}{4}$ as that of the non-BISR device. This leads to reduced maximum stress in the beam. As a result, the fatigue life of the unpowered EFC BISR device is increased to about 10^{111} sec.

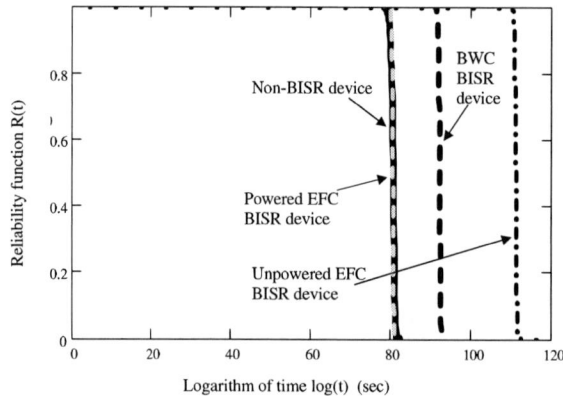

Figure 6. The reliability curves for BISR and non-BISR device ($t = 10^x$).

On the other hand, we can also see that the lifetime of both non-BISR and BISR devices are long enough for normal application. This is because in normal operation, the displacement of the movable mass is very small. Hence, the resulted maximum stress on each beam is much lower than the fracture strength ($\sigma_c = 1.10 GPa$). Although no endurance limit is found in poly-silicon material, a very low maximum stress level on the beams leads to extremely large N_f value. As a result, both non-BISR and BISR devices exhibit extremely long lifetime in material fatigue failure mode. This indicates that for normal operation of the MEMS comb accelerometer, the crack of a beam due to material fatigue in cycling loading is very rare, and it is not a major concern for device reliability. The device reliability may therefore be determined by other defect sources and failure mechanisms. However, this does not mean that the material fatigue is not a threat for MEMS device reliability. For other MEMS devices such as aluminum micromirror devices (DMD), the material fatigue of aluminum leads to much lower reliability in its working mode. Thus, for these MEMS devices, the material fatigue due to cyclic loading is a serious concern and needs to be addressed carefully. Meanwhile, MEMS material under cyclic loading develops aging at a life-time much shorter than that of fatigue. The MEMS material aging has been observed after cyclic loading of as short as 1.0×10^9 cycles. Assume this aging lead to the failure of the device, and the frequency of the cyclic loading is 1kHz, the MTTF for aging can be calculated as

$$MTTF = N_f/f_0 = 1.0 \times 10^9/(1 \times 10^3) = 10^6 sec = 11.5 days.$$

This shows that the material aging can be a serious concern for MEMS comb accelerometers. However, the aging mechanism of MEMS material has not been well understood yet. A quantitative MEMS aging model is not available at this time. As a result, we cannot quantitatively evaluate the MEMS reliability increase of BISR design in material aging. However, as we saw before, the reduce in mass in modularized design help reduce the maximum stress level in the material. This will help alleviate the aging in the material and lead to prolonged MTTF for aging failure. As a result, the device reliability of the BISR design in case of material aging will be enhanced.

3.2. Weak Device and Material Fatigue

ANSYS simulation result turns out that beam fracture due to material fatigue is not a serious threat for a defect-free MEMS comb accelerometer in its normal operation. However, some MEMS actuators (such as microresonator, vibratory gyroscope, etc.) have large vibration amplitude in their working modes. This may cause the maximum stress in a device large enough to cause the material fatigue failure in relatively short lifetime cycles. For these MEMS devices, material fatigue may become a major failure mechanism and needs thorough consideration. ANSYS can be used to guide the design optimization in order to reduce the maximum stress while maintaining the device performance requirement.

For an accelerometer device with certain defects, material fatigue can become a major concern. For example, as shown in Figure 7, the accelerometer has a point stiction on the beam, and this can greatly reduce the device sensitivity. But, if the beam section from the point stiction to the central mass is narrowed to $0.1\mu m$, it will greatly decrease the spring constant of the defective beam. Assume the length of the narrowed beam section as $6\mu m$. If both defects (beam stiction plus narrowing) occur simultaneously, they mask each other and cause the device sensitivity to be within the tolerable range. This defective device may still be accepted as a "good" one and released to the market. However, during the in-field usage, such device is very unreliable. The narrowed beam will have very large stress in it during in-field operation, and this may cause the device lifetime due to fatigue failure becomes much shorter. For a non-BISR device, this can be a serious reliability threat. However, for a BISR device, even if one or two modules have such combined defects, other modules can still be connected to guarantee the device function. Thus the reliability of the BISR device can still be very high.

We assume a cyclic loading of $360g$ acceleration is applied and simulated both the non-BISR, WBC and EFC BISR devices for the above combined defects. The simulation results are shown in Table 3. The lifetime of non-BISR device is estimated to be about 6.0×10^5sec, which is about 6.9 days. Thus material fatigue failure does become a serious threat to the reliability of the non-BISR device when such combined defects exist. The fatigue lives of BWC/unpowered-EFC/powered-EFC BISR devices becomes 5.47×10^{33}sec, 2.04×10^{38}sec and 1.08×10^{10}sec separately. However, due to the redundancy repair, the material fatigue lifetimes of BWC/unpowered-EFC/powered-EFC devices remain about 10^{92}sec, 10^{81}sec and 10^{111}sec separately, which are far more than enough for normal operation. There may be still some other combined defects cases in which the non-BISR device will have a serious reliability problem, while the BISR device can still maintain high reliability due to redundancy repair. In reality, due to the imperfection in device fabrication process, even after careful manufacturing testing, there may be still some hard-to-detect defects in the released "good" devices. Even worse, some new defects may still be developed during in-field usage. Some of these defects may be a potential threat to the device lifetime due to material fatigue. Thus, it is necessary to have the redundancy repair to ensure the high device reliability.

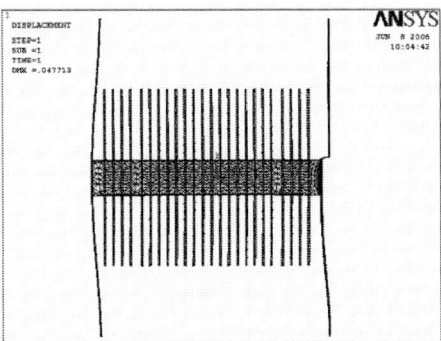

Figure 7. The accelerometer with combined defects (stiction + narrowed-beam).

Table 3. The reliability analysis for accelerometers with combined defects.

Items	non-BISR device	BWC BISR module	unpowered EFC module	powered EFC module
Displacement	$0.38\mu m$	$0.11\mu m$	$0.09\mu m$	$0.38\mu m$
Maximum stress σ_f	713.9MPa	202.6MPa	164.14MPa	693.05MPa
No. of cycles to failure N_f	2.4×10^9	5.47×10^{36}	2.04×10^{41}	1.08×10^{10}
Beam lifetime (MTTF) (sec)	2.4×10^6	5.47×10^{33}	2.04×10^{38}	1.08×10^{10}
device lifetime (MTTF) (sec)	6.0×10^5	$\sim 10^{92}$	$\sim 10^{81}$	$\sim 10^{111}$

4. Conclusion and Future Research

In this paper, we analyzed the reliability for poly-silicon MEMS comb accelerometers under material fatigue failure mechanism due to long term cyclic loading. Based on ANSYS stress simulation and reported S-N curve of poly-silicon material, the mean-time-to-failure (MTTF) lifetimes and failure rates for both BISR and non-BISR poly-silicon MEMS comb accelerometers are derived. The reliability functions for both BISR and non-BISR designs are proposed. Simulation results show that the fatigue lifetime of MEMS accelerometers made by poly-silicon material can be good enough for general purpose applications. That is, poly-silicon can be a good MEMS structure material in terms of material fatigue. However, for some "weak" devices with certain structure defects, material fatigue may become potential threats. Further, material aging may also degrade the device long-term reliability. Hence, it is necessary for introducing redundancy repair to further enhance MEMS device reliability. Compared to non-BISR design, BISR MEMS accelerometer demonstrates effective reliability improvement due to redundancy repair. In the future, the MEMS reliability under material fatigue failure mechanism for other MEMS materials, such as single-crystal-silicon, Al, SiO_2, Si_3N_4, will be further studied. The MEMS reliability under other failure mechanisms such as stiction, mechanical fracture, particle contamination, will also be explored.

References

[1] B. Stark (editor), "MEMS Reliability Assurance Guidelines for Space Applications", Jet Propulsion Laboratory Publication 99-1, Pasadena, USA, Jan. 1999.

[2] J. A. Walraven, "Failure mechanisms in MEMS", *Proceedings of IEEE International Test Conference, ITC 2003*, pp. 828-832, 2003.

[3] D. M. Tanner, J. A. Walraven, K. Helgesen, L. W. Irwin, F. Brown, N. F. Smith, and N. Masers, "MEMS reliability in shock environments", *Proceedings of IEEE International Reliability Physics Symposium*, San Jose, CA, USA, pp. 129-138, Apr. 10-13, 2000.

[4] D. M. Tanner, J. A. Walraven, K.S.Helgesen, L. W. Irwin, D. L. Gregory, J. R. Stake, and N. F. Smith, "MEMS reliability in a vibration environment", *IEEE 38th Annual International Reliability Physics Symposium*, San Jose, California, USA, pp. 139-145, 2000.

[5] W. M. V. Spengen, R. Puers, and I. D. Wolf, "On the physics of stiction and its impact on the reliability of microstructures", *Journal of Adhesion Science and Technology*, Vol. 17, No. 4, pp. 563-582, 2003.

[6] X. Xiong, Y. Wu, and W. Jone, "Reliability Analysis of Self-Repairable MEMS Accelerometer", Proc. of 21th IEEE International Symposium on Defect and Fault Tolerance in VLSI Systems (DFT'06), pp. 236-244, Arlington, VA, USA, Oct. 2006.

[7] J. M. Illston, J. M. Dinwoodie, and A. A. Smith, "Concrete, Timber and Metals", Van Nostrand Reinhold, Crystal City, VA, 1979.

[8] M. Tabib-Azar, K. Wong, and W. Ko, Aging Phenomena in heavily doped (p+) micromachined silicon cantilever beams, *Sensors and Actuators A*, Vol. 33, pp. 199-206, 1992.

[9] K. E. Peterson, "Silicon as a mechanical material", *Proceedings of IEEE*, Vol. 70, Issue 5, pp. 420-457, 1982.

[10] J. A. Connally and S. B. Brown, "Slow crack growth in single-crystal silicon", *Science*, Vol. 256, pp. 1537-1539, 1992.

[11] C. L. Muhlstein, S. B. Brown and R. O. Ritchie, "High-cycle fatigue of single-crystal silicon thin films", *Journal of Microelctromechanical Systems*, Vol. 10, pp. 593-600, 2001.

[12] H. Kahn, R. Ballarini, A. H. Heuer, "Dynamic fatigue of silicon", *Sensors and Actuators A: Physical, Current Opinion in Solid State and Material Science*, Vol. 8, pp. 71-76, 2004.

[13] C. L. Muhlstein, R. T. Howe, and R. O. Ritchie, "Fatigue of polycrystalline silicon for microelectromechanical system applications: crack growth and stability under resonant loading conditions", *Mechanics of Materials*, Vol. 36, pp. 13-33, 2004.

[14] J. Bagdahn and W. N. Sharpe Jr., "Fatigue of poly-crystalline silicon under long-term cyclic loading", *Sensors and Actuators A: Physical*, Vol. 103, pp. 9-15, 2003.

[15] X. Xiong, Y. Wu, and W. Jone, "Design and analysis of self-repairable MEMS accelerometer," *Proceedings of the 20th IEEE International Symposium on Defect and Fault Tolerance in VLSI Systems (DFT'05)*, Monterey, CA, USA, pp. 21-29, Oct. 3-5, 2005.

IEEE International Symposium on Defect and Fault Tolerance of VLSI Systems

Fault-Tolerance with Graceful Degradation in Quality: A Design Methodology and its Application to Digital Signal Processing Systems

Nilanjan Banerjee, Charles Augustine, Kaushik Roy
Purdue University
{nbanerje,caugust,kaushik}@purdue.edu

Abstract

The tremendous increase in device density of present day designs is accompanied by a corresponding increase in transistor failures (hard faults), posing a major challenge to current fault tolerant techniques and tools. We propose a novel "design-time" fault-tolerance methodology at architecture/circuit levels to improve the reliability of applications, where it is possible to classify computations into two categories- (i) those which contribute to quality degradation and, (ii) those which result in total system failure. The proposed scheme enhances system reliability by making appropriate trade-offs between area, output quality (signal to noise ratio or mean square error), and fault tolerance. This low-overhead generic methodology is suitable not only for scaled CMOS technologies, but is also applicable to future nanotechnologies (carbon nanotubes etc.) as well, where such defects are expected to be prevalent. We evaluated this technique on a widely used DSP system – Finite Impulse Response (FIR) filters (where minor degradation in quality can be tolerated). Results show that our technique achieves an improvement between 73.4%-450% (in terms of total system failure probability under iso-redundancy) compared to conventional fault tolerance techniques.

1. Introduction

The radical scaling in feature sizes and increase in integration density has been successfully achieved by CMOS technology over last few decades. However, this trend faces serious challenges, due to fundamental physical limits and increasingly complex manufacturing processes. In fact, for scaled technologies (32 nm and below), defect probability of devices (source-drain opens and shorts, gate oxide opens etc.) are increasing rapidly due to limitations in lithography. New technologies such as carbon nanotubes (CNTs), polysilicon nanowires (NWs), and molecular transistors are therefore being sought as a possible replacement to silicon. However, such technologies have their own set of limitations. Whether they are assembled lithographically (CNTs etc.) or self-assembled chemically (molecular arrays), they experience severe unreliability due to imprecision in fabrication/self-assembly processes [1]. Therefore, the necessity to cope with intrinsic hard failures (such as transistor shorts/opens) must be recognized as a key aspect of nano-scale systems design [1-4]. A likely scenario in future technologies is to design functional units with certain degree of fine-grained, built-in immunity to such hard failures. This feature would enable systems to tolerate a certain degree of defects and still function with reasonable levels of accuracy. *This kind of pervasive fault tolerance based on implicit acceptance that a certain percentage of devices may fail in a random fashion will require considerably different approaches from classical fault-tolerance methods.*

Conventional strategies for system design consist of hierarchical characterization of several levels of abstraction, from device to architecture, with intrinsic verification methods and tools for each level. Of course, the assumption for existing verification methods is that the probability of a device defect is inherently very low. With highly unreliable components (such as the ones expected in the nanotechnology regime), *the design methodologies will have to take into account the reliability of devices as a design consideration.* Adding redundancy to the devices often helps in reducing the device level uncertainties. However, only device level redundancy incurs significant area/power overhead. To guarantee a high level of

1550-5774/08 $25.00 © 2008 IEEE
DOI 10.1109/DFT.2008.43

reliability in presence of defects under a fixed area/power budget, there is a need to develop "intelligent" redundancy schemes [5, 6]. In this paper, we introduce a redundancy scheme that attacks this problem at several levels of abstraction - architecture and circuit levels.

It should also be noted that all hard faults do not necessarily result in system failures in all applications. For a large class of applications, especially in digital signal processing domain, it is possible to classify

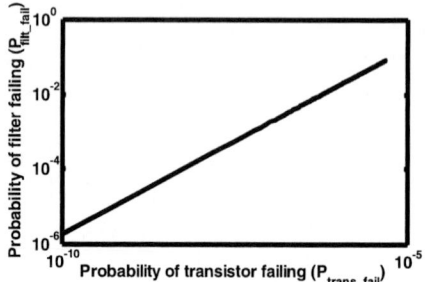

Fig.1. Catastrophic failure probability (CFP) vs. transistor defect rate

computations into "critical/less-critical" based on their contributions towards the overall output quality (say, in terms of peak-signal-to-noise-ratio (PSNR)). Faults in critical sections lead to drastic degradation in system output. We define such drastic degradation of the output by the term "catastrophic failure". Conversely, faults in the less-critical sections may only lead to degradation in output quality (say PSNR) and not cause a "catastrophic failure". We designate such failures as "quality degradation". We apply our redundancy scheme to such systems and investigate trade-offs in quality and catastrophic failure resiliency under severe defect densities for two scenarios:

 i) Fault tolerance with no quality (or PSNR) degradation
 ii) Fault tolerance with degraded output quality (within allowed limits).

To distinguish among various terms described above, we henceforth use the following definitions to denote faults at device/system levels:

a)Hard Faults/ Device Defects: open/short faults at transistor level,

b) Catastrophic Failures/Quality Degradation: System level failures

The drastic increase in catastrophic failure probability (CFP) with increase in transistor defect density for a high-pass FIR filter (Fig. 1) reinforces the necessity of developing fault-tolerant techniques which allow systems to operate under minor quality degradation without suffering catastrophic failures.

2. Fault-tolerance methodology

In this section, we provide a top-down overview (Fig.2) of the fault-tolerance scheme proposed in this paper. The basic premise behind this approach is that there exists a wide class of applications where it is possible to significantly improve the fault resilience against complete system failure (catastrophic failures (CF)) by allowing an acceptable deterioration in the output quality. This technique is well-suited for most DSP algorithms where approximate signal processing is possible by suitably demarcating computations into two classes -- critical and less-critical – based on output quality requirements. Therefore, as a first step, we obtain the specifications of the system and the quality requirements from the user. We then partition the system to separate the critical computations from the non-critical ones through a sensitivity analysis. Of course, the size of the critical/less-critical sections is solely dictated by the user-specified quality requirements. In the second step, we implement the architecture for this partitioned system. It might be possible to further improve reliability of critical sections by segregating critical parts within the critical section itself (like most significant bits of adders, multipliers etc.). After proper identification of critical sections, we selectively introduce transistor level redundancies in the system. By selective redundancy we imply that under a fixed redundant-area overhead, our scheme provides more redundant transistors for critical sections and less (or no) redundant transistors in the less-critical sections. This is in contrast to conventional fault tolerance schemes where redundancy is distributed uniformly across all sections of a design. Since fault-tolerance capability in the critical sections is increased considerably by our methodology, the possibility of preventing catastrophic failures is remarkably increased. For the case studies in this paper, we assumed all the redundancy in the critical sections, with the non-critical sections having no redundant transistors. This in fact can therefore be interpreted as a "design time" fault tolerance technique since the methodology is embedded in the design flow without any overhead in design time (as will be shown in the case studies). This scheme significantly improved circuit/system robustness in presence of manufacturing defects *(hard faults) with acceptable quality degradation (if any, due to failures in the non-critical sections).* Partitioning into critical/non-critical sections is specific to system and should be tailored accordingly. We evaluated this methodology for several designs and it is discussed in following sections.

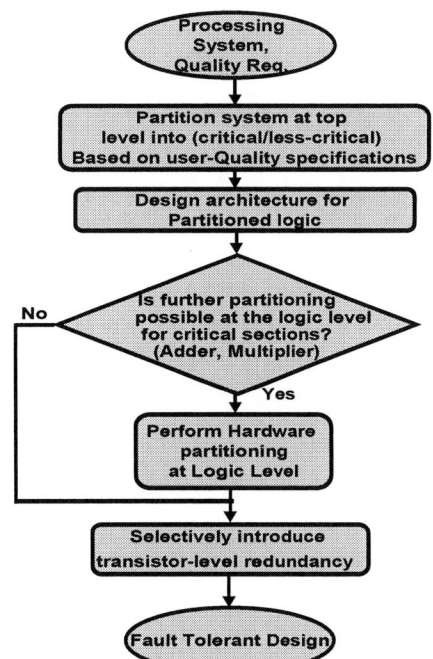

Fig. 2: Top-Down Fault Tolerance Methodology

2.1 CFP estimation in original/partitioned architectures

To justify that architecture partitioning results in minimization of CFP under a fixed area devoted to redundancy, we perform failure estimates for the original and the partitioned designs. We assume that *any hard failure in critical sections of the partitioned design has a catastrophic effect on the output response, whereas any defect in less-critical sections would result in having a minor effect on quality.* Following are notations/assumptions that constitute this probability model:

- P_s, P_o = probability of short and open, respectively, for a transistor. For simplifying analysis at architectural level, we assume that there can be either opens or shorts in a particular technology and the probability is equal to P. We also assume that since the original design is not partitioned into critical/non-critical sections, CFs can occur from failures in any transistors.

- Probability of short/open failures with two transistors in series or in parallel (one of the transistors is an added redundant transistor), respectively, is P_{red}

- Number of transistors in the original architecture: N_{orig}

- Number of transistors in the partitioned architecture: N_{part} with the critical/less-critical sections having $N_{critpart}$ and $N_{noncritpart}$ transistors respectively, so that $N_{critpart} + N_{noncritpart} = N_{part}$

CFP of the original system with no redundancy = $(1-(1-P)^{Norig})$, where $(1-P)^{Norig}$ is the probability of system working without CFs. With, say, *A%* area (uniformly distributed across the "critical section") devoted for redundancy, the number of redundant transistors is *Norig*A/100.* Therefore, the number of transistors protected by redundancy = *Norig*A/100.* For the 2-transistor system (parallel or series connected transistors as described in the second bullet), the equivalent failure probability is P_{red} *($P_{red} << P$).* CFP of the new system is given by = (1- working probability) = $(1 - (1-P_{red})^{(Norig*A/100)} \bullet (1-P)^{Norig(1-A/100)})$ (1)

With the preferential redundancy scheme, all redundant transistors are introduced in the critical sections of the partitioned architecture. The CFP for new architecture is computed as:
(1- working prob. of critical sections) = $(1 - (1 - P_{red})^{(Norig*A/100)} \bullet (1-P)^{(Ncritpart-A*Norig/100)})$ (2)
where, *A*Norig/100* redundant transistors are now present in the critical sections. The ratios of the CFPs of (2) and (1) is given by and should be much less than 1:

$$\frac{CFP_{part}}{CFP_{orig}} = \frac{(1 - (1 - P_{red})^{(Norig*A/100)} \bullet (1-P)^{(Ncritpart-A*Norig/100)})}{(1 - (1 - P_{red})^{(Norig*A/100)} \bullet (1-P)^{Norig(1-A/100)})} << 1 \quad (3)$$

From eqn. 3, we infer the following:

1) As long as $N_{critpart} < N_{orig}$, we have an improved CFP. This implies that based on quality specifications, the system should be partitioned efficiently with minimal area overhead. This is non-trivial because most conventional designs (where there exists no explicit demarcation between critical/less-critical sections) tend to maximize the optimization of hardware resources by sharing across all sections. Consequently, it is difficult to separate critical from less-critical sections without introducing significant hardware overhead, which in turn can offset CFP improvements obtained through partitioning. Also, relaxing quality constraints allow reduction in the number of transistors in the critical sections, as discussed in Section 4. It is worth mentioning that in certain instances the original architecture might have clearly demarcated critical/less-critical sections. The preferential redundancy scheme provides even more advantages in this case since no extra area is incurred in partitioning.

2) Develop transistor level techniques for obtaining $P_{red} << P$

In the following section, we discuss how redundancy at transistor level can reduce the error probabilities. In the subsequent sections, we develop low/no-overhead architecture and logic techniques for partitioning a system/circuit into critical/less-critical sections.

3. Defect mechanisms and tolerance with transistor-level redundancy

Under high defect densities, redundancy at transistor level is mandatory to achieve high confidence in terms of fault tolerance. In this section, we introduce how manufacturing defects create short/open faults. We then discuss redundant transistor configurations and their impact on fault tolerance.

3.1 Defects in Silicon Processes

We analyze defect mechanisms in Si-lithography processes and propose various transistor configurations to avoid these defects.

A. Source Drain shorts: Due to manufacturing imperfections, source/drain shorts can occur if there is a short between source and drain terminals of transistors, or if the transistor terminals in vicinity of Vdd/ground are shorted to these connections (Fig. 3(a)). Shorting to Vdd/ground is denoted by stuck at faults where terminals shorted to Vdd are said to be stuck at '1' and ground shorted terminals are called stuck at '0'. Such failure probabilities can be improved using a redundant transistor in series ("series redundancy") (Fig. 3(b)).

B. Source/drain/gate opens: Opens occur if drain, source, gate terminals or their contacts and/or polysilicon layers have an open connection. Parallel connection of two transistors (Fig.3(c)) effectively reduces the failure probability in this case. This is called "parallel redundancy".

C. Gate Oxide shorts: These defects consist of unwanted conducting paths through the gate oxide of a MOSFET from the gate material to the substrate or drain and source regions (short between diffusion and polysilicon layers). The proposed transistor level redundancy shown in Figs. 3(b) and 3(c) are not able to handle such defects.

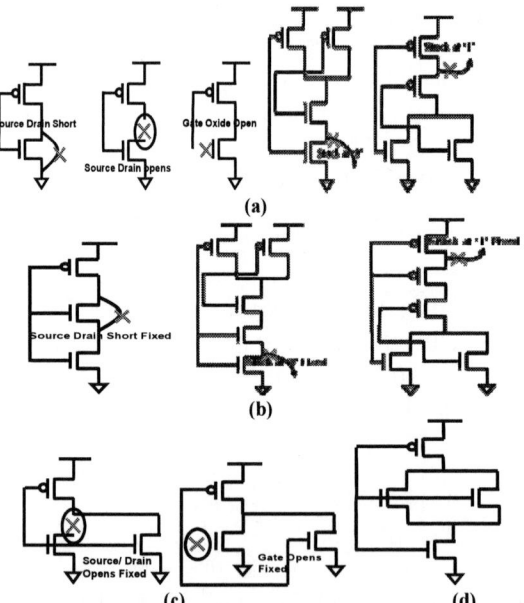

Fig. 3: Transistor defects (a) Short, Open defects (b),(c) Series and Parallel Redundancy reduces failure prob. (d) Additional redundancies improve defect resiliency at significant overhead -2-parallel and 1-series

326

D. Resistive Bridges: The bridging defects result from various reasons e.g., poor patterning, narrow metal regions near via, missing salicides, inadvertent drop of metal between wires and so forth. In extreme cases, they can represent an open or a short fault. For example, zero resistance between two connections indicates a short whereas infinite resistance denotes an open connection. Besides these two extreme cases, the proposed transistor-level redundancies are not able to overcome bridging errors.

3.2 Different redundancy configurations at the transistor level

We analyze parallel/series connections of transistors to determine their effectiveness in achieving fault tolerance. The following additional terms will be useful in doing the analyses:

P_{o_eq}, P_{s_eq}: equivalent probability of open/shorts in the modified transistor configurations

P_{work_eq}: working probability of modified configurations in presence of both opens/shorts assuming they are independent $= (1 - P_{o\_eq})(1 - P_{s\_eq})$

Series Redundancy for Short Faults: To ensure correct functionality of the shorted inverter (Fig. 3(a)), we provide a series connected transistor (Fig. 3(b)). We evaluate equivalent error probabilities for this configuration as:

$$P_{o\_eq} = 2 \times P_o - P_o^2, \quad P_{s\_eq} = P_s^2, \qquad P_{work\_eq} = (1 - P_{o\_eq})(1 - P_{s\_eq}) = (1 - P_s^2)(1 - (2 \times P_o - P_o^2))$$

In this case, we find that for shorts the probability of failure for the transistor system shown in Fig. 3(b) is reduced considerably. However, the failure probability in case of opens increase. *This configuration is therefore effective for technologies more prone to short defects.*

Parallel Redundancy for Open Faults: Parallel redundancy (Fig. 3(c)) reduces failure probability in presence of opens. Since the parallel connected transistors have equal individual probability of opens/shorts, the failure probability of the new inverter structure is:

$$P_{o\_eq} = P_o^2, \quad P_{s\_eq} = 2 \times P_s - P_s^2, \qquad P_{work\_eq} = (1 - P_{o\_eq})(1 - P_{s\_eq}) = (1 - P_o^2)(1 - (2 \times P_s - P_s^2))$$

This case is complementary to series redundancy where probability of open faults is considerably lessened but probability of short faults increase. *This configuration, therefore, is suitable for designs where open faults are more predominant.* In both series/parallel redundancy cases, failure probability in presence of both opens and shorts marginally increase than the nominal case, given by $(1 - (1 - P_{o\_eq})(1 - P_{s\_eq}) = P_{o\_eq} + P_{s\_eq} - P_{o\_eq} * P_{s\_eq})$. Therefore, to achieve fault tolerance in presence of both open and short faults, we need other configurations that are resilient to both opens and shorts, but incurs higher area overhead.

3 Tx – 2 parallel plus 1 in series: In this case (Fig. 3(d)), open fault occurs when either the lower transistor is defective, or when both the top transistors have open defects (lower transistor is defect-free). On the other hand, short occurs when i) the lower transistor and one of the top transistors (Fig. 3(d)) have shorts or, ii) all three have short defects. The equivalent fault and working probabilities are evaluated as follows:

$$P_{o\_eq} = P_o \times (1 - P_o^2 + P_o), \quad P_{s\_eq} = P_s^2 \times (2 - P_s), \quad P_{work\_eq} = (1 - P_{o\_eq})(1 - P_{s\_eq}) = (1 - P_o \times (1 - P_o^2 + P_o))(1 - (P_s^2 \times (2 - P_s)))$$

In the above case we find that fault probabilities in presence of shorts and opens are both considerably reduced. The probability of correct operation in presence of both opens/shorts improves considerably if we introduce additional redundancy but it introduces considerable area overhead.

3.3 Applicability to emerging non-Si nanotechnologies

Emerging device technologies such as CNTs, nanowires (NWs), molecular electronics etc. hold promise of dramatically higher integration densities coupled with reduced power dissipation. Researchers have successfully demonstrated several high-density electronic devices including CNTs, NWs and SETs. These nanotechnologies can be assembled lithographically (CNTs etc.) or self-assembled chemically (molecular arrays). However, they experience severe unreliabilities due to: (i) imprecision in fabrication/self-assembly processes, (ii) reduced tolerance to noise and thermal perturbation [1]. For instance, CNT-based designs often suffer from shorts due to misaligned CNTs and the inability in the current technologies to ensure 100% semiconducting tubes for channels [1]. The main component of a CNFET is a single-wall carbon nanotube (SWCNT); its conductance is determined by so-called chirality of the tube and is extremely hard to control during manufacturing. Conducting nanotubes can lead to defective CNFETs similar to source-drain shorts. Most molecular electronics consist of logic arrays of orthogonal sets of wires with molecular switches at intersection of these wires, which might result in a lot of open defects. They are usually fabricated using bottom-up techniques to obtain complex 2-D meshes of

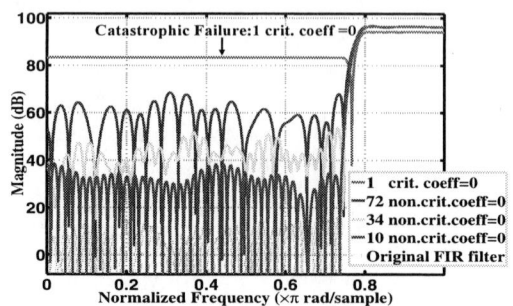

Fig. 4: Filter response with increasing no. of zero coefficients

wires. However, a fine-grained control of average spacing between wires and wire density is often not possible. This is because both spacing/density are dictated by mechanical processing variables like exposure rate/duration etc., which cannot be deterministically controlled in complex processes. In fact, it has been observed that 2-D wires are rarely equidistant from one another, and parallel wires often go askew or intersect causing open and short faults [7]. Redundancy in form of series/parallel transistors can improve failure probabilities in this case. In fact, since defect densities in such technologies are higher than CMOS, transistor level redundancies are absolutely essential.

4. Architecture partitioning: Impact on fault tolerance

The proposed fault tolerance technique is applicable to all systems where minor quality deterioration is acceptable. We demonstrate the effectiveness of above methodology by applying it to widely used DSP application FIR filter. As mentioned before, determining critical/less-critical sections based on sensitivity analysis needs to be customized for each application. Also, it is essential to define quality metric for any system, which can vary across applications. In following subsections, we explain sensitivity analysis, quality metrics and determine CFP versus quality tradeoffs for FIR filters.

4.1 Case Study: Finite Impulse Response (FIR) filter

The output of a transposed-form N-tap FIR filter [8], is given by the following expression:

$$Y[n] = c_0 * x[n-1] + c_1 * x[n-2] + c_2 * x[n-3] + .. + c_{n-1} * x[0] \qquad (1)$$

where, $c_0, c_1, ..., c_{n-1}$ denote coefficients and $x[n-1]...x[0]$ pixel inputs. Computational complexity of multipliers is reduced by Common Subexpression Elimination (CSE) methods [9], where all coefficient multiplications of transposed-form FIR filter is replaced by single multiplier block (Fig. 5(a)). The common subexpressions in coefficients are exploited to share adders mitigating hardware complexity/reducing power consumption.

In FIR designs, some coefficients play a more important role in shaping the filter response compared to others. If some of the not-so-critical coefficients are not computed (set to zero), the response is marginally affected without drastic degradation of filter characteristic. This property makes FIR filters ideal candidates for proposed fault-tolerance technique and enables us to partition the design into critical/less-critical sections based on quality requirements. The quality metric for FIR filters is defined by "normalized pass and stop band ripple" [8]. We follow a two-step approach to determine critical sections for a user-specified quality: 1) Conduct a sensitivity analysis of each coefficient on the output, where each coefficient is individually set to zero and the effect on the pass/stop band ripple is noted. Once this is tabulated, the coefficients are ordered from the least to the most critical ones based on their effect on pass (stop) band ripple. 2) Set increasing number of coefficients to zero starting from least critical and tabulate the impact on the output. This helps us identify the total effect on the pass (stop) band ripple (and hence, the quality) by setting a group of coefficients to zero. *It should be noted that retaining less-critical coefficients (in case of no defects) improves quality of the filter.*

We took a 121-tap high-pass FIR filter [10] with following specifications: Fstop=0.74 and Fpass=0.8 (of normalized sampling frequency), Passband peak-to-peak ripple = 0.080dB and Peak stopband ripple = -80.3 dB. We recorded magnitude response of original/altered filters by setting increasing number of coefficients (10 to 72) to

Table 1. Failure Probability of original/modified architectures with different coefficients in each group (crit/less-crit)

Design Spec	Passband Ripple (Norm.)	Stopband Ripple (Norm.)	Less Crit. Coeff.	Adders		P_{filt_fail} (P_{trans_fail} =0.5*10^{-6})
				Crit	Less-Crit	
Original	1.0	1.0	N/A	57	N/A	0.043
Modified (Values norma-Lized w.r.t. original Design)	1.0	1.05	9	55	2	0.038
	1.0	1.11	17	52	6	0.036
	1.0	1.17	22	48	10	0.029
	1.0	1.25	27	46	13	0.027
	1.0	1.33	28	45	15	0.014
	1.0	1.43	32	38	20	0.013
	1.0	1.54	33	37	21	0.012
	1.0	1.67	36	36	22	0.006

P_{filt_fail} = **Catastrophic Fail. Prob. of Filter, P_{trans_fail}= Tran. Fail. Prob.**

zero starting from least critical coefficient. The results show that filter quality is minimally affected when 10 least impacting coefficients are set to zero. However, the magnitude response gets gradually degraded with increasing number of zero-valued coefficients. Therefore, when less-critical coefficients are affected by hard faults, they can be set to zero and lead to minor "quality degradation" of the filter characteristic. On the other hand, if the critical coefficients suffer from hard faults and they are set to zero (even 1 single critical coefficient, Fig. 4), it results in a "catastrophic failure". Also, this analysis shows that in order to maintain a minimal passband ripple and a stopband attenuation ≥ -50dB (say, for instance this was the minimum requirements specified by the user), 87 coefficients (from Fig. 4) are critical (needed to maintain quality) and 34 coefficients are less critical (improved quality).

4.1.1 Partitioning and Probabilistic Analysis

Even after identification of critical/less critical coefficients, the challenge remains in proper partitioning of circuit with minimal area overhead for CFP improvements. This is because FIR systems are normally designed using CSE ([9], Section 4.1) where hardware is shared among all coefficients, and any random scheme for partitioning this hardware based on critical/less-critical coefficients can cause significant area increase. Fig. 5(a) shows the concept behind conventional CSE-based filter implementation, and 5(b) shows the "proposed/modified" CSE-based partitioning method. In the modified scheme, we divide the

coefficient set into two subsets {critical, less critical} based on quality requirements of the user. Each subset is henceforth referred by the term "*group*". Each group therefore, contains one or more coefficients and is denoted by G_1 and G_2. The standard CSE technique is then modified to implement critical group first (G_1 before G_2). To minimize hardware in proposed modified CSE algorithm, critical groups share resources with not-so critical groups *but not vice versa*. To elucidate this further, consider the two groups G_1 (critical) and G_2 (less critical). As shown in Fig. 5(b), first, G_1 is independently implemented using adders and shifters with maximum amount of sharing using CSE. G_2 is then implemented and some of the hardware from G_1 can be reused (through CSE again) to form the set G_2. The technique partitions hardware

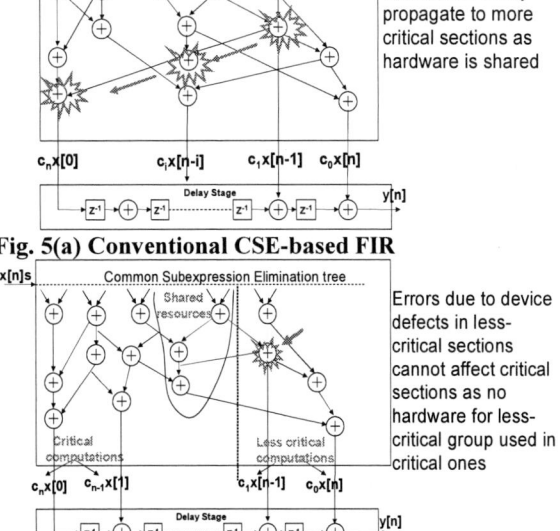

Fig. 5(a) Conventional CSE-based FIR

Fig. 5(b) Proposed HFTCSE-based FIR

responsible for "catastrophic failures" from those resulting in "quality degradation".

The "Hard Fault Tolerant CSE" (HFTCSE) algorithm is implemented using CSE. The primary difference between HFTCSE and CSE [9] is that HFTCSE implements common-subexpressions to the groups according to their priority (G_1 ahead of G_2). Also while performing CSE for G_2, it checks whether G_2 coefficients can be derived from the already computed subexpressions of G_1. HFTCSE code, developed in C++ generated optimum hardware for original/partitioned filter designs in VHDL and was then synthesized using Synopsys Design Compiler. The synthesized output was converted to Hspice netlist. We computed CFP for original /modified filter design and results are summarized in Table 1 and in Fig. 6.

One very interesting observation obtained from these experiments is the improvement of CFP of the partitioned FIR architecture compared to the original un-partitioned case even without the introduction of any redundancy (Table 1 and Fig. 6). For critical sections, only a sub-set of the total number of coefficients needs to be computed. Therefore hardware requirements and the number of transistors reduce, minimizing CFP. *Standard CSE-based FIR allows resource sharing among all coefficients and hence, hard failures in less-critical coefficients could drastically affect the filter characteristics (because of error propagation in shared hardware).* Fig. 6 plots filter failure probability (P_{filt_fail}) under different device (P_{trans_fail}) defect probabilities. As expected, HFTCSE provides higher CFP and is operable under higher defect probability with minor quality deterioration.

We also compared CFP improvement with uniformly distributed redundant transistors (conventional case) to the partitioned architecture with redundant transistors only in the critical sections. The failure probabilities under "iso-redundancy" with varying transistor defect probabilities are shown in Fig. 6. The amount of redundancy (area of the number of redundant transistors) was incremented from 10%-50% (of original system area) for both original/modified architectures, assuming the architecture to have either "open" or "short" faults. The CFP of both original and partitioned 121-tap filters improves with increase in the number of redundant transistors. However, our proposed technique significantly improves CFP compared to the original design. In fact, as the pass band ripple and stop-band attenuation requirements are further relaxed, we see a remarkable improvement in the failure probability with respect to catastrophic failures (450% improvement for 50% area devoted to

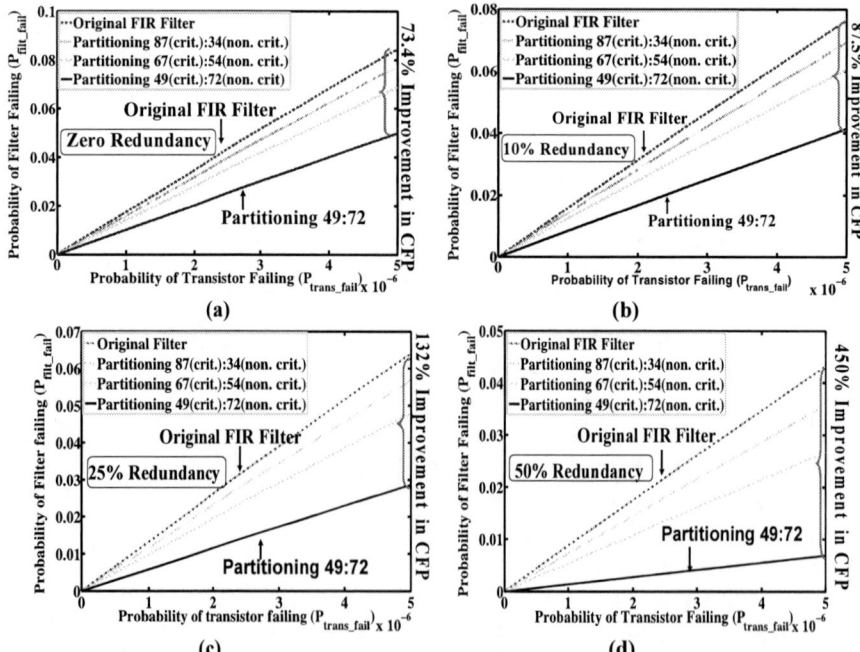

Fig.6: CFP with iso-area redundancy of original/partitioned designs (Crit: Critical, Non-crit: Non-Critical)   (a) 0% (b) 10% (c) 25% (d) 50%

redundancy, Fig. 6(d), where, *%improvement = (CFP$_{old}$-CFP$_{new}$)*100/CFP$_{new}$)*. This occurs because size of critical sections (number of important coefficients, Fig. 4) in partitioned design reduces with lower quality requirements. This in turn allows more redundant transistors for fewer components in the proposed scheme, reducing CFP. In case of high defects densities for both opens and shorts, alternative transistor configurations (2-redundant transistor case, Fig. 3(d)) can be employed along with architecture partitioning. But this is achieved under larger area overhead.

We also evaluated the overhead associated with our fault tolerance technique. The results show that number of adders required by the partitioned FIR architecture is larger than the number for the original case (Table 1). Also the number increases with relaxed pass/stop band ripple requirements. This is because the original architecture is implemented with conventional CSE, where opportunities of sharing exist across all coefficients. On the other hand, HFTCSE method is used for partitioning designs where the critical sections are implemented independently of less-critical ones, reducing sharing opportunities. As the number of less-critical coefficients become larger (relaxed quality), it further reduces sharing and furthers increment in number of adders.

Two important aspects need to be noted at this point. *Separating design into critical/non-critical sections and setting the non-critical coefficients to zero (in presence of failures/defects) can be misinterpreted as a filter with reduced number of taps. However, this is not true since designing a filter with just less number of taps will still retain the possibility of hardware sharing among the critical/less-critical computations. A failure in less-critical sections can propagate into critical coefficients resulting in complete system failure. Similar is the case when our technique is compared with coefficient truncation, a popular technique to reduce computational complexity of filter designs. As in the previous case, truncating coefficients does not prevent hardware sharing among all coefficients and retains the risk of getting affected by failures in less-critical sections.*

5. Conclusion

We provide a fault-tolerant design methodology at architecture/circuit levels to improve the reliability of applications, where it is possible to classify computations into those which result in minor quality degradation and those which result in total system failure. The basic idea behind this approach is to effectively segregate the critical/less-critical computations of a design based on pre-specified output quality requirements, and provide preferential redundancy to the critical components. We applied this method to several DSP applications, where robustness with respect to hard faults was significantly improved.

6. References

[1] Mishra et al., "Scalable Defect tolerance for Molecular Electronics", 1st NSC Workshop, 2002.

[2] Bhadhuri et al., "Using quantum model of computation for reliability evaluation of defect tolerant nano-architectures", IEEE Nano 2004, pp. 622-624

[3] Jacome et al., "Defect Tolerant Probabilistic Design paradigm for Nanotechnologies, DAC 2004, pp. 596-601.

[4] Tahoori et al., "Defects and Fault Tolerance in QCA at Nano scale", VTS 2004, pp. 291-296.

[5] Huang et al., "On the Defect Tolerance of Nano-Scale Two-Dimensional Crossbars" DFT 2004, pp. 96-104.

[6] Goldstein et al., "NanoFabrics: spatial computing using molecular electronics", ISCA 2001, pp. 178-191.

[7] Stan et al., "Molecular Electronics: From Devices and Interconnect to Circuits and Architecture" IEEE Proceedings, Vol. 91, No.11, 2003.

[8] J.G. Proakis et al., "Digital Signal Processing: Principles, Algorithms and Applications", Prentice Hall, NJ, 1996.

[9] C. Yao et al., "A Novel CSE Method for Synthesizing Fixed-Point FIR Filters", TCAS-1, 2004, pp. 2211-2215.

[10] Y.C. Lim et al., "Discrete coefficient FIR digital filter design based upon an LMS criteria," IEEE TCAS, volume 30, 1983, pp. 723-739.

[11] R.C.Gonzalez et al., "Digital Image Processing", Prentice Hall, 2002

[12] B. W. Johnson, "Design and analysis of fault tolerant digital systems," Addison Wesley Publishing, 1989.

[13] Y. M. Hsu, "Concurrent error correcting arithmetic processors," PhD dissertation, UTexas at Austin, 1995.

[14] Sirisantana et al., "Enhancing Yield at the End of the Technology Roadmap", IEEE Design and Test of Computers, 2004.

Design Space Exploration for the Design of Reliable SRAM-based FPGA Systems

Cristiana Bolchini, Antonio Miele

Dip. Elettronica e Informazione – Politecnico di Milano

P.zza L. da Vinci, 32 – 20133 Milano – Italy

{*bolchini, miele*}*@elet.polimi.it*

Abstract

1 Introduction

In the last decade, several studies have been proposed to mitigate the effects of Single Event Upset (SEU) faults in SRAM-based FPGAs, proposing periodic scrubbing of configuration bitstreams [1], triplication and voting [2, 3], information redundancy via Error Detecting And Correcting (EDAC) codes [4, 5], or duplication with comparison [6], when erroneous data can be discarded and recomputed.

These reliability-oriented techniques based on fault detection and/or correction can be efficiently coupled with partial dynamic reconfiguration [7] to achieve SEU mitigation properties [6, 8]; yet, the task is not an easy one. In fact, when designing a reliable system, the designer usually selects one of the available techniques and applies it to either the entire system (using a coarse grain granularity), or to every module in it (using a fine grain granularity); in a few cases only, a solution in between is adopted, and is designed ad hoc, based on the designer's knowledge of the various modules. In general, to the authors' knowledge, there is no available tool for exploring the solution space corresponding to intermediate granularity levels, where sets of modules are clustered together and the reliability oriented technique is applied to the entire cluster as if it were a unique component.

Thus, given the combination of available techniques and the possible granularity levels they can be applied at, the designer faces too many solutions, that can hardly be compared manually; our previous work [6, 8] was based on the generation of a large number of promising solutions, to be then analyzed with respect to a few parameters, requiring a significant time in terms of effort. Furthermore, changing the metrics taken into account to investigate different trade-offs would require the process to be rerun.

The main contribution of this paper is the introduction of an automatic design space exploration approach, based on genetic algorithms, for reliable systems implemented on SRAM-based FPGAs, exploiting classical fault tolerance techniques coupled with partial dynamic reconfiguration. The paper is focused on problem modeling, in order to define a flexible framework that can be easily extended and integrated with other mitigation techniques and metrics. In particular, the proposed methodology models the problem of reliable system design and identifies significant figures of merit for comparing different solutions; such set of parameters can be further personalized based on the designer's experience and needs in the specific situations.

It is worth noting that the particular platform here considered, introduces yet another degree of complexity, since even though these techniques are applied correctly, from a functional point of view, an opportune, ad-hoc placement and routing process is necessary to guarantee that the functional detection/correction properties hold after FPGA mapping. The problem we tackle aims at exploring the solution space and estimating costs. During the implementation phase, then, it will be necessary to adopt suitable approaches, such as the RoRA one [9]; the authors have developed a reliability-oriented place and route algorithm for guaranteeing that, in the specific case of the TMR technique, no SEU fault causes multiple modules to fail, thus invalidating the adopted reliability technique.

The paper is organized in the following way. Section 2 lays the motivations and background for the proposed methodology and framework, identifying the requirements that have led to their definition and implementation. Section 3 introduces the framework and its characterizing elements; an in-depth discussion of the adopted metrics is presented in Section 4. The proposed approach has been applied to some test circuits to evaluate its effectiveness; the results are discussed in Section 5. Concluding remarks and future work are summarized in the last section.

2 Motivations and Background

The problem of partitioning the system into blocks of logic for applying the TMR fault mitigation technique has been previously investigated in [3]. The authors observe that by using different partitioned groups of logic, here called clusters, as the basis for triplicating and voting, different levels of overheads can be achieved, in terms of area, performance and robustness. In the presented case study, the authors select three different schemas, namely "maximum logic partition", "medium logic partition" and "minimum logic partition", and compare the achieved solutions with respect to the identified parameters.

In our own experience, given a circuit such as Noekeon [10], the trivial solutions corresponding to (a) the triplication of the entire system, which we will label "minimum logic partition" to be aligned with the previous approach, or to (b) the triplication of each

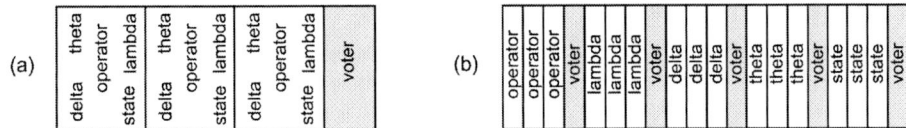

Figure 1. (a) minimum logic partition, and (b) maximum logic partition.

333

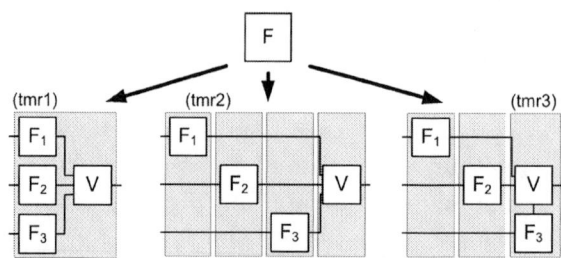

Figure 2. Application of the TMR with different types of mapping.

component in the system, labeled "maximum logic partition", do not identify the most interesting alternatives the designer has. More precisely, there are 52 possible partitionings of the five components, which to apply the TMR to, leading to different area and performance characterization, especially due to the impact of voters, whose dimension depends on the number of lines they monitor; in particular, the most interesting ones are a few solutions corresponding to clusters constituted by two components, as later discussed in Section 5. An additional aspect that comes into play in the present context is the fact that the different replicas can be mapped onto different independently reconfigurable areas in different ways (Figure 2), used by the partial dynamic reconfiguration. In particular, voters not only perform the majority vote, but also identify the faulty replica, so that it is possible to trigger a specific re-writing of the corrupted configuration bits, where necessary. The number of possible solution thus farther grows, reaching the total number of 1866 alternatives (in the specific example) to be compared and evaluated. As a result, additional aspects need be taken into account with respect to [3], and the issue of identifying the most convenient solution with respect to the interesting parameters becomes even more complex.

In this paper, the framework applies the 1-Dimensional (1D) reconfiguration possibility, supported for Xilinx's Virtex2PRO Family [11, 7], yet the problem has been modeled and the framework implemented to support also 2D reconfiguration for more recent platforms and associated design flows.

The overall scenario thus sees the need for a way to explore the solution space by evaluating different partitioning solutions, aiming at identifying the optimal application of reliable techniques coupled with partial reconfiguration on the system specification in order to obtain a hardened FPGA-based system, comparing possible alternatives from different points of view. The input information consists of the circuit to be implemented, the available set of reliability-oriented techniques and the parameters the designer wants to analyze; the expected output is the (set of) solution that best fits the given cost function. The next two sections present these aspects in detail.

3   Methodology Overview

Figure 3 presents an overview of the proposed framework; it consists of three main steps performing 1) an analysis and characterization of the nominal system specification in input, 2) the design space exploration for obtaining a reliable version of the system, and 3) the implementation and the synthesis of the identified reliable system. By following the core based design approach commonly adopted for FPGA-based system design, the circuit is specified through a structural description composed by functional units interconnected by signals; the adopted language is VHDL, restricted to the subset for structural specifi-

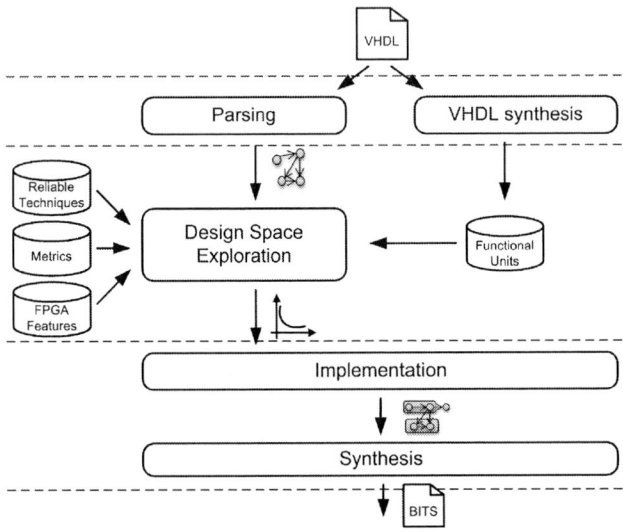

Figure 3. The proposed reconfiguration for reliability framework.

cation. Moreover, the description is organized hierarchically, where each functional unit can be modeled with another structural description or is considered as a black box, if the functionality is described with a process, and wrapped in a new component.

In the first step, the VHDL description is parsed to build an agile internal representation based on graphs where nodes represent functional units and edges interconnections. At the same time, an automatized synthesis process of the VHDL is performed (by using Xilinx's tool-chain) for determining the implementation costs and the characteristics of each type of functional unit composing the structural description; all information is stored in a •• • ••• • • • •• • •••• ••• •• ••• ••.

The second step represents the core of the proposed methodology and aims at performing the design space exploration for identifying the most promising hardened versions of the considered circuit. Together with the internal representation of the considered circuit, this step takes in input several repositories containing information necessary for performing the design space exploration; they are described in the following paragraphs.

Functional Units' Repository. It contains the descriptions of all the classes of functional units of the current design, together with those of standard units (the most relevant are TSC components such as voters and checkers) and of the FPGA components (such as tri-state buffers, I/O buffers and hardwired multipliers). Each description consists of the specification of the interface (ports) and the list of required FPGA resources (e.g., the number of slices, BRAMs and hardwired components).

Device features' repository. It stores a description of all the various characteristics and features for each considered FPGA device, including the available resources; in particular, the number of available basic reconfigurable cells, in terms of slices, is considered together with the height and width of the grid of basic cells composing the device. Moreover, the constraint on the reconfigurable areas' shape is stored: when considering the module-based 1D reconfigurability approach, each reconfigurable area has a rectangular shape with a height equal to the device grid and a variable width multiple of a fixed number of columns (usually 4) [7]; therefore, the repository entry contains the basic area size, and an indicative maximum number of reconfigurable areas.

Techniques' Repository. It contains a model of each reliable technique provided in the framework for hardening the system. The techniques are modeled in terms of the required modification on the circuit specification and binding on FPGA resources. In this work, three different implementation schemata of the TMR technique have been considered: as shown in Figure 2, the three techniques modify the specification by triplicating the part of the circuit and by introducing a voter of the correct size; they differ from each other by the imposed mapping constraints. In fact, the first version maps the three replicas and the voter onto the same reconfigurable area, the second one maps the four modules on different reconfigurable areas while the last one clusters the voter with one of the three replicas.

Metrics' Repository It contains a set of metrics the exploration phase uses for estimating in advance characteristics and costs of the hardened versions of the circuit.

The solution space is composed by all reliable implementations of the input circuit, obtained by applying the TMR (and in the future a wider set of techniques) at different granularity levels on the nominal specification. As shown in Figure 4(a), the strategy adopted for generating a reliable solution consists of two steps: first, the functional units of the circuit are partitioned into several clusters; then, on each cluster, a specific technique selected from the repository is applied. It is worth noting that not all possible combinations of clusters and techniques are feasible: in fact, it may happen that for a solution some constraints, such as the ones defined by the FPGA characterization (device dimension or number of reconfigurable areas) are not satisfied, and thus, this alternative is discarded.

A multi-objective strategy has been adopted for performing the design space exploration; the aim is to define a versatile approach to lead the exploration in several directions for finding different solutions, all interesting but with different properties. For the same reason, objective functions (usually one or two) are not predefined but can be specified as the minimization (or maximization) of a single metric or of a combination of metrics available in the repository, thus giving the designer the opportunity to specify the figures of merit to be used for expressing and comparing costs and benefits, on the basis of his/her necessities. Because of the multi-objective approach, the method adopted for comparing different solutions is the Pareto dominance: when minimizing with respect to all figures of merit $\{1, ..., n\}$, a solution vector g is better than solution vector h if and only if $\forall i \in \{1, ...n\} : g_i \leq h_i \wedge \exists j : g_j < h_j$. It is worth noting that the Pareto dominance is defined as a partial order relationship; therefore, when considering two solutions k and j, i) k may dominate j or ii) j may dominate k or iii) k and j may be incomparable (or indifferent).

A multi-objective genetic algorithm, called SPEA2 [12], has been adopted for implementing the design exploration engine. The choice of a genetic algorithm is motivated by the fact

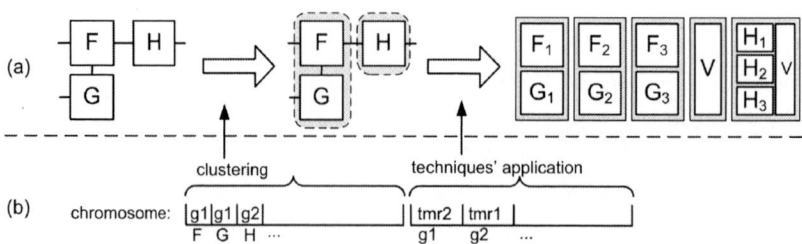

Figure 4. Strategy for generating a reliable solution: (a) the approach, and (b) the solution representation.

that this kind of evolutionary strategies is widely used for search and optimization problems due to its performance and efficiency. When considering the presented approach for generating a reliable version of the system, the solution is coded in a chromosome composed by two different parts (shown in Figure 4(b)). The first part, characterized by a number of positions equal to the number of functional units in the circuit, is used for specifying the cluster which each functional unit, considered as a black box, is assigned to. Each position of the second part of the chromosome represents a cluster of components and its value is the ID of the technique applied to it. The size of the second part of the chromosome is set as the maximum number of clusters for which there is a feasible solution. Evolution operators do not present particular aspects in their implementation:

The **selection operator** is defined by the SPEA2 algorithm with an approach based on the Pareto dominance;

The **mutation operator** has been implemented to flip the value contained in a random position of the chromosome, implying either a change in the assignment of the cluster of a functional unit or in the type of technique applied to a cluster;

The **crossover operator** has been implemented as a single point crossover which cuts in a specific point the chromosomes of the two parents and recombines subparts for generating new children.

The standard workflow of the genetic algorithm is composed by the generation of an initial population of feasible solutions and the iteration of the execution of the three operators on the population for a given number of generations; unfeasible solutions produced during the evolution are tagged with an infinite cost and are automatically discarded by the algorithm during subsequent generations. Finally, at the end of the execution, the set of solutions – computed on the final population – which are not dominated by any other one (the • • •••• •••• •) represents the set of best solutions achieved by the algorithm with respect to the specified figures of merit.

During the last step of the framework flow, the designer manually analyzes the structure and the characteristics of the best solutions, computed by the algorithm, by means of all available metrics and chooses the preferred one to be implemented. The VHDL description of the final solution is then automatically generated and synthesized by following the reconfigurable system design flow [13, 14]. In this phase, a reliability oriented place and route strategy [9] should be taken into account.

A first framework prototype has been implemented for supporting the automation of the design space exploration; it is available for download at [13]. Current efforts are devoted to integrating the proposed framework with reconfigurable system design flow proposed by the Earendil methodology [13, 14].

4 Design Metrics

We have identified a set of metrics that are usually taken into account when comparing different solutions; in particular, we have selected and refined the ones adopted in our previous works and in the Earendil methodology; nevertheless, should the designer wish to add new ones, the proposed framework allows it, through the • ••••••• •••••••••.

R_{fu_set}: it represents the number of resources required for implementing a set of functional units mapped onto the same reconfigurable area; it is computed as the sum of implementation costs obtained by the initial synthesis of the functional units. Currently, it is represented by the number of slices but it can be extended by considering other FPGA resource types.

D_{rec_area}: it represents the dimension of a reconfigurable area, estimated as the minimum rectangular area satisfying the reconfigurable systems' implementation constraints [7] and containing resources for implementing the set of functional units mapped onto.

R_{system}: it represents the sum of resources required for implementing the entire system.

D_{system}: it represents the dimension of the system, obtained as the sum of D_{rec_area} for all reconfigurable areas belonging to the system.

N_{areas}: it represents the number of reconfigurable areas contained into the system. It is constrained by an upper bound specified in the FPGA characterization.

$ro\text{-}index$: it is a fragmentation index, previously introduced in [6], aiming at estimating the degree of actual utilization of a reconfigurable area' resources. It is computed as ration between R_{fu_set} and D_{rec_area} of a specific area.

These parameters evaluate the solutions in terms of the global impact of the techniques' application on the system and from a local point of view by estimating the characteristics of each reconfigurable area of the system. The designer can use them for performing an overall evaluation of a solution, or can choose a subset of them, for defining the objective functions for the design space exploration.

5   Experimental Session

The proposed framework has been tested on two circuits, a Noekeon cipher and a FM Digital Signal Demodulator, running two sessions using different figures of merit. The aim of the first session (labeled with •) is to achieve a final system characterized by small and balanced areas, quickly reconfigurable in case of a fault, with limited internal fragmentation; the adopted objective functions maximize the average ro-index and the minimize a linear combination of the normalized average area dimension and its standard deviation. In the second session, the goal is to achieve a small system (minimization of the system dimension), still with small and balanced areas (minimization of the linear combination used in the first session). Figures 5 and 6 show the experimental results; marked points represent the optimal solutions.

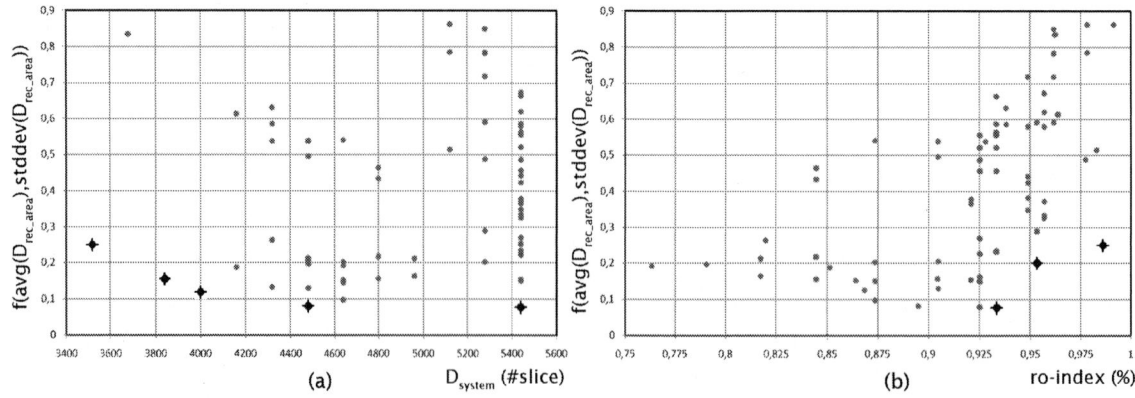

Figure 5. Experimental results on the Noekeon cipher circuit

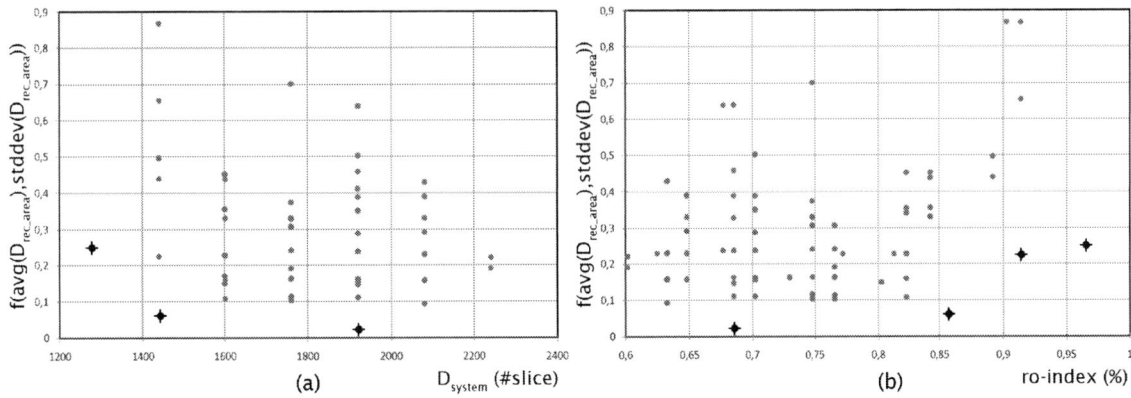

Figure 6. Experimental results on the FM Digital Signal Demodulator circuit

For the example introduced in Section 2, in session (a) on the Noekeon circuit one of the most promising solutions corresponds to the following partitioning: Cluster 0 includes only component `theta`; Cluster 1 includes components `operator`, `delta` and `state`; and finally Cluster 2 includes component `lambda`. The TMR is applied with schema •• •• (Figure 2) to Cluster 0 and also to Cluster 2, while •• •• is applied to Cluster 1. There are 6 independently reconfigurable areas, having a dimension between 800 and 1120 slices, for a total cost of 4862 slices (and a system dimension of 5440). The trivial solution of triplicating the entire system, mapping each replica on different areas (•• ••) for a combination of figures of merit falls into the set of optimal solutions, characterized by a low total area, but having bigger reconfigurable areas, such that recovery from a fault will require a wider action.

From these experiments, as well as from others here not reported, it is possible to draw some conclusions on the proposed framework.

Number of possible solutions. The application of the TMR at different levels of granularity produces a relevant number of possible solutions, differing with respect to the considered parameters; thus an automatic framework such as the proposed one, is necessary to perform the solution space exploration.

Different figures of merit identify different optimal solutions. The identification of different parameters to characterize the possible solutions is a difficult task, considering that there are some dependencies; the selected set of parameters lead to different optimal solutions thus giving the designer an instrument to drive the exploration space.

Optimal solution filtering. The designer can further reduce the number of solutions of the Pareto front (totaling at most 6 in our examples) by filtering on desired cut-off values of the objective functions; for instance, if the designer is interested in solutions with a balanced trade-off among the several objective function, s/he can delete the extreme points of the Pareto front.

6 Conclusions

The paper presented a design space exploration approach for the implementation of SRAM-based FPGAs able to mitigate the effects of SEU faults, both in their configuration and application memory cells, achieved by applying standard fault tolerance techniques and exploiting the partial, dynamic reconfiguration. We have identified a set of figures of metric that can be used to compare the various possible solutions, derived by applying the Triple Modular Redundancy at different levels of granularity, i. e., to the entire system or to every module composing the system, or to clusters of modules.

Results show that the number of available solution is quite relevant, and the solution is rarely a trivial one, thus motivating the use of such automatic tool. Among the several identified solution(s) according to the designer's selected important parameters, the framework produces the modified VHDL of the selected solution including the necessary redundancy and the additional information required for correctly mapping the system onto the FPGA, distributing it in the identified reconfigurable areas.

Present work is devoted to extend the set of available reliability oriented design techniques, and to allow the designer to the express his/her own set of figures of merit and design constraints, to explore the solution space.

References

[1] C. Carmichael, E. Fuller, P. Blain, and M. Caffrey. SEU mitigation techniques for Virtex FPGAs in space application. In • • • • • • • •••••, page 24, 1999.

[2] F. Lima Kastensmidt, G. Neuberger, R. Hentschke, L. Carro, and R. Reis. Designing Fault-Tolerant Techniques for SRAM-Based FPGAs. ••• • • ••••• • •••••••••• ••••••, 21(6):552–562, Nov./Dec. 2004.

[3] F. Lima Kastensmidt, L. Sterpone, L. Carro, and M. Sonza Reorda. On the Optimal Design of Triple Modular Redundancy Logic for SRAM-based FPGAs. In • •••• • •••• •• • ••••• • •••• •••• ••• ••• •• ••••••, pages 1290–1295, 2005.

[4] C. Bolchini, F. Salice, and D. Sciuto. Designing self-checking FPGAs through error detection codes. In • •••• •• • • •••• •••• • •••••• •••• •• ••••••• •• • ••• • •••••• •, pages 60–68, 2002.

[5] P. Kubalik and H. Kubatova. Design of self checking circuits based on FPGA. In • •••• •••• •••• • •••• • •••••••••••• •• • , pages 378–381, 2003.

[6] C. Bolchini, D. Quarta, and M. Santambrogio. SEU Mitigation for SRAM-Based FPGAs through Dynamic Partial Reconfiguration. In • •••• • • •••• • •••• •••• ••• ••••• •• • •••, pages 55–60, 2007.

[7] Xilinx Inc. • •••• • ••••• • •••••• •••••••••• • ••••. Xilinx Inc., 2006.

[8] C. Bolchini, A. Miele, and M. Santambrogio. TMR and Partial Dynamic Reconfiguration to mitigate SEU faults in FPGAs. In • •••• •• • • •••• ••• •••• • ••••• •••• •••• ••••••• •• • ••• •••••• •, pages 87–95, 2007.

[9] L. Sterpone and M. Violante. A New Reliability-Oriented Place and Route Algorithm for SRAM-Based FPGAs. ••• • ••••••• ••• ••• •••••, 55(6):732–744, 2006.

[10] J. Daemen, M. Peeters, G. Van Asshe, and V. Rijmen. Nessie proposal: Noekeon.

[11] Xilinx Inc. • •••••• ••• ••• • ••••••• ••• • •••• • •••• • ••••, 2005.

[12] E. Zitzler, M. Laumanns, and L. Thiele. SPEA2: Improving the Strength Pareto Evolutionary Algorithm for Multiobjective Optimization. In • •••••••••• • ••••••• ••• • ••••••• • •••• ••••••• ••• • ••••••• ••• •• •••••••••• ••• ••••••••• • •••••, pages 95–100, 2002.

[13] DRESD: Dynamic Reconfigurable Embedded System Design' research group. http://www.dresd.org/.

[14] V. Rana, M. Santambrogio, and D. Sciuto. Dynamic reconfigurability in embedded system design. In • •••• •• • • ••••• ••• •••• • •••••• ••• •••••• ••••• •, pages 2734–2737, 2007.

SESSION 7

ERROR DETECTION AND CORRECTION (1)

IEEE International Symposium on Defect and Fault Tolerance of VLSI Systems

A Low Cost Scheme for Reducing Silent Data Corruption in Large Arithmetic Circuits

Abhisek Pan	James W. Tschanz	Sandip Kundu
Univ. of Massachusetts	*Intel Corporation*	*Univ. of Massachusetts*
Amherst, MA	*Hillsboro, OR*	*Amherst, MA*
apan@ecs.umass.edu	*james.w.tschanz@intel.com*	*kundu@ecs.umass.edu*

Abstract

Aggressive scaling of CMOS transistors in last four decades has resulted in circuits with progressively higher packing density, increased switching speed, and higher power density. However in future, CMOS technology nodes are predicted to suffer from greater intermediate to long-term reliability and circuit marginality problems. To address these problems researchers have proposed the usage of redundant circuits to detect and, in some cases, to correct transient or permanent field failures. The proposed solutions target 100% of circuit errors but are expensive in terms of area and more importantly power overhead, some times exceeding 200%. In this paper, we investigate a flexible, lower-overhead error detection scheme that provides trade-off between reliability and circuit overhead in terms of area, performance, and power. Simulation studies on a 32-bit multiplier circuit show that this scheme can provide greater than 90% fault coverage with 15 to 20% area overhead including overhead for the comparator circuitry. Additionally, the redundant portion of the circuit can be turned off when concurrent checking is not critical, resulting in power savings. Currently, mainstream desktop processors use low-cost error detection schemes in busses, IOs and sometimes in embedded memories to reduce likelihood of silent data corruption (SDC). The discussed approach provides significant reduction in likelihood of silent data corruption in generally unprotected arithmetic circuits at a small cost.

1. Introduction

Aggressive scaling of CMOS transistors in last four decades has resulted in circuits with progressively higher packing density, increased switching speed and higher power density. Since transistors are already operating near the limits of their tolerance, this race for creating smaller and faster systems are making the systems more susceptible to failure in presence of external influences or intrinsic design and manufacturing inadequacies [1]. Therefore future CMOS technology nodes are predicted to suffer from greater intermediate to long-term reliability and circuit marginality problems [2].

A circuit, while in operation, can fail due to two kinds of faults: permanent faults, which once developed continue to manifest themselves in a consistent manner, and transient faults, which appear for a small duration and may or may not appear intermittently.

Permanent faults can occur because of conductor electro-migration, broken wires, device degradation due to NBTI and PBTI, gate-oxide charging and threshold voltage instability due to hot carrier effects [13]-[15]. A reasonable amount of security from data corruption due to permanent errors can be achieved by continually testing the circuit throughout its lifetime using testers. Transient faults in a circuit can occur due to intrinsic causes such as high on-chip temperature and supply-line noise, or because of external influences like strikes by high-energy heavy ions and neutrons from cosmic rays, low energy alpha particles from radioactive decays, low-energy thermal neutrons [16]-[20]. A strike by such an energetic external entity can cause the values of single or multiple bits in the circuit to flip temporarily

1550-5774/08 $25.00 © 2008 IEEE
DOI 10.1109/DFT.2008.42

343

(single-bit upsets or multiple-bit upsets). The error due to transient faults can severely compromise system integrity through silent data corruption (SDC) if the system does not have concurrent error-detection capabilities.

Extensive research has already been carried out in order to device novel techniques to equip the digital circuits with concurrent error detection and recovery capabilities [3]-[12]. Most of the proposed techniques are similar in their goals of achieving 100% tolerance from potential circuit errors. However the solutions often come with considerable area, performance and power cost, sometimes exceeding 200%, making the manufacturers wary of incorporating the techniques in commercial chips. In contrast, low overhead sub-optimal schemes such as the one investigated here can provide feasible alternatives if we can achieve acceptable concurrent error-detection using them. They can provide effective trade-off between reliability and circuit overhead in terms of area, performance and power.

Our goal in this paper is to achieve concurrent error detection for arithmetic circuits at the lowest cost. Frequency of arithmetic operations is high and transistor count in arithmetic circuits such as multipliers and dividers is large. Thus, protection of arithmetic circuits can significantly reduce overall SDC rate in a processor. The proposed scheme is based on modulo arithmetic. We test our proposed scheme on a multiplication unit, which forms an important part of the data path circuit of processors today, and is used extensively in signal and data processing, control, and cryptography.

The rest of the paper is organized as follows: in section II we review some of the previous work in this field, in section III we outline our scheme and the points of investigation, in section IV we describe our simulation methodology, in section V we report the simulation results, and conclude the paper in the last section.

2. Related work

Nicolaidis et al [21] presented an efficient self-checking implementation of array-based multipliers using differential logic circuits to implement each cell of the array. Each cell is shown to have a hardware overhead of about 50% with respect to standard implementations in terms of transistor count. Additional overheads for duplicating the inputs and placing a double-rail checker at the output are also considered.

Nicolaidis et al [22], [23] proposed fault secure parity predicted booth multipliers, which were improved upon by Marienfeld et al [24] to propose code-disjoint self-checking output-duplicated Booth-2 multipliers with improved error detection capabilities and lower overhead. The scheme proposed by Marienfeld et al [24] incur a hardware overhead of 38 % to 43 % for 16 bit multipliers and 28 % to 35 % for 32 bit multipliers. These schemes are fault-secure for single stuck at faults, but involve the use of considerable hardware overhead. These schemes are targeted towards specific implementation of the multiplier circuit. Moreover, the computational circuits are composed of self-checking modules that cannot be tuned on and off.

We propose an modulo-checking based error detection scheme, which (i) provides a high level of fault security instead of total fault security, but requires lower hardware overhead, (ii) is independent of the multiplier implementation and provides flexibility in terms of deciding the extent of fault security the circuit may have.

Modulo-checking has been studied previously in [27]-[28]. It was shown by Wakerly, that a low-cost modulo-n code can detect all unidirectional multiple errors that affect fewer that n bits [29]. However, our focus is on general data-path circuits that may have bidirectional errors. In [30], a similar scheme has been used as part of an integrated low-cost mechanism to protect microprocessor pipeline and memory from long–term reliability problems. By contrast, this study focuses on cost efficiency of modulo checking in terms of concurrent error detection coverage vis-à-vis overhead.

3. Proposed Scheme

For a positive integer n, two integers a and b are called congruent modulo n if a and b, when divided by n, leave the same remainder, which is denoted by

$$a \equiv b \ (\text{mod } n) \tag{1}$$

Here a becomes the remainder or the common residue of the modular division of b by n (denoted by $a = b$ (mod n)) when a is the least non-negative integer among those satisfying (1) (n is called the group length).

Relation 1: If $a_1 \equiv b_1$ (mod n) and $a_2 \equiv b_2$ (mod n), then $a_1 \times a_2 \equiv b_1 \times b_2$ (mod n)　　[Chapter 3,[26]]

The relation is valid for other arithmetic operations of addition and subtraction also, but we concentrate on multiplication for our simulation studies. Let us consider a k-bit multiplier, with k-bit inputs $i1$ and $i2$, and 2k-bit output $o1$. It follows from the above discussion that for any integer n, the product of $i1$(mod n) and $i2$(mod n) is equal to $o1$(mod n). So, for the multiplier, if the modulo operations on the inputs and the output is carried out and the above relation 1 is checked to be untrue, it can be said that the output multiplier is incorrect. However, if the equality is satisfied, all we can say is that the output belongs to the set of integers {... (p − 2n), (p − n), p, (p + n), (p + 2n) ...}, where p is the product of $i1$ and $i2$. Hence, the error detection scheme will fail if the output is erroneous but still belongs to the above set, which is referred to as *aliasing*.

Now, if the same error detection operation is done modulo m, m being co-prime to n, and an error is said to be detected if relation 1 is violated during either of the two modulo checking operations, the set of possible outputs for which the scheme fails changes to ...(p − 2mn), (p − mn), p, (p + mn), (p + 2mn)... that is, the interval between the individual elements in the set increases from m to mn, reducing the chance of aliasing. So, clearly, by deciding upon how many such modulo operations to carry out on the circuit and with what numbers, we can control the extent of error-detection the scheme can offer. From the circuit point of view, each modulo checking will be implemented through a separate checker independently, and we can employ more than one checker in parallel to increase the fault security provided by the circuit. The schematic diagram for the circuit considering a single checker module is shown in Figure 1.

The computational overhead for the scheme involves the remainder calculation for both the inputs and the outputs, modulo multiplication of the inputs and comparing the result of the modulo-multiplication with the output residue. To reduce the computational effort for the remainder calculation, we restrict the choice of moduli into the Mersenne Prime numbers, which are of the form $2^p − 1$ where p is a prime. Choosing moduli in the form $2^p − 1$ greatly simplifies the residue calculation (low-cost residue codes), and choosing them to be co-prime decreases the probability of aliasing.

For modulo n calculation, we need to consider successive chunks of p bits from the input and convert them to modulo n format (if the p bit have all ones, convert them to all zeroes, otherwise leave unchanged) and add all the bit-chunks modulo n to get the modulo n equivalent of the input. The schematic representation is shown in Figure 2. The algorithm to calculate modulo n for a k-bit binary number, where $n = 2^p − 1$ is given in Figure 3. We implemented the modulo-multiplier circuit through Wallace tree type multipliers with end-around carry. For the comparator module, we used arrays of XOR and NOR gates to arrive at the final single bit checker output, which ideally would be high if the product is correct, and should be low otherwise.

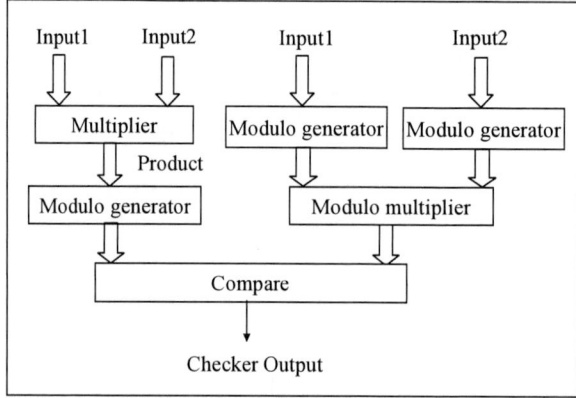

Figure 1: Self-Checking Multiplier Circuit

Although modulo checking provides a low-cost convenient checking methodology, it is not comprehensive because it does not provide 100% error detection. In [31], it was reported that for a 4X4 bit two's-complement multiplier with modulo-3 checking, error-detection probabilities were 97% for single bit error, 52% for double errors, and 74% for triple errors. However, errors in the output do not occur with uniform probability and depend upon the possible faults in the multiplier circuit implementation. Hence we perform simulation studies to estimate the fault coverage provided by different modulo schemes for our multiplier circuit. Similar studies, done for modulo-3 checking scheme on 4X4 bit to 12X12 bit multipliers estimated the fault coverage to be around 93%, much higher than the error-detection probabilities [31].

The error-detection scheme described above, by its very nature, is independent of the structure of the multiplier circuit. This allows for power savings opportunity by turning on or off of the checker on demand. Thus error detection can be dynamically traded against power savings.

4. Simulation methodology

For our studies, we have considered Wallace tree-based multipliers. Wallace multipliers form a class of parallel high-speed multipliers consisting of AND gates, tree of carry-save adders and carry-propagate adders at the output. A 4X4 bit Wallace tree multiplier is shown in Figure 4.

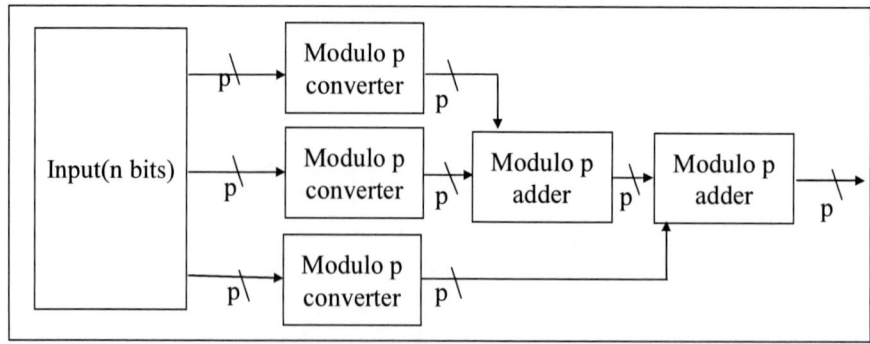

Figure 2: Modulo Calculation Scheme

```
Algorithm: modulo n calculation
Input: k bit binary string in[(k-1)...0],  integers: n, p
Output: p bit binary string

i ← 0, j ← 0
while (k−i ≥ p+1) do
  if (in[(i+k)...i] = [11...1]) do
     temp(j)←[00...0]
  else do
     temp(j)←in[(i + k)...i]
  i ← i+p
  j ← j+1

if (k − i)  do
  temp(j) = in[00..(k − 1)...i]
else do
  j ← j−1

while (j)  do
  temp (j−1)←[(temp(j)+temp(j−1))mod p]
  j ← j−1

return temp(j )
end
(temp is binary number of k bits)
```

Figure 3: Modulo Calculation Algorithm

The simulation study is described through the flow-chart of Figure 5. For simulation, a structural Verilog model of the 32X32 bit multiplier was constructed. This Verilog netlist was read into a commercial fault simulation tool [25] and a fault list generated by the tool for single stuck at fault model was applied to the netlist. Then, random sets of ten input patterns were generated at a time through a random pattern generation program and the fault simulation for the multiplier circuit was carried out. The list of the faults covered in the simulation was taken out. Next, the Verilog model for the error checking circuit was wrapped around the multiplier. This self-checking circuit was again fault simulated with the same patterns. The fault list comprised of the faults in the multiplier circuit and additional faults for

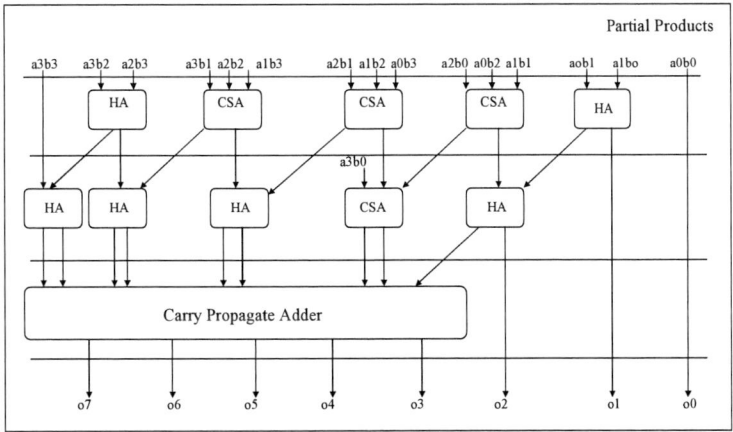

Figure 4: Wallace Multiplier

the extra circuitry added. The primary output that was observed was the single output of the checker circuit. The expected output of the checker circuit is always one, when the operation produces correct result. In the presence of a fault, if the actual product deviates from the correct value, ideally the checker output should also change to zero. We note that, this is exactly the condition required by the fault simulator to declare that a fault has been detected or covered. If the checker output remains at one even when the product changes from the correct value (which can happen due to aliasing, as was remarked before), the checker fails to perform its intended function. Correspondingly the fault simulator tool also fails to detect the fault, as the actual checker output does not deviate from the expected output. Hence, it is obvious that the fault security provided by the checker is has a direct correlation with the percentage fault coverage reported by the fault simulator tool. This process was repeated for a thousand times in total. Fault coverage for modulo 3, modulo 7, modulo 31, and modulo 127 checkers is reported.

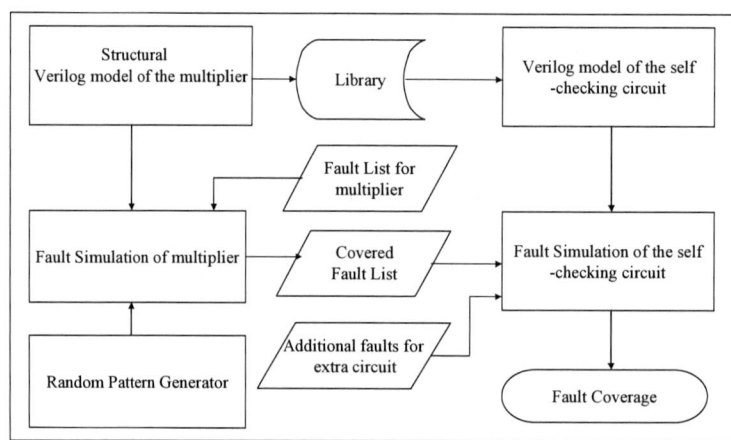

Figure 5: Simulation Flow

5. Simulation results

The area overhead for the different modulo schemes for the 32-bit multiplier, considering equivalent gate-counts is shown in Table 1. The data for the schemes presented here include the area overhead for input and output modulo-generators and the comparator.

TABLE 1: AREA OVERHEAD COMPARISON FOR 32-BIT MULTIPLIER

Detection Scheme	Mod 3	Mod 7	Mod 31	Mod 127	Nico-laidis [23]	Mari-enfeld [24]
Area Overhead %	14.5	16	18.7	21.7	46 to 76	28 to 35

Considering the fault-secure property of our scheme, we present simulation results for the individual modulo schemes in Figures 6-9. The relative distribution of fault coverage values is plotted in Figure 10. We find that for modulo-3, the minimum fault coverage for any pattern-set is 88%, and the mean is around 93%. For modulo-7, the figures improve marginally to 89% and 94% respectively. For mod-31 the mean improves considerably to 97% and remains increases slightly for mod-127. Hence, there is a marked improvement in fault-coverage while going from mod-7 to mod-31, which makes it worthwhile to bear the slight extra overhead for using up to mod-31 checkers at-least. The improvement does not rise appreciably for higher moduli, as seen in the mod-127 scheme that we report.

6. Conclusion

Silent data corruption due to transient errors is expected to rise due to growing transistor count, reduced supply voltage, increased noise levels and higher soft-error rate in highly scaled CMOS technologies. Processor designers are reluctant to use fault-tolerant design

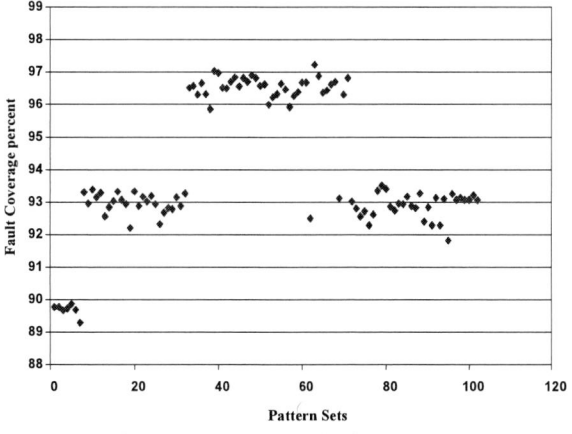

Figure 6: Modulo-3 fault coverage

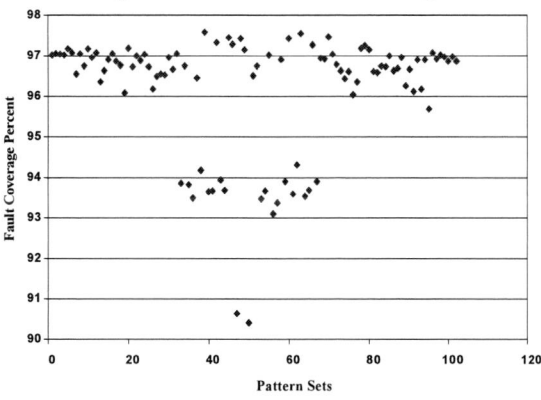

Figure 7: Modulo-7 Fault Coverage

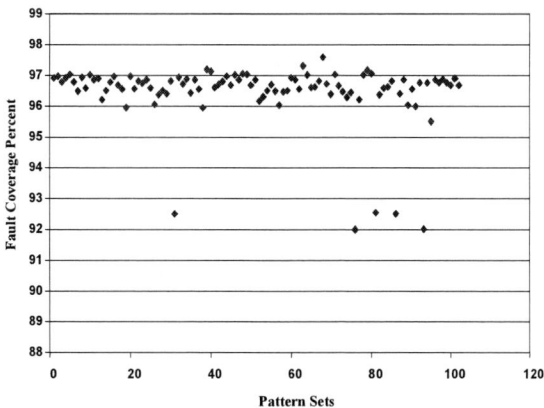

Figure 8: Modulo-31 Fault Coverage

techniques due to high area and power overhead. In this paper, we demonstrated that a modulo arithmetic technique can reduce SDC by over 90% on an average while adding less than 15% hardware overhead that also includes the comparator circuitry. Increasing the overhead by further 5% can improve detection capabilities by more that 5%. The power overhead of the proposed scheme can be further reduced due to separable nature of the checker whereby the checking can be turned off when error detection is not needed. The extra circuitry does not affect the actual implementation of the multiplier, freeing up designers from any constraints arising out of the checker. The simulation results show fault coverage improvement does not scale well for higher-overhead schemes. The separable nature of the scheme ensures that the error-checking operation does not impact performance of the main multiplier output. This scheme is also amenable as a BIST checker. In summary, the proposed scheme improves protection against SDC without any adverse effect on design or performance, making it a practical choice for arithmetic circuits.

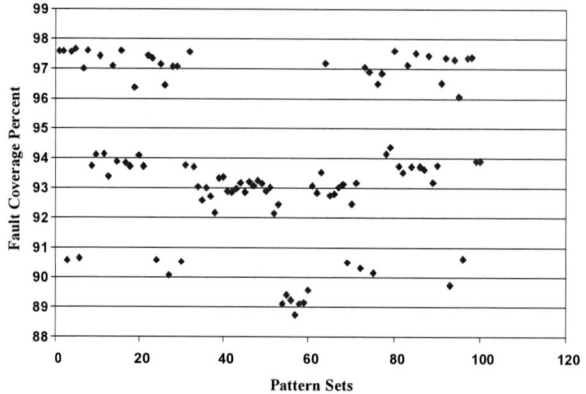

Figure 9: Modulo-127 Fault Coverage

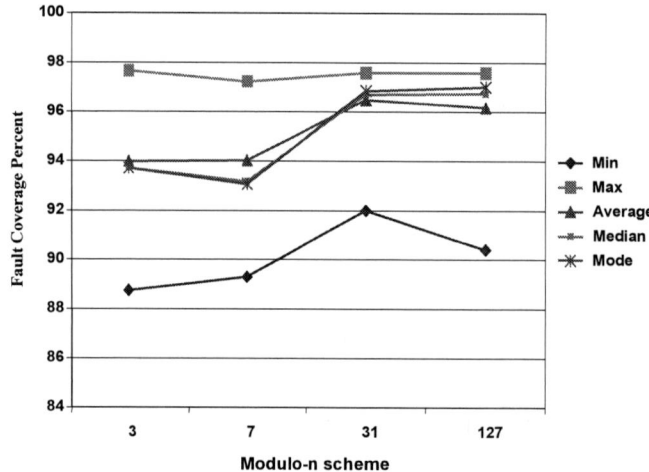

Figure 10: Relative Distribution of Results

7. References

[1] S. Y. Borkar, "Designing Reliable Systems from Unreliable Components: The Challenges of Transistor Variability and Degradation," *IEEE Micro*, Vol. 25, Issue 6, pp. 10-16, Nov.-Dec. 2005.

[2] Shekhar Borkar, Tanay Karnik, Vivek, "Reliable system-on-a-chip design in the nanometer era," Design Automation Conference (DAC'04), 2004

[3] D. A. Anderson and G. Metze, "Design of Totally Self-Checking Check Circuits for m-out-of-n Codes," IEEE Transactions on Computers, vol. C-22, no. 3, pp. 263-269, March 1973.

[4] J. E. Smith and G. Metze, "Strongly Fault Secure Logic Networks," IEEE Transactions on Computers, vol. C-27, no. 6, pp. 491-499, June 1978.

[5] M. Nicolaidis, I. Jansch and B. Courtois, "Strongly Code Disjoint Checkers," Proc. 14th International Symposium on Fault-Tolerant Computing, Orlando, FL, pp. 16-21, June 1984.

[6] N. K. Jha, "Fault Detection in CVS Parity Trees with Application to Strongly Self-Checking Parity and Two-Rail Checkers," IEEE Transactions on Computers, vol. 42, no. 2, pp. 179-189, Feb. 1993.

[7] S. Tarnick, "Controllable Self-Checking Checkers for Conditional Concurrent Checking," Proc. 12th IEEE VLSI Test Symposium, Cherry Hill, NJ, pp. 144-150, April 1994.

[8] J. Khakbaz, "Totally Self-Checking Checker for 1-out-of-n Code using Two-Rail Codes," IEEE Transactions on Computers, vol. C-31, no. 7, pp. 677-681, July 1982.

[9] J. Khakbaz and E. J. McCluskey, "Self-Testing Embedded Parity Checkers," IEEE Transactions on Computers, vol. C-31, no. 8, pp. 753-756, August 1984.

[10] M. Nicolaidis, "Fault Secure Property Versus Strongly Code Disjoint Checkers," IEEE Transactions on Computer-Aided Design, vol. 13, no. 5, pp. 651-658, May 1994.

[11] E. Fujiwara and K. Matsuoka, "A Self-Checking Generalized Prediction Checker and Its Use for Built-In Testing," IEEE Transactions on Computers, vol. C-36, no. 1, pp. 86-93, January 1987.

[12] S. Kundu and S. M. Reddy, "Embedded Totally Self-Checking Checkers: A Practical Design," IEEE Design and Test of Computers, pp. 5-12, August 1990.

[13] M. Agostinelli, S. Pae, W. Yang, C. Prasad, D. Kencke, S. Ramey, E. Snyder, S. Kashyap, M. Jones, "Random charge effects for PMOS NBTI in ultra-small gate area devices," Proc. Intl. Reliability Physics Symp., pp. 529–532, 2005.

[14] Mickael Denais, Vincent Huard, Chittoor Parthasarathy, Guillaume Ribes, Franck Perrier, Nathalie Revil, and Alain Bravaix, "Interface Trap Generation and Hole Trapping Under NBTI and PBTI in Advanced CMOS Technology With a 2-nm Gate Oxide," IEEE Transactions on Device And Materials Reliability, vol. 4, no. 4, December 2004

[15] G. Groseneken, R. Degraeve, B. Kaczer, and P. Rousel, "Recent Trends in Reliability Assessment of Advanced CMOS Technology," *Proceedings of IEEE 2005 International Microelectronics Test Structure,* vol. 18, April 2005.

[16] R. C. Baumann, "The impact of technology scaling on soft error rate performance and limits to the efficacy of error correction," in *Digest of International Electron Devices Meeting,* pp. 329-332, 2002

[17] P. Shivakumar, M. Kistler, S. W. Keckler, D. Burger, and L. Alvisi, "Modeling the effect of technology trends on the soft error rate of combinational logic," in *Proc. Int'l conf. Dependable Systems and Networks,* pp. 389-398, 2002

[18] I. Polian, J. P. Hayes, S. Kundu, and B. Becker, "Transient fault characterization in dynamic noisy environments," in *Proc. Int'l Test Conf.,* 2005

[19] H. Kobayashi, H. Usuki, K. shiraishi, H. Hiroo Tsuchiya, N. Kawamoto, G. Merchant, and J. Kase, "Comparison between neutron-induced system-SER and accelerated-SER in SRAMs," in *Proc. Int'l Reliability Phys. Symp.,* pp. 288-293, 2004

[20] N. Seifert, X. Zhu, and L. W. Massengill, "Impact of scaling on soft-error rates in commercial microprocessors," *IEEE Trans. On Nuclear Science,* Vol. 49, No. 6, pp. 3100-3106, 2002

[21] H. Nicolaidis, M. Bederr, "Efficient implementations of self-checking multiply and divide arrays," European Design and Test Conference, 1994. EDAC, The European Conference on Design Automation. ETC European Test Conference. EUROASIC, The European Event in ASIC Design, Proceedings, pp. 574–579, 28 Feb-3 Mar 1994.

[22] M. Nicolaidis and R. Duarte, "Design of fault-secure parity-prediction booth multipliers," DATE '98: Proceedings of the conference on Design, automation and test in Europe, pp. 7-14, 23-26 Feb 1998.

[23] M. Nicolaidis, R. O. Duarte, S. Manich, and J. Figueras, "Fault-secure parity prediction arithmetic operators," IEEE Des. Test, vol. 14, no. 2, pp. 60–71, 1997.

[24] D. Marienfeld, E. S. Sogomonyan, V. Ocheretnij, and M. Gossel, "New self-checking output-duplicated booth multiplier with high fault coverage for soft errors," in ATS '05: Proceedings of the 14th Asian Test Symposium on Asian Test Symposium. Washington, DC, USA: IEEE Computer Society, 2005, pp. 76–81.

[25] Synopsys Inc., TetraMAX® ATPG, www.synopsys.com.

[26] I. Koren and C. M. Krishna, Fault-Tolerant Systems, Morgan-Kaufman, San Francisco, CA, 2007.

[27] A. Avizienis, "A set of algorithms for a diagnosable arithmetic unit," Tech. Rep. 32-546, Jet Propulsion Lab., Calif. Inst. Technol., Pasadena, CA, Mar. 1964.

[28] Avizienis, A., "Arithmetic Error Codes: Cost and Effectiveness Studies for Application in Digital System Design," *Computers, IEEE Transactions on*, vol.C-20, no.11, pp. 1322-1331, Nov. 1971.

[29] Wakerly, J.F., "Detection of Unidirectional Multiple Errors Using Low-Cost Arithmetic Codes," *Computers, IEEE Transactions on*, vol.C-24, no.2, pp. 210-212, Feb. 1975.

[30] Shyam, S., Constantinides, K., Phadke, S., Bertacco, V., and Austin, T. 2006. Ultra low-cost defect protection for microprocessor pipelines. *SIGPLAN Not.* 41, 11 (Nov. 2006), 73-82.

[31] Debany, W.H.; Macera, A.R.; Daskiewich, D.E.; Gorniak, M.J.; Kwiat, K.A.; Dussault, H.B., "Effective concurrent test for a parallel-input multiplier using modulo 3," VLSI Test Symposium, 1992. '10th Anniversary. Design, Test and Application: ASICs and Systems-on-a-Chip', Digest of Papers., 1992 IEEE , vol., no., pp.280-285, 7-9 Apr 1992.

IEEE International Symposium on Defect and Fault Tolerance of VLSI Systems

Adaptive Error Control for NoC Switch-to-Switch Links in a Variable Noise Environment

Qiaoyan Yu and Paul Ampadu

ECE Dept., University of Rochester, Rochester, NY 14627
<qiaoyan, ampadu>@ece.rochester.edu

Abstract

We present an adaptive error control method for switch-to-switch links in a variable noise environment, to meet reliability requirements and achieve energy-efficiency. Unlike worst-case error correction coding (ECC), the proposed method is capable of selecting the most effective ECC scheme based on predicted link quality at runtime. Our method configures the ECC codec to obtain the desired error correction strength using a set of single-error correction (SEC) codes combined with interleaving. The method can effectively handle multi-cycle and adjacent multi-wire errors existing on switch-to-switch links. An experimental case study shows that the adaptive method can increase the energy-efficiency by up to 19% over a fixed ECC scheme, for a given residual flit error rate. Furthermore, energy reduction affected by different multi-wire and multi-cycle noise scenarios are compared.

1. Introduction

Networks-on-chip (NoCs) have been proposed to manage communication among the increasing resources in the systems-on-chips (SoCs) [1, 2]. Unfortunately, reliability of the global links challenges NoC design. Single error correction codes (SECs) have been used to address independent single-bit errors in on-chip communication links [3-6]. As technology scales, however, the links become more vulnerable to increased crosstalk, cosmic noise interference and spurious voltage swings; thus, multiple-bit errors existing on the parallel bus (multi-wire errors) and errors lasting multiple cycles (multi-cycle or burst errors) have become more important [7-10]. Consequently, more powerful error correction coding (ECC) is required to recover errors. Unfortunately, the increased codec complexity usually implies increased hardware and energy costs.

To trade off performance, energy and reliability, several approaches have been investigated [6-8, 11-14]. One method combines various codes in a general way to tackle coupling noise [6]. A self-calibrating on-chip link has been proposed to achieve high-performance and low-power consumption by dynamically adjusting the operating frequency and voltage swing [11], while error detection combined with retransmission is used to ensure reliability. To reduce the energy wasted by error detection in favorable noise conditions, the redundancy of error detection schemes can be adapted based on the number of errors found over a fixed time window by a victim line operating at half supply voltage [12]. The reconfigurable approach proposed in [8] can adapt to different fault types at runtime: predetermined Hamming coding and interleaving combined with stop-and-wait ARQ to handle transient errors; multi-cycle redundant flit transmission or spare wires to protect the flit from intermittent and permanent errors, respectively. To achieve the required QoS levels, different error control schemes are provided for different data types in [7] and [13].

In this paper, an energy-efficient, flexible and powerful error control scheme to protect switch-to-switch links from multi-wire and multi-cycle errors is proposed. Unlike previous schemes, our adaptive error control mechanism can fix a variable number of erroneous bits, by configuring a set of SEC codecs combined with interleaving. The remainder of this paper is organized as follows: in Section 2, the framework of the

1550-5774/08 $25.00 © 2008 IEEE
DOI 10.1109/DFT.2008.40

proposed adaptive error control scheme is described. In Section 3, our proposed configurable approach combining SECs with interleaving is presented in detail. The proposed ECC scheme is evaluated in Section 4. Conclusions are presented in Section 5.

2. Adaptive error control scheme

2.1 Overview of the proposed adaptive scheme

Switch-to-switch communication at the physical and data link layers is the focus of this paper; routing-related issues like deadlock and starvation are outside the scope here. We propose an error control method for flit-level transmission through global interconnects between switches. As shown in Fig. 1, a flit from a network interface is first stored in flit buffers; error detection codes are then attached to these flits to assist in channel quality evaluation at the receiver. By using the result of a channel evaluation module, the most effective ECC scheme is selected in the transmitter to protect the flit, and a matched decoding scheme is used in the receiver.

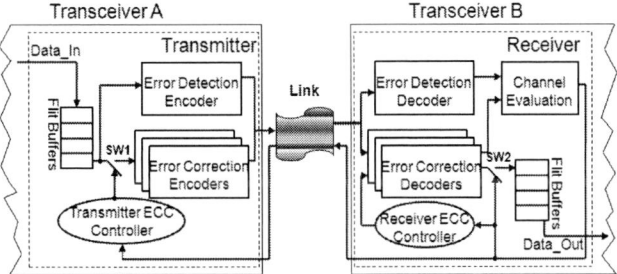

Fig.1. Adaptive switch-to-switch error control architecture

2.2 Transceiver design

The internal structure of the transmitter switch is shown in Fig. 2 (a). Flits are stored in the flit buffers. If the *Send_New* signal is enabled, the flit buffer pointer is incremented, and a new flit is fetched from the flit buffers. This flit is appended with an error detection code and delivered to the network. Subsequently, if the *Resend* signal is received from the ECC controller, the last transmitted flit is encoded with an appropriate error correction scheme based on the channel quality grading. The appropriate error correction coding is obtained by configuring the configurable ECC encoder. The *Coding Selection* block generates the *Send_New* and *Resend* signals, and also selects an efficient ECC approach for the last flit retransmission, according to the feedback from the receiver.

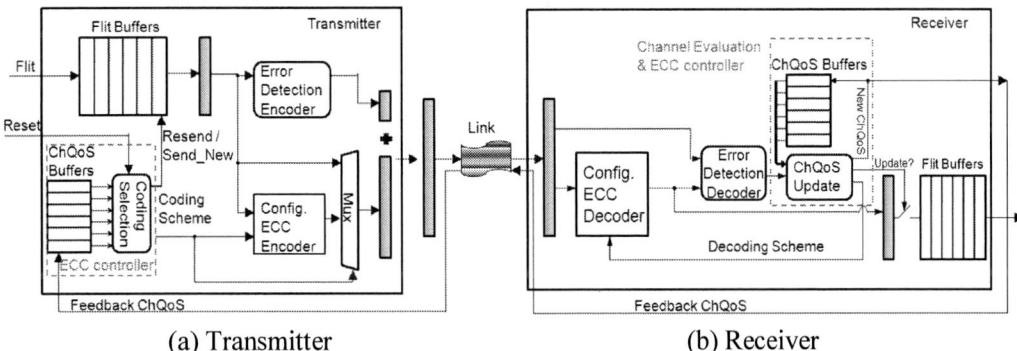

(a) Transmitter   (b) Receiver

Fig.2. Transceiver diagram for the adaptive error control scheme

The operation of the transmitter is demonstrated by the state machine of Fig. 3. The transmitter starts in the *Normal send* state, in which a flit with error detection check bits is transmitted. As soon as a retransmission request is received, the transmitter enters the

ECC resend state, and retransmits the last flit with appropriate error correction coding. If the transmitter consecutively switches between *Normal send* and *ECC resend*, a burst error is detected by the ECC controller and the state is changed to the *Burst Error Prediction* mode. In this state, subsequent flits are directly transmitted with the most recent ECC scheme, without waiting for the ChQoS feedback from the receiver. Doing so effectively eliminates the latency associated with the ChQoS feedback signal; thus, the overall system throughput is improved. This *Burst Error Prediction* mode is maintained until the burst error counter reaches its maximum value (set by the designer), at which point the state machine returns to the *Normal send* mode.

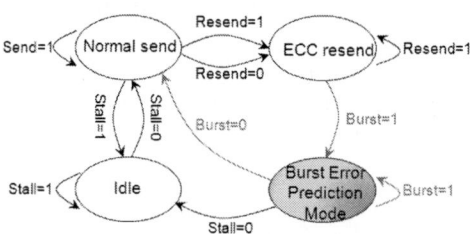

Fig.3. State machine of the switch

The relevant components of the receiver switch are shown in Fig. 2 (b). Error detection is first performed on the received flit. If an error is detected, a retransmission will be requested. Unlike conventional retransmission methods, the receiver of this adaptive architecture provides a channel quality ChQoS to the transmitter, instead of a single NACK (not acknowledged) signal. Whereas a previous approach [12] has combined error detection with channel quality feedback to improve retransmission and detection, our approach explicitly uses the channel quality to assist in selecting an appropriate error correction scheme for the next flit transmission.

2.3 Channel grading

To achieve the expected reliability and energy-efficiency, effective channel evaluation is necessary for the adaptive error control scheme. We propose a channel quality grading scheme shown in Fig. 4. The channel quality is initialized as the first grade, *G0* (no error occurs in the channel). This grade is maintained until an error is detected by the *Error Detection* block from Fig. 2. For simplicity, any detected error increases the channel quality grade, which is then fed back to the transmitter. The channel grade remains at the same level in the *Burst Error Prediction* mode and decreases to a lower level when *Burst Error Prediction* terminates or no error is detected in *Normal Send* mode. The number of channel quality grades is limited by the number of error correction choices. If the maximum error correction capability is exceeded, the receiver informs the transmitter to wait *m* cycles and then retransmit the flit.

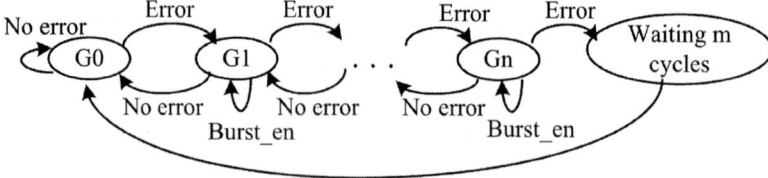

Fig.4. Channel (link) quality grading algorithm

3. Hardware-efficient implementation

3.1 Configurable error control coding

Different error correction strengths can be achieved by selecting different codes. However, the codec cost rises quickly as the number of error bits increases. Thus, direct implementation of a bank of complex codes is not feasible. Here, we propose a configurable error control coding method, employing a set of single error correction (SEC) encoders and an interleaver to recover one- to m-bit errors. Consider a K-bit information flit encoded and transmitted over an N-bit bus. Assume that j-bit adjacent error correction capability is required by the configurable control signal. As shown in Fig. 5, the K-bit information is first separated into j groups, and then encoded in parallel using multiple SEC encoders. All the n_j-bit codewords from selected encoders are regarded as sub-codewords, which are interleaved to create the N-bit codeword C_j. Here, each $(n_j,\ k_j)$ Hamming or shortened Hamming code can be used to correct a single-bit error.

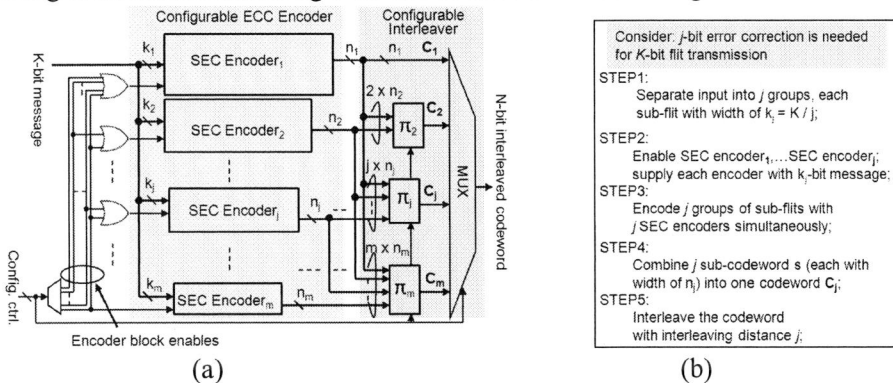

(a)   (b)

Fig.5. (a) Schematic and (b) algorithm for the configurable ECC encoder

Selection of SEC encoders for the configurable ECC scheme is important to accomplish the configuration process. Consider a configurable encoder capable of correcting up to $m = 3$ adjacent errors for a $K = 32$-bit message wire. To correct three-bit errors, three SEC encoders are needed. SEC encoder$_1$ employs a shortened Hamming code HM (38, 32); SEC encoder$_2$ uses HM (21, 16); SEC encoder$_3$ makes use of HM (15, 11). For convenience, we use the following systematic generator matrices for the three encoders below

$$G_j = [I(k_j) \mid P_j] \tag{1}$$

$$P_1^T = \begin{bmatrix} 1101101010101101010101010101010101010101 \\ 1011011001101100110011001101101011001 \\ 0111000111100011110000111100000111000111 \\ 0000111111100000000111111111000000 \\ 0000000000011111111111111110000000 \\ 0000000000000000000000000000111111 \end{bmatrix} \tag{2}$$

$$C_j = \text{message} \times G_j \tag{3}$$

Here, $j = 1, 2, 3$ and $I(k_j)$ is an identity matrix, $k_1 = 32$, $k_2 = 16$, and $k_3 = 11$. The transpose of the parity matrices are shown in equation (2), and the codeword C_j is given by equation (3). As highlighted in equation (2), P_3 is part of P_2, and P_2 is further a segment of P_1. Consequently, the computation of C_1 invariably includes computation of C_2 and C_3. This fact facilitates a hardware-efficient implementation of the configurable ECC, through XOR-tree-sharing.

As shown in Fig. 6, SEC encoder$_1$ uses XOR-tree-sharing to generate the check bits for C_2 and C_3, because of the relationship among the parity matrices. When single-bit error correction is requested, only C_1 is used. If three-bit error correction is desired, three sub-codewords with identical widths from the three SEC encoders are interleaved to create a complete C_3. In this case, some outputs of SEC encoder$_1$ and SEC encoder$_2$ are excluded during the C_3 calculation. Similarly, SEC encoder$_2$ can compute the check bits for C_3 using this XOR-tree-sharing approach. Consequently, the hardware cost for computing C can be reduced.

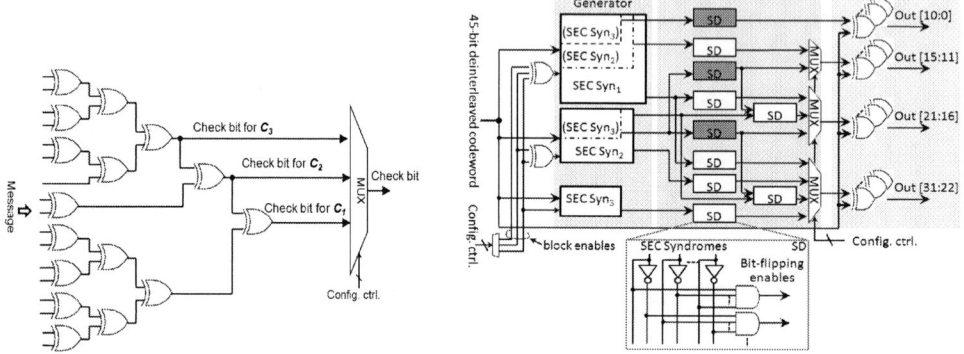

Fig.6. XOR-tree-sharing in SEC encoder　Fig.7. Schematic of configurable ECC decoder

In the configurable ECC decoder, both syndrome computation and error correction circuits of the SEC decoders can be shared. Syndrome computation using equation (4) is implemented with XOR trees. Consequently, the XOR-tree-sharing approach shown in Fig. 6 can also be applied.

$$\text{Syndrome}_j = C_j \times H_j^{\,T} = C_j \times \begin{bmatrix} P_j \\ I\bigl(n_j - k_j\bigr) \end{bmatrix}$$

(4)

Here, H_j (j = 1, 2, 3) is the parity-check matrix.

As shown in Fig. 7, the syndrome computation circuit SEC Syn_1 (for HM (38, 32)) can also be used to compute part of SEC Syn_2 (for HM (21, 16)) or SEC Syn_3 (for HM (15, 11)). The configurable control determines the necessary syndrome computation blocks. Syndromes are decoded to correct the error bits. As shown in Fig. 7, syndrome decoders (SD) are used to produce the bit-flipping enable signals, indicating which bit should be flipped. Since the syndromes for different SEC decoders are overlapped, some SDs (gray SDs in Fig. 7) can be shared to correct multiple bits.

3.2 Configurable interleaving

The multiple groups of SEC codecs allow correcting multiple errors if they are not in the same SEC codeword. To make this approach useful for adjacent multi-wire errors, an interleaver is needed to distribute the adjacent errors among the available groups. As shown in Fig. 8, the adjacent two-bit errors (indicated by the shaded bits) in the interleaved codeword $\Pi(C)$ is spread out to different sub-codewords after deinterleaving, so that each error bit can be corrected by different SEC decoders.

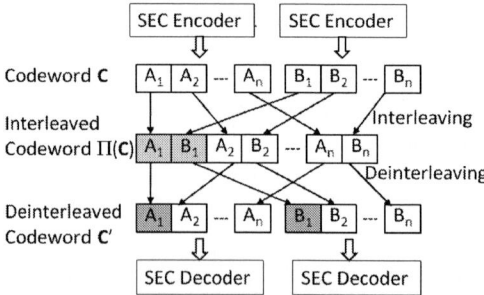

Fig.8. Interleaver breaks adjacent multi-wire error pattern

In order to match the configurable coding, it is necessary to change the interleaving distance (the minimum distance between the bits from the same sub-codeword). The configurable control signals are employed to determine the interleaving distance at runtime. For the $K = 32$ and $m = 3$ example discussed previously, Fig. 9 (a) shows the internal architecture for the configurable interleaver. The configurable control determines the ON/OFF status of switch1, switch2 and switch3, which control the inputs for different

356

interleavers. For instance, assume the configurable control closes switch2. Each set of 21 bits is interleaved with interleaving distance two. Because the link width is 45 bits, three redundant zero bits are appended to the interleaved codeword. The schematic for the configurable interleaver is shown in Fig. 9 (b). The received codeword will be deinterleaved with the same interleaving distance as the interleaver used, determined by the configurable control signal. Switch1, switch2 and switch3 allow the appropriate codeword pass through the matched deinterleaver. Finally, the codeword will be decoded by HM (38, 32) decoder, HM (21, 16) or HM (15, 11) in parallel.

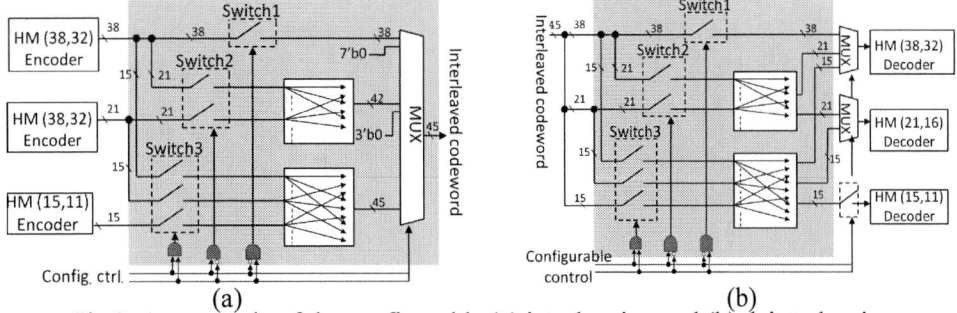

Fig.9. An example of the configurable (a) interleaving and (b) deinterleaving

4. Evaluation of proposed ECC scheme

4.1 Area reduction by XOR-tree-sharing

To evaluate hardware-efficiency and performance, we implemented the proposed architecture with and without XOR-tree-sharing. Single error correction (SEC) codes, HM (38, 32), HM (21, 16) and HM (15, 11), were combined with interleaving to protect against adjacent multi-wire errors. The maximum link width was set to 45 to allow transmission of a 32-bit message using three groups of HM (15, 11). The ECC codecs were implemented with VerilogHDL and synthesized by Synopsys Design Vision using a 180 nm TSMC typical library at 1.8 V. A maximum input-to-output delay constraint of 2 ns was set to accomplish encoding and decoding within one clock cycle of a 400 MHz clock.

As shown in Table 1, the proposed XOR-tree-sharing approach reduces the encoder and decoder area by 24% and 32%, respectively, compared to the case without XOR-sharing. Thus, the total codec area decreases by about 30%. As can be seen, the XOR-tree-sharing approach is more effective in the decoder, because more XOR gates can be shared in both syndrome computation and the error correction circuit. A limitation of this method is an increase (up to 18%) in total power, because of configuration overhead. However, this overhead is negligible compared to the link switching power, which will be shown in the next section.

Table 1: Area and power comparison of ECC codecs without and with XOR-tree-sharing

	Without XOR sharing	With XOR sharing
Encoder Area (μm^2)	15101 (100%)	11444 (76%)
Decoder Area (μm^2)	26221 (100%)	17951 (68%)
Total Area (μm^2)	41322 (100%)	28266 (71%)
Encoder Power (mW)	6.6 (100%)	8.8 (130%)
Decoder Power (mW)	10.5 (100%)	11.3 (107%)
Total Power (mW)	17.1 (100%)	20.1 (118%)

Table 2: Equivalent gates and dynamic power comparison between two ECC schemes

	Equivalent Gate Count	Dynamic Power (mW)
Fixed ECC	1048	11.8 (100%)
Adaptive ECC	2755	35.6 (300%)

4.2 Energy-efficiency over fixed ECC Scheme

To evaluate the proposed method, we compared the energy-efficiency per useful flit of the adaptive ECC to a fixed ECC scheme. Here, energy-efficiency per useful flit refers to the average energy to *successfully* transfer a flit via the NoC links (includes retransmission, codec energy, etc). Consider a 32-bit flit encoded and transmitted. Assume that three-bit adjacent error correction capability is required against the worst-case noise scenario. The fixed ECC scheme is implemented using three groups of HM (15, 11) combined with interleaving. In the proposed adaptive ECC scheme, SEC codes, HM (38, 32), HM (21, 16) and HM (15, 11), are combined with interleaving to protect against adjacent three-bit errors, and CRC-4(Circular Redundancy Code) is used to detect errors.

Table 2 shows the equivalent gate count and dynamic power of the synthesized netlists for the fixed and adaptive ECC schemes in a 180 nm TSMC technology at 1.8 V. Since the increased area overhead caused by error control coding is a small fraction of the overall switch area (e.g. 2.8% [8]), the 1.7X increased gate count required by the adaptive ECC scheme is still acceptable. Although the dynamic power of the adaptive ECC is three times that of the fixed ECC, this limitation is quickly overcome as the link power dominates the total power in deep submicron technology.

Link switching and ECC basic module power were simulated in Cadence Spectre with Predictive Technology Model CMOS 65 nm technology [15]. Supply voltage and clock frequency are set to 1.0 V and 1 GHz, respectively. As shown in Fig. 10, the dynamic power of ECC codec is significantly less than the link power.

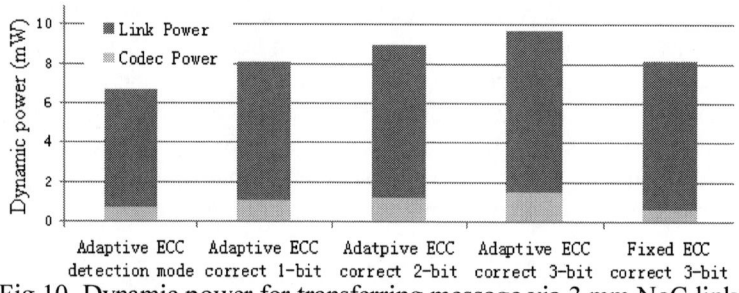

Fig.10. Dynamic power for transferring message via 3 mm NoC link

The Gaussian pulse function (5) is widely accepted as a model for the fault occurrence probability ε affecting a single wire [3, 5, 12].

$$\varepsilon = Q\left(\frac{V_{dd}}{2\sigma_N}\right) = \int_{\frac{V_{dd}}{2\sigma_N}}^{\infty} \frac{1}{\sqrt{2\pi}} e^{-y^2/2} dy \tag{5}$$

Here, the noise voltage is a normal distribution with deviation σ_N, and V_{dd} is the supply voltage. In previous works [3, 5, 12], faults are typically assumed to be statistically independent. Based on the multi-wire and multi-cycle fault model in [7], a simplified fault model (6) for transient errors has been used to simulate the noise on NoC links here.

$$Fault\ Matrix = \varepsilon \times P = \varepsilon \times \begin{bmatrix} p(1,1) & \dots & p(1,\tau_{max}) \\ p(2,1) & \dots & p(2,\tau_{max}) \\ \vdots & \ddots & \vdots \\ p(w_{max},1) & \dots & p(w_{max},\tau_{max}) \end{bmatrix} \tag{6}$$

where ω_{max} is the maximum bus width, and τ_{max} is the maximum transfer cycle during which the fault propagates. The element $p(\omega, \tau)$ of the matrix P is the probability of ω wires retaining faults for τ cycles. We assume that the occurrence probabilities (p_ω and p_τ) for both ω-wire and τ-cycle faults are independent, and they follow a Gaussian function with standard deviations σ_ω and σ_τ, respectively.

$$p_\omega = \frac{2^i \times \exp\left(-0.5\left(\frac{(\omega-1)/(2\omega_{max}-1)}{\sigma_\omega}\right)^2\right)}{\sum\limits_{n=-(\omega_{max}-1)}^{(\omega_{max}-1)} \exp\left(-0.5\left(\frac{n/(2\omega_{max}-1)}{\sigma_\omega}\right)^2\right)} \qquad \left(i = \begin{cases} 0, & (\omega = 1) \\ 1, & (1 < \omega \le \omega_{max}) \end{cases}\right) \tag{7}$$

$$p_\tau = \frac{\exp\left(-0.5\left(\frac{\tau/\tau_{max}}{\sigma_\tau}\right)^2\right)}{\sum\limits_{n=1}^{\tau_{max}} \exp\left(-0.5\left(\frac{n/\tau_{max}}{\sigma_\tau}\right)^2\right)} \tag{8}$$

$$p(\omega,\tau) = \frac{p_\omega \times p_\tau}{\sum\limits_{i=1}^{\omega_{max}}\sum\limits_{j=1}^{\tau_{max}} \left(p_i \times p_j\right)} \tag{9}$$

Using this model, we obtained the residual flit error rates for different error correction codes, as shown in Fig.11 (a). Assume that the target residual flit error rate is below 10^{-9}. Because the adaptive ECC scheme can dynamically select the most energy-efficient coding method in a noise variable environment, the average energy per useful flit estimated by (10) is 19% less than that of the fixed ECC scheme, shown in Fig. 11(b).

$$E_{avg} = E_{detection} + \left(\sum_N \sum_{\omega,\tau} p(\omega,\tau) \times \left(E_{correction}(\omega,\tau) + E_{link}(\omega) + E_{mode\_switching}(\omega,\tau)\right)\right) \tag{10}$$

Here, E_{avg}, $E_{detection}$, $E_{correction}$, E_{link} and $E_{mode\_switching}$ represent the respective energies for successfully transferring a flit via a 3 mm NoC link. When the noise voltage increases beyond maximum ECC capability, the benefit on energy achieved by the adaptive scheme starts to degrade.

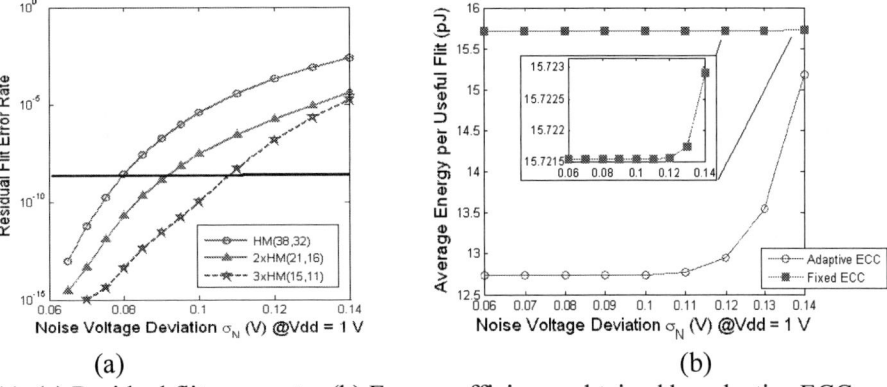

Fig.11. (a) Residual flit error rates (b) Energy-efficiency obtained by adaptive ECC

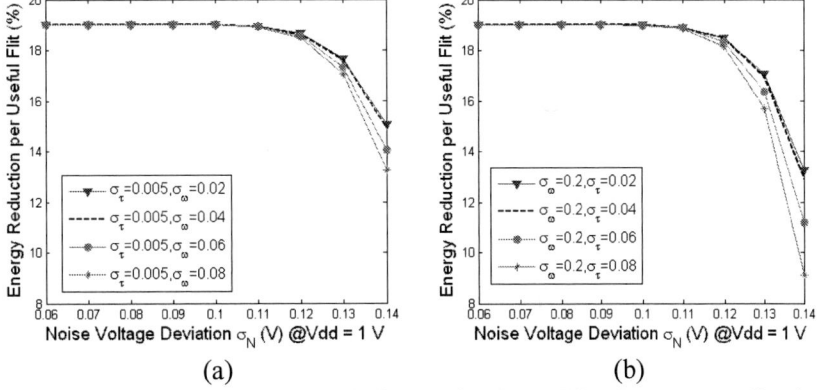

Fig.12. Energy reduction per useful flit by adaptive ECC schemes over fixed ECC schemes (a) Multi-wire (b) Multi-cycle effects

For different probability characteristics of multi-wire and multi-cycle errors, the average energy reduction by the adaptive ECC varies, particularly in a large noise voltage region. As shown in Fig. 12(a), the increasing multi-wire error (i.e. increasing σ_ω) does not significantly degrade the benefit achieved by the adaptive ECC, because of appropriate code selection. In contrast, variations in the number of cycles errors persist (because of large σ_τ) result in frequently switching between different error control modes. This affects the energy-efficiency of the adaptive ECC, as shown in Fig. 12(b).

5. Conclusion

To achieve high reliability and energy-efficiency, an adaptive error control scheme is proposed to address the reliability issue of NoC links in a noise variable environment. Because of technology scaling, adjacent multi-wire multi-cycle errors, rather than single-wire single-cycle errors, have been modeled in the proposed ECC scheme. Taking advantage of the adjacent error characteristics, a set of SEC codecs and simple block interleaving are configured to obtain different error correction capabilities. For global interconnect links, the adaptive ECC scheme can achieve a 19% better energy efficiency than a fixed ECC scheme, while maintaining the same reliability in a noise variable environment. Moreover, the energy-efficiency degradation caused by multi-wire and multi-cycle errors has been demonstrated. Using an XOR-tree-sharing approach, the codec area overhead can be reduced by 30%, compared to without XOR-tree-sharing.

Reference

[1]   L. Benini and G. De Micheli, "Networks on chips: a new SoC paradigm," *Computer*, Vol. 35, Issue 1, pp. 70–78, Jan. 2002.

[2]   T. Dumitras and R. Marculescu, "Towards on-chip fault-tolerant communication," in *Proc. ASP-DAC 2003*, pp. 225–232, Jan. 2003.

[3]   S. Murali, L. Benini, M. J. Irwin and G. De Micheli, "Analysis of error recovery schemes for networks on chips," *IEEE Design & Test of Computer*, Vol. 22, Issue 5, pp. 434–442, Sept.-Oct. 2005.

[4]   M. Ali, M. Welzl, S. Hessler and S. Hellebrand, "A fault tolerant mechanism for handling permanent and transient failures in a Network on Chip," in *Proc. ITNG'07*, pp. 1027–1032, Apr. 2007.

[5]   D. Bertozzi, L. Benini, and G. De Micheli, "Error control scheme for on-chip communication links: the energy-reliability tradeoff," *IEEE Trans. Computer-Aided Design of Integrated Circuits and Syst.*, Vol. 24, No. 6, pp. 818–831, June 2005.

[6]   Srinivasa R. Sridhara, and Naresh R. Shanbhag, "Coding for system-on-chip networks: A unified framework," *IEEE Trans. Very Large Scale Integr.(VLSI) Syst.*, Vol. 13, pp. 655–667, June 2006.

[7]   H. Zimmer and A. Jantsch, "A fault model notation and error-control scheme for switch-to-switch buses in a Network-on-Chip," in *Proc. CODES+ISSS'03*, pp. 188–193, Oct. 2003.

[8]   T. Lehtonen, P. Liljeberg, and J. Plosila, "Online reconfigurable self-timed links for fault tolerant NoC," *VLSI Design*, Vol 2007, Article ID 94676, pp. 1–13, 2007.

[9]   G. De Micheli and L. Benini, *Networks On Chips*, Morgan Kaufmann, Chapter 4, pp. 81–90, 2007.

[10] A. Ganguly, P. P. Pande, B. Belzer and C. Grecu, "Design of low power & reliable networks on chip through joint crosstalk avoidance and multiple error correction coding," *J. Electronic Testing*, Vol. 24, No. 1, pp. 67–81, Jan. 2008.

[11] F. Worm et al, "Self-calibrating networks-on-chip," in *Proc. ISCAS'05*, pp. 2361–2364, May 2005.

[12] L. Li, N. Vijaykrishnan, M. Kandemir and M. J. Irwin, "Adaptive error protection for energy efficiency," in *Proc. Intl. Conf.e on Computer Aided Design (ICCAD'03)*, pp. 2–7, Nov. 2003.

[13] D. Rossi, P. Angelini and C. Metra, "Configurable error control scheme for NoC signal integrity," in *Proc. IOLTS 2007*, pp. 43–48, July 2007.

[14] Q. Yu and P. Ampadu, "Adaptive error control for reliable systems-on-chip", in *Proc. ISCAS'08*, pp.832–835, May 2008.

[15] Arizona State Univ., Predictive Technology Model [Online]. Available: http://www.eas.asu.edu/~ptm/

Arbitrary Error Detection in Combinational Circuits by using Partitioning

Osnat Keren,
Bar-Ilan University,
•••••• •• •••••••••••

Ilya Levin,
Tel-Aviv University,
•••••• •••••••••••••

Vladimir Ostrovsky,
Tel-Aviv University,
••••••• ••••••••••••

Beni Abramov,
Tel-Aviv University,
••••• •••••••• •• ••••••••

Abstract

• •• •••••• •••••• • ••• ••••••• ••• ••••••••• • •••••••••••• ••••••• ••• ••• ••••••
•••••••• • •• ••••••• •• ••••• •• ••••••••• ••• ••• •••••• •• •• ••• • •••••••••••• •••••••••• ••••
•• ••••• ••• •••• ••• ••• •••••••••• ••••• •••••••• ••• •••••• •• • •••••••• • • •••••• •• ••••
•••••••••• • •••• ••••••••• ••• ••••••••• •••••• •• •• • ••••••• ••••••• ••• •••••••••
••••• • •••••••• • •• ••••• ••••• •••••••• ••• ••••••• • • ••••• ••• ••••••••• •• ••••• •••• •••••••
•• ••••• •••••••••• • ••••• ••• •••••• •••• •• ••••• •• ••• •••••••• •••••••••

1 Introduction

The majority of the known concurrent checking schemes assumes that a set of output words of the functional circuit to be checked is complete, i.e., any binary vector may occur. However, it is often reasonable to construct a so-called context-oriented concurrent checking scheme, where: a) the number M of possible output vectors, is much smaller than 2^k, where k is the width of the output vector, and b) the set of possible outputs is known in advance. The context-orientation has some advantages in comparison with the universality. Namely, it allows utilizing the redundancy of the circuit's output codewords, which is an intrinsic feature of such circuits. One of the ways of utilizing the redundancy is partitioning the functional circuit into a number of separate independent sub-circuits. Each of these sub-circuits implements its own subset of output signals. Since the sub-circuits have no common elements, any single fault may result in errors only in a subset of the output signals.

The context-orientation was studied in [3], where a Sum-Of-Minterms (SOM) checker was proposed. The SOM checker tests whether an output word belongs to the set of possible code-words of the circuit to be checked. Experimental results show that for $M < 2^k/3$, the system in [3] has smaller implementation cost than a solution based on duplication of the functional unit. In [4], the authors developed a specific architecture for checking sequential circuits without introducing any redundant coding variables. This architecture uses signals of logic products of the functional unit for providing the self-checking property. Signals of these products form additional inputs of the checker.

Partitioning a functional circuit for mutually checking components was proposed in [5] as an alternative way for exploring the context-orientation. The authors examined a two-block partition, minimizing the number of encoded variables in a concurrent checking scheme that detects any arbitrary errors.

In the present paper, we propose a new technique that allows to detect •••••••• errors in a combinatorial circuit. That is, all single faults (that may cause any binary vector on the

circuit's output) are detectable by the suggested scheme. Similarly with [4], we don't use any redundant coding for the output vectors. We propose to transfer the input variables (and not the products) into the checker. Actually, these input variables are used instead of the coding variables. A method for designing such a circuit is the main contribution of the paper. We show that the partitioning of the initial circuit into independent sub-circuits followed by choosing an optimized set of input variables is efficient for detecting both unidirectional and arbitrary errors. The proposed partitioning algorithm is heuristic and does not provide the optimal partition. Nevertheless, it is simple and provides good solutions.

The paper is organized as follows: Section 2 includes basic definitions, and recalls related work on on-line testing for arbitrary errors. Section 3 presents the suggested structure, and Section 4 contains experimental results. Section 5 concludes the paper.

2 Preliminaries

Consider a functional unit that has m inputs and k outputs. The logic unit can be represented as a multi-output function $Y = f(X)$ where $X = (x_{m-1}, \ldots, x_0)$ and $Y = (y_{k-1}, \ldots, y_0)$. In this paper, the binary output vectors are referred to as •• •••• ••••• • •••••. Assume that the logic unit can produce only M distinct information words out of the 2^k possible combinations, that is, $M < 2^k$.

In order to detect a single fault in the system, conventional methods encode the information words by adding redundancy bits. Namely, each information word $Y_i = (y_{k-1}^{(i)}, \ldots, y_0^{(i)})$ is encoded to a codeword $Z_i = (z_{n-1}^{(i)}, \ldots, z_0^{(i)})$ of length $n \geq k$. The set of codewords $\{Z_i\}_{i=1}^M$ forms a code.

Definition 1 • •••• •• •••••• •••••• •••• • •• ••••• ••• k •••• •••••••• $\{j_s\}_{s=0}^{k-1}$ •••• •••• $z_{j_s}^{(i)} = y_s^{(i)}$ ••• ••• $1 \leq i \leq M$ ••• $0 \leq s < k.$ ••• ••• •••••• $r = n - k$ •••••• ••••• ••• ••••••••••

Clearly, systematic codes are preferable since they allow extracting an information word Y_i out of a codeword Z_i without additional processing.

The assumption that a fault causes unidirectional errors allows to implement the functional unit as a single circuit. However, in cases where a fault may cause an arbitrary error (that is, not necessarily unidirectional), the fault may not be detected in functional units that are implemented as a single circuit. In order to detect any fault the functional unit should be implemented by at least two independent circuits [6].

Coding schemes for such a case, were discussed in [1, 6, 7, 8]. In this paper we assume that the functional unit is implemented as two independent circuits. Without loss of generality, we assume that the first circuit realizes the first n_1 bits of Z, that is, $c_1 = (z_{n_1-1}, \ldots, z_0)$, and the second circuit realizes the remaining $n_2 = n - n_1$ bits, $c_2 = (z_{n-1}, \ldots, z_{n_1})$. The output of the functional unit is denoted by $\hat{Y} = (c_2, c_1)$. Obviously, there is a one-to-one mapping between each Y_i and $\hat{Y}_i$.

The overall system is ••• •••••••• in respect to a single fault in one of the circuits, if either, a fault maps a codeword on itself, or it maps a codeword to a non codeword [2]. In [6] it was shown, that iff $c_2 = f_1(c_1)$, and, $c_1 = f_2(c_2)$, then the functional unit is fault secure in respect to arbitrary errors in one of the two circuits.

There are several ways to partition the set of information bits $\{y_j\}_{j=0}^{k-1}$ into two sets. In this paper, we follow the approach introduced in [6], where partitioning is done with the aim to minimize the number of y's in c_1 that completely specify the information words; namely,

$$Y = \hat{f}(y_{j_{k_1-1}}, \dots, y_{j_0}) = \hat{f}(c_1). \tag{1}$$

The remaining information bits (if any) form c_2. Note that this partition may not be the optimal in terms of the implementation cost.

Definition 2 (Distance) $\cdots$ $\cdots\cdots$ $d(\hat{Y}_i, \hat{Y}_j)$ $\cdots\cdots$ $\cdots\cdots$ $\cdots\cdots$ $\hat{Y}_i = (c_2^{(i)}, c_1^{(i)})$ $\cdots$ $\hat{Y}_j = (c_2^{(j)}, c_1^{(j)})$ $\cdots$ $d(\hat{Y}_i, \hat{Y}_j) = |\{k | c_k^{(i)} \neq c_k^{(j)}), \; k = 1, 2\}|.$

Definition 3 (Minimum distance) $\cdots$ $\cdots\cdots$ $\cdots\cdots$ $\cdots\cdots$ $\cdots\cdots$ $\cdots\cdots$ $\mathcal{Y} = \{\hat{Y}_i\}_{i=1}^M$ $\cdots$ $d_{min} = min_{\hat{Y}_i, \hat{Y}_j \in \mathcal{Y}, \; i \neq j} d(\hat{Y}_i, \hat{Y}_j).$

Clearly, any fault in one of the circuits implementing c_1 or c_2 is detectable if and only if all codewords are of distance two from each other. Nevertheless, in some cases, it is impossible to find a partition that leads to a d_{min} that equals two ($d_{min} = 2$). The following example illustrates such a case.

Example 1 $\cdots\cdots$ $\cdots$ $\cdots\cdots\cdots$ $\cdots\cdots$ $\cdots\cdots$ $\cdots\cdots$ $X = (x_4, \dots x_0)$ $\cdots$ $\cdots$ $\cdots$ $\cdots$ $Y = (y_4, \dots y_0).$ $\cdots$ $\cdots$ $\cdots\cdots$ $\cdots\cdots$ $\cdots\cdots$ $\cdots\cdots$ $M = 6$ $\cdots\cdots$ $\cdots\cdots$ $\cdots$ $\{Y_i\}_{i=1}^6$ $\cdots$ $\cdots\cdots$ $\cdots\cdots$ 2^5 $\cdots$ $\cdots\cdots\cdots$

$$Y_1 = (01011), \; Y_2 = (00001), \; Y_3 = (00101), \; Y_4 = (10111), \; Y_5 = (11010), \; Y_6 = (11111).$$

$\cdots$ $\cdots\cdots\cdots$ $\cdots\cdots$ $\cdots\cdots$ $\cdots\cdots$ $\cdots\cdots$ $\cdots\cdots$

$\cdots$ $c_1 = (y_4, y_3, y_2)$ $\cdots$ $c_2 = (y_1, y_0).$ $\cdots$ $\cdots\cdots$ $\cdots\cdots$ $\cdots\cdots$ $\cdots\cdots$ $\cdots\cdots$ $c_1 \times c_2$ $\cdots\cdots$ $\cdots\cdots$ $\cdots\cdots$ $\cdots\cdots$ c_1 $\cdots$ c_2 $\cdots$ $\cdots\cdots\cdots$ $\cdots\cdots$ $\cdots$ c_1 $\cdots$ c_2 $\cdots$ $\cdots\cdots$ $\cdots\cdots$ c_1 $\cdots$ $\cdots\cdots$ $\cdots\cdots$ $\cdots\cdots$ c_2 $\cdots\cdots$ $\cdots\cdots$ $\cdots\cdots$ $c_1;$ $\cdots\cdots$ c_1 $\cdots\cdots$ $\cdots\cdots$ Y_6 $\cdots$ $\cdots\cdots$ $Y_1.$

$\cdots$ $\cdots\cdots$ $\cdots\cdots$ $\hat{Y}_1$ $\cdots$ $\hat{Y}_2$ $\cdots$ $d((010, 11), (000, 01)) = 2.$ $\cdots\cdots$ $\cdots$ $\cdots\cdots$ $\cdots\cdots$ $\hat{Y}_1$ $\cdots$ $\hat{Y}_4$ $\cdots$ $d((010, 11), (101, 11)) = 1.$ $\cdots\cdots$ $\cdots\cdots$ $\cdots\cdots$ $\cdots\cdots$ $\cdots\cdots$ $\cdots\cdots$ $k = 5$ $\cdots\cdots$ $\cdots\cdots$ $\cdots\cdots$ d_{min} $\cdots\cdots$ $\cdots\cdots$ $\cdots\cdots$ $\cdots\cdots$ c_1 $\cdots\cdots$ $\cdots\cdots$

	000	001	011	010	110	111	101	100
00	Y4	Y1	Y5	Y6	Y3	Y3	Y5	Y2
01	Y5	Y6	Y6	Y5	Y1	Y5	Y5	Y1
11	Y2	Y2	Y5	Y5	Y5	Y6	Y5	Y5
10	Y5	Y1	Y5	Y5	Y5	Y6	Y2	Y5

(a) K-map

(b) The $c_1 \times c_2$ plane

	000	001	011	010	110	111	101	100
00					Y3	Y3		Y2
01								
11	Y2	Y2						
10								Y3

(c) K-map for Y_2 and Y_3

Figure 1: K-map for the functional unit in Ex. 1 (left), K-map for the characteristic functions of Y_2 and Y_3 in Ex. 2 (center), and the location of the information words on the $c_1 \times c_2$ plane (right).

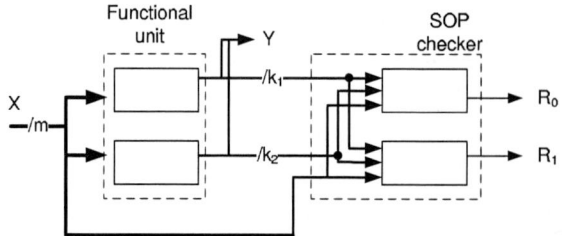

Figure 2: Proposed architecture of a non redundant functional unit with SOP checker

3 Arbitrary error detecting architecture

3.1 General error detection architecture

In this paper, we suggest a new approach for detecting an arbitrary single fault in the functional unit. Instead of encoding an information word Y_i into a codeword Z_i by adding r redundancy bits, we suggest to use the (output) information bits as they are (uncoded), and use the x's as additional inputs to the checker. We will show that this solution is simpler and has a lower implementation cost than the duplication based solutions.

The suggested architecture is shown in Figure 2. The functional unit is implemented as two independent circuits. The first circuit realizes k_1 bits of Y, that is, $c_1 = (y_{j_{k_1-1}}, \ldots, y_{j_0})$; The second circuit realizes the remaining $k_2 = k - k_1$ bits, that is, $c_2 = (y_{j_{k-1}}, \ldots, y_{j_{k_1}})$. The input variables X together with $\hat{Y} = (c_2, c_1)$ enter a (Sum-Of-Products) SOP based checker. While the conventional SOM checker cannot be minimized (since it comprises minterms of distance two), the suggested checker comprises: a) products and not minterms, and, b) this products may further minimized.

3.2 Characteristic functions and SOP checker

In this subsection we present a checker that is based on products rather than on minterms. We assume that the functional unit is implemented as two independent circuits, and that the partition to c_1 and c_2 fulfills Eq. 1. The inputs to the checker are X and the output $\hat{Y}$ of the functional unit.

Recall that the input lines are routed directly to the checker. The inputs are used to provide 'side-information' on the 'problematic', that is, information words that are of distance one from each other. In this paper we consider two cost functions:

- • • •• •• •• • • • ••• • • • •• • •••• That is, use of the minimal number of x's that are sufficient to provide desired property.

- • • •• •• • • • • • • •••• That is, use of a set of x's that reduce the number of literals in the SOP representation of the checker's function. In other words, the complexity criteria is the implementation cost of the checker (e.g. two-input AND-OR gates).

We start by defining a Characteristic function in respect to a given partition:

Definition 4 (Characteristic function) $\cdots \mathcal{Y} = \{\hat{Y}_i\}_{i=1}^{M}$ •• ••• • •••• M • •••• • •••••• •• ••• •• • •••••• • • • •• ••• • • •• •••••••••••••••• •• •••••• $g_i(X)$ •• Y_i, •• •••••••• •• $\mathcal{Y}$, ••

$$g_i(X) = \begin{cases} 1 & f(X) = Y_i \\ 0 & f(X) = Y_j \text{ and } d(\hat{Y}_i, \hat{Y}_j) = 1 \\ \phi & otherwise \end{cases} \tag{2}$$

$\bullet \bullet \bullet \bullet \bullet$ ϕ $\bullet \bullet \bullet \bullet \bullet \bullet$ $\bullet \bullet \bullet$ $\bullet \bullet \bullet \bullet \bullet$ $\bullet \bullet \bullet \bullet$

Note that the characteristic functions are not necessarily pairwise orthogonal (disjoint). That is, $\sum_X g_i(X)g_j(X) \geq 0$.

Example 2 $\cdots$ $\bullet \bullet \bullet \bullet \bullet \bullet \bullet \bullet \bullet$ $\hat{Y}_2$ $\bullet \bullet$ $\bullet \bullet \bullet \bullet \bullet \bullet$ $\bullet$ $\bullet \bullet \bullet \bullet \bullet \bullet \bullet \bullet \bullet \bullet$ $\bullet \bullet \bullet$ $\bullet \bullet \bullet \bullet$ $\bullet \bullet \bullet \bullet \bullet \bullet$ $\hat{Y}_3,$ $\bullet \bullet \bullet$ $\bullet \bullet$ $\bullet \bullet$
$\bullet \bullet$ $\bullet \bullet \bullet \bullet \bullet \bullet \bullet \bullet$ $\bullet \bullet$ $\bullet$ $\bullet \bullet \bullet \bullet$ $\bullet$ $\bullet \bullet \bullet \bullet \bullet$ $\bullet \bullet \bullet$ $\bullet \bullet \bullet \bullet \bullet$ $\bullet$ $\bullet \bullet \bullet \bullet \bullet$ $\bullet \bullet \bullet$ $\bullet$ $\bullet \bullet \bullet \bullet \bullet \bullet \bullet \bullet \bullet \bullet \bullet$ $\bullet \bullet \bullet$ $\bullet \bullet \bullet \bullet \bullet$ $\bullet \bullet$ $\bullet \bullet \bullet \bullet \bullet$ $\bullet \bullet \bullet \bullet$
$\bullet \bullet \bullet \bullet \bullet$ $\bullet \bullet \bullet$ $\bullet \bullet \bullet$ $\bullet \bullet \bullet \bullet \bullet \bullet$ $\bullet \bullet$ $X \bullet \bullet$ $\bullet \bullet \bullet \bullet \bullet \bullet \bullet \bullet \bullet$ Y_2 $\bullet \bullet \bullet$ $Y_3.$ $\bullet \bullet \bullet \bullet$ $\bullet \bullet \bullet$ $\bullet \bullet \bullet \bullet$ $\bullet \bullet$ $\bullet \bullet \bullet$ $\bullet \bullet \bullet \bullet \bullet \bullet \bullet \bullet \bullet \bullet$
$\bullet \bullet$ $\bullet \bullet \bullet$ $\bullet \bullet \bullet \bullet \bullet \bullet$ $\bullet \bullet \bullet$ $\bullet \bullet \bullet \bullet \bullet$ $\bullet \bullet \bullet$ $\bullet$ $\bullet \bullet \bullet \bullet$ $\bullet \bullet \bullet \bullet \bullet \bullet \bullet \bullet \bullet \bullet \bullet \bullet \bullet \bullet$ $\bullet \bullet \bullet \bullet \bullet \bullet \bullet$ $g_2(X)$ $\bullet \bullet \bullet$ $g_3(X)$ $\bullet \bullet \bullet$ $\bullet \bullet \bullet$
$\bullet \bullet \bullet \bullet \bullet \bullet \bullet$ $\bullet \bullet \bullet$ $\bullet$ $\bullet \bullet \bullet \bullet \bullet \bullet \bullet \bullet \bullet \bullet \bullet$ $\bullet \bullet \bullet \bullet \bullet \bullet \bullet$ $\bullet \bullet \bullet$ $g_2(X)$ $\bullet \bullet \bullet \bullet$ $\bullet \bullet$ $g_2(x) = x_4' x_3' x_1 x_0 + x_4 x_3'.$ $\bullet$ $\bullet \bullet \bullet \bullet \bullet \bullet$ $\bullet \bullet \bullet$
$\bullet \bullet \bullet$ $\bullet \bullet \bullet \bullet \bullet$ $\bullet \bullet \bullet \bullet \bullet \bullet \bullet \bullet \bullet$ $\bullet \bullet \bullet$ $\bullet \bullet \bullet$ $\bullet \bullet \bullet \bullet \bullet \bullet \bullet \bullet \bullet \bullet$ $\bullet \bullet \bullet \bullet \bullet \bullet \bullet$ $\bullet \bullet \bullet$ $g_2(X) = x_3'$ $\bullet \bullet \bullet$ $g_3(X) = x_3.$

A SOP checker is based on dividing the set $\mathcal{Y}$ into two non-empty and disjoint sets Π_0 and Π_1. The SOP checker consists of two independent circuits. Each circuit implements the function

$$R_i = \vee_{\hat{Y}_j \in \Pi_i} m(Y_j)g_j(X) \quad, \quad i = 0, 1,$$

where $m(Y_j)$ is the minterm in the variables $y_0, \ldots y_{k-1}$ that represents the word $\hat{Y}_j$. Indeed $m(Y_j)$ can be written as a product of two minterms $m(c_1^{(j)})$ and $m(c_2^{(j)})$ that represent the sub-words $c_1^{(j)}$ and $c_2^{(j)}$ which compose the word $\hat{Y}_j$, $m(Y_j) = m(c_1^{(j)})m(c_2^{(j)})$.

3.3 Characteristic functions construction

In this subsection we present two greedy approaches for generating the characteristic function in respect to the complexity criteria mentioned above.

Let $\mathcal{G} = \{0, 1, *\}$, and, $p \in \mathcal{G}$. Let a be a Boolean variable. We define a^p as

$$a^p = \begin{cases} a & if \quad p = 1 \\ \bar{a} & if \quad p = 0 \\ 1 & if \quad p = * \end{cases}.$$

Definition 5 (cube) $\bullet$ $\bullet \bullet \bullet \bullet$ $P = (p_{m-1}, \ldots, p_0) \in \mathcal{G}^m,$ $\bullet \bullet \bullet \bullet \bullet \bullet$ r $\bullet \bullet$ $\bullet$ $\bullet \bullet \bullet \bullet \bullet$ $\bullet \bullet \bullet$ $\bullet \bullet \bullet \bullet \bullet \bullet$
$\bullet \bullet \bullet$ 2^r $\bullet \bullet \bullet \bullet \bullet \bullet \bullet$ $\bullet \bullet \bullet \bullet$ $\bullet \bullet$ $X = (x_{m-1}, \ldots, x_0) \in \{0, 1\}^m,$ $\bullet \bullet \bullet$ $\bullet \bullet \bullet \bullet \bullet$ $\bullet \bullet \bullet$ $\bullet \bullet \bullet \bullet \bullet \bullet \bullet \bullet \bullet \bullet \bullet \bullet \bullet$ $\bullet$ $\bullet \bullet \bullet \bullet \bullet$
$\bullet \bullet \bullet \bullet \bullet \bullet \bullet$ $f_P(X) = \Pi_{i=0}^{m-1} x_i^{p_i}$ $\bullet \bullet \bullet \bullet \bullet \bullet$ $\bullet \bullet \bullet \bullet$ $\bullet \bullet \bullet$ $\bullet \bullet \bullet \bullet \bullet$ $\bullet \bullet$ r $\bullet \bullet \bullet \bullet \bullet \bullet$ $\bullet \bullet$ $\bullet \bullet \bullet$ $\bullet \bullet \bullet$ $\bullet \bullet \bullet \bullet$ $* \bullet \bullet$ $\bullet \bullet$ $p.$

An intersection between two cubes P_i and P_j comprises elements in the intersection of the cosets, or equivalently, the assignments of X for which $f_{P_i}(X) \cdot f_{P_j}(X) = 1$. Two cubes are called disjoint if their intersection is empty.

Let F_i be the set of cubes $\{P_j^{(i)}\}_{j=1}^{N_i}$ that are associated with the information word Y_i. Each element X in the union of the cosets defined by F_i satisfies: $f(X) = Y_i$. Clearly, $F_i \cap F_j = \Phi$ for $i \neq j$.

Example 3 $\bullet \bullet \bullet$ $\bullet \bullet \bullet \bullet \bullet \bullet$ $\bullet \bullet \bullet \bullet \bullet \bullet \bullet \bullet \bullet$ $\bullet$ $\bullet \bullet \bullet$ $\bullet \bullet \bullet$ $\bullet \bullet \bullet \bullet \bullet$ $\bullet \bullet \bullet \bullet \bullet$ $\bullet$ $\bullet \bullet \bullet \bullet$ $\bullet \bullet$ $\bullet$ $\bullet \bullet$ $\bullet \bullet \bullet$
$\bullet \bullet$ $\bullet \bullet \bullet \bullet$ $\bullet \bullet$ $\bullet \bullet$

$$
\begin{aligned}
F_1 &= \{(001*0), (1*001)\}, & F_2 &= \{(10000), (00*11), (10110)\}, \\
F_3 &= \{(11*00)\}, & F_4 &= \{(00000)\}, \\
F_6 &= \{(1111*), (01000), (0*101)\}.
\end{aligned}
$$

$\bullet$ $\bullet \bullet \bullet \bullet$ $\bullet \bullet \bullet$ $\bullet \bullet \bullet \bullet \bullet$ $\bullet \bullet \bullet$ $\bullet \bullet \bullet \bullet \bullet \bullet$ $\bullet \bullet \bullet$ $\bullet \bullet \bullet \bullet \bullet \bullet \bullet \bullet$ $\bullet$ $\bullet \bullet \bullet$ $\bullet \bullet \bullet$ $\bullet \bullet \bullet \bullet \bullet$ $\bullet \bullet \bullet \bullet$ $\bullet$ $\bullet \bullet \bullet \bullet$ Y_5:

$$F_5 = \{ (011*0), (0*001), (1010*), (1*101), (10*11), (**010), (01*1*), (*101*) \}.$$

Let $h_i(X)$ be the Boolean function defined by F_i, that is, $h_i(X) = \vee_j f_{P_j^{(i)}}(X)$. The characteristic function $g_i(X)$ covers $h_i(X)$, $g_i(X) \geq h_i(X)$.

3.4 Characteristic functions minimizing the number of literals

In order to reduce the number of literals, each original set of cubes F_i has to be covered by a new set of cubes $\hat{F}_i$, that are of larger order than the original cubes. Namely, let $P = (p_{m-1}, \ldots, p_0) = P_j^{(i)} \in F_i$ be a cube in the set associated with the information word Y_i. Changing one symbol of P from 0 (or 1) to $*$ defines a cube $\hat{P}$ that has larger order and covers P. The w'th symbol of P, symbol p_w, $0 \leq w \leq m-1$, can be changed to $*$ if after the change the modified cube $\hat{P}$ and the set $A_i = \cup_{s|d(\hat{Y}_i, \hat{Y}_s)=1} F_s$ remain disjoint, that is $\hat{P} \cap A = \Phi$.

Example 4 $\cdots$ $\cdots$ F_1 $\cdots$ $\cdots$ $\cdots$ $P_1 = (001 * 0)$ $\cdots$ $P_2 = (1 * 001)$. $\cdots$ $\cdots$ A_1 $\cdots$ $A_1 = F_4 \cup F_6$. $\cdots$ $P_1 \cap A_1 = \Phi$. $\cdots$ $\cdots$ $\cdots$ P_1 $\cdots$ $*$ $\cdots$ $\cdots$ $\cdots$ $(*01 * 0) \cap A_1 = \Phi$. $\cdots$ $\cdots$ P_2 $\cdots$ $\cdots$ $*$ $\cdots$ $\cdots$ $\cdots$ $\cdots$ $\hat{F}_1 = \{(*01 * 0), (* * 0 * 1)\}$ $\cdots$ F_1 $\cdots$ $\cdots$ $\cdots$ $\cdots$ $\cdots$ $\cdots$ $\cdots$ $\cdots$ $\cdots$ $\cdots$ $\cdots$

$$\hat{F}_1 = \{(*01 * 0), (* * 0 * 1)\}, \quad \hat{F}_2 = \{(*0 * **)\}, \quad \hat{F}_3 = \{(*1 * **)\},$$
$$\hat{F}_4 = \{(*00 * 0)\}, \quad \hat{F}_5 = \{(* * * **)\}, \quad \hat{F}_6 = \{(*1 * 1*), (*1 * *0), (* * 1 * 1)\}.$$

$\cdots$ $\cdots$ $\cdots$ $\hat{Y}_5$ $\cdots$ $\cdots$ $\cdots$ $\cdots$ $\cdots$ $\cdots$ A $\cdots$ $\cdots$ $\cdots$ $\cdots$ $\hat{F}_5$ $\cdots$ $(* * * * *)$. $\cdots$ $\cdots$ $\cdots$ $\cdots$ $\cdots$ Y_5 $\cdots$ $g_5(X) = 1$.

Let F be the set that comprises all the cubes: $F = \cup_{i=1}^M F_i$. Denote by $|F|$ the number of products in F, $|F| = \sum_{i=1}^M N_i$. Let $W(P)$ be the number of literals in the product that corresponds to the cube $P = (p_{m-1}, \ldots, p_0)$, $W(P) = |\{w|p_w \neq *, 0 \leq w < m\}|$. We define the $\cdots$ of F as

$$\mathcal{D}(F) = \frac{\sum_{P \in F} W(P)}{m|F|}.$$

In Example 3, the density of the original set is $\mathcal{D}(F) = 73/(5 * 18) = 81\%$, while the density of the encoded set is $\mathcal{D}(\hat{F}) = 16/(5 * 9) = 36\%$. Although the $\cdots$ is not a measure of the implementation's complexity, it can be used as an indicator to the simplification that the suggested approach can provide.

3.5 Characteristic functions with minimized number of inputs

Let $h_i(X)$ be the Boolean function defined by a set of cubes F_i, that is associated with the information word Y_i : $h_i(X) = \vee_j f_{P_j^{(i)}}(X)$. By its definition, $h_i(X)$ is a characteristic function of Y_i. Denote by $\hat{X}$ a subset of the input variables. Let $\hat{F}_i$ be a set of cubes that is constructed from F_i by assigning $*$ at positions that correspond to x's that are not in $\hat{X}$, that is, $\hat{F}_i = \{\hat{P}_j^{(i)}\}_{j=1}^{N_i}$ where $\hat{P} = (\hat{p}_{m-1}, \ldots, \hat{p}_0) \in \mathcal{G}^m$, and

$$\hat{p}_i = \begin{cases} p_i & x_i \in \hat{X} \\ * & x_i \notin \hat{X} \end{cases}.$$

Clearly the Boolean function $\hat{h}_i(\hat{X})$ that corresponds to $\hat{F}_i$ covers $h_i(X)$, that is, $\hat{h}_i(\hat{X}) \geq h_i(X)$. Nevertheless, $\hat{h}_i(\hat{X})$ may not be a characteristic function.

Example 5 • • • •• ••• ••• •• •••• •• •• •••• ••• •• ••• $\hat{X} = \{x_1, x_0\}$, •• • ••••• $\hat{F}_3 = \{(***00)\}$. • •• ••••••••••• ••• •••••••• $\hat{h}_3(\hat{X}) = x_1'x_0'$, •• •••• • •••••••••••• ••• •••••• • • •••••••• ••• $\hat{X} = \{x_3, x_0\}$, ••• ••••••• $\hat{h}_3(\hat{X}) = x_3 x_0'$ •• • ••••••••••••• ••• •••••• •

Denote by $\hat{X}_{opt}$ the the minimal subset of x's for which all the corresponding functions $\{\hat{h}_i(\hat{X}_{opt})\}_{i=1}^M$ are characteristic functions. Then, the functional unit combined with a SOP checker that uses these characteristic functions is fault secure.

Example 6 • • • • ••• •••••••• • •••• ••• ••• ••••••• •• • •••• ••• • •• $\hat{X}_{opt} = \{x_3, x_2, x_0\}$ ••• ••• •••••••••••• • ••• •• •••••• •••

$$\hat{F}_1 = \{(*01*0), (**0*1)\}, \qquad \hat{F}_2 = \{(*00*0), (*0**1), (*01*0)\},$$
$$\hat{F}_3 = \{(*1**0)\}, \qquad \hat{F}_4 = \{(*00*0)\},$$
$$\hat{F}_5 = \left\{ \begin{array}{l} (**0*1), (*01**), (**1*1), (*0**1), \\ (**0*0), (*1***) \end{array} \right\}, \qquad \hat{F}_6 = \{(*11**), (*10*0), (**1*1)\}.$$

Clearly, the two approaches for reducing the checker's complexity, reducing the number of inputs and reducing the density, can be combined.

4 Experimental results

In this section we present results obtained from experiments with a number of ISCAS89 benchmarks; we used the combinatorial part of these sequential circuits which inherently have the context orientation property. The results of the experiments are presented in Tables 1 and 2.

Table 1 compares the suggested structures with duplication solutions and with strict encoding (i.e. coding the information using $\lceil log_2(M) \rceil$ bits). The comparison is done in terms of the density measure. The first column of the Table contains the benchmark name; columns 2, 3, 4 contain the number of inputs m, the number of codewords M, and the number of information bits k. The number of information bits, k_1 and k_2, which are implemented as c_1 and c_2, are written in the 5'th column. The number r_d of x's in a checker having minimal number of literals (minimal density) is is given in column 6, and the minimal number r_i of x's that can be used to construct the checker is given in column 7. The density of the original set $\mathcal{D}(F)$ that indicates the complexity of the duplication and the strict encoding, and the density of the encoded sets are given in columns 8 to 11. The density of a checker designed for minimal number of literals and the density of a checkers designed to minimal number of inputs are denoted by $\mathcal{D}(\hat{F}_d)$ and $\mathcal{D}(\hat{F}_i)$, respectively. The density of a checker that combines both properties is denoted by $\mathcal{D}(\hat{F}_c)$. The last row of the table refers to normalized values of the parameters. The CPU-time of both encoding procedures is given in the two last columns.

Notice that in some cases (e.g. ••••), no partitioning is possible; in such cases $k_2 = 0$.

On average, the encoding scheme that is based on minimizing the number of literals, improves the density by a factor of 0.54 and about 66% of the AND matrix of the PLA that specifiys the characteristic functions contains don't care values.

The simulation results show that the number of inputs r_i cannot be reduces significantly in respect to r_d. The overall improvement in the density obtained by combining the two

Table 1: Density and CPU-time

	m	M	k	(k_1, k_2)	r_d	r_i	$\mathcal{D}(F)$	$\mathcal{D}(\hat{F}_d)$	$\mathcal{D}(\hat{F}_i)$	$\mathcal{D}(\hat{F}_c)$	sec_d	sec_i
s27	7	6	4	(3,1)	6	4	81	38	57	34	0.078	0.141
s298	17	332	20	(16,4)	13	13	98	44	76	44	3.078	75.484
s386	13	23	13	(11,2)	13	12	67	27	92	26	0.047	3.312
s420	35	36	18	(18,0)	17	17	55	42	42	42	0.000	0.031
s510	25	73	13	(9,4)	25	-	27	18	-	-	0.063	$> 5E + 4$
s832	23	70	24	(22,2)	22	22	39	30	59	30	0.563	1.192 E+4
s1494	14	168	25	(21,4)	13	13	64	37	93	37	0.297	15.39
average							1	0.5476	1.0371	0.5272		

Table 2: Number of LUTs in the overall system

	single circuit	circuit c_1 and c_2	suggested scheme	ref. [6] + SOM ch.	strict enc. + SOM ch.
s27	12	(7,5)	22	25	27
s298	2410	(2312,253)	3937	5310	5733
s386	63	(56,10)	163	188	194
s420	104	(104,133)	341	430	383
s510	81	(70,11)	303	401	389
s832	346	(345,4)	725	921	1032
s1494	674	(545,135)	1322	1720	1832
total	1	1.08	1.85	2.44	2.60

approaches, is negligible in respect to the density of the checker designed for minimal number of literals.

Note, that the complexity (in terms of computation time) of constructing a checker that has minimal number of inputs is much larger than the complexity of obtaining a checker with minimal number of literals. In light of the small difference between the overall densities, we find the large computational time consumed for minimizing the number of inputs, unjustified.

Table 2 shows the complexity of the overall system in terms of the number of Look-Up-Tables (*LUTs*). We used •••••••• ••••••••••• and LeonardoSpectrum. The number of LUTs required for implementation of the functional unit as one circuit is written in the second column of the table; the number of LUTs required to implement the functional unit as two independent circuits is written in the third column. Columns 4, 5, and 6 show the number of LUTs required for implementing the overall system, that is, the functional unit circuits plus a SOM (or SOP) checker: the 4'th column corresponds to the proposed scheme with reduced density, the 5'th column corresponds to the coding scheme presented in [6] combined with a SOM checker, and the 6'th column corresponds to a system based on the strict encoding. The table clearly demonstrates the efficiency of the suggested structure. On average, the presented scheme allows detection of any arbitrary fault by increasing the implementation cost by 85%. This is better than the conventional method of duplication or strict encoding, or the method presented in [6].

5   Conclusions

Known techniques for designing concurrently checking circuits are usually based on introducing a significant redundant portion into the original scheme. Reducing this redundant portion is one the main concerns in designing concurrently checking circuits. In a case when all possible circuit's output vectors are known in advance the designing of the such circuits may be simplified by using so-called context-oriented techniques. One of the context-oriented techniques that is in the focus of our study is based on the partitioning of the initial circuit for two independent sub-circuits. The partitioning utilizes the context-orientation property of the original circuit by using correlation between its output variables.

In our paper, we proposed a technique that being based on the partitioning avoids the necessity to introduce any additional redundancy into the initial scheme to be checked. It uses already existing input variables to achieve the required effect of detecting an arbitrary fault. Some of the input variables are used as additional inputs entered into a checker, in addition to the original output variables of the circuit to be checked.

We have formulated theoretical fundamentals of the proposed design method. Based on the fundamentals, we proposed a solution of the problem of selecting the optimized set of input variables to be added to the output vector.

The proposed approach has been implemented and investigated. Experimental results, obtained using a number of standard benchmarks, indicate a significant improvement in detection of arbitrary errors, in comparison with the conventional methods and in terms of the required hardware overhead.

References

[1] Kaushik De., Chitra Natarajan, Devi Nair, Prithviraj Banerjee, "RSYN: A System for Automated Synthesis of Reliable Multilevel Circuits," •••• •••••••••••• •• • •••• ••••• •••••••••• •• •••• •••••• •,vol. 2, no. 2, pp. 186-195, June 1994.

[2] P. Lala, •••••••••• ••• ••••••• •••••• • •••••• • •••••, Morgan Kaufmann Publishers, San-Francisco / San-Diego / New-York/ Boston/ London/ Sydney/ Tokyo, 2000.

[3] I. Levin, M. Karpovsky, "On-line Self-Checking of Microprogram Control Units", ••• •••• •• • • ••••••••••••• • •••••• • •••••••, Capri, pp. 153 - 159, 1998.

[4] I. Levin and V. Sinelnikov, " Self-checking of FPGA based Control Units," • ••••• •• •• •••• •••••• ••• •••••• •• • ••••, pp. 292-295, 1999.

[5] V. Ostrovsky and I. Levin, "Implementation of Concurrent Checking Circuits by Independent Sub-circuits," • ••••••••••• •• ••••• •• • • ••••••••••••• •••••• •• • ••••• ••• ••••••••••••• •• • ••• ••••• • •• • ••••, pp. 343-351, 2005.

[6] V. Ostrovsky, I. Levin, O. Keren, B. Abramov, "Designing Concurrent Checking Circuits by using Partitioning," accepted for publication in the •••••••••••••••••• • •••• • •••••• • •••••••• •••••• • ••••• on Aug-2007.

[7] V.V. Saposhnikov, A. Morosov, Vl.V. Saposhnikov, M. Gossel, "A New Design Method for Self-Checking Unidirectional Combinational Circuits," ••••••••••• ••••••••• • •••• ••• • ••••• ••• • •••••••••, vol. 12, pp. 41-53, Feb. 1998.

[8] E.S. Sogomonyan, "Design of Built-in Self-Checking Monitoring Circuits for Combinational Devices," • •••• ••••• ••• ••• ••• •••••••, vol. 35, no. 2, pp. 280-289, 1974.

Error Detect Logic Resulting in Faster Address Generate and Decode for Caches

Prashant D. Joshi
Intel Corporation
Prashant.d.joshi@intel.com

Abstract

With the complexity of the integrated circuit designs increasing along with the shrinking of the technologies with each passing generation, the vulnerability of these components increases. The sensitivity to transient faults caused by noise, extra terrestrial rays etc. are a cause of concern and require mitigating circuits in any cutting edge of technology. These problems can affect the storage as well as logic elements. The use of arithmetic residues for error detection is well known. This paper explores the use of residues in 1-hot representations of the base and displacement of the address, which are used to generate the word line in a cache. It reduces the delay in comparison with the traditional decode which has no error detection capabilities. It was a welcome case where the addition of error detection capabilities reduced the delays, in comparison with the traditional way of address base-plus-offset summing and decoding in cache accesses.

1. Introduction

The first level cache RAMs of most microprocessors are indexed with the base plus offset addresses. To access such a cache, the address addition must be done first, before the decode can start. This adds an unnecessary delay into the cache access, since the address itself just serves an intermediate stage to be used to be decoded into a 1-hot word line into the cache. Also, to add some logic error checking capabilities into this operation requires more hardware to cater to the indexed part of the address sources only. The read cycle of the cache is one of the most critical paths and often determines the operating frequency of the microprocessor. As such, adding any delay to this path while obtaining error detection is not practical. Residue arithmetic has been widely used in the past for error checking of logical and arithmetic functions. As such, the use of residue arithmetic was explored in the case of error checking in the address generation. Since the address itself was just an intermediate step towards obtaining the decoded 1-hot word lines, combining the 1-hot residues for the base and displacement to generate the final 1-hot word line was explored. In this case, the adoption of this method decreased the delay of the decoded words lines from the original addition indexed decode methods in use.

The next section captures how traditional residue arithmetic has been used for error checking, and how it is applicable to the current problem of decoding. Section III describes how the 1-hot residue combination takes place. Section IV does a comparison of other methods, followed by the conclusion.

As an example the typical addition indexed address generation and decode is show in Figure 1. It shows that bits 9 through 7 of the address are used for the decoding, and hence the addition of the base and address needs to complete till bit 9, before the decode part can start. The carry propagation chain from bit 0 to bit 9, dictates when the decode can begin. For the simple case of 8 word lines the decoder is shown as a set of AND gates that decode the 3 address lines. A typical cache will have many more address bits to be decoded, based on the RAM size, and the number of words lines.

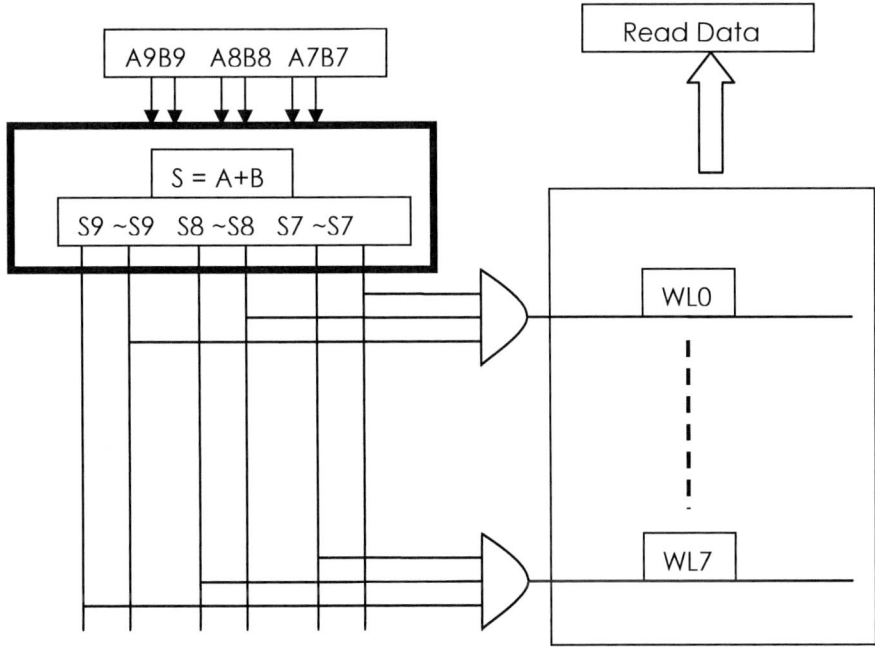

Figure 1. Addition Indexed Memory address generate and decode.

2. Decode using residue concepts to enable error correction

A residue of a number x, is the remainder of x divided by an integer m. This residue is also called x modulo m, and the terms modulo m and residue will be used interchangeably in this paper. Use of residues in arithmetic circuits for error detection is well known. A typical scenario of residue based error detection is shown in Figure 2. A logical or an arithmetic operation like adding is well suited for checking with residues. By applying the same operation on the residues of the operands and then comparing with the residue of the result of the original operands, one can spot errors. By appropriately increasing the value of the number 'm' one can get increasingly greater error coverage, and possibly error detection. There is a cost associated with all this, in the form of added area and power.

In the current problem of generating the sum followed by the decode, one can use the residues in the following way. If the two numbers to be added are the base (B) and displacement (D), one could find the residues of B and D modulo 3 separately. Once the address is added up, the result of the Address modulo 3 is compared with the modulo 3 result of the addition of the modulo 3 Base + modulo 3 Displacement. If they are not equal then an error took place somewhere in the address computation and this can be flagged.

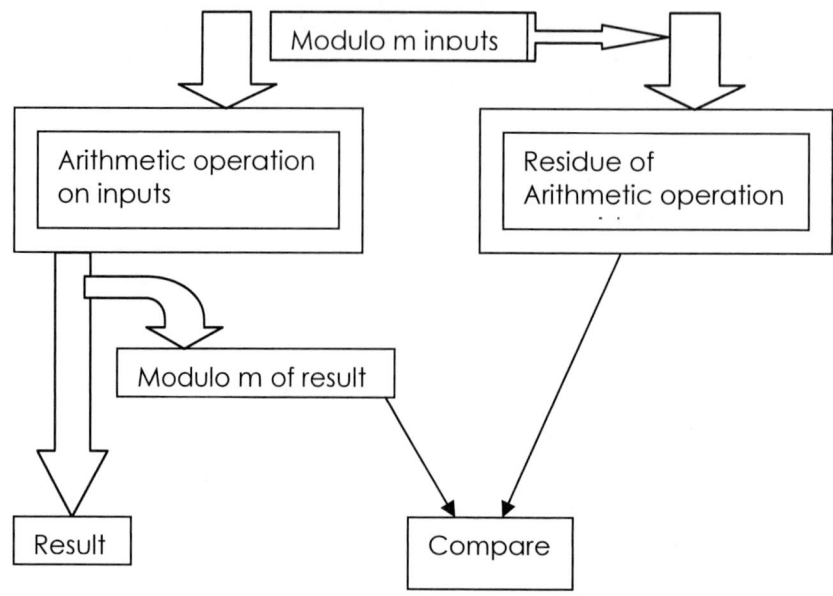

Figure 2. Typical Residue arithmetic/logical checker

To introduce error detection in the address generate and decode, in Figure 3, the arithmetic operation is that of addition and the inputs are the base and displacement numbers. The addition is checked by adding the residue of the inputs, and comparing the residue of this addition with the residue of the original inputs' sum. This is then followed by a traditional decode, which generates the 1-hot word lines.

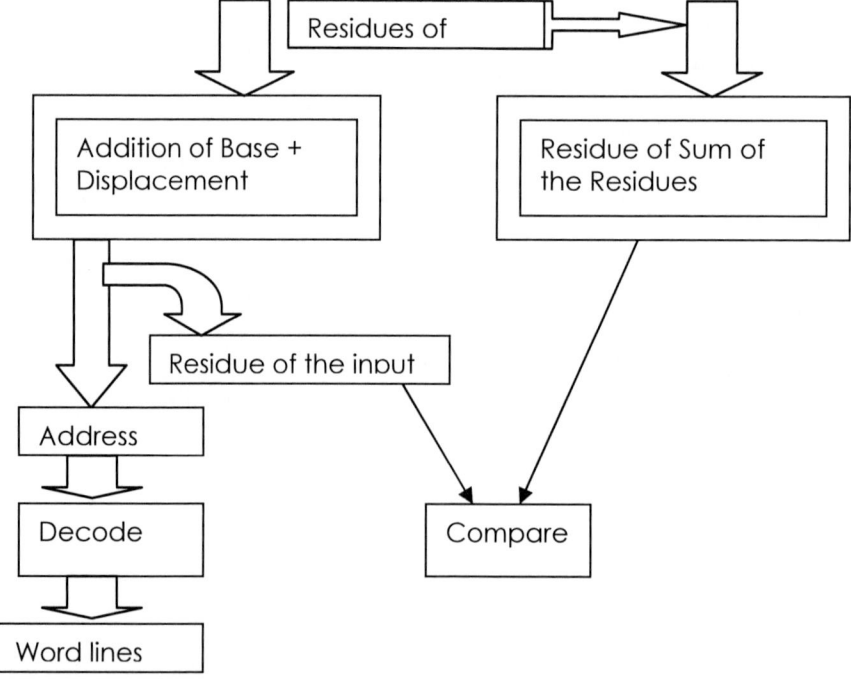

Figure 3. Use of residue arithmetic to check for correctness of address generation

To eliminate the need for the intermediate result of the Address generation, one can think of coming up with a slightly different Decode block that uses its input as the residues of the Base and Address instead. This would require us to be able to represent the numbers Base and Displacement, in the form of the residues of more than one number. If we can find the residue of the Base and Displacement, against 3 numbers say, m_1, m_2 and m_3, such that all are relatively prime, then the Chinese Remainder Theorem tells us that we can obtain a unique representation of the Base or Displacement, only if these numbers are smaller than the product of the three m_1, m_2 and m_3.

Formally, the Chinese Remainder theorem can be stated as follows:

Theorem: Let m_1, m_2, ..., m_k be pairwise relatively prime integers. If a_1, a_2, ..., a_k are any integers, then

. There exists an integer a such that $a = a_i \pmod{m_i}$ for each $i = 1, 2, ..., k$; and

. If $b = a_i \pmod{m_i}$ for each $i = 1, 2, ..., k$, then $b = a \pmod{m_1 m_2 ... m_k}$

Figure 4 shows how this would eliminate the need for the addition, and instead create the need of residue calculations of various relatively prime numbers.

Lemma1 : Let m_1, m_2, ..., m_k be numbers such that pairwise each has a GCD of p, a prime integer. If a_1, a_2, ..., a_k is the quotient of dividing m_1, m_2, ..., m_k with p, then the product p m_1 m_2 ... m_k will be the number of unique numbers represented.

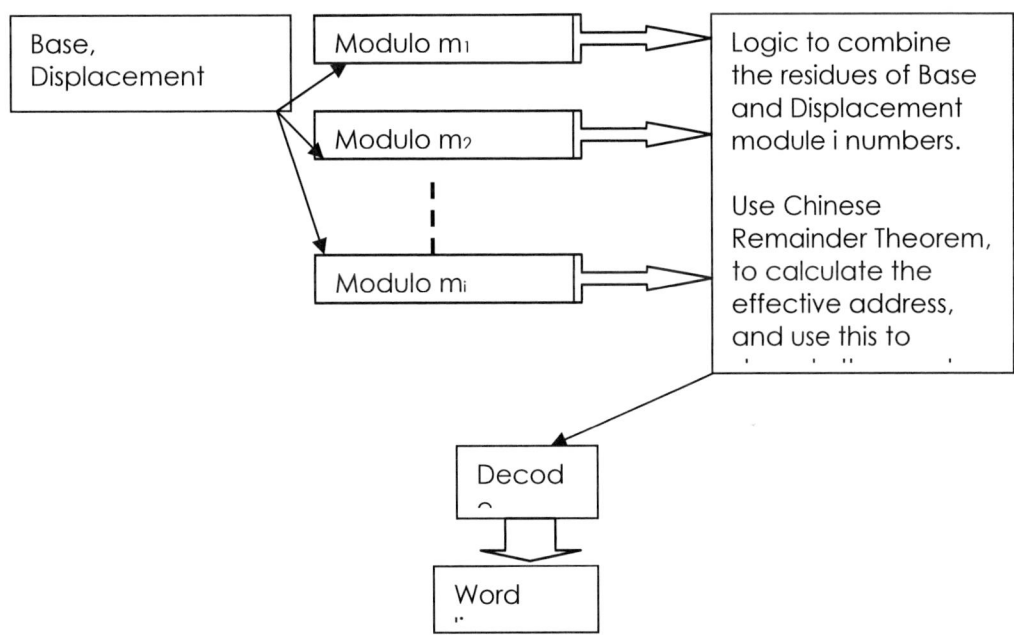

Figure 4. Using Chinese Remainder Theorem to determine the sum

This requires the logic to perform the combination of the residues of Base and Displacement to each of the m_i's and come up with a new set of i residues representing the result of adding the Base and Displacement. This however, has lost the ability to do any error detection. To enable the error detection, a slightly modified version of the Chinese Remainder theorem is used where, the numbers m_i are not mutually prime, but in fact have a common factor[1]. This has some interesting properties. If each of the m_is is a multiple of a prime number p, such that each of the multiples are relatively prime, then the Chinese Theorem now would need to be modified to state that the distinct numbers that are represented by the i residues of the m_i's would be bounded by $m_1 m_2 ... m_i / p^{(i-1)}$ based on Lemma 1.

Lemma 2: Let p be the GCD of two integers m1, m2, and let N be an integer. (N modulo m1) modulo p is equal to (N modulo m2) modulo p.

Using Lemma 2, now we have a method of modifying the steps of Figure 4, to enable error detection. Figure 5 shows how one could use these results to enable the decode using the residues, while enabling error detection using m_is such that each has a pairwise GCD of p.

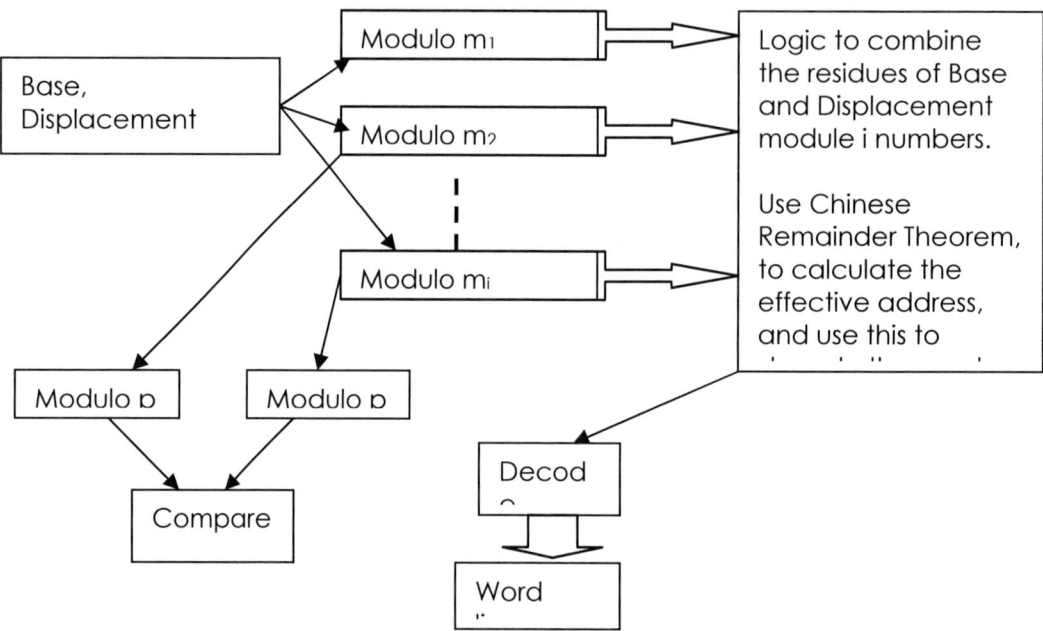

Figure 5. Use of relatively non prime numbers to enable error detection.

In the current case it is useful to use p=2, since the residue against 2 is easy to calculate (LSB). Also, if the number of word lines is 2^j, then m_1, m_2, ..., m_i needs to be such that $2\,m_1\,m_2\,...m_i > 2^{j+i}$. As an example, if the number of word lines needs to be 256, then finding the values of the Base and Displacement modulo 10, modulo 14 and modulo 16 would suffice since the GCD of each pair is 2, $_{and}$ $2*5*7*8 > 256$.

Notice that instead of 16, using 8 as one of our modulo numbers would have been sufficient to satisfy the equation. However, since we are not planning on doing the actual addition of the base and displacement, any carry out of the MSB would unfortunately get counted in our residue decode. As such, we need to ensure that the number of unique numbers represented is $2^{(j+1)}$. This means that we need to ensure that $2\,m_1\,m_2\,...m_i > 2^{j+i+1}$.

Another fact to consider is that since the end product has to be a 1-hot vector, it might be desirable to start proceeding to this format in the intermediate stages, like the residues against the m_is. Also, the assumption is that the number of word lines is a power of 2.

3. Word Line generation using 1-hot residue combination

This section deals with the generation of the residues as 1-hot vectors and combining them to obtain the word line vector. Instead of keeping the residues in binary format, the idea is to represent them in 1-hot notation right from the start, and keep combining these 1-hot vectors to end up with the final 1-hot decoded word line. To generate the residue of the base or the displacement of the address field of these registers, there is no need to wait for an addition to complete on these two. For the typical case of generating the residues in 1-hot notation, each output is simply a sum of products of $S=2^j/m_i$ terms. In the above case, where the number of word lines is 256, the number of sum of product terms is 256/10 or 26 terms. Each minterm has j variables. So for 256 word lines, the number of variables in each minterm is 8. Hence the number of stages in generating the 1-hot residue numbers is $\log_3 j + \log_3 S$. In the above example, this corresponds to $\log_3 26 + \log_3 8$, which corresponds to about 5 stages. To find the 1 hot residue of the case of modulo 10, the logic that needs to be incorporated is as follows. This uses 3 input gates in the logic minimizations.

The generation of the minterms themselves is done in 3 stages. For example the minterm for the value of the base (or displacement) to be 0 is simply the AND of 8 bits: 00000000.

The generation of the 1-hot vector which shows the residue of the Base (or displacement) is simply the sum of at most 26 terms as shown. This can be done in 3 stages:

0 = 0+10+20+30+40+50+60+70+80+90+100+110+120+130+140+150+160+170+180+190+200+210+220+230+240+250
1 = 1+11+21+31+41+51+61+71+81+91+101+111+121+131+141+151+161+171+181+191+201+211+221+231+241+251
...
9 = 9+19+29+39+49+59+69+79+89+99+109+119+129+139+149+159+169+179+189+199+209+219+229+239+249

Now to combine the modulo m_i of the base and displacement requires m_i terms. For example to combine the modulo 10 of the base and displacement into one 1-hot vector requires the following evaluation (where one variable is from base and the other from displacement in the minterms):

0 = 00 + 19 + 28 + 37 + 46 + 55 + 64 + 73 + 82 + 91
1 = 01 + 10 + 29 + 38 + 47 + 56 + 65 + 74 + 83 + 92
...
9 = 09 + 18 + 27 + 36 + 45 + 54 + 36 + 27 + 18 + 09

This would require 3 stages, and all the residues would be calculated in parallel. The worst case would occur for the largest m_i, so 14 would be the worst case, however, the number of stages does not change. Hence the total number of stages up to this point is now $\log_3 m_i + \log_3 j + \log_3 S$. Since the computation of the residues for all the m_is is ongoing in parallel, the total delay to obtain the residue of the addition of base + displacement has taken this time. The end result is i 1-hot vectors which now need to be decoded into one vector of 2^j terms which serves as the word line.

The i-tuple that represents the residues, represents a distinct number. Since we have used residues of relatively non-prime, the same number can have multiple representations. By judicious use of the m_is, we can minimize this duplication to less than 4-5 terms. For example, in the above case, since the total number of unique word lines is 256, but the product of 10.14.8 = 1120, the number of terms that need to be OR'ed is 5, which can be done in 2 stages. The total number of stages up to now is $\log_3((m_1 m_2 \ldots m_i)/2^j) + \log_3 m_i + \log_3 j + \log_3 S$.

The above description covers the decode part of the problem. To see how this would give us error correction, consider that one of the residue calculation for both the base, displacement is being done in the address generation unit. This will ensure that there is at least one calculation of the residue of the sum with p, prior to any data corruption that may occur. Now the base and displacement is sent along with the one hot decoded value of the m_i, as well as the residue of the sum with respect to p. From lemma 2, once we calculate the residue of the residues of sum from the other m_i values and compare that sent along with the address from the address generation unit, we have error detection. The word line decode is done in parallel with the error detection. Since this verification is going on in parallel with the combination of the residues of the base and displacement, this does not add to the critical path. Also since p is a factor of each of the m_is, the number of stages for determining the 1-hot modulo p for each is simply $\log_3(m_i/p)$, followed by the comparison of equality of i 1-hot vectors of length p. If we assume p to be 2, as in the example, this just checks to see if all the residues are even, or all the residues are odd. This can be done in 5 stages, which is in time to determine if an error occurred and request a replay.

The use of buffers for long wires from the address generate to the cache are not considered in the stage count comparison since this applies to all types of solutions, and would be determined by the actual floorplan of the chip.

4. Comparison of regular addition indexed address decode and residue address decode

For the typical case we can estimate what good values of the m_is to use, as well as the corresponding delays this would mean. For 256 word lines, the use of the residues against 16, 10 and 14, the number of stages this would probably mean is as follows. The value of S is 26, i is 3, j is 8. The number of stages hence comes to about $2 + 3 + 2 + 3$, or 10 stages. In the last half of this computation, the error detection is ongoing in parallel.

For a typical adder followed by a decoder we would have 11 stages. In the circuits that were designed for the two approaches, based on 256 word lines and 64 byte cache lines, it was seen that the delay of the residue based decode was better than that of the regular adder-decoder method. The delays have been described entirely in the number of stages rather than actual delays in ps, as these numbers will depend on the technology that these circuits are drawn in. The SAM decode described in [3] also would result in about 10-11 stage delays for this, but has no error detection capabilities.

There is a price to pay however in terms of area and the number of wires to be routed from the address generation unit to the cache. In the number of wires to be routed from the address generating unit is 8 for the regular case, while this number is $8+8+10+10 + 1 = 37$ (base + displacement + residue of base + residue of displacement + odd/even residue) wires for the residue address decode described in this paper .

As far as the number of devices are concerned, the regular decode and the part that combines the i 1-hot numbers into generating the final word lines is about the same. The number of devices that are used to generate the adder are about half those used in the residue calculations for the various m_is, and it depends on the number and values of the m_is.

5. Conclusion

This paper has described an error detection method for address generate and decode. This method uses residues of multiple numbers of the base and displacement, to avoid doing the actual addition of the base and displacement. Instead using Chinese Remainder Theorem the residues of the two numbers are generated and merged, and then used to create the 1-hot word line fed to the cache. This method reduces the number of stages of logic as required in a traditional adder-decoder based word line generation of the cache, while at the same time enabling error detection.

6. Acknowledgements

The author would like to thank his colleague, Dr. Leigang Kou for many fruitful discussions on this and related topics.

References
1. Rajendra S. Katti, "A New Residue Arithmetic Error Correction Scheme", *IEEE Transactions on Computers*, vol. 45, pp. 13-19, January 1996
2. D. Mandelbaun, "Error correction in residue arithmetic", IEEE Transactions on Computers, vol. 21, pp. 538-545, June 1972

3. Raymond Heald et al., "64-Kbyte Sum-Addressed-Memory Cache with 1.6ns Cycle and 2.6-ns Latency", IEEE Journal of Solid-State Circuits, vol. 33, pp. 1682-1689, November 1998.

INVITED TALK

A Case Study of ATPG Delay Path Performance Based on Measured Power Rail Integrity

Zahi Abuhamdeh, Robert Hannagan, *TranSwitch Corporation*

It is a well known problem that power rail integrity can affect the performance of a chip. This degradation of performance can produce failures in extreme chips built closer to Worst Case (WC) process and operated under slowest environmental conditions. ATPG Delay Path can be used to measure the performance of the most vulnerable delay paths of the chip in a production environment and operate as an affective screen for such collective defects in performance and power rail integrity. However, it does not offer diagnostic capability into weather the failure was due to IR Drop or actual process delay on the chip. Process Monitors have been proven to provide an accurate reading into the operating conditions of the power rails. As such, they can be used as a measure of when the total WC conditions have been reached, due to fabrication or operating environment. This presentation will correlate the Process Monitor readings to actual performance degradation and possible failure of chips. By performing an IR Drop measurement in a production environment on several hundred chips, and then grading the performance of the ATPG Delay Path performance, a relationship between the Process Monitor reading and the maximum expected performance of the chip can be reached. This correlation can serve as a diagnostic screen into any power rail integrity issues encountered during ATPG Delay Path testing in the production of a chip.

Speaker Bio – Zahi Abuhamdeh has been the Director of DFT & Diagnostics at TranSwitch Corporation for the past 7 years. His group is responsible for all the DFT work done at TranSwitch. Before this, Zahi founded and operated a DFT consultancy group for companies like Sun Microsystems, Texas Instruments & Lucent Technology to name a few. Before this, he held positions with Toshiba, Texas Instruments & Thinking Machines Corporation. Zahi has received his BS & MS in EE from George Mason University in 1987 & 1989, respectively.

SESSION 8
TESTING TECHNIQUES

IEEE International Symposium on Defect and Fault Tolerance of VLSI Systems

ATPG Heuristics Dependant Observation Point Insertion for Enhanced Compaction and Data Volume Reduction

Santiago Remersaro* (sremersa@mentor.com), Janusz Rajski**, Thomas
Rinderknecht**, Sudhakar M. Reddy*, Irith Pomeranz***
* ECE Department, The University of Iowa, Iowa City, Iowa
** Mentor Graphics Corporation, Wilsonville, Oregon
*** School of ECE, Purdue University, West Lafayette, Indiana

Abstract

As digital circuits grow in gate count so does the data volume required for manufacturing test. To address this problem several test compression techniques have been developed. This paper presents a novel and scalable technique for inserting observation points to aid compression by reducing pattern count and data volume. Experimental results presented for industrial circuits demonstrate the effectiveness of the method.

1. Introduction

Test point insertion has been studied in the literature for a long period of time. Most of the previous work done on test point insertion (TPI) has been in the area of logic Built-In Self Test (BIST) [1][2]. The goal of BIST oriented TPI is to decrease the pattern count needed to achieve the desired fault coverage by increasing the testability of hard to detect random pattern resistant faults. Mainly, two techniques have been used to identify target sites for test point insertion: exact fault simulation [3][4] and approximate testability measures [5][6][7].

In recent years, with the invention and commercialization of test compression techniques such as EDT [8] and Broadcast Scan [9], logic BIST has become less utilized and deterministic test gained more importance.

When compression, along with ATPG, is utilized the relevant variables that determine its performance are the fill rate (percent of specified bits in test patterns) of the patterns created by the ATPG engine, the number of patterns and the total data volume, defined here as the sum of the specified positions before X-fill in every test cube in the test set. These parameters determine to what extent the scan chains should be segmented for compression and this in turn determines the maximum achievable compression ratio. The usual result after applying compression is that the pattern count is similar to ATPG without compression, but the data volume is reduced and hence tester time needed for the circuit under test (CUT) decreases.

Thus, two objectives gain relevance for test point insertion in deterministic test: enhanced compaction (pattern count reduction) and data volume reduction. Pattern count reduction impacts test cost by directly reducing test time. Each pattern takes a fixed amount of time to be loaded from the tester to the CUT, depending on the maximum length of the scan chains. If fewer test vectors are needed, less tester time will be used. Fill rate reduction, when pattern count remains fixed, allows higher compression ratios by properly adjusting the maximum scan chain length. This in turn results in shorter test application time per pattern and, again, a reduction in tester time utilization. The objective of our work is to develop a technique that will identify locations for observation points (OPs) that enhance pattern compaction, i.e. reduce the pattern count, and reduce the specified bits prior to X-fill.

The remainder of the paper is organized in the following manner: Section II reviews the related

1550-5774/08 $25.00 © 2008 IEEE
DOI 10.1109/DFT.2008.39

385

existing literature in TPI. Section III explains how to find the possible locations for OPs and describes the proposed method. Section IV gives experimental results based on industrial designs and Section V concludes the paper.

2. Previous works

Previously, some papers have been published that insert test points (TP) with the objective of enhancing compaction to reduce test application time [10]-[14]. They will be described in this section.

In [10], a method was proposed that combines approximate testability measures from Controllability/Observability Program (COP) and Sandia Controllability/Observability analysis program (SCOAP) and Test Counts, which is the number of times a line must become zero (one) during the application of a test set. Three methods are derived and compared, one based on COP, another on SCOAP and a third that integrates Test Counts. The type of test point inserted is a transparent scan cell at the output of a gate. A transparent scan cell behaves as a buffer during the circuit normal mode and as a scan cell in test mode. The authors ran experiments for the three methods on ISCAS circuits and for the COP and Test Count methods on small industrial designs using the single stuck-at fault model, inserting in industrial designs around one TP per thousand gates (more TPs in the smaller circuits). For the industrial designs an average reduction in pattern count of 36.39% for the COP method and 39.26% for the Test Count method was obtained. The average time required to compute the Test Count test points was 59.49% of the ATPG time required for the baseline (pre-TP insertion) ATPG run, but the ATPG time after TP was reduced to 43.58% of the original. In [11] an extension of the methods in [10] to work with transition faults is given by the same authors. Results are obtained for ISCAS circuits and a subset of the industrial designs of [10].

In [12], a method for inserting OPs to enhance test compaction for the single stuck-at fault model is presented. The method starts from a compact test set and reduces it by combining two of its test vectors (τ_0, τ_1) into one. It achieves this by targeting the faults detected by both τ_0 and τ_1 with a new test pattern $\tau_{0,1}$ and adding OPs to detect the activated but unobserved faults. If every fault detected by τ_0 and τ_1 is detected by $\tau_{0,1}$ after inserting observation points, the two initial tests are replaced by $\tau_{0,1}$. Results are given for some ISCAS benchmark circuits.

In [13] and [14] the authors introduce a method based on fault detection probabilities and value assignment probabilities to enhance compaction of patterns. The authors propose algorithms to search the circuit lines for test point locations, both control and observe, based on these metrics. In both works, a set of three industrial designs are used to prove the effectiveness of the method. In [14] the designs are broken into their constituent blocks and test points are inserted in the blocks with the highest pattern count. The main purpose was to reduce tester time, measured as a function of test pattern count, achieving a reduction of 33.62% in this metric.

Another work [15], does not share the same objective of improving compaction in deterministic test, it is designed for BIST but it introduces several concepts relevant for TPI. Also, it is implemented in a commercial tool which is available and will be used for comparison purposes. Given these facts, and for the sake of completeness, some of its key new concepts are discussed here. Multi-Phase Test Point Insertion (MTPI) utilizes the AND-OR type of control point. It introduces the control points at the outputs of gates to control the value of a stem and in this way improve its controllability. In MTPI several control points are controlled by the same input. This reduces the extra logic that needs to be inserted. The test patterns are divided into a small number of phases. During each phase a subset of control points are enabled, increasing the testability of certain faults (possibly blocking others), while the rest are disabled. OPs on the other hand, are always enabled. To reduce the number of scan cells needed to capture the OPs values, several observation points are wired together through XOR gates (this concept is not novel to MTPI). To find the test point locations MTPI simulates a small number of random patterns and based on them it calculates signals with low controllability as potential control points and signals with low observability as potential OPs.

3. The proposed method

The principal step in OP insertion is the addition of a branch to the output of a gate. This branch in turn feeds an extra output of the circuit, preferably a scan cell and not a primary output. The addition of a scan cell to the circuit introduces an area penalty. When the number of scan cells added to a circuit becomes large this area penalty may become unacceptable due to design constraints. Due to this there is a limit on the number of OPs that can be inserted, and not all gates can be chosen. This calls for careful selection of the observation points. Since an XOR gate takes less area than a scan cell, two OPs can be observed together using an XOR gate. In this way more observation points can be inserted with similar area overhead. We will not discuss the problem of how many OPs can be inserted because it depends on design constraints. Instead, we will focus on finding a heuristic that successfully identifies the best observation points for a given circuit in an orderly manner. The first identified OPs should give the greatest benefit in compaction and data volume reduction. Progressively adding more OPs would add diminishing returns.

OPs cannot be used to reduce pattern count for the path delay or the transition fault models. For path delay faults, an OP will observe only the first part of the path. The industry currently uses the transition fault model to also detect small delay defects. This requires propagating the fault effect (FE) through a path that has the smallest slack. An OP will reduce the combined delay of the fault and the gates that the FE traverses, increasing in this way the minimum delay that the test set can detect. This makes the insertion of OPs risky for these fault models, therefore we will not analyze the effectiveness of OPs for them.

In this work we will focus on identifying OPs for the single stuck-at fault model. We will target the SAF fault list of a circuit using combinational patterns. The remainder of this paper assumes stuck-at faults and combinational test patterns in the discussion.

3.1. Overview of the proposed procedure

The proposed procedure first runs ATPG on the circuit. When a cube is created for a target fault it is immediately analyzed to extract one of the FE propagation paths. Then the input assignments needed to propagate the FE through the propagation path are identified. These input assignments are divided into two categories: essential and propagation, defined later. With this information a database is created. After the ATPG is done with test generation a post-processing step on the database is done to select OPs. After each OP selection the database is updated to reflect the presence of the newly identified OP. When the desired number of OPs is identified they are inserted into the circuit and ATPG is run again to create the final test set. Figure 1 shows an overview of the flow of this process.

If the ATPG heuristics that choose the propagation path and the gate input to justify a required value are changed, the information contained in the database will change and the OPs inserted will be different.

3.2. Creating a database from the ATPG run

The goal for this step is to create a database of input assignments for each test cube that is created when ATPG is targeting a fault as a primary objective. Each input assignment will be associated with a gate in the FE propagation path. This database will be used to select the OPs.

Every time a fault is targeted by ATPG as a primary target and a cube is successfully created for it, the propagation path of the fault is traced backward from the detection gate, PO or pseudo-primary output (PPO) of a scan cell, to the fault site. Only the first encountered propagation path is recorded, even if the fault has multiple paths. The tracing operation is linear in the circuit depth.

Once one FE propagation path is found we record for every gate in it the input, meaning both primary inputs and pseudo-primary inputs, assignments necessary to propagate the fault effect to the gate output. In the case of the fault site, and if the fault is at one of the gate inputs, we combine the activation assignments with the assignments made to propagate the fault to the gate output. The reason behind this is that a fault should not be observed at its fault site. The earliest point of observation is the output of g. This is because the only way to insert an OP is at the output of a gate, if we try to insert an OP at a branch a new fault will appear between the OP and the gate input which is fed by the branch, which is equivalent to say that the OP is at the stem. We will call the set consisting in the union of the activation input assignments and the propagation input assignments needed to propagate the FE to the output of the fault site *essential* input assignments. The rest of the input assignments will be called *propagation* input assignments and the will be associated with a gate in the FE path.

A. For each fault targeted as a primary objective do
    1. *Compute Propagation Path from PO to fault site*
    2. *Compute Essential Input Assignments*
    3. *Compute Propagation Input Assignments per gate in the Propagation Path*
    4. *Store in database*
When ATPG/data gathering phase is complete:
B. For the desired number of observation points do
    1. *Check each gate cost function*
    2. *Select OP as the gate output with higher cost function*
    3. *Update database removing assignments after OP*

Figure 1: OP selection process flow

Observation 1: Since we need to propagate a FE to a gate output to observe it, *essential* input assignments cannot be avoided with OPs. On the other hand, *propagation* input assignments can be eliminated with OPs, since the OP eliminates the need to sensitize part of the observation path.

To find the input assignments that a gate needs to propagate a FE, we trace every input of a gate in the identified propagation path backwards. If a gate has a controlling value at its output one or more of its inputs is set to a controlling value. In this case, we will mark as required the first input that is set to a controlling value. If during this process we encounter a gate g that was already transited to justify a value for a gate that appears earlier in the propagation path, we stop tracing. In this case the inputs assigned to justify a value in g are needed to propagate the FE through multiple gates but we only assign those inputs to the gate in the propagation path of the FE that has the lower gate level. We do this because if an observation point is added in the propagation path to remove some of the input assignments the ones belonging to a gate with a lower level will still be necessary even if they are not needed in the upper level gate.

Observation 2: By ending the tracing at a previously traced gate, the operation becomes linear in the number of gates of the input cone of the detection gate.

Observation 3: If a fault has multiple propagation paths, every propagation path after the first can be treated as assignments needed in the faulty circuit that will map into assignments needed in the good circuit eventually.

In Figure 2 we illustrate the process of mapping a faulty circuit value into required good circuit values. In this figure, required values are underlined and the first propagation path of the fault effect is in italics. First it is important to observe that to activate f we need a zero in one of the inputs of $g1$, this is the *essential* assignment for f. We can consider this a case in which a faulty circuit value (0/1 at the output of $g1$) maps into a required good circuit value (0 at one of its inputs). We consider the propagation path *g1, g3* as the primary and the other (*g1, g2, g3*) as assignments in faulty circuit that will map by tracing back into assignments in good circuit or paths that will end up in the fault site. If we analyze Figure 2 we see that to propagate the fault effect through *g3* only the one in the fault circuit is needed at the output of *g2*, and that the zero in the good circuit is unnecessary. Then if we trace back the one in the faulty circuit at the output of *g2* we find the fault site and a one in good circuit. This process can always be applied since the presence of any faulty circuit value can be explained by the presence of the fault and the input assignments, which are good circuit assignments.

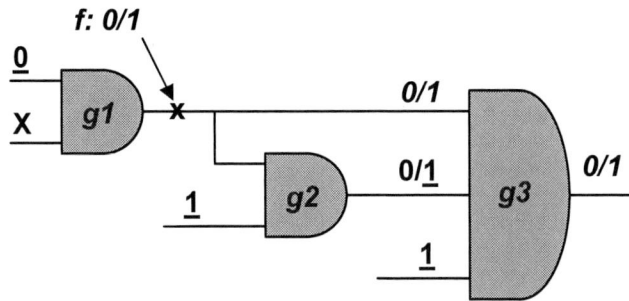

Figure 2: Mapping faulty circuit to good circuit

After this process is complete no faults remain in the fault list, because either they were targeted by ATPG or dropped by fault simulation. For every fault that was targeted as a primary objective and a cube was created for it we have one of its propagation paths and for every gate *g* in the FE propagation path we have the inputs that were assigned to a logic value to propagate through *g*.

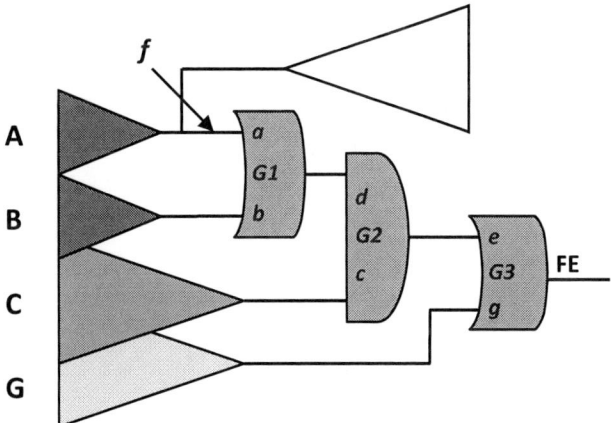

Figure 3: Input assignment identification process

Figure 3 illustrates the process of creating a database from the cubes created by the ATPG. Fault *f* is located at input *a* of gate *G1*. The sets of input assignments A and B of the cones feeding *a* and *b* cannot be avoided by inserting OPs. Thus the sets of inputs assignments A and B will be identified as *essential* to *f* and the sets C and G will be classified as *propagation* input assignments. The set of assignments $C - A \cup B$ will be associated with *G2* since it is the required set of *propagation* input assignments to propagate through *G2*. The set of assignments $G - A \cup B \cup C$ is associated to *G3* for the same reason. If we observe the output of gate *G2*, we will not

389

need $G - A \cup B \cup C$, the set of inputs assignments to set g to a zero that are not required for setting a value in a gate of lower level in the FE path. Also, if we observe the output of gate $G1$ the input assignments $G \cup C - A \cup B$ are not needed to detect the fault. In this example the best OP will be the output of gate $G1$ but given the presence of multiple faults and reconvergent fanout in the circuit several contributions coming from faults in different parts of the circuit changes the outcome.

Next we introduce two methods to locate OPs. One is designed to reduce pattern count and the other to reduce data volume, defined as the sum of the specified inputs in all test cubes before random fill.

3.3. Conflict reduction (CR) oriented method

Two faults cannot be targeted with a single test when a conflict exists in the assignments needed to activate and propagate them to an output. This inter-fault conflict differs from the usual meaning of the term conflict, which is between assignments to create a test for a target fault. Here the word conflict is used to refer to the inter-fault conflict.

Given a circuit input and a set of test cubes for every circuit fault (the database), the number of compatibility conflicts created by the cubes in it is min $\{E_0+P_0, E_1+P_1\}$, where E_0 is the number of times the input was set to 0 to excite the fault and to propagate it to the output of the gate which the fault is associated with i.e. the number of the cubes with zero *essential* input assignments in that input, and P_0 is the number of times the input was set to 0 to propagate through a gate other than the one at which the fault is located i.e. the number of cubes with zero *propagation* input assignments. All avoidable conflicts can be eliminated by observing every gate output in the circuit, and then the remaining conflicts will be min $\{E_0, E_1\}$. Then for a given input the number of avoidable conflicts will be min $\{E_0+P_0, E_1+P_1\}$ − min $\{E_0, E_1\}$.

The database records the extra input assignments needed to propagate the FE through every gate in the FE propagation path. If an observation point is inserted at the output of a gate g, all the input assignments that are needed to propagate a FE in every cube through an upper level gate are not needed any more. We will consider that a conflict is removed only if the input assignment removed is of the same logic value as min $\{E_0, E_1\}$ because this is the minimum achievable conflicts for that input.

Now we can compute for every gate in the circuit how many conflicts will be removed if its output is observed. We will select the output of a gate g that will remove the most conflicts for every circuit input as our first OP. After this we will search through the FE propagation path of every cube and remove all the input assignments of gates in the propagation paths that come after g. Then we will repeat this in the actualized list of FE propagation paths and inputs assignments until we have identified the desired number of OPs. In this work we identified up to one OP per five hundred gates in the circuit.

3.4. Data volume reduction (DVR) oriented method

Using the same information from the FE propagation paths we derived a second technique oriented at reducing the number of specified positions in the test cubes before random fill. After collecting the information from the cubes we will assign to each gate output a cost function. This cost function is simply the sum over all the cubes of the number of input assignments that will become unnecessary if that gate output were to be observed. The gate output with the highest cost function will be selected as the first OP and the input assignments that come after that gate will be removed from the cubes. After that the second iteration will select the second best OP and so on.

Both methods yielded very similar results due to the high correlation in the sets of OPs identified. The more input assignments that an observation point will remove the more likely that those removed input assignments will remove input conflicts. The data volume reduction (DVR) method runs faster (~4X) than the conflict reduction (CR) method. This is the only significant difference in the results for the two methods.

4. Experimental results

All the results in this section are using the single stuck-at fault model tested with combinational patterns. In all instances Test Coverage (TC) was the same, or slightly higher, than without TPs. The insertion of OPs never reduced TC and any increase, when it happened, although beneficial, is not reported because it is not the target in this work. Only relative run times are given since absolute times vary from one computer to another. Also the relative run time results have some noise due to the varying loads on the computers running the experiments. All results shown were obtained using a commercial ATPG tool.

In Table 1, the circuit name is given in the first column. The gate count in the circuit under test rounded to the nearest thousand is given in the second column. After this we show the ratio between time spent in observation point identification for the CR method and the ATPG time. The other two columns show respectively the original pattern count and data volume, measured in bits before filling the unspecified values, needed to test the circuits. It can be seen from Table 1 that the OP identification process will take approximately twice the time of a regular ATPG run. In the last row N/A means non-applicable.

In Table 2, one of the circuits for which the ATPG runs faster (due to its small size) is used to analyze the effectiveness of the method and how much additional benefit we can obtain from inserting progressively more OPs. In Table 2, the first column has the number of observation points, and then under the test case name there are two columns: the first is the number of patterns that the ATPG produced and the second is the specified bits before X-filling of the entire test set. Here we employed the CR method to identify OPs. It can be seen from Table 2 that as more OPs are inserted further reduction in pattern count and data volume (sum of specified bits) is achieved. There are very few exceptions in which the curve is not monotonic. This behavior can be attributed to the ATPG heuristics that order the fault list producing noise when a few new faults (due to the observation points) are added to the fault list.

In Table 3, a comparison between the CR and DVR methods is conducted for the eight circuits in the experiment. In Table 3, the first column shows the number of OPs inserted relative to the circuit gate count, the next column *Pattern Count Red %* shows the average percentage pattern count reduction for the eight circuits in the experiment for both the CR and DVR methods. Last, under the column labeled *Data Volume Red %*, the same information is shown for the average percentage data volume reduction. It can be seen from the experimental data that both methods perform similarly for the same number of OPs. The difference is minimal and can be attributed to the fault order.

In Table 4, a comparison between our method and MTPI [15] is done for the circuits whose netlists were available in Verilog format. This was necessary to run MTPI on them. While MTPI was devised for BIST, we used if for comparison in deterministic test due to its availability. In Table 4, after the circuit name we show the pattern count reduction both for the DVR method and MTPI. In the last column we show the data volume reduction for both methods. It can be seen from the data that DVR clearly outperforms MTPI when applied to deterministic test.

In Table 5, results obtained using an industrial compression tool based on EDT [8] are shown for the circuits using the DVR method to compute OPs. In Table 5, after the circuit name, under the column *Original*, we show the original pattern count using EDT. Under the column *Red %* we show the percentage pattern count reduction when inserting one OP per 500 circuit gates after the tool was done generating patterns. Then, under the column *@ Same TC*, we show the percentage pattern count reduction at the same test coverage that was achieved without OPs. Last, under the column *Chains*, we show the number of scan chains that each circuit has. Since the number of scan chains of every circuit in the sample was small we only use one scan channel for each circuit. Due to the fact that we were not interested in benchmarking compression but to compare results with and without OPs we did not reconfigure the scan chains to obtain a better compression ratio. Since compression is being used, we do not report data volume which is proportional to pattern count. From the large reductions obtained it can be seen that the insertion of the OPs under compression greatly reduces test set length.

5. Conclusions

In this paper we presented a new observation point insertion technique to enhance compaction and reduce data volume for scan based test. The method analyzes the cubes created via ATPG and based on them inserts the test points. Because of this the method is ATPG heuristics dependant. Since the method run times do not exceed those of test pattern creation the technique is scalable to large industrial designs. Experimental results with industrial designs showed substantial reductions in pattern count and total specified positions before X-filling. Comparisons with other test point insertion methods demonstrate the advantages of the method. Experiments with industrial test compression tools show that under compression environments the technique is applicable and achieves large benefits in test duration and therefore test cost.

References

[1] V.D. Agrawal, C.R. Kime and K.K. Saluja, "A Tutorial on Built-In Self Test, Part 1: Principles" IEEE Design and Test of Computers March 1993, Volume 10, Issue 1, pp. 73-82.

[2] V.D. Agrawal, C.R. Kime and K.K. Saluja, "A Tutorial on Built-In Self Test, Part 2: Applications" IEEE Design and Test of Computers April 1993, Volume 10, Issue 2, pp. 69-77.

[3] A.J. Briers and K.A.E. Totton, "Random Pattern Testability by Fast Fault Simulation", Proc. ITC 1986, pp. 274-281.

[4] V.S Iyengar and D. Brand, "Synthesis of Pseudo-Random Pattern Testable Designs", Proc. ITC 1989, pp. 501-508.

[5] K.-T. Cheng and C.-J. Lin, "Timing-Driven Test Point Insertion for Full-Scan and Partial-Scan BIST", Proc. ITC 1995, pp. 506-514.

[6] Y. Savaria, M Yousef, B. Kaminska and M. Koudil, "Automatic Test Point Insertion for Pseudo-Random Testing", Proc. International Symposium on Circuits and Systems 1991, pp. 1960-1963.

[7] B.H. Seiss, P.M. Trouborst and M.H. Schulz, "Test Point Insertion for Scan-Based BIST", Proc. of 2nd European Test Conference 1991, pp. 253-262.

[8] J. Rajski, J. Tyszer, M. Kassab and N. Mukherjee, "Embedded Deterministic Test", IEEE Transactions on CAD of Integrated Circuits and Systems, Vol. 23, No. 5, May 2004, pp. 776-792.

[9] K.-J. Lee, J.-J. Chen and C.-H. Huang, "Using a single input to support multiple scan chains", Proc. International Conference CAD 1998, pp. 74-78.

[10] M.J. Geuzebroek, J.Th. van der Linden and A.J van de Goor, "Test Point Insertion for Compact Test Sets", Proc. ITC 2000, pp. 292-301.

[11] M.J. Geuzebroek, J.Th. van der Linden and A.J van de Goor, "Test Point Insertion that facilitates ATPG in reducing test time and data volume", Proc. ITC 2002, pp. 138-147.

[12] I. Pomeranz and S.M. Reddy, "Test-Point Insertion to Enhance Test Compaction for Scan Designs", Dependable Systems and Networks, 2000. Proceedings International Conference on, 25-28 June 2000, pp. 375 – 381.

[13] M. Yoshimura, T. Hosokawa and M. Ohta, "A Test Point Insertion Method to Reduce the Number of Test Patterns", ATS 2002, pp. 298-304.

[14] M. Yoshimura, T. Hosokawa and M. Ohta, "Design for Testability Strategies Using Full/Partial Scan Designs and Test Point Insertions to Reduce Test Application Times", ASP-DAC 2001, pp. 485-491.

[15] N. Tamarapalli and J. Rajski, "Constructive Multi-Phase Test Point Insertion for Scan-Based BIST", Proc. ITC 1996, pp. 649-658.

Table 1: Observation Point Selection time

Circuit	K Gates	OPI time / ATPG time	Pattern Count	Data Volume
C_210	210	0.49	1002	733330
C_260	260	0.37	8745	4357725
C_419	419	0.15	4613	3082955
C_845	845	0.47	1029	749927
C_2007	2007	1.11	2325	2766406
C_2225	2225	1.49	28038	9843423
C_2500	2500	1.07	7525	5457533
C_2508	2508	1.52	4729	7621035
Average	N/A	0.83	N/A	N/A

Table 2: Results for C_260

# of OP Inserted	C_260	
	# Patterns	Data Volume
None	8745	4357725
1	8130	4156080
2	7454	3948161
3	6889	3768812
4	6101	3536561
5	5532	3364807
6	4774	3144804
7	4133	2967506
8	3727	2782390
9	3726	2602537
10	3681	2415063
11	3649	2253949
12	3661	2173013
1 / 40000 Gates	4774	3144804
1 / 20000 Gates	3660	2170960
1 / 10000 Gates	3654	2135085
1 / 5000 Gates	3666	2074099
1 / 2000 Gates	3673	1929533
1 / 1000 Gates	3675	1734588
1 / 500 Gates	3667	1534205

Table 3: CR vs. DVR

# of OPs / # Gates	Pattern Count Red %		Data Volume Red %	
	CR	DVR	CR	DVR
1/ 10000	28.06	30.38	17.90	19.44
1/ 5000	32.22	32.42	21.18	22.53
1/ 1000	37.84	37.47	29.51	30.74
1/ 500	41.44	41.55	34.50	34.40

Table 4: DVR vs. MTPI

Circuit	Pattern Count Red %		Data Volume Red %	
	DVR	MTPI	DVR	MTPI
C_260	58.50	5.22	62.29	5.60
C_419	68.01	53.00	54.31	13.39
C_2007	16.42	7.52	56.73	18.31
C_2225	13.45	6.69	32.58	2.91
C_2508	57.04	37.97	9.07	13.11
Average	**42.68**	22.08	**43.00**	10.66

Table 5 Results with EDT

Circuit	Original	Red %	@ Same TC	Chains
C_210	1122	47.27	47.27	5
C_260	9051	57.65	60.74	4
C_419	7509	51.44	67.16	19
C_845	1096	30.02	30.02	16
C_2007	2997	22.06	42.01	8
C_2225	27825	14.34	15.42	14
C_2500	6359	12.19	14.62	64
C_2508	5750	53.55	57.60	32
Average	N/A	36.06	41.85	N/A

IEEE International Symposium on Defect and Fault Tolerance of VLSI Systems

Detection of Transistor Stuck-open Faults in Asynchronous Inputs of Scan Cells*

Fan Yang[1], Sreejit Chakravarty[2], Narendra Devta-Prasanna[2], Sudhakar M. Reddy[1] and Irith Pomeranz[3]

[1]*ECE Department*
University of Iowa
Iowa City, Iowa, USA
{fyang\reddy@engineering.uiowa.edu}

[2]*LSI Corporation*
Milpitas, California, USA
{narendra.devta-prasanna\sreejit.chakravarty@lsi.com}

[3]*School of ECE*
Purdue University
West Lafayette, Indiana, USA
{pomeranz@ecn.purdue.edu}

Abstract

Typically faults in gates internal to scan cells are assumed to be detected using what are known as flush tests. Recently we have shown that several scan cell internal faults are not detected either by the existing flush tests or by ATPG tests. A new set of tests to detect such faults were proposed. These works considered faults in fully-synchronous flip-flops. In many designs asynchronous inputs are used to set and/or reset flip-flops. Considering a scan cell implementation used in an industrial design we show that stuck-open faults in some transistors driven by asynchronous inputs require two new flush tests. Such faults, if left undetected, cause functional failures. The two new tests increase the overall stuck-open fault coverage of each scan cell by approximately 5%. This will significantly improve the overall test quality due to the large number of scan cells contained in large industrial designs.

1. INTRODUCTION

Scan chains are universally used in VLSI designs to facilitate loading the tests applied to detect defects as well as observing the circuit responses to the tests. Faults in the scan chains may affect the functional mode of operation if left undetected. This is because scan cells contain the functional flip-flops. It was noted earlier that a large percentage of transistors in large industrial designs reside in scan cells [10]. Flush tests that scan sequences containing 1s and 0s into the scan chains are typically used to ascertain the integrity of the scan chains. Additionally, automatic test pattern generation (ATPG) tools are used to generate tests targeting stuck-at faults (SAFs) and transition delay faults (TDFs) at inputs and outputs of scan cells. We call the latter tests ATPG SAF/TDF boundary tests in this paper. In [1-9] it is shown that these tests miss many internal faults in the scan cells.

---

*Work of Fan Yang and S.M. Reddy was supported in part by SRC Grant No. 2004-TJ-1583 and the work of I. Pomeranz was supported in part by SRC Grant No. 2004-TJ-1244.

1550-5774/08 $25.00 © 2008 IEEE
DOI 10.1109/DFT.2008.11

Much work has been done to detect faults internal to the flip-flops. Makar and McCluskey developed checking sequence based functional tests to detect internal faults in flip-flops and latches [4-7]. In [8], Makar and McCluskey presented checking sequence based IDDQ testing to detect some faults internal to the scan cells. However, checking sequence based tests may not be practical due to the large number of tests generated and the difficulty of generating the tests. Recently, in [10, 17], we investigated the detection of SAFs, stuck-on faults (SONs) and stuck-open faults (SOPs) internal to the scan chains. These investigations lead to the discovery that standard test methods using flush tests and ATPG boundary tests achieve similar fault coverage as checking sequence based tests. However, additional flush tests applied at lower speed have to be added in order to detect additional faults. Design methods to minimize the occurrence of undetectable SOPs were proposed in [16]. Methods to augment scan cells to detect all stuck-open faults were proposed in [3]. However, the area overhead of these methods is high. All the earlier works consider fully-synchronous scan flip-flops only. In this work, we consider open faults in transistors driven by asynchronous set/reset inputs.

The resistive open fault model is used in this study by injecting a large resistance between the drain, source or gate of the faulty transistor and the node to which they would otherwise be connected. Two pattern tests are required for the detection of open faults [11]. If the resistance introduced by the open is not large, the open will cause a small delay defect [15]. In this paper, we consider large resistance opens only which cause gross delay defects.

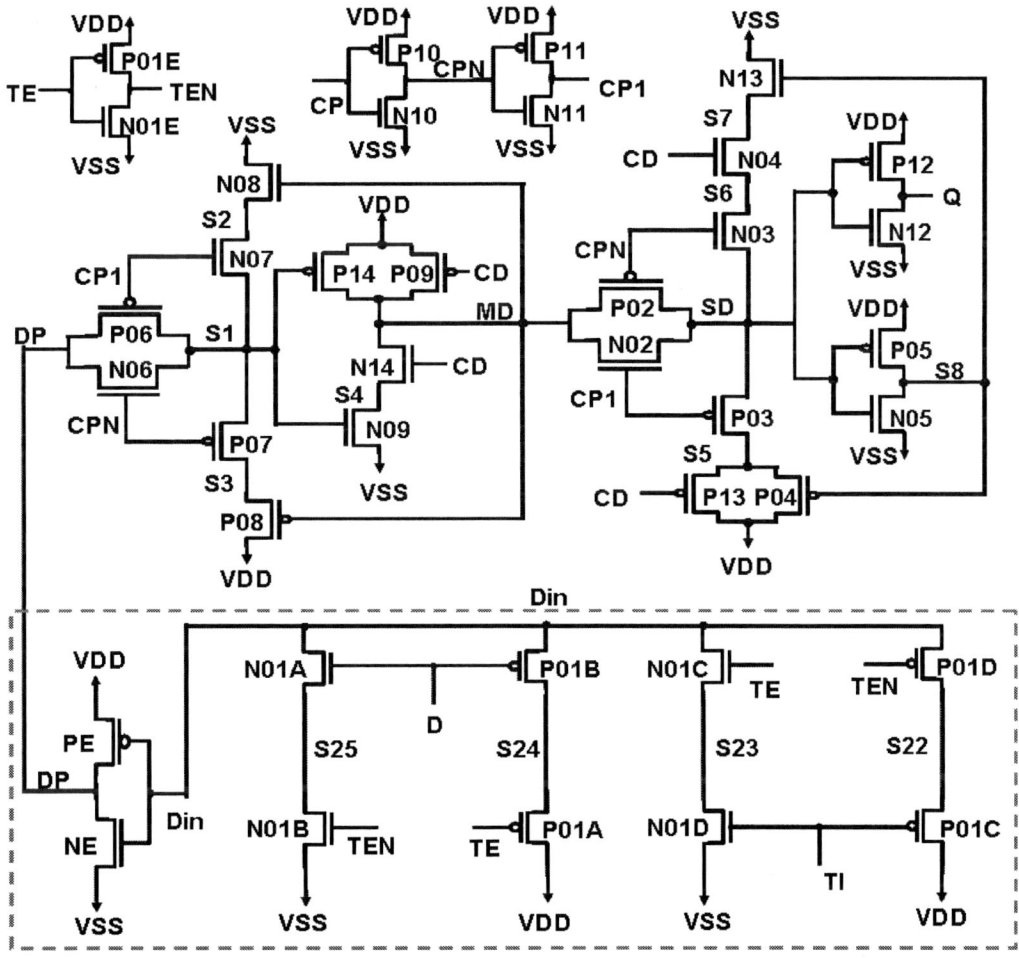

Figure 1: Scan cell with asynchronous reset input CD

In Figure 1 we give the transistor level circuit diagram of a scan cell with asynchronous reset input used in a 90nm industrial design. In Figure 1, node D is the functional (data-in) input driven by the combinational logic and TI is the test input (scan-in) applied to the scan cell. The circuit in the dashed rectangle is the multiplexer used for selecting between D and TI. The value at node TI is selected to propagate to the scan cell output (Q) if the test enable signal (TE) is 1 and the value at D is propagated to the output by setting TE to 0. The circuit between nodes DP and MD is the master stage. The slave stage is the circuit between the output (MD) of the master stage and the scan cell output Q. Input pin CD is used to asynchronously reset the scan cell (i.e. CD asynchronously sets the output Q to 0 when CD changes to 0). We will use this scan cell for our study. However, the conclusions drawn are expected to hold for other scan cells which use asynchronous inputs.

In the scan cell of Figure 1, there are a total of 38 transistors. Four transistors (10.5%) N14, N04, P09 and P13 are driven by the asynchronous reset input CD. Removing these four transistors results in a fully-synchronous flip-flop. To detect all the detectable open faults in the fully-synchronous scan cell, we have proposed a set of tests in [17]. No earlier work has considered the opens in the transistors driven by the asynchronous inputs to scan cells. In this paper, we consider detection of stuck-open faults in the four transistors driven by the asynchronous reset signal CD.

Table 1 summarizes the results obtained in this work. The transistors driven by signal CD are of two types: (i) the P transistors are on when CD = 0 (P09 and P13); (ii) the N transistors are on when CD = 1 (N04 and N14). In Table 1, column 3 indicates that open faults in the transistors will affect the functional mode of operations. Columns 4 and 5 indicate that except for opens in N14 open faults in other transistors will not be detected by the standard flush tests and ATPG boundary tests. Column 6 shows that open faults in N14 are detected by the existing flush test 0011...0011 and opens in N04 can be detected by the new flush test proposed in [17]. In this paper, we propose two new flush tests to cover the remaining open faults in transistors P09 and P13 in Figure 1. Thus the two new flush tests improve the overall stuck-open fault coverage of a scan cell by 5.26%. Considering that the number of scan cells in a design can be very high [10], additionally detecting open faults in two transistors per scan cell can have a significant impact on the overall test quality. The same two flush tests detect the faults in all the scan cells of the design.

Table 1: Summary of the transistor SOPs considered

Fault Class	Faulty Transistor	Functional Failure	Existing Flush test	ATPG SAF/TDF Boundary test	Tests for Detection	% Weight
CD = 1	N14	Yes	Yes	No	0011..0011	2.63%
	N04	Yes	No	No	[17]	2.63%
CD = 0	P09	Yes	No	No	New flush test	2.63%
	P13	Yes	No	No	New flush test	2.63%

The remainder of the paper is organized as follow. In Section 2, the proposed new flush tests for the detection of open faults in P09 and P13 are presented. In Section 3, we discuss the detection of stuck-open faults in N04 and N14. Finally, the paper is concluded in Section 4.

2. DETECTION OF OPENS IN TRANSISTORS TURNED ON WHEN CD = 0

Transistors P09 and P13 are turned on when CD = 0. P09 is located in the master stage and P13 is in the slave stage. In the following subsections, we present the proposed new flush tests for the detection of opens in P09 and P13. We also discuss why standard flush tests and ATPG SAF/TDF boundary tests do not detect these faults. The impact of these faults, if left undetected, on the functional operation is also discussed. Detection of faults was ascertained using HSPICE simulation of the proposed tests on a scan chain segment of three scan cells with a fault injected into the second cell.

2.1. Detection of stuck-open faults in P09

Consider the open faults of transistor P09 denoted by arrows in Figure 2. To detect the fault, we first need to initialize Q to 1 in clock cycle i. In the clock cycle i+1 we set CD to 0 and change CD back to 1 when CP = 1. This is shown in Figure 3(a).

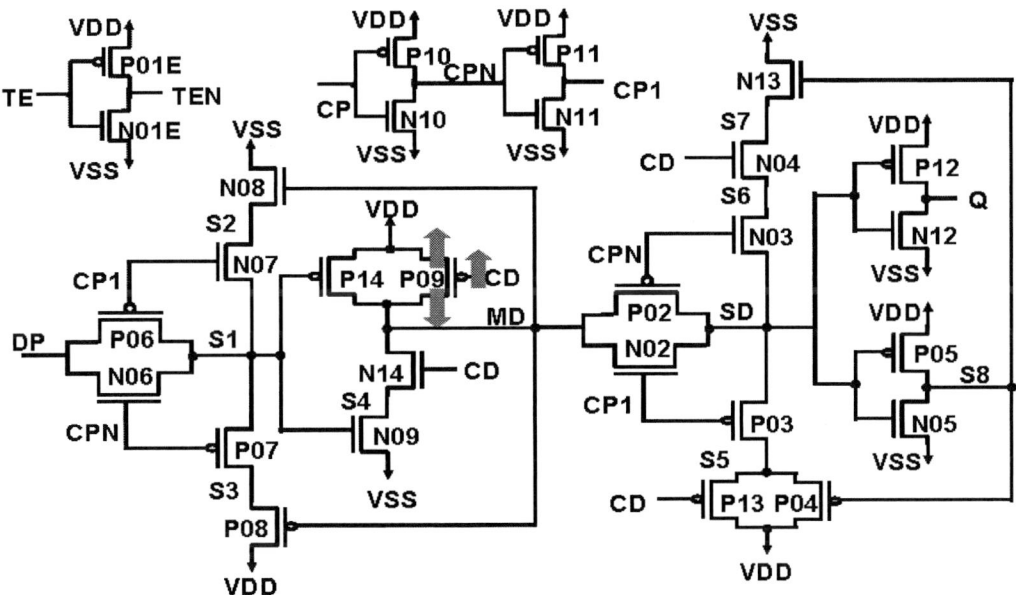

Figure 2: SOPs in transistor P09

We refer to Figures 2 and 3 in the following discussion. In clock cycle i, output Q is initialized to 1 by setting TI = 1, TE = 1 or D = 1, TE = 0. This sets DP and S1 to 1. Since CD = 1, N14 is on. Hence, node MD is driven by VSS. When clock (CP) changes to 1, P02 and N02 are on and the value 0 at node MD is propagated into the slave stage. Q is thus set to 1 at the first rising edge of clock cycle i+1.

Next, we consider clock cycle i+1. When CP = 1 we change CD to 0 as shown in Figure 3(a). Note that in order to avoid holding time violation CD needs to be changed to 0 after the required hold time. This will turn on P09 and turn off N14 in the fault-free circuit. The node MD is then driven by VDD through transistor P09. The value 1 propagates to the slave stage since N02 and P02 are on when CP = 1. Thus the output Q changes to 0 in the fault-free circuit. Meanwhile, node S1 is set to 0 since MD = 1 turns on N08 and N07 when CP = 1. CD is then changed back to 1 before CP changes to 0 (cf. Figure 3(a)). This turns on N14 and turns P09 off. In the fault-free circuit, the node MD retains the value 1 since S1= 0 turns P14 on. Therefore nodes SD and Q retain the values 1 and 0 within CP = 1, respectively. When CP changes to 0, P02 and N02 are off. Although CD = 1 turns off P13, SD = 1 sets the node S8 to 1 and turns on P04. Therefore the output Q still retains the value 0 in the rest of the clock cycle i+1. The solid lines in Figure 3(a) show the waveforms of CP, CD and Q in the fault-free circuit. However, in the faulty circuit, changing CD to 0 when CP = 1 turns off N14 and node MD will be floating due to the open fault in P09. Besides, P13 is on and N04 is off since CD = 0. Since

CP = 1 turns off P03, CD = 0 will not change the value at node SD. Hence, in the faulty circuit, nodes MD, SD, S8 and Q retain the values 0, 0, 1 and 1, respectively. Since CD is changed back to 1, which turns off P13 before CP changes to 0, the output Q retains the value 1 which is faulty at the next rising edge of the clock. The waveform of Q for the faulty circuit is shown in dashed lines in Figure 3(a).

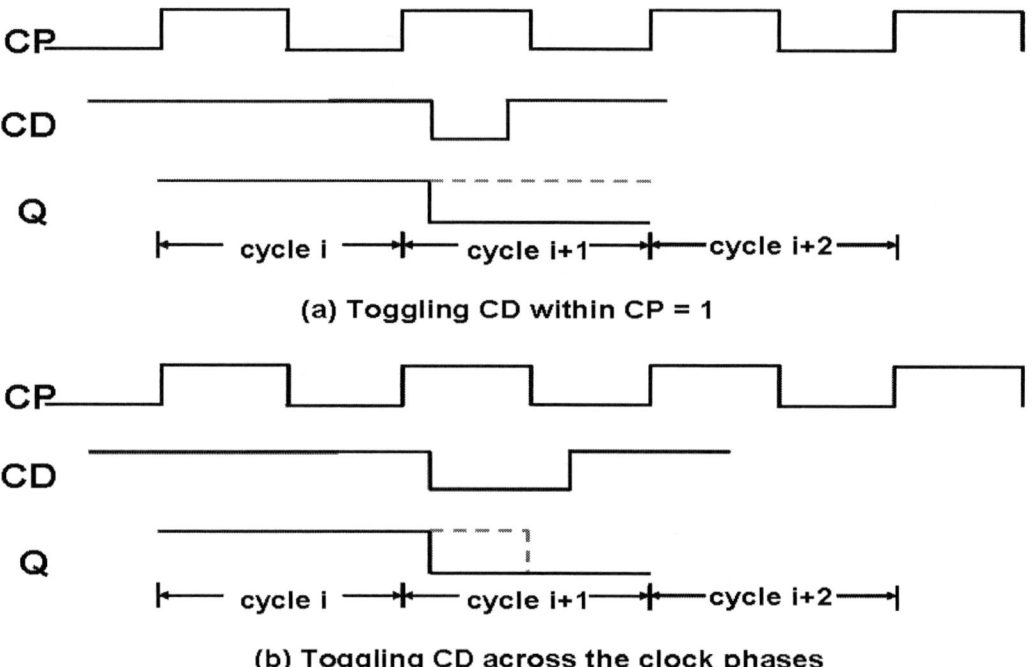

(a) Toggling CD within CP = 1

(b) Toggling CD across the clock phases

Figure 3: Detection of SOPs in transistor P09

It is important to note that if CD is changed back to 1 after the falling edge in cycle i+1, as shown in Figure 3(b), the faulty value 1 will not be observed at the rising edge of the subsequent cycle. This is explained next. When CP changes to 0, transistors P03 and N03 are turned on while P02 and N02 are turned off. The node SD is driven by VDD through transistors P03 and P13 since CD is 0. Hence, the output Q changes to 0 at the falling edge instead of changing simultaneously when CD changes to 0. Thus, we will not observe the faulty value at the rising edge of the next clock cycle. Therefore, the open fault in transistor P09 will not be detected. The fault-free and faulty waveforms are shown in Figure 3(b) in solid and dashed lines, respectively.

To accomplish the detection of this fault, we propose a new flush test called *master reset transistor flush test* discussed next. The test scans in all ones (11...1) into the scan chain with CD = 1. This helps to initialize the output of the scan cells in the scan chain to 1. Then the test applies the values TI = 1, D = X, TE = 1 for one clock cycle toggling CD to 0 and back to 1 within CP = 1 duration. This is shown in the waveform of CD in clock cycle i+1 in Figure 3(a). Finally, the output values stored in the scan cells are shifted out to compare with the expected values. Note that the open resistance is assumed large enough such that P09 is not turned on in the duration of toggling CD to 0 shown in Figure 3.

2.2. Detection of stuck-open faults in P13

Next, we discuss the detection of open faults in P13 shown in Figure 4. We propose another flush test called *slave reset transistor flush test*. Again, the 11...1 sequence is scanned into the scan chain to initialize the outputs of the scan cells. This is followed by the values TI = 1, D =

X, TE = 1 and toggling CD to 0 and back to 1 within CP = 0 phase as shown in the waveform of CD of Figure 5(a). The detection of this fault is shown in Figure 5(a) by the faulty value 1 observed at the rising edge of clock cycle i+2. From Figure 5(b), we can also see that changing CD back to 1 across the clock phases (changing CD back to 1 after the next rising edge) still detects the fault. However, this will introduce additional test cycles. Therefore, we toggle CD within CP = 0 in order to minimize the testing time.

Again, we note that the open resistance is assumed to be large enough such that P13 is not turned on during CD = 0 shown in Figure 5.

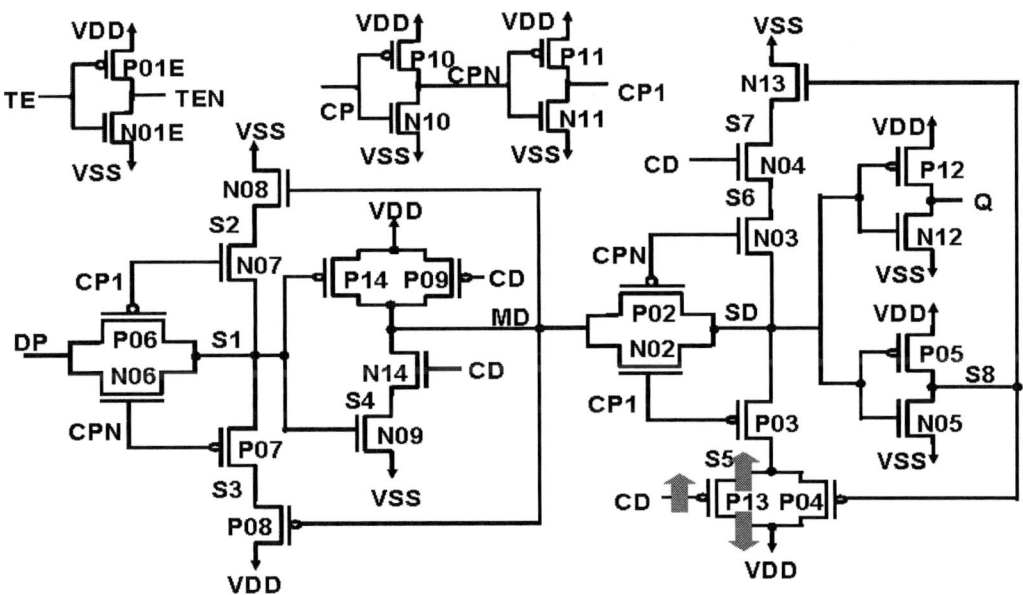

Figure 4: SOPs in transistor P13

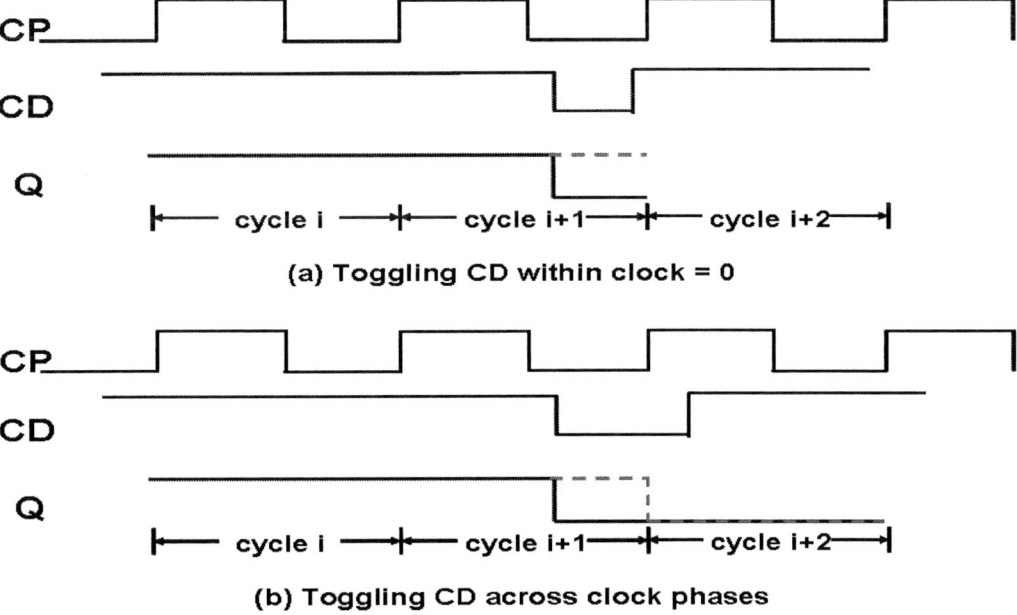

(a) Toggling CD within clock = 0

(b) Toggling CD across clock phases

Figure 5: Detection of SOPs in transistor P13

2.3. Test escape using standard flush tests and ATPG SAF/TDF boundary tests

We have seen that in order to detect the open faults of transistors P09 and P13, we need to set CD to 0. However, when applying the standard flush tests and ATPG boundary tests targeting TDFs at inputs TI, TE, D and output Q, CD is held at 1.

Since detection of open faults requires two pattern tests, the boundary TDF test at CD is the only other possible choice. We investigated this option. We found that ATPG boundary tests targeting transition delay faults at CD do not provide the detection of these faults. The ATPG tools model the scan cell as a black box. The test patterns targeting CD slow-to-fall can initialize the scan cell to 1 by applying TE = 0, D = 1, TI = X. However, the slow-to-fall fault at CD does not model the SOPs in transistors P09 and P13 properly. The test computed by existing ATPG tools holds CD at 0 for both clock phases (CP = 1 and CP = 0) in the second test pattern. Hence, they do not ensure that CD changes within CP = 1 or CP = 0 as discussed above. Therefore, the ATPG tests do not detect these SOPs.

To have the ATPG generate tests to detect these faults, controlling the timing of the change in signal CD is required. This can potentially lead to additional tests for each scan cell. The proposed flush tests have the advantage that all the SOPs in transistors driven by CD in the scan cells are detected by the same two flush tests.

2.4. Impact of open faults in P09 and P13 on functional mode of operation

It is important to note that the open faults in the transistor P09 will result in errors during the functional mode of operation. Five possible cases may happen when trying to asynchronously reset the scan cell in clock cycle i+1. They are shown in Figure 6 as waveforms CD1, CD2, CD3, CD4 and CD5, respectively. In case 1 the asynchronous reset pulse occurs when CP = 1. The output of the scan cell will not be set to 0 in the faulty circuit. Thus, it results in functional failure. When the asynchronous reset pulse is over a time greater than half the clock cycle as in CD2, the scan cell output will not set to 0 until the falling edge. This will lead to a delay and a functional failure may occur if the scan cell is connected to a negative edge triggered flip-flop. In the third case CD3 changes back to 1 in clock cycle i+2. This results in the same errors as in the case of CD2. Finally, no functional error occurs in the case of CD4 and CD5 where CD changes when CP = 0 since the output will be reset using P13.

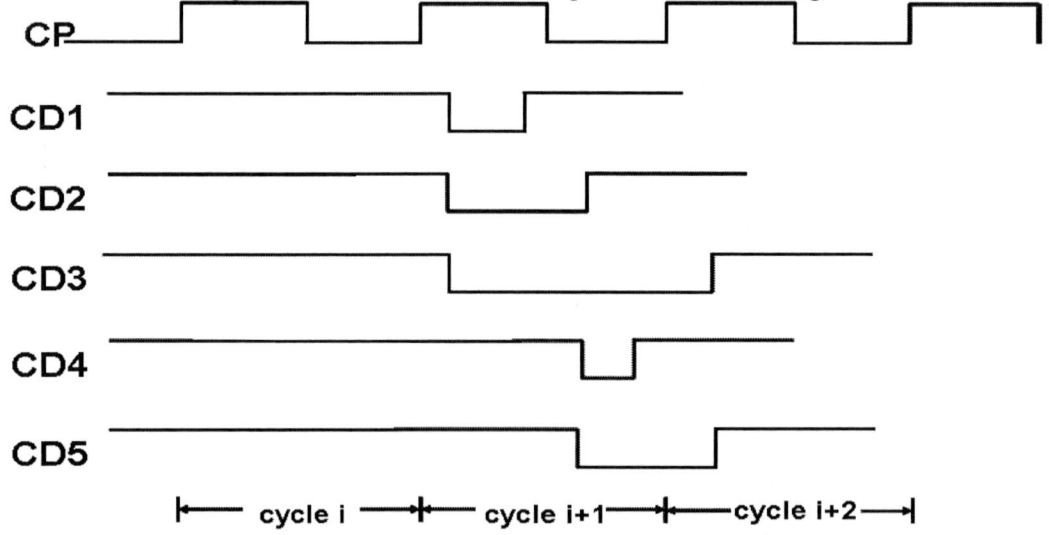

Figure 6: Cases of CD reset in functional mode of operation

Similarly, for the stuck-open faults in P13, if left undetected, it will introduce a delay in the affected scan cell, which may lead to functional failure as in CD4 and CD5 shown in Figure 6.

In the cases of CD1, CD2 and CD3 no functional errors occur since the output will be reset using P09.

Therefore, it is important to detect the open faults in P09 and P13 as they may lead to functional failures if left undetected.

3. DETECTION OF OPENS IN TRANSISTORS TURNED ON WHEN CD = 1

Similar to the case of transistors P09 and P13 discussed in Section 2, one of the transistors in this case, N14, is in the master stage and the other N04 is located in the slave stage.

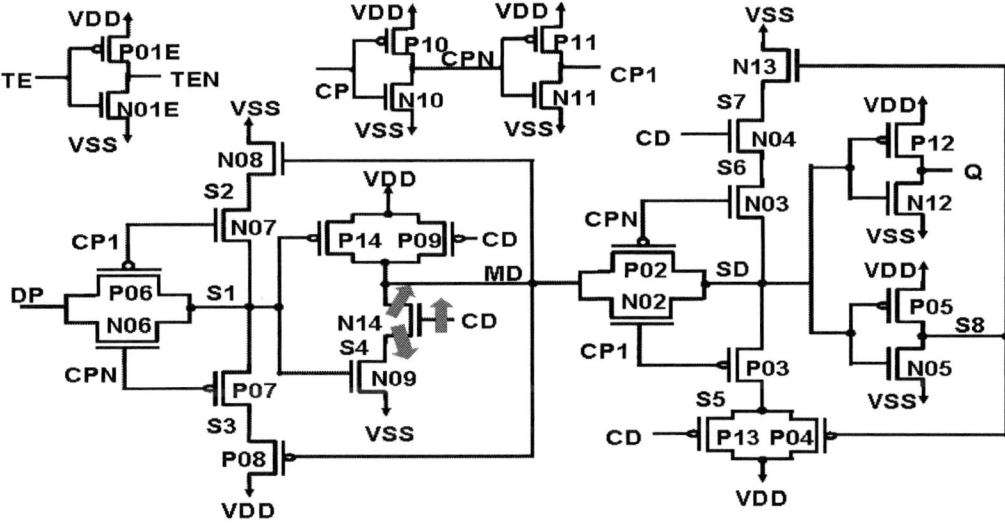

Figure 7: SOPs in transistor N14

Figure 7 shows the master and slave stages with stuck-open faults in transistor N14 denoted by the arrows. Assume that DP = 1 is propagated into the master stage when CP = 0. Node S1 is set to 1 and turns on transistor N09. In the fault-free circuit, MD is driven by VSS through N14 with CD = 1. However, MD will not be set to 0 due to the open fault in transistor N14. Therefore, the output Q can never be set to 1 in the faulty circuit. These open faults are detected by the standard flush test 0011...0011 with CD = 1.

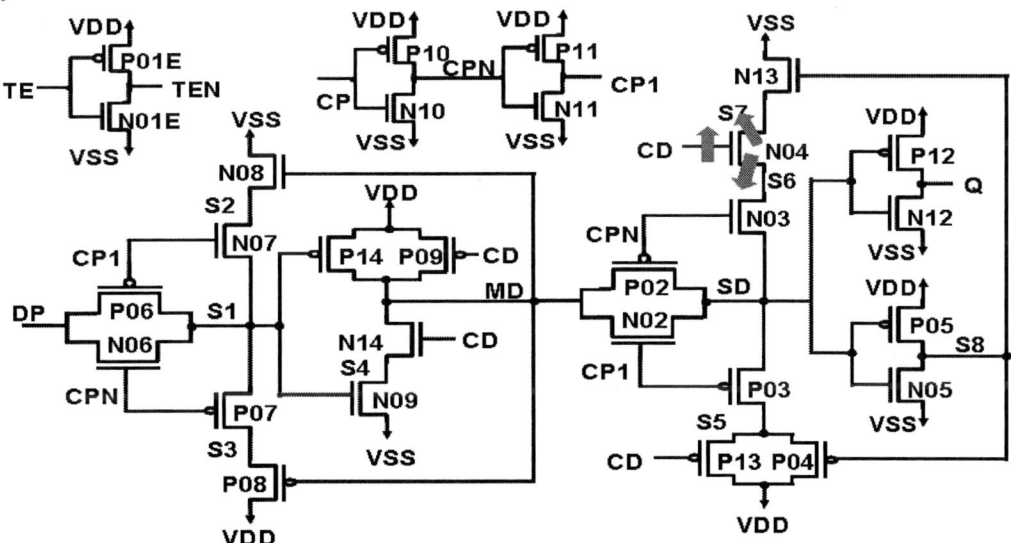

Figure 8: SOPs in transistor N04

In functional mode of operation, if left undetected, these faults will result in errors when the value 1 on D attempts to propagate to output Q.

Next, consider the open faults in the transistor N04 highlighted by arrows in Figure 8. To detect these faults, we apply TI = 1. The node MD is set 0 when CP = 0 and the value 0 propagates to the node SD when CP changes to 1. Hence, the transistors P05 and P12 are turned on. When CP changes to 0 SD is driven by VSS through the transistors N04 and N13. Thus, in the fault-free circuit the output Q is set to 1 at the rising edge and retains the value for one clock cycle. However, in the faulty circuit when CP changes to 0 the node SD is floating due to the open fault in N04. If we wait for a long enough time, the node S8 will discharge to 0 and turn on P04. The output is thus set to 0 which is faulty. These faults can be detected by applying the flush test at very slow speed. We proposed such tests in [17].

4. CONCLUSIONS

We investigated the detection of stuck-open faults in transistors driven by an asynchronous reset signal in a scan flip-flop. We found that the existing flush tests and ATPG tests targeting inputs and outputs of the scan cell do not detect the open faults in the transistors which are used to reset the scan cell. To detect these faults, we proposed two new flush tests which improve the total open fault coverage by over 5%. In addition, we showed that these faults can lead to failures in the functional mode of operation if left undetected. Therefore, it is important to detect them. The two new flush tests together with the tests we proposed earlier in [17] detect all the detectable large resistance stuck-open faults in a scan cell with asynchronous inputs.

5. REFERENCES

[1] M.K. Reddy and S.M. Reddy, "Detecting FET Stuck-Open Faults in CMOS Latches and Flip-flops," *IEEE Design and Test of Computers*, Vol. 3, No. 5, pp. 17-26, October, 1986.

[2] K, J. Lee and M.A. Breuer, "A Universal Test Sequences for CMOS Scan Registers," *IEEE Custom Integrated Circuits Conference*, pp. 28.5.1-28.5.4, 1990.

[3] W.K. Al-Assadi, "Faulty Behavior of Storage Elements and Its Effects on Sequential Circuits," *IEEE Transactions on VLSI*, Vol. 1, No. 4, December, 1993.

[4] S.R. Makar and E.J. McCluskey, "Functional Tests for Scan Chain Latches," *Proc. ITC*, pp. 606-615, 1995.

[5] S.R. Makar and E.J. McCluskey, "Checking Experiments To Test Latches," *Proc. VLSI Test Symp.*, pp. 196-201, 1995.

[6] S.R. Makar and E.J. McCluskey, "Checking Experiments for Scan Chain Latches and Flip-Flops," CRC Technical Report 96-5, August 1996.

[7] S.R. Makar and E.J. McCluskey, "ATPG for Scan Chain Latches and Flip-Flops," *Proc. VLSI Test Symp.*, pp. 364-369, 1997.

[8] S.R. Makar and E.J. McCluskey, "Iddq Test Pattern Generation for Scan Chain Latches and Flip-Flops," *IEEE IDDQ Testing*, pp. 2-6, November 1997.

[9] C. Aissi and J. Olaniyan, "Design and Implementation of A Full Testable CMOS D-Latches," *IEEE 5th IPFA*, pp. 194-199, 1995.

[10] F. Yang, S. Chakravarty, N. Devtaprasanna, S.M. Reddy, I. Pomeranz, "On the Detectability of Scan Chain Internal Faults – An Industrial Case Study," *Proc. VLSI Test Symp.*, pp. 79-84, April 2008.

[11] R.L. Wadsack, "Fault Modeling and Logic Simulation of CMOS and NMOS Integrated Circuits," *Bell Syst. Tech. J.*, Vol. 57, pp.1449-1474, May-June 1978.

[12] T. M. Storey and W. Maly, "CMOS Bridging Fault detection," *Int. Test Conf.*, pp. 842-851, 1991

[13] T. M. Storey, W. Maly, J. Andrews, and M. Miske, "Stuck Fault and Current Testing Comparison Using CMOS Chip Test," *Proc. Int. Test Conf.*, pp. 311-318, 1991.

[14] V.C. Champac and J. Figueras, "Testability of Floating Gate defects in Sequential Circuits," *Proc. VLSI Test Symp.*, pp. 202-207, 1995.

[15] V.C. Champac, A. Zenteno, J.L. Garcia, "Testing of Resistive Opens in CMOS Latches and Flip-flops," *Proc. European Test Symp.*, pp. 34-40, 2005.

[16] J. Xu, R. Kundu, F.J. Ferguson, "A Systematic DFT Procedure for Library Cells", *Proc. VLSI Test Symp.*, pp.460-466, 1999.

[17] F. Yang, S. Chakravarty, N. Devtaprasanna, S.M. Reddy, I. Pomeranz, "Detection of Internal Stuck-open Faults in Scan Chains," *Proc. Int. Test Conf.*, October 2008.

IEEE International Symposium on Defect and Fault Tolerance of VLSI Systems

Efficient Determination of Fault Criticality for Manufacturing Test Set Optimization

Yiwen Shi, Kellie DiPalma, and Jennifer Dworak
Brown University

Abstract

Defective part levels of zero are almost impossible to achieve in an era of complex defects, process variations, and limited testing resources. It is important to ensure that any defects missed during test will impact the end user as little as possible. However, optimizing test sets for superior detection of critical defects requires an understanding of relative fault criticality, and this determination may be very expensive. In this paper, we will present a method which efficiently estimates this criticality under realistic usage conditions using a translation between combinational and sequential fault analysis. We will show that the resulting criticalities are sufficiently accurate to select an optimized test set which is as effective as one created with perfect criticality information.

1. Introduction

Scaling of devices to smaller feature sizes has introduced a host of significant issues for quality, reliability, and test. These circuits are less tolerant of process variations, and many of the defects that occur are exhibiting more complex behavior that may only be detected under very specific logical or environmental conditions. As a result, the number of patterns and testing conditions required to fully test these devices and achieve defect levels of zero is prohibitively expensive. Thus, test sets must be optimized with a two-prong strategy in mind—they should detect as many defective parts as possible while simultaneously ensuring that any defects that are unavoidably missed are functionally unimportant.

It is well-known that some applications, such as graphics applications, are inherently tolerant of a small number of errors in the output stream, and this realization has lead to a number of advances—including lossy compression. Similarly, with respect to test, some internal errors in a design may have negligible impact on the quality of the result from the view of the end user. For example, consider the two figures shown in Figure 1. In each case, the picture was created using the output from a discrete cosine transform circuit that was simulated with a single stuck-at fault inserted. The stuck-at fault inserted into the circuit used to produce the picture on the left is obviously much less critical for acceptable behavior than the stuck-at fault that led to the picture on the right. While we would ideally detect both defects, it is obviously much more important to ensure that defects similar to the fault on the right are detected during testing. This type of criticality comparison is especially well-matched to embedded systems with well-defined applications.

In the past, we have proposed a greedy algorithm that can be used to choose a near-optimal subset from a superset of test

Figure 1: Comparison of the impact of two different stuck-at faults

patterns. We showed that an order of magnitude reduction in failure rate could be achieved without increasing the defect level [1]. However, implementing this algorithm requires an estimate of the criticality of different faults from the perspective of the end user. Determining this criticality is non-trivial and has the potential to be exceedingly expensive. In this paper, we will present a method for efficiently estimating this criticality in sequential circuits, and we will show that it can be used to develop test sets that are as effective as a test set chosen with exact criticality numbers.

2. Previous Work

Multiple researchers have previously investigated the relationship between errors within integrated circuits and their ultimate effect on the overall behavior of the device or system. For example, previous work has addressed the presence of soft errors in microprocessors and the need to identify architectural structures in the processor that are most likely to lead to uncorrected errors so that fault tolerance techniques can be cost-effectively implemented [2], [3].

Other researchers have dealt specifically with manufacturing defects and the problem of decreasing yields. These researchers have proposed allowing chips that are defective to be sold—possibly at a reduced cost—so that yields can be maintained [4] [5] [6] [7] [8] [9]. Some of these techniques identify faults as being acceptable because the magnitude of the error difference between the good and defective circuit does not exceed a chosen threshold [4]. Thus, ATPG techniques need to ensure that the magnitude of the error difference is greater than the threshold when unacceptable faults are being tested. Other work has identified faults in combinational circuits that are expected to have acceptably low failure rates using pseudo-random pattern sampling and ignored them during ATPG [7]. Still other research has focused on estimating the failure rates of chips which have already failed tests to determine whether they should be sold anyway [8]. Once again, this error rate determination is accomplished through a sampling of pseudo-random patterns. Similarly, the authors of [9] apply pseudo-random patterns in a BIST methodology to a circuit under test and then estimate the error rate from a compressed test response composed of one's counting. The ultimate goal is to determine whether a known defective part should be sold.

[10] noted that achieving high fault coverage may be overkill—especially in the case of transition faults. Although some of these transitions may be tested for in the artificial world of scan, they are redundant during normal operation. Thus, transitions and paths that can never be sensitized or observed during normal operation should not necessarily be included in a test set even if they can be detected with AC scan

In the past, we have also considered the effect of failure rates on the quality of test and the resulting circuit reliability. In [11], we showed that the presence of scan chains during testing can cause significant differences in the time to failure for defects during manufacturing test as opposed to during normal sequential operation because of differences in the immediate observability of circuit sites. We proposed a new metric: ELF-MD (for *Expected Latency to Failure due to Manufacturing Defects*) that could be used to optimize test sets. In [1], we showed that random or ATPG patterns are insufficient for truly capturing fault criticality, but that if the criticalities of different faults are known, it is possible to achieve an order of magnitude reduction in failure rates, without increasing the defect level. However, we did not provide an efficient strategy for obtaining true, functional criticality data.

Previous researchers have attempted to find effective ways of estimating fault coverage for functional test patterns. Determining fault criticality is a related, but distinct problem. Unlike fault coverage questions which look at faults in aggregate, optimizing test sets for fault criticality requires that the criticality of each fault be determined individually. Furthermore, fault coverage questions are generally binary in nature—a fault is either covered by the test set, or it isn't. In contrast, fault criticality requires a probabilistic estimate of the effect that a given fault will have on the overall circuit behavior.

In this paper, we will propose a method of estimating criticality that takes into account both the combinational and sequential behavior of the circuit. We will show that the effectiveness of the patterns chosen with these criticality estimates is highly correlated with the effectiveness of patterns chosen with the considerably more expensive full sequential fault simulation.

3. Functional Environment for Estimating Fault Criticality

Three main issues must be addressed to effectively estimate criticality:
- Determination of the metric used to characterize "degree of correctness"
- Determination of the input sequences which characterize normal operation
- Analysis of the interaction of those input sequences with potential faults

3.1 Criticality Metrics

Optimizing test sets to ensure that the most egregious faults are detected requires a determination of the relative importance of different faults as experienced by the end user. We call this importance "fault criticality." Fault criticality may be expressed as a failure rate, the mean-squared-error, PSNR, or some other metric that captures the degree to which the circuit is operating acceptably/unacceptably. In this paper, we will use a form of failure rate that also takes into account the importance of different outputs through a weighting scheme.

3.2 Choice of Input Sequences to Represent Normal Operation

Many previous researchers who have investigated applications for error tolerance have estimated the importance of an error by finding the detection profile of a fault when random or ATPG patterns have been applied. However, we have previously shown that the detection profile of faults under such conditions is significantly different from that which occurs when the circuit is operating in normal functional mode—even for combinational circuits [1]. Thus, some way of effectively capturing the true functional environment and constraints of the design are necessary.

If the design is part of an embedded system with a well-defined application, the actual application that will eventually be run may be used to determine criticality. However, in many cases, an application may not be available, and we will need an alternative way of defining the functional input space. A good choice would be the use of an appropriate subset of the design verification patterns, which are written specifically to capture the entire functionality of the design—including corner cases. A weighting system could be used to determine the relative importance of the various verification test sequences analyzed based upon the expected frequency with which the tested functionality is expected to appear in normal circuit operation. A verification sequence that is representative of the functionality 80% of the time could be given proportionally more influence than a verification sequence that is representative of only 1% of the normal operation.

4. Combinational and Sequential Criticality Translation

Once appropriate input sequences have been chosen to represent the functional environment, an efficient methodology must be developed for estimating the actual criticality of each fault. This criticality estimation is complicated by the fact that most of the circuits we

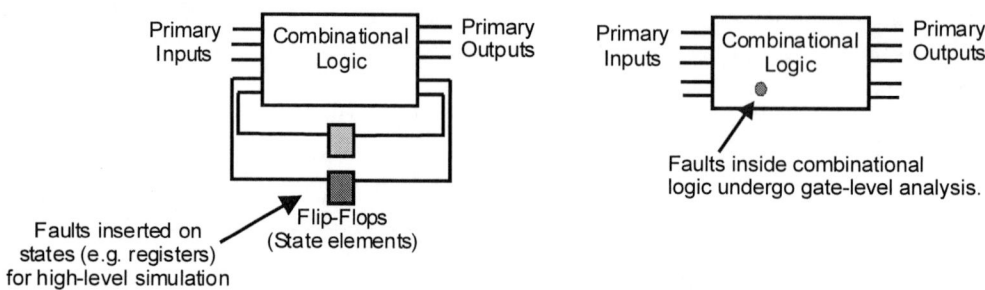

Figure 2: Huffman Model of a sequential circuit and combinational version for gate-level simulation

are interested in are sequential in nature. Thus, criticality must be estimated across multiple time cycles, and the behavior of different faults could vary widely. For example, one error could propagate to a flip-flop and persist in the circuit across multiple time cycles—causing multiple errors. A different error could propagate to a flip-flop initially but die out in subsequent cycles—never even reaching an output visible by the end user. Thus, this sequential behavior means that an estimate of criticality through combinational circuit analysis alone (as occurs in scan-based ATPG) is highly inadequate [11]. Unfortunately, while an understanding of the sequential behavior of the circuit in the presence of a fault is mandatory for accurately estimating criticality, sequential simulation of long test sequences is prohibitively expensive.

4.1 Division of Circuit Analysis between the Gate Level and the RTL

One potential solution to the determination of criticality across multiple clock cycles is to perform the simulation on a high level representation of the design, such as the RTL. However, many of the signals that will become fault sites at the gate level are not present in the RTL. In this paper, we propose a methodology which combines combinational gate level analysis with high level sequential simulation for different sets of faults—ultimately translating between both levels of the simulation to find the criticality of gate level faults internal to the circuit logic.

For an overview of this technique, consider Figure 2. In this figure, the circuit shown on the left represents the Huffman model for a sequential circuit. The state elements (flip-flops) are shown below the box representing the combinational logic. These state elements are present in both the RTL and gate level versions of the design. Thus, the criticality of gate level faults inserted on any of these state elements can be determined through RTL circuit simulation using a commercially available simulator. Furthermore, it is possible to insert the faults at different error-insertion rates.

In contrast, the circuit on the right shows only the combinational logic. Flip-flops have become pseudo primary outputs (PPO's) and pseudo primary inputs (PPI's) just as is done in scan chain analysis. Functionally valid input patterns for the true and pseudo PIs can be extracted by saving a sample of the states from high level good-circuit simulation performed previously. The combinational logic is realized at the gate level, and on any given clock cycle, it is possible to determine whether a single error that appeared at an internal fault site propagated to either a true primary output or to a state-holding element by the end of the cycle. However, this doesn't indicate the output failure rate over subsequent clock cycles. Thus, to determine the true criticality of a fault internal to the combinational logic, we combine two different probabilities:

- The probability that an error at internal fault site i propagates to a flip-flop (or true PO) in the current cycle (as determined from combinational analysis)

- The failure rate at a PO in subsequent cycles provided that an error appears at the flip-flop in the current cycle (as determined by sequential analysis).

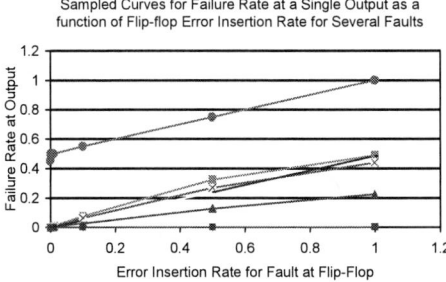

Figure 3

4.2 Failure Rate Translation

In the simplest estimation of overall fault criticality, we could assume that failure rates can be translated from the combinational to the sequential regime in a proportional manner. For example, assume that a fault propagates through the combinational logic to a given flip-flop 50% of the time. Further assume that the corresponding stuck-at fault at the flip-flop in question causes a failure rate of 70% at a PO. We could then estimate the ultimate failure rate at the output due to this internal fault as 35%.

However, it is generally not this simple. For example, if we insert errors at a flip-flop with different error insertion rates, we will often find that the slope depends upon the flip-flop involved. Furthermore, the resulting curve may not be linear—especially for low failure rates. These differences in failure rates for faults inserted at several different flip-flops as a function of insertion rate is shown in Figure 3.

Thus, we characterize the slope for each flip-flop/fault as a function of its fault insertion rate. A sequence of simulations is done. In this paper, we chose fault insertion rates of 1, 0.5, 0.1, 0.01, 0.001 and 0.0001 for each stuck-at fault inserted on a flip-flop (although fewer could be used). Although the entire curve is often nonlinear as a whole, linear interpolation can be used to estimate the shape of curve between the simulated values, and in our experience this does not introduce a significant amount of additional error. Thus, once the probability of an internal fault reaching a flip-flop is known, curves such as those shown in Figure 3 can be used to extrapolate the corresponding error rate at a true output.

(Note: Because each curve represents either a stuck-at 1 or a stuck-at 0 fault, the probability that an inserted error is actually excited is dependent upon the one's probability of that flip-flop during normal sequential simulation. For example, if a flip-flop is equally likely to be a logic one as a logic zero, and we randomly insert a stuck-at 0 fault at that flip-flop on 10% of the patterns, only about 5% of the patterns will actually contain an error. This must be taken into account when using the curve to extrapolate fault criticalities for internal faults.)

Unfortunately, this analysis may still be too simplistic for many faults. It works well whenever an internal fault propagates along a single path to a single flip-flop. However, many faults have the potential to propagate along many paths to many flip-flops. Thus, an internal fault may arrive at one or more flip-flops as either a faulty one or a faulty zero depending upon the pattern applied and the number of inversions along the taken path(s). As a result, the corresponding failure rates at different flip flops and with different polarities must be *combined* in some way to produce a single, accurate estimate of the failure rate of each output due to the presence of the internal fault. Two cases must be considered:

- Case 1: An internal error propagates to a single flip-flop as both a faulty 1 and a faulty 0 during combinational simulation (obviously on different patterns.)
- Case 2: An internal error propagates to multiple flip-flops.

Case 1: In this situation, the internal fault appears as an error at single flip-flop, behaving as both a stuck-at one and a stuck-at zero fault depending upon the pattern. Thus, two propagation probabilities exist for this fault's propagation to flip-flop i:

$$P_{sao-ff\text{-}i}, P_{sal-ff-i.}$$

This leads to two estimated criticalities at a given output due to the error at this flip-flop extracted from a chart such as that in Figure 3.

$$C_{sao-ff\text{-}i}, C_{sal-ff-i.}$$

Because the patterns on which these two different types of error can occur are disjoint, to find the combined estimated criticality of the internal fault, we add these two criticalities and get:

$$C_{ff-i} = C_{sao-ff-i} + C_{sal-ff-i}$$

Case 2: In this case, we have a fault propagating to multiple flip-flops—each of which can individually cause an error at the output in question. We must now combine these criticalities across flip-flops. First, assume that all necessary analysis described in Case 1 has already been completed, and thus, for this fault, propagation to each flip-flop i is represented by a single criticality. C_{ff-i}. Because a fault may propagate to multiple flip-flops on a single pattern, a simple addition is not generally appropriate. Instead, we take a weighted average of the individual flip-flop criticalities. Each weight corresponds to the total probability that the internal error propagates to that flip-flop.

Thus, the individual weight corresponding to flip-flop i for this internal fault is defined as:

$$W_i = P_{sao-ff-i} + P_{sal-ff-i}$$

The combined criticality at the output is then determined with the following equation:

$$C = \frac{\sum_{i=1}^{num\_ffs} W_i C_{ff-i}}{\sum_{i=1}^{num\_ffs} W_i}$$

While Case 1 and Case 2 are sufficient to obtain a good estimate for many internal faults, they do not take into account the correlation that may exist when faults propagate to multiple outputs/flip-flops. Simultaneous failures on multiple flip-flops can magnify (or cancel) the effect of the individual failures. Simulating all possible combinations of flip-flop failures is obviously impossible. Fortunately, only a few combinations actually need to be simulated. We can identify likely groups of correlated flip-flops as those with identical propagation probabilities for a given fault.

Thus, for a given internal fault, if $P_{sao-ff-i} = P_{sao-ff-j} = P_{sal-ff-k}$ then flip-flops i, j, and k should be simulated together with the appropriate failure polarities. (In some cases, groups with extremely low propagation probabilities may also be ignored without significantly impacting the results.) For a circuit with almost 59000 faults, we found that only 146 flip-flop combinations actually needed to be simulated together and many of these combinations could be used to improve the estimate of multiple faults—in some cases several hundred. Furthermore, the reduction in percent error of the estimate was dramatic—changing in some cases a 4000% error to 0.04%.

5. Experimental Results for Criticality Estimation

The previous criticality estimation procedure was applied to a benchmark circuit obtained from opencores.org. This circuit implements a color conversion between several different specification systems, including, for example, the transformation of RGB to and from YCbCr. The flattened gate-level netlist contains 11,186 gates and 995 flip-flops. The circuit's data outputs consist of three 8-bit integers: y1[7:0], y2[7:0] and y3[7:0]. To obtain a final criticality for each fault, we take the weighted average of the criticality of each output such that the weight is determined by the output's bit position in its respective integer. Thus, we assign the

MSB of each of the three integers weight 1, the next bit weight 0.5, the third bit weight 0.25, etc. A functional testbench was provided with the circuit on opencores, and this was used as an indicator of the true functional behavior of the circuit under normal operation.

Whenever possible, we used commercial tools to implement our algorithm. Mentor Graphics ModelSim was used to perform the high level RTL fault analysis. A random ten percent of the total cycles from good circuit sequential simulation were saved, and the corresponding state and input values were used to as inputs to the combinational form of the circuit. Mentor Graphics DFTAdvisor and FastScan were used for the corresponding fault simulation and combinational gate level analysis. Finally, Mentor Graphics FlexTest was used to determine the true criticality of each internal fault through actual gate-level sequential simulation.

Our initial analysis consisted of comparing the criticality results estimated with our methodology to those obtained from the much longer FlexTest simulation. Results are shown in Figures 4 and 5. Both of these figures present histograms which show the percent error distribution for each fault when the estimated criticality is compared to the true criticality. The difference between the two figures merely consists of which faults are included in the graphs. Figure 4 shows the data for all faults. It is apparent that a significant number of the faults have very low percent error when the estimated and true criticalities are compared. However, in some cases this is due to the fact that a particular fault was never detected during our functional simulation. Thus, the true and estimated criticalities are both zero and the percent error is necessarily zero. Once these faults are removed, we obtain Figure 5 for the faults that have nonzero true and estimated criticalities. The fraction of faults with high percent errors is obviously much larger now. Fortunately, most of these high percent errors correspond to cases where the actual criticality is very small, and thus, even small absolute differences in criticality can lead to very high percent errors. Furthermore, our estimate is often pessimistic—giving additional criticality to a fault instead of underestimating criticality.

6. Validity for Test Set Optimization

While obtaining accurate estimates of fault criticality is satisfying for its own sake, our ultimate goal is to use these criticalities to choose an optimal subset from a superset of test patterns. In the past, we have proposed a greedy algorithm for this purpose. However, our previous analysis was done only on a combinational circuit. We now wish to determine the effectiveness of the criticality estimation methodology described above for sequential circuits when choosing a test set.

In our greedy test selection algorithm, each test pattern is assigned an "ELF quality" value, where "ELF" stands for "Expected Latency to Failure." The goal is to maximize the expected

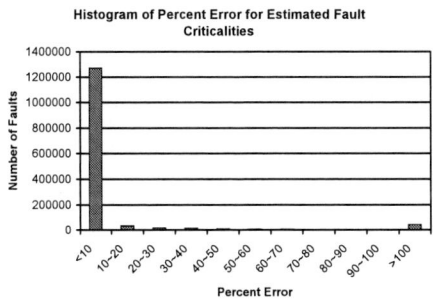

Figure 4: Histogram of Fault Criticalities for All Faults

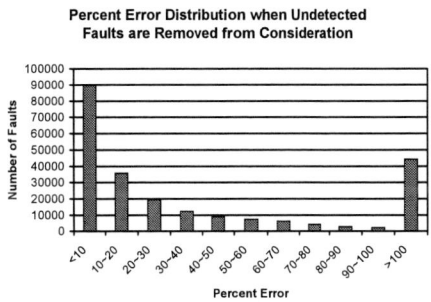

Figure 5: Histogram of Fault Criticalities for Detectable Faults

latency to failure by detecting faults (and therefore defects) with high criticality. The ELF Quality of a test pattern is defined in the following manner:

$$ELF\ Quality = \sum_{i=1}^{num\ faults} C_i * D_i$$

where C_i is the criticality of fault i and D_i is 1 if fault i is detected by this pattern and 0 otherwise. Patterns with the highest ELF Quality are included in the test set. We need to determine if our errors in C_i severely alter the effectiveness of our set.

To compare the effectiveness of our two "optimized" test pattern sets, we will find the correlation between the number of observations of the least detected faults in both test sets. Each additional observation increases the chances of having detected an unmodeled defect at the fault site, but the effectiveness of an additional observation decreases exponentially as the number of previous observations increases [12]. Thus, we will compare the *log* of the number of observations in both test sets. Test sets with similar log(number of detections) for the least detected faults can be considered similarly effective.

Each test set was chosen from the same superset of test patterns. The superset consisted of a 15-detect test set that contained 5008 patterns and was generated with FastScan. Subsets were chosen to be some percentage of the length of the original test set. For example, consider Figure 6. This figure compares two optimized test sets that are 5% of the length of the superset. The x-axis represents the $\log_2$ of the (number of detections + 1) for an optimized set created with *true* criticality values. The y-axis represents an optimized test set created with our *estimated* criticality values. All faults detected 31 times or less are included in the graph. The fault detections for the two sets are highly correlated with an R^2 value of 0.98. In contrast, if we simply take the first 5% of the vectors generated by Fastscan, the correlation drops to 0.87.

Table 1 Presents the R^2 values for test sets of different lengths when the test sets were chosen based on 1) estimated criticalities, 2) the first x% of the superset, 3) a random sample of the superset. The set of columns on the left calculates R^2 based on all faults detected less than or equal to 31 times. The set of columns on the right filters those faults so that the faults with very low criticality are not considered in the calculation.

It's apparent from this chart that even with *estimated* criticality values, the overall fault detection profile and thus unmodeled defect detectability is highly correlated to that which occurs when the criticality is known exactly. This correlation is much higher than would occur with an alternative method that does not consider criticality. It is also pronounced for highly critical faults that are only rarely detected by a test set and thus are likely to be the source of undetected defects that cause catastrophic fails for the end user. Thus, our criticality estimates are sufficiently accurate for producing highly effective test sets optimized for fault criticality.

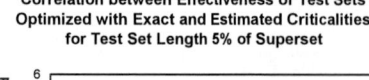

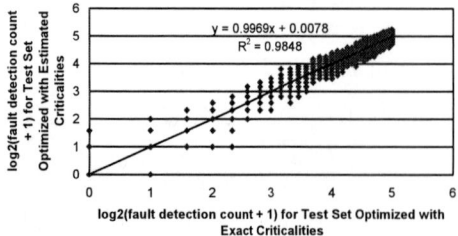

Figure 6: Correlation between $\log_2$(detections + 1) for test sets 5% as long as the superset (generated with estimated or true criticality)

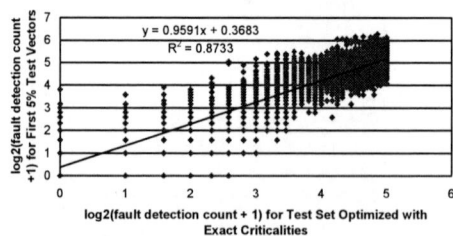

Figure 7: Correlation between $\log_2$(detections + 1) for test sets 5% as long as the superset (generated with true criticality or by truncation)

Table 1: R^2 Values in Comparison to "True" Optimized Set

Test Length	Unfiltered Faults			Filtered Faults		
	Est	First	Random	Est	First	Random
2%	0.98	0.83	0.83	0.95	0.56	0.61
5%	0.98	0.87	0.85	0.98	0.80	0.80
10%	0.96	0.81	0.73	0.97	0.75	0.66
30%	0.99	0.89	0.82	0.96	0.80	0.76

7. Conclusions and Future Work

In this paper, we have presented a methodology for estimating the criticality of gate-level faults using a combination of combinational and sequential analysis. We have shown that these estimated criticalities can be used to successfully optimize test sets based on ELF-Quality. The overall time required for our criticality analysis is approximately 53% of the time required for a full analysis. However, additional optimizations could make this analysis even shorter. In the future, we will investigate the impact of reducing the number of patterns considered for combinational simulation as well as an optimization of the failure rate simulations performed at the RTL. Furthermore, we anticipate that creating a specialized tool for extracting the combinational data from simulation would also greatly enhance the speed of our analysis. Our current implementation, which uses commercial tools, requires significant file I/O and subsequent parsing of large output files to extract the exact data needed. A dedicated tool should be able to reduce this time overhead dramatically.

8. References

[1] J. Dworak "Which Defects Are Most Critical? Optimizing Test Sets to Minimize Failures due to Test Escapes," *Proceedings of the 2007 IEEE International Test Symposium (ITC'07)*, Santa Clara, California, October 23-25, 2007.

[2] S. Mukherjee, C. T. Weaver, J. Emer, and S. K. Reinhardt, "Measuring Architectural Vulnerability Factors," *IEEE Micro*, Nov.-Dec. 2003, pp. 70-75.

[3] A. Biswas, P. Racunas, R. Cheveresan, J. Emer, S. S. Mukherjee and Ram Rangan, "Computing Architectural Vulnerability Factors for Address-Based Structures," *Proceedings of the 32nd International Symposium on Computer Architecture (ISCA'05)*, 2005.

[4] "An ATPG for Threshold Testing: Obtaining Acceptable Yield in Future Processes," *Proc. International Test Conference*, 2002, pp. 824-833.

[5] M. Breuer, S. Gupta, and TM Mak, "Defect and Error Tolerance in the Presence of Massive Numbers of Defects," *IEEE Design and Test of Computers,* May-June 2004, pp. 216-227.

[6] M.A. Breuer ,"Intelligible test techniques to support error tolerance," *13th Asian Test Symposium*, 2004, pp. 386-393.

[7] K.J. Lee, T.Y. Hsieh, and M.A. Breuer, "A novel test methodology based on error-rate to support error tolerance," *Proc. International Test Conference,* 2005, pp. 1136-1144.

[8] T-Y. Hsieh, K-J. Lee, and M. A. Breuer, "An Error-Oriented Test Methodology to Improve Yield with Error Tolerance," *Proceedings 24th IEEE VLSI Test Symposium (VTS)*, 2006.

[9] S. Shahidi and S. K. Gupta, "Estimating Error Rate during Self-Test via One's Counting," *Proc. International Test Conference,* 2006, Paper 15.3.

[10] J. Rearick, "Too Much Delay Fault Coverage is a Bad Thing," *Proc. International Test Conference,* 2001, pp. 624-633.

[11] J. Dworak, D. Dorsey, A. Wang, and M.R. Mercer, "Excitation, Observation, and ELF-MD: Optimization Criteria for High Quality Test Sets," *Proceedings of the 2004 IEEE VLSI Test Symposium (VTS'04)*, Napa Valley, California, April 25-29, 2004, pp. 9-15

[12] J. Dworak, J. D. Wicker, S. Lee, M. R. Grimaila, K. M. Butler, B. Stewart, L-C. Wang, and M. R. Mercer, "Defect-Oriented Testing and Defective-Part-Level Prediction," *IEEE Design and Test of Computers*, January-February, 2001, Vol. 18, No. 1, pp. 31 - 41.

IEEE International Symposium on Defect and Fault Tolerance of VLSI Systems

Core Test Wrapper Design to Reduce Test Application Time for Modular SoC Testing

Hyunbean Yi and Sandip Kundu
University of Massachusetts, Amherst, USA
bean@engin.umass.edu, kundu@ecs.umass.edu

Abstract

Conventional test access mechanism (TAM) and test wrappers of complex System-on-Chip (SoC) designs do not adequately utilize the system resources available in the functional mode of operation. With the advent of Network-on-Chip (NoC), the internal data transaction bandwidth has risen dramatically. This increase does not automatically translate to benefits during test. In this paper, we present a core test wrapper which takes advantages of the functional interconnect bandwidth to improve test application efficiency. Experimental results clearly demonstrate the benefit of the proposed approach in improving test application time.

1. Introduction

The number of Intellectual Property (IP) cores in a System-on-Chip (SoC) has been increasing over time, in commensurate with the growth in number of applications it performs. However, as the complexity of a SoC grows, on-chip bus becomes a bottleneck for overall throughput. To overcome these limitations, Network-on-Chip (NoC) has emerged as a new communication architecture [1,2].

Modular testing is one of efficient methods to test complex SoCs or NoCs that feature many IP cores [3,4]. In modular testing, a test access mechanism (TAM) enables the exchange of test data between external pins of a chip and its embedded cores. A test wrapper designed around an embedded core enables core isolation and provides an interface between a TAM and the embedded cores. By applying predetermined test patterns supplied by IP core providers, test generation time and test data volume can be reduced [3]. However, the area overhead due to the insertion of TAMs such as Daisychain [5], TestRail [6], and TestBus [7], and test wrappers is an issue in cost sensitive SoC designs. Accordingly, several researchers have focused on the reuse of a functional interconnect as a TAM and the design of a test wrapper fitted to the functional interconnects [8-12]. When a functional interconnect is utilized as a TAM, the test architecture models for an SoC and an NoC associated with Automated Test Equipment (ATE) are as shown in Fig. 1. Since the ATE interface between an ATE and a chip may be different from the functional interconnect protocol, a Test Interface Unit (TIU) module is required to convert the ATE interface into the functional interconnect protocol. Moreover, the number of ATE channels, the data width of a functional interconnects, and the number of scan chains of a core under test (CUT) may also be different. This may necessitate serial-to-parallel/parallel-to-serial conversion schemes in the TIU or in the test wrapper for bandwidth matching [11,12].

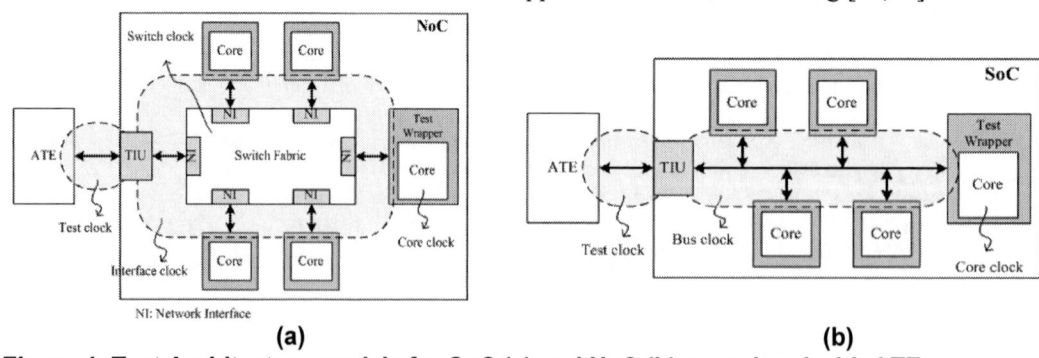

(a) **(b)**

Figure 1. Test Architecture models for SoC (a) and NoC (b) associated with ATE

1550-5774/08 $25.00 © 2008 IEEE
DOI 10.1109/DFT.2008.13

This paper presents a test wrapper design approach to optimize test bandwidth and reduce test time *without adding any serial-to-parallel or parallel-to-serial conversion* that tends to make the test inefficient. This point will be explained in detail in section 3. The rest of the paper is organized as follows. Section 2 reviews related work. In Section 3, a detailed analysis of representative existing test wrappers and the precise scope of our problem are described. We list our assumptions in Section 4 and describe our core test wrapper architecture in detail in Section 5. Experimental results are given in Section 6 and we conclude the paper in Section 7.

2. Related work

There have been several investigations on implementation of TAMs and test wrappers based on the Advanced Microcontroller Bus Architecture (AMBA) [13]. For structural testing, a combination of Test Interface Controller (TIC) and External Bus Interface (EBI) can be used as a TIU and the TH as a test wrapper. Test stimuli and responses are transferred through the external bidirectional test bus (TBUS) of the EBI. Feige et al. [8] presented a Scan-Test Harness (STH) to enable a scan test through the TIC and AMBA bus. Lin and Liang [9] added a separate observation-dedicated TAM to simultaneously perform scan-in and scan-out operations. Song et al. [10] illustrated a more detailed STH architecture including some control logics. They used the TBUS as a test stimuli application dedicated path and modified the EBI so that the address bus of the EBI can be used as a test responses observation dedicated path.

In [14], Marinissen et al. presented a test wrapper for embedded core test which consists of wrapper input cells, wrapper output cells, test ports, and wrapper control signals. They introduced an efficient configuration between wrapper cells and scan chains, named as TAM chain as shown in Fig. 2 in which wrapper input cells, core-internal scan chains, and wrapper output cells are concatenated in a row. For multiple TAM chains, some algorithms such as Largest Processing Time (LPT), First Fit Decreasing (FFD), MultiFit and COMBINE are used to minimize the maximum sum of scan lengths assigned to a TAM chain [14]. Amory et al. [11] presented a test wrapper based on the IEEE 1500 wrapper [15] for the reuse of a functional interconnect as a TAM. They also used a TAM chain named as wrapper chain as shown in Fig. 3 and used DTL [16] protocol for network interface. By enabling some wrapper cells at each end of a wrapper chain to perform parallel-to-serial/serial-to-parallel conversion, they reduced test length. However, if the test bandwidth is not a multiple of the test clock frequency, the test bandwidth utilization cannot be optimized. Hussin et al. [12] also used the wrapper chain proposed in [11] and added the bandwidth matching registers to the test wrapper to increase the test bandwidth utilization. However, the area overhead of the test wrapper due to the bandwidth matching registers is too high and test control scheme is too complex.

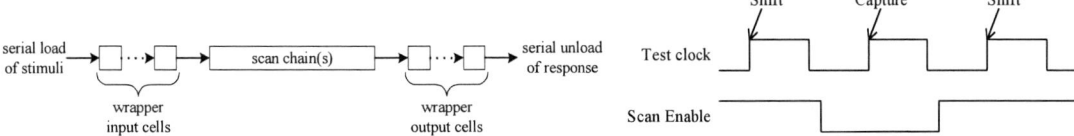

(a) TAM chain configuration　　　　　　**(b) Timing diagram for scan testing**
Figure 2. TAM chain and its operation

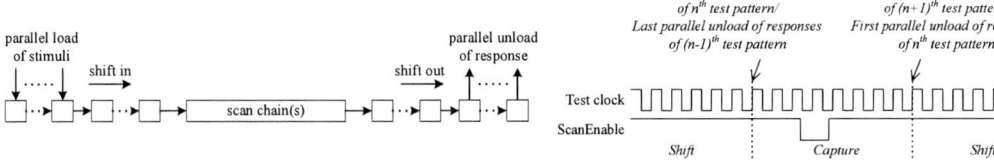

(a) Wrapper chain　　　　　　**(b) Timing diagram for scan testing**
Figure 3. Wrapper chain and its operation

3. Analyses of Previous Wrappers and Problem Statement

The test wrapper presented in [14] is based on the TAM chain in Fig. 2. In order to compare with other test wrappers, let us assume that there is no bidirectional terminal. For a core, if we denote the number of test patterns, the number of primary inputs, the number of primary outputs, and the number of scan flip-flops by p_c, i_c, o_c, and r_c, respectively, then test stimulus bits to be loaded $s_i = i_c + r_c$ and test response bits to be unloaded $s_o = o_c + r_c$. The unload of the responses of a test pattern and the load of the stimuli of next test pattern are overlapped in time and p_c capture operations are conducted. Therefore, if, for the TAM width w, w balanced TAM chains are configured, the test time of a core T_C can be defined as equation (1) [17].

$$T_C = \left\lceil \frac{\max(s_i, s_o) \cdot p_c + \min(s_i, s_o)}{w} \right\rceil + p_c \qquad (1)$$

The test wrapper proposed in [11] is based on the wrapper chain in Fig. 3. If the functional interconnect provides the guaranteed bandwidth and latency for testing, the test bandwidth b_{test} = $\min(b_{in}, b_{out})$ for ATE input bandwidth b_{in} and ATE output bandwidth b_{out}. Then, the number of wrapper chains $wc = \lfloor b_{test} / f_{test} \rfloor$ for the test frequency f_{test} and the first k wrapper input cells and the last k wrapper output cells of each wrapper chain are connected to the data inputs and the data outputs of the functional interconnect protocol, respectively. Test stimuli are loaded in parallel k-bit quanta from input data bus at regular intervals of k test clock cycles and shifted into a wrapper chain in serial at test clock. Likewise, test responses are shifted in serial at test clock by the end of a wrapper chain and unloaded in k-bit parallel onto output data bus at regular intervals of k test clock cycles. Accordingly, for each test pattern, the last k-bit word of test stimuli is directly consumed in parallel and the first word of test responses is directly transported away in parallel. Hence, the number of test clock cycles for load and shift of test stimuli t_i and the number of test clock cycles for shift and unload of test response t_o are calculated by

$$t_o = \left(\left\lceil \frac{wc_o}{k} \right\rceil - 1 \right) \cdot k + 1 \text{ and } t_i = \left(\left\lceil \frac{wc_i}{k} \right\rceil - 1 \right) \cdot k + 1 \qquad (2)$$

where wc_i and wc_o denote respectively the test data input bit length and the test data output bit length. Since scan chains and wrapper cells are configured with wc wrapper chains, wc_i $= \lceil s_i / wc \rceil$ and $wc_o = \lceil s_o / wc \rceil$, respectively.

Fig. 3(b) shows a timing diagram example of shift and capture operations for the case $k = 8$. For each test pattern, the capture operation is conducted within the *Capture* period between the last parallel loading of stimuli and the first parallel unloading of responses. Although the last word of stimuli can be directly consumed and the first word of responses can be directly transported away, the shift operations cannot be conducted within the *Capture* period until the first stimuli of next test pattern arrive. Therefore, the number of test clock cycles required for a capture operation t_{cap} is not '1' but $(k-1)$ where "−1" denotes the first test clock in the *Capture* period consumed to load the last word of stimuli. As a result, the test time of a core T_A can be defined as equation (3).

$$T_A = \{t_{cap} + \max(t_i, t_o)\} \cdot p_c + \min(t_i, t_o) \qquad (3)$$

However, since $wc = \lfloor b_{test} / f_{test} \rfloor$, the test wrapper has to be redesigned according to the test environment. Moreover, if b_{test} is a multiple of f_{test}, the test bandwidth utilization is maximized because all the w-bit data bus wires can be used for testing, but if b_{test} is not a multiple of f_{test}, the test bandwidth cannot be optimized. For example, if $w = 32$, $b_{test} = 3200$ Mbps, and $f_{test} = 500$

MHz, then $wc = 6$, $k = 5$, and only thirty wires out of 32-bit data bus wires can be used for testing and remaining two bits cannot be used.

In order to compare the test wrappers, let us take a closer look at them. In the TAM chained test wrapper [14], the number of TAM chains and the TAM width are the same by nature. Let us apply this TAM chained test wrapper to the environment where it uses a functional interconnects as a TAM. If we use a test wrapper configured with w TAM chains and create a new scan test clock so that test data can be shifted at the interval of k test clock cycles and captured within the interval as shown in Fig. 4, then the test time T_C (see Eq. (1)) can be reduced by p_c. Here, we observe that there is no difference between this operation (see Fig. 4) and the operation of the test wrapper in a wrapper chain (see Fig. 3) in terms of test time. This is due to the shifts that are performed just to catch up to the test bandwidth with small number of wrapper chains. Thus, if we generate the Scan Test Clock shown in Fig. 4, we obtain

$$T_A \geq (T_C - p_c). \qquad (4)$$

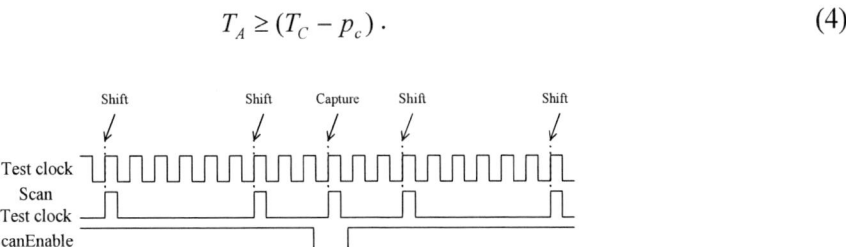

Figure 4. Waveform of the Scan Test Clock to shift and capture in *w*-bit parallel

4. Assumptions

Following assumptions are used in our test wrapper design:

- As with [11] and [12], it is assumed that a functional interconnect provides a guaranteed bandwidth and latency for testing and supports full-duplex and burst data transfer with separate read and write data buses. This allows scan-in and scan-out operations to be conducted at the same time.
- We assume that the number of scan chains in a core equal data width of the functional interconnects.
- If a functional interconnect is used as a TAM, the scan enable signal must be generated in each test wrapper because functional interconnect protocols may not support scan enable signal. Therefore, in this paper, we assume that in each core scan enable signal is carefully routed like clock signal or is locally regenerated at each flip-flop using a technique such as Fast Scan Enable Generator [18] or Delay Test Scan Flip-Flop (DTSFF) [19] developed for scan-based delay testing so that capture operations can be performed at the full test clock speed.

5. Core Test Wrapper Design

Each end of a TAM chain or a wrapper chain is connected to a data bus. Thus, this paper envisions a test wrapper which can load/unload in parallel. In order to do so without breaking the serial configuration of the IEEE 1500 wrapper, some extra logic is needed. The proposed test wrapper architecture is shown in Fig. 5. According to our assumption, the number of scan chains is equal to the data width of functional interconnect w. Some of the wrapper input/output cells are assigned to short scan chains to minimize scan-in and scan-out lengths and the others are grouped in w-bit. If wrapper input cells or wrapper output cells are divided by j groups for a positive integer j, the size of jth group which is the last group can be equal to or less than w. Let us assume that there are two test modes; one is "serial test mode" in which the test wrapper is

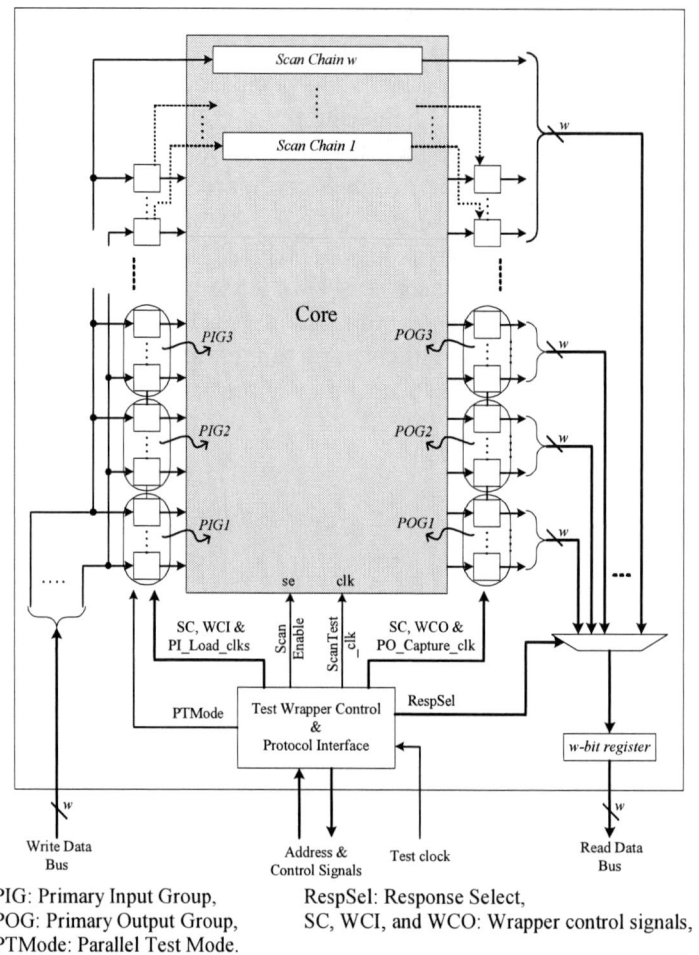

PIG: Primary Input Group,   RespSel: Response Select,
POG: Primary Output Group,   SC, WCI, and WCO: Wrapper control signals,
PTMode: Parallel Test Mode.

Figure 5. Proposed Core Test Wrapper Architecture

controlled by the IEEE Std. 1500, and the other is "parallel test mode" in which the Test Wrapper Control & Protocol Interface logic enables scan chains or primary input groups (PIGs), selects scan chains or primary output groups (POGs), and generates test clock pulses in time so that test data can be applied and observed in w-bit parallel. The wrapper cells are designed based on the IEEE 1500 wrapper cell. The IEEE 1500 standard wrapper cell can be used for the wrapper output cells, but the wrapper input cells need to be slightly modified to load test data from Write Data Bus.

5.1. Assigning Wrapper Cells to Short Scan Chains

The length of scan shift is determined by a longest scan chain out of w scan chains, and the numbers of PIGs and POGs are determined by wrapper input cells and wrapper output cells, respectively. Modern scan insertion tools can easily create well-balanced multiple scan chains. If w balanced scan chains are created, the difference between the length of a longest one and the length of a shortest one becomes 1. Therefore, at most $(w - 1)$ wrapper cells are assigned to short scan chains.

For a core, the number of PIs PIn, the number of POs POn, the number of PIGs $PIGn$, and the number of POGs $POGn$ are given. Furthermore, to illustrate our wrapper cell assignment algorithm, we define the number of short scan chains $SSCn = \{w - (r_c \bmod w)\}$ for the number of scan flip-flops r_c, $PIr = (PIn \bmod w)$, and $POr = (POn \bmod w)$. Then, the number of the wrapper input cells selected to be assigned $SWICn$ and the number of the wrapper output cells selected to be assigned $SWOCn$ are determined by Algorithm 1. $PIGn$ and $POGn$ are determined by the remaining wrapper input/output cells, and thus we can get $PIGn = \lceil (PIn \cdot SWICn)/w \rceil$ and $POGn = \lceil (POn \cdot SWOCn)/w \rceil$.

Algorithm 1 shows that the test bandwidth utilization can be maximized by finding a combination of $SWICn$ and $SWOCn$ so that the sum of $PIGn$ and $POGn$ and the difference between $PIGn$ and $POGn$ become a minimum. In Line 1-3, since $PIr \leq SCCn$, $PIGn$ is reduced by 1 if PIr wrapper input cells are used to be assigned. However, since $POr > SCCn$, $POGn$ cannot be reduced even if $SCCn$ wrapper output cells are used to be assigned. Therefore, $SWICn = PIr$ and $SWOCn = 0$. In case of Line 8, wrapper cells in the bigger one of PI and PO are used. Since PI application and PO observation are conducted simultaneously, the test time can be optimized by reducing the difference between $PIGn$ and $POGn$. Line 16 is the best case for test time reduction because each of $PIGn$ and $POGn$ can be reduced by 1. On the contrary, in case both PIr and POr are greater than $SCCn$ like Line 20, $PIGn$ or $POGn$ cannot be reduced, even though wrapper cell assignment is performed. Therefore, in this case we do not need to assign wrapper cells.

Algorithm 1: Wrapper Cell Assignment Algorithm
Input: $SCCn$, PIr, POr, PIn, POn
Output: $SWICn$, $SWOCn$
1 **if** $(PIr \leq SCCn \parallel POr > SCCn)$ {
2        $SWICn = PIr$;
3        $SWOCn = 0$;
4 } **else if** $(PIr > SCCn \parallel POr \leq SCCn)$ {
5        $SWICn = 0$;
6        $SWOCn = POr$;
7 } **else if** $(PIr \leq SCCn \parallel POr \leq SCCn)$ {
8   **if** $((PIr + POr) > SCCn)$ {
9   **if** $(PIn > POn)$ {
10     $SWICn = PIr$;
11      $SWOCn = 0$;
12   } **else** {
13      $SWICn = 0$;
14      $SWOCn = POr$;
15   }
16   } **else** {   // $(PIr + POr) \leq SCCn$): Best case
17     $SWICn = PIr$;
18     $SWOCn = SCCn - PIr$;
19   }
20 } **else** {   // $(PIr > SCCn \parallel POr > SCCn)$: Worst Case
21     $SWICn = 0$;
22     $SWOCn = 0$;
23 }

5.2. Test Wrapper Control and Protocol Interface

Test Wrapper Control & Protocol Interface logic generates control signals and clock pulses necessary to perform scan testing. To generate the signals in time, address of functional interconnect [8-10] or a counter [11,12] can be used. In accordance with each scan test step, Address Decoder or Counter generates several control signals such as the scan shift enable (ScanShift_En), the capture enable (Capture_En), and the response select (RespSel) and the PI_Load_clks (PIG1_clk, PIG2_clk, ..., and PIGj_clk) to load stimuli onto one PIG at a time. The RespSel signal is generated so that the scan response can be selected during the scan shift in operation and the response of a POG can be selected when stimuli is loaded onto a PIG. In this way, scan shift in and scan shift out, and PI application and PO observation are conducted simultaneously. The scan enable signal (ScanEnable), the scan test clock (ScanTest_clk), and the PO_Capture_clk to capture test response onto wrapper output cells are generated. To generate scan test clock pulses, some clock gating logics are required. We can get glitch-free clock pulses by using clock gating cells (CGCs) [20,21].

If $PIGn = 3$, $POGn = 3$, and test data are transferred at regular intervals of 4 Test clock cycles, then the timing diagram for scan testing is displayed as shown in Fig. 6. By using a scan test procedure (*Scan Shift In → PI Application → Scan & PO Capture → Scan Shift Out → PO Observation*), scan shift in and scan shift out can be overlapped and PI application and PO observation also can be overlapped. In the *Scan Shift In/Out* period, scan chains operate with the ScanTest_clk at regular intervals of 4 Test clock cycles and the RespSel keeps selecting scan chains. In the *PI Application/PO Observation* period, stimuli are loaded into each of PIGs with each PIG clk (PIG1_clk, PIG2_clk, and PIG3_clk) at regular intervals of 4 Test clock cycles and RespSel selects one POG at a time. In the *Scan & PO Capture* period, test results are captured into scan chains with the ScanTest_clk and wrapper output cells with the PO_Capture_clk at the same time. This capture operation is performed within an interval for a PI application/PO observation. Therefore, no additional interval is consumed for a capture operation, and thus, the test time can be reduced.

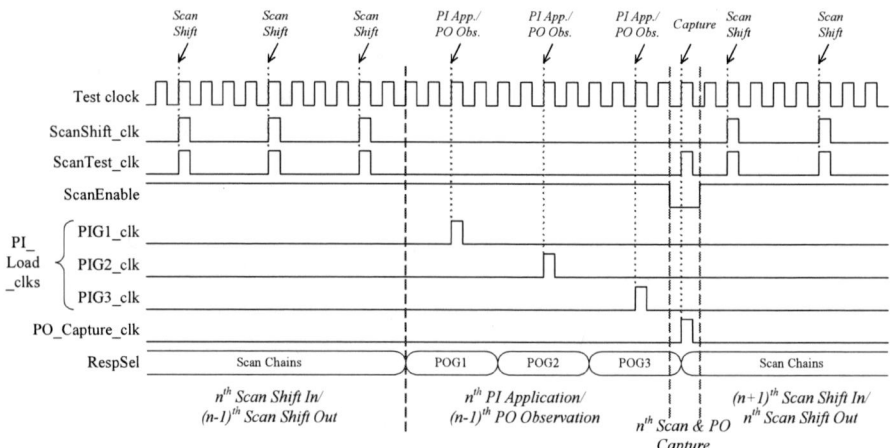

Figure 6. Timing Diagram Example for Scan Testing with Proposed Test Wrapper

5.3. Test Time Analysis

In the proposed test wrapper, all the scan shift in/out and PI application/PO observation operations are performed in w-bit parallel. Accordingly, scan shift in/out or PI/PO application/observation operations are conducted at regular intervals of m Test clock cycles, with $m = \{(f_{test} \cdot w)/b_{test}\}$. The length of the maximum scan chain $l_s = \lceil r_c/w \rceil$ for the number of scan flip-flops r_c, and $PIGn$ (the number of PI groups) and $POGn$ (the number of PO groups) are obtained from the results of Algorithm 1 in subsection 5.1. The capture operation does not need to be counted because it is performed within an interval for a PI application/PO observation. Also, the last PO observation operation for the last test pattern consumes not m Test clock cycles but only 1 Test clock cycle. Therefore, the test time of the proposed test wrapper T_P is

$$T_P = \{\max(c_i, c_o) \cdot p_c + \max(c_i, c_o) - 1\} \cdot m + 1 \qquad (5)$$

with $c_i = (PIGn + l_s)$ and $c_o = (POGn + l_s)$.

6. Experimental Results

In order to evaluate the proposed wrapper, we used AMBA-based IP cores supplied by Gaisler Research [22]. The IP cores have 32-bit separate read and write data buses and are supplied with Verilog HDL models and thus are suitable for our experiment. The cores are

Table 1. Comparison of Test Time

Core	Test Time (# of Test clock cycles)									
	Case 1: b_{test} = 1600 Mbps and f_{test} = 400 MHz					*Case 2: b_{test} = 3200 Mbps and f_{test} = 500 MHz*				
	Conv. [14]	*Amory* [11]	*Prop.*	Red.(%) (/Conv.)	Red.(%) (/Amory)	*Conv.* [14]	*Amory* [11]	*Prop.*	Red.(%) (/Conv.)	Red.(%) (/Amory)
Leon3 Processor	66289	65513	65513	1.2	0	41431	43866	40946	1.2	6.7
SDRAM Controller	14137	13713	13713	3.0	0	8836	9076	8571	3.0	5.6
Ethernet MAC	3649	3489	3489	4.4	0	2281	2186	2181	4.4	0.2
VGA	125241	125161	125161	0.1	0	78276	83116	78226	0.1	5.9
GPIO	1513	1489	1489	1.6	0	946	931	931	1.6	0
PS/2	10497	10073	10073	4.0	0	6561	6746	6296	4.0	6.7
Timer	3121	2785	2785	10.8	0	1951	1991	1741	10.8	12.6
UART	2369	2337	2337	1.4	0	1481	1466	1461	1.4	0.3

designed to interface with AMBA 2.0 bus. Hence, we assume that each of them has Advanced eXtensible Interface (AXI) port [23], which is a widely used network interface (NI) in NoC, and we implemented a simple AXI logic as a protocol interface in test wrapper.

Table 1 shows a test time comparison of the proposed wrapper (*Prop.*) with existing test wrappers, *Conv.* with TAM chains [14] and *Amory* with wrapper chains [11]. The test times for the *Case 1* when b_{test} = 1600 Mbps and f_{test} = 400 MHz and the *Case 2* when b_{test} = 3200 Mbps and f_{test} = 500 MHz are presented. For *Conv.*, in order to compare with other test wrappers, we assumed that the test wrapper is configured with 32 TAM chains and all shift and capture operations are conducted at regular interval of m Test clock cycles for $m = \{(f_{test} \cdot w)/b_{test}\}$. For *Amory*, in the *Case 1*, b_{test} is a multiple of f_{test}. Then, the number of wrapper chains wc $= \lfloor 1600/400 \rfloor = 4$ and (32 mod 4) = 0 and thus, the test bandwidth utilization is maximized because all data bus wires can be utilized as test data carriers. In the *Case 2*, however, the test bandwidth utilization becomes low because only 30 ($wc = \lfloor 3200/500 \rfloor = 6$ and (32 mod 6) = 2) out of 32 bus wires can be used. Therefore, the test times for all the cores are slower than those of *Prop.* and some of them are slower than those of *Conv.*. The proposed wrapper (*Prop.*) does not need to be reconfigured according to the cases and can always optimize the test bandwidth. As a result, the test time of *Prop.*, compared to *Conv.*, is reduced 3.3% on average for all the cases, and compared to *Amory*, is the same for the *Case 1* and is reduced 4.8% on average for the *Case 2*.

An area comparison of test wrappers is presented in Table 2. The area of proposed wrappers increased on an average by 19.4% compared to IEEE 1500 wrappers due to modification of the wrapper input cells with an additional MUX, a control input and AXI protocol logic. This represents an increase of 4.9% compared to the result (14.5%) of the experiment which Amory [11] carried out with ITC'02 test benchmarks [24]. This area increase is relatively small when compared to the total chip area at an SoC or an NoC level and hence can be easily justified by the corresponding test time reduction.

Table 2. Comparison of Area

Core	Area (# of equivalent gates (2-input NAND))		
	IEEE 1500 wrapper	*Prop.* (with AXI)	Area Inc. (%)
Leon3 Processor	5276	6279	19.0
SDRAM Cont.	9004	10523	16.9
Ethernet MAC	5788	6847	20.7
VGA	3068	3703	18.3
GPIO	5692	6703	17.8
PS/2	2716	3327	22.5
Timer	3324	3959	19.1
UART	3020	3659	21.2

7. Conclusions

In this paper we have presented a core test wrapper to utilize a functional interconnect as a TAM and to improve the test performance. The proposed test wrapper optimizes the test bandwidth by enabling all data bus lines to be used as parallel test data without any need for serial-to-parallel/parallel-to-serial conversion. Moreover, by generating scan test control signals so that a capture operation can be interleaved with shift operation, additional test time reduction is achieved. Also, our test wrapper design is independent of tester bandwidth and test clock frequency. We designed the test wrapper based on IEEE 1500 wrapper and added AXI interface logic for experimental purposes. Our experimental results with AMBA based cores show that we always achieve the fastest test time at the expense of a small but acceptable area overhead.

References

[1] L. Benini and G. De Micheli, "Networks on Chips: A New SoC Paradigm," *IEEE Computer*, 2002, pp. 70-80.

[2] P. P. Pande et al., "Design, Synthesis, and Test of Networks on Chips," *IEEE Design & Test of Computers*, 2005, pp. 404-412.

[3] E. J. Marinissen and T. Waayers et al., "Infrastructure for modular SOC testing," *Proceedings IEEE Custom Integrated Circuits Conf.*, 2004, pp. 671-678.

[4] T. Waayers et al., "Definition of a robust Modular SOC Test Architecture; Resurrection of the single TAM daisy-chain," *Proceedings IEEE Int. Test Conf.*, 2005, Paper 25.3, pp. 1-10.

[5] J. Aerts and E. J. Marinissen, "Scan Chain Design for Test Time Reduction in Core-Based ICs," *Proceedings IEEE Int. Test Conf.*, 1998, pp. 448-457.

[6] E. J. Marinissen et al., "A Structured and Scalable Mechanism for Test Access to Embedded Reusable Cores," *Proceedings IEEE Int. Test Conf.*, 1998, pp. 284-293.

[7] P. Varma and S. Bhatia, "A Structured Test Re-use Methodology for Core-Based System Chips," *Proceedings IEEE Int. Test Conf.*, 1998, pp. 294-302.

[8] C. Feige et al, "Integration of the Scan-Test Method into an Architecture Specific Core-Test Approach," *J. Electronic Testing: Theory and Applications*, 1998, pp. 125-131.

[9] C. Lin and H. Liang, "Bus-Oriented DFT Design for Embedded Cores," *IEEE Asia-Pacific Conf.*, 2004, pp. 561-563.

[10] J. Song et al., "Design of Test Access Mechanism for AMBA Based System-on-a-Chip," *IEEE VLSI Test Symp.*, 2007. pp. 375-380.

[11] A. M. Amory et al., "Wrapper Design for the Reuse of a Bus, Network-on-Chip, or Other Functional Interconnect as Test Access Mechanism," *IET Comput. Digit. Tech.*, 2007, pp. 197-206.

[12] F. A. Hussin et al., "Optimization of NoC Wrapper Design Under Bandwidth and Test Time Constraints," *Proceedings IEEE Europ. Test Symp.*, 2007, pp. 35-42.

[13] ARM, "AMBA Specification (Rev 2.0)," May 1999.

[14] E. J. Marinissen et al., "Wrapper design for embedded core test," *Proceedings IEEE Int. Test Conf.*, 2000, pp. 911 – 920.

[15] IEEE Computer Society, "IEEE Standard Testability Method for Embedded Core-based Integrated Circuits," Aug. 2005.

[16] Philips Semiconductors, "Device Transaction Level (DTL) Protocol Specification," Ver. 2.2, 2002.

[17] S. K. Goel and E. J. Marinissen, "SOC Test Architecture Design for Efficient Utilization of Test Bandwidth," *ACM Trans. on Design Automation of Electronic Systems*, 2003, pp. 399-429.

[18] S. Wang et al., "Hybrid Delay Scan: A Low Hardware Overhead Scan-based Delay Test Technique for High Fault Coverage and Compact Test Sets," *Proceedings of the Design, Automation and Test in Europe Conf. and Exhibition*, 2004, pp. 1296-1301.

[19] G. Xu and A. D. Singh, "Delay Test Scan Flip-Flop: DFT for High Coverage Delay Testing," *IEEE Int. Conf. on VLSI Design*, 2007, pp. 763-768.

[20] M. Beck et al., "Logic Design for On-Chip Test Clock Generation – Implementation Details and Impact on Delay Test Quality," *Proceedings of the Design, Automation and Test in Europe*, 2005, pp. 56-61.

[21] Synopsys On-Line Documentation, Vol. 1, 2003.

[22] J. Gaisler et al., "Gaisler Research IP Core's Manual," Ver. 1.0.16, 2007.

[23] ARM, "AMBA AXI Protocol Specification," V. 1.0, 2003.

[24] E. J. Marinissen et al., "A Set of Benchmarks For Modular Testing of SOCs," *Proceedings IEEE Int. Test Conf.*, 2002, pp. 519-528.

INVITED TALK

Computing at the Nanoscale

John E. Savage, *Brown University*

Advances have been been made recently in assembling nanoscale devices without using photolithography. This important development, which offers the potential for greatly increasing the density of memory cells and logic gates, introduces a new model of computation and new analytical challenges. In this talk we provide an introduction to this new area.

The difficulty of assembling irregularly placed nanoscale devices has caused the research community to focus on the crossbar. All known methods for controlling individual nanowires (NWs) in a crossbar by mesoscale wires (MWs) introduces randomness in the connections. This introduces several questions. First, which methods of controlling NWs with MWs devotes the smallest amount of area for this purpose? Second, how can stochastically assembled chips be configured after assembly? Third, since errors will occur during assembly, how can chips be designed to minimize the effect of such errors? Finally, what computational limitations do stochastically assembled, crossbar-based computers introduce? We will address these and other questions.

Speaker Bio – John Savage is Professor of Computer Science at Brown University. He earned his PhD in Electrical Engineering at MIT in 1965 and his bachelor's and Master's degrees, also at MIT, in 1962. He was employed by Bell Laboratories from 1965 until 1967 when he joined the faculty at Brown University. He is a founder of the Department of Computer Science and was its chair from 1985 to 1991.

Savage's early research was in information theory and communication theory. His work on the complexity of decoders for error correcting codes in the 1960s led him into theoretical computer science and to the introduction of circuit complexity into the field. His first book, *The Complexity of Computing*, published in 1976, became the standard reference on circuits. He 1998 book *Models of Computation* provides a comprehensive treatment of theoretical computer science. He has contributed to research on space-time tradeoffs, area-time tradeoffs in VLSI, I/O time-space tradeoffs, silicon compilers, and parallel algorithms for VLSI and the finite-element method. His current research focus is computational nanotechnology and coded computation. He is a Fellow of AAAS and ACM, a Life Fellow of IEEE, and a Guggenheim Fellow.

1550-5774/08 $25.00 © 2008 IEEE
DOI 10.1109/DFT.2008.70

SESSION 10

ERROR DETECTION AND CORRECTION (2)

A Generalized Approach for the Use of Convolutional Coding in SEU Mitigation

L. Frigerio, M. A. Radaelli, F. Salice
Dipartimento di Elettronica e Informazione
Politecnico di Milano, Milano, Italy
Email: laura.frigerio@polimi.it

Abstract

1 Introduction

In recent years the continuous evolution in the fabrication technology process allowed the scaling of components size. The reduction of devices physical size induces also a proportional decrement of supply and threshold voltages. As a side effect, devices are becoming more vulnerable to radiation and this means that particles with small charge, which were once negligible, are now able to produce upsets [1]. The most common effect induced by a charged particle that hits a memory cell is referred to as SEU (Single Event Upset). A SEU provokes the reverse of the transistors state, causing a bit-flip in the memory cell. This error is usually defined as soft-error, since the normal functioning of the component is restored after the hit. However the memory content can be altered and therefore data can be corrupted. In order to preserve data integrity the problem of SEU must be addressed and suitable solutions must be introduced.

Several techniques can be used to protect a digital design, like space, time or information redundancy [2]. Among them, the most common way to protect memory content is to use information redundancy techniques [3]. Error Detection And Correction coding (EDAC) schemes can be exploited to produce a transparent architecture, that is able to mitigate possible errors that occur during the functioning. These schemes are usually composed of an encoder used to introduce information redundancy and a decoder that reconstructs the original data and corrects possible errors.

Convolutional codes are among the most used EDAC techniques and are usually exploited to protect data streams, especially during transmissions [4]. In this case the decoder is usually quite complex, in order to protect data from multiple or burst errors that can occur

1550-5774/08 $25.00 © 2008 IEEE
DOI 10.1109/DFT.2008.36

during the transmission. Typical schemes of decoding are based on trellis and maximum likelihood (e.g. Viterbi algorithm [5]). When dealing with Single Event Upsets, these decoding schemes are too complex with respect to the problem under consideration, therefore convolutional encoding appears less appealing, especially when compared to simpler schemes, like light block coding (e.g. Hamming coding). However, the development of light decoding schemes based on convolutional codes, can enable the use of this type of coding with the aim of protecting data from SEU.

The use of convolutional codes has been proposed for error detection purposes in FSM designs. In [7] an error detection with latency is proposed in which the FSM is realized such that all sequences of states are code sequences in a convolutional code. An error-detecting decoder for the convolutional code is used to monitor the sequencing of states. The approach is extended in [8] by considering sequential circuits and microprogrammed control units. Several improvements of this approach have also been presented in [9], by reducing the needed hardware and adding protection also to output logic.

In this paper, we propose a scheme for both error detection and correction, based on convolutional encoding, that can be exploited to protect data from SEU. This work generalizes our previous work [10] that proposes two light convolutional decoders for combinatorial and sequential circuits. The presented architectures can be applied to applications which present a need for reduced coding and decoding times, such as fast, synchronous, memory-based systems. In this paper a methodology is developed to allow considering all the possible combinations of generator polynomials of order three. The automatic generation of synthetizable VHDL code for all the possible architectures is also considered, by exploiting a framework for cores generation. The comparison with state-of-the art solutions shows that the use of convolutional codes is advantageous in terms of performance and error coverage.

The rest of the paper is organized as follows: Section 2 introduces the convolutional codes, Section 3 presents the proposed approach, Section 4 illustrates the architecture of the light decoder, Section 5 describes how the automate the generation of the VHDL code, Section 6 reports experimental results and Section 7 concludes the paper.

2 Convolutional codes

A convolutional code is an error-correction code where the source information is processed as a stream and the transmitted bits are a linear function of the past source bits. It produces a n-bits coded word starting from a m-bits word ($m \leq n$). The quantity m/n is called •• •• and represents the number of generated bits for each input bit. For example, with a rate of 1/2, the encoder produces two output bits for each input bit.

Usually, the rule for generating the transmitted bits involves feeding the present source bit into a linear shift-register (•••• •• ••••••••) of length (•• ••••) l, and transmitting one or more linear functions of the state of the shift register at each iteration [6]. The way in which the original bits are combined to generate the output is described by •• •• •• •• •• •• •• •• ••.

Let us consider, for example, a convolutional code with rate 1/2 and generator polynomials $G_1 = (1,1,1)$ and $G_2 = (1,0,1)$. The corresponding encoder is depicted in Figure 1a. Each coded bit is generated by combining the bits of the original data through modulo-2 additions, as described by the generator polynomials or equivalently by the following

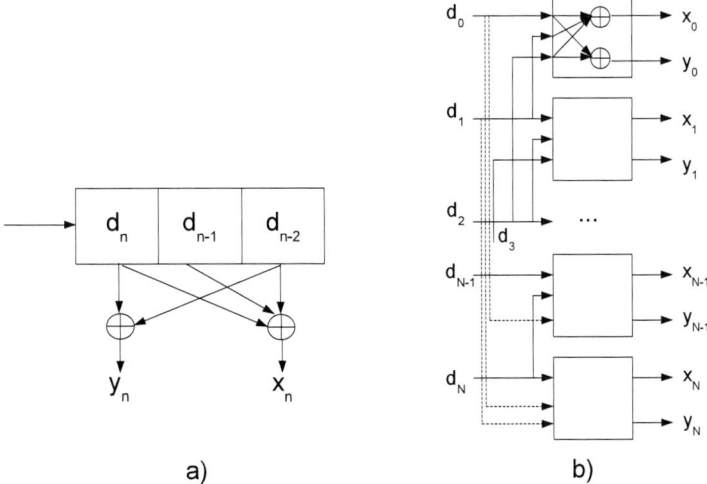

Figure 1. a) Sequential encoder with rate $1/2$, generator polynomials $G_1 = (1,1,1)$ and $G_2 = (1,0,1)$, b) Correspondent combinatorial encoder.

equations:

$$\begin{aligned} x_n &= d_n \oplus d_{n-1} \oplus d_{n-2} \\ y_n &= d_n \oplus d_{n-2} \end{aligned} \tag{1}$$

where d_i represents the i-th bit of the input sequence and x_i and y_i are the corresponding outputs. Bits of the input sequence are fed sequentially into the register, that is usually initialized at zero.

Equivalently we can obtain a combinatorial architecture by unfolding the sequential coding scheme over the entire data word. Figure 1b represents this situation, with a circular structure for the input data (Tail Biting technique).

One of the most common ways to decode the produced bits is by the analysis of the trellis that describes all the possible sequences that can be generated by the encoder. Compared with the encoding process, that is simple, the decoding phase is usually very complex and requires considerable time and resources when implemented in hardware. This approach, however, guarantees a high correcting power and is suited, for example, to protect disturbed transmission over channels. When operating to protect memory data from SEU, the required correcting power is limited, and usually the mitigation architecture is based on the hypothesis of single-error. In the following we discuss how to obtain a light decoder that can be used to protect data from SEU.

3 Proposed approach

The generic equations for the generator polynomials of a convolutional coding scheme, with order $l = 3$ are the following:

$$x_n = \alpha_1 d_n \oplus \beta_1 d_{n-1} \oplus \gamma_1 d_{n-2} \tag{2}$$

$$y_n = \alpha_2 d_n \oplus \beta_2 d_{n-1} \oplus \gamma_2 d_{n-2} \tag{3}$$

with $G_1 = (\alpha_1, \beta_1, \gamma_1)$, $G_2 = (\alpha_2, \beta_2, \gamma_2)$ the generator polynomials. Bits d are the original data, while bits x, y are the coded data.

Not all the combinations of coefficients of G_1 and G_2 generate a valid convolutional code. In particular the following situations generate non-valid polynomials:

- All the coefficients of one of the two polynomials are equal to zero.
- Polynomials G_1 and G_2 are equal.
- Both polynomials contains only a value different from zero. In this case the input values are not convolved but only delayed and transmitted in output.
- Polynomial G_1 and G_2 are linearly dependent. This happens when both the polynomials have two coefficients equal to one.

These configurations are referred as non-valid (NV) in Table 1 listing all the possible configurations.

Table 1. Configurations of Generator Polynomials

$\alpha_1\beta_1\gamma_1$	$\alpha_2\beta_2\gamma_2$	$\alpha_\oplus\beta_\oplus\gamma_\oplus$	Cat.	$\alpha_1\beta_1\gamma_1$	$\alpha_2\beta_2\gamma_2$	$\alpha_\oplus\beta_\oplus\gamma_\oplus$	Cat.	$\alpha_1\beta_1\gamma_1$	$\alpha_2\beta_2\gamma_2$	$\alpha_\oplus\beta_\oplus\gamma_\oplus$	Cat.	$\alpha_1\beta_1\gamma_1$	$\alpha_2\beta_2\gamma_2$	$\alpha_\oplus\beta_\oplus\gamma_\oplus$	Cat.
000	000	000	NV	010	000	010	NV	100	000	100	NV	110	000	110	NV
000	001	001	NV	010	001	011	NV	100	001	101	NV	110	001	111	C_1
000	010	010	NV	010	010	000	NV	100	010	110	NV	110	010	100	C_1
000	011	011	NV	010	011	001	C_1	100	011	111	C_1	110	011	101	NV
000	100	100	NV	010	100	110	NV	100	100	000	NV	110	100	010	C_1
000	101	101	NV	010	101	111	C_1	100	101	001	C_1	110	101	011	NV
000	110	110	NV	010	110	100	C_1	100	110	010	C_1	110	110	000	NV
000	111	111	NV	010	111	101	C_1	100	111	011	C_1	110	111	001	NA
001	000	001	NV	011	000	011	NV	101	000	101	NV	111	000	111	NV
001	001	000	NV	011	001	010	C_1	101	001	100	C_1	111	001	110	C_1
001	010	011	NV	011	010	001	C_1	101	010	111	C_1	111	010	101	C_1
001	011	010	C_1	011	011	000	NV	101	011	110	NV	111	011	100	NA
001	100	101	NV	011	100	111	C_1	101	100	001	C_1	111	100	011	C_1
001	101	100	C_1	011	101	110	NV	101	101	000	NV	111	101	010	C_2
001	110	111	C_1	011	110	101	NV	101	110	011	NV	111	110	001	NA
001	111	110	C_1	011	111	100	NA	101	111	010	C_2	111	111	000	NV

The proposed approach for the convolutional decoding is based on two phases:

- Double Decoding: for each original bit d_n two decoding options are generated: d_{n_a}, d_{n_b}. The equations that generate the two options are referred respectively as E_a^n and E_b^n and are XOR chains of coded bits.
- Forward Correction: when an error is present, the decision on which value transmit in output is taken by comparing the values of a forward couple d_{n+k_a}, d_{n+k_b}, with $1 \leq k \leq 2$ for polynomial order equal to 3. k is referred as ••••• ••••• .

In order to be effective for the decoding and correcting scheme, equations E_a^n and E_b^n must satisfy the following properties:

- Disjoint Functions: E_a^n and E_b^n depend on different set of coded bits. With A_n, B_n the support sets of E_a^n and E_b^n we obtain: $A_n \cap B_n = 0$.
- Forward inclusion: $A_n \subset B_{n+k}$, $B_n \cap (A_{n+k} \cup B_{n+k}) = \emptyset$, $1 \leq k \leq 2$, so that an error on the support set of E_a^n causes an error on E_b^{n+k}, but an error on the support set of E_b^n does not affect neither E_a^{n+k} nor E_b^{n+k}.

To give an example of the proposed approach, let us consider the generator polynomials $G_1 = (1, 1, 1)$ and $G_2 = (1, 0, 1)$. By manipulating Equation 1 we obtain the following couple of decoding functions:

$$
\begin{aligned}
E_a^n : d_{n_a} &= x_{n+1} \oplus y_{n+1} \\
E_b^n : d_{n_b} &= y_n \oplus x_{n-1} \oplus y_{n-1}
\end{aligned}
\tag{4}
$$

with support $A_n = \{x_{n+1}, y_{n+1}\}$, $B_n = \{y_n, x_{n-1}, y_{n-1}\}$. With check index equal to 2 we consider the following couple to perform the correction:

$$
\begin{aligned}
E_a^{n+2} : d_{n+2_a} &= x_{n+3} \oplus y_{n+3} \\
E_b^{n+2} : d_{n+2_b} &= y_{n+2} \oplus x_{n+1} \oplus y_{n+1}
\end{aligned}
\tag{5}
$$

The equations satisfy the above properties and can be used for decoding and correction purposes. If d_{n_a} and d_{n_b} are different only one of the two is correct. If d_{n_a} is correct, that means that x_{n+1} and y_{n+1} are also correct and d_{n+2_b} is equal to d_{n+2_a} (under the hypothesis of single fault). Otherwise, if d_{n_a} is incorrect, either x_{n+1} or y_{n+1} are corrupted causing also d_{n+2_b} to be wrong and to differ from d_{n+2_a}. We can therefore decide which is the correct value to be used to reconstruct the word.

In the following we explain how to derive equations E_a^n and E_b^n for all the couple of valid generator polynomials. Let us consider the combination of Equation 2 and 3 with a XOR operation:

$$
x_n \oplus y_n = (\alpha_1 \oplus \alpha_2)d_n \oplus (\beta_1 \oplus \beta_2)d_{n-1} \oplus (\gamma_1 \oplus \gamma_2)d_{n-2}
\tag{6}
$$

Coefficients of 6 are referred as: $\alpha_\oplus = \alpha_1 \oplus \alpha_2, \beta_\oplus = \beta_1 \oplus \beta_2, \gamma_\oplus = \gamma_1 \oplus \gamma_2$.

For all the valid configurations of generator polynomials, at least one among the equations 2, 3 and 6 contains only one term d in the right side. Equation E_a^n is therefore computed by deriving the term d in function of x and y and eventually shifting the equation to obtain d_n in the left side. Equation E_b^n is computed by substituting equation E_a^n in Equations 2 or 3.

More in detail, we can classify all the possible valid cases as follows:

- Equation 2 or Equation 3 has only one coefficient equal to one (case C_1). Without loss of generality, let us consider that this property is satisfied by Equation 2. Equation E_a^n is computed by $d_{n_a} = x_{n+v}$, $0 \leq v \leq 2$. Equation E_b^n is computed by making explicit the first term d in the equation 3 and substituting the other terms d by using E_a^n:

  - $\alpha_2 = 1$: $E_b^n : d_{n_b} = y_n \oplus \beta_2 x_{n+v-1} \oplus \gamma_2 x_{n+v-2}$. The check index is 2 if $\gamma_2 = 1$, 1 otherwise.

  - $\alpha_2 = 0 \rightarrow \beta_2 = \gamma_2 = 1$: $E_b^n : d_{n_b} = y_{n+1} \oplus x_{n+v-1}$. The check index is 1.

- One between Equation 2 and Equation 3 has two coefficients equal to one, and the other has all the coefficients equal to one. Equation 6 has only one coefficient equal to one and can be used to derive E_n^a. To derive E_n^b, let us consider without loss of generality, that Equation 3 has two coefficients equal to one. We obtain:

  - $\alpha_2 = 0$: $E_n^a = x_n \oplus y_n$. It is not possible to identify a E_n^b that satisfies the above properties (case NA).

  - $\beta_2 = 0$: $E_n^a = x_{n+1} \oplus y_{n+1}$, $E_n^b : d_{n_b} = y_n \oplus x_{n-1} \oplus y_{n-1}$. The check index is 2 (case C_2).

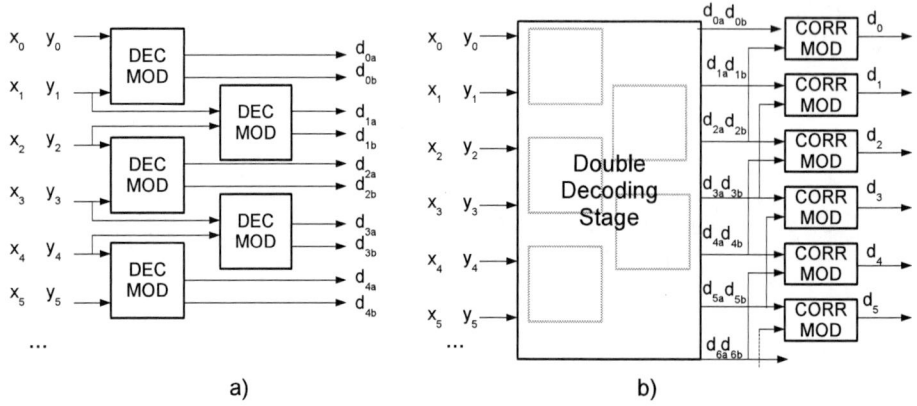

Figure 2. a) Decoder Stage b) Decoder Architecture

- $\gamma_2 = 0$: $E_n^a = x_{n+2} \oplus y_{n+2}$. It is not possible to identify a E_n^b that satisfies the above properties (case NA).

It is worth noting that the proposed approach can correct multiple errors in the same word if they are separated by a minimum distance. The distance depends on the generator polynomials and can be easily computed by considering the support sets of Equations E_a^n and E_b^n and the support sets of the correspondent Equations referred to the check index k: E_a^{n+k} and E_b^{n+k}. One error can be corrected in the elements composing the support sets, therefore in order to correct two errors they have to be separated by a number of bit-couples q equal to the differences between the greater and the smaller index in the considered support sets. Analyzing all the configurations we obtain that the • •• •• •• •• •• •• q can assume values in $3 \leq q \leq 5$.

4 Architecture

The decoder architecture is organized in two stages correspondent to the decoding phases: Double Decoding and Forward Correction.

The Double Decoding stage is composed of a set of modules `DEC_MOD`, each one taking in input a subset of coded bits and generating in output two decoding options for each original bit. Each module `DEC_MOD` implements equations E_a^n and E_b^n for a bit d_n, therefore its input bits are the elements of the support sets A_n and B_n and the output are bits d_{n_a} and d_{n_b}. An example of the decoding stage is represented in Figure 2a.

The Forward Correction stage chooses the bit to transmit in output in presence of corrupted bits in the coded word, by analyzing couples of decoding options provided by the previous stage and deciding which bit to propagate as output. The Forward Correction stage is composed of `CORR_MOD` modules, each taking as input d_{n_a}, d_{n_b}, d_{n+k_a} and d_{n+k_b} (with k the check index) and generating the decoded value d_n. Considering for example ••••• ••••• $= 2$ we obtain the architecture depicted in Figure 2b.

5 Automatic generation of the VHDL architecture

When generating a VHDL description for one specific decoder architecture, we exploit its modular structure and instantiate the simple components DEC_MOD and CORR_MOD along the whole word that has to be coded. For a defined couple of generator polynomials, each component DEC_MOD and CORR_MOD has a fixed architecture and the connections among them follow a regular pattern. The structure is therefore well suited for an automatic implementation.

In order to automate the VHDL description for each possible couple of generator polynomials, we take advantage of the RoadRunner framework for core generations [11]. The aim of the framework is to support the development and integration of parametric VHDL cores in complex designs. RoadRunner allows to create cores that are not directly described in VHDL, but rather produced by core generators written in C++. A generator is a C++ executable that exposes a set of parameters that are used for the configuration of the modules. The structure of the RR library also allows easy integration of already designed cores into more complex ones.

We exploit the RoadRunner framework to create a parametric core able to generate the VHDL description of a decoder architecture for all the valid generator polynomials and for each dataword length. The module is organized in two stages: the first generates the circuit equations by analyzing the input generator polynomials, and the second produces the VHDL description for the decoder, by considering the dataword length.

With this approach the final user can generate all the possible architectures by simply specifying the dataword length and the couple of generator polynomials.

6 Experimental Results

This section reports the experimental results related to the implementation of the proposed architectures. The solutions have been validated by randomly generating one or more errors in the coded word and verifying the effects on the decoded word.

In the following, we report synthesis results in terms of area and timing obtained for an FPGA implementation (in particular a Xilinx Spartan3 XC3S50 device), using the Xilinx XST engine. We remind that, when the implementation is performed on FPGA, suitable mechanisms should be used to protect also the decoder combinatorial logic [2], while when it is performed on ASIC only the sequential parts should be additionally protected from SEU.

All the valid configurations of generator polynomials have been considered. By implementing only once the symmetric couples of generator polynomials (same G_1 and G_2 in inverted order), we obtain 15 possible decoder architectures. Table 2 reports the number of Slices, LUTs and the path delay obtained for a dataword lenght of respectively 8, 16, 24, 32 bits.

As it can be noticed, the area figures (Slices and LUTs) increase linearly with the dimension of the dataword length, while the delay remains practically the same for all the dataword length. This is due to the fact that the increase in the dataword length results in the instantiation of further modules to cover the whole word, increasing the decoder size, but not affecting the logic depth. The proposed correction scheme allows to maintain a fixed latency and provide a correction power of one error every 3, 4 or 5 coded couples,

Table 2. Synthesis results for the all valid generator polynomials

Dimension		001 011	001 101	001 110	001 111	010 011	010 101	010 110	010 111	011 100	100 101	100 110	100 111	101 111
8	Slices	8	8	8	12	8	8	8	13	8	8	8	12	14
	LUT	16	16	16	20	16	16	16	24	16	16	16	21	24
	Delay(ns)	8.245	8.245	8.245	9.313	8.245	8.245	8.245	10.561	8.245	8.245	8.245	9.378	9.378
16	Slices	16	16	16	25	16	16	16	25	16	16	16	25	25
	LUT	32	32	32	43	32	32	32	45	32	32	32	46	48
	Delay(ns)	8.245	8.245	8.245	9.317	8.245	8.245	8.245	10.646	8.245	8.245	8.245	9.378	9.378
24	Slices	24	24	24	37	24	24	24	41	24	24	24	35	37
	LUT	48	48	48	65	48	48	48	72	48	48	48	67	72
	Delay(ns)	8.245	8.245	8.245	9.313	8.245	8.245	8.245	8.245	9.369	8.245	8.245	9.386	9.378
32	Slices	32	32	32	53	32	32	32	53	32	32	32	50	55
	LUT	64	64	64	93	64	64	64	93	64	64	64	90	96
	Delay(ns)	8.245	8.245	8.245	9.386	8.245	8.245	8.245	9.709	8.245	8.245	8.245	9.386	9.378

depending on the chosen polynomials.

This behaviour differentiate this approach from the classical decoder schemes based on block codes. For example, the Hamming Code, one of the most used codes in SEU protection, produces decoders that increase both in size and in logic depth with respect to the size of the data, and allow to correct a single fault regardless of the dimension of the dataword. The proposed scheme can be used alternatively to the classical ones, when a greater correction power is required and short latencies have to be maintained.

For further evaluation on the competitiveness of the proposed solution, we tested it against a Hamming decoding circuit. In order to obtain consistent data, circuits with the same correction power have been compared. Since the proposed decoder architectures have a correction power of one error every q (• •• •• •• •• •• •• ••) coded couples depending on the generator polynomials, this dimension is considered for the comparison with the Hamming decoders. q can assume values 3, 4, 5 for the considered polynomials, therefore the coded word for the proposed approach are respectively of length 6, 8, and 10 bits, while the coded word with Hamming approach are respectively of length 6, 7 and 9 bits (considering both data and check bits).

Figure 3a reports the obtained results referring to area occupation and Figure 3b reports the result in term of delay. The numbers in parenthesis represent the minimum distance q for each generator polynomials couple.

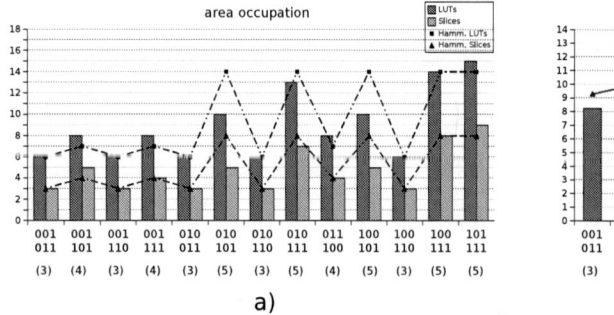
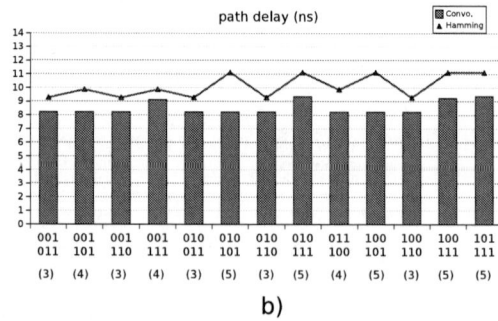

Figure 3. Area occupation and delay for the comparison with Hamming coding

Results highlight similarities in area occupation between the proposed architectures and Hamming decoders. Some configurations of the generator polynomials produce circuits requiring slightly greater resources with respect to the Hamming decoders, while others generate circuits presenting much more competitive results. Area occupation is therefore comparable with Hamming solutions, making the convolutional approach an effective alternative. The timing analysis shows that we obtain advantages in applying the proposed architectures with respect to an Hamming one, since every configuration of generator polynomials produces faster combinatorial circuits.

7   Conclusion

The paper presents an approach for the protection of memory data from Single Event Upsets, based on convolutional codes. A method is presented to derive light decoder architectures for all the valid combinations of Generator Polynomials of order three. A regular architecture is proposed, with a fixed latency for each configuration and a correction power of one error every 3,4 or 5 couples of coded bits. By exploiting a parametric core generator framework, the VHDL description of the each decoder is automatically generated, requiring to the final user only the specification of the generator polynomials and the data-word length. Experimental results show that the proposed solution is comparable for the area occupation and competitive for performance in relation to state of the art solutions (Hamming Coding). Moreover, the main advantage of proposed approach is the increase of the correction power, since multiple errors can be corrected in the same word, when they are separated by a minimum distance that depends on the generator polynomials (3, 4 or 5 couple bits). The proposed solution represents therefore a valid alternative for the protection of data from SEU.

References

[1] Allan. H. Johnston, *Scaling and Technology Issues for Soft Error Rates*, 4th Annual Research Conference on Reliability, Stanford University, Oct. 2000

[2] F. Lima Kanstensmidt, L. Carro, R. Reis, *Fault tolerant techniques for SRAM-based FPGAs*, Springer, 2006.

[3] Chen, C.L., and M.Y. Hsiao, *Error-Correcting Codes for Semiconductor Memory Applications: A State-of-the-Art Review*, IBM J. Res. Develop., Vol. 28, pp. 124-134, Mar. 1984.

[4] Ajay Dholakia, *Introduction to convolutional codes with applications*, Kluwer Academic Publishers, 1994

[5] G. D. Forney. *The Viterbi algorithm*, Proceedings of the IEEE, Vol. 61, no. 3, pp. 268278, Mar 1973.

[6] D.J.C. MacKay, *Information Theory, Inference, and Learning Algorithms*, Cambridge University Press, 2003.

[7] L.P. Holmquist, L.L. Kinney, *Error detection with latency in sequential circuits*, Proceedings of the IEEE international Test Conference, pp. 926-933, Washington, DC, 1988.

[8] L. P. Holmquist, L. L. Kinney, *Concurrent Error Detection for Restricted Fault Sets in Sequential Circuits and Microprogrammed Control Units Using Convolutional Codes*.Proceedings of the IEEE international Test Conference, pp. 926-935, Washington DC, 1991.

[9] K. Rokas, Y. Makris, D. Gizopoulos, *Low Cost Convolutional Code Based Concurrent Error Detection in FSMs*, p. 344, 18th IEEE International Symposium on Defect and Fault Tolerance in VLSI Systems (DFT'03), 2003.

[10] L. Frigerio, M. A. Radaelli, F. Salice, *Convolutional Coding for SEU mitigation*, European Test Symposium, pp.191-196, 2008.

[11] C. Bolchini, C. Brandolese, L. Frigerio, V. Rana, F. Salice, M. Santambrogio, *RoadRunner and IP-Gen: A Combined Solution to Speedup Configurable Systems Design*, IEEE Southern Conference on Programmable Logic, SPL'2007 Mar del Plata, Argentina, 2007, pages 75-78.

IEEE International Symposium on Defect and Fault Tolerance of VLSI Systems

A Novel Error Detection And Correction Technique for RNS based FIR Filters

S. Pontarelli‡, G.C. Cardarilli†, M. Re†, A. Salsano†

{pontarelli, salsano}@ing.uniroma2.it,

{marco.re, g.cardarilli}@ieee.org

†University of Rome "Tor Vergata", Department of Electronic Engineering

Via del Politecnico 1, 00191, Rome, ITALY

‡(ASI) Italian Space Agency, Viale Liegi, 26 00198 Rome, ITALY

Abstract

In this paper a novel technique for detecting and correcting errors in the RNS representation is presented. It is based on the selection of a particular subset of the legitimate range of the RNS representation characterized by the property that each element is a multiple of a suitable integer number m. This method allows to detect and correct any single error in the modular processors of the RNS based computational unit. This subset of the legitimate range can be used to perform addition and multiplication in the RNS domain allowing the design of complex arithmetic structures like FIR filters. In the paper, the architecture of a FIR filter with error detection and correction capabilities is presented showing the advantages with respect to filters in which the error detection and correction are obtained by using the traditional RNS technique.

I. INTRODUCTION

Nowadays digital systems are very often used to implement high complexity Digital Signal Processing (DSP) algorithms working in real time. The requirements in terms of speed and circuit complexity of these applications are stringent, and the mandatory use of technologies with the best available feature size increases the probability of the occurrence of faults.

To face these problems much research work has been published on fault detection in DSP architectures. In particular, with respect to the basic arithmetic operations, self-checking adders based on residue codes [1], [2], parity codes [3], or Berger codes [4] have been proposed. With respect to basic DSP building blocks, the RRNS (Redundant Residue Number System) representation has been used in the implementation of FIR filters [5], [6], [7] allowing fault detection and correction.

The use of RRNS gives to the designer, advantages both in terms of error detection and correction capabilities and in terms of maximum operating frequency. In fact, the operations on each residue digit are independent and so the addition, subtraction, and multiplication on the full dynamic range are split in different channels and performed in parallel on each of the moduli over a smaller dynamic range. The error detection in RRNS is based on the following consideration: the representation range is divided in two intervals: the *legitimate range* and the *illegitimate range*. An error in a single module is detected if, after the conversion from the RNS to the two's complement representation, the result belongs to the illegitimate range.

In this work the RNS representation range defined by the chosen moduli set is divided in two subsets: the legitimate subset is composed by any element that is exactly divisible by an additional modulus, i.e. the integer number m, while the other elements belong to the illegitimate subset. The paper shows how elements of the legitimate subset can be used to perform addition and multiplication and the architecture of a FIR filter with fault detection and correction capabilities is presented.

The paper is organized as follows: in Section II a background on RNS and RRNS arithmetic is given. In Section III, the new method to obtain error detection is presented, and the implementation of a FIR filter based on this method is described. In Section IV the extension of this method to achieve error correction is presented, while in Section V the implementation of a set of example FIR filter with error correction capabilities is shown, and the advantages with respect to the traditional RRNS representation are discussed. The Conclusions are drawn in Section VI.

II. RNS-RRNS BACKGROUND

In this section a short background on RNS and RRNS is presented, and the traditional techniques to implement FIR filters with error detection and correction capabilities based on the RRNS are shown.

A. Background on Residue Number System

A Residue Number System (RNS) is defined by a set of relatively prime integers $\{m_1, m_2, \cdots, m_P\}$.

The dynamic range of the system is given by the product of the moduli m_i:

$$M = \prod_{i=1}^{P} m_i$$

1550-5774/08 $25.00 © 2008 IEEE

DOI 10.1109/DFT.2008.32

Any integer $X \in [0, M - 1]$ has a unique RNS representation given by

$$X \xrightarrow{RNS} (<X>_{m_1}, <X>_{m_2}, \cdots, <X>_{m_P}) \tag{1}$$

where $<X>_{m_i} = X \bmod m_i$

A comprehensive description of the RNS theory and its applications to computer systems and DSP can be found in [8], [9], and [10]. In RNS, operations such as addition and multiplication, are executed in parallel as shown in the following formula

$$Z = X \text{ op } Y \xrightarrow{RNS} \begin{cases} <Z>_{m_1} = <X_{m_1} \text{ op } Y_{m_1}>_{m_1} \\ \cdots \\ <Z>_{m_P} = <X_{m_P} \text{ op } Y_{m_P}>_{m_P} \end{cases} \tag{2}$$

As a consequence, arithmetic operations are split into several modular operations with reduced word length.

The conversion of the RNS representation of Z is accomplished by using the Chinese Remainder Theorem (CRT)

$$Z = CRT(Z_{m_1}, \cdots, Z_{m_P}) = \left\langle \sum_{i=1}^{P} <Z_{m_i} \cdot k_i>_{m_i} \cdot M_i \right\rangle_M \tag{3}$$

where $M_i = \frac{M}{m_i}$ and k_i are obtained by $<M_i \cdot k_i>_{m_i} = 1$, *i.e.* they are the multiplicative inverse of M_i modulo m_i.

Clearly, the input and output conversions, constitute a significant overhead in systems implemented in RNS. However, efficient methods to perform those conversions are presented in [11], [12], and [13].

B. Background on Redundant Residue Number System for error detection and correction

A Redundant Residue Number System (RRNS) is defined as a residue number system added with r additional moduli. The first k moduli form a set of non redundant moduli, and their product represents the legitimate range, M that is,

$$M = \prod_{i=1}^{k} m_i$$

The remaining $P - k = r$ moduli form the set of redundant moduli that allows error detection and correction where M_R is defined as

$$M_R = \prod_{i=k+1}^{P} m_i$$

Given a residue vector $(x_{m_1}, \ldots, x_{m_P})$, the corresponding integer X belongs to the interval $[0, M_T - 1]$, where $M_T = M \cdot M_R$.

This interval, usually called total range, can be divided into two adjacent intervals by considering the ranges defined by the non redundant and redundant moduli. The interval $[0, M - 1]$ is called the *legitimate range* and the interval $[M, M_T - 1]$ is the *illegitimate range*.

In RRNS error detection and correction is obtained by constraining the operands and the results in the legitimate range. This restriction defines the dynamic range of the system. The m_i-projection of X, denoted X_i, is defined as the residue vector $(x_{m_1}, \cdots, x_{m_{i-1}}, x_{m_{i+1}}, \cdots, x_{m_P})$, *i.e.* the representation of X with the i-th residue digit deleted.

A single error occurs when a legitimate vector $(x_{m_1}, \cdots, x_{m_i}, \cdots, x_{m_P})$ is changed into a different residue vector, $(x_{m_1}, \cdots, \overline{x}_{m_i}, \cdots, x_{m_P})$ by the occurrence of an error in the i-th digit, the number corresponding to this vector is $\overline{X}$.

In [6] has been proved that, under the hypothesis of ordered moduli (i.e. $m_i < m_{i+1} \; \forall i$), in a RRNS representation with $r = 2$ any error in a single module produces an illegitimate number $\overline{X}$. Moreover, $\overline{X}_i$ is legitimate, where i is the residue affected by an error, while the other projections $\overline{X}_j$, for $j \neq i, i = 1, \ldots, P$ are all illegitimate. From these considerations is straightforward to detect and correct an error in the RRNS representation. The erroneous module is that characterized by his m_i-projection belonging to the legitimate range, while the correct value of the integer can be obtained by performing the reverse conversion of the X_i projection.

C. Implementation of FIR Filters in RNS

A N taps FIR filter is expressed by

$$y(n) = \sum_{k=0}^{N-1} a_k x(n - k) \tag{4}$$

The RNS implementation of the FIR filter is a direct consequence of equation (2),

$$y(n) = \sum_{k=0}^{N-1} a_k x(n-k) \xrightarrow{RNS} \begin{cases} Y_{m_1}(n) = \left\langle \sum_{k=0}^{N-1} \left\langle < a_k >_{m_1} \cdot x_{m_1}(n-k) \right\rangle_{m_1} \right\rangle_{m_1} \\ \cdots \\ Y_{m_P}(n) = \left\langle \sum_{k=0}^{N-1} \left\langle < a_k >_{m_P} \cdot x_{m_P}(n-k) \right\rangle_{m_P} \right\rangle_{m_P} \end{cases} \tag{5}$$

Adding error detection capabilities to the FIR filter requires the use of an additional modulus m_{P+1} and a comparator after the RNS to binary converter. A well know architecture of a RNS FIR filter capable of error detection using the RRNS representation is given in Fig. 1 ([6]).

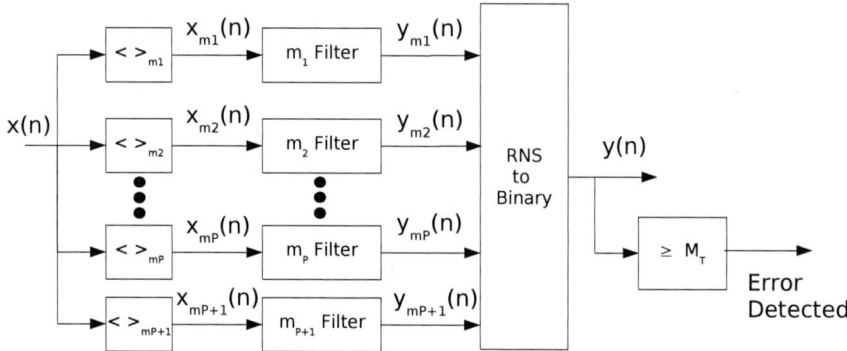

Fig. 1.   RNS implementation of a FIR filter with error detection

The binary to RNS conversion is performed by reducing modulo m_i the input sequence $x(n)$, providing the residue digits x_{m_i}. The parallel filters mod m_i compute the residues Y_{m_i} (eq. (5)), while the the standard binary representation of the result $y(n)$ is obtained by the RNS to the binary conversion block. An error inside one of the modulo m_i filters produces a result belonging in the *illegitimate range* and the comparator after the RNS to the binary converter allows the detection this error.

An RRNS FIR filter with error correction capabilities is shown in Fig. 2 in the case of P=3.

It requires two additional moduli with respect to the RNS representation, a reverse converter for each m_i projection, and a block choosing the value of the m_i projection that is in the legitimate range. The blocks called CRT (Chinese Remainder Theorem) performs the reverse conversion for the m_i projections, while the block called "choose legitimate" selects which input belong in the legitimate range.

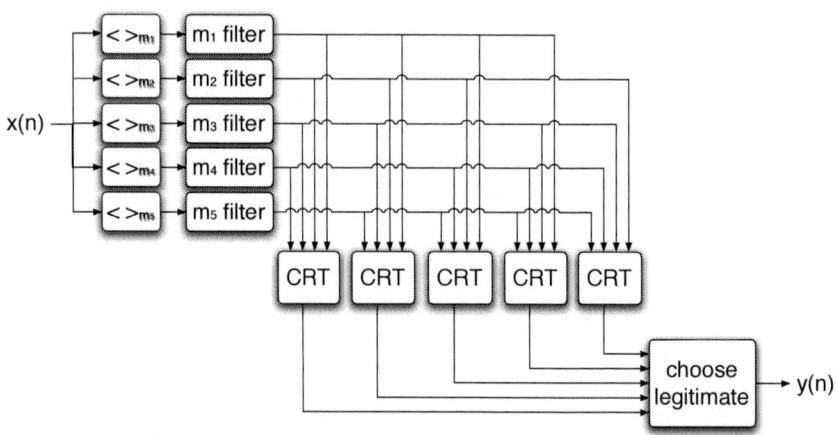

Fig. 2.   RNS implementation of a FIR filter with error correction

The main overhead of this architecture is related to the implementation of a CRT block for each module. For a RNS with a moduli set of i element we need i CRT blocks. Each CRT block is composed by:
1) $i-1$ modulo m_i constant multipliers to compute $< Z_{m_i} \cdot k_i >_{m_i}$
2) $i-1$ constant multipliers for M_i
3) $i-2$ modulo M adders

Therefore, the overall overhead due to the i CRTs grows quadratically with the number of element i of the moduli set.

III. ERROR DETECTION IN RNS BY USING SCALED VALUES

To define the method for error detection in RNS using scaled values we restrict the discussion to a moduli set composed only by prime numbers. The choice is not reductive because in RNS the use of prime moduli gives to the designer the possibility to exploit the isomorphism technique [8] to perform multiplication modulo a prime avoiding the use of multipliers.

Given the moduli set $\{m_1, m_2, \cdots, m_P\}$ the range of the traditional RNS representation is $[0, M-1]$, with $M = \prod_{i=1}^{P} m_i$. Without loss of generality, we suppose that the moduli set is ordered. The subset $\mathcal{C}$ of the range constitute the legitimate subset and is formed by the numbers $x \in [0, M-1]$ with $m \mid x$ (i.e. x is exactly divisible by m), with the constrains that m is relatively prime with respect to each element of the moduli set and $m > m_P$. The illegitimate subset is composed by all the numbers not belonging to $\mathcal{C}$. The mapping between the integer representation and the Scaled RNS (SRNS) representation is defined by

$$X \xrightarrow{Scaled\ RNS} (< m \cdot X >_{m_1}, < m \cdot X >_{m_2}, \cdots, < m \cdot X >_{m_P}) \tag{6}$$

The SRNS representation range is $[0, M_s - 1]$, with $M_s = \lfloor \prod_{i=1}^{n} m_i / m \rfloor$. It must be noticed that if $m \approx m_P$, the range representable with this mapping is similar to the range of an RRNS representation with $\{m_1, m_2, \cdots, m_P\}$ as the set of non redundant moduli and $m = m_{P+1}$ as the redundant modulus.

For the SRNS representation a relationship similar to equation (2) can be used for the addition operation, i.e.

$$Z = X + Y \xrightarrow{Scaled\ RNS} \begin{cases} < m \cdot Z >_{m_1} = << m \cdot X >_{m_1} + < m \cdot Y >_{m_1} >_{m_1} \\ \cdots \\ < m \cdot Z >_{m_P} = << m \cdot X >_{m_P} + < m \cdot Y >_{m_P} >_{m_P} \end{cases} \tag{7}$$

that is valid until the operand and the results belong to the range $[0, M_s - 1]$.

Instead, for multiplication the following relationship holds:

$$Z = X \cdot Y \xrightarrow{Scaled\ RNS} \begin{cases} < m \cdot Z >_{m_1} = << X >_{m_1} \cdot < m \cdot Y >_{m_1} >_{m_1} \\ \cdots \\ < m \cdot Z >_{m_P} = << X >_{m_P} \cdot < m \cdot Y >_{m_P} >_{m_P} \end{cases} \tag{8}$$

that is valid until the operand and the results belong to the range $[0, M_s - 1]$. We invite the reader to notice that, differently from addition, multiplication is performing scaling only one of the two operands.

Now let us suppose that a single module error occurs to a number represented in the scaled RNS. If the original value is $X = (x_{m_1}, \cdots, x_{m_i}, \cdots, x_{m_P})_{SRNS}$, the corresponding erroneous value can be expressed as $\overline{X} = (x_{m_1}, \cdots, \overline{x}_{m_i}, \cdots, x_{m_P})_{SRNS}$ and the value of the single module error is defined as $E = (0, \cdots, e_i, \cdots, 0)$, with $e_i = x_{m_i} - \overline{x}_{m_i}$ and $e_i \in [0, m_i - 1]$. The error is detected if the value $\overline{X}$ is an element of the illegitimate range, i.e. $m \nmid X$ (m not divide X). This condition occurs if the integer representation of the error E in not divisible by m.

Now, we define the condition under which all possible single module errors E respect the condition $m \nmid E$.

Applying the Chinese Remainder Theorem to E we obtain

$$E = CRT((0, \cdots, e_i, \cdots, 0)) = < e_i \cdot k_i >_{m_i} \cdot \prod_{j=1, j \neq i}^{n} m_j$$

The constrain that m is relative prime with respect to m_i have as a consequence that

$$m \mid E = e_i \cdot k_i \cdot \prod_{j=1, j \neq i}^{n} m_j \iff m \mid < e_i \cdot k_i >_{m_i} \tag{9}$$

where k_i are the multiplicative inverse of M_i modulo m_i.

The value of $< e_i \cdot k_i >_{m_i}$ is less than m_i and therefore a sufficient condition for equation (9) is:

$$m > m_i \ \forall i \tag{10}$$

For the set of moduli that respect the condition expressed in eq. (9) the single module error detection can be performed easily by checking the remainder modulo m of the result after the RNS to binary conversion. From the above discussion is possible to detect an error in the scaled RNS representation computing $< E >_m$. If no errors occur $< E >_m = 0$, otherwise $< E >_m \neq 0$.

scaled RNS value	integer value	RNS representation
0	0	$(0,0,0)$
1	11	$(2,1,4)$
2	22	$(1,2,1)$
3	33	$(0,3,5)$
4	44	$(2,4,2)$
5	55	$(1,0,6)$
6	66	$(0,1,3)$
7	77	$(2,2,0)$
8	88	$(1,3,4)$
9	99	$(0,4,1)$

TABLE I

RNS REPRESENTATIONS FOR THE CODEWORD DIVISIBLE BY 11

A. A numerical example

Given the set of moduli 3, 5, 7, $M = 105$, if $m = 11$ is chosen, the codeword values in the RNS range are {0,11,22,33,44, 55,66,77,88,99}. For the chosen moduli set the k_i values are 2, 1, 1 and equation (9) is respected for all the possible e_i values. The correct codewords represented in RNS are shown in Table I. In the first column the value used in the proposed coding is reported, while in the second column the corresponding integer value is shown.

All the values that a single module error can assume are reported in Table II. It is easy to see that for all the possible errors, the residue modulo 11 is always different from zero, allowing the detection of any possible single module error.

error RNS coordinates	Integer value	value mod 11
$(1,0,0)$	70	4
$(2,0,0)$	35	2
$(0,1,0)$	21	10
$(0,2,0)$	42	9
$(0,3,0)$	63	8
$(0,4,0)$	84	7
$(0,0,1)$	15	4
$(0,0,2)$	30	8
$(0,0,3)$	45	1
$(0,0,4)$	60	5
$(0,0,5)$	75	9
$(0,0,6)$	90	2

TABLE II

VALUE OF ERRORS IN AN RNS MODULE

An arithmetic computation in the SRNS representation is illustrated in the following example. Given the expression $2 \cdot 3 + 1 = 7$, using the SRNS representation the operation without error can be expressed as

$$(2,2,2) \cdot (0,3,5) + (2,1,4) = (2,2,0)$$

The multiplication step is performed between the scled value of 3, i.e. (0,3,5) and the unscaled value of 2, i.e. (2,2,2). Now we suppose that an error occurs in the second residue changing the result from 2 to 4 (the error is $(0,2,0)$). The erroneous result is $(2,4,0)$, that corresponds to the integer value 14, that is not divisible by 11.

B. FIR filter with error detection capability

Using the result presented in the previous section the design of a FIR filter with error detection capability that exploits the characteristic of the SRNS representation is obtained by using the following additional hardware resources:

1) a constant multiplier before the conversion of the input sequence into the RNS representation,
2) a constant divider after the conversion of the output sequence from the RNS representation to the binary one,
3) a modulo m reduction block to check if the result is a multiple of m.

The architecture of the FIR filter with error detection is shown in Fig. 3.

The $< a_k >_{m_i}$ coefficients of the FIR filter are computed in standard RNS representation to respect the equivalence defined in equation (8) for multiplication in scaled RNS. The overhead of this schema is strictly dependent by the choice of the scaling factor m. With the right choice this overhead can be drastically reduced. For example, if we choose $m = 2^i$, the multiplication is performed by shifting left the input sequence samples, the division by shifting right discarding the i less significant bits of the result after the RNS to binary conversion, the modulo m reduction of the result consists in taking the i less significant bits of the result, and the congruence to zero is checked by the logic OR of the i less significant bits. In this case, the overhead of the additional hardware resources, is negligible. In fact, the additional hardware resources for $m = 2^i$ are:

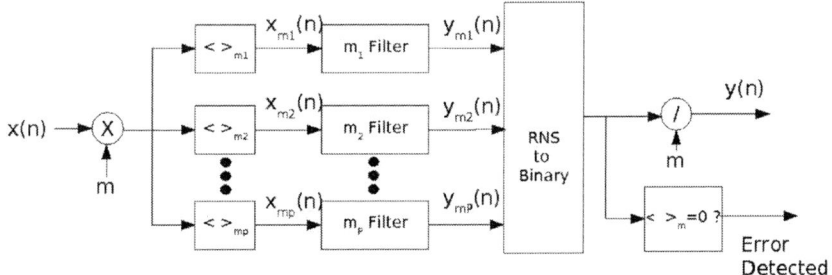

Fig. 3. SRNS implementation of a FIR filter with error detection capability

1) the constant multiplier simply is a shift of the input bits,
2) the result of the division by 2^i is obtained discarding the i less significant bits.
3) the bit composing the result of the modulo 2^i operation are the i less significant bits of the RNS to binary conversion.
4) the block for error detection is an i-input OR.

IV. ERROR CORRECTION IN RNS BY USING SCALED VALUES

To add error correction capabilities to the scaled RNS representation the sufficient and necessary condition is that two different errors correspond to two different values of $< E >_m$. We call this condition as no aliasing condition. The no aliasing condition allows to define a one to one map between the value of $< E >_m$ and the value of E that is the value of the error occurred in the scaled RNS representation. Subtracting E to the output of the RNS to binary conversion of Fig. 3 we can obtain the correct value of the output. To identify the relationship between the moduli set $\{m_1, m_2, \cdots, m_P\}$ and m that satisfy the no aliasing condition we take two different errors $E' = (0, \cdots, e'_i, \cdots, 0)$ and $E'' = (0, \cdots, e''_j, \cdots, 0)$.

First of all we define the no aliasing condition $< E' >_m \neq < E'' >_m$ as $< E' - E'' >_m \neq 0$.

If $i = j$ the errors E' and E'' correspond to errors occurred in the same m_i module. In this case $E' - E'' = (0, \cdots, e'_i - e''_i, \cdots, 0)$ and the no aliasing condition correspond to the condition defined in the previous section for error detection. Instead, if the errors occur in different moduli the value $E' - E''$ is:

$$E' - E'' = (0, \cdots, e'_i, 0, \cdots 0, e''_j, \cdots, 0)$$

To show the condition that must be satisfied in this case, we remark that the moduli set $\{m_1, \cdots, m_i, \cdots, m_j, \cdots, m_P\}$ is equivalent to the moduli set $\{m_1, \cdots, m^*, \cdots, m_P\}$, where $m^* = m_i \cdot m_j$ substitute the two moduli m_i and m_j. Using this moduli set equivalence the value of $E' - E''$ can be expressed as:

$$E' - E'' = (0, \cdots, e^*, \cdots, 0)$$

where $e^* = CRT(e'_i, e''_j)$ and the CRT is computed on the moduli set composed by m_i and m_j. After this manipulation the case of $i \neq j$ is re-conducted to the first case and the no aliasing condition is the same one defined for error detection. The condition $< E' - E'' >_m \neq 0$ is therefore equivalent to $m^* < m$. This condition is verified for all the value of i and j if

$$m > m_i \cdot m_j \; \forall i, j \tag{11}$$

If the condition expressed in eq. (11) is satisfied all possible errors occurring in a single module correspond to one and only one value of $< E >_m$ and therefore from the value of $< E >_m$ we can derive the error value E.

Let us suppose that an error e_i occurs in the i module changing the value of the output from Y (the correct value) to $\bar{Y} = Y + E$. The value of E is:

$$E = e_i \cdot M_i \tag{12}$$

where e_i is in the range $[-m_i + 1, m_i - 1]$.

The corresponding $< E >_m$ value is:

$$< E >_m = < e_i \cdot M_i >_m$$

Now we can multiply modulo m for the inverse modulo m of M_i. This inverse, M_i^{-1} exist because m is relative prime with respect to all the m_i, and therefore is prime with respect to M_i. After this multiplication the equation becomes:

$$< e_i >_m = < M_i^{-1} \cdot E >_m = << M_i^{-1} >_m \cdot < E >_m >_m$$

441

Defining

$$\bar{e}_i = <<M_i^{-1}>_m \cdot <E>_m>_m$$

and using the inequality $m_i < m$ the following equation holds:

$$e_i = \begin{cases} \bar{e}_i & if \quad \bar{e}_i < m_i \\ \bar{e}_i - m & if \quad \bar{e}_i > m_i \end{cases} \tag{13}$$

The e_i values computed from equation (13) can be used to compute some candidate erroneous values E_i defined as $E_i = e_i \cdot M_i$.

The correct value E is one of the E_i candidate values. The choice of the right E_i value can be performed computing $<\bar{Y} - E_i>_M$. The only value of $<\bar{Y} - E_i>_M$ that is exactly divisible by m is the corrected value of output. The uniqueness and the existence of an E_i that satisfy the condition $<\bar{Y} - E_i>_M | m$ is assured by the no aliasing condition previously discussed.

A. A numerical example

Now we give a numerical example of the scaled RNS representation with error correction capabilities. The moduli set is chosen as $\{11, 13, 15\}$, $M = 2145$, and $m = 256$.

The M_i values for the chosen moduli set are $M_1 = 195$, $M_2 = 165$, $M_3 = 143$, while $k_1 = 7$, $k_2 = 3$, $k_3 = 2$.

The values of $<M_i^{-1}>_{256}$ are $<M_1^{-1}>_{256} = 235$, $<M_2^{-1}>_{256} = 45$ and $<M_3^{-1}>_{256} = 111$.

The codeword values in the RNS range are $0, 256, 512, 768, 1024, 1280, 1536, 1792, 2048$ that are all divisible by 256. The correct codewords represented in RNS are shown in Table III. In the first column the value used in the proposed coding is reported, in the second column the corresponding integer value is shown and in the last column the RNS representation is given.

scaled RNS value	integer value	RNS representation
0	0	$(0, 0, 0)$
1	256	$(3, 9, 1)$
2	512	$(6, 5, 2)$
3	768	$(9, 1, 3)$
4	1024	$(1, 10, 4)$
5	1280	$(4, 6, 5)$
6	1536	$(7, 2, 6)$
7	1792	$(10, 11, 7)$
8	2048	$(2, 7, 8)$

TABLE III

ERROR CORRECTING SCALED RNS REPRESENTATIONS CODEWORDS

All the values that a single module error can assume are reported in Table IV. It is easy to see that for all the possible errors, the residue modulo 256 is always different from zero. The no aliasing condition is verified, and all modulo 256 values the are different.

error RNS coordinates	Integer value	value mod 256	error RNS coordinates	Integer value	value mod 256	error RNS coordinates	Integer value	value mod 256	error RNS coordinates	Integer value	value mod 256
$(1, 0, 0)$	1365	85	$(10, 0, 0)$	780	12	$(0, 9, 0)$	165	165	$(0, 0, 6)$	1716	180
$(2, 0, 0)$	585	73	$(0, 1, 0)$	495	239	$(0, 10, 0)$	660	148	$(0, 0, 7)$	2002	210
$(3, 0, 0)$	1950	158	$(0, 2, 0)$	990	222	$(0, 11, 0)$	1155	131	$(0, 0, 8)$	143	143
$(4, 0, 0)$	1170	146	$(0, 3, 0)$	1485	205	$(0, 12, 0)$	1650	114	$(0, 0, 9)$	429	173
$(5, 0, 0)$	390	134	$(0, 4, 0)$	1980	188	$(0, 0, 1)$	286	30	$(0, 0, 10)$	715	203
$(6, 0, 0)$	1755	219	$(0, 5, 0)$	330	74	$(0, 0, 2)$	572	60	$(0, 0, 11)$	1001	233
$(7, 0, 0)$	975	207	$(0, 6, 0)$	825	57	$(0, 0, 3)$	858	90	$(0, 0, 12)$	1287	7
$(8, 0, 0)$	195	195	$(0, 7, 0)$	1320	40	$(0, 0, 4)$	1144	120	$(0, 0, 13)$	1573	37
$(9, 0, 0)$	1560	24	$(0, 8, 0)$	1815	23	$(0, 0, 5)$	1430	150	$(0, 0, 14)$	1859	67

TABLE IV

VALUE OF ERRORS IN AN RNS MODULE

In the following example, an arithmetic computation in the SRNS representation is illustrated. Given the expression $2 \cdot 3 + 1 = 7$, using the SRNS representation the operation without error can be expressed as

$$(2, 2, 2) \cdot (9, 1, 3) + (3, 9, 1) = (10, 11, 7)$$

Now we suppose that an error occurs in the third residue changing the result from 7 to 8 (i.e. the error is $(0, 0, 1)$). The erroneous result is $(10, 11, 8)$, that corresponds to the integer value 2078, that is not divisible by 256. Computing the values

442

of $\bar{e}_i$ from the value $< E >_{256}= 30$ we obtain $\bar{e}_1 =< 30 \cdot 235 >_{256}= 138$, $\bar{e}_2 =< 30 \cdot 45 >_{256}= 70$, $\bar{e}_3 =< 30 \cdot 111 >_{256}= 2$. e_1 and e_2 are out of the range $[-m_i + 1, m_i - 1]$ and therefore do not represent a possible candidate value.

However, the candidate values for E_i are: $E_1 = 138 \cdot 195 = 26910$, $E_2 = 70 \cdot 165 = 11550$ and $E_3 = 2 \cdot 143 = 286$. The value of $< \bar{Y} - E_i >_M$ are:

$$< \bar{Y} - E_1 >_M =< 2078 - 26910 >_M= 908$$

$$< \bar{Y} - E_2 >_M =< 2078 - 11550 >_M= 1253$$

$$< \bar{Y} - E_3 >_M =< 2078 - 286 >_M= 1792$$

$< \bar{Y} - E_3 >_M$ identify the error in the third digit, and the correct value is 1792, that is the value corresponding to the SRNS value of 7.

B. FIR filter with error correction capability

Now we present the architecture of a FIR filter that use scaled RNS representation to achieve error correction. Like in the case of error detection we need some additional hardware resources. For the case of a SRNS representation with a moduli set of i elements we have:

1) a constant multiplier before the conversion of the input sequence into the RNS representation,
2) a constant divider after the conversion of the output sequence from the RNS representation to the binary one
3) a modulo m reduction block to obtain $< E >_m$
4) i modulo m constant multiplier to obtain $\bar{e}_i$
5) i constant multiplier to obtain the different candidate value of E_i
6) i comparator between $\bar{e}_i$ and m_i
7) i modulo M adder to compute the candidate correct value
8) i modulo m reduction block to select the correct value
9) a mux selecting the right E_i value

The architecture of the FIR filter with error correction capabilities is shown in Fig. 4

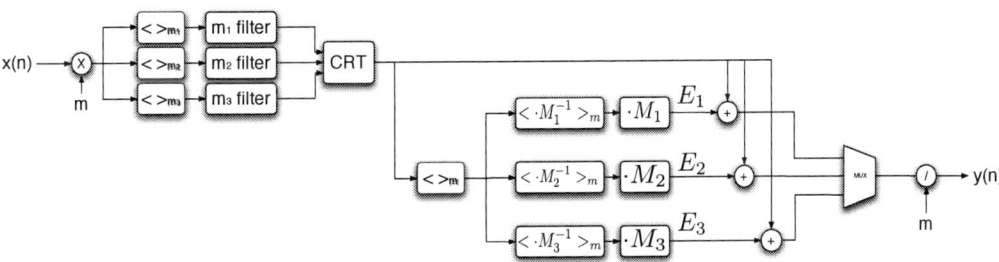

Fig. 4.   SRNS implementation of a FIR filter with error correction capability

Also in this case the overhead of this schema is strictly dependent by the choice of the scaling factor m and the use of $m = 2^n$ allows to reduce the resource occupation of the additional blocks.

V. IMPLEMENTATION OF SCALED RNS FILTERS

In this section some FIR filter has been implemented to carry out a comparison between our technique and the standard RRNS one.

The dynamic range of a filter depends on the dynamic range of coefficient, input data and number of taps, therefore the choice of moduli sets is dependent by these parameters. We choose two cases, one with 8 bits input data and coefficients and another with 12 bits input data and coefficients. The number of taps used to implement the filters vary from 32 to 128 taps and we restricted our analysis to constant coefficient filters. The implementation has been carried out on the Xilinx Virtex V FPGA and the moduli has been chosen to satisfy the relationship $m_i \leq 2^6$. In table V the different parameters the dynamic range of the filter and the chosen moduli set are presented. The moduli set has been defined for the RNS representation, without error detection and correction capability, for the standard RRNS and for the proposed scaled RNS.

For dynamic ranges up to 23 bits the unredundant representation use 4 moduli, the RRNS use six moduli, while the SRNS use six or seven moduli depending on the dynamic range of the filter. For the dynamic range between 29 and 31 both the RRNS and the SRNS require 8 moduli. The dynamic range of the SRNS is computed as $M_s = \lfloor log_2(\prod_{i=1}^{n} m_i/m) \rfloor$, where $m = 2^{12}$ is chosen to satisfy equation (11). The table shows how for most of the example filter the number of moduli of the RRNS is the same of the SRNS representation. The resource usages of the filters taken into account are summarized in table

443

Name	input/coefficient length	number of taps	dynamic range	unredundant moduli set	RRNS moduli set	SRNS moduli set
FIR1	8	32	21	61,59,53,47	64,63,61,59,53,47	63,61,59,53,47,43
FIR2	8	64	22	61,59,53,47	64,63,61,59,53,47	63,61,59,53,47,43
FIR3	8	128	23	61,59,53,47	64,63,61,59,53,47	63,61,59,53,47,43
FIR4	12	32	29	61,59,53,47,43,41	64,63,61,59,53,47,43,41	63,61,59,53,47,43,41,37
FIR5	12	64	30	61,59,53,47,43,41	64,63,61,59,53,47,43,41	63,61,59,53,47,43,41,37
FIR6	12	128	31	61,59,53,47,43,41	64,63,61,59,53,47,43,41	63,61,59,53,47,43,41,37

TABLE V

PARAMETERS AND DYNAMIC RANGE OF THE EXAMPLES FILTERS

VI both for RRNS and for SRNS. In the last column of table VI the ratio between resource occupation of RRNS and SRNS is given. The resource usage has been divided in three parts:

1) resource occupation for the forward converters (modulo m_i reduction)
2) resource occupation for all the modulo m_i filters
3) resource occupation for the CRT and the error correction blocks

Filter Name	RRNS				SRNS				ratio between RRNS and SRNS
	forward (#LUTs)	m_i filters (#LUTs)	CRT and error correction (#LUTs)	total usage (#LUTs)	forward (#LUTs)	m_i filters (#LUTs)	CRT and error correction (#LUTs)	total usage (#LUTs)	
FIR1	70	1760	1344	**3174**	84	1920	650	**2654**	83%
FIR2	70	3520	1344	**4934**	84	3840	650	**4574**	92%
FIR3	70	7040	1344	**8454**	98	8960	737	**9795**	115%
FIR4	98	2400	2472	**4970**	112	2560	853	**3525**	71%
FIR5	98	4800	2472	**7370**	112	5120	853	**6085**	82%
FIR6	98	9600	2472	**12170**	112	10240	853	**11205**	92%

TABLE VI

RESOURCE OCCUPATION OF EXAMPLES FILTERS

It can be noticed that, when the area overhead due to the error correction of RRNS a significant part of the resource occupation, the use of the SRNS allows to reduce this overhead up to 30% with respect to the RRNS method. Instead, when the number of moduli used for SRNS is greater than the one used for RRNS the RRNS is more convenient than SRNS.

VI. CONCLUSIONS

In this paper a novel error detection and correction technique for RNS representation is proposed. It is based on the use of a subset of the RNS legitimate range in which each element of the subset is a multiple of a number m with the constrain that $m > m_i$, where m_i are the elements of the set of moduli used in the RNS representation. This method allows to detect and correct any single error in the modular processors of the RNS based computational unit. The paper also shows how this method is suitable for the design of FIR filters with error detection and correction capabilities. Comparing the proposed technique with respect to the traditional one the use of scaled RNS allows reducing overhead avoiding the use of different CRT modules for error correction.

REFERENCES

[1] W. W. Peterson, "On Checking an Adder", I.B.M. J. Res. Develop., vol 2, pp. 166-168, Apr 1958.

[2] D. Nikolos, A.M. Paschalis, G. Philokyprou, "Efficient Design of Totally Self-Checking Checkers for All Low-cost Arithmetic Codes", IEEE Transactions on Computers, vol. 37, N. 7, pp. 807 - 814, July 1988.

[3] M. Nicolaidis, "Carry Checking/Parity Prediction Adders and ALUs", IEEE Transactions on Very Large Scale Integration (VLSI) Systems, Vol. 11, N. 1, Feb 2003 pp. 121-128.

[4] J.-C. Lo, S. Thanawastien, T. R. N. Rao, M. Nicolaidis, "An SFS Berger Check Prediction ALU and its Application to Self-Checking Processors Design", IEEE Transactions on Computer Aided Design, pp. 525-540, Mar. 1992.

[5] S. Bandyopadhyay, G.A. Jullien, A. Sengupta, "A Systolic Array for Fault Tolerant Digital Signal Processing using a Residue Number System Approach", Proceedings of the International Conference on Systolic Arrays, 25-27 May 1988, Page(s):577 - 586

[6] Mark H. Etzel and W. K. Jenkins, "Redundant Residue Number Systems for Error Detection and Correction in Digital Filters", IEEE Transactions on Acoustics, Speech and Signal Processing, vol. ASS-28, No 5, pp. 538-544, October 1980.

[7] W. K. Jenkins, "The Design of Error Checkers for Self-Checking Residue Number Arithmetic", IEEE Transactions on Computers, Volume C-32, Issue 4, Apr 1983 Page(s):388 - 396

[8] I. Vinogradov, "An Introduction to the Theory of Numbers", New York: Pergamon Press, 1955.

[9] N. Szabo and R. Tanaka, "Residue Arithmetic and its Applications in Computer Technology", New York: McGraw-Hill, 1967.

[10] M. Sodestrand, W. Jenkins, G. A. Jullien, and F. J. Taylor, "Residue Number System Arithmetic: Modern Applications in Digital Signal Processing", New York: IEEE Press, 1986.

[11] T. V. Vu, "Efficient Implementation of the Chinese Remainder Theorem for Sign Detection and Residue Decoding", IEEE Transactions Circuits Systems-I, vol. 45, pp. 667-669, June 1985.

[12] S.Piestrak, "A High-Speed Realization of a Residue to Binary Number System Converter", IEEE Transactions Circuits Systems-II Analog and Digital Signal Processing, vol. 42, pp. 661-663, Oct. 1995.

[13] G. Cardarilli, M. Re, and R. Lojacono, "A Residue to Binary Conversion Algorithm for Signed Numbers", European Conference on Circuit Theory and Design (ECCTD97), vol. 3, pp. 1456-1459, 1997.

[14] F. Barsi and P. Maestrini, "Error Correcting Properties of Redundant Residue Number Systems", IEEE Transactions Compututer, vol. C-22, pp. 307-315, Mar. 1973.

IEEE International Symposium on Defect and Fault Tolerance of VLSI Systems

An Asymmetric Checkpointing and Rollback Error Recovery Scheme for Embedded Processors

Hamed Tabkhi, Seyed Ghassem Miremadi, and Alireza Ejlali
Dependable Systems Laboratory (DSL)
Department of computer engineering
Sharif University of Technology, Tehran, Iran
tabkhi@ce.sharif.edu, miremadi@sharif.edu, ejlali@sharif.edu

Abstract

This paper presents a checkpointing scheme for rollback error recovery, called Asymmetric Checkpointing and Rollback Recovery (ACRR) which stores the processor states in an asymmetric manner. In this way, error recovery latency and the number of checkpoints are reduced to increase the probability of timely task completion for soft real-time applications. To evaluate the ACRR, this scheme was studied analytically. The analytical results show that the recovery latency is reduced as non-uniformity of the checkpoint increases. As a case study, the ACRR is implemented and simulated on a behavioral VHDL model of LEON2 processor. The simulation results follow the results obtained in the analytical study.

1. Introduction

Embedded processors are widely used in safety-critical real-time systems such as flight control systems, automotive electronics and fabrication equipment systems [9], where the occurrence of failures in such systems can cause catastrophic consequences. Fault-tolerant and error recovery methods are used to prevent these systems to fail [16]. The dominant error recovery technique for the uniprocessor systems is the rollback error recovery [1, 11, 12, 14, 18, 19, 20] that is based on re-execution of specific instructions, when an error occurs during execution of these instructions. Two main measures to evaluate the rollback error recovery are the recovery latency and the number of checkpoints inserted in an application program [8]. The optimal recovery latency and the number of checkpoints have been analytically investigated by several researchers [7, 10, 17, 21, 22]. These investigations are based on one or more of the following assumptions:

- A zero error detection latency is assumed [7, 10, 21, 22],
- The interval between checkpoints in an application program is uniform [7, 21, 22],
- The interval between checkpoints in an application program is non-uniform [10].

The main drawback of the above studies is the assumption of the zero error detection latency. However, many practical systems are involved with error detection latency [3, 13], which in turn may impose delay in the error recovery latency. The assumption to zero error detection latency certifies that the optimal rollback should be to the most recent checkpoint [10, 21]. However, it has been shown that rollback to the most recent checkpoint is not a valid choice, because of the error detection latency which in practical systems, is not zero [6, 15, 19, 20].

1550-5774/08 $25.00 © 2008 IEEE
DOI 10.1109/DFT.2008.27

This paper presents a checkpointing and error recovery scheme that supports the non-zero error detection latency. This scheme called Asymmetric Checkpointing and Rollback Recovery (ACRR). The ACRR scheme is based on a static non-uniform checkpoints placement that stores the processor states in an asymmetric manner. The analytical results show that the ACRR has less average recovery latency than the uniform checkpointing methods. Consequently, the ACRR scheme increases the probability of timely task completion for soft real-time systems. As a case study, the ACRR scheme is implemented and simulated on a behavioral VHDL model of LEON2 processor. In implementation of the ACRR scheme, a checkpointing mechanism that stores a lower amount of information of the processor state with deterministic checkpointing latency is presented.

The organization of this paper is as follow: Section 2 introduces some definitions and restrictions in checkpointing and rollback error-recovery. The analytical study of the ACRR scheme is presented in Section 3. Section 4 presents an experimental implementation of the ACRR scheme on LEON2 processor. The ACRR simulation results are presented in Section 5, and finally Section 6 concludes this paper.

2. Checkpoint and Rollback Policies

The following five definitions are necessary in the discussion:
- Checkpointing interval: the duration of time between two consecutive checkpoints (speculative epoch).
- Checkpointing latency: the duration of required time that the execution of a program is stalled to store the state of a processor.
- State restoration latency: the duration of required time for restoring the last fault free state of a processor
- Recovery latency: the duration of required time to re-executing faulty instructions.
- Total checkpointing and recovery latency: the duration of required time for checkpointing, state restoration, and error-recovery that is added to the real execution time of an application program.

The error detection latency should not be neglected in the checkpointing and error recovery methods, especially in embedded systems where cost and power consumption are important factors. The error detection mechanisms with low detection latency such as module redundancy impose a high overhead which cannot be acceptable for a varied range of the embedded systems like automotive electronics [14].

If an error occurs before a checkpoint and be detected afterward, the stored data in this checkpoint may be corrupted. As a result, two most recent checkpoints should have been stored previously to prevent the processor from rollback to an invalid checkpoint. It means that the processor should be returned to the second recent checkpoint which in this paper is called penultimate checkpoint (Figure 1).

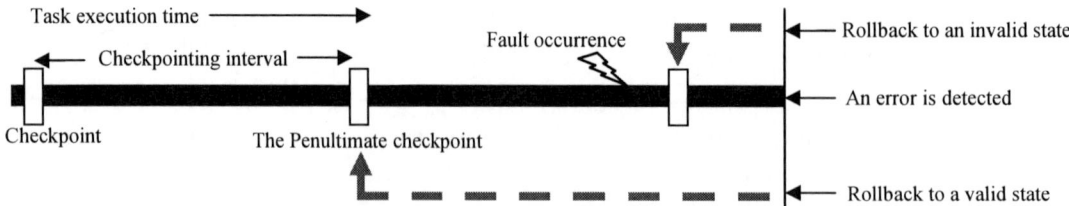

Figure 1. Rollback to the penultimate checkpoint

Rollback to the penultimate checkpoint is a general constraint which is imposed by the error detection latency and is independent of the checkpointing interval. Also, the checkpointing interval must be greater than the worst-case latency of the error detection mechanism. Otherwise, the error may corrupt the stored data of both recent checkpoints. The analytical studies which neglect error detection latency, relinquish the above constraints [7, 10, 21, 22]. In contrast to the analytical studies, most of the practical rollback error recovery methods store two most recent checkpoints and rollback to the penultimate checkpoint in case of error occurrence [6, 15, 19, 20].

3. Optimal Checkpoints Placement

This study is performed for uniprocessor and aperiodic task which assumes K errors would occur during task execution time and the probability of error occurrence is uniform during task execution time. Below notations are used through the remainder of this paper:
- E: Real execution time of the task without checkpointing and error recovery.
- I: Total execution time of the task with checkpointing and error recovery.
- N: Number of checkpoints.
- N_{op}: Optimal number of checkpoints.
- T: Checkpointing interval that equaled to $E/(N+1)$.
- R: Recovery latency.
- S: State restoration latency.
- C: Checkpointing latency.
- K: Number of errors which occur during task execution time.
- P: Probability of timely task completion.

3.1 Uniform Checkpointing

As it mentioned in Section 2, the processor rolls back to the penultimate checkpoint when an error is detected. Therefore, the two recent intervals affect the recovery latency and must be considered in the analysis of the optimal checkpoints placement. When the checkpoints are uniform and the checkpointing intervals are equal (Figure 2(a)), the average recovery latency is given by:

$$R_{average} = T + \frac{\int_T^{2T} t\,dt}{T} = T + \frac{\frac{1}{2}T^2}{T} = \frac{3}{2}T = \frac{3}{2}\frac{E}{N+1} \qquad (4\text{-}1)$$

The average execution time of a program is given by:

$$I_{average} = E + (N+K)C + KS + KR_{average} = E + (N+K)C + KS + K\frac{3}{2}\frac{E}{N+1} \qquad (4\text{-}2)$$

In equation (4-2), the number of times that checkpointing latency has been considered is $N+K$. because, when a processor rollbacks to the penultimate checkpoint, one extra checkpoint must be repeated each time. We can derive the value of N which minimizes I by differentiating (4-2) with respect to N. This gives a cubic equation in N, given by:

$$\frac{dI_{average}}{dN} = 0 \quad \Rightarrow \quad C - K\frac{3}{2}\frac{E}{(N+1)^{-2}} \quad \Rightarrow N_{op} = \sqrt{\frac{3}{2}\frac{KE}{C}} - 1 \qquad (4\text{-}3)$$

In equation (4-3), N_{op} should not be less than zero and $E/(N_{op}+1)$ should be greater than the error detection latency. Otherwise as described before, error may corrupt two consecutive checkpoints.

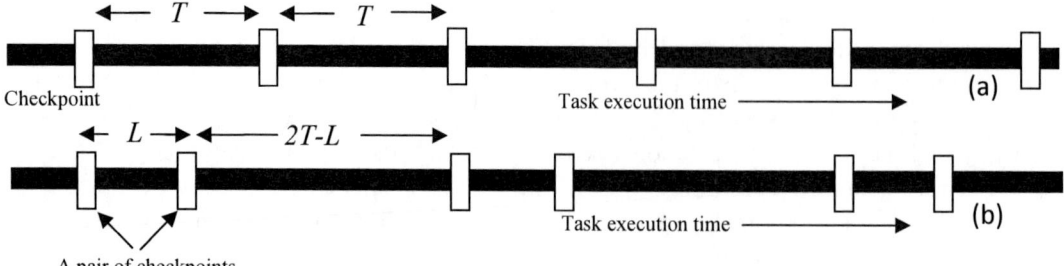

Figure 2. Checkpoints placement

3.2 The ACRR: Asymmetric checkpointing and rollback recovery

Instead of placing N checkpoints in the task execution time uniformly, the ACRR scheme inserts $N/2$ pairs of checkpoints with interval L inside the pairs, and $2T-L$ between two consecutive pairs. In such approach, sum of two consecutive intervals always equals to $2T$ which is similar to the uniform checkpointing (Figure 2(b)). In fact, the uniform checkpointing is a special case of the ACRR scheme where L is equal to T. The value of T depends on the number of checkpoints. It should be noticed that both L and $2T-L$ are greater than the error detection latency. In this case, with assumption that the probability of error occurrence is uniform during the task execution time, the average recovery latency is given by:

$$R_{average} = \frac{L}{2T}(2T-L+\frac{\int_0^L tdt}{L}) + \frac{2T-L}{2T}(L+\frac{\int_0^{2T-L} tdt}{2T-L}) = T+L-\frac{L^2}{2T} \qquad (4\text{-}4)$$

In comparison with the uniform checkpointing, it can be shown that for all values of L, R_{averag} from the equation (4-4) are equal or less than R_{averag} from the equation (4-1). Therefore, the R_{averag} in uniform checkpointing is the worst-case of the R_{averag} in the ACRR scheme with the same value of N. The average recovery time (R_{averag}) for different values of parameter L is shown in Figure 3. The E values are obtained from real execution of four benchmarks on the LEON2 processor. Figure 3 illustrates that in the ACRR scheme, the R_{averag} decreases as the parameter L decreases.

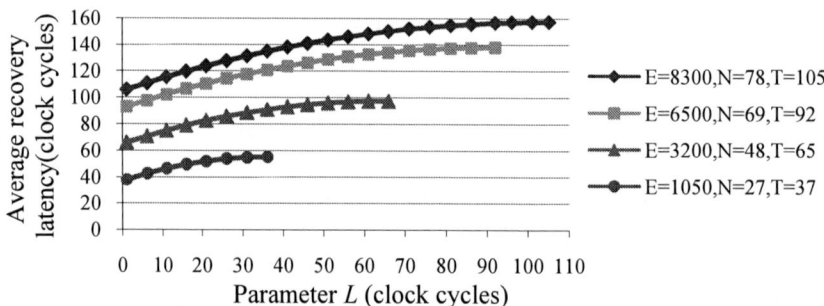

Figure 3. The average recovery latency for the same *N* when *K*=1

By substituting the values of R_{averag} derived from (4-4) in the equation (4-2), we obtain:

$$I_{average} = E+(N+K)C+KS+K(T+L-\frac{L^2}{2T}) = E+(N+K)C+KS+K(\frac{E}{N+1}+L-\frac{L^2(N+1)}{2E})$$

$$(4\text{-}5)$$

We can derive a new optimal value of N which minimizes $I_{average}$ by differentiating (4-5) with respect to N. This gives a cubic equation in N, given by:

$$\frac{dI_{average}}{dN} = 0 \quad \Rightarrow \quad C + K(-\frac{E}{(N+1)^{-2}} - \frac{L^2}{2E}) = 0 \quad \Rightarrow \quad N_{op} = \sqrt{\frac{2KE^2}{2EC - L^2 K} - 1} \quad (4\text{-}6)$$

Based on the experimental implementation in Section 5, the value of C is chosen 2 clock cycle. The N_{op} for the different ranges of parameter L is shown in Figure 4 (when C=2, and K=1). This figure illustrates that reduction of L value will decrease the amount of N_{op}.

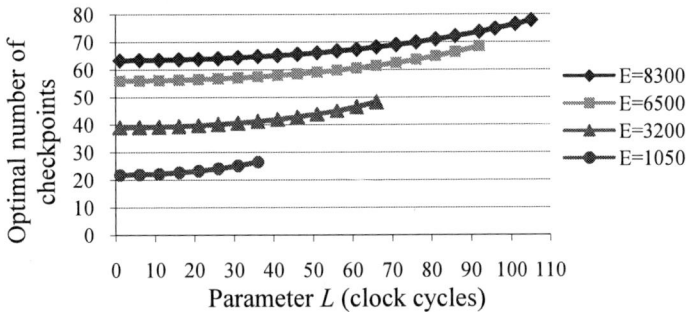

Figure 4. The optimal number of checkpoints for different ranges of parameter L

The total latency is equal to $(N+K) \times C + K(R+S)$. Figure 5 illustrates the Total latency of the ACRR scheme for the different values of parameter L (when E=6500, C=2, K=1). According to the equation (4-6), decreasing parameter L leads to reduction in N_{op} value. This means that the amount of N_{op} in the ACRR scheme is lower than the amount of N_{op} in the uniform checkpointing with the same task execution time.

The fact that the N_{op} decreasing can lead to an increase in R_{averag} for several values of L is not important, because the total latency, which is the sum of total checkpointing latency and recovery latency, always decreases as N decreases. The relations between N_{op}, L, R_{averag}, and the total latency in ACRR scheme, are illustrated in Figure 5.

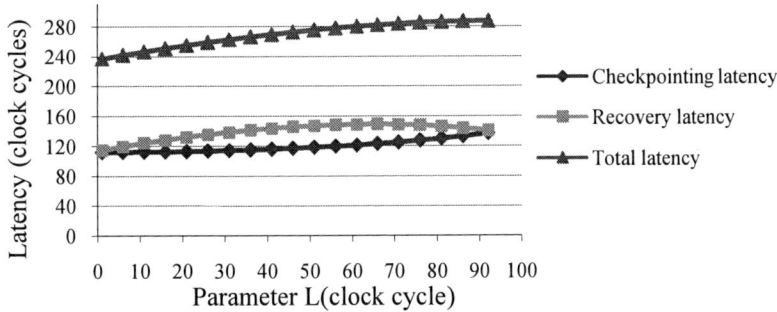

Figure 5. Total checkpoint and recovery latency for Matrix multiply when K=1

Both of (4-5) and (4-6) equations show that the optimal number of checkpoints decreases as the parameter L decreases. In the ACRR scheme the lower bound of L is error detection latency which seems to be the best value for L. Although we explain in Section 5 that determining a precise value for error detection latency is difficult in some cases. The analytical results show that the total latency is reduced as the non-uniformity of the checkpointing intervals increase.

The reduction of total checkpointing and recovery latency can increase the probability of timely task completion for soft real-time applications. As an example, for a soft real-time task with parameters E=8300, C=2, S=20, and soft deadline equals to 8750 clock cycle, both uniform checkpointing and ACRR schemes can tolerate a single error (K=1) without missing the soft deadline. In case of $K=2$, the figures for the uniform checkpointing scheme are: $Nop=110$, $T=74$, $P=16\%$. For the ACRR scheme, the figures for $K=2$, and $L=40$ are: $Nop=95$, $T=40$ and 132, $P=40\%$, and the figures for $K=2$, and $L=15$ are: $Nop=90$, $T=15$ and 165, $P=61\%$.

4. The ACRR implementation

Fault occurrences for any reason such as cosmic ray radiations may cause to control or data errors. These errors finally lead to incomplete execution, data violation or processor crash. So, the aim of error recovery is returning processor to a valid state that the wrong manipulations on data and control become ineffective. The state of processor includes the memory elements values, which hold data or control signals like registers file, control registers, and cache memory. There is no need to restore all memory elements if we need to rollback to a valid state. Checkpoints can accomplish in a manner that only manipulated data during checkpointing interval have been stored (incremental checkpointing). In this section, the ACRR scheme in micro architecture-level is implemented on the VHDL model of LEON2 processor. LEON2 is a 32 bit processor that conforms to the SPARC V8 architecture [4].

4.1 Recovery unit controller (RUC)

In order to implement ACRR, LEON2 processor is extended by adding an extra unit, called RUC (Recovery Unit Controller). RUC has been connected to execution unit, registers file and data cache memory. RUC schedules and controls all related checkpointing and recovery activities. As mentioned before, two most recent checkpoints should have been stored to prevent from returning to an invalid checkpoint. RUC manages this activity and determines the checkpointing intervals with respect to error detection latency. All error signals from different parts of the processor are sent to the RUC. Regarding to the detection latency of this error signals, RUC determines the valid checkpoint for error recovery.

4.2 Checkpointing mechanism

The applied checkpointing mechanisms in each part of a processor may be different, but the checkpointing consistency between these mechanisms can leads to a successful checkpointing and error recovery. For implementing the ACRR scheme, two different policies have been applied: 1) for global, status, and control registers and register file, 2) for data memory and cache.

Two backup registers are allotted to each control and status registers. The backup registers hold the content of the control and status registers in two most recent checkpoints. In each checkpoint, RUC stalls pipeline and the content of control and global registers is copied to the backup registers. In case of error occurrence, RUC stalls pipeline and the content of the backup registers are restored to the original registers. Both store and restoration can be done in just one clock cycle. For register file the same scenario is applied. Two extra register files are allotted for holding backup values of the register file.

For data memory, the same scenario cannot be used. Memory, both in size and access frequency is not comparable with register file. Most of the pervious works have implemented memory checkpointing in cache-level [2, 15, 19]. These works try to avoid writing the manipulated data in the cache memory to the main memory until be assured of the data correctness. The weakness point of these methods is undeterministic checkpointing latency. In these methods, the required time for checkpointing is unpredictable and depends on the number of write instructions in each checkpointing interval. Also in these methods the performance of the data cache degrades as the checkpointing interval increases. As a result, the total performance of the systems that using the cache-level error recovery is unpredictable.

In the ACRR scheme, instead of holding manipulated data in cache memory until the next checkpoint, the cache writes are updated in the main memory and the previous values of the written addresses is stored, simultaneously. Two backup buffers are allotted to the data cache. The backup buffers hold previous data in two most recent checkpointing intervals. If any write instruction executes during the checkpointing interval, the previous data and its address are stored in the backup buffers. Therefore, all of the memory checkpointing activity can be done parallel with real execution of the application program. In case of error occurrence, the old data in the backup buffers restore to their original places either in the data cache or main memory. In this case, the state restoration latency contain the duration of required time to update the data cache and main memory with the valid data from the backup buffers. In order to have shorter and predictable state restoration latency, copy to the main memory can be done in parallel with re-execution. According to the equation (3-5), reducing the checkpointing latency, causes an increase in the number of checkpoints and decrease in the checkpointing interval. Note that the probability of write instruction in each interval is reduced as the checkpointing interval decreases. This leads to low and more predictable state restoration latency.

In the proposed scheme, the checkpointing latency equals to 2 clock cycle which is a constant value and is independent of checkpointing interval. Since the main goal of this experimental work is implementing the ACRR scheme on the LEON2 processor as a case study, we neglect interrupt or I/O action during the runtime of benchmarks. However it is possible to remove this limitation without any negative effects on the proposed scheme. Also we assume the checkpoint memory elements are robust and fault-tolerant techniques like error correction codes are applied to them.

5. Experimental Results

To carry out the experiments to evaluate the ACRR scheme, four benchmarks have been run on the LEON2 (Bitcount and basicmath of automotive benchmarks from MiBench suit [5], bubble sort and matrix multiply). A controllable random signal generator is used for sending error detection signals to the Recovery Unit Controller (RUC). To determine the optimal number of checkpoints, the amount of C (checkpointing latency) and L (error detection latency) in each benchmark is required. Following to the practical implementation in Section5, the amount of C equals to 2 clock cycle that is constant value in different benchmarks and different checkpointing intervals. The total latency and overhead of uniform checkpointing are shown in Table 1 for each benchmark. The numbers of checkpoints are obtained analytically by formulas which have been presented in Section 2.

Table 1. The uniform checkpointing overhead

Benchmark	Total execution Time [cycles]	number of checkpoints	Checkpointing interval [cycles]	Total overhead [%]
Bubble sort	1171	27	38	11.5
Basicmath	3409	48	65	6.5
Matrix multiply	6793	69	92	4.5
Bitcount	8633	78	105	4.0

In practical applications, error detection latency differs depending on the fault nature and applied error detection mechanism. For example, in [3] and [13] control flow checking methods for embedded and COTS processors are presented which the maximum amount of error detection latency is about 50 instructions and 60 clock cycles in these methods, respectively. However, values of L are assumed to be greater than error detection latency and the benchmarks are executed for two cases, L=15 and L=30 on the LEON2 processor. Totally, more than 1500 times, the error signal has been activated. The total latency improvement ratio is the proportion of latency decrease in the ACRR scheme to uniform checkpointing. The obtained results are presented in Table 2.

Table 2. The ACRR scheme overhead (when *K*=1, *L*=15 *and L*=35)

Benchmark	Total execution time [cycles]	Number of checkpoints	Ch. P. Interval [cycles]		Overhead [%]	Late. Imp.* ratio [%]
			Interval1	Interval2		
Bubble sort, L=15	1158	22	73	15	10	12
Bubble sort, L=35	1165	26	40	35	10.9	5
Basicmath, L=15	3379	39	143	15	5.5	16
Basicmath, L=35	3391	41	115	35	5,9	9
Matrix multiply, L=15	6748	56	211	15	3.8	18
Matrix multiply, L=35	6763	57	187	35	4.0	11
Bitcount, L=15	8578	63	241	15	3.3	19
Bitcount. L=35	8594	64	217	35	3.5	13

* Total latency improvement ratio relative to uniform checkpointing

Table 3. The ACRR scheme for Bitcount benchmark (when k=2)

Benchmark	Total execution time [cycles]	Number of checkpoints	Ch. P. Interval [cycles]		Overhead [%]	Prob. * D=8750	Prob. D=8800
			Interval1	Interval2			
Uniform	8767	110	74	74	5.6	16	76
ACRR, L=15	8719	90	165	15	5.0	59	82
ACRR, L=35	8738	93	140	35	5.2	45	77

* Probability of timely task completion with soft deadline D

Table 2 shows that the latency improvement ratio in the ACRR scheme increases as the execution time of the benchmarks grows. Note that average execution time increases as the parameter L increases. Also, the optimal number of checkpoints in the ACRR scheme decreases relative to uniform checkpointing.

The experiments are repeated for the Bitcount benchmark and the results are averaged out over these runs. We are interested here in the probability of timely task completion P which the task completes before the soft deadline D. To illustrate more advantages of the ACRR scheme relative to the uniform checkpointing, we note that if we set K=2 (using the value of L as before), The proposed ACRR scheme provides higher value of P relative to the uniform checkpointing as the slack time decreases. In some cases, up to 40% increase is obtained in the probability of timely task completion; the results are shown in Table 3.

6. Conclusions

A checkpointing and rollback recovery scheme, called Asymmetric Checkpointing and Rollback Recovery (ACRR) was presented which store the states of a processor in an asymmetric manner. It was shown that the ACRR scheme reduced the error recovery latency and the optimal number of checkpoints to increase the probability of timely task completion. The ACRR scheme was evaluated analytically and practically. The evaluation results showed that the ACRR scheme reduced the average task execution time and was more likely to meet soft deadlines. The amount of these improvements depended on the error detection latency. The proposed approach could be extended to a set of multiple periodic tasks with considering other important parameters in embedded systems such as power consumption.

7. References

[1] M. Bashiri, S. G. Miremadi, and M. Fazeli, "A Checkpointing Technique for Rollback Error Recovery in Embedded Systems," *Proceeding of the 18th IEEE International Conference on Microelectronics (ICM 06),* 16-19 Dec. 2006, pp. 174-177,

[2] N. S. Bowen, and D.K. Pradhan, "Processor and Memory-Based Checkpoint and Rollback Recovery," *Proceeding of Computer,* Vol. 26, Feb. 1993, pp. 22-31.

[3] M. Fazeli, R. Farivar, and S. G. Miremadi, "A software-based concurrent error detection technique for power PC processor-based embedded systems," *Proceeding of the 20th IEEE International Symposium on Defect and Fault Tolerance in VLSI Systems,* Oct. 2005, pp. 266- 274.

[4] J. Gaisler, *Leon2 Processor,* www.gaisler.com

[5] M. Guthaus, J. Ringenberg, D. Ernst, T. Austin, T. Mudge, and R.Brown, "MiBench: A Free, Commercially Representative Embedded Benchmark Suite," *IEEE International Workshop on Workload Characterization* 2 dec. 2001, pp. 3-14.

[6] M. J. Iacoponi, "Hardware Assisted Real-Time Rollback in the Advanced Fault-Tolerant Data Processor," *Proceeding of the 10th IEEE Digital Avionics Systems Conference,* 1991, pp 169-274.

[7] V. Izosimov, P. Pop, P. Eles, and Z. Peng, "Scheduling of Fault-Tolerant Embedded Systems with Soft and Hard Timing Constraints," *Design, Automation, and Test in Europe (DATE 2008),* Munich, Germany, 10-14 Mar. 2008, pp. 915-920.

[8] I. Koren, and C. M. Krishna, *Fault Tolerant Systems,* Morgan Kaufmann, USA, 2007.

[9] P. Marwedel, *Embedded System Design,* Springer, Netherlands, 2006.

[10] R. Melhem, and E. Elnozahy, "The Interplay of Power-Management and Fault-Recovery in Real-Time Systems," *IEEE Transaction on Computers 2004,* Vol. 53, Issue 2, pp. 217-231, Feb. 2004.

[11] N. Nakka, K. Pattabiraman, and R. Iyer, "Processor-Level selective Replication," *37th Annual IEEE/IFIP International Conference on Dependable Systems and Networks,* june 2007, pp. 544-553.

[12] M. Pflanz, and H. T. Vierhaus, "On-line Check and Recovery Techniques for Dependable Embedded Processors," *IEEE MICRO,* Vol. 21, Issue 5, Sep/Oct. 2001, pp. 24-40.

[13] F. Rota, S. Dutt, and S. Krishna, "Off-Chip Control Flow Checking of On-Chip Processor-Cache Instruction Stream", *Proceeding of the 21th IEEE International Symposium on Defect and Fault Tolerance in VLSI Systems,* Oct. 2006, pp. 507-515.

[14] T. Sakata, T. Hirotsu, H. Yamada, and T. Kataoka, "A Cost-Effective Dependable Microcontroller Architecture with Instruction-Level Rollback Recovery," *37th Annual IEEE/IFIP International Conference on Dependable Systems and Networks,* June 2007, pp. 256-265.

[15] S. Shyam, K. Constantinides, S. Phadke, V. Bertacco, and T. Austin, "Ultra Low-Cost Defect Protection for Microprocessor Pipelines," *Proceedings of the 2006 ASPLOS Conference,* 2006, pp. 73-82.

[16] L. Spainhower, and T. A. Gregg, "G4: A Fault tolerant CMOS mainframe," *proceeding of 28th fault tolerant computing,* June 1998, pp. 432-440.

[17] N. H. Vaidya, "A Case for Two-Level Recovery Schemes," *IEEE Transaction on Computer,* Vol. 47, Issue 6, June 1998, pp. 656-666.

[18] Y. Tamir, and M. Tremblay, "High Performance VLSI Systems Using Micro rollback," *IEEE Transaction on Computers,* Vol. 39, Issue 4, Apr. 1990, pp. 548-554.

[19] R. Teodorescu, J. Nakano, and J. Torrellas, "SWITCH: A Prototype for Efficient Cache-Level Checkpointing and Rollback," *IEEE MICRO,* Vol. 26, Issue 1, Jan/Feb. 2006, pp. 28-39.

[20] H. Wang, S. Rodriguez, C. Dirik, A. Gole, V. Chan, and B. Jacob, "TERPS: The Embedded Reliable Processing System," *Proceedings of the ASP-DAC, IEEE Design Automation Conference,* Vol. 2, Jan. 2005, pp. d/1-d/2.

[21] Y. Zhang, and K. Chakrabarty, "Energy-Aware Adaptive Checkpointing in Embedded Real-Time Systems," *Proceedings of the Design, Automation and Test in Europe Conference and Exhibition,* 2003, pp. 918-923.

[22] Y. Zhang, and K. Chakrabarty, "Fault Recovery Based on Checkpointing for Hard Real-Time embedded systems," *Proceedings of the 18th IEEE International Symposium on Defect and Fault Tolerance in VLSI systems,* 3-5 Nov. 2003 pp. 320-327.

IEEE International Symposium on Defect and Fault Tolerance of VLSI Systems

Design and Evaluation of a Timestamp-Based Concurrent Error Detection Method (CED) in a Modern Microprocessor Controller

Michail Maniatakos[1], Naghmeh Karimi[2], Yiorgos Makris[1],
Abhijit Jas[3], Chandra Tirumurti[4]

[1] EE Department - Yale University
[2] ECE Department - University of Tehran
[3] Validation & Test Solutions - Intel Corporation
[4] Strategic CAD Labs - Intel Corporation
{michail.maniatakos, naghmeh.karimi, yiorgos.makris}@yale.edu
{abhijit.jas, chandra.tirumurti}@intel.com

Abstract

This paper presents a concurrent error detection technique for the control logic of a modern microprocessor. Our method is based on execution time prediction for each instruction executing in the processor. To evaluate the proposed method, we use a superscalar, dynamically-scheduled, out-of-order, Alpha-like microprocessor, on which we execute SPEC2000 integer benchmarks and we consider the coverage and the detection latency for faults in the scheduler module of the microprocessor controller. Experimental results show that through this method, a large percentage of control logic faults can be detected with low latency during normal operation of the processor.

1. Introduction

The rapidly shrinking feature sizes of semiconductor fabrication, along with the corresponding physical challenges that they incur, continue to give rise to various design robustness concerns. For example, the frequent occurrence of transient errors has, once again, surfaced as a problem of contemporary interest. While soft errors, occurring due to strikes by neutrons or alpha particles which potentially lead to corresponding single event upsets (SEUs) in memory bits, or single event transients (SETs) in combinational logic have received the lion's share of attention, they only constitute part of the problem. Indeed, various other issues related to design marginalities, process variations and corner operating conditions are starting to cause errors and to play an increasingly important role. Ranging in duration from single events to permanent faults, such errors have revived interest in concurrent error detection (CED) and/or correction methods that may ameliorate or resolve their effect.

CED [1, 2] has been extensively studied in the past and numerous ideas and solutions have been investigated along various directions. The simplest approach is duplication, wherein a replica of the circuit is added to the design, possibly diversely implemented to avoid common mode failures [3]. The original and the replica serve as predictors of the functionality of each other and a simple comparator indicates any discrepancy in their outputs, thus detecting potential malfunctions. While simple, this technique is prohibitively expensive. Partial duplication solutions focusing only on the most critical parts of a circuit have, therefore, also been explored [4]. Another very popular CED approach has been the use of various codes, especially within the context of finite state

1550-5774/08 $25.00 © 2008 IEEE
DOI 10.1109/DFT.2008.59

machine (FSM) controllers. Several redesign and resynthesis methods are described in [5, 6], wherein parity or various unordered codes are employed to encode the states of the circuit. Utilization of multiple parity bits is also examined in [7] within the context of FSMs. These methods guarantee latency-free error detection; on the down side, they are intrusive and expensive. Non-intrusive CED methods have also been proposed. Implementations based on Bose-Lin and Berger codes are presented in [8] and [9], respectively, while parity-based CED methods are described in [1, 10, 11]. While the aforementioned methods guarantee latency-free detection of all errors, their cost is often prohibitive. Trading-off the incurred cost by allowing a non-zero, yet bounded latency has also been investigated [12].

At a coarser level, an attempt to identify inherent invariance either at the gate-level [13] or at the RTL [14] of a design has been made. Such invariance can be monitored during the normal operation of a circuit to identify errors that cause it to be violated. In [13] such invariance is mined from the gate-level of a controller implementation in the form of assertions, which are evaluated through simulation in order to select a cost-effective appropriate subset. The same principle governs the approach in [14]; therein, however, invariance is identified through a path-construction algorithm, which exploits inherent transparency channels that exist in the RTL description of a modular design.

At an even higher architectural level, several concurrent error detection and/or correction methods have been proposed. The concept of watchdog processors, which compute control-flow signatures and compare them to expected correct values, known at compilation time, is proposed in [15]. Concepts akin to instruction-level duplication and comparison to identify erroneous results are examined in [16, 17]. In [18], the authors examine the vulnerability of different parts of a microprocessor to soft errors and recommend various strategies (including register file protection with codes, parity coding to protect instruction words, and a timeout counter to flush the pipeline when no activity occurs for prolonged periods) to detect/correct such errors. Similar analysis is performed in [19], based on the concept of Architectural Vulnerability Factor (AVF), which prioritizes microprocessor modules based on their susceptibility. Such metrics can prove very useful in guiding allocation of CED resources

This pluralism of options implies that no one solution fits the needs of every circuit or even every part of a circuit. Furthermore, it stresses the fact that generic solutions typically incur prohibitive cost, often times without providing commensurate coverage. Therefore, while developing CED methods for a circuit, it is important to tailor the solutions by leveraging the specifics of each module.

In this work, we combine several of the key ideas proposed in the above references and we develop a CED method for the scheduler module of a modern microprocessor controller. The proposed method utilizes architectural information (i.e. the functionality of the scheduler) to construct an invariant (i.e. the relation between the dispatching and starting execution time-stamps of an instruction). Consequently, monitoring of this simple invariant during normal operation enables detection of any fault or error resulting in a discrepancy between the expected and actual timestamps, with small detection latency, often even before the error corrupts the architectural state. We note that, while the microprocessor datapath is equally important, we mainly focus on control logic for two reasons. First, CED for datapath is understood much better and various residue code-based techniques have been successfully applied. Second, advanced architectural features complicate significantly the task of the controller, making it much harder to analyze or predict its behavior in the presence of errors.

The remaining of this paper is organized as follows. In Section 2, we briefly review a fault simulation infrastructure that we have previously developed around a modern microprocessor and which we use to evaluate the proposed CED method. In Section 3, we discuss the details of the timestamp-based CED method as well as the Scheduler module of the microprocessor controller, which is targeted by this method. In Section 4, we experimentally assess the coverage and the detection latency of the proposed CED method using the aforementioned infrastructure to perform extensive experiments while the target microprocessor executes typical SPEC benchmark programs. Conclusions and future directions are provided in Section 5.

2. Background

The research work described herein builds upon a previously developed infrastructure, which is presented in detail in [20]. The employed model is the Illinois Verilog Model (IVM) [18], an Alpha 21264-like microprocessor featuring superscalar, out-of-order execution. The complexity of such a model reflects most of the features of modern, high-performance microprocessors enabling accurate evaluation of CED techniques targeting the same. The developed infrastructure supports simulation of actual programs (i.e. SPEC benchmarks), injection and simulation of Register Transfer Level (RTL) faults (both stuck-at and transients), as well as extensive I/O capabilities (i.e. Machine State Dumping and Trace Dumping features).

Using the developed infrastructure, in [20] we investigated the correlation between RTL faults in the control logic and their instruction-level impact on the execution flow of typical programs. Specifically, we injected stuck-at faults at the Scheduler and the Reorder Buffer (ROB) modules of the microprocessor and we studied their impact on the execution of integer SPEC benchmarks. The arising Instruction Level Errors (ILEs) were divided into five groups reflecting the key aspects of instruction execution in a superscalar out-of-order microprocessor, namely (i) the operation that is executed, (ii) the operands that are being used, (iii) the functional unit where execution takes place, (iv) the starting and finishing time of execution, and (v) the order of commitment. These groups were further divided in 13 Types, as detailed in Table 1.

Table 1. Instruction level error types

Group 1: Operation Errors	Type 1:	Incorrect (yet valid) operation code used
	Type 2:	Invalid operation code used
Group 2: Operand Errors	Type 3:	Incorrect (yet valid) register addressed
	Type 4:	Invalid register addressed
	Type 5:	Premature use of register contents
	Type 6:	Incorrect immediate operand used
Group 3: Execution Errors	Type 7:	Incorrect functional unit type utilized
	Type 8:	Multiple functional units utilized
Group 4: Timing Errors	Type 9:	Early commencement
	Type 10:	Late or no commencement
	Type 11:	Longer duration
	Type 12:	Shorter duration
Group 5: Order Errors	Type 13:	Commitment order violation

Using RTL fault injection, 16,904 stuck-at faults were injected at the Scheduler module for each of the SPEC benchmarks executed. For every injected fault, an automated cycle-by-cycle analysis of the Scheduler traces was used to classify the fault to one of the ILE types defined in Table 1. The classification results are presented in Figure 1, averaged among the executed SPEC benchmarks.

The results of the study show that Timing Errors were the most dominant group of errors, particularly Type 10 (Late or no instruction commencement) and Type 11 (Longer instruction duration). Based on this observation, in this work we develop a timestamp-based CED technique which specifically targets these erroneous behaviors.

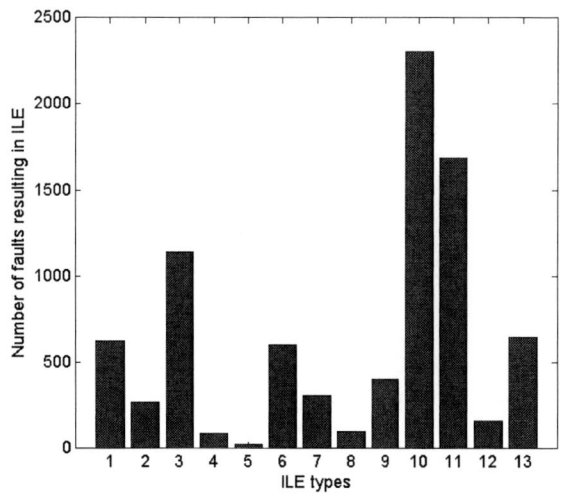

Figure 1. Classification of stuck-at faults of Scheduler module into ILE Types

3. Developed CED method

In this section, we propose a cost-effective strategy for Concurrent Error Detection (CED) of the control logic of the above processor. Using this strategy, transient faults, as well as permanent faults occurring due to gradual degradation during the lifetime of the microprocessor can be detected during its normal operation. The following subsections present the structure of the targeted Scheduler module and the CED technique.

3.1. Scheduler module

The targeted Scheduler is a dynamic module which can issue up to 6 instructions in each clock cycle. Instructions are issued out of order depending on the following factors:

- Availability of the instructions in the Scheduler module
- Avoidance of data hazards
- Avoidance of structural hazards

The Scheduler module contains an array of up to 32 instructions waiting to be issued. Each instruction coming to the Scheduler, resides in this buffer until an acknowledgement is received from the execution unit that it can start execution. At this time, the corresponding location in the scheduler list can be used for another newly arriving instruction to the scheduler module.

Structural hazards are considered by the Scheduler before issuing an instruction. The microprocessor model has 2 simple, 1 complex, 1 branch and 2 memory instruction functional units. Thus there is a limitation on the number of instructions of each type that can be issued in each clock cycle.

The microprocessor model also includes a Rename module. The renaming process removes the possibility of Write-After-Read (WAR) and Write-After-Write (WAW) hazards. However, due to the dependency of the instruction operands there can still be a Read-After-Write (RAW) hazard. To deal with such RAW hazards, the Scheduler module uses a Scoreboard technique [21].

Furthermore, for each instruction coming to the Scheduler, the Reorder Buffer module assigns an identification number, called *ROBid*. This *ROBid* follows the instruction until it commits and serves as a mechanism for ensuring in-order instruction commitment in the out-of-order execution of the microprocessor.

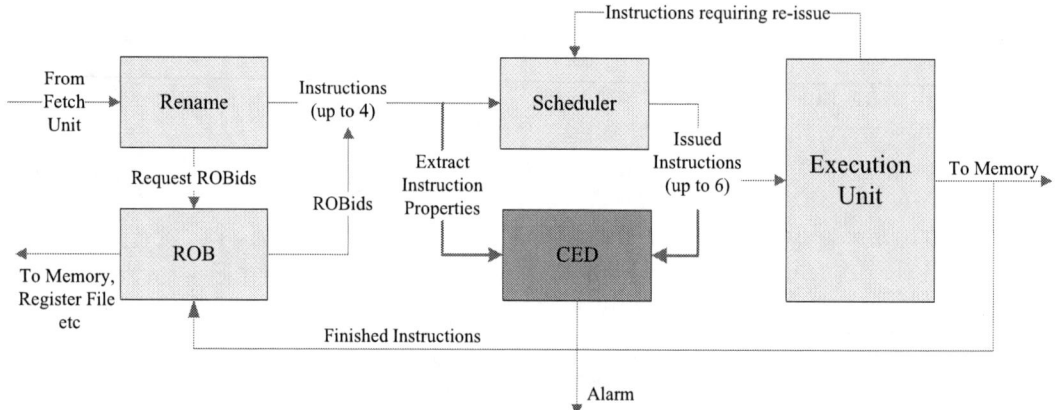

Figure 2. Block diagram of proposed CED technique

3.2. CED strategy

Based on the functionality of the Scheduler module discussed above, we implemented a CED mechanism to detect a portion of permanent as well as transient errors of this module, during the normal operation of the microprocessor. Our method is based on execution time prediction for each instruction resides in Scheduler module. The prediction is based on the fact that each coming instruction to the Scheduler starts execution after a certain number of clock cycles assuming that there is no structural or data hazard. A high-level block diagram of the proposed method is shown in Figure 2.

The CED mechanism keeps track of the incoming instructions to the Scheduler and predicts the execution time of these instructions based on the information gathered from the Scoreboard and Scheduler units. If an instruction encounters no data or structural hazards due, its execution time is predicted considering the structure of the microprocessor model. Then the *ROBid* of that instruction, which uniquely identifies it, is stored along with the predicted execution time.

The CED module checks if instructions stored in its internal buffer are correctly executed at the appropriate functional unit at the correct timestamp. An instruction may be replayed if the operands are not ready; this happens in the analyzed microprocessor model because forwarding mechanisms may provide the necessary operands directly to the execution unit, so the scheduler tries to issue instructions even if their operands are not ready yet. The developed CED technique checks the Scoreboard module, which is part of the Scheduler, for the availability of the operands and predicts the starting timestamps.

The CED algorithm can be summarized as follows:

1. Extract *ROBid*, Instruction Type and Operands information from instruction array entering the Scheduler.
2. Based on bookkeeping information of functional units utilized and operands in use, predict when an instruction should start execution.

3. Track instruction execution at the functional units. Raise alarm if any discrepancy identified.

The proposed CED method is not duplication, since it reuses parts of the Scheduler such as the Scoreboard, which can be protected easily by using techniques such as parity.

4. Experimental Results

In this section, we discuss the fault model used to evaluate the developed CED method and we present an extensive analysis of the results. Specifically, we report statistics about the fault coverage and the latency of the proposed CED method.

4.1. Experimental setup

To asses the fault coverage of the developed CED, the stuck-at fault model is used. Because our target is control logic, and more specifically the scheduler, faults are injected only in this module. A total of 16,904 s@0 and s@1 faults are injected using the RTL model fault injection technique of [20], which was briefly described in section 2.

In order to evaluate the developed CED method, six different SPEC benchmarks are utilized. Each benchmark is executed at the fault-injected RTL model for 20,000 cycles. After the end of the simulation, the architectural state of the microprocessor is compared to a fault-free (golden) run. If any discrepancies are identified, then the error is classified into one of two different sets: i) if the error propagates to the architecture register file, then the execution is marked as Erroneous, and ii) otherwise, if a discrepancy exists in the machine state but not in the architecture register file, the fault is likely masked and propagates to a part of the microprocessor that does not affect the execution – thus we classify the execution as Masked.

Besides architectural state discrepancies, an injected fault may lead to a different simulation outcome: stall of the pipeline, if the executed program uses unimplemented instructions. As explained in [20], the microprocessor model lacks certain instructions, such as system calls or floating-point operations. Even though the golden run is carefully chosen so that no such instructions are fetched, a fault may still drive the microprocessor to incorrectly call one of these instructions in the same time window. In this case, due to the described microprocessor model limitations, the execution stalls and the corresponding run is marked as Stalled. This is indeed an erroneous behavior, yet we cannot tell whether the proposed CED method will detect the fault after the unimplemented instruction. Thus, the number reported is very pessimistic and covers faults detected before the microprocessor stalls; in a full instruction set implementation the fault coverage would be higher. The complete classification process is presented in Figure. 3.

Besides the fault coverage metric, detection latency is also reported and analyzed. Latency is defined as the difference between the time that the CED error signal is activated and the time when a discrepancy first appears in the architecture register file of the microprocessor. Sometimes, the CED output signal may be activated before a discrepancy appears. We call this case an early detection. This type of detection provides an easy recovery, since the register file is unaffected. Similarly, if the CED fires after a discrepancy is identified, we call it a late detection. In this case, the microprocessor should rollback to a previous valid checkpoint.

4.2. Experiment results and discussion

The most important aspect of a CED method is the attained fault coverage. Our fault coverage analysis consists of two different parts: i) fault coverage of the cases where the architectural state is different at the end of the simulation, and ii) fault coverage of the cases where a pipeline stall occurs. As explained in section 4.1, the latter is a pessimistic estimate because there is no capability to evaluate the CED behavior after the stalling point.

Figure 4 presents the classification of the 16,904 runs for each SPEC benchmark. The Erroneous and Stalled classes are the target of our CED method. The differences between the numbers of corrupted runs for each benchmark exist because each benchmark uses a different set of instructions during the simulation window, which may or may not use the stuck-at bit. Nevertheless, this variability provides a better estimate of the CED performance, since different instructions are utilized for each SPEC benchmark.

Proceeding to the actual results, Figure 5 presents the fault coverage percentage for

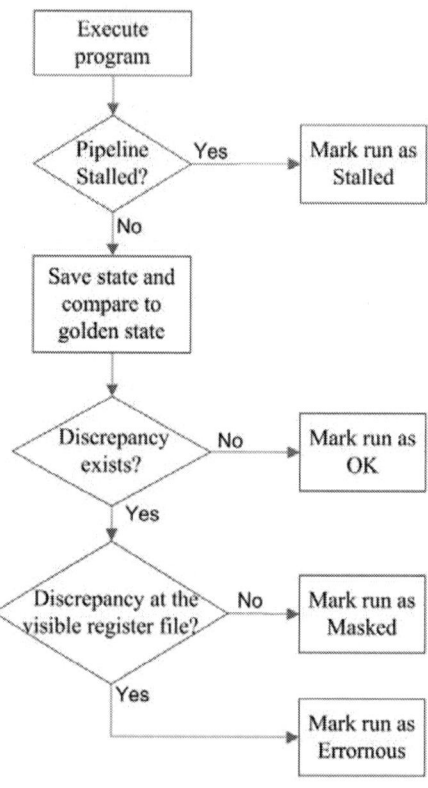

Figure 3. Simulation outcome classes

each of the aforementioned classes (Erroneous and Stalled). The CED detects 52.2% of the Erroneous cases on average. The consistency of the fault coverage among the 6 benchmarks corroborates that the CED method is independent of the program load.

Figure 6 shows the fault coverage of applying our CED method for each ILE type discussed in section 2. Even though the proposed CED method targets timing issues (Types 9 and 10), fault coverage in these groups is close to 50%. This happens because, as described in [20], errors that do not fall into one of the 13 Types usually appear as timing errors. These errors cannot be targeted by our CED method, as individual instruction timing appears to be correct.

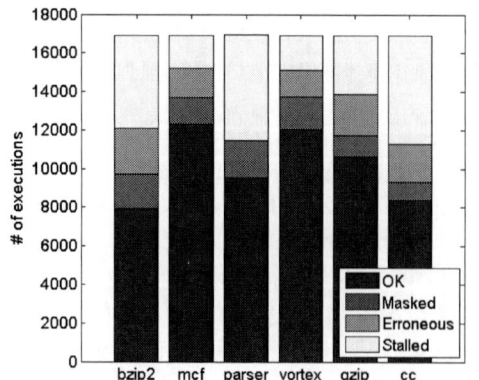

Figure 4. Simulation outcome results

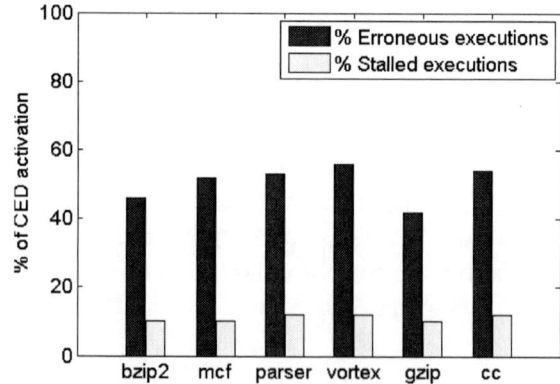

Figure 5. Fault coverage results

Another interesting conclusion drawn from Figure 6, is that the CED method excels at detecting faults concerning utilization of Functional Units (FUs) and In-Order Instruction Commitment. This is expected due to the nature of the CED method and its location between the Scheduler, the ROB and the Execution Unit. The CED checks the validity of the FUs utilization by checking the opcodes and the operands. In addition many commitment order violation (Type 13) errors are the result of timing errors.

As defined in section 4.1, detection latency is the difference between the time that the CED error signal is activated and the time when a discrepancy first appears in the architecture register file of the microprocessor. Early

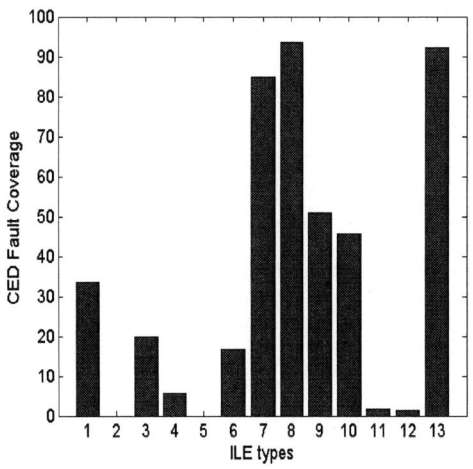

Figure 6. Correlation of CED detections to ILEs

detection implies that the CED alarm signal is triggered before the discrepancy appears; otherwise we have a late detection.

The results presented in Table 2 show that, in most cases, an early detection is reported. Latency may not be reported for all detections, because a discrepancy may not appear at the programmer-visible register file within the time window of 20,000 cycles. As can be observed, an average of 95% of the total number of reported latencies are early detections (latency <= 0).

The normalized average latency presented in the last column of Table 2 is the average of latencies, with an early detection considered as latency of 0 cycles. The early detection property of the CED indicates that, in most cases, a pipeline flush and restart is enough to correct the fault. For the late detection cases, the worst case latency does not exceed 35 cycles, according to Table 2, so a checkpoint and restore operation could be fine-tuned accordingly to correct the microprocessor state.

Table 2. Fault detection latency results

SPEC Benchmark	Early detection	Normalized Average Latency (in clock cycles)
bzip2	93%	17
mcf	98%	22
parser	96%	5
vortex	97%	22
gzip	95%	34
cc	92%	35
Average	**95%**	**23**

5. Conclusion and Future Work

The CED method proposed in this work for control logic of modern microprocessors is based on the observation that the impact of most low-level faults in a modern microprocessor is quickly visible on the starting and finishing instruction execution timestamps. Predicting these timestamps and checking against their actual execution values during normal operation results in detection of over 52% of the faults in the Scheduler module with an average detection latency of 35 cycles. We are currently extending the scope of the CED method to other control logic units. Our ultimate goal is to efficiently detect faults in the control logic of a microprocessor by adding an ensemble of small CED techniques, each targeting a different set of faults.

Acknowledgement

This research was sponsored by a generous gift by Intel Corporation and was performed while the second author was a visiting student at Yale University.

6. References

[1] M. Goessel, and S. Graf, Error Detection Circuits, McGraw-Hill, 1993.
[2] S. Mitra, and E. J. McCluskey, "Which Concurrent Error Detection Scheme to Choose?" in Proc. of the International Test Conference, 2000, pp. 985–994.
[3] A. Avizienis, and J. P. J. Kelly, "Fault Tolerance by Design Diversity: Concepts and Experiments", IEEE Transactions on Computers, vol. 17, no. 8, pp. 67-80, 1984.
[4] K. Mohanram, and N. A. Touba, "Cost-Effective Approach for Reducing Soft Error Rate in Logic Circuits," in Proc. of the International Test Conference, 2003, pp. 893–901.
[5] G. Aksenova, and E. Sogomonyan, "Design of Self-Checking Built-in Check Circuits for Automata with Memory," Automation and Remote Control, vol. 36, no. 7, pp. 1169–1177, 1975.
[6] S. Dhawan, and R. C. D. Vries, "Design of Self-Checking Sequential Machines," IEEE Transactions on Computers, vol. 37, no. 10, 1988, pp. 1280–1284.
[7] C. Zeng, N. Saxena, and E.J. McCluskey, "Finite State Machine Synthesis with Concurrent Error Detection," in Proc. of the International Test Conference, 1998, pp. 672–679.
[8] D. Das, and N. A. Touba, "Synthesis of Circuits with Low-Cost Concurrent Error Detection Based on Bose-Lin Codes," Journal of Electronic Testing: Theory and Applications, vol. 15, no. 2, 1999, pp. 145–155.
[9] N. K. Jha and S.-J. Wang, "Design and Synthesis of Self-Checking VLSI Circuits," IEEE Transactions on Computer-Aided Design of Integrated Circuits and Systems, vol. 12, no. 6, 1993, pp. 878–887.
[10] R. A. Parekhji, G. Venkatesh, and S. D. Sherlekar, "Concurrent Error Detection Using Monitoring Machines," IEEE Design and Test of Computers, vol. 12, no. 3, 1995, pp. 24–32.
[11] S. Almukhaizim, P. Drineas, and Y. Makris, "Entropy-Driven Parity-Tree Selection for Low-Overhead Concurrent Error Detection in Finite State Machines," IEEE Transactions on Computer-Aided Design of Integrated Circuits and Systems, vol. 25, no. 8, 2006, pp. 1547-1554.
[12] S. Almukhaizim, P. Drineas, and Y. Makris, "On Concurrent Error Detection with Bounded Latency in FSMs in Proc. of the IEEE Design Automation and Test in Europe Conference (DATE), 2004, pp. 596-601.
[13] R. Vemu, A. Jas, J. A. Abraham, S. Patil, and R. Galivanche, "Low-Cost Concurrent Error Detection Technique for Processor Control Logic," in Proc. of Design Automation and Test in Europe, 2008, pp. 897-902.
[14] Y. Makris, I. Bayraktaroglu, and A. Orailoglu, "Enhancing Reliability of RTL Controller-Datapath Circuits via Invariant-Based Concurrent Test," IEEE Transactions on Reliability, vol. 53, no. 2, 2004, pp. 269-278.
[15] A. Mahmood, and E. J. McCluskey, "Concurrent Error Detection Using Watchdog Processors-A Survey", IEEE Transactions on Computers, vol. 37, no. 2, 1988, pp. 160-174.
[16] A. Mendelson, and N. Suri, "Designing High-Performance and Reliable Superscalar Architectures-the Out of Order Reliable Superscalar (O3RS) Approach", In Proc. of International Conference on Dependable Systems and Networks, 2000, pp. 25-28.
[17] A. K. Somani, and J. Nickel, "REESE: a Method of Soft Error Detection in Microprocessors", in Proc. of International Conference on Dependable Systems and Networks, 2001, pp.401-410.
[18] N. J. Wang, J. Quek, T. M. Rafacz, S. J. Patel, "Characterizing the Effect of Transient Faults on a High-Performance Processor Pipeline", in Proc. of International Conference on Dependable Systems and Networks, Florence, Italy, 2004, pp.61-70.
[19] S. S. Mukherjee, C. Weaver, J. Emer, S. K. Reinhardt, and T. Austin, "A Systematic Methodology to Compute the Architectural Vulnerability Factors for a High-Performance Microprocessor," In Proc. of International. Symposium on Microarchitecture, 2003, pp. 29-40.
[20] N. Karimi, M. Maniatakos, A. Jas, and Y. Makris, "On the Correlation between Controller Faults and Instruction-Level Errors in Modern Microprocessors," in Proc. of International Test Conference, 2008.
[21] J. L. Hennessy, and D. A. Patterson, Computer Architecture: A Quantitative Approach, Third Edition, Morgan Kaufmann Publishers, 2003.

SESSION 11

TESTING FOR TIMING AND PARAMETRIC FAILURES

IEEE International Symposium on Defect and Fault Tolerance of VLSI Systems

Novel On-Chip Clock Jitter Measurement Scheme
For High Performance Microprocessors*

C. Metra	M. Omaña	TM Mak	A. Rahman	S. Tam

DEIS – Univ. of Bologna (Italy)   *Intel Corporation, Santa Clara (CA, USA)*

{cecilia.metra, martin.omana}@unibo.it   *{t.m.mak, asifur.rahman, simon.tam}@intel.com*

Abstract

In this paper we present an on-chip clock jitter digital measurement scheme for high performance microprocessors. The scheme enables in-situ jitter measurement of the clock distribution network during the test or the debug phase. It provides very high measurement resolution, despite the possible presence of power supply noise (constituting a major cause of clock jitter) affecting itself. The resolution is higher than a min sized inverter input-output delay, and can on principle be further increased, at some additional costs in terms of area overhead and power consumption. In this paper, a resolution of the 1.8% of the clock period is achieved with limited area and power costs.

1. Introduction

Clock is one of the most critical signal in any synchronous system, that has to be distributed throughout the chip by means of a complex network [1]. With the continuous scaling of technology and increase in clock frequency, it is becoming increasingly difficult to guarantee the availability of correct clock signals throughout the chip due to the increasing likelihood of manufacturing defects (most likely resulting in clock duty cycle variations [2]), clock jitter and duty cycle distortion due to process variation and power supply noise [3-4].

Jitter affecting a signal of the clock distribution network produces uncertainties on its period and rising/falling edges, thus forcing the designer to either increase the timing margins, or face the possibility of operating malfunctions. For high performance microprocessors, adopting minimum timing margin is desirable and the application of on-chip jitter measurement schemes for the clock distribution (in addition to that for the PLL output) could be used during the test or debug phase to validate the design and manufacturing assumptions.

Power supply noise modulating the delay of the clock distribution network is currently recognized as one of the main causes of clock jitter [4-5]. It is expected to keep on increasing with technology scaling, due to the increasing complexity and integration density resulting in high on-die switching activities [6]. Additionally, the use of clock gating contributes to significant di/dt power supply noise degrading the jitter of the clock network [4]. Parameter variations occurring during fabrication and possibly affecting the buffers of the clock distribution network are another likely cause of clock jitter [7], whose likelihood is expected to keep on increasing with technology scaling.

Several off-chip and on-chip schemes have been presented so far for clock jitter measurement. On-chip schemes imply some die area overhead and may require higher circuit complexity, but they generally provide higher measurement resolution than off-chip schemes [8]. To minimize timing margins, the availability of high resolution clock jitter measurement schemes is a primary concern for high performance microprocessors. Therefore, for this application on-chip approaches are generally preferred to off-chip alternates.

One of the major difficulties in accurately measuring clock jitter by an on-chip scheme is that the measurement scheme is itself affected by the Power Supply Noise (*PSN*) that mainly contributes to the clock jitter we want to measure. This way, we risk to miss the contribution to clock jitter of *PSN*, thus obtaining inaccurate time measurements.

Several on-chip jitter measurement approaches for signals of the clock distribution network

---

*Work partially supported by Intel Corporation (Santa Clara, CA) research grant.

1550-5774/08 $25.00 © 2008 IEEE

DOI 10.1109/DFT.2008.51

have been proposed so far (e.g., those recently presented in [9, 5, 10, 11, 4]).

In [9], a Vernier Delay Line (VDL) is employed. An additional Delay-Lock Loop (DLL) is used to calibrate the delay of the elements within the VDL against process parameter variations, temperature and *PSN*. This approach allows high resolution measurements (7% of the clock period), but implies non negligible area overhead and power consumption (approx. more than 3 times the area and power required by our solution).

In [5], a solution using a reference clock and two delay lines is presented. An output measurement encoded by a thermometer code is produced. The reference clock is generated by means of a Delay-Locked Loop (DLL), a digital low-pass filter and a counter, which may imply non negligible design costs. The resolution of this approach is equal to an inverter delay, and may be too low for high performance microprocessors. The effects of *PSN* on jitter measurement are compensated by the same DLL that generates the reference clock.

In [10], a circuit that under-samples the input clock signal with another sampling signal (featuring a frequency very close to that of the input clock) is presented. This allows to achieve high resolution jitter measurements, but requires the calibration of the sampling signal, which may be expensive to achieve. The scheme is robust against process variations, but no details are given concerning the possible impact of *PSN* on the achieved measurement.

In [11], the input clock is sampled using a random clock. The resolution of the approach depends on the number of performed samples. It requires the generation of a random clock (through chaos-based circuits), with consequently non negligible design complexity, area overhead and power consumption (approx. more than 2 times the area and power required by our solution when achieving the same resolution). Similarly to [5], no details are given concerning the possible impact of process variations on the achieved measurement and, as in [10], no details are provided on the impact of *PSN* on the achieved measurement.

In [4], a circuit constituted by latches and NOT chains is presented. It may require non negligible area overhead and power consumption (an increase of approx. 70% in area and power consumption compared to our solution). Moreover, resolution is given by an inverter input-output delay. In this approach, the inverters of the NOT chains are calibrated to compensate the effects of *PSN* and parameter variations on the achieved measurements.

Based on the high measurement resolution needs of high performance microprocessors and the approaches proposed so far, in this paper we present a new on-chip digital measurement scheme, that allows to measure jitter of the signals of the clock distribution network during the test or debug phase with very high resolution, despite the presence of *PSN* (up to 17% of nominal Vdd) affecting itself. Resolution is higher than a min sized NOT input-output delay, and can on principle be increased as much as desired, at some additional costs in terms of area overhead and power consumption during the measurement phase. A resolution of the 1.8% of the clock period (i.e., a resolution of 6ps for a 3GHz clock) can be easily achieved with limited costs in terms of area and power. Also design costs are limited.

Our approach can be employed to measure clock jitter of the clock distribution network of any high performance microprocessor.

Compared to the alternative on-chip measurement approach most recently proposed in [4], our scheme allows area and power consumption savings of the 41% and 40%, respectively, while featuring higher measurement resolution and lower sensitivity to *PSN*.

The effectiveness of our approach has been verified by means of electrical level simulations, performed considering also statistical variations of electrical parameters up to the 35%.

The rest of the paper is organized as follows. In Section 2, we describe the block structure of our clock jitter measurement scheme. In Section 3, we present its possible implementation. In Section 4, we show some of the results of the electrical level simulations performed to verify its effectiveness and robustness to electrical parameter variations and *PSN*. In Section 5, we evaluate the costs of our approach and compare them to those of the most recently proposed alternate measurement scheme. Finally, in Section 6, we draw some conclusive remarks.

2. Block Structure

A possible strategy to measure the jitter of an input clock signal consists in measuring the duration of its high or/and low phase/s over time, and comparing them to their expected duration for the case of jitter-free clock. This is a well assessed and widely used approach (employed also in [4]), that we will also consider as a general mean to measure clock jitter.

Our scheme can be adopted to measure either the high clock phase only, or both clock phases. Measuring both phases allows the measurement of the clock duty-cycle and a better characterization of jitter within a considered clock period, but also requires a cost increase. For the sake of brevity, we will show only our scheme for clock high phase measurement, which can be easily extended to measure both clock phases by straightforward modifications.

2.1. Basic block structure

The basic block structure of our proposed scheme is shown in Fig. 1(a).

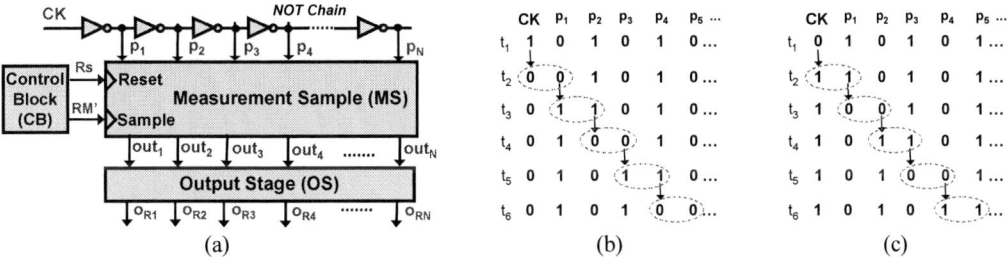

(a)   (b)   (c)

Fig. 1. Basic block structure of our scheme (a); schematic representation of the propagation of the CK falling (b) and rising (c) edges within the NOT chain.

The NOT Chain implements a delay line, which delays the input clock signal (CK), whose jitter has to be accurately measured, by a given time entity. The outputs of the inverters p_i (i=1..N) are progressively delayed and inverted (for i odd) versions of CK. Denoting by τ the input-output delay of each NOT in the chain, the global delay of the NOT chain is given by $N\tau$. Such a global delay should be long enough to cover, under jitter-free conditions, the whole period (T_{CK}) of the input clock CK (i.e., $N\tau > T_{CK}$). It could be easily verified that the measurement resolution will be equal to the input-output NOT delay (τ).

After a CK falling (rising) edge, the logic values simultaneously present at the outputs of each NOT of the chain are reported in Fig. 1 (b) ((c)), in which each row represents the snap-shots at one specific instant of time, whereas the time axis is vertical. The position of the CK falling (rising) edge is identified by the occurrence of two successive 0s, or two successive 1s, whose location moves progressively right.

The duration of the clock high phase is given by the number of NOTs within the chain that the CK rising edge has to pass through before the CK falling edge arrives to the input of the chain, which is identified by the Measurement Sample (MS) and Output Stage (OS) blocks.

MS samples the values present on signals p_i upon the falling edge of signal RM', which is generated by the Control Block (CB). This takes place immediately after the occurrence of the falling edge of the CK signal identified by CK=p_1=0 (at t_2 in Fig. 1(b)). The p_i sampled values are given to signals out$_i$ that, in our scheme, are given to OS, that encodes them into a word belonging to the thermometer code (which is given on signals o_{Ri}).

Such output thermometer encoding allows to easily and quickly derive the duration of the clock jitter. In fact, as an example, the produced encoded o_{Ri} (i=1..N) word can be compared, for instance through N parallel EXORs, with that expected for the case of jitter-free clock, thus providing an N bit string with a number of 1s equal to the difference between the number of 0s in the produced encoded word and in the expected one. Then, the number of 1s in such a produced string can be counted. Jitter measurement can then be simply obtained by multiplying such a 1s count by the scheme resolution.

Of course, after measurement sampling, and after a time long enough to allow the system to read the measurement, the measurement scheme could be automatically re-initialized (e.g., by generating Rs=1), thus making our scheme ready to be used for a following measurement.

Unfortunately, the measurement obtained with this circuit may be incorrect in presence of *PSN*. In fact, other than being a major cause of the clock jitter we want to measure, *PSN* will slow down the behavior of the NOTs in the chain. Consequently, should *PSN* be the only cause of jitter, due to the slowed down behavior of the chain NOTs, the obtained number of 0s would be the same as if *PSN* were not present, and a jitter equal to 0 would be measured. In practice, other causes of jitter are generally present together with *PSN*, but being this a major contributor, the measurement error would be anyway significant.

2.2. Basic block structure improvements

To solve the problem outlined above, we here propose to sample the values present on the p_i signals (by asserting RM'=0) when the CK falling edge arrives to the input of the second NOT of the chain (i.e., when $p_1=p_2=1$, at t_3 in Fig. 1(b)), rather than when it arrives to the input of the first NOT of the chain (i.e., when CK=p_1=0). In fact, since the *PSN* is most likely to occur upon clock edges (due to the simultaneous transition of all clock buffers, flip-flops, etc) rather than at the middle or end of its period [6, 12, 13], and being the NOTs' behavior temporarily delayed because of such a *PSN*, we propose to wait in time for the disappearance of such a temporary influence before sampling the outputs of the NOT chain. This assumption is adequate because of the duration of the power supply noise is a random variable with a maximum value in the order of the rise time of the clock signal [13]. That is, we wait for the NOTs to go back to their "normal" input-output delay, before sampling the outputs of the NOT chain. In order to still obtain a thermometer code at the output of OS, we can feed this block with out_i with i=2..N only.

Of course, if the clock signal being measured requires a buffering scheme to route it to the input of the NOT chain, we could simply use the delay of such buffering scheme to delay the sampling instant, rather than using the first NOT of the NOT chain as described above.

Therefore, by properly sampling the outputs of the NOT chain, we can achieve the goal of jitter measurement with low sensitivity to *PSN*, which mainly contributes to the clock jitter we want to measure. Of course, the measurement resolution remains equal to the NOT delay (τ) which, as introduced, is not high enough for high performance microprocessors.

In order to obtain a resolution higher than a NOT delay we could for instance replace the previously described basic block scheme of Fig. 1(a), by the one in Fig. 2(a), employing two NOT chains rather than one. NOT chain 1 consists of NOTs each with a delay equal to τ. Instead, NOT chain 2 consists of NOTs, out of which, the first NOT has a delay equal to $(1+\frac{1}{2})\tau$, while all other NOTs have a delay equal to τ. This way, as illustrated in Fig. 2(b), the outputs of the corresponding NOTs of the two chains present a phase difference equal to $\tau/2$ (we here consider only the phase difference between signal edges, without addressing existing signal inversions, that will be accounted for by the OS block).

In particular, considering as output the alternated succession of the two NOT chains' outputs (i.e., p_{11}; p_{21}; p_{12}; p_{22}; p_{13}; p_{23}; etc., in Fig. 2(b)), any two following outputs will have a phase difference equal to $\tau/2$. Therefore, the scheme in Fig. 2(a) will provide a measurement resolution of half a min sized NOT input-output delay.

As introduced previously, to achieve low sensitivity to *PSN*, the values at the outputs of the NOT chains are sampled by MS when the CK falling edge arrives at the input of the second NOT of the NOT chain 1 (i.e., when $p_{11}=p_{12}=1$).

Differently from the basic scheme in Fig. 1(a), here MS receives 2N signals from the NOT chains, and produces as output 2N signals out_i (i=1..2N). At the sampling instant, this block will give on signal out_1, the sampled value of p_{11}; on signal out_2, the sampled value of p_{21}; on signal out_3, the sampled value of p_{12}; on signal out_4, the sampled value of p_{22}; and so on.

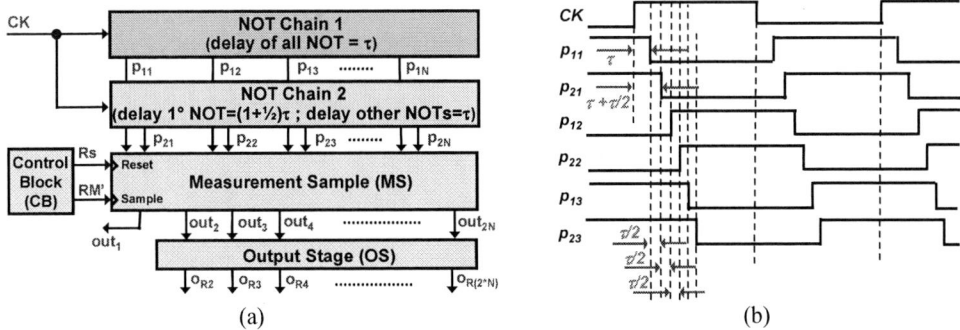

(a)   (b)

Fig. 2. Block structure of our scheme to provide high measurement resolution (a); schematic representation of the signals produced at the outputs of its NOT chains (b).

As in the previous basic scheme, we feed OS with out_i (i=2..2N), so that the outputs o_{Ri} (i=2..2N) are encoded by a thermometer code.

Therefore, by properly adding a NOT chain to the basic scheme in Fig. 1(a), and by properly sizing their NOTs, we can achieve a resolution that is half of that of the basic scheme. This is obtained at the cost of an area and power consumption increase that, however, as proven in Section 5, still compare favorably to that of the alternate approach most recently proposed.

By straightforward modifications, our scheme can achieve even higher resolutions in jitter measurement, with low sensitivity to *PSN*. For instance, we can implement the scheme in Fig. 2(a) with *n* NOT chains in which: 1) the first chain consists of NOTs each with a delay equal to τ; 2) all other *j-th* chains (j=1..n) consist of NOTs, out of which, the first NOT has a delay equal to $(1+i/n)\tau$, while all other NOTs have a delay equal to τ. This way, the outputs of the corresponding NOTs of the *n* chains present a phase difference equal to $(1/n)\tau$.

Of course, the possibility to achieve a resolution as high as possible is limited by the ability to control the NOT delays, despite possible parameter variations occurring during fabrication. To solve this problem, we could implement the NOT chains by using balanced delay lines (like those in [14, 15]), or by using inverters whose delay can be properly calibrated after fabrication to compensate for parameter variations (e.g. like those in [16]).

Finally, Fig. 3 shows schematically the timing of the control signals RM' and Rs of our scheme in Fig. 2(a) with respect to the CK signal.

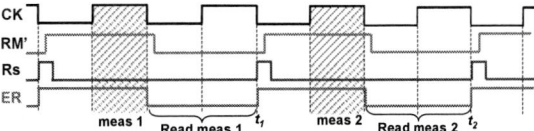

Fig. 3. Schematic representation of the timing of the control signals RM' and Rs.

After the CK falling edge following the CK high phase that was being measured (meas 1 in Fig. 3) signal RM' flips to 0, thus the logic values present at the outputs of the NOTs of the 2 NOT chains are sampled and given on signals out_i (i=1..2N). Signal RM' remains low till reset, which is automatically activated at t_1 (Fig. 3), after a time interval long enough to allow to read the measurement (Read meas 1 in Fig. 3). Afterwards, the circuit is ready to measure the duration of the following CK high phase (meas 2 in Fig. 3).

Our considered scheme measures the CK high phase. Thus, to avoid to sample incorrect measurements, the Rs pulse should be activated when CK=0, and should be deactivated before the following CK rising edge corresponding to the next CK high phase to be measured.

In order to have sufficient time to read out the measurement and to avoid the criticality associated with the generation of the Rs signal (e.g., due to process parameter variations), the Rs pulse could be activated every other CK cycle (Fig. 3). To achieve this, we need an additional signal "Enable Reset" (ER in Fig. 3) with half the CK frequency, which enables to reset the circuit only after its rising edges.

3. Possible Implementation

As an example, let us describe an implementation of our scheme for the measurement of the clock high phase, considering a standard 65 nm CMOS technology, a clock frequency of 3 GHz, and the case of n=2 NOT chains (Fig. 2(a)).

Let us start from the NOT chains. Their possible implementation is shown in Fig. 4(a).

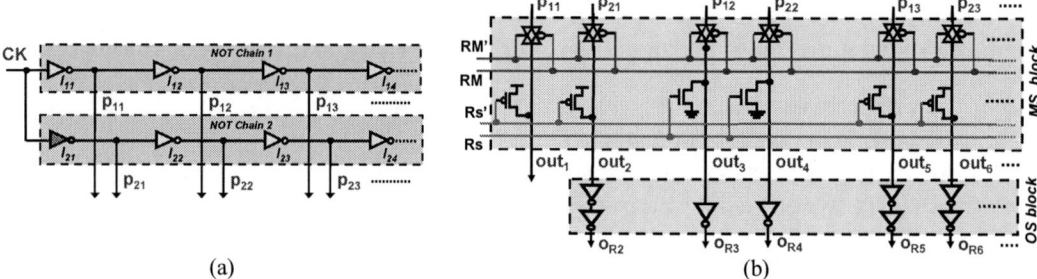

(a)   (b)

Fig. 4. Possible implementation of: (a) the NOT chains; (b) the *MS* and *OS* blocks.

It consists of 2 NOT chains out of which: in chain 1, all NOTs present a delay $\tau \cong 12ps$; in chain 2, the first NOT (I_{21}) has a delay $= (1 + \frac{1}{2})\tau \cong 18ps$, while all other NOTs present a delay $\tau \cong 12ps$. This way, as introduced previously, the outputs of the NOTs at the same level i ($i = 1..N$) within each chain k-th ($k = 1, 2$) present a phase difference of $\frac{1}{2}$ of a min sized NOT delay, thus providing a measurement resolution of $\tau/2 \cong 6ps$.

Since each chain needs to cover a full T_{CK} ($\cong 333ps$), the number N of NOTs within each chain should be $N \geq T_{CK}/\tau = 28$. Therefore, we implemented each chain with $N=29$ NOTs.

To compensate possible variations of electrical parameters occurring during manufacturing, the chain inverters have been implemented by NOTs with a programmable delay (e.g. like those in [16]).

As for MS and OS, their possible implementation is shown in Fig. 4(b).

The inputs of MS (p_{ki} k=1..2; i =1..N) are connected to its outputs (out_i i=1..2N) through transfer gates driven by RM' and by its complement RM. This way, when RM'=1, all transfer gates are conductive, thus connecting the outputs of the NOT chains to signals out_i. Instead, when RM' flips to 0, all transfer gates become OFF, thus the outputs of the NOT chains are sampled and disconnected from signals out_i. Therefore, signals out_i (i=1..2N) remain in a high impedance state, keeping latched the logic values of the outputs of the NOT chains till reset.

When reset is activated, signal RM' flips to 1 again (Fig. 3) and RM to 0. Thus, all transfer gates in Fig. 4(b) become conductive again, and the outputs of the NOT chains p_{ki} are connected again to signals out_i. In addition, while reset is activated (Rs=1 and Rs'=0), all nMOS transistors driven by Rs and all pMOS transistors driven by Rs' (Fig. 4(b)) become conductive. They force all out_i to assume a logic value making all output signals o_{Ri} become equal to 1, thus removing the previous measurement of the CK high phase.

As for OS, while RM'=0, it buffers the out_i (i=2..2N) signals and encodes them by a thermometer code. Instead, when the circuit is reinitialized (Rs=1), it forces all outputs o_{Ri} to assume a logic 1.

As for CB, Figs. 5(a) and 5(b) show a possible implementation of the circuits generating

RM' (and RM) and Rs' (and Rs), respectively, for the circuit in Fig. 2(a).

As described in the previous section, we sample the measurement when $p_{11}=p_{12}=1$. When this is the case, signal RM' should flip to 0, and it should be kept at 0 till the application of the reset pulse (Fig. 3). As shown in Fig. 5(a), this can be easily achieved by exploiting signals out_1 and out_3 generated at the output of MS (Fig. 4(b)), that remain latched at the high logic value till reset, thus keeping RM'=0 till reset.

(a)   (b)

Fig. 5. Possible implementation of circuits generating: (a) RM' and RM; (b) Rs' and Rs.

In fact, before the CK falling edge (CK=1), it is $p_{11}=out_1=0$ and $p_{12}=out_3=1$, thus RM'=1. When the CK falling edge propagates to p_{11} and out_1 as a 1, it becomes $out_1=out_3=1$, thus making RM' flip to 0. This way, all transfers gates in MS (Fig. 4(b)) become OFF, thus keeping $out_1=out_3=1$ till the application of the Rs signal. Then, while Rs=1, node out_3 is forced to 0, thus making RM' flip again to 1. This way, after the application of the Rs pulse, the circuit becomes ready to start a new measurement of the following CK high phase.

It is worth remembering that sampling takes place when the effect of PSN has disappeared (i.e., after the CK falling edge has propagated through the first inverter I_{11} of the chain). Therefore, the sampling circuitry and the sampling events present low susceptibility to PSN.

As for signal Rs (and Rs'), we showed in Section 2 that the Rs pulse should be activated every other CK cycle, and after the CK falling edge before the following measurement (Fig. 3). Since these instants coincide with the rising edges of signal ER (Fig. 3), we can simply activate the Rs pulse after the rising edge of signal ER. As an example, a possible implementation of the circuit generating Rs' and Rs is shown in Fig. 5(b). This is a pulse generation circuit, in which Rs' will assume a logic 0 only for a short time (i.e., equal to the delay of inverter I1) after the rising edge of signal ER.

Finally, the signal ER required to generate the Rs pulse after every other falling edge of CK (as shown in Fig. 3) can be obtained in many different ways. As an example, we consider its possible generation by a standard divide-by-2 circuit of the kind in [17].

4. Verification

We have verified the described behavior of our scheme by means of electrical level simulations performed by HSPICE, considering a standard 65 nm CMOS technology, a Vdd = 1V, and a clock frequency of 3 GHz. In addition, we modeled the *PSN* as a triangular pulse occurring at every CK falling/rising edge and having a peak value of 0.17V (17% Vdd).

We have also performed Monte Carlo simulations, considering statistical variations (with uniform distribution) of electrical parameters (oxide thickness, transistor conductance threshold, and electron/hole mobility) up to 35%.

As an example, we considered the implementation of our scheme for the case of n=2 (Fig. 2(a)), which (as shown previously) provides a measurement resolution of $\tau/2 \cong 6$ps.

Fig. 6(a) shows the simulation results obtained for nominal values of electrical parameters, considering the case of: i) no jitter affecting the first measured CK high phase (CK HP 1); ii) a jitter of 7ps widening the secondly measured CK high phase (CK HP 2). As we can see, when no jitter affects the CK high phase (CK HP 1), while RM'=0 (Read meas 1 in Fig. 6(a)), as expected, our scheme gives to its outputs (o_{Ri}) a word encoded by the thermometer code with 28 zeros (from o_{R2} to o_{R29}). Since with our scheme we measure jitter as difference in the number of 0s between the produced output (=28) and the one expected for the jitter-free case (= $T_{CK}/2 * 1/Res = 168 / 6 = 28$) multiplied by the resolution of our scheme (equal to 6ps), we obtain a measurement of jitter = 0ps as it is actually the case. Instead, when for instance a jitter of 7ps affects the CK high phase (CK HP 2), while RM'=0 (Read meas 2 in Fig. 6(a)) our scheme gives to its outputs (o_{Ri}) a word encoded by the thermometer code with 29 zeros (from o_{R2} to o_{R30}). Therefore we obtain a measurement of jitter = 6ps. The resolution of our scheme is 6ps, thus jitter is measured with a maximum error of 6ps. In the example above, the actual jitter of 7ps is measured as a jitter of 6ps, and thus with an error of 7ps - 6ps = 1ps.

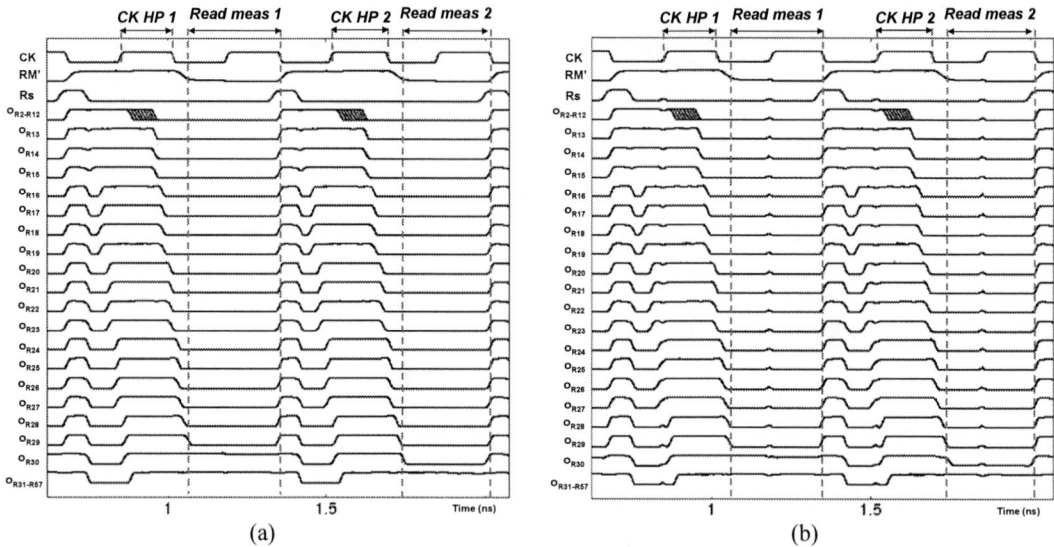

(a)   (b)

Fig. 6. Simulation results for nominal values of electrical parameters, considering (a) no power supply noise and (b) a power supply noise of 17% of Vdd.

Fig. 6(b) shows the simulation results obtained under the same conditions as for Fig. 6(a), but in the presence of *PSN* of 17% Vdd. As can be seen, the obtained results are identical to those obtained in the case of absence of *PSN*. Therefore, the presence of *PSN* does not affect the measurements produced by our scheme.

Assuming inverters properly calibrated after fabrication, we verified that our scheme behaves as described, despite statistical variations of electrical parameters up to the 35%.

Finally, regarding the susceptibility of the NOT chains to temperature variations, we expect similar results to those presented in [18]. In particular, we expect a measurement error of approx. 5% for a temperature range between -40°C and 90°C [18].

5. Costs and comparison

We have evaluated the costs in terms of area and power consumption of our proposed jitter measurement scheme and compared them to those of the approach recently proposed in [4].

Since the scheme in [4] is intended to measure both the high and low phases of CK, for comparison purpose we also consider the version of our scheme allowing to measure both CK phases. Thus, our scheme requires the addition of another MS, OS and CB block. Additionally, since the approach in [4] allows a measurement resolution equal to a min sized NOT delay, for the purpose of comparison we also consider this as the target resolution of our scheme. Thus, our scheme requires to use only one NOT chain (n=1).

As for area cost, it has been roughly estimated in squares, while as for power consumption, electrical level simulations have been performed. Both schemes have been implemented considering the same standard 65nm CMOS technology and a CK frequency of 3GHz.

The obtained results are reported in Tab. 1, showing the reductions in terms of area and power consumption allowed by our scheme.

Additionally, our scheme features the advantages of: 1) being scalable in the achievable measurement resolution (over one inverter input-output delay); 2) featuring low sensitivity to *PSN*. However, our scheme is conceived to measure jitter at every other CK cycle, while the solution in [4] allows jitter measurements at every CK cycle. As for the sensitivity to parameter variations, a similar behavior can be reasonably expected for the two approaches.

Tab. 1. Area and power costs of the compared schemes and our allowed reductions.

	Area (squares)	$\Delta A\ (\%)=$ $(A_{[4]}\text{-}A_{OUR})\,/\,A_{[4]}$	Power Comp. (μW)	$\Delta P\ (\%)=$ $(P_{[4]}\text{-}P_{OUR})\,/\,P_{[4]}$
Our scheme	1952	41%	634	40%
Scheme in [4]	3332	-	1048	-

6. Conclusions

We proposed a novel clock jitter digital measurement scheme useful for post-silicon in-situ monitoring of the jitter of clock signals of a clock distribution network during the test or debug phase. Our scheme provides high and scalable measurement resolution, despite the possible presence of power supply noise affecting itself. Resolution is higher than a min sized inverter input-output delay, and can be considerably increased at some additional costs in terms of area and power consumption during the measurement phase. We showed that a resolution of 6ps for a 3GHz clock can be achieved with a reduction in area and power of the 41% and 40%, respectively, compared to the alternate, most recently proposed approach.

7. References

[1] S. Tam, S. Rusu, U. Desai, R. Kim, J. Zhang, and I. Young, "Clock Generation and Distribution for the First IA-64 Microprocessor," *IEEE J. of Solid State Circuit*, Vol. 35, No. 11, November 2000, pp. 1545- 1552.

[2] C. Metra, D. Rossi, T. M. Mak, "Won't On-Chip Clock Calibration Guarantee Performance Boost and Product Quality?" *IEEE Trans. on Computers*, Vol. 56, Issue 3, March 2007, pp. 415-428.

[3] J.M. Cazeaux, M. Omaña, C. Metra, "Novel On-Chip Circuit for Jitter Testing in High-Speed PLLs", IEEE Trans. on Inst. and Measurement, Vol.54, No. 5, October 2005, pp.1779-1788.

[4] R. Franch, P. Restle, N. James, W. Huott, J. Friedrich, R. Dixon, S. Weitzel, K. Van Goor, G. Salem, "On-Chip timing Uncertainty Measurements on IBM Microprocessors", Int. Test Conference, 21-26 Oct. 2007, pp. 1-7.

[5] R. Kuppuswamy, K. Wong, D. Ratchen, G. Taylor, "On-die Clock Jitter Detector for High Speed Microprocessors", in IEEE Symp. on VLSI Circuits Digest of Tech. Papers, June 2001, pp. 187-190.

[6] C. Metra, L. Schiano, M. Favalli, "Concurrent Detection of Power Supply Noise", IEEE Trans. on Reliability, Vol. 52, Issue 4, Dec. 2003, pp. 469-475.

[7] Application Note 1448-2, "Finding Sources of Jitter with Real-Time Jitter Analysis", Agilent Technologies.

[8] H.C. Lin, A. Chong, E. Chan, M. Soma, H. Haggag, J. Huard, J. Braatz, "CMOS Built-In Test Architecture for High-Speed Jitter Measurement", in Proc. of IEEE Int. Test Conf, 2003, pp. 646-652.

[9] P. Dudek, S. Szczepanski, and J.V. Hatfield, "A High Resolution CMOS Time-to-Digital Converter Utilizing a Vernier Delay Line", IEEE Trans. on Solid-State Circuits, vol.35, pp. 240-247, Feb., 2000.

[10] S. Sunter, A. Roy, "On-Chip Digital Jitter Measurement, from Megahertz to Gigahertz", IEEE Design and Test of Computers, Vol. 21, Issue 4, pp. 314-321, July-August 2004.

[11] R. Z. Bhatti, M. Denneau, J. Draper, "Duty Cycle Measurement and Correction Using a Random Sampling Technique", in Proc. of Midwest Symp. On Circuits and Systems, pp. 1043-1046, Vol. 2, 2005.

[12] E. Alon, V. Stojanovic, M. Horowitz, "Circuits and Techniques for High-Resolution Measument of On-Chip Power Supply Noise", IEEE J. Solid-State Circuits, Vol. 40, No 4, pp. 820-828, April 2005.

[13] P. Heydari, M. Pedram, "Analysis of Jitter due to Power-Supply Noise in Phase-Locked Loops", in Proc. of IEEE Conf, Custom Interated Circuits, pp. 443 – 446, 2000.

[14] J. F. Genat, "High Resolution Time-to-Digital Converter", *Nuclear Instruments and Methods in Physics Research*, A315

[15] R. Datta, G. Carpenter, K. Nowka, J. A. Abraham, "A Scheme for On-Chip Timing Characterization", in Proc. of IEEE VLSI Test Symp. Pp 24-29, May 2006.

[16] M. Maymandi-Nejad, M. Sachdev, "A Digitally Programmable Delay Element: Design and Analysis", IEEE Trans. on Very Large Scale Integration (VLSI) Systems, Vol. 11, No. 5, October 2003.

[17] R. Chen, "High-speed CMOS frequency divider", Elect. Letters, vol. 33, no. 22, pp. 1864–1865, Oct. 1997.

[18] Z. Abuhamdeh, et Al., "Separating Temperature Effects from Ring-Oscillator Readings to Measure True IR-Drop on a Chip", Int. Test Conference, 21-26 Oct. 2007.

Prioritization of Paths for Diagnosis

Rajsekhar Adapa, Spyros Tragoudas
Department of Electrical and Computer Engineering
Southern Illinois University Carbondale, IL 62901
{adapa, spyros}@engr.siu.edu

Abstract

Existing techniques for path delay fault (PDF) diagnosis do not provide high diagnostic resolution. These techniques fail to prune a large number of fault free candidates from the set of possible candidate faults. This paper presents an efficient technique to prioritize faults (PDFs) among the set of possible candidate faults. A novel approach for PDF prioritization is presented and its effectiveness is demonstrated on the ISCAS benchmarks.

1. Introduction

Timing related defects in a chip can be effectively modeled using appropriate delay fault models such as the transition delay fault model and the path delay fault model (PDF). When compared to the transition delay fault model, the path delay fault model is more suitable to detect distributed delay defects caused by process variations and noise related aspects in deep sub-micron devices. These timing defects also known as delay defects are generally checked by applying a set of test patterns and comparing the observed response with the expected response. If the delay along one or more paths starting from the input to the output of the circuit exceeds the timing specification of the circuit, the observed response mismatches with the expected response. The paths which cause the delay to exceed the timing specification are said to have delay defects. The process of identifying such paths whose delay exceeds the timing specification is known as *delay fault diagnosis* [11, 12].

Fault diagnosis techniques are generally classified as cause-effect and effect-cause techniques [1, 3]. Cause-effect techniques determine the set of possible candidate faults using fault dictionaries [1, 2]. Cause-effect techniques have high space and time complexity and are not feasible to path delay diagnosis due to the exponential number of paths in the designs. On the other hand, effect-cause diagnosis techniques apply a set of patterns to the circuit under test and analyze the failing responses to determine the possible fault candidates that have caused the circuit to fail. The possible fault candidates are placed in a *suspect set*. Each fault in the suspect set is simulated with respect to all failing patterns, and the fault which explains the maximum number of the failing patterns is reported as the most likely fault candidate [3, 4].

Several effect-cause diagnosis techniques are proposed for delay fault models [6, 7, 10, 11, 12]. The techniques in [6, 7, 10] target the transition fault model, where diagnosis is done under the single fault assumption. These techniques report the most likely fault candidate but pruning depends on actual delay values. [11, 12] target the path delay fault model. The techniques in [11, 12] are more robust because they prune both single and multiple path delay faults in a delay independent manner. Even after applying the techniques of [11, 12] the suspect set still has a prohibitively large number of PDFs. Pruning the fault free candidates from the suspect set using fault prioritization is the only way to lead to diagnosis within an acceptable time limit. It is worthy to mention that even for simple fault models such as stuck-at, transition, bridge, open interconnect, this fault prioritization has already been proposed [13, 15, 18].

In traditional effect-cause techniques, the fault in the suspect set that is detected (sensitized) by the maximum number of failing patterns is typically considered as the candidate fault, i.e., it receives the highest rank (priority) when compared to other faults in the suspect set. Faults with high priority are subjected to failure analysis, where e-beam or laser probe techniques are applied to identify the exact defect location. Lower ranked paths will be examined only if failure analysis tools fail to identify the defect location.

A traditional brute-force approach to identify the PDFs which are detected by the maximum number of failing patterns is done by simulating the PDFs in a path enumerative manner, where each PDF is taken at a time and simulated with respect to all failing patterns. However, this process is involved with extremely high space and time complexity due to the exponential number of PDF's in the circuit under test. Hence existing PDF effect-cause diagnosis techniques cannot point to a single suspect fault and the diagnosis is inaccurate [11, 12]. In particular, the huge number of faults does not allow conventional path delay simulators [14, 20] to simulate a single PDF for the set of all failing patterns. Hence cannot report the number of patterns that detect each PDF. Due to the large number of PDFs in a circuit, conventional path delay fault simulators estimate the fault coverage in a non-enumerative manner. These simulators report the total number of PDFs detected by the given test set, however, they do not keep track of the number of patterns that detect each PDF. Therefore these simulators do not have the ability to rank the PDFs based on the number of patterns that detect each PDF.

The recent work proposed in [16] determines the exact PDF coverage for a given test set in a non-enumerative way. [16] uses zero suppressed binary decision diagram (ZDD) [9], a canonical data structure which stores the PDFs in an implicit manner. It is shown in [16] that ZDDs can store extremely large number of PDFs, however, they do not give any information about the patterns that detect each PDF.

This paper presents a novel technique to prioritize the PDFs in the suspect set in a fault implicit manner and solves all the above mentioned problems. The proposed method uses the circuit structure to store the pattern information, it is shown that the proposed technique overcomes the space limitations, and we also present a method to implicitly prioritize the PDFs in the suspect set. This remaining section of the paper is organized as follows. Section 2 gives necessary preliminary concepts and existing work in PDF diagnosis. Section 3 presents the proposed method. Experimental results are presented in Section 4, and Section 5 concludes.

2. Preliminaries

This section gives a brief overview of the existing path delay diagnosis techniques. Due to the exponential number of PDFs in the design, PDFs are stored implicitly and fault grading is done in a non-enumerative manner. Under these conditions, it is not feasible to report the number of failing patterns that detect each PDF. Hence traditional techniques report the set of PDFs that are detected by at least α patterns.

In effect-cause diagnosis of PDFs, a set of input patterns are applied to the circuit under test, the set of input patterns whose observed response matches the expected response is termed as the passing test set, and the set of input patterns whose observed response is different from the expected response is termed as the failing test set. The set of all PDFs sensitized by the failing test set which may likely explain the failing behavior of the circuit is termed as the *suspect set*. The set of all PDFs sensitized by the passing set and is guaranteed to be fault free is termed as a fault free set.

Some of the PDFs are sensitized by both failing patterns and passing patterns, and are present in both fault free set and the suspect set. If a PDF is sensitized robustly by a passing test, such PDFs are guaranteed to be fault free. On the other hand, some PDFs sensitized by a passing test pattern may not be fault free, if they are sensitized non-robustly, in which case the transition may be masked during the test application process. A subset of fault free candidates are pruned from the suspect set using various path sensitization rules [11, 12]. After applying the techniques of [11, 12] the suspect set has a large number of PDFs. This paper presents a novel technique to prioritize the PDFs remaining in the suspect set for diagnosis, based on the number patterns that detect each PDF.

As the proposed method stores the pattern information and PDF information in an implicit manner, we compare its effectiveness with [16], which also stores the PDF information in an implicit manner. However, [16] does not give any information on the number of patterns that detect each PDF. It is shown in [16] that all the PDFs detected by a given test set are implicitly stored in a ZDD. A straight forward approach to store the pattern information is to append one node for each pattern that detects the PDF. Figure 1(a) shows 4 PDFs and the corresponding patterns which detect each PDF. Figure 1(b) shows the corresponding ZDD where one node per pattern is associated with the PDF. For example, PDF F_1 is detected by 3 patterns P_1, P_2, P_3, this is shown by the path with nodes $P_{1a}, P_{2b}, P_{3b}, a_3, b_1, d_1, e_1$ in Figure 1(b), where nodes P_{1a}, P_{2b}, P_{3b} correspond to patterns P_1, P_2, P_3. This approach prioritizes PDFs and is used in the studied problem formulation. Unfortunately, it has high space and time complexity.

The work in [5] presents a non-enumerative ATPG technique by processing partial paths (fanout free segments) in the circuit. The proposed method utilizes the framework presented in [5] to identify the set of PDFs that have high probability of explaining the failure.

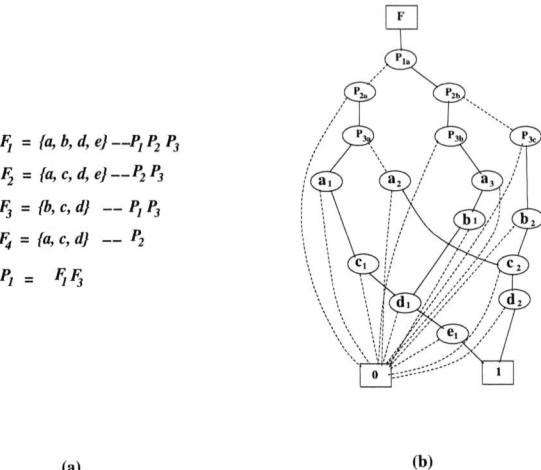

$F_1 = \{a, b, d, e\} \text{ --}P_1 P_2 P_3$

$F_2 = \{a, c, d, e\} \text{ --}P_2 P_3$

$F_3 = \{b, c, d\} \text{ -- } P_1 P_3$

$F_4 = \{a, c, d\} \text{ -- } P_2$

$P_1 = F_1 F_3$

(a)   (b)

Figure 1. PDFs in Traditional ZDD with Pattern Information

Definition 1 (Fanout-Free Segment [8]): *A partial path or a fanout-free segment is a subpath in the circuit between a primary input and a fanout stem or between two fanout stems or between a fanout stem and a primary output, with no other fanout stems in between the starting node and terminal node of the segment.*

As a path P is divided into fanout-free segments, if a pattern p_i sensitizes a path P, all the fanout-free segments are sensitized by the pattern p_i. If a segment is sensitized by a pattern, we refer to such segments as *pattern-segments*. The number of pattern-segments are polynomial in number because the pattern-segments for fanout-free segments are invariant for a given primary input transition and a given pattern-segment input transition [5]. The proposed approach uses the pattern-segments to quickly identify the number of patterns that detect a PDF. The proposed method is discussed in Section 3.

3. Segment Based Method

In this section we present the proposed method to identify a set of paths that have high probability of explaining the failure in the circuit. The proposed technique is superior in terms of time and space requirements when compared to the traditional ZDD method. The input to the proposed method is the failing test set, and the coverage factor α. The coverage factor α is a parameter which identifies a set of PDFs that are detected by at least by α patterns. The output is a set of PDFs that are detected by at least α patterns. From here after, the terms sensitized and detected are used interchangeably, also the terms paths, PDFs and faults are used interchangeably.

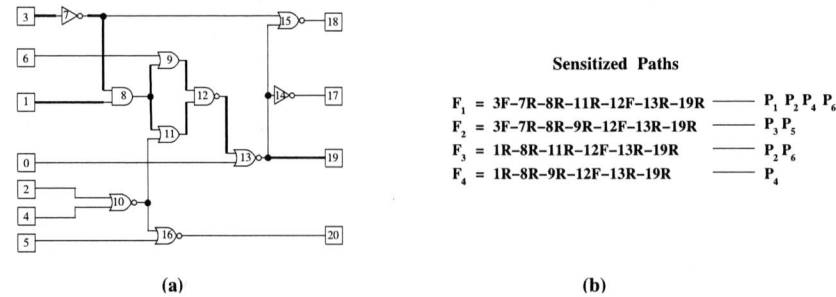

Sensitized Paths

$F_1 = 3F\text{--}7R\text{--}8R\text{--}11R\text{--}12F\text{--}13R\text{--}19R \quad\text{------}\quad P_1\ P_2\ P_4\ P_6$

$F_2 = 3F\text{--}7R\text{--}8R\text{--}9R\text{--}12F\text{--}13R\text{--}19R \quad\text{------}\quad P_3\ P_5$

$F_3 = 1R\text{--}8R\text{--}11R\text{--}12F\text{--}13R\text{--}19R \quad\text{------}\quad P_2\ P_6$

$F_4 = 1R\text{--}8R\text{--}9R\text{--}12F\text{--}13R\text{--}19R \quad\text{------}\quad P_4$

(a)   (b)

Figure 2. Circuit and Sensitized Paths

The proposed method uses the fanout-free segments in the circuit [5, 8] to store the pattern and PDF information. For each input pattern, we identify the fanout-free segments on the sensitized path. These segments are associated with the input

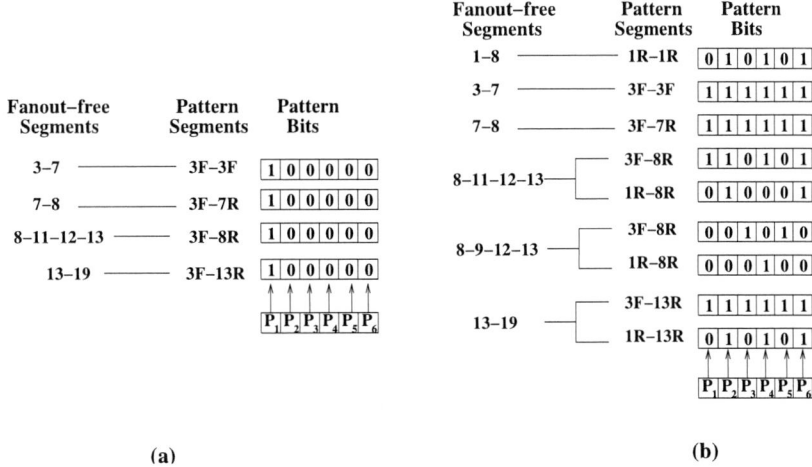

Figure 3. Pattern Segments

patterns and such segments are called pattern segments. The pattern indexes that sensitize each pattern segment are stored in a compact way. The input to the proposed method is a failing test set and the coverage factor α, and the output is a set of paths (PDFs) that are detected by at least α patterns. The following explains the proposed method.

Consider the combinational circuit shown in Figure 2(a). Let P_1, P_2, P_3, P_4, P_5 and P_6 be the failing patterns. The PDFs (paths) that are detected by the failing patterns are given in Figure 2(b). It can be seen that PDF F_1 is sensitized by four patterns P_1, P_2, P_4 and P_6, PDF's F_2, F_3 are sensitized by 2 patterns each, and PDF F_4 is sensitized by 1 pattern. Each PDF is represented in terms of the node indexes of the circuit, and each node index is tagged with the type of the transition along the path, R denoting the rising transition and F denoting the falling transition. For example, the PDF F_1 has a falling transition at the primary input (node 3) and a rising transition at the primary output (node 19).

The process of identifying the set of PDFs that are detected by at least α patterns involves several steps. The initial step invokes a grading tool where each failing pattern P_i is applied to the circuit under test(CUT) and the set of PDFs that are sensitized by P_i are are stored implicitly in a ZDD 'ζ'. It should be noted that the ZDD 'ζ' contains the PDF's sensitized by only one pattern, the ZDD 'ζ' is only used as temporary storage class for the PDFs sensitized, once the PDF's are assigned to the fanout-free segments the ZDD 'ζ' can be deleted. Instead of the ZDD 'ζ' one can use any of the data structures presented in [14, 15] to temporarily store the PDFs sensitized by a pattern P_i. For simplicity we have used the ZDDs.

Each PDF sensitized by P_i is split into fanout-free segments. For example, assume pattern P_1 is applied to the CUT, from Figure 2(b), it can be seen that pattern P_1 detects the PDF F_1. The PDF F_1 (F_1 = 3F-7R-8R-11R-12F-13R-19R) is next split into fanout-free segments. The fanout-free segments along the PDF F_1 are (3-7), (7-8), (8-11-12-13) and (13-19). More details on fanout-free segments are given in [5]. From here after we interchangeably use the terms fanout-free segments/segments.

Each fanout-free segment is further split into pattern-segments based on the transitions at the primary input and at the start of the segment. If t_{pi} and t_s denote the transitions at the primary input and at the start of the segment S, there are only four possible pattern segments since t_{pi} and t_s can either be a rising R or a falling F transition [5]. For instance, consider the segment (7-8) and primary input 3, the possible pattern-segments are (3R-7F), (3F-7R). The pattern-segments (3F-7F) and (3R-7R) are not possible due to the structure of the circuit (node 7 is the inverting output of node 3). Based on the primary input transition and the segment input transition, the PDF F_1 which is sensitized by the pattern P_1 is split into 4 pattern-segments denoted by (3F-3F), (3F-7R), (3F-8R) and (3F-13R).

If a PDF F is sensitized by a pattern P_i, all the pattern-segments associated with F are also sensitized by P_i. This information is stored for each pattern-segment and later used to identify PDFs which are detected by at least α patterns. Each pattern-segment PS is associated with a pattern-bit array to store the information regarding the patterns which sensitize the pattern-segment PS. Each bit in the pattern-bit array correspond to the pattern P_i and can have a value of 0 or 1. A value of 1 at location i in the pattern-bit array indicates that the pattern segment PS is detected by pattern P_i and a 0 indicates PS is not detected by P_i.

As PDF F_1 is sensitized by pattern P_1, its corresponding pattern-segments (3F-3F), (3F-7R), (3F-8R), (3F-13R) are also sensitized by P_1. This information is stored in the pattern-bit array of all the pattern-segments. The values associated with the pattern-bit arrays of all the pattern-segments after applying pattern P_1 is shown in Figure 3(a). The value 1 in the pattern-bit array corresponding to bit position P_1 of the pattern-segments (3F-3F), (3F-7R), (3F-8R), (3F-13R) indicate that these pattern-segments are sensitized by pattern P_1. The pattern-bit array values of all the pattern-segments corresponding to PDFs F_1, F_2, F_3, F_4 after applying all the patterns $P_1, P_2, P_3, P_4, P_5, P_6$ are given in Figure 3(b).

After applying all the failing patterns the next step identifies all the PDFs which are detected by at least α patterns. This is done by traversing the pattern-segments which are sensitized by at least α patterns. If a pattern segment PS has n bits set to 1 in the pattern bit array, this implies that the pattern-segment is sensitized by n patterns. If a pattern-segment PS is sensitized by n patterns, any PDF along the PS cannot be sensitized by more than n patterns. If n is less than α, all the PDF's along the pattern-segment PS are ignored. The set of PDFs that are sensitized by at least α patterns are obtained by traversing along the pattern-segments which are sensitized by at least α patterns.

The set of PDFs that are sensitized by at least α patterns are obtained by a set of depth first search DFS traversals done starting from the primary input to the primary output along the pattern-segments which are detected by more than α patterns. The PDFs are constructed as the pattern-segments are visited. This is explained as follows. Let α be 4, in the initial step, the method identifies all the pattern-segments which are detected by at least 4 patterns. From Figure 3(b), pattern-segments (3F-3F), (3F-7R), (3F-8R), (3F-13R) corresponding to fanout-free segments (3-7), (7-8), (8-11-12-13), (13, 19) are sensitized by at least 4 patterns. Then a DFS traversal [21] is done starting at the input pattern-segment and the DFS visits pattern-segments that are sensitized by at least α patterns (i.e., $\alpha = 4$). In this example, the DFS procedure first visits the pattern-segment (3F-3F), the only possible successor of (3F-3F) that is detected by 4 patterns is (3F-7R), the search is followed by visiting the pattern-segments (3F-8R) and (3F-13R). It can be observed that all the pattern-segments ((3F-3F), (3F-7R), (3F-8R), (3F-13R)) visited by the DFS traversal constitute a single PDF F_1. Hence the PDF F_1 is formed by traversing the pattern-segments. These PDFs can be stored in any of the data structures given in [14, 16, 20]. The PDF's are given high priority for diagnosis when compared to the PDF's that are sensitized by fewer failing patterns.

Algorithm 1 Segment-Based Method $(P,\ \alpha)$

1: P - Set of Failing Patterns
2: α - Coverage Factor
3: **for each** P_i in P **do**
4:      Apply pattern P_i and store the PDFs sensitized by P_i in a temporary ZDD 'ζ'
5:      **for each** PDF F_j in 'ζ' **do**
6:          Split PDF F_j into Fanout-free segments
7:          Based on the primary input transition and transition at the start of segment, identify the appropriate Pattern-Segment PS
8:          Update the Pattern Bit Array of Pattern-Segment PS with pattern P_i (set bit value to 1)
9:      **end for**
10:     Delete ZDD 'ζ'
11: **end for**
12: **for each** Pattern Segment PS_j in the Circuit **do**
13:     **if** PS_j is sensitized by less than α patterns **then**
14:         Mark PS_j is invalid
15:     **end if**
16: **end for**
17: **for each** Valid Pattern Segment PS_j in the Circuit **do**
18:     Do a DFS traversal starting from the Primary Input along the valid Pattern segments
19:     Form the PDFs during the DFS traversal and store them in a ZDD F_{high}
20: **end for**
21: F_{high} - Set of PDFs that have high probability of explaining the failing behavior

---

The Segment-based method is outlined in Algorithm 1. In Lines 3-11, each failing pattern P_i is applied and each PDF sensitized by pattern P_i is split into fanout-free segments and further into pattern-segments. The pattern-segments are later updated with the pattern index in the pattern bit array. Lines 12-16 check whether a pattern-segment is sensitized by at least α patterns, if not, it is marked as invalid. In lines 17-20, the DFS traversal is done along the valid pattern-segments. The PDFs are formed during the DFS traversal. These PDFs can be stored in any of the data structures given in [14, 16, 20], for simplicity we used the ZDD. The ZDD F_{high} has set of PDFs sensitized by at least α patterns.

4. Experimental Results

Benchmark	Suspect Set	PDFs $\alpha = 20$	Trad ZDD-Based		Segment-Based		
			Nodes	Mem (KB)	Mem (KB)	%Red	CPU (sec)
c880	664	20	9567	149.4	8.2	94.5	316
c1355	1171	48	13506	211.0	11.8	94.4	475
c1908	865	39	10702	167.2	12.3	92.6	321
c2670	870	29	11469	179.2	37.49	79.0	550
c3540	406	32	5880	91.8	9.2	89.9	508
c5315	956	63	14033	219.2	45.86	79.0	524
c7552	1526	76	17138	267.7	35.4	86.7	679

Table 1. Path Prioritization using Segment Based Method

The proposed technique is implemented in C language and evaluated on the ISCAS benchmark circuits using a SunWBlade-1000 Workstation with 4GB RAM. Two sets of experiments are conducted to demonstrate the effectiveness of the approach. The first one considers around 500 random patterns taken from [17] as the failing patterns. These failing patterns form our failing test set. The PDFs (faults) that are sensitized by the failing patterns is the initial suspect set. The segment-based method is evaluated against the traditional method (where PDFs along with pattern information is stored in a ZDD, Figure 1(b)) and the results of the experiments are tabulated in Table 1.

Column 2 shows the initial suspect set. These are the PDFs that are sensitized by the 500 failing patterns. A coverage factor of α equal to 20 is chosen for experimentation, i.e., the proposed techniques will identify a set of PDFs that are detected by at least 20 patterns. Columns 3 shows the number of PDFs that are detected (sensitized) by at least 20 patterns by both the proposed methods. The experiments are conducted using the traditional (brute-force) approach (Figure 1(b)), and the segment-based method. In the first experiment, the set of PDFs sensitized by at least α patterns are returned using the traditional ZDD-based method, where one variable (node) per pattern is used (Figure 1(b)). The number of nodes in the ZDD having all the PDFs in the suspect set along with the pattern information is given in column 4 and its memory requirements are given in column 5.

The next set of experiments are conducted using the segment-based method. The memory usage for the segment-based method is given in column 6. It should be noted that the memory usage corresponds to the memory associated with pattern-segments and the memory required to store the PDFs which are sensitized by at least 20 patterns. In our experiments the PDF's which are sensitized by at least 20 patterns are stored in a ZDD, other data structures proposed in [14] can also be used. In order to be memory efficient, all the pattern-bit information and path information is stored in computer words. For example, an integer word (int in C language) on a 32-bit machine is represented by 32 bits. This integer word can store information of 32 patterns. Column 7 gives the percentage reduction of the memory requirements of the proposed method when compared to the traditional method. It can be seen that on an average there is a 88% savings in the memory used by the segment-based method when compared with the traditional method. The CPU time in seconds for identifying the set of PDFs that are sensitized by at least 20 patterns is given in column 8.

The PDFs which are sensitized by at least α patterns are then fed to the failure analysis tools to identify the exact defect location. If such tools fail to identify the exact defect location, the value of α can be relaxed to a smaller value (for example $\alpha = 15$) and a new set of PDFs satisfying the requirements of the coverage factor α are returned. In this way, a guided search for identifying the faulty location is guaranteed with the proposed methods.

Additional experiments are conducted using fictitious test sets to show the scalability of the approach. A test set of 10k fictitious patterns and around 50k paths ending at various outputs of the circuit are considered for experimentation. It should be noted that these are fictitious patterns which are generated randomly and no fault simulation is performed to determine their testability characteristics. Table 2 shows the memory sizes of the traditional method and the proposed method for ISCAS 89 benchmarks. The results corresponds to PDFs which are detected by at least 500 failing patterns. Column 2 corresponds to the number of nodes in the ZDD for the traditional approach, column 3 shows the memory size of the ZDD in MB.

Column 4 shows the number of pattern segments and the segment memory is given in column 5. Column 6 gives the percentage reduction of the memory requirements of the proposed method when compared to the traditional method. It can

Ckt	Trad Nodes	Trad MB	Pattern Segs	Seg MB	%Red
c5315	7113346	108.5	772	0.92	99.1
c7552	7699212	117.4	614	0.77	99.3
s13207	7463792	113.8	2544	2.13	98.1
s15850	8466579	129.9	614	3.00	97.6
s35932	7033982	107.3	8478	9.69	90.9
s38417	7086425	108.1	4064	5.00	95.3
s38584	7060038	107.7	8144	9.46	91.1

Table 2. Memory Size for Fictitious Sets

be seen that on an average there is a 96% savings in the memory used by the segment-based method when compared with the traditional method. This shows that the proposed method is scalable to path intensive designs. By incorporating the proposed method in the current diagnosis tools we can reduce the number of faults in the suspect set and can accurately identify the defect location within acceptable time limit.

5. Conclusions

This paper presents a novel method to identify a set of PDFs that have high probability of explaining the failure in the circuit. The proposed method identifies the set of PDFs in a non-enumerative manner based on the number of failing patterns that sensitize each PDF. It is shown that the proposed method is both space and time efficient when compared to traditional method. The proposed technique can be incorporated in to existing diagnostic tools to reduce the size of suspect set and speed up the diagnosis process.

References

[1] M. Abramovici, M. A. Breuer and A. D. Friedman, *Digital Systems Testing and Testable Design*, Piscataway, New Jersey: IEEE Press, 1990.

[2] I. Pomeranz, S. M. Reddy. "On Dictionary-Based Fault Location in Digital Logic Circuits", *IEEE Transactions on Computers*, Vol 46, No 1, January 1997, pp. 48-59.

[3] J. Waicukauski, E. Lindbloom, "Failure Diagnosis of Structured VLSI", *IEEE Design & Test of computers, Aug 1989, pp49-60.*

[4] Z. Wang, K. H. Tsai, M. M. Sadowska, J. Rajski, "An Efficient and Effective Methodology on the Multiple Fault Diagnosis", *Proc International Test Conference*, Oct 2003, pp. 329-338.

[5] S. Padmanaban and S. Tragoudas, "Efficient Identification of (Critical) Testable Path Delay Faults Using Decision Diagrams" , *IEEE Trans. on Computer Aided Design* vol. 24, No. 1, January 2005, pp. 77-87.

[6] K. Yang, K. T. Cheng, "Timing-reasoning-based delay fault diagnosis". *Proc. of DATE,* 2006, pp 418-423.

[7] E. Z. Wang, M. Marek-Sadowska, K. Tsai, J. Rajski,"Delay-fault diagnosis using timing information", *IEEE Trans. on CAD of Integrated Circuits and Systems*, vol.24, no. 9, September 2005, pp. 1315-1325.

[8] Y. Shao, S. M. Reddy, S. Kajihara, and I. Pomeranz, "An efficient method to identify untestable path delay faults" *Proc. Asian Test Symp.*, 2001, pp. 233-238.

[9] S-I. Minato, "Zero-Suppressed BDDs for Set Manipulation in Combinatorial Problems", *Proc. of Design Automation Conference*, 1993, pp. 272-277.

[10] P. Girard, C.Landrault, S. Pravossoudovitch., "Delay-Fault Diagnosis by Critical-Path Tracing", *IEEE Design and Test of Computers*, vol. 9, no. 4, October/December 1992, pp. 27-32.

[11] P. Pant, Y. C. Hsu, S. K. Gupta and A. Chatterjee, "Path Delay Fault Diagnosis in Combinational Circuits With Implicit Fault Enumeration", *IEEE Trans. Computer Aided Design*, vol.20, no. 10, Oct. 2001, pp. 1226-1235.

[12] S. Padmanaban and S. Tragoudas, "An Implicit Path Delay Fault Diagnosis Methodology ", *IEEE Trans. on CAD of Integrated Circuits and Systems*, vol.22, no. 10, Nov. 2003, pp. 1399-1408.

[13] R. Adapa, S. Tragoudas, M. Michael. "Accelerating Diagnosis via Dominance Relations between Sets of Faults", *Proc VLSI Test Symposium,* 2007. pp 219-224.

[14] M. A. Gharaybeh, M. L. Bushnell and V. D. Agrawal," Path-Status Graph with Application to Delay Fault Simulation", *IEEE Trans on CAD of Integrated Circuits and Systems* vol. 17, no. 4, April 1998, pp. 324-332.

[15] I. Pomeranz, S. M. Reddy, "An Efficient Non-Enumerative Method to Estimate the Path Delay Fault Coverage in Combinational Circuits". *IEEE Trans. on CAD of Integrated Circuits and Systems* vol. 13, no. 2, Feb 1994, pp.240-250.

[16] S. Padmanaban, M. K. Michael, S. Tragoudas, "Exact path delay fault coverage with fundamental ZBDD operations", *IEEE Trans. on CAD of Integrated Circuits and Systems*, vol. 22, no. 3, March 2003, pp. 305-316.

[17] M. K. Michael and S. Tragoudas., "Function-based compact test pattern generation for path delay faults", *IEEE Trans. on VLSI Systems*, vol. 13, No. 8, August 2005, pp. 996-1001.

[18] B. Seshadri, X. Yu, S. Venkataraman, "Accelerating Diagnostic Fault Simulation Using Z-diagnosis and Concurrent Equivalence Identification", *Proc VLSI Test Symposium*, May 2006, pp. 380-385.

[19] R. Adapa, S. Tragoudas, M. K. Michael, "Accelerating Diagnosis via Dominance Relations between Sets of Faults", *Proc VLSI Test Symposium*, May 2007, pp. 219-224.

[20] I. Pomeranz, S. M. Reddy, "An Efficient Non-Enumerative Method to Estimate the Path Delay Fault Coverage in Combinational Circuits". *IEEE Trans. on CAD of Integrated Circuits and Systems* vol. 13, no. 2, Feb 1994, pp. 240-250.

[21] T. H. Cormen, C. E. Leiserson, R. L. Rivest, C. Stein, "Introduction to Algorithms". *The MIT Press* 2nd edition 2001.

Delay Fault Testability on Two-Rail Logic Circuits

Kazuteru NAMBA   and   Hideo ITO

Graduate School of Advanced Integration Science
Chiba University
1-33 Yayoi-cho, Inage-ku, Chiba 263-8522, JAPAN
{namba, h.ito}@faculty.chiba-u.jp

Abstract

The importance of redundant technologies for improving dependability and delay fault testability are growing. This paper presents properties of a class of redundant technologies, namely two-rail logic, and discusses testability of path delay faults occurring on two-rail logic circuits. The paper reveals the following characteristics of two-rail logic circuits: While the number of paths in two-rail logic circuits is just twice that in ordinary single-rail logic circuits, the number of robust testable path delay faults in two-rail logic circuits is twice or more that in the single-rail logic circuits. This suggests two-rail logic circuits are more testable than ordinary single-rail logic circuits. On two-rail logic circuits, path delay faults are always functional sensitizable and may be robust testable. Even if faults that no codeword input vectors functionally sensitize occur, the circuits are still strongly fault secure for unidirectional stuck-at faults as well as they work correctly.

1: Introduction

In the recent and future high-density and low-power VLSIs, soft and hard errors frequently occur during system operation. In addition, VLSIs become sensitive to noise and thus it becomes difficult to achieve good signal integrity[1, 2, 3, 4]. So, redundant technologies for improving dependability of VLSI systems become increasingly important. Error control coding is one of well-known redundant technologies[1]. While error control coding technologies are widely used in memory system, they are also available for logic circuits. Two-rail logic circuit design is a well-known class of dependable designs for logic circuits and it uses a class of error control codes, namely a two-rail code[5]. On two-rail logic circuits, every signal is encoded with the two-rail code. Two-rail logic circuits can be constructed without inverter elements, and inverter-free two-rail logic circuits are strongly fault secure for unidirectional stuck-at faults[6].

The recent high-density VLSIs also lead to increasing delay faults caused by manufacturing defects, and so manufacturing testing detecting delay faults is also of increasing significance[4, 7, 8]. From these, delay fault testing on two-rail logic circuits is important. In [9, 10], a universal delay fault test set for unate circuits were shown. Since inverter-free two-rail logic circuits are unate, the universal test set is available for delay fault testing on two-rail logic circuits. However, the studies[9, 10] target not two-rail logic circuits but unate circuits. There are still many important and useful but not clarified properties relating to delay fault testing on two-rail logic circuits. This paper deals with testability such as robust testability of path delay faults on inverter-free two-rail logic circuits and reveals some characteristics of two-rail logic circuits, e.g. the relationship between delay fault testability of ordinary single-rail logic circuits and that of two-rail logic circuits.

This paper is organized as follows: Section 2 is preliminary. In this paper, there are two key words, namely corresponding paths and corresponding path delay faults. They are defined in Sect. 3. Section 4 provides discussion about testability of path delay faults in two-rail logic circuits. Section 5 shows experimental results. Section 6 concludes this paper.

1550-5774/08 $25.00 © 2008 IEEE
DOI 10.1109/DFT.2008.19

Table 1. Notations.

$\overline{x}$	inverted value of x	$\Rightarrow$	logical implication	$\{,\}$	set brackets
$\neg$	logical negation	$\Leftrightarrow$	if and only if	$\in$	set membership
$\wedge$	logical conjunction	$\forall$	universal quantification		

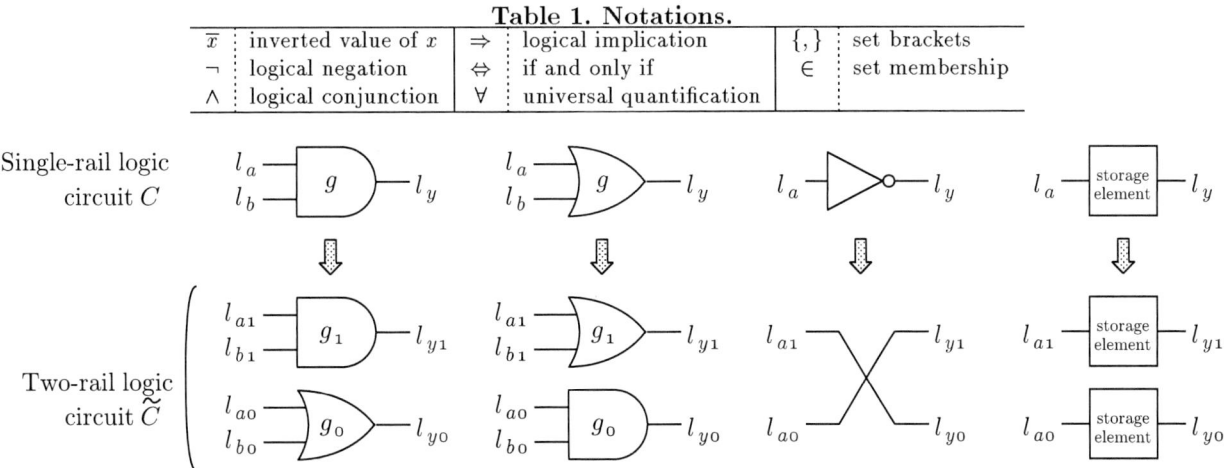

Figure 1. Construction rule of two-rail logic circuits.

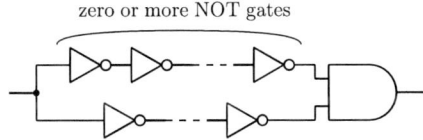

Figure 2. Example of case not discussed in this paper.

2: Preliminaries

Table 1 shows notations used in this paper. In 2.1, some terminologies and known theorems related to two-rail logic circuits are introduced. In 2.2, notations, definitions and properties related to delay fault testing are described.

2.1: Two-rail logic circuits

(1) Construction: A code which contains two codewords (0,1) and (1,0) corresponding to binary information bits '0' and '1' is called a *two-rail code*[11]. A logic circuit we can construct by encoding each value of every line in a logic circuit C with the two-rail code is called a *two-rail logic circuit* corresponding to the *single-rail logic circuit* C.

Let C be a circuit consisting of only three classes of gates, namely AND, OR and NOT gates, and storage elements. A logic circuit $\tilde{C}$ constructed by the construction rule shown in Fig. 1 is a two-rail logic circuit corresponding to C. In this paper, it is assumed that combinational circuits and the combinational parts of sequential circuits consist of only the three classes of gates. We can assume it without lose of generality. It is because any gates comprised by combinational circuits, e.g. NAND, NOR, XOR and MUX, can be equivalently converted into circuits consisting of only the three classes of gates, and path delay fault testability on the gates is equivalent to that on the converted circuits. In this paper, it is also assumed that no circuits include a redundant structure that there are some gates whose multiple inputs are connected back to an identical line directly or through one or more NOT gates as shown in Fig. 2. This paper discusses only inverter-free two-rail logic circuits constructed by the rule shown in Fig. 1. A line l in C and lines l_1, l_0 in $\tilde{C}$ are called *corresponding lines* to each other, and a gate g in C and gates g_1, g_0 in $\tilde{C}$ are called *corresponding gates* to each other. Note that this paper deals with only combinational circuits unless otherwise noted.

(2) Codeword input vectors: Let $\text{val}(v, l)$ be a function taking the value on the line l in a circuit when the input vector v is applied into the circuit. For primary input lines l_i, we can also regard $\text{val}(v, l_i)$ as a function taking the value of the element of v corresponding to l_i. If, for any primary input lines l_i of a single-rail logic circuit C, an input vector v to C and an input vector $\tilde{v}$ to the two-rail logic circuit $\tilde{C}$ satisfy the following equality, $\tilde{v}$ is called a *codeword input vector* corresponding to v: $\text{val}(v, l_i) = \text{val}(\tilde{v}, l_{i1}) = \overline{\text{val}(\tilde{v}, l_{i0})}$. If an input vector v' to a two-rail logic circuit is the codeword input vector corresponding to some input vector, v' is called a codeword input vector.

483

Table 2. Nine-valued logic.

Symbol	Meaning	Ini.	Tr.
S0	Stable logic 0	0	0
S1	Stable logic 1	1	1
T0	Falling transition	1	0
T1	Rising transition	0	1
H0	Hazard-possible 0	0	0
H1	Hazard-possible 1	1	1
U0	S0, T0 or H0	x	0
U1	S1, T1 or H1	x	1
XX	Unspecified (U0 or U1)	x	x

Ini. : initial values,   Tr. : transition values
x : unspecified value

Table 3. Truth table.
(a) AND gate.

x	S0	S1	T0	T1	H0	H1
S0	S0	S0	S0	S0	S0	S0
S1	S0	S1	T0	T1	H0	H1
T0	S0	T0	T0	H0	H0	T0
T1	S0	T1	H0	T1	H0	T1
H0	S0	H0	H0	H0	H0	H0
H1	S0	H1	T0	T1	H0	H1

(b) OR gate.

+	S0	S1	T0	T1	H0	H1
S0	S0	S1	T0	T1	H0	H1
S1	S1	S1	S1	S1	S1	S1
T0	T0	S1	T0	H1	T0	H1
T1	T1	S1	H1	T1	T1	H1
H0	H0	S1	T0	T1	H0	H1
H1	H1	S1	H1	H1	H1	H1

Theorem 1: Let C be a single-rail logic circuit and v be an input vector to C. For any lines l in C, the following equality holds: $\mathrm{val}(v, l) = \mathrm{val}(\tilde{v}, l_1) = \overline{\mathrm{val}(\tilde{v}, l_0)}$. □

(3) Strongly fault secure property: Let C be a circuit and F be a set of faults occurring in C. If, for any faults f in F, the faulty circuit C always outputs correct output vectors or vectors which the fault-free circuit C never outputs, C is referred to as *fault secure* for F. If for any faults f in F, there is an input vector v such that the circuit C outputs an output vector which the fault-free circuit C never outputs, the circuit C is referred to as *self-testing* for F. If C is fault secure and self-testing for F, the circuit C is referred to as *totally self-checking (TSC)* for F[11]. If for any faults f in F, either of the following two is true, C is referred to as *strongly fault secure (SFS)* for F[12]: 1) the circuit C is TSC for $\{f\}$; or 2) the circuit C is fault secure but not TSC for $\{f\}$, and if f occurs in C, the resultant circuit is SFS for $F - \{f\}$. (The circuit is deemed as to be SFS, if $F - \{f\}$ is empty.) It is well-known that two-rail logic sequential circuits are SFS for unidirectional stuck-at faults[6].

2.2: Path delay fault testing

(1) Nine-valued logic system: *Nine-valued logic system*[8] consists of nine symbols; S0, S1, T0, T1, H0, H1, U0, U1 and XX. Every symbol represents a pair of initial and transition values of a signal. Table 2 illustrates the meaning of the symbols. Table 3 is truth table for AND and OR gates on nine-valued logic system.

(2) Path delay faults: In general, a path is an alternate connection of lines and gates from an input to an output. A *path delay* represents the sum of delays on all lines and gates along the path. Path delays for rising and falling transitions on a path are usually different. In this paper, path delays for rising (falling) transitions on a path mean ones in the case that a rising (falling) transition is assigned into the primary input line of the path. If a path delay on a path p exceeds some specified period, it is deemed that a *path delay fault* occurs on p.

(3) Testabilities of path delay faults: There are three classes of testabilities of path delay faults, namely *robust testability, non-robust testability* and *functional sensitizability*.

Let f be a path delay fault occurring on a path p and v be an input vector (a pair of initial and transition vectors). If f can be detected by v regardless of occurrence of delay faults outside p, the fault f is referred to as robust testable for v. If f can be detected by v when no faults occur outside p, the fault f is referred to as non-robust testable for v. If there is a possibility that f affects the output value at the output of p when v is applied, f is referred to as functional sensitizable for v.

If there is an input vector v such that a fault f is robust testable for v, the fault f is referred to as robust testable. In a similar fashion, we can define non-robust testable and functional sensitizable faults. Let V be a set of input vectors. If there is an input vector v in V such that f is robust testable for v, the fault f is referred to as robust testable for V. In a similar way, we can define non-robust testable and functional sensitizable faults for V.

(4) On-input and Off-input: If an input line of a gate is on a path p, it is called an *on-input* of p. If a gate g is on p and an input line of the gate g is not on p, the line is called an *off-input* of p. The testability of a fault f occurring on a path p for an input vector v depends upon the values at on- and off-inputs of p when v is applied. The fault f has the respective testabilities for an input vector v, if and only if the following two are true: 1) the fault f is for rising (falling) transitions and the value in v corresponding to the primary input line of p is T1 (T0), and 2) the values on all on- and off-inputs of p become ones shown in Table 4 when v is applied.

Table 4. Testabilities vs. values at on- and off-inputs.

Testability	Gate	On-input	Off-input
Robust testability	AND	T0	S1
		T1	U1
	OR	T0	U0
		T1	S0
Non-robust testability	AND	T0 or H0	U1
		T1 or H1	U1
	OR	T0 or H0	U0
		T1 or H1	U0
Functional sensitizability	AND	T0 or H0	U1, T0 or H0
		T1 or H1	U1
	OR	T0 or H0	U0
		T1 or H1	U0, T1 or H1

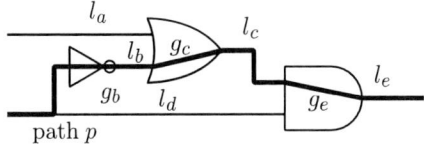

(a) Path p in single-rail logic circuit C.

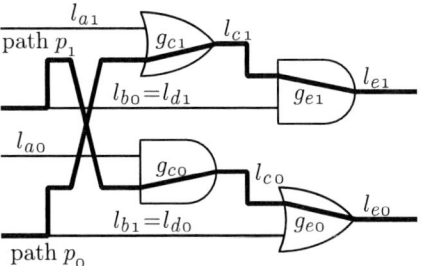

(b) Paths p_1, p_0 in two-rail logic circuit $\tilde{C}$.

Figure 3. Example of corresponding paths.

3: Corresponding paths and path delay faults

The previous section presents the definition of corresponding lines and gates in two-rail logic circuits. In a similar fashion, we can suppose corresponding paths and path delay faults. They are defined in this section.

Definition 1: Let C be a single-rail logic circuit, p be a path in C and p_1, p_0 be paths in $\tilde{C}$. If p, p_1, p_0 satisfy the following four conditions, the three paths are called *corresponding paths* to each other:
1. If a line l_i is the primary input line of p, the corresponding lines l_{i1} and l_{i0} are those of p_1 and p_0, respectively.
2. Suppose the path p includes an AND (OR) gate g, an input line l_a of g and the output line l_y of g. If l_{a1} is on p_1 (p_0) and l_{a0} is on p_0 (p_1), the gate g_1 and the line l_{y1} are on p_1 (p_0), and g_0 and l_{y0} are on p_0 (p_1).
3. Suppose the path p includes a NOT gate g, the input line l_a of g and the output line l_y of g. The output line l_{y1} (l_{y0}) is the same as l_{a0} (l_{a1}) and is necessarily on the path that includes l_{a0} (l_{a1}).
4. The paths p_1, p_0 do not include any lines and gates except those that are recursively regarded as on the paths according to the above conditions 1.-3.

□

Example 1: Figure 3 (a) shows an example of a single-rail logic circuit C and a path p in C. Figure 3 (b) illustrates the corresponding two-rail logic circuit $\tilde{C}$ and corresponding paths p_1, p_0. The path p contains lines and gates l_d, g_b, l_b, g_c, l_c, g_e, l_e. Meanwhile, p_1 contains $l_{d1} = l_{b0}$, g_{c0}, l_{c0}, g_{e0}, l_{e0}, and p_0 contains $l_{d0} = l_{b1}$, g_{c1}, l_{c1}, g_{e1}, l_{e1}.

□

Theorem 2: Let C be a single-rail logic circuit and p be a path in C. For any gates g in C and input lines l of g, if l is an on-input (off-input) of p, the corresponding input line l_1 of g_1 is an on-input (off-input) of either path p_1 or p_0 that includes g_1, and the input line l_0 of g_0 is an on-input (off-input) of either p_1 or p_0 that includes g_0. □

From the Definition 1, Theorem 2 can be easily proved. So, the proof is omitted here.

Theorem 3: For any paths p' in a two-rail logic circuit $\tilde{C}$, there is exactly one path corresponding to p' in the single-rail logic circuit C corresponding to $\tilde{C}$.

Proof: Let p' be a path in a two-rail logic circuit $\tilde{C}$. The path p' is composed of the following subpaths:
1) subpaths from the gate g_a' to g_y' for any lines l' on p' except primary input and primary output lines, where g_a' is the gate whose output line is l' and g_y' is the gate whose input line is l';
2) a subpath from l_i' to g_i', where l_i' is a primary input line of p' and g_i' is the gate whose input line is l_i'; and
3) a subpath from g_o' to l_o', where l_o' is a primary output line of p' and g_o' is the gate whose output line is l_o'.

For every subpath 1), there is a gate g_a in C corresponding to g_a' as well as a gate g_y corresponding to g_y'. The output line of g_a corresponds to l'. The input line of g_y that is the on-input of the subpath also corresponds to l'. If $(g_a' = g_{a1}) \wedge (g_y' = g_{y1})$, or $(g_a' = g_{a0}) \wedge (g_y' = g_{y0})$; from the construction rule of two-rail logic circuits, l_a and l_y are the same, or there is a subpath consisting of an even number of NOT gates from l_a to l_y. Alternatively, if

$(g'_a = g_{a1}) \wedge (g'_y = g_{y0})$, or $(g'_a = g_{a0}) \wedge (g'_y = g_{y1})$; there is a subpath consisting of an odd number of NOT gates from l_a to l_y. Consequently, there is a subpath consisting of zero or more number of NOT gates from g_a to g_y. Moreover, there is exactly one subpath from g_a to g_y because it is assumed that no gates whose multiple inputs are connected back to the same line through zero or more number of NOT gates.

In a similar fashion, for the subpath 2), there is exactly one subpath from l_i to g_i, where l_i is the primary input line of C corresponding to l_i' and g_i is the gate in C corresponding to g_i'. For the subpath 3), there is also exactly one subpath from g_o to l_o, where g_o is the gate in C corresponding to g_o' and l_o is the primary output line of C corresponding to l_o'.

From the construction rule of two-rail logic circuits, for every gate in $\tilde{C}$, there is exactly one corresponding gate in C. Consequently, there is exactly one path p from l_i to l_o in C consisting of the above subpaths. From Definition 1, the input line l_i' of p' is the same as either l_{i1} or l_{i0}. Moreover, we can recursively show that if $l_i' = l_{i1}$, then $p' = p_1$; and if $l_i' = l_{i0}$, then $p' = p_0$. In sum, the path p corresponds to p'.

From Definition 1, there are no paths corresponding to p' except paths consisting of the above subpaths. Therefore, there is exactly one path corresponding to p' in C. $\hfill Q.E.D.$

Corollary 1: If the number of paths in a single-rail logic circuit C is n, that in the corresponding two-rail logic circuit $\tilde{C}$ is $2n$. $\hfill \square$

Definition 2: Let C be a single-rail logic circuit and p be a path in C. The following three path delay faults f, f_1, f_0 are called *corresponding path delay faults* to each other:
- the path delay fault f for rising (falling) transitions on p,
- the path delay fault f_1 for rising (falling) transitions on p_1, and
- the path delay fault f_0 for falling (rising) transitions on p_0.

$\hfill \square$

4: Testability of path delay faults on two-rail logic circuits

4.1: Relation of path delay fault testability

Theorem 4: For any path delay faults f in a single-rail logic circuit and the corresponding faults f_1, f_0 in the corresponding two-rail logic circuit, the following three are true:

$$\mathrm{R}(f) \Leftrightarrow \mathrm{R}(f_1, \tilde{\mathbf{V}}) \Leftrightarrow \mathrm{R}(f_0, \tilde{\mathbf{V}}),$$

$$\mathrm{NR}(f) \Leftrightarrow \mathrm{NR}(f_1, \tilde{\mathbf{V}}) \Leftrightarrow \mathrm{NR}(f_0, \tilde{\mathbf{V}}), \quad \text{and} \quad \mathrm{FS}(f) \Leftrightarrow \mathrm{FS}(f_1, \tilde{\mathbf{V}}) \Leftrightarrow \mathrm{FS}(f_0, \tilde{\mathbf{V}}),$$

where, $\mathrm{R}(f)$ / $\mathrm{NR}(f)$ / $\mathrm{FS}(f)$ are predicates meaning "a fault f is robust testable / non-robust testable / functional sensitizable", $\mathrm{R}(f, V)$ / $\mathrm{NR}(f, V)$ / $\mathrm{FS}(f, V)$ are those meaning "a fault f is robust testable / non-robust testable / functional sensitizable for a set V of input vectors", and $\tilde{\mathbf{V}}$ is the set of all codeword input vectors to the two-rail logic circuit. $\hfill \square$

In order to prove Theorem 4, the following Lemma is needed.

Lemma 1: For any path delay faults f in a single-rail logic circuit C and the corresponding faults f_1, f_0 in the corresponding two-rail logic circuit $\tilde{C}$; and for any input vectors v to C; the following three are true:

$$\mathrm{R}(f, \{v\}) \Leftrightarrow \mathrm{R}(f_1, \{\tilde{v}\}) \Leftrightarrow \mathrm{R}(f_0, \{\tilde{v}\}),$$

$$\mathrm{NR}(f, \{v\}) \Leftrightarrow \mathrm{NR}(f_1, \{\tilde{v}\}) \Leftrightarrow \mathrm{NR}(f_0, \{\tilde{v}\}), \quad \text{and} \quad \mathrm{FS}(f, \{v\}) \Leftrightarrow \mathrm{FS}(f_1, \{\tilde{v}\}) \Leftrightarrow \mathrm{FS}(f_0, \{\tilde{v}\}).$$

Proof: Let v be an input vector to C, p be a path in C, l_i be the input line of p and f be a path delay fault occurring on p such that $\mathrm{R}(f, \{v\})$.

If $f \in \mathbf{F}_r(C)$, we have $\mathrm{val}(v, l_i) = \mathrm{T1}$, where $\mathbf{F}_r(C)$ is the set of all path delay faults for rising transitions on paths in C. From the definition of codeword input vectors, for the input lines l_{i1}, l_{i0} of the paths p_1, p_0 corresponding to p, it follows that $\mathrm{val}(\tilde{v}, l_{i1}) = \mathrm{T1}$ and $\mathrm{val}(\tilde{v}, l_{i0}) = \mathrm{T0}$. Likewise, if $f \in \mathbf{F}_f(C)$, we have $\mathrm{val}(v, l_i) = \mathrm{T0}$, where $\mathbf{F}_f(C)$ is the set of all path delay faults for falling transitions on paths in C. From the definition, it follows that $\mathrm{val}(\tilde{v}, l_{i1}) = \mathrm{T0}$ and $\mathrm{val}(\tilde{v}, l_{i0}) = \mathrm{T1}$.

Suppose v and $\tilde{v}$ are applied into C and $\tilde{C}$, respectively. For any gates g except NOT gates in C, the values on the input lines of g become ones shown in "Robust testability" row in Table 4. From the construction rules of two-rail logic circuits, if g is an AND gate, a gate g_1 corresponding to g is also an AND gate, and if g is an OR gate, g_1 is also an OR gate. From Theorems 1 and 2, the values at the on- and off-inputs of p are the same as those of either path p_1 or p_0 that includes g_1. Consequently, for any g_1, the values on the input lines of g_1 become

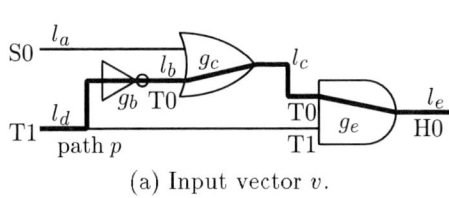

(a) Input vector v.

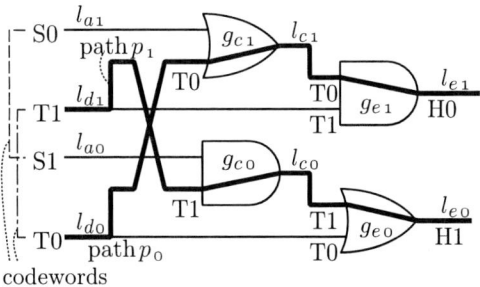

(b) Codeword input vector $\tilde{v}$.

Figure 4. Example of codeword input vectors.

ones shown in "Robust testability" row in Table 4. From the construction rules of two-rail logic circuits, if g is an AND gate, the other corresponding gate g_0 is an OR gate, and if g is an OR gate, g_0 is an AND gate. From Theorems 1 and 2, the values at the on- and off-inputs of p are the same as the inverted values of those of either p_1 or p_0 that includes g_0. Consequently, for any g_0, the values on the input lines of g_0 become ones shown in "Robust testability" row in Table 4, too. From Definition 1, the paths p_1, p_0 include only gates corresponding to gates on p. Consequently, for any gates on p_1 and p_0, the values of the input lines of the gates become ones shown in "Robust testability" row in Table 4. From those, $\mathrm{R}(f_1, \{\tilde{v}\})$ and $\mathrm{R}(f_0, \{\tilde{v}\})$ are true.

Therefore, we can establish that $\mathrm{R}(f, \{v\}) \Rightarrow \mathrm{R}(f_1, \{\tilde{v}\})$ and $\mathrm{R}(f, \{v\}) \Rightarrow \mathrm{R}(f_0, \{\tilde{v}\})$. In a similar way, the other implications can be proved. *Q.E.D.*

Example 2: In the example shown in Fig. 4 (a), an input vector v is applied into the single-rail logic circuit C shown in Fig. 3 (a). As shown in the figure, the fault f for rising transitions on p is robust testable for v. Meanwhile, in the example shown in Fig. 4 (b), the codeword input vector $\tilde{v}$ is applied into the corresponding two-rail logic circuit $\tilde{C}$ shown in Fig. 3 (b). The corresponding faults f_1, f_0 are robust testable for $\tilde{v}$. □

Proof of Theorem 4: Suppose a path delay fault f in C is robust testable. There is an input vector v such that $\mathrm{R}(f, \{v\})$. From Lemma 1, for the fault f_1 corresponding to f in $\tilde{C}$, it follows that $\mathrm{R}(f_1, \{\tilde{v}\})$, Therefore, $\mathrm{R}(f) \Rightarrow \mathrm{R}(f_1, \tilde{\mathbf{V}})$ is true.

In a similar fashion, the other implications can be proved from Lemma 1. *Q.E.D.*

Theorem 5: For any path delay faults f occurring in two-rail logic circuits, the following is true: $\mathrm{NR}(f) \Rightarrow \mathrm{R}(f)$.

Proof: Let $\tilde{C}$ be a two-rail logic circuit, p be a path in $\tilde{C}$, l_i be the input line of p and f be a path delay fault on p such that $\mathrm{NR}(f, \tilde{\mathbf{V}})$ and $f \in \mathbf{F}_\mathrm{r}(\tilde{C})$.

When v is applied into $\tilde{C}$, the values on the inputs of the gates on p become ones shown in "Non-robust testability" row in Table 4. That is to say, the following is true:

$$\forall l \in \mathbf{L}_{\mathrm{on}}(p),\ \mathrm{val}(v, l) = \mathrm{T0},\ \mathrm{T1},\ \mathrm{H0}\ \text{or}\ \mathrm{H1},$$
$$\forall l \in \mathbf{L}_{\mathrm{offA}}(p),\ \mathrm{val}(v, l) = \mathrm{S1},\ \mathrm{T1}\ \text{or}\ \mathrm{H1}, \quad \text{and} \quad \forall l \in \mathbf{L}_{\mathrm{offO}}(p),\ \mathrm{val}(v, l) = \mathrm{S0},\ \mathrm{T0}\ \text{or}\ \mathrm{H0},$$

where $\mathbf{L}_{\mathrm{on}}(p)$ is the set of all on-inputs of p and where $\mathbf{L}_{\mathrm{offA}}(p)$ and $\mathbf{L}_{\mathrm{offO}}(p)$ are the sets of all off-inputs of p on AND and OR gates, respectively. Since $\mathrm{val}(v, l_\mathrm{i}) = \mathrm{T1}$, from Table 3, we can recursively show that the following is true: $\forall l \in \mathbf{L}_{\mathrm{on}}(p),\ \mathrm{val}(v, l) = \mathrm{T1}\ \text{or}\ \mathrm{H1}$.

Let v' be an input vector which we can obtain by replacing T0, H0 in v with S0 and S1, H1 with T1. The input vector v' includes only S0 and T1. The transition vector of v' becomes the same as that of v. Since $\mathrm{val}(v, l_\mathrm{i}) = \mathrm{T1}$, we have $\mathrm{val}(v', l_\mathrm{i}) = \mathrm{T1}$.

Since v' includes only S0 and T1, and $\tilde{C}$ consists of only AND and OR gates, from Table 3, the values of all lines in $\tilde{C}$ become S0 or T1 when v' is applied into $\tilde{C}$. Since the transition vectors of v and v' are the same, for any lines in $\tilde{C}$, the transition values for v are the same as those for v'. Hence we can establish the following equality for any lines l:

$$\mathrm{val}(v', l) = \begin{cases} \mathrm{S0} & (\text{if } \mathrm{val}(v, l) = \mathrm{S0},\ \mathrm{T0}\ \text{or}\ \mathrm{H0}) \\ \mathrm{T1} & (\text{if } \mathrm{val}(v, l) = \mathrm{S1},\ \mathrm{T1}\ \text{or}\ \mathrm{H1}) \end{cases}$$

From those, the following is true:

$$\forall l \in \mathbf{L}_{\mathrm{on}}(p),\ \mathrm{val}(v', l) = \mathrm{T1}, \quad \forall l \in \mathbf{L}_{\mathrm{offA}}(p),\ \mathrm{val}(v', l) = \mathrm{T1}, \quad \text{and} \quad \forall l \in \mathbf{L}_{\mathrm{offO}}(p),\ \mathrm{val}(v', l) = \mathrm{S0}.$$

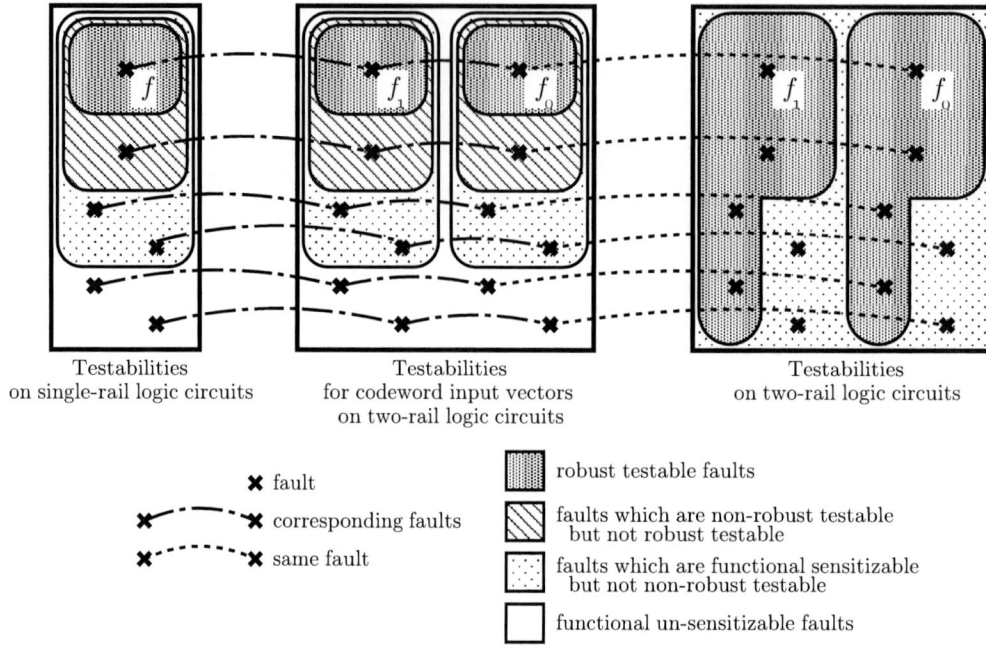

Figure 5. Relation of testabilities.

Consequently, the values at all on- and off-inputs of p become ones shown in "Robust testability" row in Table 4 when v' is applied. Therefore, it follows that $NR(f) \Rightarrow R(f)$ for any $f \in \mathbf{F}_r(\tilde{C})$.

In a similar fashion, we can prove the theorem for faults f such that $f \in \mathbf{F}_f(\tilde{C})$ by supposing an input vector which can be obtained by replacing T1, H1 in v with S1 and S0, H0 with T0. $\hspace{1cm} Q.E.D.$

From Theorems 4 and 5, for any path delay faults in single-rail logic circuits which are non-robust testable, the corresponding faults in two-rail logic circuits are robust testable. It suggests that two-rail logic circuits are more testable than the corresponding single-rail logic circuits.

Theorem 6: For any path delay faults f in two-rail logic circuits, $FS(f)$ is true.

Proof: Let $\tilde{C}$ be a two-rail logic circuit and f be a path delay fault on a path p for rising transitions. Suppose v is an input vector whose every element is T1. Since $\tilde{C}$ consists of only AND and OR gates, from Table 3, we can obtain that $\text{val}(v, l) = \text{T1}$ for any lines l including on- and off-inputs of p. Consequently, when v' is applied, the values at all on- and off-inputs of p become ones shown in "Functional sensitizability" row in Table 4, and thus $FS(f, \{v\})$ is true. Therefore, for any faults $f \in \mathbf{F}_r(\tilde{C})$, it follows that $FS(f)$.

In a similar way, we can prove $FS(f)$ for any faults $f \in \mathbf{F}_f(\tilde{C})$, by supposing an input vector whose every element is T0. $\hspace{1cm} Q.E.D.$

From Theorem 6, path delay faults which are functional un-sensitizable for codeword input vectors are functional sensitizable. Additionally, as experimentally shown in Sect 5, some of them are robust testable.

Figure 5 summarizes the relation of testabilities disclosed above. For example, Theorem 4 gives that for any faults f, we have $R(f) \Leftrightarrow R(f_1, \tilde{\mathbf{V}}) \Leftrightarrow R(f_0, \tilde{\mathbf{V}})$. Moreover, $R(f_0, \tilde{\mathbf{V}}) \Rightarrow R(f_0)$ is obviously true.

4.2: Relation between testability and SFS property on two-rail logic circuits

This subsection discusses SFS property of two-rail logic sequential circuits. Note that the discussed SFS property is for only unidirectional stuck-at faults. So, it is assumed that path delay faults occur only when circuits are manufactured. In other words, for any line and gate delays, the delay times are unchangeable during system operation.

Theorem 7: Let $\tilde{C}$ be a two-rail logic circuit which is sequential circuit. Even if n (≥ 1) path delay faults $f_{d,0}, \cdots, f_{d,n-1}$ such that $\neg FS(f_{d,i}, \tilde{\mathbf{V}})$ ($0 \leq i < n$) occur in the combinational part of $\tilde{C}$, the circuit $\tilde{C}$ can work correctly and SFS for unidirectional stuck-at faults. $\hspace{1cm}\square$

Proof: The theorem is proved by the inductive method.

Suppose path delay faults $f_{d,i}$'s occur and no unidirectional stuck-at faults occur in $\tilde{C}$. If a codeword input vector is applied into the combinational part, $f_{d,i}$'s are not sensitized and the combinational part outputs correct output vectors. Consequently, if the storage elements in $\tilde{C}$ are initialized to a correct vector consisting of codewords, and correct codeword input vectors are always applied into $\tilde{C}$; then vectors stored in the storage elements and output vectors of $\tilde{C}$ are always correct vectors consisting of codewords. Therefore, $\tilde{C}$ where $f_{d,i}$'s occur can work correctly.

Suppose path delay faults $f_{d,i}$'s and m (≥ 0) unidirectional stuck-at faults $f_{u,0}, \cdots, f_{u,m-1}$ occur in $\tilde{C}$; and the faulty circuit $\tilde{C}$ always outputs correct output vectors. Let $f_{u,m}$ be an unidirectional stuck-at-1 (-0) fault other than $f_{u,j}$'s ($0 \leq j < m$). Imagine the following two cases: $f_{u,m}$ occurs in $\tilde{C}$ at the time t, and $f_{u,m}$ does not occur in $\tilde{C}$. The circuit $\tilde{C}$ in the former and later cases is designated by $\tilde{C}_f$ and $\tilde{C}_n$, respectively. The circuits $\tilde{C}_f$ and $\tilde{C}_n$ are always applied the same input vectors. Additionally, at any times before t, i.e. before the occurrence of $f_{u,m}$, the values on any lines and storage elements in $\tilde{C}_f$ are the same as those in $\tilde{C}_n$. Hence the values at each input of every gate, line and storage element in $\tilde{C}_f$ are equal to those in $\tilde{C}_n$ before t. The circuits $\tilde{C}_f$ and $\tilde{C}_n$ are inverter-free because the fault-free circuit $\tilde{C}$ is inverter-free and unidirectional stuck-at faults do not make inverter elements. In addition, any line and gate delays in $\tilde{C}_f$ are the same as those in $\tilde{C}_n$ because unidirectional stuck-at faults do not change them. Hence for any gates, lines and storage elements such that the values at the outputs are not affected by $f_{u,m}$ at t, the values in $\tilde{C}_f$ become equal to those in $\tilde{C}_n$ at t. Meanwhile, for ones such that the values are affected by $f_{u,m}$, the values in $\tilde{C}_f$ become 1's (0's) and thus they become greater (less) than or equal to those in $\tilde{C}_n$ at t. From those, regardless of whether $f_{u,m}$ affects, the values at each output in $\tilde{C}_f$ become greater (less) than or equal to those in $\tilde{C}_n$ at t. Furthermore, we can recursively show that the values at each input of every gate, line and storage element in $\tilde{C}_f$ are greater (less) than or equal to those in $\tilde{C}_n$, and those at each output in $\tilde{C}_f$ are also greater (less) than or equal to those in $\tilde{C}_n$ at any times after t. Accordingly, the values on pairs of primary output lines of $\tilde{C}_f$ corresponding to each other are always correct codewords or non-codewords and never incorrect codewords. Consequently, the circuit $\tilde{C}$ where path delay faults $f_{d,i}$'s and unidirectional stuck-at faults $f_{u,j}$'s ($0 \leq j < m$) occur is fault secure for any unidirectional stuck-at-1 (-0) faults other than $f_{u,j}$'s.

Therefore, two-rail logic circuits where path delay faults $f_{d,i}$'s such that $\neg\text{FS}(f_{d,i}, \tilde{\mathbf{V}})$ occur in the combinational parts are SFS for unidirectional stuck-at faults. $\quad Q.E.D.$

As mentioned in 4.1, path delay faults which are functional un-sensitizable for codeword input vectors may be robust testable. From Theorem 7, testing for such faults can be regarded as over-testing[13].

5: Experimental results

Table 5 demonstrates experimental results indicating a relationship between path delay fault testabilities on two-rail logic circuits and those for codeword input vectors. In the experimentation, the two-rail logic circuits corresponding to the combinational parts of ISCAS89 benchmark circuits are used. The columns whose elements in "Code" row are "R", "NR-R", "FS-NR" and "FU" show the numbers of faults which have testabilities for codeword input vectors such as shown at the bottom of the table. The columns whose elements in "Any" row are "R", "NR-R", "FS-NR" and "FU" show the numbers of faults which have the respective testabilities. For example, the column whose element in "Code" row is "NR-R" and whose one in "Any" is "R" shows the number of faults f such that $\text{R}(f) \wedge \text{NR}(f, \tilde{\mathbf{V}}) \wedge \neg\text{R}(f, \tilde{\mathbf{V}})$.

For any circuits, there are no faults f such that $\text{NR}(f) \wedge \neg\text{R}(f)$ or $\neg\text{FS}(f)$ as proved in Theorems 5 and 6. For 8 out of 10 circuits, there are faults f such that $\text{FS}(f) \wedge \neg\text{FS}(f, \tilde{\mathbf{V}})$. For all of the 8 circuits, there exist faults f such that $\text{R}(f) \wedge \neg\text{FS}(f, \tilde{\mathbf{V}})$.

6: Conclusion

This paper has clarified path delay fault testabilities on inverter-free two-rail logic circuits. Two-rail logic circuits have the following characteristics:

- There are two corresponding path delay faults in a two-rail logic circuit for every fault in the corresponding single-rail logic circuit. For every fault in two-rail logic circuits, there is exactly one corresponding fault in the single-rail logic circuit.
- For any path delay faults which are non-robust testable but not robust testable in a single-rail logic circuits, the corresponding faults in the two-rail logic circuits are always robust testable. This suggests that two-rail logic circuits are more testable than ordinary single-rail logic circuits.

Table 5. Testabilities vs. those for codeword input vectors

Code	R	NR-R		FS-NR			FU			
Any	R	R	NR-R	R	NR-R	FS-NR	R	NR-R	FS-NR	FU
s526n	1,390	46	0	48	0	86	60	0	2	0
s641	3,958	582	0	396	0	1,374	470	0	196	0
s713	2,368	7,476	0	660	0	3,624	804	0	72,316	0
s820	1,960	8	0	0	0	0	0	0	0	0
s832	1,968	24	0	32	0	0	0	0	0	0
s953	4,604	20	0	0	0	0	0	0	0	0
s1196	7,162	356	0	1,646	0	1,600	1,052	0	576	0
s1238	7,178	190	0	1,238	0	3,582	456	0	1,592	0
s1488	3,750	82	0	16	0	0	0	0	0	0
s1494	3,764	90	0	48	0	0	2	0	0	0

Code : testabilities for codeword input vectors.

Any : testabilities (for any input vectors).

R : number of robust testable path delay faults.

NR-R : number of path delay faults which are non-robust testable but not robust testable.

FS-NR : number of path delay faults which are functional sensitizable but not non-robust testable.

FU : number of functional un-sensitizable path delay faults.

- There may be some robust testable faults which no codeword input vectors functionally sensitize in a two-rail logic circuit. Even if such faults occur, the circuit is still strongly fault secure for unidirectional stuck-at faults as well as it can work correctly.

Future works include development of path delay fault test generation for two-rail logic circuits based on this study.

Acknowledgment

This research was partially supported by a grant from CASIO Science Promotion Foundation and the Grant-in-Aid for Scientific Research (C) No.19560335.

References

[1] E. Fujiwara, Code design for dependable systems: theory and practical applications, Wiley-Interscience, 2006.

[2] S. Mitra, N. Seifert, M. Zhang, Q. Sbi and K.S. Kim, "Robust system design with built-in soft-error resilience," IEEE Des. & Test Comput., pp.43–52, Feb. 2005.

[3] T. Karnik, P. Hazucha and J.Patel, "Characterization of soft errors caused by single event upsets in CMOS processes," IEEE Trans., Dependable & Secure Comput., vol.1, No.2, pp.128–143, 2004.

[4] L.-T. Wang, C.-W. Wu and X. Wen, VLSI test principles and architectures: design for testability, Morgan Kaufmann, 2006.

[5] F.F. Sellers, M.-Y. Hsiao, and L.W. Bearnson, Error Detecting Logic for Digital Computers, McGraw-Hill, 1968.

[6] T. Nanya and T. Kawamura, "A note on strongly fault-secure sequential circuits," IEEE Trans. Comp., Vol.C-36, No.9, pp.1121–1123, Sep., 1987.

[7] A. Krstic and K.-T. Cheng, Delay fault testing for VLSI circuits, Kluwer Academic Publishers, 1998.

[8] A.K. Pramanick and S.M. Reddy, "On the detection of delay faults," Proc. IEEE Int'l Test Conf., pp.845–856, 1988.

[9] U. Sparmann, H. Muller and S.M. Reddy, "Universal delay test sets for logice networks," IEEE Trans. Very Large Scale Integr. Systems, vol.7, no.2, pp.156–165, Jun., 1999.

[10] U. Sparmann, H. Muller and S.M. Reddy, "Minimal delay test sets for unate gate networks," Proc. IEEE Asian Test Symp., pp.155–163, 1996.

[11] J.F. Wakerly, Error detecting codes, self-checking circuits and applications, North-Holland, 1978.

[12] J.E. Smith and G. Metze, "Strongly fault secure logic networks," IEEE Trans. Comp., vol.C-27, No.6, pp.491–499, Jun., 1978.

[13] J. Rearick, "Too much delay fault coverage is a bad thing," Proc. IEEE Int'l Test Conf., pp.624–633, 2001.

IEEE International Symposium on Defect and Fault Tolerance of VLSI Systems

Diagnosis of Analog Circuits by Using Multiple Transistors and Data Sampling

Yukiya Miura   Jiro Kato
Graduate School of System Design, Tokyo Metropolitan University
miura@tmu.ac.jp   katou-jirou@ed.tmu.ac.jp

Abstract

We have proposed a method for diagnosing analog circuits that is realized by combining the operation-region model and the X-Y zoning method. In this paper, we propose two improved methods for diagnosing analog circuits by using multiple transistors and data sampling. Diagnosis by multiple transistors gives results of diagnostic resolution that is higher than that by a single transistor. Diagnosis by multiple data sampling gives results of diagnostic sequence length and processing time that are shorter than that by a fixed data sampling. We demonstrate the effectiveness of the proposed methods by applying them to ITC'97 benchmark circuits with hard faults and soft faults. The proposed method can elevate diagnostic resolution by 6.4%, reduce diagnostic sequence length to 5.7%, and reduce processing time to 2.06%, in the maximum compared with our previous method.

1. Introduction

Analog and mixed-signal circuits have become key components for processing analog data. In order to produce highly reliable products in a short turn around time, fault diagnosis is an important technique in the circuit development and fault analysis stages [1]. In order to reduce the testing cost of analog circuits, a hierarchical circuit modeling and its application to circuit testing are proposed [2]. This method is effective for testing parametric faults. However, the application to hard faults and fault diagnosis is unknown. Sensitivity-based methods have been proposed for diagnosing analog circuits, which are based on the relationship between circuit parametric variations and performance variations [3]-[7]. These methods maximize the effect of parametric faults and yield accurate results. Although they have been suitable for diagnosing parametric faults, their efficiency in circuit testing and diagnosing hard faults is unknown. Thus, we need to develop another approach for diagnosing faults that can be applied to hard and soft faults in analog circuits.

To deal with this problem, we have proposed a diagnosis method of analog circuits by combining an X-Y zoning method and an operation-region (OR) model [8]-[10]. The X-Y zoning method uses the characteristics of circuit input and output and is very simple and easy to implement to detect faults [11], [12]. The operation-region (OR) model can be used to model circuit behaviors by utilizing changes in the operation regions of MOS transistors [13]-[15]. The operation regions of MOS transistors give us internal information about the circuit. These two methods are easily combined because both are achieved by applying the same input values. By using these methods, we implemented diagnostic procedures by the way similar to diagnosis for digital circuits [16]-[18]. We demonstrated the effectiveness of our diagnosis methods by applying them to ITC'97 benchmark circuits with hard and soft faults.

In this paper, we propose two diagnosis methods that are improved methods of our previous ones to obtain higher diagnostic resolution and shorter diagnostic sequence length and processing time. For these purposes, we use multiple transistors and data sampling. We demonstrate the effectiveness of proposed methods by applying them to ITC'97 benchmark

1550-5774/08 $25.00 © 2008 IEEE
DOI 10.1109/DFT.2008.31

circuits. We show that proposed methods give more efficient results without losing diagnostic performance compared with the previous methods.

The rest of the paper is organized as follows. Section 2 briefly introduces the OR model, the X-Y zoning method and our previous diagnosis methods. A diagnosis method by multiple transistors is presented in Section 3. Section 4 presents a diagnosis method by multiple data sampling. We show the effectiveness of both methods through simulation results for fault diagnosis using ITC'97 benchmark circuits. Finally, we conclude this work in Section 5.

2. Prior works

2.1 Operation-region model and X-Y zoning method

A MOS transistor in a fault-free circuit operates in one of three regions: the saturation (S), the linear (L), or the cut-off region (C). Utilizing the operation regions of MOS transistors can enable us to model the behavior of a circuit consisting of MOS transistors. We call such a modeling method an *operation-region model* (*OR model*) (see Fig. 1). For generating the OR model, (1) apply input signals to the circuit under test, (2) sample node voltages, Vd, Vg, Vs, of each transistor, and (3) calculate ORs of each transistor. A transistor's OR arranged in time series is called an *OR sequence*, which consists of *OR data* and *time data*. The length of the OR sequence is called *OR length*. Since the OR model can represent circuit behavior by discrete data, it has features of easy handling of the model and easy data processing. Based on this model, we investigated the relationship between circuit behaviors and transistor's ORs and showed that the OR model can be applied to circuit testing [13]-[15]. The OR of a transistor in a faulty circuit may become another operation region (O) which does not belong to any in three regions; e.g., if a reverse voltage is applied to a transistor terminal due to a bridging fault, the transistor's OR is calculated as "O". Thus, we must consider four ORs for the OR model. Note that we assume to use simulation data for generating the OR model in this research because circuit simulation is usually carried out at the design phase. This approach is the same as the sensitivity-based approach.

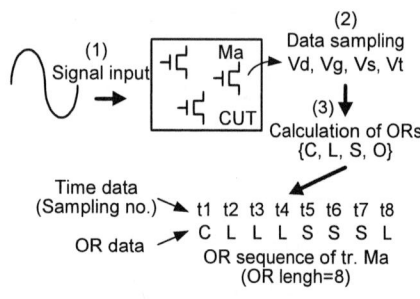

Fig. 1. Operation-region model.

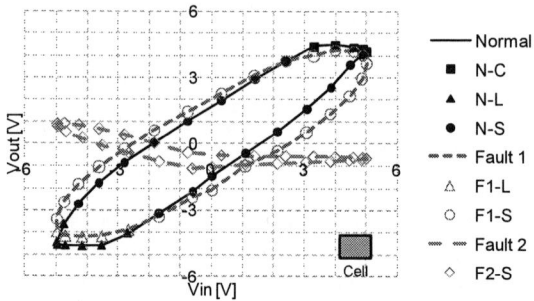

Fig. 2. X-Y zoning method and OR map.

The X-Y zoning method is proposed for testing analog circuits, which uses the characteristics of the input-output voltages of the circuit [11], [12]. An ellipse (i.e., a Lissajous curve) is drawn by plotting the relationship between input and output voltages. If there is a fault in a circuit, the shape of the ellipse for a faulty circuit is different to that for a fault-free circuit. Therefore, faults can be detected by extracting the differences between these shapes. Figure 2 shows an example of ellipses for fault-free and faulty circuits for the ITC'97 benchmark, a continuous-time state-variable filter. In this example, ellipses for faulty circuits are different to that for the fault-free circuit; as a result, these faults can be detected with the X-Y zoning method.

The OR model and the X-Y zoning method are easily merged because both can be achieved using the same input values. In Fig. 2, the dots represent the ORs of one transistor. Each dot on the X-Y plane corresponds to information on an input value, an output value, and a transistor's

492

OR. Therefore, since the location on the X-Y plane represents the existence of a fault and the transistor's OR represents the characteristics of the fault (e.g., fault type or fault location), the combination of both methods can be used to diagnose faults. The X-Y curve plotting the transistor's OR is called an *OR map*. Since we can use multiple OR maps such as a fault dictionary, based on this motivation, we have proposed a fault diagnosis method [8]-[10].

2.2 Basis of our diagnosis methods [8]-[10]

We have proposed diagnostic methods for analog circuits based on diagnostic methods for digital circuits. The *preset test* and the *adaptive test* use pass/fail information on all faults for a given test set, and then a *fault table* is generated [16]-[18]. The fault set in the diagnostic procedure is classified into two subsets using the pass/fail information by applying test vectors in a certain order. This procedure is recursively repeated. If all faults are identified (i.e., the *diagnostic resolution* becomes one.) or all test vectors are applied, the procedure is finished, and a *diagnostic tree* is formed. The preset test uses test vectors that are applied by a predetermined order, while the adaptive test uses test vectors that are applied by the order based on circuit responses.

The X-Y plane is divided into small areas that contain only one plot (i.e., one OR of a utilizing transistor) to handle data on the OR map as indicated by the dashed lines in Fig. 2. We can then obtain discrete data. We call the small areas *cells*. In OR maps generated by several faults and a fault-free circuit, all the cells have the following *cell data*.

{(cell coordinate), (fault set of OR=C), (fault set of OR=L), (fault set of OR=S), (fault set of OR=O)}

The set of all cell data for the OR maps is called *OR map data*. When we apply the preset test to an analog circuit, the test vectors for the digital circuit correspond to the cell coordinates of the OR map data for the analog circuit, and the circuit responses correspond to the OR data contained in the cell data. A fault set for an analog circuit is divided into a maximum of five subsets because cell data have five kinds of OR information, i.e., C, L, S, O, and Φ, where Φ means no data on faults is plotted in the cell.

In order to select cells by an appropriate order, we used the following heuristic measurements for selecting cells to diagnose circuits using fewer cells.

(M1) First, select cells with many kinds of OR data because faults can be efficiently diagnosed if many faults are distinguished from fewer cells.

(M2) If several cells have the same number of OR kinds, we used the following criterion to select one cell from these:

Distinguishability criterion (d.c.)=Π(# of OR),

where (# of OR), OR$\in${C, L, S, O, Φ}, denotes the number of each OR plotted in the cell. This means that if we select the cell with the greatest value of the distinguishability criterion, the cell can divide into the greatest number of subsets for faults.

Fault diagnosis based on the preset test is carried out after cells are arranged using the above measurements. By using one fault table arranged by heuristic measurements (e.g., Table 1(b)), given faults are identified forming the diagnostic tree. This procedure is recursively repeated until diagnosis resolution became one or all cells are checked. Note that if the diagnostic resolution of any subset becomes one, this procedure is halted for this subset (i.e., *early termination*).

In the adaptive test, we recalculate the table of ordered cells (e.g., Table 1 (b)) consisting of only faults that are included in a fault set at each node in the diagnostic tree. We always extract the highest priority cell among them and divide the present fault set into subsets by using the cell. The following is the procedure of the above.

(a) For the fault set at any node in the diagnostic tree, calculate the order of cells that have information of faults included in the set using heuristic measurements M1 and M2.

(b) Extract the cell with the highest selection priority, and divide the present fault set into new subsets by the cell.

(c) Repeat (a) and (b) recursively until resolution becomes one or division is impossible.

As shown in step (c), if a cell cannot divide the present fault set into new subsets, this procedure is halted for that node because the highest priority cell have only one OR kind. Therefore, we can easy achieve the early termination by using the adaptive test.

We explain this procedure by using Table 1 and Fig. 3. We assume that OR map data are obtained from nine faults, f1 to f9, and these fault information are plotted in six cells, A to F (Table 1(a)). These cells are arranged using two heuristic measurements above to select cells, and we obtain the fault table of Table 1(b). Its difference with the digital circuit is that each column has OR information for each fault. After the initial fault set is diagnosed by cell B that has the highest priority in Table 1(b), we obtain three subsets, {f1, f2, f3}, {f4, f5}, and {f6, f7, f8} that do not reach a resolution of one (Fig. 3). We recalculate the cell order for each subset and obtain the highest priority cells A for {f1, f2, f3}, E for {f4, f5}, and F for {f6, f7, f8} (Table 1(c)). All faults are finally identified by repeating this calculation and we obtain the diagnostic tree of Fig. 3. When diagnostic resolution becomes one or maximum resolution is obtained, the number of observed cells is called a *diagnostic sequence length* (*diagnostic length* for short). For example, the diagnostic length for faults f2 and f3 is three. The diagnostic tree generated with the proposed method becomes a quinary tree (i.e., 5-ary tree).

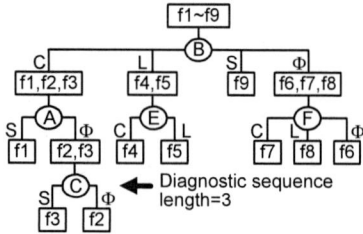

Fig. 3. Diagnostic tree of adaptive test.

Table 1. OR map data & fault table.

(a)

Cell	OR=C	OR=L	OR=S	OR kind	d.c.
A	-	f5, f6, f7	f1	3	15
B	f1, f2, f3	f4, f5	f9	4	18
C	-	-	f3	2	8
D	-	f4, f5, f6	f7, f8	3	24
E	f4, f6	f5, f9	-	3	20
F	f7	f8	-	3	7

(b)

Cell	f1	f2	f3	f4	f5	f6	f7	f8	f9
B	C	C	C	L	L	Φ	Φ	Φ	S
D	Φ	Φ	Φ	L	L	L	S	S	Φ
E	Φ	Φ	Φ	C	L	C	Φ	Φ	L
A	S	Φ	Φ	Φ	L	L	L	Φ	Φ
F	Φ	Φ	Φ	Φ	Φ	Φ	C	L	Φ
C	Φ	Φ	S	Φ	Φ	Φ	Φ	Φ	Φ

(c)

Cell	{f1, f2, f3} OR, d.c.	{f4, f5} OR, d.c.	{f6, f7, f8} OR, d.c.
D	-	1, 0	2, 2
E	-	2, 1	2, 2
A	2, 2	2, 1	2, 2
F	-	-	3, 1
C	2, 2	-	-

2.3 Objective of this paper

Our previous methods are based on the preset test and the adaptive test for digital circuits. Based on these methods, in this paper, we improve our previous methods to elevate diagnostic resolution. In addition, we also improve the previous methods to shorten diagnostic sequence length and CPU time. For these purposes, we use multiple transistors and multiple data sampling.

3. Fault diagnosis by multiple transistors

In our previous method, we used OR data of only one transistor for fault diagnosis. For elevating diagnostic resolution, we first propose a diagnosis method by observing ORs of multiple transistors in a circuit. If we use OR data of two or more transistors simultaneously for generating the OR map, we can extract minute change in circuit behavior by a fault. The basis of the diagnostic procedure is the preset test and the adaptive test described in Sect. 2, and in addition, we use cell data where OR data of two or more transistors are plotted. One transistor functions one of four states and each transistor functions independently. If we use n transistors, the maximum number of kinds of ORs is 4^n+1 including the non-plotting state, Φ. Thus, if we use multiple transistors for fault diagnosis, one node of the diagnostic tree can produce many branches more. For example, when two transistors are used, the maximum number of kinds of ORs is 17, {CC, CL, CS, CO, LC, LL, LS, LO, SC, SL, SS, SO, OC, OL, OS, OO, Φ}. One fault set is divided into 17 subsets at maximum, and the diagnostic tree becomes a 17-ary tree. If we use OR data of multiple transistors for fault diagnosis, it is possible to identify faults that are not identified by using single transistor.

We applied the proposed method to two ITC'97 benchmark circuits, the continuous-time state-variable filter (CT filter) and the leap-frog filter (LF filter) [19]. Table 2(a) summarizes simulation conditions. We used transistors with long OR sequences to diagnose faults because they showed good diagnostic resolution [9]. The number of transistors to use simultaneously is three at maximum. We used a sine-wave input of 800 Hz for the CT filter and 500 Hz for the LF filter. We determined the cell size for the number of sampling with equations in the table. We inserted 234 faults and 195 faults into random locations of the CT filter and the LF filter, respectively (Table 2(b)). Soft faults (i.e., parametric faults) mean variations in capacitances and resistances whose values were obtained from Ref. [20]. Open faults were modeled by inserting high resistance into the signal line and transistor short faults were modeled by connecting the source-drain terminals of a transistor with small resistance [20]. Line short faults were modeled by connecting two signal lines with small resistance.

Table 2. Summary of simulation.

(a)

	CT filter		LF filter	
Output	LPO (L), BPO (B), HPO (H)		Out	
Transistor	M19, M29	(OR length=9)	M14, M17, M34	(OR length=5)
	M39	(OR length=7)	M37, M54, M57	(OR length=5)
Cell size	{10mV*10mV}*(200/(# sampling))		{30mV*30mV}*(200/(# sampling))	

(b)

Fault	Value	CT filter	LF filter
Soft fault	$\pm 1\sigma$, $\pm 6\sigma$	36	36
Open	100MΩ	48	40
Tr. short	5Ω, 500Ω, 10kΩ	126	81
Line short	5Ω, 500Ω, 10kΩ	24	38
Fault-free		1	1
Total		235	196

Table 3 summarizes a part of diagnostic results for changes in the number of transistors, where we show the minimum, maximum, mean, and median values for the diagnostic length for each combination of transistors. The table also shows the best and worst diagnostic resolutions and the number of faults of resolution one. In the table, areas (a) and (b) show results of our previous methods and results by multiple transistors diagnosis, respectively. Due to the limited space, we show a part of results for the CT filter in this paper. Figure 4(a) shows the number of circuits of resolution one by changing the number of transistors. Figure 4(b) shows the distribution of circuits of resolution two or more, where the value in the figure means diagnostic resolution of the circuit. For example, there are eight circuits having resolution two when three transistors, M19, M29, and M39, are used at diagnosis of 200 sampling.

(1) The diagnostic resolution of the preset test is the same as that of the adaptive test because both tests use the same cell data and the difference between these tests are the calculation method for ordering the priority for cell selection.

(2) The number of circuits of resolution one increases as the number of transistors to use diagnosis increases. If we use three transistors, the ratio of resolution one at 200 (10) sampling increases 2.1% (6.4%) on average compared with resolution by a single transistor (Fig. 4(a)).

(3) By using multiple transistors, diagnostic performance of circuits having resolution two or more is also improved. Figure 4(b) shows the distribution of circuits having resolution two or more. The number of circuits having lower resolution decreases as the number of

transistors for diagnosis increases.

(4) Diagnostic resolution of a circuit by multiple transistors never deteriorates than that of diagnosis by their individual transistor. The number of kinds of ORs in one cell increases by using multiple transistors, and a node in the diagnostic tree (i.e., a fault set) can be divided into several nodes by the cell that cannot divide into several nodes when a single transistor is used.

Table 3. Diagnosis results of CT filter (LPO).

# sampling		Preset test					Adaptive test				
		M19	M29	M39	M19 M29	M19 M29 M39	M19	M29	M39	M19 M29	M19 M29 M39
200	min.	(a) 2	3	3	(b) 1	2	(a) 2	2	3	(b) 1	2
	max	4814	4769	4496	5359	5615	51	52	53	49	48
	mean	442.4	516.3	534.6	488.8	493.6	17.0	17.7	21.1	16.0	15.2
	median	163	190	177	164	109	12	12	17	11	10
	resolution	(1, 5)	(1, 5)	(1, 5)	(1, 3)	(1, 2)	(1, 5)	(1, 5)	(1, 5)	(1, 3)	(1, 2)
	res.=1	94.9%	93.6%	93.6%	96.2%	96.6%	94.9%	93.6%	93.6%	96.2%	96.6%
10	min.	3	2	2	1	1	1	1	2	1	1
	max	261	262	259	263	263	22	21	24	22	22
	mean	116.4	126.9	122.5	105.0	101.6	7.4	7.6	9.8	7.0	7.4
	median	41	48	64	36	35	7	8	9	6	7
	resolution	(1, 9)	(1,17)	(1,30)	(1, 9)	(1, 6)	(1, 9)	(1, 17)	(1, 30)	(1, 9)	(1, 6)
	res.=1	64.3%	60.0%	62.6%	68.9%	70.6%	64.3%	60.0%	62.6%	68.9%	70.6%
Multiple sampling {10, 20, 40,100, 200}	min.	(c) 3	2	2	(d) 1	1	(c) 2	2	2	(d) 1	1
	max	280	285	291	286	283	22	21	24	22	22
	mean	118.6	129.7	126.7	107.1	103.5	8.0	8.6	11.0	7.5	7.1
	median	42	49	65	36	35	8	8	12	6	6
	resolution	(1, 5)	(1, 5)	(1, 5)	(1, 3)	(1, 2)	(1, 5)	(1, 5)	(1, 5)	(1, 3)	(1, 2)
	res.=1	94.9%	93.6%	93.6%	96.2%	96.6%	94.9%	93.6%	93.6%	96.2%	96.6%

resolution: (a, b)=(best diagnostic resolution, worst diagnostic resolution)
(a): our previous results [9], [10] (b): diagnosis results by multiple transistors
(c): diagnosis results by multiple sampling
(d): diagnosis results by both multiple transistors and multiple sampling

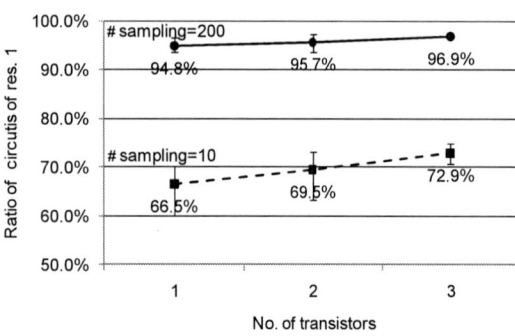

Fig. 4(a). The number of circuits of resolution 1.

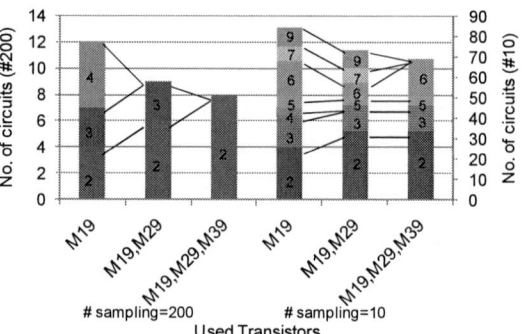

Fig. 4(b). The number of circuits of resolution 2 or more.

(5) Distributions of diagnostic length by multiple transistors diagnosis are almost the same with diagnosis by a single transistor. The priority of cell selection and diagnostic length of each circuit depend on transistors used at diagnosis. Diagnostic length of some circuits become shorter and others become longer compared with diagnosis by a single transistor, and then, statistical distribution of diagnostic length does not very change compared with single transistor diagnosis. On the other hand, the preset test uses all of cells in the fault test if diagnostic resolution of a circuit is two or more. Since the number of cells in the fault table increases as the number of transistors used at diagnosis increases, the maximum diagnostic length become longer than that of single transistor diagnosis. Note that mean and median values of the diagnostic length by multiple transistors diagnosis are slightly improved than

diagnosis by a single transistor. This is because multiple transistors diagnosis elevates diagnostic resolution.

4. Fault diagnosis by multiple data sampling

The number of sampling in our previous diagnosis method is fixed during diagnostic processing; however, it is possible to identify a circuit by diagnosis of small sampling number if the circuit gives a large change in its behavior. Figure 5 shows the distribution of diagnostic resolution of one and CPU time, which are results of the preset test for the CT filter [9]. From the results, circuits of 94% becomes resolution one by diagnosis of 200 sampling, while circuits of 66% becomes resolution one even if the number of sampling is 10. One the other hand, CPU time increases dramatically as the number of sampling increases, this is because the size of the fault table becomes large together with the increment of sampling number (Note that diagnostic length shown in Table 3(a) shows the same trend). From these characteristics, we expect a shorter CPU time and diagnostic length if the number of two or more data sampling is used.

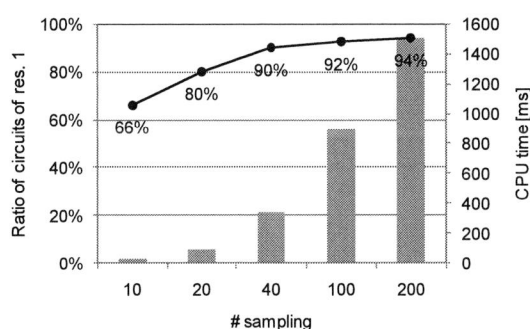

Fig. 5. Circuits of resolution 1 vs. CPU time.   Fig. 6. Diagnosis by multiple sampling.

In this section, we propose a diagnosis method by using multiple data sampling to reduce diagnostic length and CPU time (Fig. 6). In the method, we prepare several kinds of data sampling number. Firstly, the smallest number of sampling is used and diagnosis is carried out for all of given faults. For each set of faults that do not reach to resolution one, fault diagnosis is repeated by using the second small number of sampling. This procedure is recursively repeated until diagnosis resolution of all given faults became one or all given data sampling are used.

In order to show the effective of the proposed method, we prepare five kinds of sampling number, {10, 20, 40, 100, and 200}. Simulation conditions are the same as described in Sect.3. Part (c) of table 3 summarizes a part of results for multiple sampling diagnosis. Part (d) shows results of the combination of multiple transistors diagnosis with multiple sampling diagnosis. Figure 7 compares diagnostic length between single sampling and multiple sampling. Table 4 summarizes CPU time of diagnosis simulation, where we use the following conditions; Windows XP, 3.2 GHz-Pentium 4, and 1GB-RAM, Visual C++. Figure 8 compares CPU time.

(1) The diagnostic resolution of multiple sampling is the same as that of a single 200 sampling because the proposed method uses 200 sampling for diagnosis finally.

(2) The maximum diagnostic length at 200 sampling of the preset test and the adaptive test reduce respectively to 5.5% and 43.8%, and the mean and median values of the preset test (the adaptive test) also reduce respectively to 23.2% and 32.5% (48.6% and 63.7%) (see Table 3(c) and Fig. 7). Comparing results of multiple sampling with those of 10 sampling, diagnostic length does not increase so much. At the initial diagnosis (i.e., the number of sampling is 10), the ratio of circuits having resolution one by single transistor diagnosis is

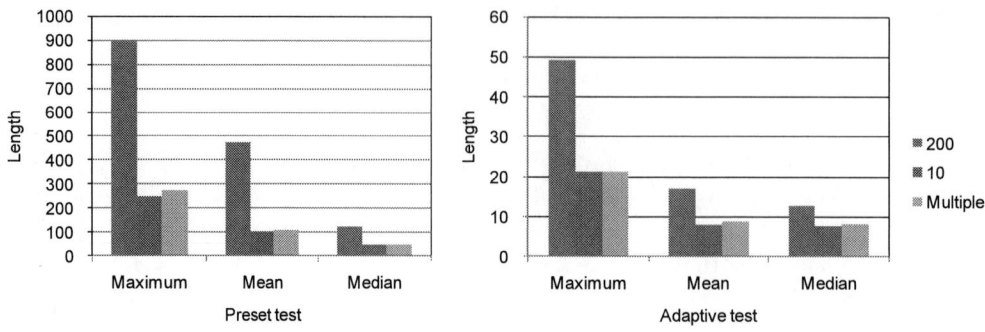

Fig. 7. Comparison of diagnostic length.

approximately 62.3%. The rest of circuits are diagnosed by using a larger sampling number; however, the number of faults in the root of the diagnostic tree (i.e., initial fault set) is 30 at most. As the number of faults decrease, the size of the fault table becomes very small, and then, diagnosis is carried out by a short diagnostic length. If we use the combination of multiple sampling diagnosis with multiple transistors diagnosis (Table 3(d)), there are a few improvements for diagnostic length compared with that of single sampling diagnosis.

(3) Table 4 and Fig. 8 compare CPU time of single sampling diagnosis of 200 and 10 sampling with that of the proposed method. In the table, improvement ratio of CPU time

Table 4. CPU time. [ms]

# sampling	Preset test				Adaptive test			
	#200	#10	Multiple	Ratio	#200	#10	Multiple	Ratio
M19	3636	35	35	0.96%	5604	118	127	2.27%
M29	3613	37	36	1.00%	5684	109	115	2.02%
M39	3425	30	32	0.93%	5513	115	118	2.14%
M19, M29	4108	72	79	1.92%	7249	192	199	2.75%
M19,M39	3896	74	81	2.08%	7290	201	207	2.84%
M29,M39	3906	70	75	1.92%	7175	187	196	2.73%
M19,M29,M39	4918	219	229	4.66%	13552	482	486	3.59%
ave.	3929	77	81	2.06%	7438	201	207	2.78%

Ratio=(CPU time of multiple data sampling)/(CPU time of 200 sampling)

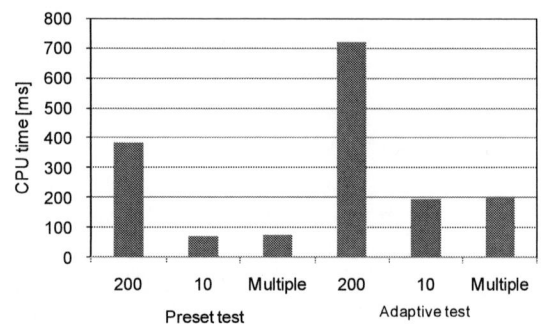

Fig. 8. Comparison of CPU time.

between 200 sampling and the proposed method is shown. CPU times of the preset test and the adaptive test by the proposed method reduce respectively to 2.06% and 2.78% of those of 200 sampling diagnosis. Increment of CPU time of the proposed method against for the 10 sampling method is small. This is because initial diagnosis of the proposed method is carried out by a small sampling number and its CPU time is small. Since second diagnosis, the size of the fault set becomes small and then increment of CPU time is small. Therefore, total CPU time of the proposed method dramatically reduces.

5. Conclusion

We have proposed a method for diagnosing analog circuits with the OR model and the X-Y zoning method. In order to improve diagnostic length and resolution, we improved our previous methods by using multiple transistors and multiple data sampling. We verified their effectiveness by applying them to two ITC'97 benchmark circuits. The ratio of resolution one by the diagnosis method of three transistors at 200 (10) sampling increased 2.1% (6.4%) compared with results of a single transistor. Diagnostic resolution can be elevated by using multiple transistors diagnosis. Since the number of ORs for the use of three transistors in the CT filter is eight at most, computational complexity does not increase so much. The maximum diagnostic length by the multiple data sampling method reduced respectively to 5.7% and 43.2% for the preset test and the adaptive test. CPU time also reduced respectively to 2.06% and 2.78% for the preset test and the adaptive test. We could dramatically reduce the diagnostic length and CPU time without degrading diagnostic resolution by using multiple sampling diagnosis. We could obtain the same results for the LF filter.

In future work, we need to investigate the optimal numbers of transistors and data sampling to obtain better resolution and CPU time. We also need to apply our methods to other circuits including board circuits to demonstrate their efficiency.

Acknowledgments

The authors would like to thank Professor Joan Figueras of Universitat Politècnica de Catalunya for his helpful comments for implementing the X-Y zoning method.

References

[1] Kabisatpathy, A. Barua and S. Sinha, Fault Diagnosis of Analog Integrated Circuits, Springer, Berlin, 2005.

[2] F. Liu and S. Ozev, "Fast Hierarchical Process Variability Analysis and Parametric Test Development for Analog/RF Circuits," Proc. Int. Conf. Computer Design, pp.161-168, 2005.

[3] L.S. Milor, "A Tutorial Introduction to Research on Analog and Mixed-Signal Circuit Testing," IEEE Trans. Circuits and Systems-II: Analog and Digital Signal Processing, vol. 45, no. 10, pp.1389-1407, October 1998.

[4] M. Slamani and B. Kaminska, "Analog Circuit Fault Diagnosis Based on Sensitivity Computation and Functional Testing," IEEE Design & Test of Computers, vol. 9, no. 1, pp.30-39, March 1992.

[5] M. Slamani and B. Kaminska, "Fault Observability Analysis of Analog Circuits in Frequency Domain," IEEE Trans. Circuits and Systems-II: Analog and Digital Signal Processing, vol. 43, no. 2, pp.134-139, February 1996.

[6] É.F. Cota, L. Carro and M. Lubaszewski, "A Method to Diagnose Faults in Linear Analog Circuits Using an Adaptive Testing," Proc. Design, Automation and Test in Europe, pp.184-188, 1999.

[7] S. Cherubal and A. Chatterjee, "Parametric Fault Diagnosis for Analog Systems Using Functional Mapping," Proc. Design, Automation and Test in Europe, pp.195-200, 1999.

[8] Y. Miura, "Fault Diagnosis of Analog Circuits by Operation-Region Model and X-Y Zoning Method," Proc. 19th IEEE Int. Symp. Defect and Fault Tolerance in VLSI Systems, pp230-238, 2004.

[9] Y. Miura, "Characteristics of Fault Diagnosis for Analog Circuits Based on Preset Test," Proc. 20th IEEE Int. Symp. Defect and Fault Tolerance in VLSI Systems, pp573-581, 2005.

[10] Y. Miura and J. Kato, "Fault Diagnosis of Analog Circuits Based on Adaptive Test and Output Characteristics," Proc. 21th IEEE Int. Symp. Defect and Fault Tolerance in VLSI Systems, pp.410-418, 2006.

[11] A.M. Brosa and J. Figueras, "Digital Signature Proposal for Mixed-Signal Circuits," Proc. Int. Test Conf., pp.1041-1050, 2000.

[12] R. Sanahuja, V. Barcons, L. Balado and J. Figueras, "X-Y Zoning BIST: An FPAA Experiment," 9th IEEE Int. Mixed-Signal Test Workshop, 2002.

[13] Y. Miura, "Fault Behavior and Change in Internal Condition of Mixed-Signal Circuits," IEICE Trans. Inf. & Syst., vol. E83-D, no. 4, pp.943-945, April 2000.

[14] Y. Miura, "A Novel Approach for Modeling Behavior of Analog and Mixed-Signal Circuits Based on an Operation-Region Model," Proc. European Test Workshop 2002, pp.215-217, 2002.

[15] Y. Miura and D. Kato, "Analysis and Testing of Analog and Mixed-Signal Circuits by an Operation-Region Model: A Case Study of Application and Implementation," Proc. 18th IEEE Int. Symp. Defect and Fault Tolerance in VLSI Systems, pp.279-286, 2003.

[16] H.Y. Chang, E.G. Manning and G. Metze, Fault Diagnosis of Digital Systems, John Wily & Sons, New York, 1970.

[17] M.A. Breuer, A.D. Friedman, Diagnosis & Reliable Design of Digital Systems, Computer Science Press, Maryland, 1976.

[18] Koren and Z. Kohavi, "Sequential Fault Diagnosis in Combinational Networks," IEEE Trans. Comput., vol.C-26, no.4, pp.334-342, April 1977.

[19] B. Kaminska, K. Arabi, I. Bell, P. Goteti, J.L. Huertas, B. Kim, A. Rueda and M. Soma, "Analog and Mixed-Signal Benchmark Circuits - First Release," Proc. Int. Test Conf., pp.183-190, 1997.

[20] "Benchmark Circuits for Analog and Mixed Signal Testing," http://www.coe.uncc.edu/~cestroud/analgobc/mixtest.html.

INVITED TALK

Design for Test Challenges of High Performance/Low Power Microprocessors

Kamran Zarrineh, *AMD*

In this presentation, we are going to describe some of the architectural features of high performance/low power microprocessors. We then explain their unique test challenges due to their architecture and why these devices are different from ASICs designs from a test perspective. In addition, we present some of the commonly used DFT techniques, e.g., scan-based, memory BIST, IEEE 1149.1, and describe why these test techniques need to accommodate the architecture uniqueness of these microprocessors. Furthermore, we explain why the presented structural DfT techniques are necessary, yet not sufficient for a complete test of these devices. We continue the presentation with describing the architecture support requirement for using functional test vectors.

Speaker Bio − Kamran Zarrineh started his career at IBM in 1991 working on different aspects of testing digital designs ranging from ATPG algorithm development to architecture of programmable memory BIST for high performance embedded memories. From 2000 to 2004, he was in charge of DFT issues for the UltraSparc V microprocessor at Sun Microelectronics. He is currently responsible for DFT of the high performance/low power microprocessors at Advanced Micro Devices in Boxborough, MA. His research interests include design and test of VLSI systems.

Zarrineh has a BS in Electrical Engineering from the University of Utah, an MS in Computer Science from the State University of New York at Binghamton, and a PhD in Electrical Engineering from the University at Buffalo. He is a member of the IEEE.

Defect-Tolerant Hybrid CMOS/Nanoelectronic Circuits

Konstantin K. Likharev, *Stony Brook University*

This talk reviews recent work on devices, circuits and defect-tolerant architectures for hybrid semiconductor/nanodevice integrated circuits. Such a circuit is essentially a CMOS stack with a simple add-on in the form of a nanowire crossbar, with similar two-terminal devices (with the functionality of programmable diodes) formed at each crosspoint. Special attention will be given to the so-called "CMOL" variety of the hybrids, in which the crossbar is connected to the CMOS circuit with an area-distributed interface. Such interface allows the CMOS subsystem to address each and every of the crosspoint devices, even with no nanoscale alignment between the CMOS and crossbar subsystems.

The recent detailed studies have shown CMOL may enable (at least) the following applications: (i) terabit-scale memories with access time below 100 ns and defect tolerance up to 10%, (ii) FPGA-like reconfigurable logic circuits with even higher (~20%) defect tolerance, and the density at least two orders of magnitude lower than that of CMOS FPGAs fabricated with similar design rules and power per unit area, and (iii) mixed-signal neuromorphic networks ("CrossNets") which may exhibit extraordinary defect tolerance (up to 90%) and provide unparalleled performance for some important information processing tasks.

Recently, the hybrid circuit concept has received a strong boost from the experimental demonstrations of reproducible crosspoint devices (programmable diodes) based on amorphous silicon, and nanowire crossbars with 15-nm-scale half-pitch. However, the transfer of semiconductor IC industry to the hybrid technology will still require a very substantial effort, and I will describe the most significant challenges on that way. References may be found at http://rsfq1.physics.sunysb.edu/~likharev/nano/.

Speaker Bio – Konstantin K. Likharev received the Candidate (Ph.D.) degree in Physics from the Lomonosov Moscow State University, Russia in 1969, and the habilitation degree of Doctor of Sciences from the Higher Attestation Committee of the U.S.S.R. in 1979. From 1969 to 1991 Dr. Likharev was a Staff Scientist of Moscow State University, and in 1991 he assumed a Professorship at Stony Brook University (Distinguished Professor since 2002). During his research career, Dr. Likharev worked in the fields of nonlinear classical and dissipative quantum dynamics, and solid-state physics and electronics, notably including superconductor electronics, single-electronics, and nanoelectronics. He is an author of more than 250 original publications, 75 review papers and book chapters, 2 monographs, and several patents. Dr. Likharev is a Fellow of the American Physical Society and a Senior Member of the IEEE.

SESSION 12
EMERGING TECHNOLOGIES

IEEE International Symposium on Defect and Fault Tolerance of VLSI Systems

A Statistical Model for Assessing the Fault Tolerance of Variable Switching Currents for a 1Gb Spin Transfer Torque Magnetoresistive Random Access Memory

Y. Asao, M. Iwayama, K. Tsuchida,
A. Nitayama, and H. Yoda
Center for Semiconductor R & D,
Semiconductor Company,
Toshiba Corporation
yoshiaki.asao@toshiba.co.jp

H. Aikawa, S. Ikegawa, and T. Kishi
Corporate R & D Center,
Toshiba Corporation

Abstract

A comprehensive statistical model of the switching probability was proposed for a 1Gb spin transfer torque magnetoresistive random access memory (STT-MRAM). Since the switching current varies with every write cycle owing to the thermal instability, the read disturbance and the write error are critical issues in the STT-MRAM. In this paper, the operating condition of read and write was designed so as not to cause the read disturbance or the write error. The effect of an error correcting code (ECC) on the read disturbance was also calculated. Finally, it was demonstrated that the 1Gb STT-MRAM could be realized with the optimal bit line voltages and the ECC.

1. Introduction

A spin transfer torque magnetoresistive random access memory (STT-MRAM) features high density, high speed, non-volatility, and low power consumption [1][2].

However, the read disturbance can occur in the STT-MRAM because the MOSFET drives both the read current I_{READ} and the write current I_{WRITE}. In addition, a switching current varies with every write cycle because a spin reversal obeys the Poisson probability process [3]–[5] as shown in Figure 1. This variable switching current (VSC) can degrade the read disturbance and also can cause the write error [6]. Thus, the I_{READ} and the I_{WRITE} should be designed so as not to cause the read disturbance or the write error as illustrated in Figure 2. Only a few papers, however, have been reported on the read disturbance and the write error [2][7].

In the present work, the spin reversal probability is expanded by implementing the distributions of the spin reversal energy, the switching current, and the MOSFET current. Further, an effect of an error correcting code (ECC) on data retention and the read disturbance is calculated. By using those, the I_{READ} and the I_{WRITE} are designed so as not to generate the read disturbance or the write error. Finally, it is demonstrated that the 1Gb STT-MRAM could be realized with the ECC.

2. Spin reversal probability

2.1. Poisson probability process

The probability that a data of the STT-MRAM is switched for a given time t at least unit time is expressed by using the Poisson distribution

$$F(t) = 1 - \exp(-\lambda t), \tag{1}$$

where λ is the average number of the spin reversal per time t given by

1550-5774/08 $25.00 © 2008 IEEE
DOI 10.1109/DFT.2008.18

$$\lambda = \tau_0^{-1} \exp\left[-\Delta(1 - I/I_{C0})\right]. \tag{2}$$

Here τ_0 is the thermally activated reversal time of 1ns, I is an applied current, I_{C0} is the switching current for 1ns, and Δ is a coefficient defined by $\Delta = E_a/k_B T$, where E_a is a spin reversal energy, k_B is the Boltzmann constant, and T is the temperature.

Assume that I_{C0} and Δ obey the two-dimensional normal distribution because they depend on a size of a magnetic tunnel junction (MTJ). The MTJ is a memory element of the STT-MRAM. Thus, (1) can be expressed as a mixed Poisson distribution

$$F(t) = 1 - \int_0^\infty \int_0^\infty \exp\left\{-t \tau_0^{-1} \exp\left[-\Delta(1 - I/I_{C0})\right]\right\} w(I_{C0}, \Delta) dI_{C0} d\Delta, \tag{3}$$

where $w(I_{C0}, \Delta)$ is the joint probability density function of I_{C0} and Δ. Assume that $w(I_{C0}, \Delta)$ obeys the bivariate normal distribution given by

$$w(I_{C0}, \Delta) = \frac{1}{2\pi \sigma_{I_{C0}} \sigma_\Delta \sqrt{1 - \rho_{I_{C0}\Delta}^2}} \exp\left\{-\frac{0.5}{1 - \rho_{I_{C0}\Delta}^2}\left[\frac{(I_{C0} - \mu_{I_{C0}})^2}{\sigma_{I_{C0}}^2} - 2\rho_{I_{C0}\Delta}\left(\frac{I_{C0} - \mu_{I_{C0}}}{\sigma_{I_{C0}}}\right)\left(\frac{\Delta - \mu_\Delta}{\sigma_\Delta}\right) + \frac{(\Delta - \mu_\Delta)^2}{\sigma_\Delta^2}\right]\right\}. \tag{4}$$

In (4), $\mu_{I_{C0}}$ and $\sigma_{I_{C0}}$ are the mean and the standard deviation of I_{C0}, respectively. In the same manner, μ_Δ and σ_Δ are the mean and the standard deviation of Δ, respectively. The correlation of I_{C0} and Δ is determined by the correlation coefficient $\rho_{I_{C0}\Delta}$.

2.2. Error correcting code (ECC)

An ECC with a code length of $n=38$ and an information length of 32 will be introduced for the 1Gb STT-MRAM. The ECC is able to correct unit bit out of 38 bits in the code length.

Assume that the 1Gb STT-MRAM is divided into $M=33{,}554{,}432$ segments. Each segment consists of the $n=38$ bits code. Thus, the number of the failing bits in the 1Gb STT-MRAM for the time t is obtained by $1024^3 \times (38/32) \times \lambda t = Mn\lambda t$.

The problem of determining the probability is identical to the classical statistics problem of placing $Mn\lambda t$ balls in M boxes, and then calculating the probability that a given box contains k balls. If $Mn\lambda t$ failing bits are distributed randomly among M segments, the probability that a given segment contains k failing bits is given by the binominal distribution

$$P_k = {}_{Mn\lambda t}C_k (1/M)^k (1 - 1/M)^{Mn\lambda t - k}. \tag{5}$$

When $Mn\lambda t$ is large and $n\lambda t$ remains finite, the binomial distribution (5) can be approximated by the Poisson distribution

$$P_k \cong (n\lambda t)^k \exp(-n\lambda t)/k!. \tag{6}$$

Since the ECC is able to correct unit bit out of 38 bits, the probability that the ECC does not save the data is given by

$$F_{\text{ECC}}(t) = 1 - P_0 - P_1. \tag{7}$$

When I_{C0} and Δ obey the two-dimensional normal distribution, (7) is expressed by

$$F_{\text{ECC}}(t) = 1 - \int_0^\infty \int_0^\infty \sum_{k=0}^1 \left\{nt\tau_0^{-1} \exp\left[-\Delta(1 - I/I_{C0})\right]\right\}^k \exp\left\{-nt\tau_0^{-1} \exp\left[-\Delta(1 - I/I_{C0})\right]\right\} w(I_{C0}, \Delta) dI_{C0} d\Delta. \tag{8}$$

3. Data retention

3.1. Probability of data retention

By substitution of $I=0$ into (3), we have the probability for data retention that is switched at least unit time for the time t

$$F(t) = 1 - \frac{1}{\sqrt{2\pi}\sigma_\Delta} \int_0^\infty \exp\left[-\frac{t}{\tau_0}\exp(-\Delta) - \frac{(\Delta - \mu_\Delta)^2}{2\sigma_\Delta^2}\right] d\Delta. \tag{9}$$

When N chips of the 1Gb STT-MRAM have 10 years data retention with a failure rate of 1ppm, $F(t)$ is smaller than $1/(N\times1024^3\times10^6)$ at $t=10$ years.

3.2. Probability of data retention with ECC

By substitution of $I=0$ into (8), we have the probability for data retention saved by the ECC

$$F_{\mathrm{ECC}}(t)=1-\sum_{k=0}^{1}\frac{1}{\sqrt{2\pi}\sigma_\Delta}\left(\frac{nt}{\tau_0}\right)^k\int_0^\infty\exp\left[-tk\Delta-\frac{nt}{\tau_0}\exp(-\Delta)-\frac{(\Delta-\mu_\Delta)^2}{2\sigma_\Delta^2}\right]d\Delta. \tag{10}$$

When N chips of the 1Gb STT-MRAM have 10 years data retention, $F_{\mathrm{ECC}}(t)$ is smaller than $1/(NM\times10^6)$ at $t=10$ years. By a numerical calculation of (9) and (10), required μ_Δ and σ_Δ with and without the ECC were obtained and are plotted in Figure 3.

4. Variable switching current

4.1. Probability density function of switching voltage

As described previously, the switching current varies with every write event owing to magnetic thermal instability. This VSC can cause the read disturbance. In this section, the probability density function of the spin reversal considering the VSC is obtained for estimating the read disturbance and the write error.

An MRAM cell consists of an MTJ and a MOSFET as illustrated in Figure 4. When the spin of the free layer is parallel to that of a pinned layer (P-state), the MTJ has low resistance R_P. On the other hand, when the spin of the free layer is antiparallel to that of the pinned layer (AP-state), the MTJ has high resistance R_{AP}.

When the free layer is on the tunnel barrier, the I_{WRITE} flows from a source line to a bit line so as to write the data from P-state to AP-state as shown in Figure 5(a), and further, the I_{WRITE} flows from the bit line to the source line so as to write the data from AP-state to P-state as shown in Figure 5(b). We have designed both the I_{READ} and the I_{WRITE} to be driven with a constant voltage source in the 1Gb STT-MRAM.

When the I_{WRITE} is large, the Joule heating increases T. Thus, $\Delta=E_a/k_BT$ decreases [8][9]. For this reason, we introduced a term βI^2 [8]–[11] to adjust Δ as

$$\Delta=E_a/\left(k_BT+\beta I^2\right). \tag{11}$$

From (1), (2), and (11),

$$I_{C0}=\frac{k_BT\Delta I}{A\left(k_BT+\beta I^2\right)+k_BT\Delta}, \tag{12}$$

where

$$A\equiv\ln\left\{-\tau_0t^{-1}\ln[1-F(t)]\right\} \tag{13}$$

Let R_{MTJ} and R_{TR} be the MTJ resistance and the MOSFET resistance, respectively. Assume that I_{C0}, Δ, R_{MTJ}, and R_{TR} obey the normal distribution. In those, I_{C0}, Δ, and R_{MTJ} are correlated because they depend on the MTJ size.

Thus, let $g(I_{C0},\Delta,R_{\mathrm{MTJ}},R_{\mathrm{TR}})$ be the joint probability density function of I_{C0}, Δ, R_{MTJ}, and R_{TR}. As mentioned before, we have designed both the I_{READ} and the I_{WRITE} to be driven with the constant voltage source. If the applied voltage between the bit line and the source line is denoted by V, the joint probability function $g(I_{C0},\Delta,R_{\mathrm{MTJ}},R_{\mathrm{TR}})$ is transformed into

$$l(V,\Delta,R_{\mathrm{MTJ}},R_{\mathrm{TR}})=g\left(\frac{k_BT\Delta V/(R_{\mathrm{MTJ}}+R_{\mathrm{TR}})}{Ak_BT+A\beta[V/(R_{\mathrm{MTJ}}+R_{\mathrm{TR}})]^2+k_BT\Delta},\Delta,R_{\mathrm{MTJ}},R_{\mathrm{TR}}\right)J, \tag{14}$$

where Jacobian is

$$J = k_B T \Delta \left| \frac{A k_B T - A\beta [V/(R_{\mathrm{MTJ}} + R_{\mathrm{TR}})]^2 + k_B T \Delta}{\left\{ A k_B T + A\beta [V/(R_{\mathrm{MTJ}} + R_{\mathrm{TR}})]^2 + k_B T \Delta \right\}^2 (R_{\mathrm{MTJ}} + R_{\mathrm{TR}})} \right|. \tag{15}$$

Consequently, the probability density function of V is expressed by

$$f_{\mathrm{SWITCH}}(V) = \int_0^\infty \int_0^\infty \int_0^\infty l(V, \Delta, R_{\mathrm{MTJ}}, R_{\mathrm{TR}}) d\Delta dR_{\mathrm{MTJ}} dR_{\mathrm{TR}}. \tag{16}$$

The probability that the read disturbance occurs within x times reading can be obtained by substitution of $1/x$ into $F(t)$ of (13). In the same manner, the probability that write error occurs within x times writing can be obtained by substitution of $1-1/x$ into $F(t)$.

4.2. Probability of read disturbance

As shown in Figure 6, the probability that the read disturbance occurs at the read voltage V_{READ} is given by

$$P(V < V_{\mathrm{READ}}) = \int_0^V f_{\mathrm{SWITCH}}(x) dx. \tag{17}$$

When N chips of 1Gb-MRAM have no read disturbance, $P(V < V_{\mathrm{READ}})$ is smaller than $1/(N \times 1024^3 \times 10^6)$.

4.3. Probability of read disturbance with ECC

After a read cycle, the ECC saves the data by rewriting a corrective one. By taking account of distributions of I_{C0}, Δ, R_{MTJ}, and R_{TR} in (8), the probability that no read disturbance occurs with the ECC is expanded as

$$F_{\mathrm{ECC}}(V) = 1 - \int_0^\infty \int_0^\infty \int_0^\infty \int_0^\infty \sum_{k=0}^1 \left\{ nt\tau_0^{-1} \exp\left[-\frac{k_B T \Delta \left(1 - \dfrac{V}{I_{C0}(R_{\mathrm{MTJ}} + R_{\mathrm{TR}})}\right)}{k_B T + \beta \left(\dfrac{V}{R_{\mathrm{MTJ}} + R_{\mathrm{TR}}}\right)^2} \right] \right\}^k$$

$$\cdot \exp\left\{ -nt\tau_0^{-1} \exp\left[-\frac{k_B T \Delta \left(1 - \dfrac{V}{I_{C0}(R_{\mathrm{MTJ}} + R_{\mathrm{TR}})}\right)}{k_B T + \beta \left(\dfrac{V}{R_{\mathrm{MTJ}} + R_{\mathrm{TR}}}\right)^2} \right] \right\} g(I_{C0}, \Delta, R_{\mathrm{MTJ}}, R_{\mathrm{TR}}) dI_{C0} d\Delta dR_{\mathrm{MTJ}} dR_{\mathrm{TR}}. \tag{18}$$

When N chips of 1Gb-MRAM have no read disturbance, $F_{ECC}(V)$ is smaller than $1/(NM \times 10^6)$.

4.4. Probability of write error

Since a write cycle is also adversely affected by the VSC, the probability that the write error occurs at the write voltage V_{WRITE} is given by

$$P(V > V_{\mathrm{WRITE}}) = \int_V^\infty f_{\mathrm{SWITCH}}(x) dx. \tag{19}$$

When N chips of 1Gb-MRAM have no write error, $P(V > V_{\mathrm{WRITE}})$ is smaller than $1/(N \times 1024^3 \times 10^6)$ as shown in Figure 7.

In calculating the read disturbance using (17) and (18), R_{MTJ} and R_{TR} are independent of the applied voltage V because V during the read cycle is lower than that during the write cycle, that is, the MOSFET is operated in the linear region. In this section, therefore, voltage dependences on R_{MTJ} and R_{TR} are introduced for estimating the write error.

As illustrated in Figure 8, R_{MTJ} takes two values, namely, R_{P} for P-state and R_{AP} for AP-state. They are related with magnetoresistance (MR) defined as $MR = (R_{\mathrm{AP}} - R_{\mathrm{P}})/R_{\mathrm{P}}$. The MTJ resistance in writing the data from AP-state to P-state is expressed by $R_{\mathrm{MTJ}}{}^{\mathrm{AP}\to\mathrm{P}}=R_{\mathrm{AP}}$. R_{AP} can be approximated by a linear function of the applied voltage V_{MTJ} on the MTJ as

$$R_{\mathrm{MTJ}}^{\mathrm{AP}\to\mathrm{P}} = R_{\mathrm{AP}} = aV_{\mathrm{MTJ}} + b, \tag{20}$$

where a and b are constant. On the other hand, the MTJ resistance $R_{\mathrm{MTJ}}{}^{\mathrm{P}\to\mathrm{AP}}$ in writing the data from P-state to AP-state is not simple because $R_{\mathrm{MTJ}}{}^{\mathrm{P}\to\mathrm{AP}}$ is increased during the spin reversal in the constant voltage writing. Assume that the spin reversal is established when the free layer spin is rotated at $90°$ with the pinned layer spin. Thus, $R_{\mathrm{MTJ}}{}^{\mathrm{P}\to\mathrm{AP}}$ is defined as the average of R_{P} and R_{AP}

$$R_{\mathrm{MTJ}}^{\mathrm{P}\to\mathrm{AP}} = \left(R_{\mathrm{P}} + R_{\mathrm{AP}}\right)/2 = aV_{\mathrm{MTJ}} + b. \tag{21}$$

Next, voltage dependence is introduced for the MOSFET resistance R_{TR}. The MOSFET characteristics are obtained by SPICE. Figure 9 shows I-V and R-V curves of the MOSFET, in which the I_{WRITE} flows from the source line to the bit line. The horizontal axis is the applied voltage V_{TR} between the source and the drain of the MOSFET. For simplicity, the I-V curve was approximated by a linear function as

$$I = cV_{\mathrm{TR}} + d, \tag{22}$$

where c and d are constant. On the other hand, Figure 10 shows I-V and R-V curves, in which the I_{WRITE} flows from the bit line to the source line. Similarly, the R-V curve was approximated by a linear function as

$$R_{\mathrm{TR}} = eV_{\mathrm{TR}} + f, \tag{23}$$

where e and f are constant.

From (21) and (22), the write current from the source lien to the bit line is solved as

$$I_{\mathrm{WRITE}}^{-}(V) = \frac{acV + bc + 1 + ad - \sqrt{(acV + bc + 1 + ad)^2 - 4a(cV + d)}}{2a}. \tag{24}$$

Thus, the joint probability density function (14) is rewritten by

$$l(V, \Delta, R_{\mathrm{MTJ}}, R_{\mathrm{TR}}) = g\!\left(\frac{k_B T\Delta \cdot I_{\mathrm{WRITE}}^{-}(V)}{Ak_B T + A\beta I_{\mathrm{WRITE}}^{-}(V)^2 + k_B T\Delta}, \Delta, R_{\mathrm{MTJ}}, R_{\mathrm{TR}}\right)J. \tag{25}$$

Further, Jacobian (15) is modified as

$$J = \left|\frac{k_B T\Delta\left(Ak_B T - A\beta \cdot I_{\mathrm{WRITE}}^{-}(V)^2 + k_B T\Delta\right)}{\left(Ak_B T + A\beta \cdot I_{\mathrm{WRITE}}^{-}(V)^2 + k_B T\Delta\right)^2} \cdot \frac{c\left(1 - aI_{\mathrm{WRITE}}^{-}(V)\right)^2}{bc + \left(1 - aI_{\mathrm{WRITE}}^{-}(V)\right)^2}\right|. \tag{26}$$

In the same manner, from (20) and (23), the write current from the bit line to the source line is solved as

$$I_{\mathrm{WRITE}}^{+}(V) = \frac{(a+e)V + b + f - \sqrt{(eV + aV + b + f)^2 - 4a(aeV + be + af)V}}{2(aeV + bf + af)}. \tag{27}$$

Thus, the joint probability density function (14) is rewritten by

$$l(V, \Delta, R_{\mathrm{MTJ}}, R_{\mathrm{TR}}) = g\!\left(\frac{k_B T\Delta \cdot I_{\mathrm{WRITE}}^{+}(V)}{Ak_B T + A\beta \cdot I_{\mathrm{WRITE}}^{+}(V)^2 + k_B T\Delta}, \Delta, R_{\mathrm{MTJ}}, R_{\mathrm{TR}}\right)J. \tag{28}$$

Further, Jacobian (15) is modified as

$$J = \left|\frac{k_B T\Delta\left(Ak_B T - A\beta \cdot I_{\mathrm{WRITE}}^{+}(V)^2 + k_B T\Delta\right)}{\left(Ak_B T + A\beta \cdot I_{\mathrm{WRITE}}^{+}(V)^2 + k_B T\Delta\right)^2} \cdot \frac{\left(1 - aI_{\mathrm{WRITE}}^{+}(V)\right)^2\left(1 - eI_{\mathrm{WRITE}}^{+}(V)\right)^2}{b\left(1 - eI_{\mathrm{WRITE}}^{+}(V)\right)^2 + f\left(1 - aI_{\mathrm{WRITE}}^{+}(V)\right)^2}\right|. \tag{29}$$

The probability of the write error (19) can be solved by using the Monte-Carlo simulation. Random numbers of the MTJ resistance and the MOSFET resistance are obtained by multiplying (20), (21), (22), and (23) by random numbers generated from the standard normal distribution.

5. Calculation results

By using (17), (18), and (19), the read voltage V_{READ} and the write voltage V_{WRITE} are plotted as a function of endurance in Figure 11. The I_{READ} and the I_{WRITE} are also plotted in Figure 12. Here, the I_{READ} and the I_{WRITE} are mean currents among the 1Gb STT-MRAM. Parameters used in the calculation are listed in Table I. The V_{READ} and the I_{READ} were calculated both with and without the ECC. A specification for V_{WRITE} is less than 1.2V as indicated in Figure 11. Both $V_{WRITE}^{AP \to P}$ and $V_{WRITE}^{P \to AP}$ satisfy the specification with the parameters in Table I.

It was also found that the MOSFET current of 50μA was required for write operating the 1Gb STT-MRAM. As indicated in Figure 12, I_{READ} of 7μA was required for our sense circuits with the ECC. The I_{READ} with ECC satisfied the specification although I_{READ} without ECC did not for 10^5 endurances. The I_{READ} without the ECC did not satisfy the specification for 10^5 endurances.

6. Conclusion

A comprehensive statistical model was proposed for the read disturbance and the write error. It was demonstrated that the 1Gb STT-MRAM could be realized by using the ECC with the code length of 38.

7. References

[1] M. Hosomi, H. Yamagishi, T. Yamamoto, K. Bessho, Y. Higo, K. Yamane, H. Yamada, M. Shoji, H. Hachino, C. Fukumoto, H. Nagao, and H. Kano, "A novel nonvolatile memory with spin torque transfer magnetization switching: Spin-RAM", IEDM Tech. Dig., 2005, pp. 473-476.
[2] K. Miura, T. Kawahara, R. Takemura, J. Hayakawa, S. Ikeda, R. Sasaki, H. Takahashi, H. Matsuoka, and H. Ohno, "A novel SPRAM (Spin-Transfer Torque RAM) with a synthetic ferromagnetic free layer for higher immunity to read disturbance and reducing write-current dispersion", Symposium on VLSI Tech. Dig., 2007, pp. 234-235.
[3] N. D. Rizzo, M. DeHerrera, J, Janesky, B. Engel, J. Slaughter, and S. Tehrani, "Thermally activated magnetization reversal in submicron magnetic tunnel junctions for magnetoresistive random access memory", Appl. Phys. Lett., Apr. 2002, Vol. 80, No. 13, pp. 2335-2337.
[4] M. P. Sharrock, "Measurement and interpretation of magnetic time effects in recording media", IEEE Trans. Magn., Nov. 1999, vol. 35, pp. 4414-4422.
[5] Y. Higo, K. Yamane, K. Ohba, H. Narisawa, K. Bessho, M. Hosomi, and H. Kano, "Thermal activation effect on spin transfer switching in magnetic tunnel junctions", Appl. Phys. Lett., Aug. 2005, Vol. 87, pp. 082502-082504.
[6] M. Iwayama, T. Kai, H. Aikawa, Y. Asao, T. Kajiyama, S. Ikegawa, H. Yoda, and A. Nitayama, "Reduction of switching current distribution in spin transfer magnetic random access memories", J. Appl. Phys., 2008, Vol. 103, 07A720.
[7] Y. Huai, M. Pakala, Z. Diao, and Y. Ding, "Spin-transfer switching current distribution and reduction in magnetic tunnel junction-based structures," *IEEE Trans. Magn.*, vol. 41, pp. 2621-2625, Oct. 2005.
[8] H. Aikawa, *et al.*, "Effect of joule heating on estimating thermal stability factor", Spintech, Hawaii, Jun. 2007, Poster 88.
[9] S. Ikegawa, H. Aikawa, N. Shimomura, T. Ueda, T. Kai, M. Nakayama, M. Iwayama, H. Yoda, M. Yoshikawa, and Y. Asao, "Pulse duration dependence and effective energy barrier for current-induced magnetization switching in Mg-based tunnel junctions", 6th International Symposium on Metallic Multilayers (IEEE MML2007), Perth, Australia, Oct. 2007, TUE-16.
[10] G. D. Fuchs, I. N. Krivorotov, P. M. Braganca, N. C. Emley, A. G. F. Garcia, D. C. Ralpha, and R. A. Buhrman, "Adjustable spin torque in magnetic tunnel junctions with two fixed layers", Appl. Phys. Lett., 2005, vol. 86, 12509.
[11] M. Yoshikawa, T. Ueda, H. Aikawa, N. Shimomura, E. Kitagawa, M. Nakayama, T. Kai, K. Nishiyama, T. Nagase, T. Kishi, S. Ikegawa, and H. Yoda, "Estimation of spin transfer torque effect and thermal activation effect on magnetization reversal in CoFeB/MgO/CoFeB magnetoresistive tunneling junctions", J. Appl. Phys., 2007, vol. 101, 09A511.

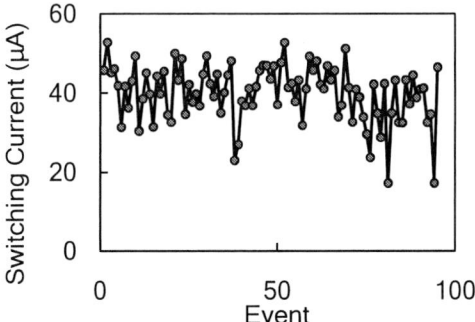

Figure 1. Variable switching currents of the magnetic tunnel junction (MTJ) measured at room temperature. The MTJ has a size of 85nm×110nm. The pulse width of the write current was 5ms. The free layer of the MTJ was 2nm CoFeB.

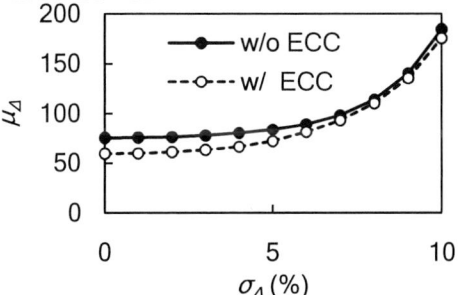

Figure 3. Required statistical parameters of the spin reversal energy of 1Gb STT-MRAM for 10 years data retention. The failure rate is 1ppm. The vertical line is the mean of Δ, and the horizontal line is the ratio of standard deviation and the mean of Δ. Open and closed circles are calculated points with and without ECC, respectively.

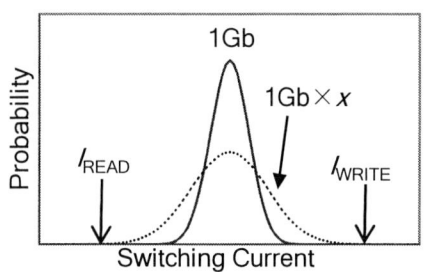

Figure 2. Design of the read current and the write current. A solid line is the probability density function of the switching current of a 1Gb STT-MRAM. A dashed line is the probability of the switching current with x times endurance, dispersion of which is caused by the variable switching current.

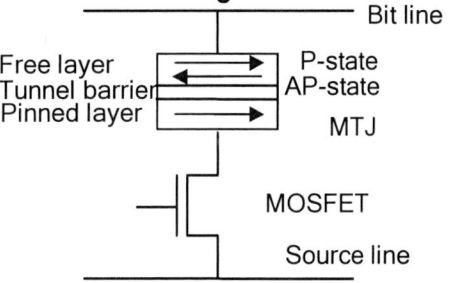

Figure 4. MRAM cell consisting of an MTJ and a MOSFET between a bit line and a source line. When the spin of the free layer is parallel to that of a pinned layer (P-state), the MTJ has low resistance. When the spin of the free layer is antiparallel to that of the pinned layer (AP-state), the MTJ has high resistance.

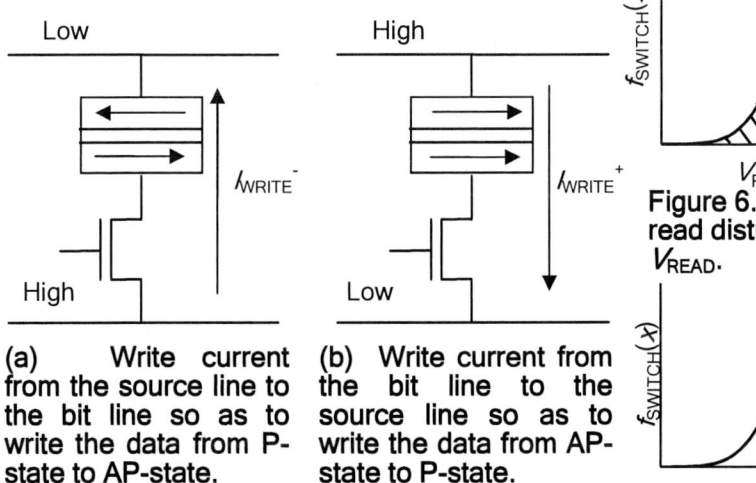

(a) Write current from the source line to the bit line so as to write the data from P-state to AP-state.

(b) Write current from the bit line to the source line so as to write the data from AP-state to P-state.

Figure 5. Directions of write current when the free layer is on the tunnel barrier.

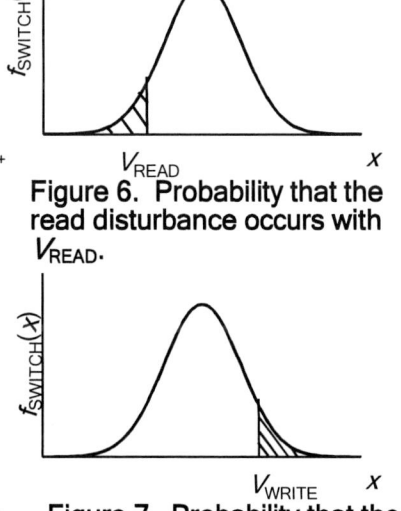

Figure 6. Probability that the read disturbance occurs with V_{READ}.

Figure 7. Probability that the write error occurs with V_{WRITE}.

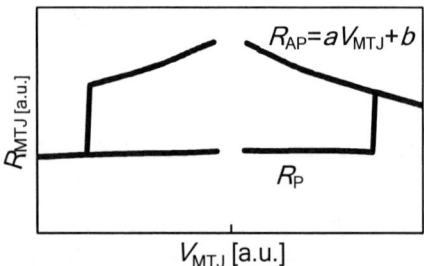

Figure 8. Resistance versus voltage curves of the MTJ. Resistance in AP-state is approximated by a linear equation as $R_{AP}=aV_{MTJ}+b$. Resistance in P-state is independent of the applied voltage because of coherent tunneling current.

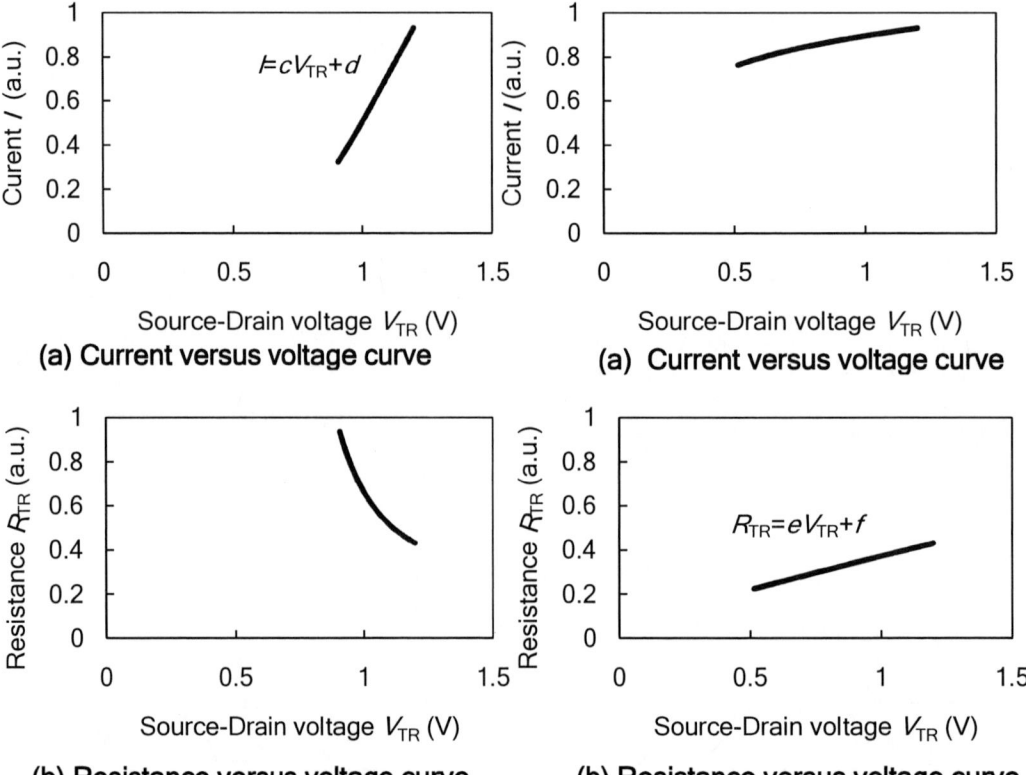

(a) Current versus voltage curve

(a) Current versus voltage curve

(b) Resistance versus voltage curve

(b) Resistance versus voltage curve

Figure 9. Current versus voltage (a) and resistance versus voltage (b) curves of the MOSFET when the write current flows from the source line to the bit line at -10 ℃. Current versus voltage curve (a) is approximated by a linear function as $I=cV_{TR}+d$.

Figure 10. Current versus voltage (a) and resistance versus voltage (b) curves of the MOSFET when the write current flows from the bit line to the source line at -10℃. Resistance versus voltage curve (b) is approximated by a linear function as $R_{TR}=eV_{TR}+f$.

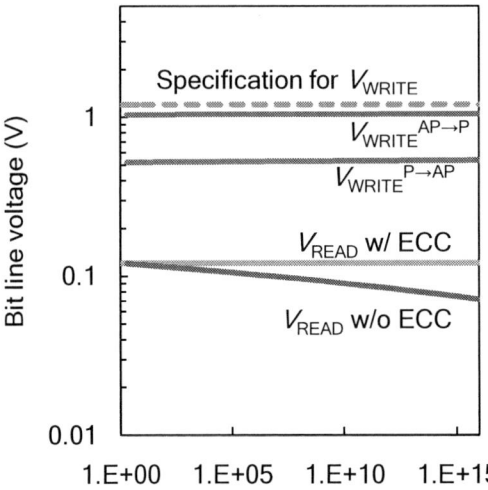

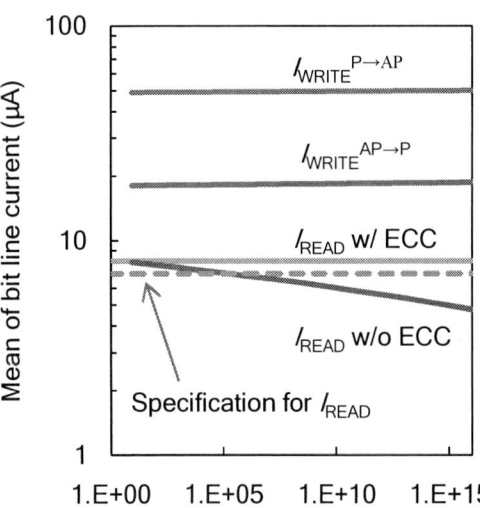

Figure 11. Design of bit line voltages during read and write cycle. A specification for V_{WRITE} is less than 1.2V and is indicated by a dashed line. Both $V_{WRITE}^{AP \to P}$ and $V_{WRITE}^{P \to AP}$ satisfy the specification. The ECC improves V_{READ}.

Figure 12. Design of bit line currents during read and write cycles. The MOSFET current of 50µA was required for operating the 1Gb STT-MRAM. A specification for I_{READ} is greater than 7µA and is indicated by a dashed line. I_{READ} with ECC satisfied the specification although I_{READ} without ECC did not for 10^5 endurances.

Table 1. Parameters used in calculation of Figure 11 and Figure 12.

Symbol	Description	Value
T	Temperature in reading / writing	85 °C / -10°C
t	Read time / Write time	10 ns / 20ns
β	Coefficient to adjust spin reversal energy in reading/writing	0.123/0.091eV/A^2
μ_{IC0}	Mean of I_{C0} in reading	13 µA
μ_{IC0}	Mean of I_{C0} in writing the data from AP-state to P-state	13 µA
μ_{IC0}	Mean of I_{C0} in writing the data from P-state to AP-state	35 µA
σ_{IC0}	Standard deviation of I_{C0}	4 %
μ_{Δ}	Mean of Δ at 85°C	100
σ_{Δ}	Standard deviation of Δ	4 %
μ_{RMTJ}	Mean of MTJ resistance in reading	11 000 Ω
σ_{RMTJ}	Standard deviation of MTJ resistance	4 %
μ_{RTR}	Mean of MOSFET resistance in reading	4 000 Ω
σ_{RTR}	Standard deviation of MOSFET resistance	4 %
$\rho_{IC0\Delta}$	Correlation coefficient of I_{C0} and Δ	0.5
$\rho_{\Delta RMTJ}$	Correlation coefficient of Δ and R_{MTJ}	0.5
$\rho_{IC0RMTJ}$	Correlation coefficient of I_{C0} and R_{MTJ}	0.5
a	Coefficient of MTJ I-V curve in writing from AP to P-state	-9 231
b	↑	12 000
a	Coefficient of MTJ I-V curve in writing from P to AP-state	-4 615
b	↑	7 500
c	Coefficient of MOSFET I-V curve when the I_{WRITE} flows from the source line to the bit line	0.000 1891
d	↑	-0.000 142 8
e	↑ when the I_{WRITE} flows from the bit line to the source line	9 846
f	↑	2 347

A Tile-Based Error Model for Forward Growth of DNA Self-assembly

Masoud Hashempour, Zahra Mashreghian Arani, Fabrizio Lombardi
Northeastern University, Dept. of ECE, Boston MA 02115
Tel: 617-373-4854, Fax: 617-373-8970
Email :{masoud, zahra, lombardi}@ece.neu.edu

Abstract

. . •• ••••• •••• •• • ••• ••• ••••• •••••••• •• ••• ••• •••• •• ••• • • • •••••••• ••• • • ••••
••••• •• • ••• •• •••• • ••••• •• •• •••••••• • ••• • •••• ••••• ••• •••••• •• ••• ••••••••
••••• • ••• •••••• • •• ••• • ••••• •• •••• ••••• ••• • ••• • ••• •••• • ••••• ••••••
•••••••• •• • •••••• • • •••• ••••• •• ••••••• ••• •• •••••• •• •••••• ••••• •• •• •••• •••
•••• • ••• ••••• •• ••••••• • ••• •••• ••• • •••• • •••••• ••• •••••• • • • ••••• ••
• ••• •••••• ••• •••• •• •••••••• • ••••• •• ••••• •••• ••••••• ••• ••••••• • ••• •••••• ••
•• •••••••• ••• •• • •••••• ••• • ••• •• •• ••••• •••• • • •••••• ••• ••• • • • •••••••• • •••
••• ••••••• • •• •••••••• • ••• ••• • •••• •• ••••• ••• • ••••• •••••••• •• •• •••• ••• •••
• •••••• •••• •••••••• ••••• •• ••••••••••• •••

••••• •••• •• Error Tolerance, DNA Self-assembly, Growth, Tiling, Nano Manufacturing.

1 Introduction

Self-assembly is a very promising approach for nano-fabrication at very high density [9]. Assembly by DNA provides a framework to process simple blocks (i.e. the tiles) into complex systems (crystals). Self-assembly relies on the utilization of a tile set in which a so-called seed tile initiates the growth process and proceeds by repeated addition of other tiles (referred to as rule and boundary tiles) to built a pattern along a specified direction of growth. During this process, however several phenomena have been observed to lead to the presence of incorrect (erroneous) tiles. Consider for example the widely used Sierpinski Triangle pattern [4] for DNA self-assembly in nano manufacturing. Figure 1 shows the Xgrow-simulated and AFM images (with no error) of a crystal using DNA self-assembly. Figure 2 shows the AFM image of a crystal grown using the Sierpinski Triangle tile set in which errors (identified by red crosses) are now present.

Tile assembly by DNA tends to be reversible, i.e. tiles can fall off from an existing aggregate [7]. Moreover, incorrect tiles can add and lock into a growing pattern, thus causing the assembly to generate an erroneous pattern as final structure. [2] has reported error rates in a range between 1 to 10 percent; so, a key challenge in DNA self-assembly is to find methods to retain a high level of error tolerance to ensure a correct final assembly [16]. An approach to error tolerance is based on redundancy in the tile set. [2] has proposed the "proofreading" tile scheme; proofreading tile sets utilize massive redundancy at block level to reduce error rates. To decrease this rather large overhead, [7] has proposed an error-resilient tiling method based on 2-way overlay redundancy, however speed of the growth process could be affected. [15] has presented an error-correction scheme referred to as "snake proofreading". A different snake proofreading approach was proposed in [8]; this approach incurs in a smaller redundancy overhead than [15]. [12] has proposed a technique by which error correction/detection can be accomplished for algorithmic self-assembly by modifying the tile set. Through the use of a tile set that allows errors to propagate to an edge of 2D (two-dimensional) assemblies, this technique permits errors to be detected and corrected.

1550-5774/08 $25.00 © 2008 IEEE
DOI 10.1109/DFT.2008.15

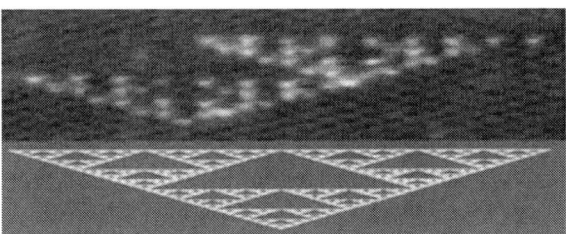

Figure 1. AFM image of error free crystal and simulated pattern for self-assembly based on Sierpinski Triangle [14]

Figure 2. AFM image of DAE E-crystal with errors based on Sierpinski Triangle self-assembly [4]

A previous paper [5] has presented an approach for assessing the error tolerance of tile sets for DNA self-assembly when a healing process (such as punctures) is intentionally induced in the fluid solution for nano-manufacturing. Initially, it has been shown that the combined and optimized solution to the process of tile binding results in a so-called Ideal Tile Set (ITS) [5] that ensures always correct aggregation for healing. Unfortunately, most tile sets for nano-manufacturing are not ITS, hence corrective actions (such as sequentially induced punctures to restart growth in areas of the pattern with erroneous tiles) are often required [10]. Moreover, this analysis is very restrictive because it does not clearly identify the causes of errors as related to the tiles and their bonds.

The objective of this paper is to analyze forward growth from the novel prospective of a single tile attachment to an existing (error-free) aggregate. In this paper, differently from previous manuscripts, the bonding characteristics of the tiles are investigated to analyze forward growth and possible errors. The proposed model is simple and relies on the matching of two sticky ends as labels (bonds) of the tile to be attached; the analysis of the cardinality of the label set is pursued to establish its upper and lower bounds such that only the possible errors are considered. Based on the bonding characteristics of the attachment, error tolerance metrics are proposed to assess the impact of single and double mismatches. A detailed evaluation is then pursued with respect to tile sets with and without redundancy in the generation of the pattern.

This paper is organized as follows. Setion 2 deals with the preliminaries inclusive of a brief review of DNA self-assembly. Forward growth is analyzed in Section 3 through binding and the tile set. This analysis is used in Section 4 to formalize a novel model. In Section 5, the detailed evaluation of the Sieripinski tile set is initially presented as an example; the error tolerance of other tile sets is also pursued. The last section provides the conclusion of this work.

2 Preliminaries

In this section, a brief review of different aspects of DNA self-assembly is presented. The basic principle is to utilize the programmability of DNA tiles to self-assemble into lattice structures (i.e. crystals) on which it is then possible for other molecular electronic devices to selectively attach. This technique has been advocated as one of the most promising methods for "bottom-up" nanofabrication.

The DNA tile is the basic element of this process. Each tile has four DNA branched junctions and each junction has four arms with sticky ends. Tiles can be designed using different sticky-ends to control self-assembly. The •••••••• • ••• • •••• ••• • •••••••• • • provides the basis for self-assembly in ideal cases. A •••• ••• consists of a finite set of unique •••• that are used to self-assemble into a DNA •••••••. A tile is assumed to be square and tiles can not rotate by assumption. Each of the four sides of a tile has a •••• ••••. The bond types of a tile determine the uniqueness of the tile. Each bond type has an associated •••• ••••••••. Bond types can be null (strength of 0), single (strength of 1) or double (strength of 2). Two bonds of the same type can glue (i.e. bond) together, with a corresponding bond strength. It is assumed that the strength between different bond types is always 0. In the ideal process assumed by the aTAM model, self-assembly always begins with a •••• ••••. A tile can be added to the existing crystal when its bond strength to the crystal is greater than or equal to 2.

In practice, a • ••• •••• (as an error) may occur during this process. Additionally, a tile can also fall off from a crystal. The •••••• • ••• • •••• ••• • •••• •••• • • provides a framework to analyze and simulate the non-ideal scenario of self-assembly. The kTAM model includes rates for both ••••••• ••• • ••• ••••••••••• of tiles from the crystal. In this model, it is assumed that the on-rate (association) r_{on} is determined only by the tile concentration, that is represented by the parameter G_{mc}. The off-rate r_{off} (dis-association) is determined by the total bond strength b and the parameter G_{se}. These rates are given by $r_{on} = k \cdot e^{-G_{mc}}$ and $r_{off,b} = k \cdot e^{-bG_{se}}$ where k is a constant, $G_{mc} > 0$ measures the entropic cost of fixing the location of a monomer unit (and thus is dependent upon monomer concentration) and $G_{se} > 0$ measures the free energy cost of breaking a single sticky-end bond; both are expressed with respect to the thermal energy. b is the total bond strength that hold the tile to the crystal, measured by multiples of the unit bond strength. While error tolerance has been extensively dealt in the technical literature [13] [2] [5] [6], there is no comprehensive model to assess it with respect to growth. This aspect will be dealt in the next few sections.

3 Growth

In this section, two important aspects of growth (namely the binding model and the tile set) are considered to establish novel conditions by which errors can be modelled.

3.1 Binding Model

This paper deals with the generic instance of forward growth (from south-east to north-west) for a single tile attachment (backward growth is usually associated with healing of DNA self-assembly [11]; healing is not considered in this manuscript). This type of growth occurs in the so-called south-east corner (Fig. 3): a tile (white box) attaches to an existing error free aggregate (made of grey colored tiles) through binding of the bonds in its sticky ends. The attachment scenario depicted in Fig. 3 is applicable to growth of the interior part of the pattern, i.e. the boundary and the seed are excluded (error tolerance of the boundary is not important in most cases of nano manufacturing and seed errors are usually fatal [6]). In the literature [10], the south-east corner of an aggregate is also referred to as an •• ••• ••••.

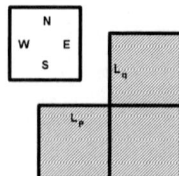

Figure 3. south-east corner in forward growth from south-east to north-west

The following types of attachment are possible: (1) •••••••••••••••••• both the south and east sides of the empty corner will be binding to a tile with matching ends; Fig. 4 shows such an attachement for the empty site of Fig. 3. (2) •••••••••••••••••••••••••••• only one side (south or east) of the empty corner will be binding to a tile with a matching end (i.e. the other side is mismatched); Fig. 5 shows such an attachement for the empty site in Fig. 3. (3) •• ••••••••••••••••••• both sides (south and east) of the empty corner will be binding with a tile with mismatched ends.

During growth in the self-assembly process, a tile will attach to the aggregate at the corner (empty site). Based on the tile to be attached to a corner, different bindings may occur: a ••••••• ••••••• occurs when both the east and south sides of the tile will match the east and south sides of the corner; a •••••••••••• is said to occur when at least one of both sides of the tile will not match its corresponding side at the corner. It has been proved in the literature [6], that an attachment with weak binding by two mismatches will likely disassociate as it is not stable. Hence, the scenario of an erroneous tile attachment will likely result from a weak binding with only one mismatch [5] [6].

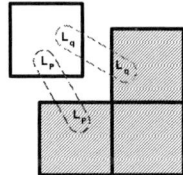

Figure 4. Attachment with two matched sticky ends

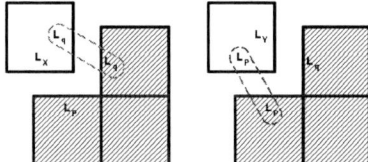

Figure 5. Attachments with one matched sticky end

3.2 Tile Set Analysis

Consider a generic assembly and its tile set (made of N_A completely different tiles). This generic tile set has $4 \times N_A$ different labels (4 labels for each tile) and can be easily generated using existing approaches. Such tile set is usually referred to as a trivial tile set due to its proven existence (for any pattern) and its construction (i.e. with a large number of tiles and bonds). The following sets should be considered for an assembly (target) and the generation of a specific pattern (crystal).

- •••••••••••• This set contains all tiles (rule, boundary and seed tiles) that are used for the target self assembly; its cardinality is related to the cardinality of the generic tile set as $1 \leq N_T \leq N_A$.

- •••••••••••••• This set contains all labels that are used for each sticky end of the tiles in the set (K denotes the set cardinality i.e. the total number of labels in this set). Therefore, $1 \leq K \leq 4 \times N_T \leq 4 \times N_A$.

A generic tile set can be synthesized [13] to generate the set for the desired pattern. Let a tile with a N1 label at north, a E1 label at east, a S1 label at south and a W1 label at west be denoted by (N1,E1,S1,W1). For forward growth (from south-east toward north-west), a tile set must generate a unique assembly [13]; hence, the south-east ordered pair (-,E,S,-) should be also unique. This yields the following Lemma to establish the range of K.

$\bullet\bullet\bullet\bullet\bullet\bullet$ $\left\lceil\sqrt{N_T}\right\rceil \leq K \leq 4 \times N_T$.

$\bullet\bullet\bullet\bullet\bullet$ The proof of the upper bound of K is trivial, i.e. each side (bond) of a tile is unique, each tile has four sides and all tiles are different. Therefore, $K \leq 4 \times N_T$ as corresponding to the so-called trivial tile set (as equivalent to the generic tile set of maximum cardinality).

For the lower bound, let $K = 1$, then only one tile ($N_T = 1$) is required as a rule tile, i.e. $L = \{a\}$ then $T = \{(a, a, a, a)\}$. In general when $K > 1$ there are at most K^2 different combinations of east and south sides of a tile (as in forward growth from south-east to north-west, these sides should be unique). So the maximum value of N_T is K^2 and $N_T \leq K^2 \implies \left\lceil\sqrt{N_T}\right\rceil \leq K$.

A tile set (T) with a label set L can be represented as a 4-tuple, i.e. N_i, E_i, S_i, and W_i represent the north, east, south, and west labels of a tile T_i in T. Therefore, $T = \{T_1, ..., T_{N_T}\}$, $L = \{L_1, ..., L_K\}$ and $T_1 = (N_1, E_1, S_1, W_1)$, $T_2 = (N_2, E_2, S_2, W_2)$, ..., $T_{N_T} = (N_{N_T}, E_{N_T}, S_{N_T}, W_{N_T})$.

In general, the number of labels (K) is significantly less than $4 \times N_T$, therefore some labels appear in multiple tiles and sides. So a K-tuple can be defined to address this label property for each side, i.e. $No = (No_1, ..., No_K)$, $Ea = (Ea_1, ..., Ea_K)$, $So = (So_1, ..., So_K)$ and $We = (We_1, ..., We_K)$. where No_J in the above K-tuple denotes the number of tiles that have the label L_J at their north side. Ea_J denotes number of tiles which have the label L_J at their east side; similar definitions apply to So_J and We_J for south and west. This yields the following conditions:

$$0 \leq No_i \leq N_T \qquad i \in \{1, ..., K\}$$
$$0 \leq Ea_i \leq N_T \qquad i \in \{1, ..., K\}$$
$$0 \leq So_i \leq N_T \qquad i \in \{1, ..., K\}$$
$$0 \leq We_i \leq N_T \qquad i \in \{1, ..., K\}$$
$$\sum_{i=1}^{K} No_i = \sum_{i=1}^{K} Ea_i = \sum_{i=1}^{K} So_i = \sum_{i=1}^{K} We_i = N_T$$

The characteristics of the label set (and in particular a label appearing on multiple tiles and sides) strongly influence the process of tile attachment, as discussed next.

4 Attachment Model

Consider the scenario depicted in Fig. 3 for a tile set T of cardinality N_T. This is the south-east corner for a single tile attachment, i.e. the east label of this empty site is L_q and the south label is L_p. For a unique assembly, there must exist a unique tile with an orderd pair for the south-east corner given by $(-, L_q, L_p, -)$; this corresponds to the correct (0 mismatch) attachment. So if n_2'' denotes the number of combinations with two matching sticky ends for this empty site, then for having a unique assembly, $n_2'' = 1$. Using the previously described K-tuple for the east and south labels (E and S respectively) the number of combinations with one and zero matching ends are given by $n_1'' = (Ea_{L_q} + So_{L_p} - 2)$ and $n_0'' = N_T - (Ea_{L_q} + So_{L_p} - 1)$, and $n_2'' + n_1'' + n_0'' = N_T$.

For a pattern, only some south-east corners are possible as result of an existing fault-free aggregate; let C denote the number of possible south-east corners (empty sites) as shown in Fig. 3. In this case, the number of possible corners is not the exhaustive combination of the east and south labels; it is equal to at most the cardinality of the tile set (i.e. N_T).

The proposed model relates the number of combinations of sticky end pairs (and labels) in the matching process for an attachment to the errors in south-east corners, i.e. not the whole self-assembly process. The following two extreme cases are possible:

$\bullet\bullet\bullet\bullet\bullet\bullet$ there is only one label (i,e. a) in the label set ($K=1$); hence, there is only one rule tile in the set ($N_T = 1$) and $No = (1), Ea = (1), So = (1), We = (1)$. The number of combinations are given by $n_2'' = 1$, $n_1'' = (Ea_a + So_a - 2) = 1 + 1 - 2 = 0$, $n_0'' = N_T - (Ea_a + So_a - 1) = 1 - (1 + 1 - 1) = 0$. This is equivalent of the scenario of an ideal tile set as analyzed in a previous paper [5]; in this case, only a very simple assembly can be generated with this tile set.

• • • • • • all labels are unique ($K = 4 \times N_T$), then $No = (L_1 = 1, ..., L_{4N_T} = 1)$, $Ea = (L_1 = 1, ..., L_{4N_T} = 1)$, $So = (L_1 = 1, ..., L_{4N_T} = 1)$ and $We = (L_1 = 1, ..., L_{4N_T} = 1)$; moreover, $n_2'' = 1$, $n_1'' = (Ea_a + So_a - 2) = 1 + 1 - 2 = 0$ for all south-east corners. and $n_0'' = N_T - (Ea_a + So_a - 1) = N_T - (1 + 1 - 1) = N_T - 1$. Also, this corresponds to an ideal tile set; however, the cardinality of the tile set is very large and impractical due to the large laboratory work required for generating the target assembly. Note that through synthesis [13], the cardinality of the tile set can be possibly reduced; unfortunately, the synthesized tile set will also show different number of combinations for the attachment.

As based on the tile attachment of Fig. 3 the proposed model has the following features: (1) It is not an exhaustive model as it considers only the possible south-east corners, as defined by the target pattern. (2) The number of attachments per corner is equal to the cardinality of the tile set, not the exhaustive combinations in the bond sets for the sides of a tile. (3) The proposed model reflects a realistic scenario of DNA self-assembly; it assumes that given a sufficient growth rate, an attachment with two mismatches will fall off the aggregate (and not appearing in the final pattern) as highly unstable [6].

Let C_j (j=0,1,2) denote the number of different tiles in the set T that can be attached to a specific corner with j mismatches, i.e. $C_0 + C_1 + C_2 = N_T$. Hence, for a specific pattern (with N_T tiles), C_0=1 and then, at a possible south-east corner $C_1 + C_2 = N_T - 1$. Let $R_C = (C_2 - C_1)/(C_2 + C_1)$ be referred to as the • • • • • • • • • • • • • • • • • •, i.e. if C_2 (i.e. the number of combinations with 2 mismatches) increases and C_1 is small, then the considered corner will be tolerant to erroneous attachments as double mismatches will fall off. If C_1=0 and no single mismatch error can occur at the corner, R_C=1; such corner is said to be a "safe corner". This metric can be extended to a tile set, i.e. let $R(x)$ be referred to as the • x and given by $R(x) = N_1/C$, where N_1 is the number of safe corners with 1-mismatch and C is the number of possible corners.

5 Evaluation

Initially in this section, different Sierpinski tile sets (i.e. with no redundancy and with proofreading redundancy) are analyzed using the proposed model.

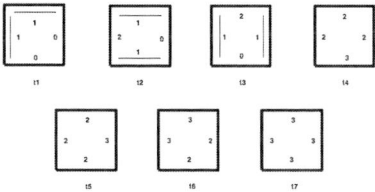

Figure 6. Sierpinski Tile Set (no redundancy)

(1) • • • • • • • • • The Sierpinski tile set (with no redundancy) is shown in Fig. 6, N_T=7. There are four possible corners (C=4) and the results are shown in Table 1. As R_C is equal for all possible corners (constant), then corner errors are independent of the aggregation and $R(\text{Sier})$=0.

(2) • The 3×3 proofreading Sierpinski tile set is shown in Fig. 7, N_T=43. In this set, a 3×3 block is utilized for each rule tile of the Sierpinski tile set; as per construction, unique bonds appear inside each block. The attachments at an empty corner of a block are given in Table 2 (note that the analysis of the other blocks is exactly the same as the one in Table 2).

Two significant differences are evident when comparing 3×3 proofreading (3P) with the Sierpinski tile set with no redundancy: (a) As there are more rule tiles, N_T and C_2 of 3P have larger values and in most cases C_1 is reduced; (b) For some corners, C_1=0 as unique bonds appear inside each block of a rule tile (N_1=4, C=9).

521

Table 1. Attachements for the Sierpinski tile set

empty site	C_0	C_1	C_2
•••• •• •••	1	2	4
•••• •• •••	1	2	4
•••• •• •••	1	2	4
•••• •• •••	1	2	4

For error tolerance of these tile sets, $R(\text{Sier})=0$ while $R(\text{Sier-3P})=4/9$; this proves that the introduction of redundancy in the proofreading tile set results in an improvement of error tolerance (albeit at an increase in size and tile set cardinality). By increasing the size of the blocks in the proofreading tile set, the following effects will occur; (a) For each possible corner, C_2 increases; (b) The number of corners with $C_1=0$ increases. Hence for a $N \times N$ proofreading approach, $R(\text{Sier-}NP)$ will be greater than $R(\text{Sier-}(N\text{-}1)P)$, as reflecting a higher tolerance due to a larger amount of redundancy.

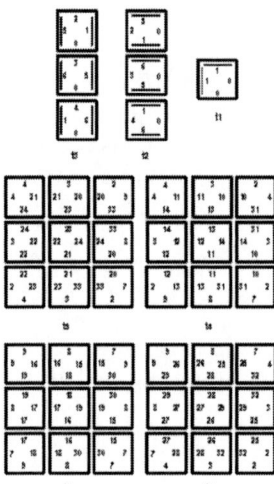

Figure 7. 3x3 Proofreading Sierpinski Tile Set

The proposed model has also been applyed to other tile sets, with and without redundancy. The results are given in Table 3. The AB tile set is ITS, hence there is no need to add redundancy. In all other cases, the addition of proofreading redundancy (3P) results in an improvement of the error tolerance as more safe corners appear. Moreover, the error tolerances of the tile sets reported in Table 3 confirm the simulation results of [5] and [13] in which an extensive experimental evaluation was reported.

6 Conclusion

This paper has presented a model based on a single tile attachment for DNA self-assembly. This model is applicable to forward growth and considers the binding features of the two bonds to match an existing (error free) aggregate. This model has captured the strong and weak bindings (for both single and double mismatches) that may occur in an aggregate; by considering the two bonds in an attachment (as occurring at a south-east corner for forward growth), it has been shown that the proposed model is capable to consider only the errors that are possible at an empty site. This

Table 2. Attachements for the 3x3 Proofreading Sierpinski tile set

empty site	C_0	C_1	C_2
•••• •• •••	1	2	40
•••• •• • •••	1	1	41
•••• •• • •••	1	1	41
•••• • •• •••	1	1	41
•••• • ••• •••	1	0	42
•••• • ••• •••	1	0	42
•••• • •• •••	1	1	41
•••• • ••• ••	1	0	42
•••• • ••• •••	1	0	42

feature considerably reduces the complexity of the evaluation process of error tolerance as it relates the two labels in the attachment to the tiles in the tile set. The upper and lower bounds on the cardinality of the label set for a tile set of a given size have been proved in this manuscript; based on the label set, it has been shown that safe corners (i.e. empty sites for attachments that can not be resulting in one mismatch) are possible in the self-assembly process. An extensive evaluation on existing tile sets (with and without redundancy) has been presented; it has been shown that error tolerance as established using the proposed model is effective in assessing the advantages of proofreading as an error tolerant method.

Current work is being pursued to extend the proposed model to backward growth (as applicable to healing) as well as defining a probabilistic analysis which would take into account association and disassociation rates in the self-assembly process.

Table 3. Error Tolerance of Tile Sets

Tile Set	R(Tile Set)
AB	1
Barcode	0
Barcode-3P	0.66
Barseed	0.33
Barssed-3P	0.66
Binarycounter	0
Binarycounter-3P	0.44
Line1	0
Line1-3P	0.66
Line2	0
Line2-3P	0.66

References

[1] E. Winfree, "Algorithmic Self-Assembly of DNA," Doctoral Thesis, Pasadena, California Institute of Technology, May, 1998.

[2] E. Winfree and R. Bekbolatov, "Proofreading tile sets : Error correction for algorithmic self-assembly," Proceedings of the Ninth International Meeting on DNA Based Computers. Madison, Wisconsin, June 2003, also in Lecture Notes in Computer Science, Vol. 2943, pp. 126-144, Springer, Berlin 2004.

[3] E. Winfree, "Self Healing Tile Sets " Nanotechnology: Science and Computation, pp. 55-78, Springer, Berlin, 2006.

[4] Rothemund, P.W.K, N. Pakadakis and E. Winfree, "Algorithmic self assembly of DNA Sierpinski Triangles," *PLoS Biology*, 2(12) e424, 2004.

[5] M. Hashempour, Z. Mashereghian Arani and F. Lombardi, "A Metric for Assessing the error Tolerance of Tile Sets for Punctured DNA Self-assemblies," *Proc. IEEE VTS*, San Diego, May 2008.

[6] X. Ma, J. Huang and F. Lombardi, "Modeling Facet Roughening Errors in Self-Assembly by Snake Tile Sets," *Proc. IEEE International Test Conference*, Santa Clara, October 2007 CD, IEEE number: 07CH37892C, ISBN: 1-4244-1128-9, paper 27.3 (10 pages).

[7] J. H. Reif, S. Sahu, and P. Yin. "Compact error-resilient computational DNA tiling assemblies," In DNA Computing 10, Berlin Heidelberg, 2004.

[8] X. Ma, J. Huang and F. Lombardi, "Error Tolerant DNA Self-Assembly Using $(2k-1) \times (2k-1)$ Snake Tile Sets " *IEEE Transactions on Nano-Bio Science*, vol. 7, no. 1, pp. 56-64, 2008.

[9] S. Zhang, D.M. Marini, W. Huang and S. Santoro, "Design of Nanostructured Biological Materials through Self-Assembly of Peptides and Proteins," Current Opinions in Chem. Biology, Vol. 6, pp. 865-871, Elsevier, 2002.

[10] M.Hashempour, Z.Mashreghian A., and F. Lombardi, "Error Tolerance of DNA Self-Healing Assemblies by Puncturing," 22nd IEEE International Symposium on Defect and Fault Tolerance in VLSI Systems (DFT'07), pp. i400-408, Rome, Italy, September 2007.

[11] M.Hashempour, Z.Mashreghian A., and F. Lombardi, "Robust Self-Assembly of Interconnects by Parallel DNA Growth ," IEEE/ACM Symposium on Nanoscale Architectures (NANOARCH'07), pp. 70-76, San Jose, USA, October 2007.

[12] Frechette S. and F. Lombardi, "Error Detection/Correction in DNA Algorithmic Self-assembly" *Proc. IEEE DATE08,* Munich, April 2008.

[13] X. Ma, and F. Lombardi, "Synthesis of Tile Sets for DNA Self-assembly," *IEEE Trans. on CAD of ICAS,* (accepted).

[14] E. Winfree, Interview in Technology Research News, "Programmed DNA Forms Fractal," by K. Patch, Technology Research News, May 4, 2006.

[15] H.-L. Chen and A. Goel, "Error free self-assembly using error prone tiles", DNA Computing 10, LNCS volume 3384:62-75, 2005

[16] J. H. Reif, S. Sahu and P. Yin, "Compact Error-Resilient Computational DNA Tiling Assemblies," Tenth International Meeting on DNA Based Computers (DNA10), Milano, Italy, June 7-10, 2004. Lecture Notes in Computer Science (Edited by C Ferretti, G. Mauri and C. Zandron), Vol. 3384, Springer-Verlag, New York, (2005), pp 293-307. Published as an invited chapter in "Nanotechnology: Science and Computation, Springer Verlag series in Natural Computing (edited by Natasha Jonoska), Springer-Verlag Berlin, Germany, pages 79-104, 2006.

IEEE International Symposium on Defect and Fault Tolerance of VLSI Systems

Checkpointing of Rectilinear Growth in DNA Self-assembly

S. Frechette+, Y. B. Kim++ and F. Lombardi++
+Riverside Research Institute
Lexington, MA USA
++Northeastern University,
ECE Department
Boston, MA 02115

Abstract

checkpointing, error tolerance, DNA self-assembly, tiling.

1 Introduction

Self-assembly has been widely advocated as an efficient manufacturing environment for nano-technologies; self-assembly can be used to avoid expensive nano-lithography in building structures (such as scaffolds), to which other molecular-based electronic devices can be attached. This process uses complexes commonly referred to as ••••; for example, in rectangular structures a tile has four types of DNA strands. For self-assemblying into specified patterns, a large number of tiles (made of DNA strands) are programmed through their bond types. Previous works have reported that the assembly of incorrect tiles occurs with error rates between 1 to 10 percent [2]; as millions of molecules are usually involved, the presence of these errors represents a serious challenge for efficient manufacturing in the nano ranges. Therefore •••••••• ••• ••••••••• •••••••• are needed.

Several techniques have been proposed. Error correction techniques that require additional matching ends, have been proposed in the literature; in [15], pads between tiles decrease the error rate by requiring more bonds to match for each assembled tile. The use of an additional layer for error rate reduction requires two layers of tiles to match perfectly (as opposed to a single layer) [16]. Two proofreading methods for error correction include a tile set proposed in [2] (in which a so-called block consists of four proofreading tiles) and the snake tile set [10] (that requires a block of sixteen proofreading tiles). The use of checkpoints for DNA-self assembly has been investigated in previous works; [14] has proposed, simulated, and evaluated a checkpoint scheme based on ••• •••••••• •••• ••• during crystalline array growth. Periodic pulses remove defective, and adversely affect some of the correct (error-free) sections of the crystalline arrays. In this process, it is assumed that a temperature pulse (i.e., a temporary increase in temperature) removes all defective regions of the structure. Although a portion of the removed tiles are correct, incorrect tiles are removed at a higher rate. Additionally in [14], the number of temperature pulses required for an error-free assembly, and the optimal temperature pulse checkpointing interval, have been established. This paper presents a detailed analysis to extend the process of [20] to tile checkpointing for rectilinear growth in DNA self-assembly. This analysis utilizes the Error Isolation Tiles of [20] while introducing a physical

1550-5774/08 $25.00 © 2008 IEEE
DOI 10.1109/DFT.2008.10

framework by which erroneoues sections of the aggregate can be removed. Checkpointing is then analyzed under a novel Markov model to establish the optimal rate. This technique is compared to existing redundancy based techniques. It is shown to exhibit superior performance by solving the proposed Markov model and simulation; a substantial reduction of error rate is also achieved due to the periodic execution of the checkpointing process.

2 Review of Error-Tolerance for DNA

In algorithmic self-assembly, the growth of a DNA crystal is used for information processing. A set of DNA tiles is used to execute an algorithm. The ·········· · ··· · ···· ··· · ·········· · · [18] provides the basis for analysis of algorithmic self-assembly in an ideal case. A ···· ··· consists of a finite set of unique ···· that is used to self-assemble into a DNA ········. A tile is assumed to be square and tiles can not rotate by assumption. Each of the four sides of a tile has a ···· ····. The bond types of a tile determines the uniqueness of the tile. Each bond type has an associated ···· ········. Bond types can be null (strength of 0), single (strength of 1), or double (strength of 2). Two bonds of the same type can glue (i.e. bond) together, with a corresponding strength. It is assumed that the strength between different bond types is always 0. In an ideal process, self-assembly always begins with a ···· ···. A tile can be added to the existing crystal when its total bond strength to the crystal is greater than or equal to 2. The crystal generated from the seed tile via a series of legal (correct) tile additions is referred to as the ········ ····· ··. The ····· in the tile denotes the ···· ····. In practice, a · ··· ···· may occur during this process, such that a tile with a total bound less than 2 is attached. Additionally, a tile can also fall off from a crystal.

The ··· ···· · ··· · ···· ··· · ········· · · [18] provides a framework to analyze and simulate the non-ideal self-assembly. The kTAM model includes rates for both ·········· ··· ·············· of tiles from the crystal. In this model, it is assumed that the on-rate (association) r_{on} is determined only by the tile concentration, that is determined by the parameter G_{mc}. The off-rate r_{off} (dis-association) is determined by the total bound strength b that holds the tile to the crystal and the parameter G_{se}. These rates are given as follows: $r_{on} = k \times e^{-G_{mc}}$ and $r_{off,b} = k \times e^{-bG_{se}}$, where k is a constant, G_{mc} is the physical parameter measuring the tile concentration, while G_{se} is the physical parameter measuring the unit bond strength.

This paper specifically addresses checkpointing for detection and correction of ···· ·· ·······, which are defined as weakly-bonded tile attachments at a location where another tile could and should attach. Growth errors occur if there exist at least two incorrect (erroneous) attachments upon the same four-sided tile. The error correction technique presented in [20] is commonly referred to as the · ···· ··· ·· ··· · ··· · ·····. · ····· ···· are defined as any tile on an edge of an assembly, in which growth has ceased. In the model [20], border tiles assemble at the same rate as the entire assembly, thus a tile mismatch is allowed to ········· ·· ·· ····, and causes a ······ ··· ··· · to occur in the border tiles, i.e., the initial error results in a disjointed line in the border tiles, that should otherwise be straight. At the break in the border line using the proposed method, a so-called · ···· ······· · ·· attaches and thus, it effectively *tags* that section of the assembly for removal. A growth error is illustrated in Figure 1; the disjoint line (as effect of this error) is evident.

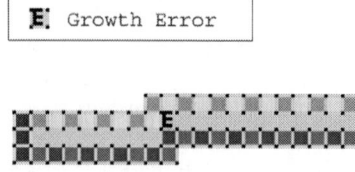

Figure 1. Initial Growth Error.

Although a set of Error Isolation Tiles that attach to every possible defective structure of two adjacent tiles could be constructed, the number of tiles in such an Error Isolation Tile set would be prohibitively large, thus very inefficient in terms of overhead in the final assembly. In addition to a tile mismatch, it is possible that a growth error is caused by weak bonds. The bond strength must not allow for shifting a subsequent column up or down by one row; if this occurs, then it is considered a growth error. The approach of [20] requires only two additional tiles, which are referred to as the Error Isolation Tile set. The rectilinear growth of the assembly results in the need for two tiles in the tile set, as opposed to just one tile. (1) The first Error Isolation Tile attaches to the two disjoint border tiles from the southeast. (2) The second Error Isolation Tile attaches from the southwest. Errors are detected when an Error Isolation Tile attaches to the assembly at the location of two disjoint and adjacent border tiles, which can only be adjacent in error. Any growth error results in a disjoint line at the southern border tiles, so only two Error Isolation Tiles are required to detect a large number of possible growth errors [20]. Border tiles are employed in most tile sets for algorithmic self-assembly [9]. In most tile sets these tiles assembly prior to the interior tiles. Thus, propagation of an error within the interior tiles will be blocked by the border tiles and prevented from reaching an edge of the assembly. In the proposed model, border tiles in the tile set may only assemble if an interior tile next to the border is assembled. The growth error in the interior will cause a break in the straight line of the border tiles. The border tiles make up the disjoint bottom row. The set employed to demonstrate the Error Isolation Tile correction technique of [20] is a modified version of the linear tile set presented in [1]. The method proposed in [20] can be applied to a modified version of any tile set that exhibits algorithmic self-assembly, such as the Sierpinski triangle or the binary counter tile sets of [9].

3 Physical Framework

In the method of [20], all Error Isolation Tiles hold a metal, i.e. • •••••••••• •••••. Thus, the assemblies with an attached Error Isolation Tile can be attracted to a charged metal and a section of the assembly (within a given radius of the Error Isolation Tile) can be raised in temperature, thus causing bonds to break and a partition of the assembly to occur. This partition divides the assembly into error-free and erroneous (defective) sections. This is accomplished as follows. (1) The device that removes the defective section of the assembly is referred to as the • ••••••••• • ••, and is shown in Figure 2a. An array of partitioner cells is illustrated in Figure 2b. The Error Isolation Tiles are attached to the charged metal strip, known as the Error Isolation Tile • ••••••••, this is shown as (a) in Figure 2a. The portion of the assembly to be removed contains the initial error and all tiles to the right of the error, so the assembly attaches to the Partitioner Cell with the Error Isolation Tile on the south side. (2) To ensure a proper orientation of the assembly, a 3-D • ••••••••• • ••• (shown as (b) in Figure 2a) is employed to prevent the assembly from attaching to the Partitioner Cell at an incorrect orientation. (3) Lastly, the • ••••••••• • •••• (shown as (c) in Figure 2a) is a 2-D strip of metal that heats the local portion of the assembly directly above it, thus causing the bonds (also directly above it) to break. Figure 3a illustrates an occupied Partitioner Cell. After an assembly attaches to a Partitioner Cell and the portion of the assembly (that is located above the Partitioner Strip) breaks, the assembly is partitioned into error-free and erroneous sections. This is illustrated in Figure 3b. The erroneous section of the assembly remains connected to the charged strip, and the error-free sections of the assembly return to the solution. The array of Partitioner Cells continues to fill with defective portions of the assembly, so there must be sufficient arrays of Partition Cells for all assemblies with Error Isolation Tiles attached. Arrays of Partitioner Cells are always immersed in the solution; they attract only erroneous assemblies when the self-assembly is in a checkpointing state. The checkpoint process is therefore complete, and it may repeat at the next checkpoint for a set interval.

The section of the assssembly located above the Partitioner Strip ((c) in Figure 2a), is removed, as well as the subsequent tile growth that could have been affected by the initial error. The Orientation Bar (given by (b) in Figure 2a) ensures that the erroneous assembly attaches at the proper orientation with respect to the Partioner Cell. A proper orientation is required, such that only tiles near

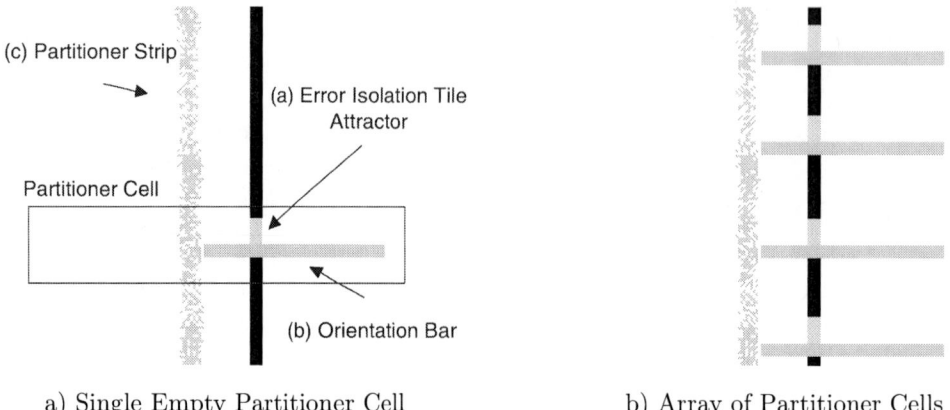

a) Single Empty Partitioner Cell   b) Array of Partitioner Cells

Figure 2. Partition Cells.

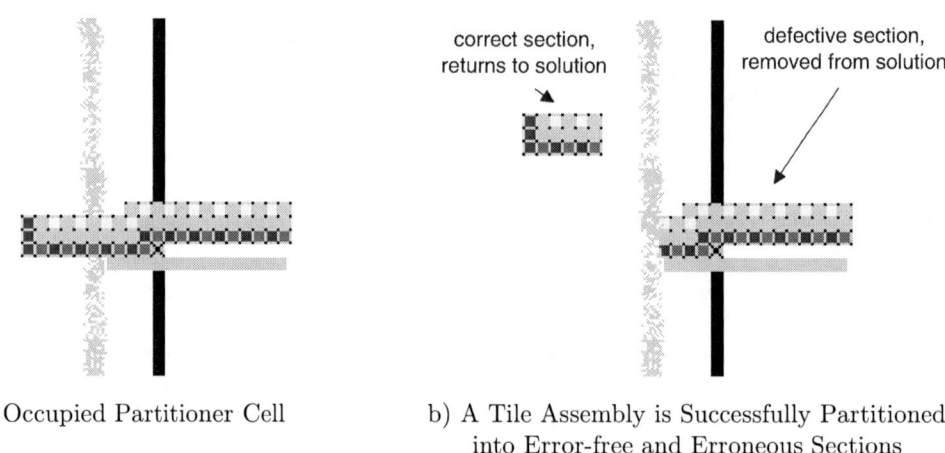

a) Occupied Partitioner Cell   b) A Tile Assembly is Successfully Partitioned
into Error-free and Erroneous Sections

Figure 3. Example of an Occupied Partitioner Cell Before and After Partitioning an Erroneous Tile Assembly.

and to the east of the error are removed. The checkpointing process occurs over a period of time; in the current state of self-assembly, ten seconds is an estimate of the length of the time required for this process. Since the items are in a solution, it is estimated that ten seconds are required for all assemblies in a solution to pass close enough to the charged metal strip and to become attached.

4 Optimal Checkpointing

The optimal checkpoint interval is determined in terms of the error rate per tile (denoted by λ). λ is given by $2 \times N \times e^{-Gse}$ in [2], where N is the number of tiles in the set ($N = 10$ for the $3 \times Y$ rectangular tile set of [20]). The rate at which an erroneous assembly attaches to a Partitioner Cell and returns to the solution, is given on a per assembled tile basis, and is represented by μ. The optimal checkpoint interval for the removal of the erroneous (defective) sections of the assembly can be determined. Given an error rate $\lambda = 2 \times N \times e^{-Gse}$, and the rate at which an erroneous tile attaches and then partitions using a Partitioner Cell (given by μ) the optimal checkpoint rate for error detection and correction is determined in this subsection.

A Markov model is employed to represent the states that an assembly is classified. A discrete-time Markov model is used; in this model, transitions occur at fixed time intervals of Δt. The

Markov diagram consists of the three possible states of an assembly and is shown in Figure 4. The value of a directed edge in the Markov diagram is the probability of moving to the respective state during the time step Δt, i.e. Δt represents the time required for a tile to attach to the assembly.

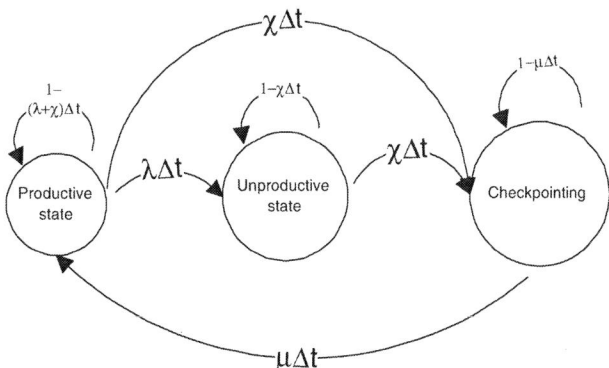

Figure 4. Markov Diagram for the Error Isolation Tile Correction Method.

The Markov diagram of Figure 4 has the three possible states of an assembly. (1) When an assembly is in the • •••• ••••• •••••, tiles can attach to the assembly, thus an Error Isolation Tile cannot be attached to the assembly. (2) If the assembly is in the • ••••••••••• •••••, then a growth error has occured. In the model [20], it is estimated that an Error Isolation Tile will always attach to an erroneous assembly prior to the occurrence of checkpointing. Since an Error Isolation Tile will eventually attach near the growth error, at the next checkpoint all subsequently attached tiles will be removed along with the entire erroneous section of the assembly. (3) When a checkpoint occurs, all assemblies are moved into the • ••••••••••••• •••••; in this state, the mechanism that partitions the assembly for containment of errors, is active. Due to the disruption incurred by the array of Partitioner Cells, it is estimated that no tile attaches to any assemblies while in the checkpointing state.

As it is estimated that all state changes in the Markov diagram exhibit an exponential distribution, then let $P_{Sn}(t)$ represent the probability of an assembly being in a given state n in the Markov diagram of Figure 4. On the assumption of an error-free initial condition and non-negative continuous random variables, in the Laplace transform domain, the first-order differential equations are solved via algebraic manipulation. Using the Final-value theorem, the steady state availability is approximated as follows:

$$
\begin{aligned}
A_{checkpoint}(steady-state) &= \lim_{s\to 0} s(P_{S0}(s)) \\
&= \lim_{s\to 0} s \frac{\frac{1}{s}((\chi+s)(\mu+s))}{\lambda\mu+\mu\chi+\mu s+\lambda\chi+\lambda s+\chi^2+2\chi s+s^2} \\
&= \frac{\chi\mu}{\lambda\mu+\mu\chi+\lambda\chi+\chi^2} \\
&= \frac{\mu}{\frac{\lambda\mu}{\chi}+\mu+\lambda+\chi}
\end{aligned}
\tag{1}
$$

Thus, the optimal value of χ, which results in a maximum of $A_{checkpoint}(steady-state)$, is given below in terms of λ and μ

$$
Optimal\ checkpoint\ rate\ (\chi) = \sqrt{\lambda\mu}
\tag{2}
$$

As an example of the convergence of χ towards a maximum $A_{checkpoint}(steady-state)$, a plot of χ versus $A_{checkpoint}(steady-state)$ for $\lambda=\mu=0.1$ is shown in Figure 5. As the tile set employed exhibits rectilinear growth (from west to east), then once an error occurs all subsequent tiles that assemble will be removed at the next checkpoint, thus guaranteeing the assembly's correctness.

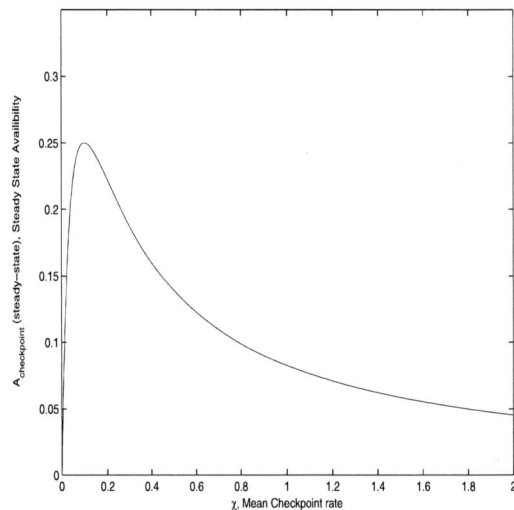

Figure 5. χ **versus** $A_{checkpoint}(steady-state)$ **for** $\lambda = \mu = 0.1$

5   Simulation Results and Analysis

Two error correction techniques are compared with the Error Fault Isolation Tile method [20]. Results were obtained using the Xgrow simulator [1]; these three techniques are compared using the probability of error and size of the final assembly as criteria. In the simulation, the following two parameters represent the physical conditions in which the crystalline arrays assemble: Gse is defined as the ••• •••••••• of the solution, and Gmc is the •••• •••••••••••. The time required for a checkpoint is fixed at ten (virtual)seconds. The optimal checkpointing rate for a given error rate and mean checkpoint length have been determined in (2). For Gse=9.1, the asymptotic error rate is given as $O(e^{-Gse})$ [2], and the mean checkpoint time is estimated as 10 (virtual) seconds. Moreover, it is also estimated that the entire checkpoint process takes 10 (virtual) seconds (as observed in the execution of the Xgrow simulation environment). Ten ••••••• •••••••• is the time spent at a checkpoint in which no tiles can assemble. In the simulations, the $3 \times Y$ rectangle assembles at a rate of approximately 0.2 assembled tile per virtual second for $Gmc = 2Gse$. Once the assembly is in the checkpointing state, the rate at which the erroneous tiles attach to the Partitioner Cell and successfully partition, is $\mu = \frac{1}{0.2 \times 10} = 0.5 \frac{partitioned\ erroneous\ tile}{assembled\ tile}$. When in the checkpointing state, the assembly ceases, and only partitioning occurs.

The optimal checkpoint rate, for $Gse = 9.1$, is given by $\sqrt{2 \times 10 \times e^{-Gse} \times 0.5} = 0.033$ checkpoints per assembled tile . Thus, the optimal checkpoint interval (or period between checkpoints) in this case is given by 161 virtual second/checkpoint. Therefore if the Error Isolation Tile method is employed with a checkpoint interval of $151 + 10 = 161$ virtual seconds, and an assembly is completed in 1,058 seconds, then $\lceil 1058/161 \rceil \times 10 = 70$ virtual seconds are spent in the checkpointing state. The optimal checkpoint rate in term of $\frac{assembled\ tile}{checkpoint}$ is simply $151 \frac{virtual\ seconds}{checkpoint} \times 0.2 \frac{assembled\ tile}{virtual\ seconds} = 30.2 \frac{assembled\ tile}{checkpoint}$, for $Gse = 9.1$.

This section presents the performance evaluation for three error correction techniques, namely 2×2 Proofreading [2], 4×4 snake proofreading [10], and the Error Isolation Tile method [20] for the assembly of a 3×20 rectangular structure. In the two error correction techniques used for comparison, each of the three tiles marked either southern border or interior, were subdivided into 2×2 proofreading or 4×4 snake proofreading tile sets.

530

[10] has proved that the 4×4 snake proofreading method results in no error for $Gmc \approx 2Gse$ (optimal growth). To compare the performance of the error correction techniques, simulation must assemble tiles at a cooler temperature, thus resulting in a greater error percentage, i.e. the cooler the temperature, the stronger the bonds and the greater the error rate due to the faster speed of assembly. Hence, the values of Gse and Gmc that were used in the presented simulations are not optimal in terms of a minimum error rate for $Gmc \approx 2Gse$ [2], so to better evaluate the error correction techniques and to increase the error rates of the original $3 \times Y$ tile set, the error correction techniques were compared in cooler and non-optimal temperature conditions.

Winfree has noted that for DNA-based self-assembly, Xgrow [18] accurately models the quantities of all error types observed in physical assemblies [5]. The •••• •• ••••••••••• •• ••••••• for the conditions of the simulation can then be determined. The optimal checkpoint interval for $Gse = 8.5$ is 98 virtual (Xgrow time) seconds, and 156 virtuals seconds for $Gse = 9.5$. Checkpointing intervals within this range were used for all simulations.

A ••••• is categorized as an assembly error if the assembly is not error-free, i.e. all columns match previous columns and the border tiles are not disjoint. The ••••• •• ••••• •••••••••• is defined as the measured proportional probability that an assembly is not error-free. In these experiments, the assembly of a 3×20 rectangular is considered. Figure 6 plots the assembly error probability versus Gse, (Gmc is held constant at 15). The •••• • ••• •••• •••••••••• is the number of mismatched tiles divided by the number of tiles used, i.e. 60, 240, and 960 for the original tile, 2×2 proofreading, and the 4×4 snake proofreading tile sets respectively. Figure 7 plots the tile mismatch probability versus Gse. The large assembly error probability and the small tile mismatch probability of the original $3 \times Y$ rectangular tile set are caused by an incorrect formation due to tiles with weak bonds, such that other tiles attach and strengthen before they fall off.

Figures 6 and 7 show a comparison of the original 3×20 tile set, 2×2 proofreading tile set, 4×4 snake tile set and the Error Isolation Tile Method (averaged over 11 trials. The Error Isolation Tile correction method [20] exhibits less errors than the 2×2 proofreading and 4×4 snake proofreading methods. The technique of [20] requires significantly less tiles than the other error correction techniques listed in Figure 6 (that require an increase in the number of tiles of the final assembly by a factor of four or sixteen fold due to the massive redundancy required), i.e. if an assembly requires either four or sixteen times as many tiles, then the final assembly is four or sixteen times as large.

6 Conclusion

A novel method for error correction in nano-scale DNA based self-assembly of crystalline arrays was presented in [20], it relies on so-called Error Isolation Tiles. In this paper. a detailed treatment of error Isolation Tiles has been persued. Initially, a physical framework has been proposed for implementing tile removal using the technique of [20]. Through the use of the border tiles in conjunction with Isolation Tiles, it has been shown that this physical framework with checkpointing can be used to remove defective (erroneous) sections. A novel Markov model has also been presented to assess the optimal checkpointing rate of self-assembly; it has been shown that [20] with checkpointing results in a smaller final assembly than all previous methods known to the authors because no redundancy in the tiles is utilized.

References

[1] E. Winfree et al.. The xgrow simulator. http://www.dna.caltech.edu/Xgrow/

[2] E. Winfree and R. Bekbolatov. Proofreading tile sets: Error correction for algorithmic self-assembly. • ••••• ••• ••••• • ••••••• •• • ••• • •••• ••• ••• •••••••• Madison, Wisconsin, June 2003.

[3] C. Mao, T. H. LaBean, J. H. Reif, and N. C. Seeman. "Logical computation using algorithmic self-assembly of DNA triple-crossover molecules," • •••••, vol. 407(6803), pp. 493-496, 2000.

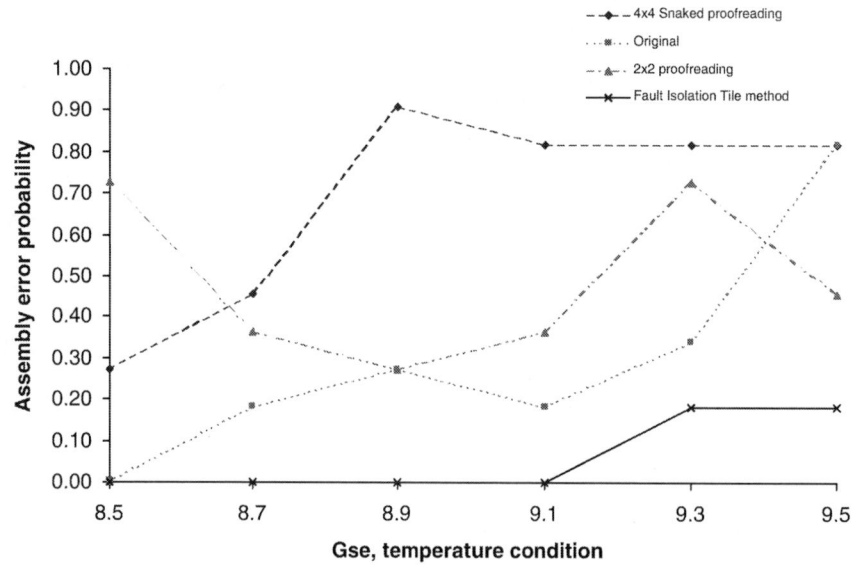

Figure 6. Assembly error probability vs. Gse**,** Gmc**=15 for a 3×20 rectangular assembly.**

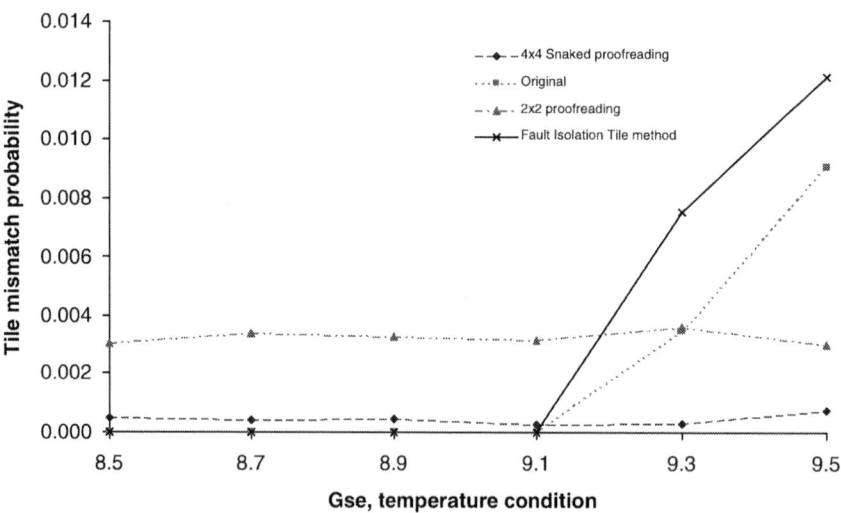

Figure 7. Tile mismatch probability vs. Gse**,** Gmc**=15 for a 3×20 rectangular assembly.**

[4] P. Rothemund, N. Papadakis, and E. Winfree "Algorithmic Self-Assembly of DNA Sierpinski Triangles," • • • • • • •••••, 2 (12). pp. 2041-2053. ISSN 1544-9173 2004.

[5] E. Winfree. Interviewed in: Technology Research News. • ••••••• • •• • • • • •••• • • •••••• by Kimberly Patch, Technology Research News, May 4, 2006

[6] E. Winfree, • •••• •• ••• • • ••• ••••• Foundations of Nanoscience: Self-Assembled Architectures and Devices, 2005, pp. 21-22.

[7] H. Yan, S. H. Park, G. Finkelstein, J. H. Reif, and T. H. LaBean. "DNA-templated self-assembly of protein arrays and highly conductive nanowires," • •••• ••• 301:1882:1884, 2003.

[8] B.H. Robinson and N.C. Seeman, "The Design of a Biochip: A Self-Assembling Molecular-scale Memory Device," • ••••••• • ••• ••••••••, vol. 4, pp. 295-300, 1987.

[9] M. Cook, PWK Rothemund, and E. Winfree, "Self-Assembled Circuit Patterns," • • • • •••• • •• ••••• •, 2004, pp. 91107.

[10] H.L. Chen and A. Goel, "Error Free Self-Assembly using Error Prone Tiles," • •••• •• ••• •••• •• ••• • •••••• • •• • • • •• • ••••• •• • • •••, Univ. of Milano-Bicocca, June 2004.

[11] "DNA damage checkpoint," Technical Document, European Bioinformatices Institute, Wellcome Trust Genome Campus, Cambridge, UK, March. 2006. [Online]. Available: http://www.ebi.ac.uk/ego /DisplayGoTerm?id=GO:0000077&selected=GO:0031571

[12] A. Carbone and N.C. Seeman, Circuits and Programmable Self-Assembling DNA Structures, • •••••• • ••••••• ••••• • ••••••••• • •• 99:12577-12582, Sept. 13, 2002. [Online]. Available: http://www.pnas.org/cgi/content/full/99/20/12577

[13] H. Yan, X. Zhang, Z. Shen and N.C. Seeman, "A Robust DNA Mechanical Device Controlled by Hybridization Topology," • ••••, vol. 415, pp. 62-65, 2002.

[14] Y. Baryshnikov, E.G. Coffman, N. Seeman and B. Yimwadsana, "Self-Correcting Self-Assembly: Growth Models and the Hammersley Process," • ••••• •• ••• •••• •• ••• • ••••• •• • •• • •• •••••, London, Ontario, 2005.

[15] J.H Reif, S. Sahu, and P. Yin, "Compact error-resilient computational dna tiling assemblies," •••• ••••• • ••••• •• • ••• • •••• ••• ••••••, Lecture Notes in Computer Science, Springer-Verlag: New York, 2004.

[16] K. Fujibayashi and S. Murata, "A Method of Error Suppression for Self-assembling DNA Tiles," • •••• ••• •• •••• •••• •• • • • •••• ••• ••••••, Lecture Notes in Computer Science, Springer Verlag: New York pp. 284-293, 2004.

[17] Kishor Trivedi, • •••••••• ••• • ••••••• • ••• • •••••••• • ••••• ••• • •• •••• ••••• • •• ••••••••, John Wiley and Sons, Inc. USA 2002, pp. 173, 673.

[18] E. Winfree, "Simulations of Computing by Self-assembly," Tech. Report CS-TR:1998.22, Caltech, 1998.

[19] K.F. Wong and M. Franklin, "Checkpointing in Distributed Computing Systems," • •••• • •••••• ••• • ••••••••• ••• • •••••• Vol. 35, pp. 67-75, 1996.

[20] S Frechette and F. Lombardi, "Error Detection/Correction in DNA Algorithmic Self-assembly" • •••• •• • • ••••• •••• Munich, April 2008.

IEEE International Symposium on Defect and Fault Tolerance of VLSI Systems

Fabrication Variations and Defect Tolerance for Nanomagnet-based QCA

Michael Niemier, Michael Crocker, and X. Sharon Hu
Department of Computer Science and Engineering
University of Notre Dame, Notre Dame, IN 46556, USA
Email: {mniemier,mcrocker,shu}@nd.edu

Abstract

1 Introduction

In this paper, we consider how fabrication variations will affect the logical correctness of circuit elements realized from a nanomagnet-based implementation of the Quantum-dot Cellular Automata (MQCA) device architecture. For MQCA, wires, gates, and inverters have all been realized, they operate at room temperature [11], and it is estimated that if 10^{10} magnets switch 10^8 times/second, they would only dissipate about 0.1 W of power [5]. When the drive circuitry is included, [14] predicts that circuits could provide performance wins over state-of-the-art, low power CMOS when considering energy delay product. Devices can scale and remain non-volatile provided their size/shape remains above the superparamagnetic limit. However, binary state in nanomagnets with feature sizes below the superparamagnetic limit can be stable for around 1 ms [16] – long enough to perform logical operations. Scaling can also decrease switching times [16].

Still, like any device with nanometer feature sizes, MQCA based circuits could suffer from defect rates that are much higher than those for today's CMOS-based circuits. Thus, an MQCA-based circuit must not only perform better for some computational task of interest (to justify a technology transition), but will need to do so in the presence of more faulty components. Fabrication processes envisioned for MQCA are similar to those for CMOS and fabrication variations should be similar as well. However, because MQCA devices process information in different ways than CMOS devices, defect tolerance mechanisms will be different. We study these issues here. We pay particular attention to how defect tolerance mechanisms affect performance – as an understanding of correctness/performance tradeoffs early in the device design process can facilitate experimental "mid-course corrections" to ensure that a new technology can ultimately best CMOS for a given application.

2 Background

Figs. 1(a)-(b) illustrate two important building blocks that would be used to construct MQCA circuits. A wire (Fig. 1(a)) is just a line of magnets that are antiferromagnetically coupled with each other. The basic logic gate in MQCA is based on the majority voting function. By setting one input of a majority gate to a logic '0' or '1', the gate will execute an AND or OR function, respectively. In MQCA, the gate performs an •••••••• majority function (Fig. 1(b)). These structures have all been experimentally demonstrated at room temperature (see Fig. 1(c),(d) [11]).

The structures illustrated in Fig. 1(c),(d) were tested with a clock that took the form of a periodically oscillating •••••••• magnetic field that drove a system to an initial state, and

1550-5774/08 $25.00 © 2008 IEEE
DOI 10.1109/DFT.2008.54

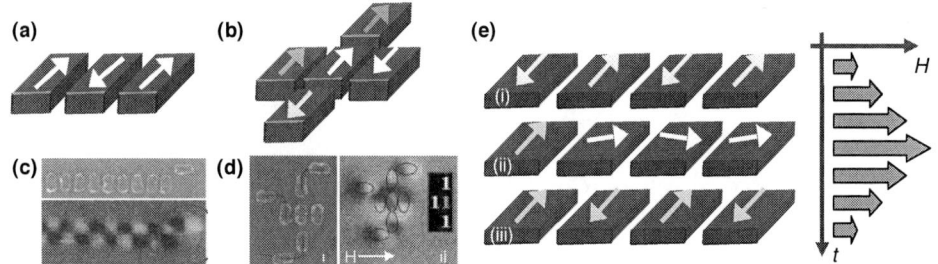

Figure 1. Schematic/experimental representations of (a)/(c) a wire segment and (b)/(d) a majority gate. Operating scheme of a wire: (e) (i) initial configuration, (ii) high-field ("null") state, (iii) after the application of the input, and the final ordered state.

then controlled the relaxation of the system to a ground state. For example, Fig. 1(e)(i), illustrates a line (or "wire") of nanomagnets that has relaxed to a logically correct, antiferromagnetically coupled ground state. In Fig. 1(e)(ii), the external field turns the magnetic moments of all magnets horizontally into a neutral logic state against the preferred magnetic anisotropy (i.e. along the hard axes of the magnets). This is an unstable state of the system, and when the field is removed, the nanomagnets relax into a new antiferromagnetically ordered ground state in accordance with the new input (Fig. 1(e)(iii)). [14] explored the use of copper wires wrapped by ferrite on the sides and bottom to provide ••••control of MQCA-based circuits. Nanomagnets would reside on the wire surface.

3   Process Variations and Defects

MQCA devices are simply three-dimensional blocks made from one layer of a magnetic material (e.g., permalloy or supermalloy). Conventional, imprint, or electron-beam lithography (EBL) are all candidate fabrication mechanisms. EBL was used to make the nanomagnets for the MQCA majority gate and wire experiments discussed in [11]. Although EBL is well accepted for patterning at the nanometer regime, it can introduce many variations that may effect the functionality of an MQCA-based circuit: (a) Over-exposure and under-exposure could lead to incorrect shapes. (b) Irregularities in the EBL gun can lead to "shot noise" that can cause rough edges. (c) The electrons that make up the beam are subject to scattering effects that can cause exposure of the resist outside of the desired shapes [10]. This can lead to bulging around the edges. (d) If the spacing between shapes is small, the resist in the gap may fail completely during development, leading to the merger of two or more shapes. The thin resist walls can also bend, creating nanomagnets with bulges and indentations [9]. (e) With liftoff, a pattern is made in resist and metal is evaporated over it. When the resist is dissolved, the metal that previously resided on the top surface is washed away, or lifted off. However, the deposited material can clump, and the liftoff process might remove a large chunk from the edges of the rounded rectangle shapes.

Scanning Electron Microscopy (SEM) images of nanomagnet lines from [12] (see Fig. 2(a)) illustrate said variations. While it is difficut to get precise information about the entire 3D nanomagnet shape from this top-down view, the images indicate that the edges of the nanomagnets are missing magnetic material. The SEM image in Fig. 2(b) shows a horizontal wire next to a vertical wire. The magnetic material bulges between the two wires. Similar behavior has been observed for nanomagnets that are placed close together.

4   Relating Defects to Faulty Behavior

Here, we consider how missing magnetic material, bulges and "shifts", and edge roughness affect how well a magnet represents and propagates binary state. Important to this discussion is how the magnetization (binary state) associated with a previous computation is "removed," by nulling the nanomagnets. This is essential as it allows the nanomagnets that make up MQCA circuit elements to be re-evaluated with new inputs (see Fig. 1(e)).

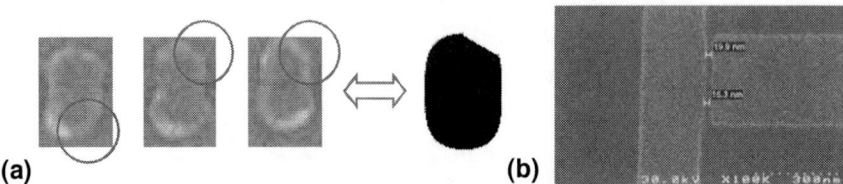

Figure 2. (a) Experimentally realized magnets can have slanted edges. (b) This SEM image shows wires made using EBL. Proximity effects and other variations have led to bulging between the two wires. The lithography design called for 25nm between wires, but the fabricated wire are much closer. Also, note that the vertical wire bulges most significantly where it is adjacent to the horizontal wire.

Fabrication variations can make this process more difficult – which can in turn lead to faulty behavior and undesirable logical states. We leverage micromagnetic simulations of supermalloy nanomagnets (the material used in [2]) performed with the OOMMF simulation suite [7] to study these issues. As in [6], we assumed a saturation magnetization of 8.0×10^5 A/m, an exchange stiffness constant of 1.05×10^{-11} J/m, an anisotropy constant K_1 of 3 J/m^3, and the default damping constant of 0.5. Each simulation stage (when the magnitude or direction of the applied field changes) was considered complete when the maximum $| d\mathbf{m}/dt |$ dropped below a preset number of degrees per nanosecond. We note that OOMMF is widely used and there is excellent correlation between simulation and experimental results (see [15, 13] for examples).

4.1 A Base Case

Before looking at defective systems, we first consider a simulation of a non-defective magnet to provide a basis for comparison – a 60x90x30nm magnet (as in [14]), in isolation, where the y-component of magnetization was initially strongly negative (see bottom inset in Fig. 3(a)). We then applied an external magnetic field (H_{clock}) along the magnet's hard axis that continuously increased in magnitude to simulate the effects of a clock. A 1.5×10^4 A/m biasing field was simultaneously applied along the magnet's easy axis. The biasing field is necessary as, after we null the magnet ($M_y = 0$), we want to "tip" it to the opposite polarization. Without a biasing field, the magnet would randomly tip up or down as the field applied along its hard axis was removed. This simulation determines the external field required to null this magnet given this local bias.

The "No Defects" curve in Fig. 3(a) illustrates how the y-component of magnetization changes as a function of the magnitude of the external field. As the magnitude of the external field (H_{clock}) increases, the magnitude of its y-component of magnetization decreases – eventually reaching 0 when the external field is approximately 0.5×10^5 A/m. The nanomagnet is now • • •••• and is tipped up by the biasing field. Note that if the magnitude of H_{clock} continues to increase, it will work to keep the magnet nulled (see the "tail" of the "No Defects" curve). This explains why after the magnet initially changes its polarization, the y-component of its magnetization trends back toward 0 – the external field can be too strong and works to re-null the magnet. When H_{clock} returns to 0, the y-component of magnetization is strongly positive. After the magnet has changed polarizations, we again applied an external field along the magnet's hard axis and a biasing field of the same magnitude (but in the opposite direction) to return the magnet to its original state. The • •••••• of the nulling field should not impact the • •••••••• of the field required to null the magnet. For a magnet with no fabrication variations, this is in fact the case as seen in Fig. 3(a).

4.2 Missing Magnetic Material

Now consider a simulation similar to the one above, but this time with magnetic material removed from the top-right of the magnet – a possible fabrication variation as discussed in Sec. 3 and seen in Fig. 2(a). While the same biasing field was used in this simulation, the

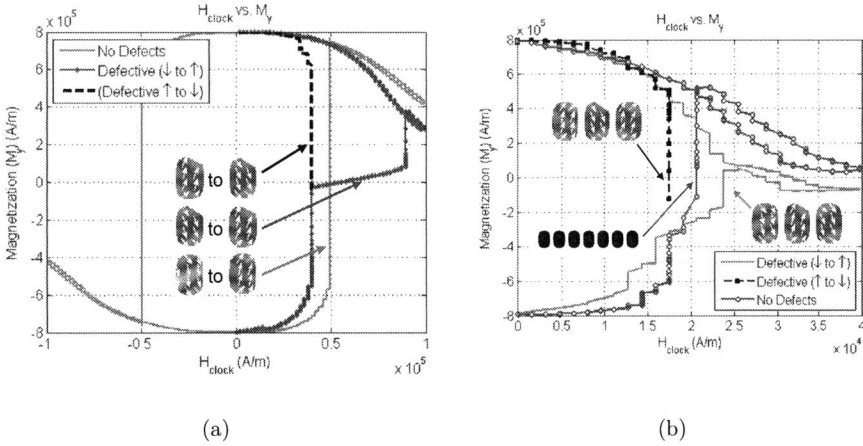

(a)   (b)

Figure 3. (a) H_{clock} vs. M_y for a 60x90nm magnet with no fabrication variation and a 60x90nm magnet with a "slanted" edge. H_{clock} required to null the slanted magnet is greater than that for the perfect magnet. (b) H_{clock} vs. M_y for the 4th magnet in a 7-cell line. A larger H_{clock} is required to facilitate a $\downarrow$ to $\uparrow$ transition than a $\uparrow$ to $\downarrow$ transition when compared to a wire with no missing magnetic material.

external field was applied in the same direction for both state transitions. These simulation results are illustrated in Fig. 3(a). We consider the down-to-up transition first (see middle inset in Fig. 3(a)). Note that a stronger external field is required to null this magnet (approximately 0.6×10^5 A/m instead of 0.5×10^5 A/m). Magnetic moments tend to align along a magnet's edge. In this simulation, the placement of the slant and the direction of the applied external field help to reinforce the initial downward polarization ($\downarrow$). For this same reason, the up-to-down transition (see top inset in Fig. 3(a)) can be accomplished when the magnitude of H_{clock} is lower (approximately 0.4×10^5 A/m). (Only the first portion of this curve is shown – i.e. until the magnet is nulled – to improve graph readability.)

We now consider the above results in the context of a simple circuit – in this case a line of 7, 60x90x30nm magnets spaced 16nm apart (see inset at lower left in Fig. 3(b)). The line was initialized to a logically correct, antiferromagnetically coupled ground state and the y-component of magnetization in the first magnet was switched in order to simulate a new input to this line (as in Fig. 1(e)). H_{clock} was then applied along the magnets' hard axes and increased from 0 A/m to approximately 4.0×10^4 A/m in approximately 1.5×10^4 A/m increments[1] The field was then allowed to relax back to 0 A/m in the same manner.

The results of three simulations are summarized in Fig. 3(b). We plot the magnitude of H_{clock} versus M_y for the 4th magnet in the line – as M_y represents binary state. We first considered a wire with no defects. In this plot, one can see that H_{clock} was approximately 2.0×10^4 A/m when the 4th magnet was nulled. When a misshapen magnet is introduced into the line (insets show magnets 3, 4, and 5), as before, the external field required to null the misshapen magnet can increase or decrease •••••••• •• ••• • ••••••••• ••••••• •••••• As seen from the inset at the top left of Fig. 3(b), the field required to null the defective magnet is actually about 20% lower if magnets 3 and 4 are initially $\downarrow$ and $\uparrow$. However, if magnets 3 and 4 are intitially $\uparrow$ and $\downarrow$, the magnitude of H_{clock} must be about 14% higher in order to "reuse" this system to correctly evaluate a new input.

The above simulations were repeated with magnets that became progressively more misshapen. These results are summarized in Fig. 4(a). As fabrication variation gets worse, the

---

[1]4.0×10^4 A/m is the magnitude of H_{clock} that a clock wire in [14] might produce assuming a current density of 10^7 A/cm^2 – and was considered to be to a practical upper limit. Realistically, placing nano-magnets closer to one another can increase coupling and lower the magnitude of H_{clock} required to null the magnets.

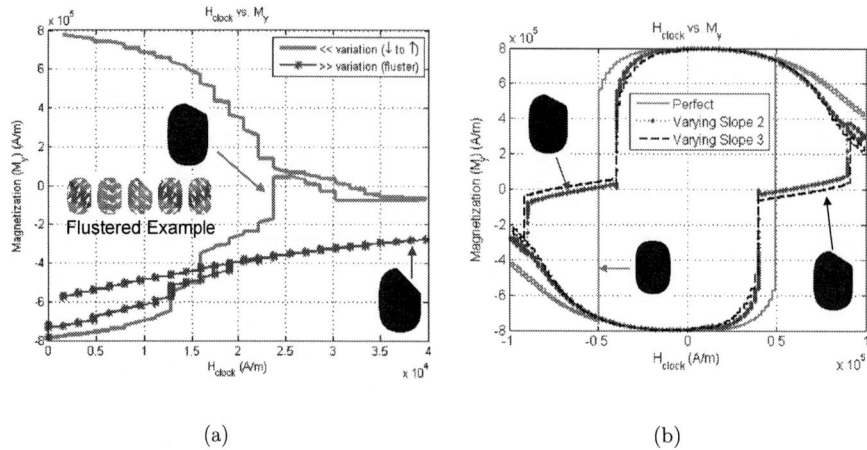

Figure 4. (a) With more missing material, it becomes more difficult to null the nanomagnet as evidenced by the y-component of the defective magnet's magnetization. (b) If H_{clock} is applied to a magnet with missing magnetic material from right- to-left, a ↑ to ↓ transition becomes more difficult.

misshapen magnet is in fact never nulled. While H_{clock} begins to cause the 4th magnet's y-component of magnetization to trend toward 0, the missing magnetic material/slant helps to re-enforce its initial state. Moreover, as the amount of magnetic material missing from the edge of the 4th magnet increases, it becomes more difficult to null when H_{clock} has a similar magnitude. After H_{clock} reaches 4.0×10^4 A/m and relaxes back to 0 A/m, the 4th magnet settles back into its old (and now logically incorrect) polarization.

This suggests that we can leverage the external drive circuitry to help tolerate the effects of fabrication variation. Increasing the magnitude of H_{clock} might increase the likelihood that the nanomagnets will behave in a logically correct manner. A stronger external field makes it easier to remove the state associated with a previous computation. However, more reliability comes at the expense of an increase in system energy. If H_{clock} is never strong enough to remove the state associated with the previous computation, missing magnetic material can induce stuck at faults. That said, whether or not a wire (for example) is stuck at 0 or is stuck at 1 can depend on (a) • •••• on the magnet material is missing from, and (b) what polarization state a group of magnets was initially in. For the example discussed above, if the misshapen magnet initially has a positive component of M_y, it can easily make a ↑ to ↓ transition. However, a ↓ to ↑ transition is more difficult suggesting that a "stuck-at-down" is most likely.

The simulations discussed thus far have assumed that H_{clock} is applied uni-directionally. However, as noted in [14], H_{clock} could also be sinusoidal. This could be advantageous from the perspective of minimizing overall system energy as the clock generation circuitry could be derived from the designs for adiabatic clocks [1, 17]. Moreover, as seen in Fig. 3(a), for a magnet with no fabrication variations, the magnitude of H_{clock} (not the direction) is of most importance when working to null a given magnet. However, if the magnet is missing magnetic material, this is not necessarily the case.

In Fig. 4(b), we illustrate the results of 2 simulations designed to determine how an external parameter – the direction of H_{clock} – might affect the logical correctness of some physical structure – a configuration of nanomagnets with fabrication variation. A setup is used that is similar to that in Fig. 3(a). The first half of this simulation is essentially identical to the ↓ to ↑ transition captured in Fig. 3(a). However, when considering the ↑ to ↓ transition here, the external field was this time applied from right-to-left If H_{clock} is applied in this manner, the magnitude of the external field must increase ($0.6 \times 10^5 A/m$

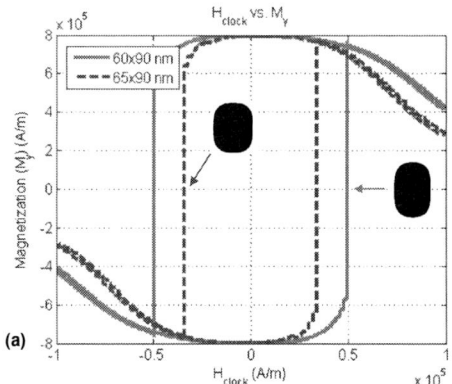

Magnet	M_y	H_y
60x90	0.965	-1.61 x 10^5 A/m
65x90	0.938	-1.57 x 10^5 A/m
70x90	0.903	-1.39 x 10^5 A/m

Figure 5. (a) M_y vs. H_{clock} for a 60x90nm magnet and a 65x90nm magnet. As the aspect ratio decreases, the magnet is easier to null. (b) Fields generated by various magnets.

vs. $0.4 \times 10^5 A/m$) to support the $\uparrow$ to $\downarrow$ transition.

These results are significant in MQCA circuit design. Consider a line of 3 nanomagnets where the middle magnet has missing magnetic material (i.e. as in Fig. 4(b)). If the initial state of this line is $\uparrow\downarrow\uparrow$ and H_{clock} is applied from left-to-right, the third magnet will be more likely to end up in a "stuck at up" state per the discussion above. However, if the initial state of the line is $\downarrow\uparrow\downarrow$ and H_{clock} is applied from right-to-left (with data still flowing from left-to-right), the third magnet is likely to end up in a "stuck at down" state. Essentially, a stuck at fault can occur, but it may change between stuck-at-up or stuck-at-down. • •• •••••••• •• •• •• •••••• •• •• •• ••• •••••••• ••••• •• ••• • ••••••• ••• •••••• ••• • ••••• • • •••••• •• ••• • •••••••• •• H_{clock}• ••• ••• ••••••• ••••• •• •••• •••••• ••• •••••••

4.3 Bulging Nanomagnets

Above we looked at how missing edges might affect a magnet's logical state and dataflow. We now consider what might happen if magnets "bulge." Because bulging usually occurs where lithographically defined shapes are adjacent and close to each other, most nanomagnets in a wire will see similar bulging from neighbors on their left and right. To consider the effects of fabrication variations that manifest themselves as bulges, we simulated lines of 60x90x30nm, 65x90x30nm, and 70x90x30nm magnets. For sufficiently strong nulling fields, both up and down inputs were successfully propagated down each of the aforementioned lines. However, it is more revealing to examine how bulges affect the quality of a signal's binary state by considering an M-H plot analogous to Fig. 3(a).

Specifically, Fig. 5 plots the y-component of magnetization against H_{clock}. (Again, a 1.5×10^4 A/m bias is used.) As one can see, the magnet with a smaller aspect ratio is actually easier to null. For example, the magnitude of H_{clock} when $M_y = 0$ for the 65x90nm magnet is lower than that for the 60x90nm magnet. This would seemingly indicate that the lower aspect ratio magnets resulted from bulging are an asset as the current density in a clock wire could decrease. While in some sense this is true, there are consequences when moving toward magnets with smaller aspect ratios.

When studying a line or "wire" of nanomagnets, as a magnet's aspect ratio decreases, the magnitude of the y-component of magnetization for each individual magnet in the line decreases as well. Magnets tended to couple with one another in the x-direction more readily. From the perspective of binary logic, this represents a "weaker" 1 or 0. While one might look at Fig. 5(a) and note that when H_{clock} was 0, M_y was strong, this is only because the global bias field was applied continuously throughout this simulation. To better quantify the y-component of magnetization as a function of aspect ratio, we initialized 60x90, 65x90, and 70x90nm magnets in isolation such that M_y was strongly negative and then let each

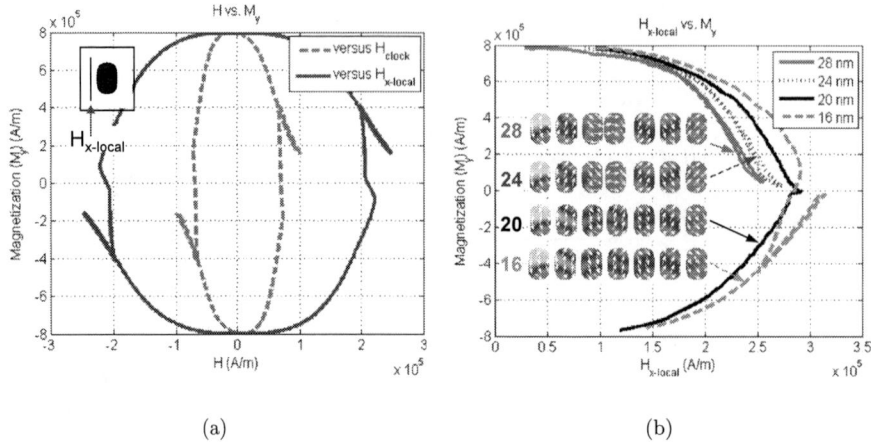

(a)   (b)

Figure 6. (a) M_y as a function of H_{clock} and M_y as a function of H_{local} in the presence of a 1.5×10^4 A/m biasing field. Note that the field seen at the field near the surface of the magnet is greater than the applied field. This will be an important measure when considering why a circuit does or does not show logically correct behavior. (b) Local field vs. M_y for magnets in a wire. Note that magnet needs to see a similar local field to null but at the expense of a larger global field.

magnet relax to a ground state with **no** applied biasing field.

We report the average Y component of their magnetization in the presence of zero bias (normalized to ± 1) in Fig. 5(b). We averaged M_y over all of the simulation mesh points and also averaged H_y 4nm away from the magnet. As the aspect ratio decreases, the magnitude of the y-component of the field below the magnet decreases. This makes sense as the magnetization in the x-direction increases as aspect ratio decreases. Logically, as aspect ratio decreases, the y-component of magnetization decreases – resulting in a potentially "weaker" 0 or 1 and a "weaker" driver due to a lower biasing field. While further discussion is beyond the scope of this paper, a strong component of M_y is desirable as it can help facilitate other circuit structures (i.e. the driver to a crossover structure).

4.4 Shifted Nanomagnets

We now consider what might happen if the space between adjacent nanomagnets in a line is non-uniform. This will bring 2 nanomagnets closer together and move two others farther apart as seen in the insets of Fig. 6(b). (With no variations, each nanomagnet would be 16nm away from its neighbors.) In these simulations, we will consider M_y as a function of the x-component of the •••• magnetic field seen in the gap near the surface of a nanomagnet (see inset in Fig. 6(a)). This measurement ($H_{x-local}$) is more insightful as it also incorporates the contribution of the magnetic material itself. One way to interpret this data is that in order to reuse a magnet to evaluate new inputs, the 60x90x30nm magnet needs to "see" this value of $H_{x-local}$. Depending on the size and shape of the surrounding magnetic material, a given value of $H_{x-local}$ may be generated with a higher or lower •••••••• field. (Note the difference in Fig. 6(a).)

We leverage H_{local} to consider how spacing variations between nanomagnets might affect dataflow in a 7 cell line with no missing magnetic material. Simulation results are illustrated in Fig. 6(b) where we plot M_y of the 5th magnet from the left versus the local field 8nm to the left of the 5th magnet. We begin by examining the 16nm curve – which is essentially a wire with •• spacing variation between the magnets. Note that when the magnet is nulled, $H_{x-local} = 2.8 \times 10^5$ A/m. We next examine the curve for the simulation with a 20nm spacing between the 4th and 5th magnets. Again, the 4th magnet is successfully nulled and that the wire functions correctly. However, while H_{local} **or** $H_{x-local}$ is nearly identical to the 16nm case, the corresponding value of H_{clock} when magnet 5 was nulled was greater than the magnitude of H_{clock} for the 16nm case – indicating that a stronger clock/more energy is needed to null magnet 5 due to weaker coupling with the adjacent magnetic material. Looking at the curves for 24 and 28nm spacing between the 4th and 5th magnets, the magnitude of H_{clock} is not sufficient to null all of the magnets in a line and a stuck-at state ensues.

4.5   Edge Roughness and Magnetic Material

A magnet's edge roughness and the magnetic material used to make it can also affect circuit functionality. While the details are beyond the scope of this paper, we do highlight two observations obtained from simulation results. First, as a magnet's edge roughness increases, the magnitude of H_{clock} required to null it generally increases as well (when compared to simulation results described by Figs. 3(a) and 4(b)). Second, simulation results indicate that magnets made with a harder magnetic material (i.e. permalloy instead of supermalloy) may be more difficult to null, but also provide a stronger, local biasing field. Depending on the type of defect, sometimes a stronger local biasing field can allow fault free operation with a lower nulling field (as an individual magnet can provide more local control over nearby devices). This will be studied further in future work.

5   Defect Tolerance at the Architectural Level

Reconfigurable logic offers another way to avoid faults via architectural-level redundancy. For example, the PLA structure in [8] can be expanded to include more rows and columns such that defective crosspoints and/or interconnect can be avoided – increasing the probability that the desired set of logic functions can be mapped onto the faulty PLA. However, for MQCA, a larger PLA not only means a larger chip area, but also more/longer clock wires to operate the logic and interconnect. Redundancy in an MQCA PLA provides a way to trade power consumption and area for fault tolerance.

Consider the yield vs. fault rate study in [3]. This study indicates that a yield of 90% is possible given a fault rate of 10^{-3} and 10% redundancy. However, if the fault rate increases to 10^{-2}, 400% redundancy is required. As seen in Sec. 4, increasing the magnitude of H_{clock} provides another level of flexibility in terms of fault tolerance. According to [14], the magnitude of H_{clock} increases linearly with the current density and current in a clock wire. However, the power increases with the square of the current. Therefore, doubling the magnitude of H_{clock} implies that current will double while the power will quadruple. Thus, from the standpoint of performance and logical correctness, it is an interesting optimization problem to determine the most effective usage of the above mechanisms for fault tolerance. Together these techniques could allow for tolerance of a higher fault rate than can be achieved by either individually. However, one technique might be sufficient to provide the required fault tolerance with the smallest increase in power.

To further quantify the tradeoffs discussed above, we worked with ALU4 from the Toronto 20 benchmark suite. Assume that the fault rate in a given MQCA PLA is normally 5×10^{-4} with a clocking field of 2.0×10^4 A/m, and an increase in the clocking field to 2.4×10^4 A/m leads to a lower fault rate of 2.5×10^{-4}. We used the mapping tools proposed in [4] to map the ALU4 benchmark to a PLA. For the higher fault rate (5×10^{-4}), the area needed to successfully map the ALU4 benchmark to the PLA with a yield of 96% is approximately 68,500 μm^2. For the lower fault rate (2.5×10^{-4}), a PLA with an area of only 24,500 μm^2 is sufficient to map the benchmark with 96% yield. These area estimates are based on 60x90nm magnets with 20nm spacings and wires crossings as described in [14].

To estimate the power consumption, the clocking wire design given in [14] is adopted, where the wire has a width of 2 μm and a length of 4 μm for an area of 8 μm^2. This means that the larger PLA will require 8,563 wires while the smaller PLA will require 3,063 wires. However, for the smaller PLA, the clocking field needs to be increased by 20%

from 2.0×10^4 A/m to 2.4×10^4 A/m. This 20% increase in field strength means a 20% increase in current, and a 44% increase in power consumption. Therefore, if the power consumed by one clocking wire region is P, then the power consumption for the larger PLA with the lower clocking field is $8563P$. The power consumption for the smaller PLA with the higher clocking field is $4411P$. In this case, a 96% yield can be achieved with less power consumption if H_{clock} is increased so that a smaller PLA can be used to map the logic. A systematic study of this tradeoff will be done in our future work.

6   Conclusions and Future Work

We have illustrated how fabrication variations might induce logical faults in MQCA-based circuits, and identified faulty behavior unique to MQCA. We have also demonstrated three different ways to mitigate the effects of faulty behavior, at the device, circuit, and architecture level, that can all be applied simultaneously to lower the overall fault rate. We will continue to leverage this work to determine what fabrication variations are most problematic in terms of both logical correctness and overall system performance to shape experimental work for this promising technology.

The authors gratefully acknowledge the support of the National Science Foundation under grant numbers CCF06-21990, CCF05-41324, and CCF-0702705, as well as the SRC NRI funded MIND center.

References

[1] A. Chandrakasan and B. Brodersen, "Low power digital cmos design," *Kluwer Academic Publishers*, 1996.

[2] R. Cowburn and M. Welland, "Room temperature Magnetic Quantum Cellular Automata," *Science*, vol. 287(5457), pp. 1466–1468, 2000.

[3] M. Crocker, X. S. Hu, and M. Niemier, "Fault models and yield analysis for QCA-based PLAs," *Int. Sym. on FPL*, pp. 435–440, 2007.

[4] M. Crocker, M. Niemier, and X. S. Hu, "Defect tolerance in QCA-based PLAs," *Int. Sym. on Nanoscale Architectures*, 2008.

[5] G. Csaba, P. Lugli, A. Csurgay, and W. Porod, "Simulation of power gain and dissipation in field-coupled nanomagnet," *J. of Comp. Elec.*, vol. 4(1-2), 2005.

[6] N. Dao, S. Whittenburg, and R. Cowburn, "Micromagnetics simulation of deep-submicron supermalloy disks," *J.of Appl. Phys.*, vol. 90(10), pp. 5235–7, 2001.

[7] M. Donahue and D. Porter, "OOMMF user's guide, version 1.0, interagency report NISTIR 6367," http://math.nist.gov/oommf.

[8] X. S. Hu, M. Crocker, M. Niemier, M. Yan, and G. Bernstein, "PLAs in Quantum-dot Cellular Automata," *Int. Sym. on VLSI*, 2006.

[9] X. Huang, G. Bazan, D. A. Hill, and G. H. Bernstein, "Stability of thin resist walls," *Journal of Electrochemical Society*, vol. 139(10), pp. 2952–2956, 1992.

[10] X. Huang, G. H. Bernstein, G. Bazan, and D. A. Hill, "Spatial density of lines exposed in poly(methylmethacrylate) by electron beam lithography," *J. of Vacuum Sci. and Tech.*, vol. 11(4), pp. 1739–1744, 1993.

[11] A. Imre, G. Csaba, L. Ji, A. Orlov, G. Bernstein, and W. Porod, "Majority logic gate for Magnetic Quantum-dot Cellular Automata," *Science*, vol. 311 no. 5758, pp. 205–208, January 13, 2006.

[12] A. Imre, "Experimental study of nanomagnets for Magnetic Quantum-dot Cellular Automata (MQCA) logic applications," *Disseration, Univ. of Notre Dame*, April 2005.

[13] S. McVitie, G. White, J. Scott, P. Warin, and J. Chapman, "Quantitative imaging of magnetic domain walls in thin films using lorentz and magnetic force microscopies," *J. of Appl. Phys.*, vol. 90(10), pp. 5220–7, 2001.

[14] M. Niemier, M. Alam, X. S. Hu, G. Bernstein, W. Porod, M. Putney, and J. DeAngelis, "Clocking structures and power analysis for nanomagnet-based logic devices," *Proc. of Int. Symp. on Low Power Electronics and Design*, pp. 26–31, 2007.

[15] L. Verma and V. Ng, "Magnetic domain patterns in a zigzag nanowire," *J. of Magnetism and Magnetic Materials*, vol. 313(2), pp. 317–321, 2007.

[16] X. Wu, C. Liu, L.Li, P. Jones, R. Chantrell, and D. Weller, "Nonmagnetic shell in surfactant-coated FePt nanoparticles," *J. Appl. Phys.*, vol. 95, pp. 6810–6812, 2004.

[17] C. Ziesler, S. Kim, and M. Papaefthymiou, "A resonant clock generator for single-phase adiabatic systems," *Int. Symp. on Low Power Elec. and Design*, 2001.

Author Index

Abate, Francesco 24
Abramov, Beni 361
Abuhamdeh, Zahi 381
Adapa, Rajsekhar 474
Aikawa, Hisanori 507
Aitken, Robert 114
Al-Assadi, Waleed K. 167
Ampadu, Paul 33, 352
Apte, Ravi 143
Armbrust, David 152
Asao, Yoshiaki 507
Augustine, Charles 323
Babu, Hafiz Md. Hasan 290
Banerjee, Nilanjan 323
Becker, Bernd 245
Bolchini, Cristiana 332
Bosio, A. 7
Breveglieri, Luca 202
Cardarilli, G.C. 436
Carro, Luigi 281
Chakravarty, Sreejit 394
Chandra, Vikas 114
Chapman, Glenn H. 220, 305
Choi, Jae-Young 229
Choi, Yoon-Hwa 16, 229
Crocker, Michael 534
Daasch, W. Robert 152
Dai, Kui 184
de Veciana, Gustavo 105
Devta-Prasanna, Narendra 394
DiPalma, Kellie 403
Dworak, Jennifer 403
Dysart, Timothy J. 72
Eisenbarth, Thomas 202
Ejlali, Alireza 445
Fazeli, Mahdi 193
Frechette, Stephen 525

Frigerio, Laura 427
Fummi, Franco 54
Ghassem Miremadi, Seyed 193, 445
Girard, P. 7
Gong, Rui 184
Hashempour, Masoud 516
Hsiao, Michael S. 143
Hu, Jiang 96
Hu, X. Sharon 534
Huang, Jiun-Lang 143
Huh, Yoonjae 16
Ibrahim, Muhammad 290
Ienne, Paolo 202
Ikegawa, Sumio 507
Ito, Hideo 272, 482
Iwayama, Masayoshi 507
Jain, Saurabh 152
Jain, Vijay K. 220
Jas, Abhijit 454
Jiang, Zhigang 143
Jone, Wen-Ben 314
Joshi, Prashant D. 370
Kakarla, Sindhu 167
Karimi, Naghmeh 454
Kato, Jiro 491
Keren, Osnat 361
Kerkhoff, Hans G. 45
Kim, Yong Bin 525
Kishi, Tatsuya 507
Kogge, Peter M. 72
Koren, Israel 202, 305
Koren, Zahava 305
Kuiken, Oscar J. 45
Kumar Goparaju, Manoj 176
Kumar Palaniswamy, Ashok 176
Kundu, Sandip 343, 412
Landrault, C. 7

543

Leung, Jenny	305	Remersaro, Santiago	385
Levin, Ilya	361	Rinderknecht, Thomas	385
Li, James C.-M.	143	Roy, Kaushik	323
Likharev, Konstantin K.	504	Ruan, Shuangyu	272
Lisboa, Carlos Arthur Lang	281	Salice, Fabio	427
Lombardi, Fabrizio	236, 516, 525	Salmani, Hassan	87
Ma, Xiaojun	236	Salsano, A.	436
Mak, TM	465	Samanta, Rupak	96
Makris, Yiorgos	454	Savage, John E.	423
Maniatakos, Michail	454	Shah, Nimay	96
Margala, Martin	134	Shazli, Syed Z.	63
Mashreghian Arani, Zahra	516	Sheu, Boryau	143
Metra, Cecilia	465	Shi, Yiwen	403
Miele, Antonio	332	Sliech, Kevin	134
Miura, Yukiya	491	Song, Jiayong	143
Mohanram, Kartik	83	Stefanni, Francesco	54
Mukherjee, Nilanjan	163	Sun, Hongbin	254
Mukherjee, Shubu	301	Tabkhi, Hamed	445
Namba, Kazuteru	272, 482	Tahoori, Mehdi B.	63
Niemier, Michael	534	Tam, Simon	465
Nigh, Phil	3	Tang, Xun	245
Nitayama, Akihiro	507	Tehranipoor, Mohammad	87
Omaña, Martin	465	Tirumurti, Chandra	454
Ostrovsky, Vladimir	361	Touba, Nur A.	125
Pan, Abhisek	343	Tragoudas, Spyros	176, 474
Park, Nohpill	211	Tschanz, James W.	343
Patitz, Zachary	211	Tsuchida, Kenji	507
Plusquellic, Jim	87	Vial, J.	7
Polian, Ilia	245, 263	Violante, Massimo	24
Pomeranz, Irith	245, 385, 394	Virazel, A.	7
Pontarelli, S.	436	Walker, Duncan	96
Pravossoudovitch, S.	7	Wang, Laung-Terng	143
Quaglia, Davide	54	Wang, Xiaoxiao	87
Radaelli, Matteo Alan	427	Wang, Zhiying	184
Rahman, Asifur	465	Wen, Xiaoqing	143
Raja Chowdhury, Ahsan	290	Wolpert, David	33
Rajski, Janusz	385	Wu, Shianling	143
Rao, Wenjing	263	Wu, Yu-Liang	314
Re, M.	436	Xiong, Xingguo	314
Reddy, Sudhakar M.	245, 385, 394	Yang, Fan	394
Regazzoni, Francesco	202	Yang, Joon-Sung	125

Yi, Hyunbean.............................. 412
Yoda, Hiroaki 507
Yu, Qiaoyan.............................. 352
Zarrineh, Kamran 503
Zhang, Ming 96

Zhang, Tong............................. 254
Zhang, Xiao 45
Zheng, Nanning 254
Zykov, Andrey....................... 105

IEEE Computer Society
Conference Publications
Operations Committee

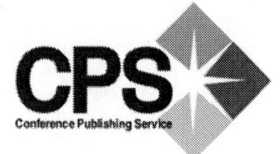

CPOC Chair
Chita R. Das
Professor, Penn State University

Board Members
Mike Hinchey, *Director, Software Engineering Lab, NASA Goddard*
Paolo Montuschi, *Professor, Politecnico di Torino*
Jeffrey Voas, *Director, Systems Assurance Technologies, SAIC*
Suzanne A. Wagner, *Manager, Conference Business Operations*
Wenping Wang, *Associate Professor, University of Hong Kong*

IEEE Computer Society Executive Staff
Angela Burgess, *Executive Director*
Alicia Stickley, *Senior Manager, Publishing Services*
Thomas Baldwin, *Senior Manager, Meetings & Conferences*

IEEE Computer Society Publications
The world-renowned IEEE Computer Society publishes, promotes, and distributes a wide variety of authoritative computer science and engineering texts. These books are available from most retail outlets. Visit the CS Store at *http://www.computer.org/portal/site/store/index.jsp* for a list of products.

IEEE Computer Society *Conference Publishing Services* (CPS)
The IEEE Computer Society produces conference publications for more than 250 acclaimed international conferences each year in a variety of formats, including books, CD-ROMs, USB Drives, and on-line publications. For information about the IEEE Computer Society's *Conference Publishing Services* (CPS), please e-mail: cps@computer.org or telephone +1-714-821-8380. Fax +1-714-761-1784. Additional information about *Conference Publishing Services* (CPS) can be accessed from our web site at: *http://www.computer.org/cps*

IEEE Computer Society / Wiley Partnership
The IEEE Computer Society and Wiley partnership allows the CS Press *Authored Book* program to produce a number of exciting new titles in areas of computer science and engineering with a special focus on software engineering. IEEE Computer Society members continue to receive a 15% discount on these titles when purchased through Wiley or at: *http://wiley.com/ieeecs*. To submit questions about the program or send proposals, please e-mail jwilson@computer.org or telephone +1-714-816-2112. Additional information regarding the Computer Society's authored book program can also be accessed from our web site at:
http://www.computer.org/portal/pages/ieeecs/publications/books/about.html

---

Revised: 21 January 2008

CPS Online is our innovative online collaborative conference publishing system designed to speed the delivery of price quotations and provide conferences with real-time access to all of a project's publication materials during production, including the final papers. The **CPS Online** workspace gives a conference the opportunity to upload files through any Web browser, check status and scheduling on their project, make changes to the Table of Contents and Front Matter, approve editorial changes and proofs, and communicate with their CPS editor through discussion forums, chat tools, commenting tools and e-mail.

The following is the URL link to the **CPS Online** Publishing Inquiry Form:
http://www.ieeeconfpublishing.org/cpir/inquiry/cps_inquiry.html

IEEE
445 Hoes Lane
Piscataway, NJ 08854-4141

ISBN 978-0-7695-3365-0